Springer Series on
SIGNALS AND COMMUNICATION TECHNOLOGY

Signals and Communication Technology

Passive Eye Monitoring
Algorithms, Applications and Experiments
R.I. Hammoud (Ed.)   ISBN 978-3-540-75411-4

Multimodal User Interfaces
From Signals to Interaction
D. Tzovaras   ISBN 978-3-540-78344-2

Human Factors and Voice Interactive Systems
D. Gardner-Bonneau, H.E. Blanchard (Eds.)
ISBN: 978-0-387-25482-1

Wireless Communications
2007 CNIT Thyrrenian Symposium
S. Pupolin (Ed.)   ISBN: 978-0-387-73824-6

**Satellite Communications
and Navigation Systems**
E. Del Re, M Ruggieri (Eds.)
ISBN: 978-0-387-47522-6

Digital Signal Processing
An Experimental Approach
S. Engelberg   ISBN: 978-1-84800-118-3

**Digital Video and Audio
Broadcasting Technology**
A Practical Engineering Guide
W. Fischer   ISBN: 978-3-540-76357-4

Three-Dimensional Television
Capture, Transmission, Display
H.M. Ozaktas, L. Onural (Eds.)
ISBN 978-3-540-72531-2

**Foundations and Applications
of Sensor Management**
A.O. Hero, D. Castañón, D. Cochran,
K. Kastella (Eds.)   ISBN 978-0-387-27892-6

**Digital Signal Processing
with Field Programmable Gate Arrays**
U. Meyer-Baese   ISBN 978-3-540-72612-8

Adaptive Nonlinear System Identification
The Volterra and Wiener Model Approaches
T. Ogunfunmi   ISBN 978-0-387-26328-1

Continuous-Time Systems
Y.S. Shmaliy   ISBN 978-1-4020-6271-1

Blind Speech Separation
S. Makino, T.-W. Lee, H. Sawada (Eds.)
ISBN 978-1-4020-6478-4

**Cognitive Radio, Software Defined Radio,
and Adaptive Wireless Systems**
H. Arslan (Ed.)   ISBN 978-1-4020-5541-6

Wireless Network Security
Y. Xiao, D.-Z. Du, X. Shen
ISBN 978-0-387-28040-0

Terrestrial Trunked Radio – TETRA
A Global Security Tool
P. Stavroulakis   ISBN 978-3-540-71190-2

Multirate Statistical Signal Processing
O.S. Jahromi   ISBN 978-1-4020-5316-0

Wireless Ad Hoc and Sensor Networks
A Cross-Layer Design Perspective
R. Jurdak   ISBN 978-0-387-39022-2

**Positive Trigonometric Polynomials
and Signal Processing Applications**
B. Dumitrescu   ISBN 978-1-4020-5124-1

Face Biometrics for Personal Identification
Multi-Sensory Multi-Modal Systems
R.I. Hammoud, B.R. Abidi, M.A. Abidi (Eds.)
ISBN 978-3-540-49344-0

**Cryptographic Algorithms
on Reconfigurable Hardware**
F. Rodríguez-Henríquez
ISBN 978-0-387-33883-5

**Ad-Hoc Networking
Towards Seamless Communications**
L. Gavrilovska   ISBN 978-1-4020-5065-7

Multimedia Database Retrieval
A Human-Centered Approach
P. Muneesawang, L. Guan
ISBN 978-0-387-25627-6

Broadband Fixed Wireless Access
A System Perspective
M. Engels; F. Petre
ISBN 978-0-387-33956-6

Acoustic MIMO Signal Processing
Y. Huang, J. Benesty, J. Chen
ISBN 978-3-540-37630-9

Algorithmic Information Theory
Mathematics of Digital Information
Processing
P. Seibt   ISBN 978-3-540-33218-3

Continuous-Time Signals
Y.S. Shmaliy   ISBN 978-1-4020-4817-3

Interactive Video
Algorithms and Technologies
R.I. Hammoud (Ed.)   ISBN 978-3-540-33214-5

Handover in DVB-H
Investigation and Analysis
X. Yang   ISBN 978-3-540-78629-0

**Speech and Audio Processing in
Adverse Environments**
E. Hänsler, G. Schmidt (Eds.)
ISBN 978-3-540-70601-4

Eberhard Hänsler · Gerhard Schmidt
Editors

Speech and Audio Processing in Adverse Environments

 Springer

Editors
Eberhard Hänsler
Technische Universität
Darmstadt
Germany

Gerhard Schmidt
Harman/Becker Automotive Systems
Ulm
Germany

ISBN 978-3-540-70601-4   e-ISBN 978-3-540-70602-1
DOI 10.1007/978-3-540-70602-1

Springer Series on Signals and Communication Technology ISSN 1860-4862

Library of Congress Control Number: 2008931047

© 2008 Springer-Verlag Berlin Heidelberg
This work is subject to copyright. All rights are reserved, whether the whole or part of the material is concerned, specifically the rights of translation, reprinting, reuse of illustrations, recitation, broadcasting, reproduction on microfilm or in any other way, and storage in data banks. Duplication of this publication or parts thereof is permitted only under the provisions of the German Copyright Law of September 9, 1965, in its current version, and permission for use must always be obtained from Springer. Violations are liable for prosecution under the German Copyright Law.

The use of general descriptive names, registered names, trademarks, etc. in this publication does not imply, even in the absence of a specific statement, that such names are exempt from the relevant protective laws and regulations and therefore free for general use.

Cover design: WMXDesign GmbH, Heidelberg

Printed on acid-free paper

9 8 7 6 5 4 3 2 1

springer.com

Preface

Users of signal processing systems are never satisfied with the system they currently use. They are constantly asking for higher quality, faster performance, more comfort and lower prices. Researchers and developers should be appreciative for this attitude. It justifies their constant effort for improved systems. Better knowledge about biological and physical interrelations coming along with more powerful technologies are their engines on the endless road to perfect systems.

This book is an impressive image of this process. After "Acoustic Echo and Noise Control" published in 2004[1] many new results lead to "Topics in Acoustic Echo and Noise Control" edited in 2006[2]. Today – in 2008 – even more new findings and systems could be collected in this book. Comparing the contributions in both edited volumes progress in knowledge and technology becomes clearly visible: Blind methods and multi input systems replace "humble" low complexity systems. The functionality of new systems is less and less limited by the processing power available under economic constraints.

The editors have to thank all the authors for their contributions. They cooperated readily in our effort to unify the layout of the chapters, the terminology, and the symbols used. It was a pleasure to work with all of them.

Furthermore, it is the editors concern to thank Christoph Baumann and the Springer Publishing Company for the encouragement and help in publishing this book.

Darmstadt and Ulm, Germany   *Eberhard Hänsler*
                                                              Gerhard Schmidt

---

[1] Eberhard Hänsler and Gerhard Schmidt: *Acoustic Echo and Noise Control,* New York, NY: Wiley, 2004

[2] Eberhard Hänsler and Gerhard Schmidt (Eds.): *Topics in Acoustic Echo and Noise Control,* Berlin: Springer, 2006

Contents

Abbreviations and Acronyms 1

1 Introduction
    E. Hänsler, G. Schmidt
1.1   Overview about the Book 8

Part I Speech Enhancement

2 Low Delay Filter-Banks for Speech and Audio Processing
    H. W. Löllmann, P. Vary
2.1   Introduction 13
2.2   Analysis-Synthesis Filter-Banks 15
    2.2.1   General Structure 15
    2.2.2   Tree-Structured Filter-Banks 16
    2.2.3   Modulated Filter-Banks 17
    2.2.4   Frequency Warped Filter-Banks 20
    2.2.5   Low Delay Filter-Banks 26
2.3   The Filter-Bank Equalizer 29
    2.3.1   Concept 29
    2.3.2   Prototype Filter Design 31
    2.3.3   Relation between GDFT and GDCT 33
    2.3.4   Realization for Different Filter Structures 35
    2.3.5   Polyphase Network Implementation 37
    2.3.6   The Non-Uniform Filter-Bank Equalizer 41
    2.3.7   Comparison between FBE and AS FB 43
    2.3.8   Algorithmic Complexity 43
2.4   Further Measures for Signal Delay Reduction 44
    2.4.1   Concept 45

	2.4.2	Approximation by a Moving-Average Filter	45
	2.4.3	Approximation by an Auto-Regressive Filter	46
	2.4.4	Algorithmic Complexity	47
	2.4.5	Warped Filter Approximation	48
2.5	Application to Noise Reduction		49
	2.5.1	System Configurations	49
	2.5.2	Instrumental Quality Measures	50
	2.5.3	Simulation Results for the Uniform Filter-Banks	51
	2.5.4	Simulation Results for the Warped Filter-Banks	53
2.6	Conclusions		55
References			56

3 A Pre-Filter for Hands-Free Car Phone Noise Reduction: Suppression of Harmonic Engine Noise Components

H. Puder

3.1	Introduction		63
3.2	Analysis of the Different Car Noise Components		64
	3.2.1	Wind Noise	65
	3.2.2	Tire Noise	65
	3.2.3	Engine Noise	66
3.3	Engine Noise Removal Based on Notch Filters		68
3.4	Compensation of Engine Harmonics with Adaptive Filters		73
	3.4.1	Step-Size Control	75
	3.4.2	Calculating the Optimal Step-Size	78
	3.4.3	Results of the Compensation Approach	80
3.5	Evaluation and Comparison of the Results Obtained by the Notch Filter and the Compensation Approach		84
3.6	Conclusions and Summary		85
	3.6.1	Conclusion	85
	3.6.2	Summary	86
References			87

4 Model-Based Speech Enhancement

M. Krini, G. Schmidt

4.1	Introduction		89
4.2	Conventional Speech Enhancement Schemes		91
4.3	Speech Enhancement Schemes Based on Nonlinearities		93
4.4	Speech Enhancement Schemes Based on Speech Reconstruction		97
	4.4.1	Feature Extraction and Control	99
	4.4.2	Reconstruction of Speech Signals	110
4.5	Combining the Reconstructed and the Noise Suppressed Signal		124
	4.5.1	Adding the Fully Reconstructed Signal	125
	4.5.2	Adding only the Voiced Part of the Reconstructed Signal	129

4.6	Summary and Outlook		133
References			133

5 Bandwidth Extension of Telephony Speech
B. Iser, G. Schmidt

5.1	Introduction		135
5.2	Organization of the Chapter		137
5.3	Basics		138
	5.3.1	Human Speech Generation	139
	5.3.2	Source-Filter Model	141
	5.3.3	Parametric Representations of the Spectral Envelope	143
	5.3.4	Distance Measures	147
5.4	Non-Model-Based Algorithms for Bandwidth Extension		149
	5.4.1	Oversampling with Imaging	149
	5.4.2	Spectral Shifting	151
	5.4.3	Application of Non-Linear Characteristics	153
5.5	Model-Based Algorithms for Bandwidth Extension		153
	5.5.1	Generation of the Excitation Signal	155
	5.5.2	Vocal Tract Transfer Function Estimation	159
5.6	Evaluation of Bandwidth Extension Algorithms		176
	5.6.1	Objective Distance Measures	177
	5.6.2	Subjective Measures	180
5.7	Conclusions		181
References			182

6 Dereverberation and Residual Echo Suppression in Noisy Environments
E. A. P. Habets, S. Gannot, I. Cohen

6.1	Introduction		186
6.2	Problem Formulation		188
6.3	OM-LSA Estimator for Multiple Interferences		191
	6.3.1	OM-LSA Estimator	191
	6.3.2	A priori SIR Estimator	193
6.4	Dereverberation of Noisy Speech Signals		195
	6.4.1	Short Introduction to Speech Dereverberation	195
	6.4.2	Problem Formulation	197
	6.4.3	Statistical Reverberation Model	199
	6.4.4	Late Reverberant Spectral Variance Estimator	200
	6.4.5	Summary and Discussion	203
6.5	Residual Echo Suppression		203
	6.5.1	Problem Formulation	204
	6.5.2	Late Residual Echo Spectral Variance Estimator	206
	6.5.3	Parameter Estimation	208
	6.5.4	Summary	210

6.6	Joint Suppression of Reverberation, Residual Echo, and Noise	210
6.7	Experimental Results	212
	6.7.1 Experimental Setup	214
	6.7.2 Joint Suppression of Reverberation and Noise	214
	6.7.3 Suppression of Residual Echo	216
	6.7.4 Joint Suppression of Reverberation, Residual Echo, and Noise	221
6.8	Summary and Outlook	223
References		224

7 Low Distortion Noise Cancellers – Revival of a Classical Technique

A. Sugiyama

7.1	Introduction	229
7.2	Distortions in Widrow's Adaptive Noise Canceller	230
	7.2.1 Distortion by Interference	230
	7.2.2 Distortion by Crosstalk	232
7.3	Paired Filter (PF) Structure	233
	7.3.1 Algorithm	233
	7.3.2 Evaluations	235
7.4	Crosstalk Resistant ANC and Cross-Coupled Structure	239
	7.4.1 Crosstalk Resistant ANC	240
	7.4.2 Cross-Coupled Structure	241
7.5	Cross-Coupled Paired Filter (CCPF) Structure	242
	7.5.1 Algorithm	242
	7.5.2 Evaluations	245
7.6	Generalized Cross-Coupled Paired Filter (GCCPF) Structure	247
	7.6.1 Algorithm	250
	7.6.2 Evaluation by Recorded Signals	251
7.7	Demonstration in a Personal Robot	261
7.8	Conclusions	261
References		263

Part II Echo Cancellation

8 Nonlinear Echo Cancellation Based on Spectral Shaping

O. Hoshuyama, A. Sugiyama

8.1	Introduction	267
8.2	Frequency-Domain Model of Highly Nonlinear Residual Echo	268
	8.2.1 Spectral Correlation Between Residual Echo and Echo Replica	269

		8.2.2 Model of Residual Echo Based on Spectral Correlation 273

- 8.3 Echo Canceller Based on the New Residual Echo Model.......... 274
  - 8.3.1 Overall Structure 274
  - 8.3.2 Estimation of Near-End Speech 275
  - 8.3.3 Spectral Gain Control 276
- 8.4 Evaluations 277
  - 8.4.1 Objective Evaluations 277
  - 8.4.2 Subjective Evaluation 279
- 8.5 DSP Implementation and Real-Time Evaluation 280
- 8.6 Conclusions 280
- References 281

Part III Signal and System Quality Evaluation

9 Telephone-Speech Quality
U. Heute

- 9.1 Telephone-Speech Signals 287
  - 9.1.1 Telephone Scenario 287
  - 9.1.2 Telephone-Scenario Model 287
- 9.2 Speech-Signal Quality 289
  - 9.2.1 Intelligibility 289
  - 9.2.2 Speech-Sound Quality 290
- 9.3 Speech-Quality Assessment 292
  - 9.3.1 Auditory Quality Assessment 292
  - 9.3.2 Aims 292
  - 9.3.3 Instrumental Quality Assessment 293
- 9.4 Compound-System Quality Prediction 293
  - 9.4.1 The System-Planning Task 293
  - 9.4.2 ETSI Network-Planning Model (E-Model) 293
- 9.5 Auditory Total-Quality Assessment.............................. 294
  - 9.5.1 Conversation Tests....................................... 294
  - 9.5.2 Listening Tests 296
  - 9.5.3 LOTs with Pair Comparisons 296
  - 9.5.4 Absolute-Category Rating (ACR) LOTs 297
- 9.6 Auditory Quality-Attribute Analysis 298
  - 9.6.1 Quality Attributes 298
  - 9.6.2 Attribute-Oriented LOTs 298
  - 9.6.3 Search for Suitable Attributes 302
  - 9.6.4 Integral-Quality Estimation from Attributes 305
- 9.7 Instrumental Total-Quality Measurement 306
  - 9.7.1 Signal Comparisons...................................... 306
  - 9.7.2 Evaluation Approaches 306
  - 9.7.3 Psychoacoustically Motivated Measures 312

9.8	Instrumental Attribute-Based Quality Measurements	320
	9.8.1 Basic Ideas	320
	9.8.2 Loudness	322
	9.8.3 Sharpness	323
	9.8.4 Roughness	323
	9.8.5 Directness/Frequency Content (DFC)	324
	9.8.6 Continuity	326
	9.8.7 Noisiness	329
	9.8.8 Combined Direct and Attribute-Based Total Quality Determination	331
9.9	Conclusions, Outlook, and Final Remarks	331
References		332

10 Evaluation of Hands-free Terminals

F. Kettler, H.-W. Gierlich

10.1	Introduction	339
10.2	Quality Assessment of Hands-free Terminals	340
10.3	Subjective Methods for Determining the Communicational Quality	342
	10.3.1 General Setup and Opinion Scales Used for Subjective Performance Evaluation	343
	10.3.2 Conversation Tests	345
	10.3.3 Double Talk Tests	346
	10.3.4 Talking and Listening Tests	347
	10.3.5 Listening-only Tests (LOT) and Third Party Listening Tests	348
	10.3.6 Experts Tests for Assessing Real Life Situations	349
10.4	Test Environment	350
	10.4.1 The Acoustical Environment	351
	10.4.2 Background Noise Simulation Techniques	351
	10.4.3 Positioning of the Hands-Free Terminal	352
	10.4.4 Positioning of the Artificial Head	352
	10.4.5 Influence of the Transmission System	354
10.5	Test Signals and Analysis Methods	354
	10.5.1 Speech and Perceptual Speech Quality Measures	356
	10.5.2 Speech-like Test Signals	356
	10.5.3 Background Noise	360
	10.5.4 Applications	363
10.6	Result Representation	365
	10.6.1 Interpretation of HFT "Quality Pies"	366
	10.6.2 Examples	368
10.7	Related Aspects	368
	10.7.1 The Lombard Effect	368
	10.7.2 Intelligibility Outside Vehicles	372

References 375

Part IV Multi-Channel Processing

11 Correlation-Based TDOA-Estimation for Multiple Sources in Reverberant Environments

J. Scheuing, B. Yang

11.1 Introduction.. 381
11.2 Analysis of TDOA Ambiguities 383
    11.2.1 Signal Model.. 383
    11.2.2 Multipath Ambiguity 384
    11.2.3 Multiple Source Ambiguity 384
    11.2.4 Ambiguity due to Periodic Signals 386
    11.2.5 Principles of TDOA Disambiguation 386
11.3 Estimation of Direct Path TDOAs 390
    11.3.1 Correlation and Extremum Positions 390
    11.3.2 Raster Matching 392
11.4 Consistent TDOA Graphs..................................... 397
    11.4.1 TDOA Graph 397
    11.4.2 Strategies of Consistency Check 398
    11.4.3 Properties of TDOA Graphs 399
    11.4.4 Efficient Synthesis Algorithm 402
    11.4.5 Initialization and Termination......................... 404
    11.4.6 Estimating the Number of Active Sources............... 405
11.5 Experimental Results... 406
    11.5.1 Localization System.................................... 406
    11.5.2 TDOA Estimation of a Single Signal Block 408
    11.5.3 Source Position Estimation 412
    11.5.4 Evaluation of Continuous Measurements................ 412
11.6 Summary 414
References 415

12 Microphone Calibration for Multi-Channel Signal Processing

M. Buck, T. Haulick, H.-J. Pfleiderer

12.1 Introduction.. 417
12.2 Beamforming with Ideal Microphones.......................... 418
    12.2.1 Principle of Beamforming 418
    12.2.2 Evaluation of Beamformers 421
    12.2.3 Statistically Optimum Beamformers 424
12.3 Microphone Mismatch and its Effect
    on Beamforming 427
    12.3.1 Model for Non-Ideal Microphone Characteristics......... 428
    12.3.2 Effect of Microphone Mismatch on Fixed Beamformers 429

         12.3.3 Effect of Microphone Mismatch on Adaptive Beamformers . 430
         12.3.4 Comparison of Fixed and Adaptive Beamformers 432
  12.4 Calibration Techniques and their Limits
         for Real-World Applications 432
         12.4.1 Calibration of Single Microphones 432
         12.4.2 Analysis of Fixed Beamformers 440
         12.4.3 Analysis of Adaptive Beamformers 444
         12.4.4 Comparison of Fixed and Adaptive Beamformers 448
  12.5 Self-Calibration Techniques 449
         12.5.1 Basic Unit 451
         12.5.2 Configurations for Array Processing 452
         12.5.3 Recursive Configuration 455
         12.5.4 Adaptation Control 457
         12.5.5 Experimental Results 458
  12.6 Summary 459
  12.A Experimental Determination of the Directivity Index 460
         12.A.1 Numerical Integration over a Spherical Surface 460
         12.A.2 Definition of the Coordinate System 462
         12.A.3 Determination of the Directivity
                 using the Normalized Cross Power Spectral
                 Densities of Microphone Signals 464
  References 465

13 Convolutive Blind Source Separation for Noisy Mixtures

   R. Aichner, H. Buchner, W. Kellermann

13.1 Introduction.. 469
13.2 Blind Source Separation for Acoustic Mixtures
       Based on the TRINICON Framework 473
       13.2.1 Matrix Formulation 473
       13.2.2 Optimization Criterion and Coefficient Update 474
       13.2.3 Approximations Leading to Special Cases 478
       13.2.4 Estimation of the Correlation Matrices
               and an Efficient Normalization Strategy 484
       13.2.5 On Broadband and Narrowband BSS Algorithms
               in the DFT Domain................................... 485
       13.2.6 Experimental Results for Reverberant Environments 488
13.3 Extensions for Blind Source Separation in Noisy Environments ... 490
       13.3.1 Model for Background Noise in Realistic Environments 491
       13.3.2 Pre-Processing for Noise-Robust Adaptation 493
       13.3.3 Post-Processing for Suppression of Residual
               Crosstalk and Background Noise........................ 498
13.4 Conclusions 518
References 519

14 Binaural Speech Segregation

N. Roman, D.L. Wang

14.1	Introduction	525
14.2	T–F Masks for CASA	528
14.3	Anechoic Binaural Segregation	529
14.4	Reverberant Binaural Segregation	533
14.5	Evaluation	536
14.6	Concluding Remarks	543
14.7	Acknowledgments	546
References		546

15 Spatio-Temporal Adaptive Inverse Filtering in the Wave Domain

S. Spors, H. Buchner, R. Rabenstein

15.1	Introduction		551
15.2	Problem Description		553
	15.2.1	Nomenclature	553
	15.2.2	Massive Multichannel Sound Reproduction	553
	15.2.3	Multichannel Active Listening Room Compensation	556
	15.2.4	Multichannel Active Noise Control	559
	15.2.5	Unified Representation of Spatio-Temporal Adaptive Filtering Problems	561
	15.2.6	Frequency-Domain Notation	563
15.3	Computation of Compensation Filters		563
	15.3.1	Classification of Algorithms	564
	15.3.2	Ideal Solution	564
	15.3.3	Adaptive Solution	565
	15.3.4	Problems of the Adaptive Solution	567
15.4	Eigenspace Adaptive Filtering		568
	15.4.1	Generalized Singular Value Decomposition	568
	15.4.2	Eigenspace Adaptive Filtering	569
	15.4.3	Problems	570
15.5	Wave-Domain Adaptive Filtering		570
	15.5.1	Concept	571
	15.5.2	The Circular Harmonics Decomposition	573
	15.5.3	The Circular Harmonics Expansion Using Boundary Measurements	574
15.6	Application of WDAF to Adaptive Inverse Filtering Problems		576
	15.6.1	Application of WDAF to Active Listening Room Compensation	577
	15.6.2	Application of WDAF to Active Noise Control	578
15.7	Conclusions		578
References			580

Part V Selected Applications

16 Virtual Hearing
K. Wiklund, S. Haykin

- 16.1 Previous Work 588
- 16.2 VirtualHearing 592
- 16.3 Room Acoustic Model 593
- 16.4 HRTF Simulation 597
- 16.5 Neural Model 600
- 16.6 The Software and Interface 605
- 16.7 Software Testing 609
- 16.8 Future Work and Conclusions 610
- References 613

17 Dynamic Sound Control Algorithms in Automobiles
M. Christoph

- 17.1 Introduction 615
  - 17.1.1 Introduction of Dynamic Volume Control Systems 616
  - 17.1.2 Introduction of Dynamic Equalization Control Systems 618
- 17.2 Previous Systems – Description and Analysis 619
  - 17.2.1 Speed Dependant Sound systems 619
  - 17.2.2 Microphone Based Dynamic Volume Control Sound Systems 621
  - 17.2.3 Non-Acoustic Sensor Based Sound Systems 643
- 17.3 Spectrum-Based Dynamic Equalization Control 645
  - 17.3.1 Frequency Domain Adaptive Filter 646
  - 17.3.2 Generalized Multidelay Adaptive Filter 649
  - 17.3.3 Step-Size Control 652
  - 17.3.4 Multi-Channel Systems 653
  - 17.3.5 Estimating the Power Spectral Density of the Background Noise 654
  - 17.3.6 Psychoacoustic Basics 658
  - 17.3.7 The Psychoacoustic Masking Model According to Johnston 661
- 17.4 Conclusion and Outlook 670
- 17.5 Acknowledgement 673
- References 674

18 Towards Robust Distant-Talking Automatic Speech Recognition in Reverberant Environments
A. Sehr, W. Kellermann

- 18.1 Introduction 679
- 18.2 The Distant-Talking ASR Scenario 680

18.3 How to Deal with Reverberation in ASR Systems? 683
18.4 Effect of Reverberation in the Feature Domain 691
18.5 Signal Dereverberation and Beamforming 695
18.6 Robust Features 699
18.7 Model Training and Adaptation 700
18.8 Reverberation Modeling for Speech Recognition 702
     18.8.1 Feature Production Model 703
     18.8.2 Reverberation Model 704
     18.8.3 Training of the Reverberation Model 705
     18.8.4 Decoding 708
     18.8.5 Inner Optimization 713
     18.8.6 Solution of the Inner Optimization Problem
           in the Melspec Domain for Single Gaussian Densities 714
     18.8.7 Simulations 717
18.9 Summary and Conclusions 722
References 723

Index 729

List of Contributors

R. Aichner
Microsoft Corporation
Redmond, WA, USA

H. Buchner
Deutsche Telekom Laboratories
Germany

M. Buck
Harman/Becker
Germany

M. Christoph
Harman/Becker
Germany

I. Cohen
Israel Institute of Technology
Israel

S. Gannot
Bar-Ilan University
Israel

H. W. Gierlich
HEAD acoustics
Germany

E. Habets
Bar-Ilan University
Israel

E. Hänsler
Technische Universität Darmstadt
Germany

T. Haulick
Harman/Becker
Germany

S. Haykin
McMaster University
Canada

U. Heute
University of Kiel
Germany

O. Hoshuyama
NEC Corporation
Japan

B. Iser
Harman/Becker
Germany

W. Kellermann
University Erlangen-Nuremberg
Germany

F. Kettler
HEAD acoustics
Germany

M. Krini
Harman/Becker
Germany

H. W. Löllmann
RWTH Aachen University
Germany

H.-J. Pfleiderer
University Ulm
Germany

H. Puder
Siemens Audiological Engineering
Group
Germany

R. Rabenstein
University Erlangen-Nuremberg
Germany

N. Roman
Ohio State University at Lima
Lima, USA

J. Scheuing
University of Stuttgart
Germany

G. Schmidt
Harman/Becker
Germany

A. Sehr
University Erlangen-Nuremberg
Germany

S. Spors
Deutsche Telekom Laboratories
Germany

A. Sugiyama
NEC Corporation
Japan

P. Vary
RWTH Aachen University
Germany

D. Wang
Ohio State University
Columbus, USA

K. Wiklund
McMaster University
Canada

B. Yang
University of Stuttgart
Germany

Abbreviations and Acronyms

AA-LP	Anti-aliasing lowpass
AC3	Adaptive transform coder 3
ACF	Autocorrelation function
ACR	Absolute-category rating
ADC	Analog-to-digital converter
ADPCM	Adaptive differential pulse-code modulation
AEC	Acoustic echo cancellation/canceller
AGC	Automatic gain control
AIR	Acoustic impulse response
ALP	Adaptive lattice predictor
AM	Amplitude modulation
AMR	Adaptive multi-rate (codec)
ANC	Active noise control or adaptive noise canceller
AP	Affine projection
APC	Adaptive predictive coding
AR	Auto-regressive
ARC	Active room compensation
AS	Analysis synthesis
ASA	Auditory scene analysis
ASL	Active speech level
ASR	Automatic speech recognition
ATC	Adaptive transform coding
BBC	British Broadcast Corporation
BGNT	Background noise transmission
BSA	Bark-spectral approximation
BSD	Bark-spectral distance
BSS	Blind source separation
BWE	Bandwidth extension
CAN	Controller area network
CASA	Computational auditory scene analysis
CB	Codebook or critical band

CCF	Crosscorrelation function
CCPF	Cross-coupled paired filter
CCR	Comparison-category rating
CD	Cepstral distance/compact disc
CDR	Connected digit recognition
CELP	Code-excited linear predictive (coding)
CMN	Cepstral mean normalization
CMOS	Comparison mean opinion score
CMS	Cepstral mean subtraction
CPU	Central processing unit
CSS	Composite source signal
CTRANC	Crosstalk resistant adaptive noise canceller
DAC	Digital-to-analog converter
DAM	Diagnostic acceptability measure
DCR	Degradation-category rating
DCT	Discrete cosine transform
DEC	Dynamic equalization control
DECT	Digital enhanced cordless telecommunications
DFT	Discrete Fourier transform
DIVA	Digital interactive virtual acoustics
DMOS	Degradation mean opinion score
DRAM	Dynamic random access memory
DSK	Digital signal processor starter kit
DSP	Digital signal processing or digital signal processor
DT	Double talk
DFTF	Discrete time Fourier transform
DTS	Digital Theater Systems Inc.
DTX	Discontinuous transmission
DVC	Dynamic volume control
DWT	Discrete wavelet transform
EAF	Eigenspace adaptive filtering
EC	Echo canceller
EC-LAF	Echo canceller using a linear adaptive filter
EDC	Energy decay curve
ELDT	Echo level during double talk
EMDF	Extended multi-delay filter
ERB	Equivalent rectangular bandwidth
ERLE	Echo return loss enhancement
ETSI	European telecommunications standards institute
FB	Filter bank
FBE	Filter-bank equalizer
FBF	Fixed beamformer
FBSM	Filter-bank summation method
FDAF	Frequency domain adaptive filter
FFT	Fast Fourier transform

FIR	Finite impulse response
FM	Frequency modulation
GAL	Gradient adaptive lattice
GCC	Generalized cross correlation
GCCPF	Generalized cross-coupled paired filter
GDCT	Generalized discrete cosine transform
GDFT	Generalized discrete Fourier transform
GMAF	Generalized multi-delay filter
GMM	Gaussian mixture model
G-MOS	Global mean opinion score
GSC	Generalized sidelobe canceller
GSM	Global system for mobile communications
GSVD	Generalized singular value decomposition
HATS	Head and torso simulator
HERB	Harmonicity-based dereverberation
HF	High frequency
HFRP	Hands-free reference point
HFT	Hands-free terminal
HINT	Hearing in noise test
HMM	Hidden Markov model
HOA	Higher-order ambisonics
HOS	Higher-order statistics
HP	Highpass
HRIR	Head related impulse response
HRTF	Head related transfer function
HTK	Hidden Markov model toolkit
IBM	Ideal binary mask
IC	Interference canceller
ICA	Independent component analysis
ICC	In-car communication
IDEC	Individual dynamic equalization control
IDFT	Inverse discrete Fourier transform
IDVC	Individual dynamic volume control
IEC	International electrotechnical commission
IFFT	Inverse fast Fourier transform
IHC	Inner hair cell
IID	Interaural intensity difference or independent identically distributed
IIR	Infinite impulse response
IMCRA	Improved minima controlled recursive averaging
INMD	In-service non-intrusive measurement device
INR	Input-to-noise ratio
IRS	Intermediate reference system
ISDN	Integrated-services digital network
ISO	International standardization organization
ITD	Interaural time difference

ITU	International telecommunication union
IWDFT	Inverse warped discrete Fourier transform
KEMAR	Knowles electronic manikin for acoustic research
LAF	Linear adaptive filter
LAR	Log-area ratio
LBG	Linde, Buzo, and Gray
LDF	Low delay filter
LEM	Loudspeaker enclosure microphone
LMS	Least mean square
LOT	Listening-only test
LP	Linear prediction or lowpass
LPC	Linear predictive coding
LPTV	Linear periodically time-variant
LQ	Listing quality
LS	Least squares
LSA	Log spectral amplitude
LSD	Log spectral distance
LSE	Least-squares error
LSF	Line spectral frequencies
LSTR	Listener sidetone rating
LTI	Linear time-invariant
MA	Moving average
MAP	Maximum a posteriori
MDF	Multi-delay filter
MDS	Multi-dimensional scaling
MF	Main filter
MFCC	Mel-frequency cepstral coefficient
MIMO	Multiple-input multiple-output
MINT	Multiple input/output inverse theorem
MLP	Multi-layer perceptron
MMSE-STSA	Minimum mean-square error short-time spectral amplitude
MOS	Mean-opinion score
MOST	Media oriented systems transport
MRP	Mouth reference point
MSC	Magnitude-squared coherence
MVDR	Minimum variance distortionless response
NC	Noise canceller
NL-EC	Nonlinear echo canceller
NLMS	Normalized least mean square
N-MOS	Noise mean opinion score
NN	Neural network
NPR	Near-perfect reconstruction
NR	Noise reduction
NS	Noise suppression
OEM	Original equipment manufacturer

OHC	Outer hair cell
OLR	Overall loudness rating
OM-LSA	Optimally-modified log spectral amplitude
PARCOR	Partial correlation (coefficient)
PAMS	Perceptual analysis measurement system
PBFDAF	Partitioned block frequency domain adaptive filter
PC	Personal computer
PCM	Pulse code modulation
PDF	Probability density function
PESQ	Perceptual evaluation of speech quality
PF	Paired filter
PHAT	Phase transform
POTS	Plain old telephone system
PPN	Polyphase network
PR	Perfect reconstruction
PSD	Power spectral density
PSQM	Perceptual speech quality measure
QMF	Quadrature mirror filter
RAM	Random access memory
RASTA	Relative spectra
REMOS	Reverberation modeling for speech recognition
RES	Residual echo suppression
RF	Radio frequency
RIR	Room impulse response
RLR	Receiving loudness rating
RLS	Recursive least-squares
RPM	Revolutions per minute
RS	Reverberation suppression
SAEC	Stereo acoustic echo cancellation
SBC	Subband coding
SC	Sylvester constraint
SD	Semantic differential or spectral distance
SDM	Spectral distortion measure
SDRAM	Synchronized dynamic random access memory
SF	Smoothing filter or subfilter
SFM	Spectral flatness measure
SII	Speech intelligibility index
SIMO	Single-input multiple-output
SIR	Signal-to-interference ratio
SIRP	Spherically invariant random process
SLR	Sending loudness rating
S-MOS	Speech mean opinion score
SNR	Signal-to-noise ratio
SOS	Second-order statistics
SPIN	Speech perception in noise

SPL	Sound pressure level
SQET	Speech-quality evaluation tool
SRA	Statistical room acoustics
SRAM	Static or synchronized random access memory
SRR	Signal-to-reverberation ratio
STFT	Short-time Fourier transform
STMR	Sidetone masking ratio
TBQ	Total background quality
TCL	Terminal coupling loss
TCM	Target cancellation module
TDOA	Time difference of arrival
TFRM	Tolerance function of raster match
TFTM	Tolerance function of triple match
TIMIT	Texas Instruments (TI) and Massachusetts Institute of Technology (MIT)
TOSQA	Telecommunication objective speech quality assessment
TRINICON	Triple-N independent component analysis for convolutive mixtures
TSQ	Total signal quality
TWRM	Tolerance width of raster match
TWTM	Tolerance width of triple match
VAD	Voice activity detection
VDA	Verband der Automobilindustrie (German, stands for *German association of the automotive industry*)
VoIP	Voice over internet protocol
WDAF	Wave-domain adaptive filtering
WDFT	Warped discrete Fourier transform
WFA	Wave field analysis
WFS	Wave field synthesis
X-RLS	Filtered-x recursive least-squares (algorithm)

1
Introduction

Eberhard Hänsler[1] and Gerhard Schmidt[2]

[1] Technische Universität Darmstadt, Germany
[2] Harman/Becker Automotive Systems, Germany

If people would speak digitally speech processing would lack some of its most challenging problems.

One of those tasks is to provide means for a comfortable conversation with a remote partner where one of them or both are in adverse environments. By "adverse environment" we mean noisy offices, railway stations, airports, shop floors, etc. Similar problems have to be solved when a speech recognition system is used. Under "comfortable environment" we understand that a speaker does not have to be "wired", i.e. to carry or to hold a microphone very close to his mouth. He should be able "just to talk" without caring where the microphone(s) is/are located. His partner at a remote location or a speech recognition system should just receive his speech signal. In case of speech recognition background noise should be suppressed as much as possible. For a human listener it may be desirable to communicate some information about the environment his partner stays in.

Systems for speech enhancement have to perform at least three main functions: echo canceling, noise suppression and speech restoration. For all three there is no clear-cut solution. Mathematical approaches have to start with models that are simplified to a high degree and such are a very rough approximation to the reality only. Necessarily, a good deal of heuristics has to enter the solution in order to match it to the real world. Thus, there are no break throughs to final solutions in any one of the subproblems. Advanced technology and cheaper hardware will always stimulate researchers and industrial developers to come up with more sophisticated methods that promise better results. The appearance of powerful simulation tools and high-capacity personal computers over the last decades have speeded up this process. The simulation and the real-time verification of algorithms do no longer require costly dedicated soft- and hardware.

This book provides an overview of recent developments and new results reported from the key researchers in speech and audio processing.

1.1 Overview about the Book

The succeeding seventeen chapters are organized in five parts. This introduction is followed by chapters on *Speech Enhancement* in PART ONE. *H. Löllmann* and *P. Vary* review the design of uniform and non-uniform analysis-synthesis filter-banks employed sub-band processing of speech and audio signals. Their main aim is to achieve low signal delay. They introduce and analyse the concept of the filter-bank equalizer. In certain applications of noise reduction information originally generated for other purposes can be used to support the solution of a problem. In case of noise reduction in a passenger car the exact speed of the engine is provided by a bus-system in the car. *H. Puder* discusses a pre-filter to reduce harmonics that are proportional to the speed of the engine from car noise. *M. Krini* and *G. Schmidt* improve noisy speech signals by partially reconstructing the spectrum. By overlaying the conventionally noise reduced signal and the reconstructed one it is possible to avoid the robot-like sound of pure synthesized speech signals. Telephone signals are transmitted with a reduced bandwidth. *B. Iser* and *G. Schmidt* explain how the bandwidth of a transmitted speech signal can be extended at the receiver, such that a listener feels a more natural sound. Reverberations often degrade the fidelity and intelligibility of speech signals and decrease the performance of automatic speech recognition devices. *E. A. P. Habets, S. Gannot*, and *I. Cohen* develop a post-processor for the joint suppression of the residual echo, the background noise and the reverberation. *A. Sugiyama* assumes the existence of a reference microphone and extends this classical noise reduction technique for the case where the crosstalk from the primary signal source to the auxiliary source can not be neglected. He describes the application of this technique together with a speech recognition system in a human-robot communication scenario.

PART TWO of the book deals with nonlinear *Echo Cancellation*. *O. Hoshuyama* and *A. Sugiyama* address the nonlinearity of the echo path in hands-free cellphones. Since common approaches like Voltera filters are too demanding for such devices they propose a nonlinear echo canceller based on the correlation between spectral amplitudes of the residual echo and the echo replica.

In PART THREE, *Signal and System Quality Evaluation*, two chapters are concerned with diagnosing the quality of speech signals. The ultimate criterion here is the judgment of human listeners. Tests of this kind, however, are time consuming and costly and not free of problems. *U. Heute* explains these tests for various scenarios and also explains possibilities for automatic quality measurement not involving human listeners. The chapter by *F. Kettler* and *H.-W. Gierlich* primarily focuss on hands-free terminals installed in passenger cars. They describe subjective tests and the necessity of objective laboratory tests. They show how the the scores for different aspects of a hands-free system can be documented by a "quality pie".

PART FOUR deals with topics on *Multi-Channel Processing*. The use of microphone and loudspeaker arrays opens a new dimension in speech and audio signal processing. *J. Scheuring* and *B. Yang* estimate the time difference of arrival in a multi-source reverberant environment. They resolve the ambiguity by using the facts that the cross-correlation maxima from a direct path and the echo show the same distance than the extrema in the autocorrelation of the microphone signals and that the cyclic sum of the time differences of arrival of all microphone pairs is zero. The performance of microphone arrays may degrade severely due to mismatched microphones. *M. Buck, T. Haulick* and *H.-J. Pfleiderer* introduce a new model that allows to study the effect of differences of microphones in an array mathematically and by simulation. They discus methods for fixed and for adaptive calibration of the microphones where in case of mass production adaptive calibration is clearly preferable. Convolutive blind source separation for noisy mixtures is the concern of the chapter by *R. Aichner, H. Buchner* and *W. Kellermann*. In contrast to conventional procedures no a-priori knowledge about source and sensor positions are required. Their method combines pre- and postprocessing algorithms such, that residual cross-talk and background noise can be handled. The results are confirmed by experiments. In their chapter on binaural speech segregation *N. Roman* and *D. L. Wang* describe the principles of binaural processing. Their special interest is an automatic sound separation in a realistic environment. Solutions of this problem are of interest in many applications of speech and speaker recognition. Substantial improvements are achieved by utilizing only reliable target dominant features and by a target reconstruction method for unreliable features. In the final chapter of this part of the book *S. Spors*, H. Buchner and *R. Rabenstein* gives a unified description of spatio-temporal adaptive methods of sound reproduction and trace them back to the problem of inverse filtering. They introduce eigenspace adaptive filters to decouple the multichannel problem. An exact solution, however, would need data-dependant transformations. Therefore, wave-domain adaptive filtering serves as an approximate solution.

Selected Applications are described in PART FIVE. *K. Wiklund* and *S. Haykin* report on a system that allows to test algorithms designed for hearing aids. It allows to simulate the real acoustic environment and the impairments of patients. Thus, hearing aids can be tuned to the needs of patients and the amount of time consuming and costly real life tests can be reduced. Automobiles provide very undesirable acoustic environments. On the other hand, passengers request pleasant acoustic conditions for listening to audio programs, to carry hands-free telephone calls, or talking to other passengers. *M. Christoph* shows how – under a practical point of view – a vehicle dependent tuning can be accomplished. Since this is done when the engine is off and the car is not moving control procedures are required to maintain the acoustic quality during the time the car is operated. Inputs for this algorithms can be signals that are available from the car-electronics or additional microphones in the passenger compartment. *A. Sehr* and *W. Kellermann* address the problem of automatic speech recognition in reverberant environments. Reverberation disperses a signal and thus, it cannot be modelled by an additive or a multiplicative term. Conventional methods for dereverberation are described. A new concept called reverberation modelling that combines the advantages of the former methods is introduced.

Part I

Speech Enhancement

2

Low Delay Filter-Banks for Speech and Audio Processing

Heinrich W. Löllmann and Peter Vary

Institute of Communication Systems and Data Processing,
RWTH Aachen University, Germany

2.1 Introduction

Digital filter-banks are an integral part of many speech and audio processing algorithms used in today's communication systems. They are commonly employed for adaptive subband filtering, for example, to perform acoustic echo cancellation in hands-free communication devices or multi-channel dynamic-range compression in digital hearing aids, e.g., [34,81]. Another frequent task is speech enhancement by noise reduction, e.g., [4,81]. This eases the communication in adverse environments where acoustic background noise impairs the intelligibility and fidelity of the transmitted speech signal. A noise reduction system is also beneficial to improve the performance of speech coding and speech recognition systems, e.g., [41].

The choice of the filter-bank has a significant influence on the performance of such systems in terms of signal quality, computational complexity, and signal delay. Accordingly, the filter-bank design has to fulfill different, partly conflicting requirements in dependence of the considered application.

One prominent example is speech and audio processing for digital hearing aids. The restricted capacity of the battery and the small size of the chip set limit the available computational power. Moreover, a low overall processing delay is required to avoid disturbing artifacts and echo effects, e.g., [1,75]. Such distortions can occur when the hearing aid user is talking. In this case, the processed speech can interfere with the original speech signal, which reaches the cochlea with minimal delay via bone conduction or through the hearing aid vent. To prevent this, the algorithmic signal delay of the filter-bank used for the signal enhancement must be considerably lower than the tolerable processing delay, i.e., the latency between the analog input and output signal of the system. In addition, a filter-bank with non-uniform time-frequency resolution, which is similar to that of the human auditory system, is desirable to perform multi-channel dynamic-range compression and noise reduction with a small number of frequency bands.

A common choice for many applications is still the uniform DFT analysis-synthesis filter-bank (AS FB). This complex modulated filter-bank can be efficiently realized by means of a polyphase network (PPN) [77] and comprises the weighted overlap-add method as a variant hereof [11, 12]. This scheme is often used for frame-wise processing, e.g., in noise reduction systems of speech coders. In this case, a fixed buffering delay occurs and the additional delay due to overlapping frames can be reduced by an appropriate window design, e.g., [53].

However, the frequency resolution of the uniform (DFT) filter-bank is not well adapted to that of the human auditory system. The non-uniform frequency resolution of the human ear declines for an increasing frequency, which can be described by the Bark frequency scale [84]. Therefore, several authors have proposed the use of non-uniform AS FBs for speech enhancement to obtain an improved (subjective) speech quality [9, 16, 19, 26, 27, 61]. One rationale for these approaches is that a filter-bank with a non-uniform, approximately Bark-scaled frequency resolution incorporates a perceptual model of the human auditory system. Another reason is that on average most of the energy and harmonics of speech signals are located at the lower frequencies.

One approach to achieve an approximately Bark-scaled frequency resolution is to employ the discrete wavelet (packet) transform, which can be implemented by a tree-structured AS FB, e.g., [9, 19, 26, 27]. Another method is to use frequency warped AS FBs [19, 26, 27, 61]. These filter-banks possess usually a lower signal delay and a lower algorithmic complexity than comparable tree-structured filter-banks.

The allpass transformation is a well-known technique for the design of frequency warped filter-banks [6, 17, 58, 79]. These filter-banks can achieve a Bark-scaled frequency division with great accuracy [73]. This property is of interest for speech and audio processing applications alike and allows to use a lower number of frequency channels than for the uniform filter-bank. A disadvantage of this approach is that the allpass transformation of the analysis filter-bank leads to (increased) aliasing and phase distortions. The compensation of these effects results in a more complex synthesis filter-bank design as well as a higher algorithmic complexity and signal delay. These drawbacks often prevent to exploit the benefits of frequency warped filter-banks for (real-time) speech and audio processing systems.

In this chapter, we discuss alternative design concepts for uniform and frequency warped filter-banks. The aim is to devise a general filter-bank design with the same time-frequency resolution as the conventional uniform and allpass transformed AS FB, but with a considerably lower signal delay.

For these purposes, the design of uniform and non-uniform AS FBs is reviewed in Sec. 2.2, and approaches to achieve a reduced signal delay are discussed. The alternative concept of the filter-bank equalizer (FBE) is introduced in Sec. 2.3. The effects of time-varying coefficients are analyzed, and an efficient implementation of the FBE is devised. A generalization of this concept is given by the allpass transformed FBE, which is presented in Sec. 2.3.6.

Measures to achieve (nearly) perfect signal reconstruction are described, and the algorithmic complexity of different filter-bank designs is contrasted. For applications with very demanding signal delay constraints, a modification of the FBE is proposed in Sec. 2.4 to achieve a further reduced signal delay with almost no loss for the subjective speech quality. In Sec. 2.5, the discussed filter-banks are applied to noise reduction and the achieved performance is investigated. Finally, a summary of this chapter is provided by Sec. 2.6.

2.2 Analysis-Synthesis Filter-Banks

In this section, some design concepts for uniform and non-uniform analysis-synthesis filter-banks are briefly reviewed, which form the basis (and motivation) for our alternative low delay filter-bank design introduced in Sec. 2.3.

2.2.1 General Structure

The general structure of an *analysis-synthesis filter-bank* (AS FB) is shown in Fig. 2.1. The discrete, real input signal $y(n)$ is split into M subband signals $y_i(n)$ by analysis bandpass filters with impulse responses $h_i(n)$ for $i \in \{0, 1, \ldots, M-1\}$. These subband filters can have different bandwidths $\Delta\Omega_i$ to achieve a non-uniform frequency resolution. The limited bandwidth of the subband signals $y_i(n)$ allows to perform a downsampling. The subsampling rates R_i for each subband can be determined by the general rule

$$R_i \leq \frac{2\pi}{\Delta\Omega_i} \quad \text{for} \quad R_i \in \{1, 2, \ldots, M\} \quad \text{and} \quad \sum_{i=0}^{M-1} \Delta\Omega_i = 2\pi\,. \quad (2.1)$$

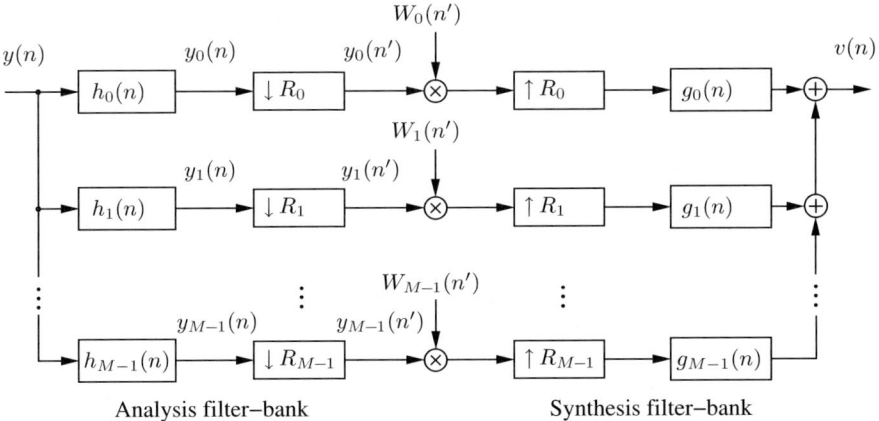

Fig. 2.1. M-channel analysis-synthesis filter-bank (AS FB) with subsampling and spectral weighting.

For a *uniform* filter-bank, the bandwidths of all subband filters are equal and the same subsampling rate $R_i = R$ is taken for each subband signal. *Critical subsampling* is performed if $R = M$. An *oversampled* filter-bank performs non-critical subsampling where $R < M$.

The signal reconstruction is accomplished by the synthesis filter-bank, which consists of the upsampling operations and interpolating bandpass filters with impulse responses $g_i(n)$. A filter-bank achieves *perfect (signal) reconstruction* (PR) with a delay of d_0 samples if

$$v(n) = y(n - d_0) \qquad (2.2)$$

for $W_i = 1 \, \forall \, i$. Accordingly, near-perfect reconstruction (NPR) is achieved if this identity is approximately fulfilled.

AS FBs are commonly used for adaptive subband processing as indicated by Fig. 2.1. The spectral gain factors $W_i(n')$ are adapted at a reduced sampling rate based on the downsampled subband signals $y_i(n')$. For example, this filtering technique is frequently used for the enhancement of noisy speech signals, e.g., [31, 80, 81].

2.2.2 Tree-Structured Filter-Banks

Tree-structured filter-banks are used to achieve a uniform or, more commonly, a non-uniform time-frequency resolution. They are mostly realized by the discrete wavelet transform (DWT) or by quadrature mirror filters (QMFs), e.g., [7, 77, 83]. Tree-structured filter-banks can realize an octave-band frequency analysis as depicted in Fig. 2.2. The input signal is split into a lowpass (LP) and highpass (HP) signal which can be each downsampled by a ratio of two. This step can be repeated successively until the desired frequency resolution is (approximately) achieved. This procedure leads to different signal delays for the subband signals which can be compensated by corresponding

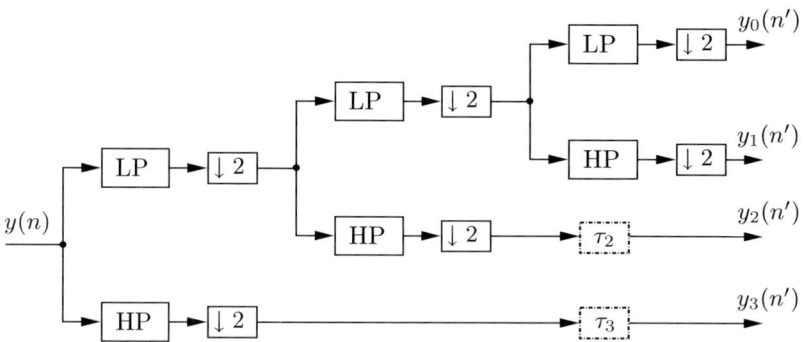

Fig. 2.2. Tree-structured filter-bank with three stages realizing an octave-band analysis.

delay elements. A more flexible adjustment of the frequency resolution can be accomplished by the wavelet packet decomposition, e.g., [7]. The signal delay of a tree-structured AS FB is equal to $(2^J - 1)\, d_\mathrm{s}$ with J marking the number of stages and d_s denoting the signal delay of the underlying two-channel AS FB. Hence, such filter-banks exhibit a high signal delay, especially if subband filters of high degrees are needed to avoid aliasing distortions owing to subband processing, cf. [9, 16, 26].

2.2.3 Modulated Filter-Banks

An important class of filter-banks constitute modulated AS FBs. The individual subband filters are derived by uniform modulation of a single prototype filter which yields a uniform time-frequency resolution, e.g., [12, 77]. The input-output relation for fixed spectral gain factors W_i and an input signal of finite energy can be written as [77]

$$V(z) = \frac{1}{R} \sum_{r=0}^{R-1} Y(z\, E_R^r) \sum_{i=0}^{M-1} H_i(z\, E_R^r)\, G_i(z)\, W_i \tag{2.3}$$

with the modulation factor defined by

$$E_R = e^{-j\frac{2\pi}{R}}. \tag{2.4}$$

Due to the subsampling operations, the AS FB is a time-variant system even for fixed gain factors W_i. To account for this behavior, we determine the overall transfer function of the filter-bank by a *series* of time-shifted unit sample sequences as input, i.e., $y(n) = \delta(n-d)$ with $d \in \mathbb{N}_0$. Inserting $Y(z) = z^{-d}$ into Eq. 2.3 leads to the transfer function

$$\begin{aligned}T_d(z) &= \frac{V(z)}{z^{-d}} \\ &= \underbrace{\frac{1}{R} \sum_{i=0}^{M-1} H_i(z)\, G_i(z)\, W_i}_{T_{\mathrm{lin}}(z)} + \underbrace{\frac{1}{R} \sum_{r=1}^{R-1} E_R^{-d\,r} \sum_{i=0}^{M-1} H_i(z\, E_R^r)\, G_i(z)\, W_i}_{\mathcal{E}_\mathrm{A}(z)}.\end{aligned} \tag{2.5}$$

The linear transfer function of the filter-bank is given by $T_{\mathrm{lin}}(z)$. The *aliasing distortions* due to the subsampling operations are represented by $\mathcal{E}_\mathrm{A}(z)$. The AS FB is a linear periodically time-variant (LPTV) system with period R since

$$T_{d+R}(z) = T_d(z) \tag{2.6}$$

according to Eq. 2.5. Therefore, a *linear time-invariant* (LTI) system is obtained if $\mathcal{E}_A(z) = 0$, so that no aliasing distortions occur. Perfect reconstruction according to Eq. 2.2 is achieved, if the transfer function of Eq. 2.5 is given by

$$T_d(z) = z^{-d_0} \quad \text{for} \quad d \in \{0, 1, \ldots, R-1\} \quad \text{and} \quad W_i = 1 \, \forall \, i \qquad (2.7)$$

where the limited set of values for d follows from Eq. 2.6. An example are paraunitary filter-banks which fulfill this condition with a delay of $d_0 = L$ samples where L denotes the degree of the FIR prototype filters [77].[1]

A filter-bank with perfect reconstruction ensures complete aliasing cancellation only if no subband processing is performed, that is, $W_i = 1$ for Eq. 2.5. Therefore, oversampled filter-banks are commonly used for adaptive subband filtering to avoid strong aliasing distortions in consequence of spectral weighting, e.g., [4, 19]. In contrast, critically subsampled AS FBs are a typical choice for subband coding systems, e.g., [77, 83].

An important realization of a (complex) modulated filter-bank is given by the *DFT filter-bank*. The subband filters have the transfer functions

$$H_i(z) = \sum_{l=0}^{L} h(l) \, E_M^{i\,l} \, z^{-l} \qquad (2.8)$$

$$G_i(z) = \sum_{l=0}^{L} g(l) \, E_M^{i\,(l+1)} \, z^{-l} \; ; \quad i \in \{0, 1, \ldots, M-1\} \qquad (2.9)$$

where $h(n)$ and $g(n)$ denote the finite impulse responses (FIRs) of the analysis and synthesis prototype filter, respectively. The use of linear-phase prototype filters leads to a signal delay of $d_0 = L$ samples, cf. [77].

A common choice for the FIR filter degree is $L = M-1$, but a higher degree can be taken to increase the frequency selectivity of the subband filters. Such a filter-bank can be efficiently implemented by a *polyphase network* (PPN). The analysis filters of Eq. 2.8 can be written as

$$H_i(z) = \sum_{\lambda=0}^{M-1} H_\lambda^{(M)}(z^M) \cdot z^{-\lambda} \cdot E_M^{\lambda \, i} \; ; \quad i \in \{0, 1, \ldots, M-1\} \qquad (2.10)$$

with the 'type 1' polyphase components defined by [77]

$$H_\lambda^{(M)}(z) = \sum_{m=0}^{l_M - 1} h(m\,M + \lambda) \, z^{-m} \; ; \quad \lambda \in \{0, 1, \ldots, M-1\} \, . \qquad (2.11)$$

It is assumed that the length of the prototype filters is an integer multiple of M according to

---

[1] One property of paraunitary filter-banks is that the sum of the subband energies is equal to the energy of the input signal.

$$L+1 = l_M M \; ; \; l_M \in \mathbb{N} \tag{2.12}$$

which can be always achieved by an appropriate zero-padding. The synthesis filters of Eq. 2.9 can be expressed by

$$G_i(z) = \sum_{\lambda=0}^{M-1} G_\lambda^{(M)}(z^M) \cdot z^{-(M-1-\lambda)} \cdot E_M^{-\lambda i} \; ; \; i \in \{0, 1, \ldots, M-1\} \tag{2.13}$$

with the 'type 2' polyphase components

$$G_\lambda^{(M)}(z) = \sum_{m=0}^{l_M-1} g\bigl((m+1)M - \lambda - 1\bigr) z^{-m} \; ; \; \lambda \in \{0, 1, \ldots, M-1\}. \tag{2.14}$$

Fig. 2.3 shows the derived PPN realization of a DFT filter-bank. The subsampling operations can be moved towards the delay elements due to the so-called 'noble identities', cf. [77]. The discrete Fourier transform (DFT) can be computed efficiently by the fast Fourier transform (FFT), e.g., [59]. Hence, this PPN filter-bank implementation possesses only a low computational complexity. The processing scheme of Fig. 2.3 can also be interpreted as weighted

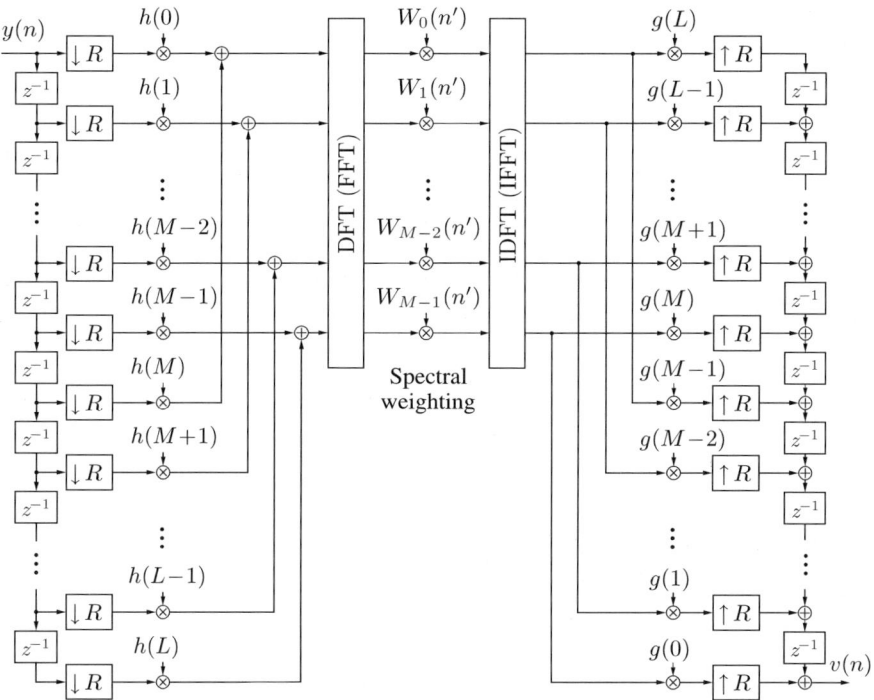

Fig. 2.3. Polyphase network (PPN) realization of a DFT AS FB for $L+1 = 2M$.

overlap-add method [11,12]. The delay chains buffer the samples of the input and output frames whose overlap is determined by the subsampling rate R.

For speech enhancement applications, the time-varying spectral gain factors $W_i(n')$ of the filter-bank might cause audible artifacts due to so-called 'block-edge effects'. This can be avoided by non-critical subsampling ($R < M$) and a dedicated prototype filter design, e.g., [31,81].

2.2.4 Frequency Warped Filter-Banks

2.2.4.1 Principle

A frequency warped digital system can be obtained by replacing the delay elements of its transfer function by allpass filters

$$z^{-1} \to H_\mathrm{A}(z) \,. \tag{2.15}$$

For this *allpass transformation*, a causal, complex allpass filter of first order is considered, whose transfer function is given by

$$H_\mathrm{A}(z) = \frac{z^{-1} - a^*}{1 - a\,z^{-1}}\,;\quad |a| < |z|\,;\quad |a| < 1\,;\quad a = \alpha\,e^{j\gamma} \in \mathbb{C}\,;\quad \alpha, \gamma \in \mathbb{R} \tag{2.16}$$

with $\mathbb{C}$ marking the set of all complex numbers and $\mathbb{R}$ marking the set of all real numbers. The asterisk denotes the complex-conjugate value. One possible implementation of this allpass filter is shown in Fig. 2.4. The frequency response reads

$$H_\mathrm{A}(z = e^{j\Omega}) = \frac{e^{-j\Omega} - a^*}{1 - a\,e^{-j\Omega}} = e^{-j\varphi_a(\Omega)} \tag{2.17a}$$

$$\varphi_a(\Omega) = 2\arctan\left(\frac{\sin\Omega - \alpha\sin\gamma}{\cos\Omega - \alpha\cos\gamma}\right) - \Omega \,. \tag{2.17b}$$

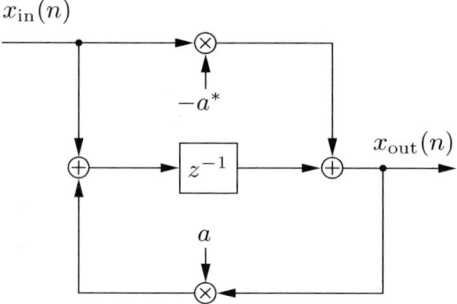

Fig. 2.4. Realization of an allpass filter of first order.

This concept can be extended to allpass filters of higher order [35, 36], but the allpass transformation of first order is of most interest here.[2]

The allpass transformation is a bilinear transformation and allows to alter the frequency characteristic of a digital system without changing its coefficients. This property is exploited for the design of variable digital filters [10, 70, 71]. The cutoff frequency of these filters is adjusted by the allpass coefficient, whereas the shape of the frequency response (e.g., number and magnitude of the ripples or the stopband attenuation) is not changed. The allpass transformation can also be employed to perform short-term spectral analysis with a non-uniform frequency resolution or to construct non-uniform digital filter-banks [6, 17, 58, 79].

The allpass transformation of the analysis filters of Eq. 2.8 yields the warped frequency responses

$$H_i(z = e^{j\varphi_a(\Omega)}) = \sum_{l=0}^{L} h(l)\, E_M^{i\,l}\, e^{-j\varphi_a(\Omega)\,l} \tag{2.18}$$

$$= \widetilde{H}_i(e^{j\Omega})\,; \quad i \in \{0, 1, \ldots, M-1\} \tag{2.19}$$

due to Eq. 2.15 and Eq. 2.17. Hence, the allpass transformed filter-bank is a generalization of the uniform filter-bank, which is included as special case for $a = 0$. The allpass transformation causes a *frequency warping*

$$\Omega \to \varphi_a(\Omega) \tag{2.20}$$

where the course of the phase response $\varphi_a(\Omega)$ is determined by the allpass coefficient a. The effect of this allpass transformation is demonstrated in Fig. 2.5. For a real and positive allpass coefficient $a = \alpha > 0$, a higher frequency resolution is achieved for the lower frequency bands and vice versa for the higher frequency bands. The opposite applies if $\alpha < 0$. Thus, the frequency resolution can be adjusted by a single coefficient without the requirement for an individual subband filter design, which is sometimes needed for the construction of non-uniform filter-banks (cf. Sec. 2.2.5). A complex allpass transformation is of interest, if a more flexible adjustment of the frequency resolution is desired, cf. [35].

The allpass transformation allows to design a non-uniform filter-bank whose frequency bands approximate the *Bark frequency scale* with great accuracy [73]. The frequency resolution of the human auditory system is determined by the so-called 'critical bands'. The mapping between frequency and critical bands can be described by the critical band rate with the unit 'Bark'. An analytical expression for the Bark frequency scale is given by [84].

---

[2] An allpass transformation of order N maps the unit circle N-times onto itself which causes a comb filter structure. This comb filter effect is undesirable for the design of filter-banks and can be avoided by additional allpass filters at the price of an increased algorithmic complexity and a higher signal delay, cf. [35, 36].

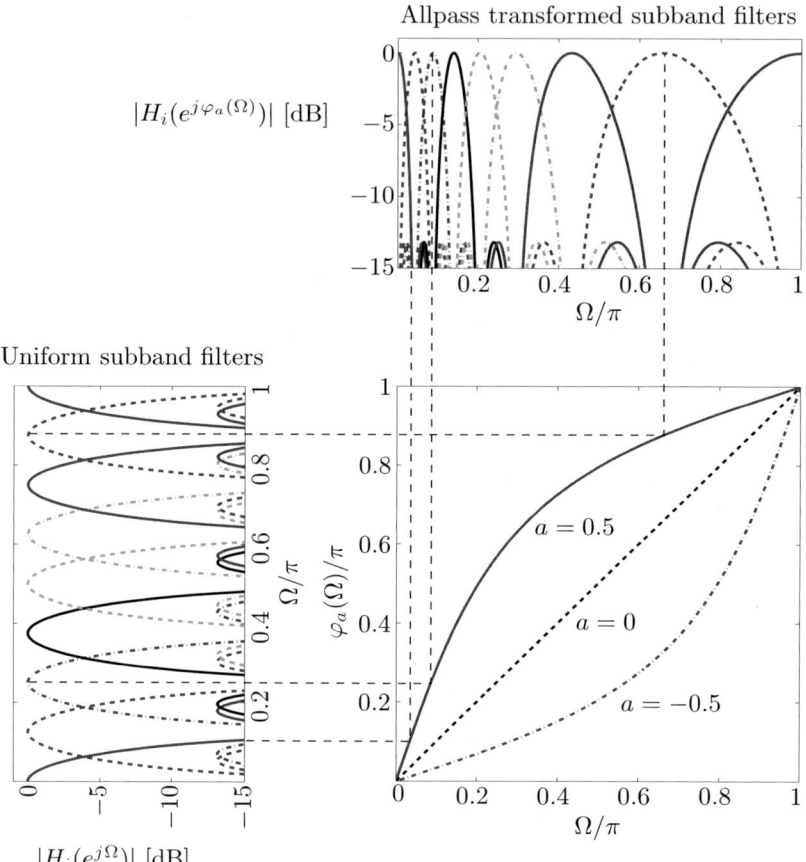

Fig. 2.5. Allpass transformation of subband filters for $M = 16$ and a rectangular prototype filter of length $L + 1 = M$.

$$\frac{\xi(f)}{\text{Bark}} = 13 \cdot \arctan\left(\frac{0.76\,f}{\text{kHz}}\right) + 3.5 \cdot \arctan\left(\left(\frac{f}{7.5\,\text{kHz}}\right)^2\right). \quad (2.21)$$

Fig. 2.6 illustrates that such a frequency division can be well approximated by means of an allpass transformed filter-bank. A filter-bank with approximately Bark-scaled frequency bands can also be realized by the wavelet packet decomposition [9]. However, the obtained tree-structured filter-bank has a significantly higher signal delay and a higher algorithmic complexity than a comparable allpass transformed filter-bank.

2.2.4.2 Signal Reconstruction

The allpass transformation of the analysis filter-bank according to Eq. 2.18 leads to phase modifications and a stronger overlap of aliasing components in

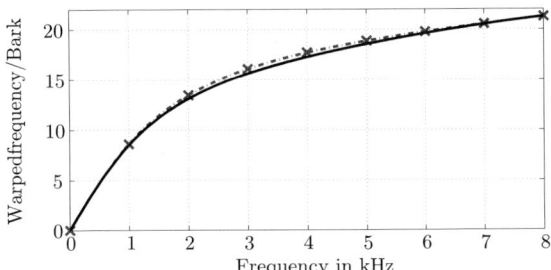

Fig. 2.6. Approximation of the Bark frequency scale: The solid line corresponds to the analytical expression of Eq. 2.21. The dashed line marks the frequency warping for an allpass transformation with $a = 0.576$ and a sampling frequency of 16 kHz.

comparison to the uniform filter-bank. These effects complicate the synthesis filter-bank design. The two main approaches to perform the signal reconstruction in this case are depicted in Fig. 2.7.

The filter-bank structure I uses $L + 1 = M$ filters with transfer functions $P(z, l)$ for $l \in \{0, 1, \ldots, L\}$ to compensate the (additional) phase and aliasing distortions caused by the allpass transformed analysis filter-bank. Perfect reconstruction can be achieved by FIR filters. Their coefficients can be determined by analytical closed-form expressions in case of a prototype filter length of $L + 1 = M$. This is shown in [72] for critical subsampling and in [21, 35] for arbitrary subsampling rates. However, the obtained synthesis subband filters show no distinctive bandpass characteristic. This causes a high reconstruction error if spectral modifications of the subband signals, such as quantization or spectral weighting, are performed [21].

Synthesis subband filters with a distinctive bandpass characteristic are obtained by the filter-bank designs proposed, e.g., in [23, 24, 49] which achieve near-perfect reconstruction. The filters with transfer functions $P(z, l)$ are designed to compensate the phase distortions due to the frequency warping.[3] The aliasing distortions are limited by the higher stopband attenuation of longer subband filters ($L + 1 \gg M$) and a lower subsampling rate R.[4]

A similar principle is used for the synthesis filter-bank structure II of Fig. 2.7, which, however, uses only a single compensation filter with transfer function $P(z, L)$. The allpass transformation is applied to the analysis and synthesis filter-bank, i.e., all delay elements of the uniform filter-bank (shown in Fig. 2.3) are replaced by allpass filters according to Eq. 2.15. If the uniform

---

[3] If not mentioned otherwise, the more general concept of frequency warping will always refer to an allpass transformation of first order so that both terms are used interchangeably.

[4] Here, the same subsampling rate R is used for each subband signal so that the DFT and IDFT can be executed at a decimated sampling rate. This is not possible for different subsampling rates according to Eq. 2.1.

(a) Synthesis filter-bank structure I

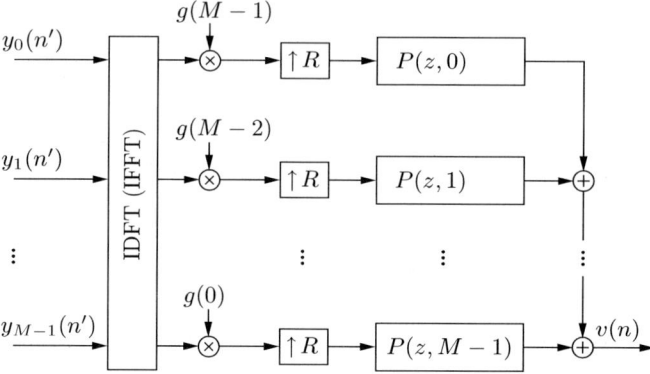

(b) Synthesis filter-bank structure II

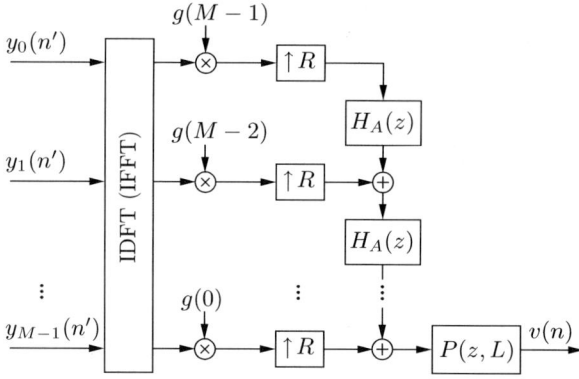

Fig. 2.7. Synthesis filter-bank structures for an allpass transformed analysis filter-bank with prototype filter of length $L + 1 = M$.

filter-bank fulfills Eq. 2.7 with $d_0 = L$, the allpass transformation leads to the frequency response

$$\widetilde{T}_d(z = e^{j\Omega}) = e^{-j\varphi_a(\Omega)L} + \mathcal{E}_A(e^{j\varphi_a(\Omega)}). \tag{2.22}$$

The aliasing distortions $\mathcal{E}_A(e^{j\varphi_a(\Omega)})$ emerge due to the non-uniform bandwidths of the allpass transformed subband filters. They can be reduced by a lower subsampling rate R and the use of subband filters of higher degrees having narrow transition bands and high stopband attenuations.

The non-linear phase term $\varphi_a(\Omega)\,L$ can cause audible distortions especially for a high prototype filter degree L. Eq. 2.17 reveals that this term corresponds to the frequency response of a cascade of L identical allpass filters termed as allpass (filter) chain. The task of the fixed *phase equalizer* at the filter-bank output is to compensate these phase distortions (see Fig. 2.7-b). The frequency response of the phase equalizer has to fulfill the general requirement, cf. [48]

$$e^{-j d_P \varphi_a(\Omega)} \cdot P_{\text{gen}}(e^{j\Omega}, d_P) \stackrel{!}{=} e^{-j \tau_P \Omega} ; \quad \tau_P \geq 0 ; \quad \tau_P \in \mathbb{R} ; \quad d_P \in \mathbb{N} \quad (2.23)$$

where $d_P = L$ for the warped AS FB described by Eq. 2.22.

The 'ideal' phase equalizer to fulfill Eq. 2.23 with $\tau_P = 0$ is obviously given by the inverse transfer function of an allpass chain

$$P_{\text{ideal}}(z, d_P) = H_A(z)^{-d_P} ; \quad |z| < \frac{1}{|a|} . \quad (2.24)$$

However, the impulse response of this phase equalizer is infinite and anti-causal (for $a \neq 0$), i.e., $p_{\text{ideal}}(n, d_P) = 0$ for $n > 0$. An approach to realize anti-causal filters is to buffer the input samples in order to process them in time-reversed order [14, 56]. This rather complex technique requires large buffers and leads to a high signal delay.

An alternative approach is to approximate the desired anti-causal phase equalizer of Eq. 2.24 by a causal FIR filter of degree N_P. Its coefficients can be obtained by shifting and truncating the impulse response $p_{\text{ideal}}(n, d_P)$ according to

$$p_{\text{LS}}(n, d_P) = \begin{cases} p_{\text{ideal}}(n - N_P, d_P) & ; \quad n \in \{0, 1, \ldots, N_P\} \\ 0 & ; \quad \text{else.} \end{cases} \quad (2.25)$$

The transfer function of an inverse allpass chain $H_A(z)^{-d_P}$ is identical to the para-conjugate transfer function of the allpass chain where the z-variable is replaced by z^{-1} and the complex-conjugate filter coefficients are used. Thus, the impulse response of the ideal phase equalizer $p_{\text{ideal}}(n, d_P)$ can be obtained by the time-reversed impulse response of an allpass chain of length d_P with complex-conjugate allpass coefficient a^*. The FIR filter approximation of an IIR filter by truncating its impulse response leads to a least-squares error, e.g., [60]. Thus, the phase equalizer according to Eq. 2.25 is termed as least-squares (LS) phase equalizer. For a complex allpass transformation, a complex output signal $v(n)$ is obtained where the (discarded) imaginary part becomes negligible for a low signal reconstruction error.

A drawback of FIR phase equalizers is that they cause significant magnitude distortions in case of a low filter degree N_P. Hence, this filter degree of the phase equalizer must be high enough to keep phase and magnitude distortions low. Such magnitude distortions are avoided by using an allpass phase equalizer. Its filter degree is determined only by the need to keep the phase distortions due to the warping inaudible. The design of phase equalizers for warped filter-banks is treated in [48] in more detail. It should be noted that the discussed synthesis filter-bank designs for near-perfect reconstruction can also be applied if the prototype filter length $L + 1$ exceeds M.

The signal delay of the considered uniform AS FB is given by $d_0 = L$. The signal delay of the warped filter-bank with LS FIR phase equalizer according to Eq. 2.25 is approximately equal to N_P. This filter degree of the phase equalizer should be considerably higher than the value $d_P = d_0$ so that the warped AS FB with phase equalizer has a significantly higher overall signal delay than the uniform filter-bank. As shown in [49], it is also beneficial to use the LS phase equalizer of Eq. 2.25 for the filter-bank structure I in Fig. 2.7, which leads to (almost) the same signal delay as for the filter-bank structure II with LS phase equalizer.

The devised concepts for phase equalization are also effective if spectral weighting is performed, cf. [48]. For speech and audio processing, a perfect equalization of the warped phase is not required due to the insensitivity of human hearing towards minor phase distortions, cf. [80,84]. A design example for an allpass transformed AS FB is given later in Sec. 2.5.

2.2.5 Low Delay Filter-Banks

One approach for the design of uniform and non-uniform AS FBs with low delay is to use the *lifting scheme*, which has been originally proposed for the construction of 'second generation wavelets' [15,76]. A single zero-delay lifting step is shown in Fig. 2.8. The new analysis lowpass filter after one lifting step is given by

$$H_0^{(1)}(z) = H_0(z) + H_1(z) B(z^2) \,. \tag{2.26}$$

Correspondingly, this procedure can be applied to the analysis highpass filter termed as dual lifting step and so on. The lifting steps for the analysis filters are followed by inverse lifting steps at the synthesis side. By this, the degree of the subband filters is increased without increasing the overall signal delay of the filter-bank, cf. Fig. 2.8. The application of this scheme to the design of (uniform) cosine modulated AS FBs with low delay is proposed in [37,38]. In [23,24], the lifting scheme is applied to the allpass transformed AS FB. The higher aliasing distortions due to the frequency warping are reduced by improving the stopband attenuation of the subband filters. The lifting scheme is used to increase the filter degree while constraining the signal delay of the filter-bank. However, the adding of further lifting steps shows no improvement

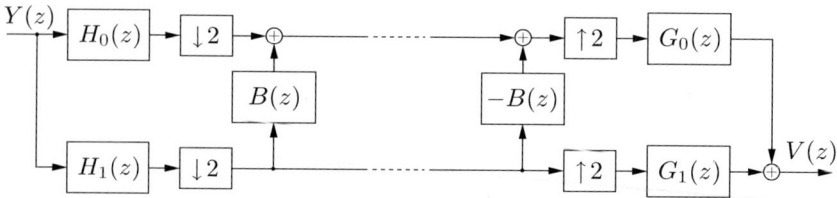

Fig. 2.8. Single zero-delay lifting step for a two-channel AS FB.

after some stages. Therefore, only a limited enhancement of the stopband attenuation and the associated aliasing cancellation can be achieved. Moreover, the analysis and synthesis filter-bank proposed in [23, 24] are operated at the non-decimated sampling rate, which causes a very high computational complexity.

The use of the *warped discrete Fourier transform* (WDFT) is an alternative approach to design a low delay filter-bank with warped frequency bands, e.g., [22, 51, 61]. The WDFT is a non-uniform DFT and calculated by the rule

$$\widetilde{Y}(i) = \sum_{n=0}^{M-1} y(n) \left(\frac{e^{-j\frac{2\pi}{M}i} - a^*}{1 - a e^{-j\frac{2\pi}{M}i}} \right)^n \; ; \; i \in \{0, 1, \ldots, M-1\}. \quad (2.27)$$

In contrast to the DFT (which is obtained for $a = 0$), the frequency points of the WDFT are non-uniformly spaced on the unit circle. The WDFT filter-bank evolves by replacing the (I)DFT in Fig. 2.3 by the (I)WDFT so that the signal delay remains the same. In this process, the center frequencies of the subband filters are shifted, but their bandwidth remain the same. This effect is illustrated in Fig. 2.9. In contrast to the allpass transformed filter-bank (see Fig. 2.5), the spectrum of the WDFT exhibits 'spectral gaps' due to the uniform bandwidths of the subband filters. This complicates the signal reconstruction, which is reflected by an ill-conditioned WDFT matrix for values of about $|a| > 0.2$ and $M > 40$, cf. [22, 61]. The numerical difficulties for the calculation of the inverse WDFT (matrix) become even more pronounced

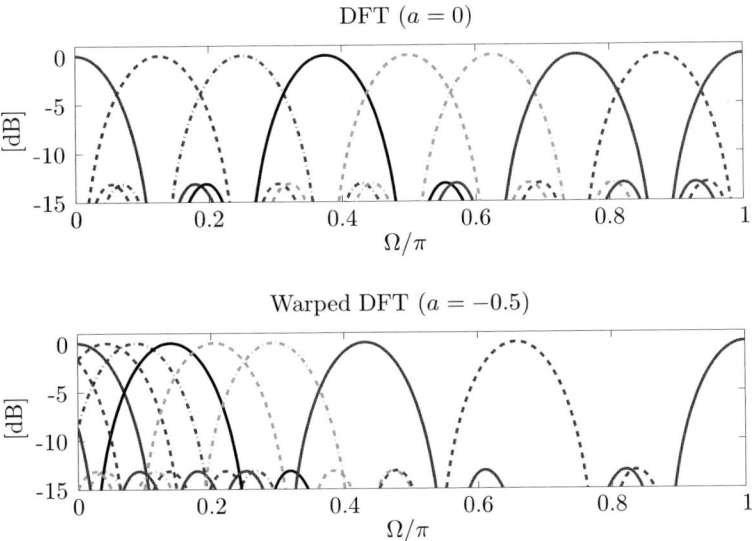

Fig. 2.9. Magnitude responses of the subband filters for a DFT and warped DFT with $M = 16$ channels.

with regard to a practical implementation with finite precision arithmetic, for instance, on a (fixed-point) digital signal processor (DSP).

There are other ways to derive a non-uniform low delay filter-bank from a uniform filter-bank. One approach is to combine an appropriate number of cosine modulated subband filters termed as 'feasible partitioning' [16, 43]. Another method is to use two different uniform filter-banks for the upper and lower frequency bands which are linked by a 'transition filter', e.g., [13, 18]. A good approximation of the Bark scale is difficult to achieve by this approach. The subband filters of these filter-banks need to have a relatively high filter degree to achieve a sufficient stopband attenuation in order to keep aliasing distortions low, especially if subband processing takes place. This causes still a comparatively high signal delay which depends, among others, on the permitted aliasing distortions.

Many designs of uniform and non-uniform filter-banks allow to prescribe an (almost) arbitrary signal delay, e.g., [16, 18, 40, 69]. However, it is problematic to achieve simultaneously a high stopband attenuation for the subband filters as well as a low signal delay. Hence, there is a trade-off between a low signal delay on the one hand and low aliasing distortions (high speech and audio quality) on the other hand, cf. [16].

A low signal delay *and* an aliasing-free signal reconstruction can be achieved by means of the *filter-bank summation method* (FBSM) depicted in Fig. 2.10. The FBSM can be derived from the filter-bank interpretation of the short-time DFT, e.g., [12]. A drawback of this filter-bank structure is its high computational complexity as no downsampling of the subband signals can be performed. Therefore, the AS FB is considered to be more suitable than the FBSM for real-world applications such as speech enhancement [19]. Moreover, the computational complexity of the FBSM is significantly increased, if we apply the allpass transformation to achieve a Bark-scaled frequency resolution.

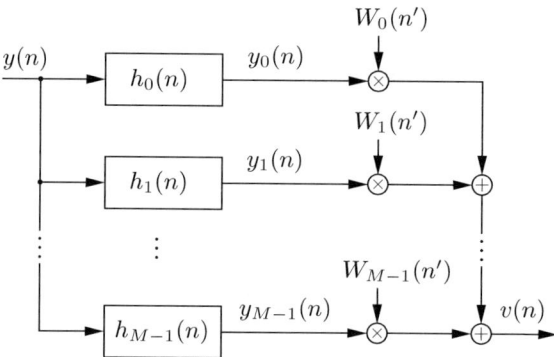

Fig. 2.10. Filter-bank summation method (FBSM) with time-varying spectral gain factors adapted at a reduced sampling rate.

In the following, we derive a uniform and warped low delay filter-bank without the high algorithmic complexity of the FBSM.

2.3 The Filter-Bank Equalizer

An alternative filter-bank concept to that of the conventional AS FB will be devised to perform adaptive subband processing with a significantly lower signal delay.

2.3.1 Concept

The FBSM of Fig. 2.10 is considered. As for the AS FB, the real input signal $y(n)$ is split into the subband signals $y_i(n)$ by means of M bandpass filters. The adaptation of the time-varying spectral gain factors $W_i(n')$ can be performed by the same algorithms as for the AS FB. This adaptation is based on the subband signals $y_i(n)$ and executed at intervals of R sample instants with n' defined by[5]

$$n' = \lfloor n/R \rfloor \cdot R \, ; \quad R \in \mathbb{N} \, . \tag{2.28}$$

The operation $\lfloor . \rfloor$ provides the greatest integer which is lower than or equal to the argument.

The impulse response $h_i(n)$ of the ith bandpass filter is obtained by modulation of a prototype lowpass filter with real impulse response $h(n)$ of length $L+1$ according to

$$h_i(n) = \begin{cases} h(n)\,\varPhi(i,n) & ; \; i \in \{0, 1, \ldots, M-1\} \, ; \; n \in \{0, 1, \ldots, L\} \\ 0 & ; \; \text{else} \, . \end{cases} \tag{2.29}$$

The choice for the general modulation sequence and the prototype filter affects the spectral selectivity and time-frequency resolution of the filter-bank. The modulation sequence $\varPhi(i,n)$ can be seen as transformation kernel of the filter-bank. In general, it has the periodicity

$$\varPhi(i, n + mM) = \varPhi(i, n)\,\rho(m) \, ; \quad m \in \mathbb{Z} \tag{2.30}$$

where $\mathbb{Z}$ denotes the set of all integer numbers. The sequence $\rho(m)$ depends on the chosen transform as shown later. For many transforms (including the DFT) it is given by $\rho(m) = 1 \; \forall m$.

The input-output relation for the FBSM of Fig. 2.10 can be written as

---

[5] This definition is more suitable for the following treatment than the common convention $n = Rn'$.

$$v(n) = \sum_{i=0}^{M-1} W_i(n') \, y_i(n) \qquad (2.31)$$

$$= \sum_{i=0}^{M-1} W_i(n') \sum_{l=0}^{L} y(n-l) \, h_i(l)$$

$$= \sum_{l=0}^{L} y(n-l) \, h(l) \sum_{i=0}^{M-1} W_i(n') \, \Phi(i,l) \qquad (2.32)$$

for the modulated bandpass filters of Eq. 2.29. The second summation is the spectral transform of the gain factors $W_i(n')$ which yields the coefficients

$$w_l(n') = \sum_{i=0}^{M-1} W_i(n') \, \Phi(i,l) \,; \quad l \in \{\, 0, 1, \ldots, L \,\} \qquad (2.33)$$

$$= \mathcal{T}\{\, W_i(n') \,\} \,. \qquad (2.34)$$

These $L+1$ *time-domain weighting factors* have the periodicity

$$w_{l+mM}(n') = w_l(n') \, \rho(m) \qquad (2.35)$$

due to Eq. 2.30 and Eq. 2.33. The input-output relation finally reads

$$v(n) = \sum_{l=0}^{L} y(n-l) \, h(l) \, w_l(n') \qquad (2.36)$$

$$= \sum_{l=0}^{L} y(n-l) \, h_s(l, n') \,. \qquad (2.37)$$

The obtained filter-bank structure is a *single* time-domain filter whose coefficients

$$h_s(l, n') = h(l) \, w_l(n') \,; \quad l \in \{\, 0, 1, \ldots, L \,\} \qquad (2.38)$$

are the product of the fixed impulse response $h(n)$ of the prototype lowpass filter and the time-varying weighting factors $w_l(n')$ adapted in the short-term spectral-domain.[6] This efficient implementation of the FBSM (which resembles a filter-bank used as equalizer) is termed as *filter-bank equalizer* (FBE) [44, 82]. A sketch of this filter-bank structure is given in Fig. 2.11. A distinctive advantage in comparison to the AS FB is that the output signal $v(n)$ is not affected by aliasing distortions. Moreover, a non-uniform (warped) frequency resolution can be achieved by means of the allpass transformation with lower efforts than for the AS FB as shown later in Sec. 2.3.6. The uniform

---

[6] For the sake of clarity, the index l instead of the discrete time index n will be used to indicate that $L+1$ filter coefficients are considered.

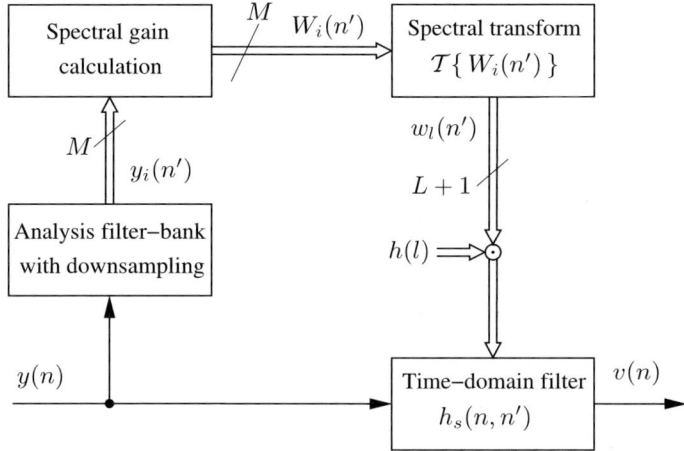

Fig. 2.11. Filter-bank equalizer (FBE) for adaptive subband filtering.

and non-uniform FBE can be used for speech and audio processing with low signal delay, e.g., [45, 68].

A similar approach has been proposed independently in [39] for dynamic-range compression in hearing aids. For acoustic echo cancellation and active noise control applications, a related time-domain filtering approach can be found in [57], where the coefficients are adapted in the uniform frequency-domain. However, the following treatment will show that the concept of the FBE is a much more general and comprehensive low delay filter-bank concept.

2.3.2 Prototype Filter Design

The objective of the prototype lowpass filter design is to achieve perfect reconstruction according to Eq. 2.2. The FBE meets this condition if the following two requirements are fulfilled [44]: Firstly, the general modulation sequence of Eq. 2.29 must have the property

$$\sum_{i=0}^{M-1} \Phi(i,n) = \begin{cases} C & ; \ C \neq 0 \,; \ n = n_0 \\ 0 & ; \ n \neq n_0 \end{cases} \quad \text{for } n, n_0 \in \{0, 1, \ldots, M-1\}. \tag{2.39}$$

Secondly, a generalized Mth-band filter with impulse response

$$h(n) = \begin{cases} \dfrac{1}{C\,\rho(m_c)} & ; \ n = n_0 + m_c M \,; \ \rho(m_c) \neq 0 \,; \ m_c \in \mathbb{Z} \\ 0 & ; \ n = n_0 + m M \,; \ m \in \mathbb{Z}\backslash\{m_c\} \\ \text{arbitrary} & ; \ \text{else} \end{cases} \tag{2.40}$$

is needed as prototype lowpass filter. Such a filter has equidistant zeros at intervals of M samples and its modulated versions add up to a delay according to

$$\sum_{i=0}^{M-1} H\left(z\, E_M^i\right) \cdot E_M^{i\, n_0} = \frac{M}{C\,\rho(m_c)}\, z^{-(n_0+m_c\, M)}. \qquad (2.41)$$

The conditions of Eq. 2.39 and Eq. 2.40 can be easily met to achieve perfect reconstruction with a delay of

$$d_0 = n_0 + m_c\, M \qquad (2.42)$$

samples. A suitable Mth-band filter according to Eq. 2.40 is given by

$$h(n) = \frac{1}{C\,\rho(m_c)}\, \frac{\sin\left(\frac{2\pi}{M}(n-d_0)\right)}{\frac{2\pi}{M}(n-d_0)}\, \mathrm{win}_L(n) \qquad (2.43)$$

with the general window sequence defined by

$$\mathrm{win}_L(n) = \begin{cases} \text{arbitrary} & ;\ 0 \le n \le L \\ 0 & ;\ \text{else}. \end{cases} \qquad (2.44)$$

A rectangular window achieves a least-squares approximation error, but other window sequences are often preferred to influence properties of the filter such as transition bandwidth or sidelobe attenuation [59]. Commonly used window sequences are the Kaiser window or the parametric window sequence

$$\mathrm{win}_L(n, \beta) = \begin{cases} \beta + (\beta-1)\cos\left(\frac{2\pi}{L} n\right) & ;\ 0 \le n \le L;\ 0.5 \le \beta \le 1 \\ 0 & ;\ \text{else}. \end{cases} \qquad (2.45)$$

The rectangular window ($\beta = 1$), the Hann window ($\beta = 0.5$), and the Hamming window ($\beta = 0.54$) are included as special cases [63].

The condition of Eq. 2.39 is met, among others, by the Walsh and Hadamard transform (cf. [2]) as well as the *generalized discrete Fourier transform* (GDFT). The transformation kernel of the GDFT reads

$$\Phi_{\mathrm{GDFT}}(i, n) = e^{-j\frac{2\pi}{M}(i-i_0)(n-n_0)} \qquad (2.46)$$

$$n, n_0 \in \mathbb{Z};\ i \in \{0, 1, \ldots, M-1\};\ i_0 \in \{0, 1/2\}$$

where Eq. 2.30 applies with $\rho(m) = (-1)^{2\, i_0\, m}$. The DFT is included as special case for $n_0 = i_0 = 0$. For $i_0 = 1/2$, a GDFT filter-bank with oddly-stacked frequency bands is obtained, cf. [12]. A value of $i_0 = 0$ leads to the evenly-stacked GDFT where the above equations apply with $\rho(m) = 1$ and $C = M$.

The Walsh and Hadamard transform are employed, among others, for image processing, cf. [25]. The evenly-stacked GDFT is of interest for speech and audio processing. Thus, this filter-bank type is considered primarily in the following without loss of generality.

2.3.3 Relation between GDFT and GDCT

For speech enhancement, the time-varying spectral gain factors $W_i(n')$ are often calculated by means of a spectral speech estimator, e.g., [5, 20, 50, 52]. For a DFT-based adaptation, the gain factors have the properties

$$\epsilon \leq W_i(n') \leq 1; \quad W_i(n') \in \mathbb{R}; \quad 0 \leq \epsilon < 1 \tag{2.47}$$

and possess the symmetry

$$W_i(n') = W_{M-i}(n'); \quad i \in \{0, 1, \ldots, M-1\}; \quad M \text{ even} \tag{2.48}$$

as the input sequence $y(n)$ is real. The limitation of the gain factors by a lower (often time-varying) threshold ϵ is favorable to avoid unnatural sounding artifacts such as musical noise, cf. [8]. The (I)DFT of the real gain factors of Eq. 2.47 yields time-domain weighting factors $w_l(n')$ corresponding to a (non-causal) zero-phase filter. A linear phase response is obtained for the considered (evenly-stacked) GDFT of Eq. 2.46 with $n_0 = L/2$ and L being even so that the coefficients exhibit the symmetry

$$w_l(n') = w_{L-l}(n'); \quad l \in \{0, 1, \ldots, L\}. \tag{2.49}$$

If the used prototype filter has the same symmetry

$$h(l) = h(L-l), \tag{2.50}$$

the time-varying FIR filter of Eq. 2.38 is given by

$$h_s(l, n') = h_s(L-l, n'); \quad l \in \{0, 1, \ldots, L\} \tag{2.51}$$

which implies a *linear phase response*. The GDFT of the gain factors $W_i(n')$ can be computed by the FFT with a subsequent cyclic shift of the obtained time-domain weighting factors by n_0 samples. Instead of the GDFT analysis filter-bank, the DFT filter-bank can be used for the FBE (see Fig. 2.11), because the magnitude of the subband signals is needed only for the calculation of the spectral gain factors.

For the considered GDFT, the weighting factors of Eq. 2.33 are given by

$$w_l(n') = \sum_{i=0}^{M-1} W_i(n') e^{-j\frac{2\pi}{M}i(l-n_0)}; \quad l \in \{0, 1, \ldots, L\}. \tag{2.52}$$

The substitution $M = 2N$ and exploiting the symmetry of Eq. 2.48 allows the following conversion

$$w_l(n') = \sum_{i=0}^{2N-1} W_i(n') e^{-j\frac{2\pi}{2N} i (l-n_0)}$$

$$= W_0(n') + \sum_{i=1}^{N-1} W_i(n') e^{-j\frac{\pi}{N} i (l-n_0)} + W_N(n') (-1)^{l-n_0}$$

$$+ \sum_{i=1}^{N-1} W_{2N-i}(n') e^{-j\frac{\pi}{N} (2N-i)(l-n_0)}$$

$$= \sum_{i=0}^{N} W_i(n') \nu(i) \cos\left(\frac{\pi}{N} i (l-n_0)\right) \quad (2.53a)$$

with

$$\nu(i) = \begin{cases} 1 & ; \; i \in \{0, N\} \\ 2 & ; \; i \in \{1, 2, \ldots, N-1\} \end{cases} \quad (2.53b)$$

Eq. 2.53 represents a FBE with $N+1$ channels and the (evenly-stacked) *generalized discrete cosine transform* (GDCT) as modulation sequence[7]

$$\Phi_{\text{GDCT}}(n,i) = \nu(i) \cos\left(\frac{\pi}{N} i (n-n_0)\right) \; ; \; i \in \{0,1,\ldots,N\} \; ; \; n, n_0 \in \mathbb{Z}. \quad (2.54)$$

For this transformation kernel, the condition of Eq. 2.39 is fulfilled with $M = N+1$ and $C = 2N$.

The relation between GDCT and GDFT FBE has been derived so far without considering the process of the spectral gain calculation. For noise reduction, the spectral gain factors are mostly calculated as (linear or non-linear) functions of the squared magnitude of the subband signals (spectral coefficients), cf. [4]. This can be expressed by the notation

$$W_i(n') = f\left(\overline{|y_i(n')|^2}\right) \; ; \; i \in \{0,1,\ldots,N\}. \quad (2.55)$$

Only $N+1$ gain factors must be calculated due to the symmetry of Eq. 2.48. The bar indicates that an averaged value (short-term expectation) is mostly taken inherently. Examples are the calculation of the *a priori* SNR by the decision-directed approach [20], or the estimation of the noise power spectral density by recursively smoothed periodograms [54]. The subband signals are complex for the (G)DFT so that

$$W_i(n') = f\left(\overline{\text{Re}\{y_i(n')\}^2} + \overline{\text{Im}\{y_i(n')\}^2}\right). \quad (2.56)$$

---

[7] Except for a normalization factor, the DCT-I is obtained for $n_0 = 0$, cf. [66]. For the oddly-stacked GDFT FBE ($i_0 = 1/2$), a similar derivation leads to a modulation sequence which includes the DCT-II for $n_0 = 0$.

It can be assumed that the real and imaginary part are uncorrelated and that both have equal variances and equal probability density functions (PDFs), e.g., [4]. Therefore, almost the same gain factors are obtained by considering the real part of the subband signals only

$$W_i(n') \approx f\left(2\,\overline{\operatorname{Re}\{y_i(n')\}^2}\right) \qquad (2.57)$$

due to the averaging process. Hence, the gain factors calculated by complex DFT values are almost equal to those computed by real DCT values. Accordingly, the replacement of the GDFT of Eq. 2.46 by the GDCT of Eq. 2.54 causes no noticeable differences for the speech enhanced by the FBE.[8]

2.3.4 Realization for Different Filter Structures

The choice of the filter structure plays an important role for digital filter implementations with finite precision arithmetic as well as for *time-varying* filters. Here, only the direct forms of a filter are of interest as they do not require an involved conversion of the time-varying filter coefficients $h_s(l, n')$ such as the parallel form or the cascade form, cf. [59].

The realization of an FIR filter by means of the direct form and transposed direct form is shown in Fig. 2.12. The input-output relations for these two filter forms are given by

$$v_{\text{df}}(n) = \sum_{l=0}^{L} y(n-l)\, h_s(l, n') \qquad (2.58)$$

$$v_{\text{tdf}}(n) = \sum_{l=0}^{L} y(n-l)\, h_s(l, n'-l)\,. \qquad (2.59)$$

Obviously, the derived FBE according to Eq. 2.37 uses a time-domain filter in the *direct form*.

The input-output relation for the transposed direct form is obtained by inserting Eq. 2.38 into Eq. 2.59 so that

$$\begin{aligned}
v_{\text{tdf}}(n) &= \sum_{l=0}^{L} y(n-l)\, h(l)\, w_l(n'-l) \\
&= \sum_{l=0}^{L} y(n-l)\, h(l) \sum_{i=0}^{M-1} W_i(n'-l)\, \Phi(i,l) \\
&= \sum_{i=0}^{M-1} \sum_{l=0}^{L} y(n-l)\, W_i(n'-l)\, h_i(l) \qquad (2.60)
\end{aligned}$$

---

[8] In [19], a different comparison between DFT AS FB and DCT-II AS FB has revealed a slightly lower noise suppression for the DCT AS FB.

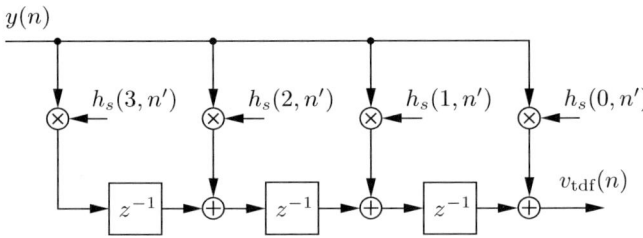

Fig. 2.12. Direct form implementations of a time-varying FIR filter with degree $L = 3$.

due to Eq. 2.33 and Eq. 2.29. The obtained relation for the *transposed direct form* corresponds to the FBSM of Fig. 2.10 with the important difference that the spectral gain factors are now applied *before* the subband filters. Fig. 2.13 shows the derived filter-bank structure. The dash-dotted boxes mark delay elements to account for the signal delay τ_a due to the analysis filter-bank and gain calculation. These delay elements might be omitted for moderately time-varying (smoothed) gain factors to avoid an additional signal delay.

Switching the coefficients of a digital filter during operation leads to transients which can cause 'filter-ringing' effects.[9] These effects might be perceived by perceptually annoying artifacts. The application to noise reduction revealed that the FBE with time-domain filter in transposed direct form yields a better perceived speech quality than the implementation with the direct form filter. This can be explained by comparing the equivalent FBSMs of Fig. 2.10 and Fig. 2.13: For the transposed direct form, the transients caused by the switching gain factors are smoothed by the following bandpass filters, which is not the case for the direct form implementation.

An alternative method to smooth the FIR filter coefficients independently of the filter form is to perform a kind of 'cross-fading' according to

---

[9] The term 'filter-ringing' is sometimes used with a slightly different meaning in the context of speech coding.

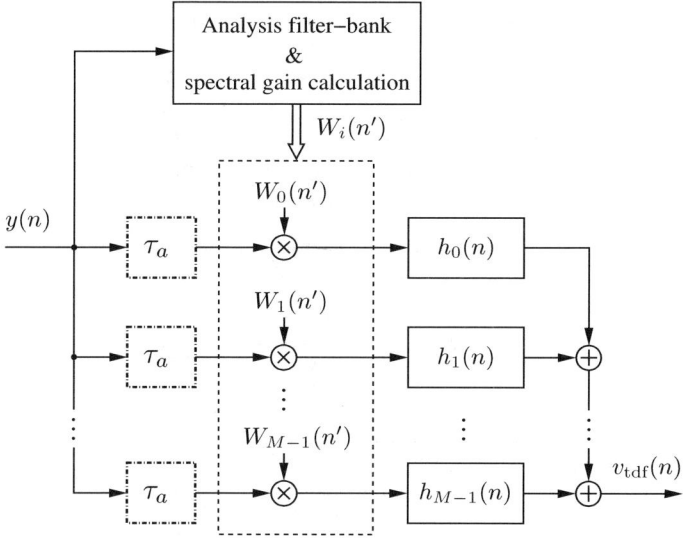

Fig. 2.13. Filter-bank summation method (FBSM) corresponding to the filter-bank equalizer (FBE) with time-domain filter in transposed direct form.

$$\bar{h}_s(l, n) = (1 - c_f(n)) \cdot h_s(l, n' - R) + c_f(n) \cdot h_s(l, n') \qquad (2.61a)$$

$$c_f(n) = \frac{n - n'}{R}; \quad l \in \{0, 1, \ldots, L\} \qquad (2.61b)$$

with n' is defined by Eq. 2.28. An existing linear-phase property is maintained. The proposed cross-fading method is very effective to avoid audible filter-ringing artifacts and is especially useful if the direct form filter is used.

It should be noted that artifacts due to time-varying spectral gain factors must also be avoided for AS FB systems, e.g., by non-critical subsampling, a dedicated prototype filter design, and smoothing of the gain factors (cf. Sec. 2.2.1).

2.3.5 Polyphase Network Implementation

An efficient polyphase network (PPN) implementation of the FBE shall be developed, which eases the utilization of prototype filters with a long or even infinite impulse response (IIR). This allows, for example, to improve the spectral selectivity of the subband filters in order to reduce the cross-talk between adjacent frequency bins. A low cross-talk can be favorable for some spectral speech estimators since most of them do not consider correlation between the frequency bands, cf. [55].

The FBE is a *time-varying* system. It can be described by the z-transform of the frozen-time impulse response which yields the so-called *frozen-time transfer function* [42]. The direct form time-domain filter of Eq. 2.38 at sample instant n' has the frozen-time transfer function

$$H_s(z, n') = \sum_{l=0}^{L} w_l(n') h(l) z^{-l} . \qquad (2.62)$$

This transfer function[10] can be expressed by means of the polyphase components of Eq. 2.11 which leads to

$$H_s(z, n') = \sum_{\lambda=0}^{M-1} w_\lambda(n') \sum_{m=0}^{l_M-1} h(\lambda + mM) z^{-(\lambda+mM)} \rho(m)$$

$$= \sum_{\lambda=0}^{M-1} w_\lambda(n') \cdot H_\lambda^{(M)}(z^M) \cdot z^{-\lambda} \qquad (2.63)$$

for $\rho(m) = 1$ and l_M defined by Eq. 2.12. The subband signals $y_i(n)$ of Eq. 2.31 are given by

$$y_i(n) = \sum_{l=0}^{L} y(n-l) h(l) \Phi(i, l) ; \quad i \in \{0, 1, \ldots, M-1\} . \qquad (2.64)$$

The z-transform leads to

$$Y_i(z) = Y(z) \sum_{l=0}^{L} h(l) z^{-l} \Phi(i, l) . \qquad (2.65)$$

Applying Eq. 2.11 and Eq. 2.30 with $\rho(m) = 1$ results in

$$Y_i(z) = Y(z) \sum_{\lambda=0}^{M-1} \sum_{m=0}^{l_M-1} h(\lambda + mM) z^{-(\lambda+mM)} \Phi(i, \lambda)$$

$$= Y(z) \sum_{\lambda=0}^{M-1} z^{-\lambda} \cdot H_\lambda^{(M)}(z^M) \cdot \Phi(i, \lambda) . \qquad (2.66)$$

The derived PPN implementation of the *direct form* FBE according to Eq. 2.63 and Eq. 2.66 is illustrated in Fig. 2.14. In contrast to the FBE realization of Fig. 2.11, the time-domain filtering and calculation of the subband signals is partly done by the same network. The PPN realization for the oddly-stacked GDFT FBE can be derived in a similar manner (cf. [44]). The same applies for an implementation with type 2 polyphase components.

The transposed direct form of a filter is derived from the direct form representation by transposition of its signal flow graph [59]: Branch nodes

---

[10] For the sake of brevity, the term transfer function refers to the frozen-time transfer function or the conventional transfer function dependent on the context.

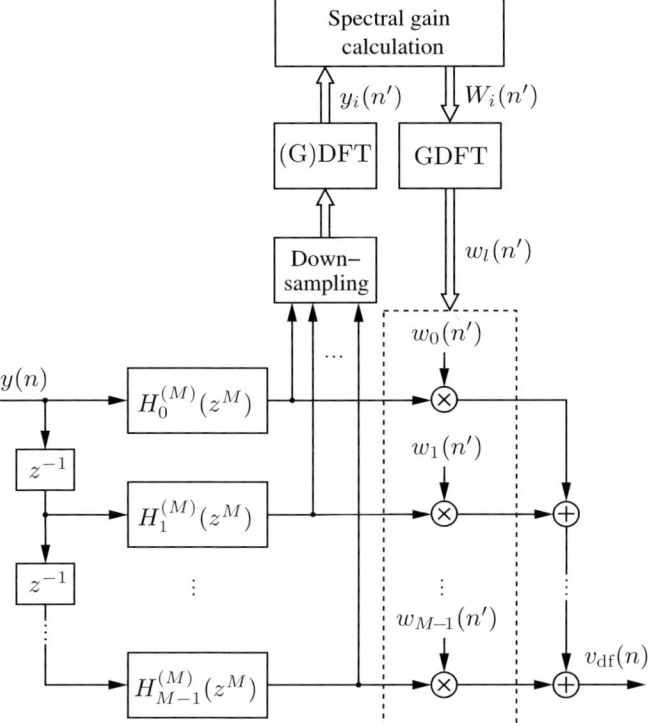

Fig. 2.14. Polyphase network (PPN) implementation of the GDFT FBE for the direct form time-domain filter.

and summations are interchanged as well as the system input and output. All signal directions are reversed. The obtained PPN implementation of the FBE for the *transposed direct form* is shown in Fig. 2.15. Delay elements might be inserted in each branch of the time-domain filter to account for the execution time τ_a to calculate the time-domain weighting factors $w_l(n')$. These weighting factors are calculated by a separate network similar to that of Fig. 2.14 but with the difference that the downsampling is performed directly after the delay elements. Thus, the PPN realization for the transposed direct form requires only a slightly higher algorithmic complexity than the direct form realization, which is discussed in Sec. 2.3.8 in more detail.

A polyphase network decomposition can be performed for FIR filters [3] as well as for IIR filters [78]. The design of IIR Mth-band filters is proposed in [67]. Hence, the developed PPN realization of the FBE enables a realization of Eq. 2.36 for L being infinite, that is, a *recursive* prototype filter.

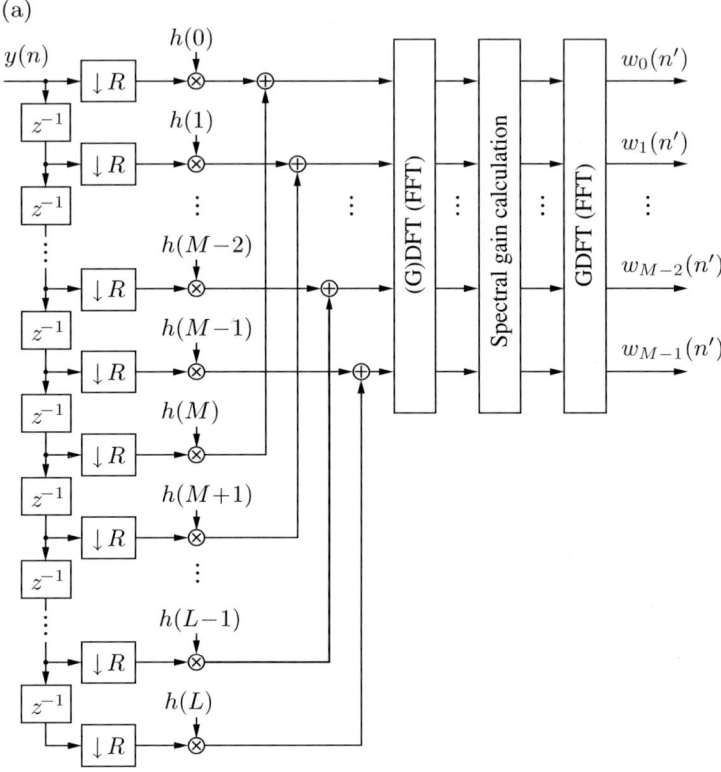

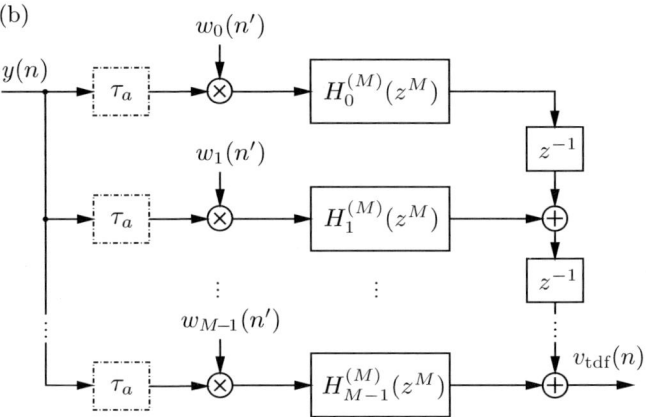

Fig. 2.15. PPN implementation of the GDFT FBE: (a) calculation of the time-domain weighting factors $w_l(n')$, (b) time-domain filter in transposed direct form.

2.3.6 The Non-Uniform Filter-Bank Equalizer

The allpass transformation can be applied to the FBE to achieve a non-uniform time-frequency resolution. The following treatment will show that the obtained allpass transformed FBE has some distinctive advantages in comparison to the corresponding allpass transformed AS FB[11] treated in Sec. 2.2.4.

2.3.6.1 Concept

The application of the allpass transformation to the bandpass filter of Eq. 2.29 yields the warped frequency responses

$$H_i(z = e^{j\varphi_a(\Omega)}) = \sum_{l=0}^{L} h(l)\,\Phi(i,l)\,e^{-jl\varphi_a(\Omega)} \qquad (2.67)$$

$$= \tilde{H}_i(e^{j\Omega})\,;\quad i \in \{0,1,\ldots,M-1\}\,. \qquad (2.68)$$

As before, the tilde-notation is used to mark quantities changed by the allpass transformation. The uniform FBE can be seen as a special case for $a = 0$ where $\varphi_a(\Omega) = \Omega$. The effect of this frequency warping on the frequency characteristic of the subband filters has been discussed already in Sec. 2.2.4. Fig. 2.16 provides a block diagram of the allpass transformed FBE. This warped FBE can be implemented efficiently by the PPN structures derived in Sec. 2.3.5 where the delay elements are substituted by allpass filters.

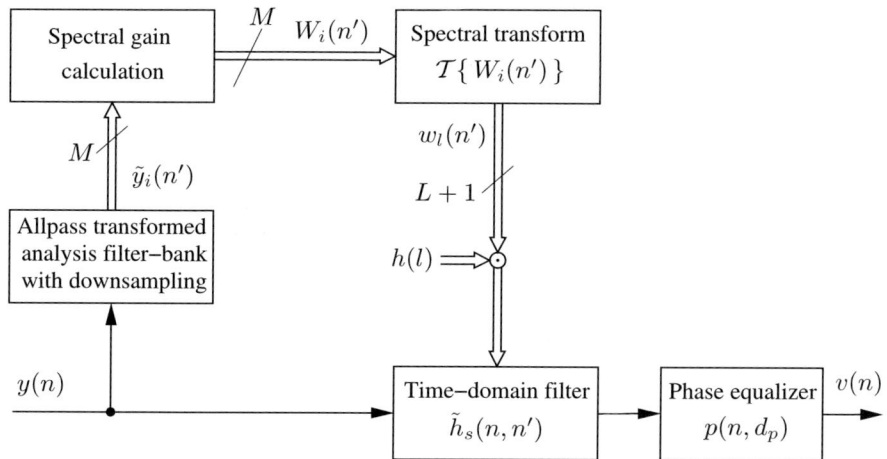

Fig. 2.16. Allpass transformed FBE for adaptive subband filtering.

---

[11] The corresponding AS FB uses the same type of analysis filter-bank as the FBE with identical values for L and M. An identical analysis filter-bank can not always be used due to different design constraints for signal reconstruction.

2.3.6.2 Compensation of Phase Distortions

The uniform FBE with perfect reconstruction fulfills Eq. 2.2, which can be expressed in the frequency-domain by the relation

$$V(e^{j\Omega}) = Y(e^{j\Omega}) \cdot e^{-j d_0 \Omega} . \tag{2.69}$$

If the allpass transformation is applied, this relation turns into

$$\widetilde{V}(e^{j\Omega}) = Y(e^{j\Omega}) \cdot e^{-j d_0 \varphi_a(\Omega)} . \tag{2.70}$$

Thus, the input signal $y(n)$ is filtered by an allpass chain of length d_0. Perfect reconstruction is achieved for a linear phase characteristic. This can be accomplished with an arbitrarily small error by means of a (fixed) phase equalizer connected to the output of the FBE as indicated in Fig. 2.16 [44, 48]. The design of the phase equalizer is similar to that for the AS FB structure II of Fig. 2.7(b). However, a phase equalizer of significantly lower degree is needed due to the lower signal delay d_0 in comparison to the AS FB.

The described phase compensation is also effective for *time-varying* filter coefficients if the symmetry of Eq. 2.48 holds. For the warped FBE with direct form filter, the (frozen-time) frequency response reads

$$\widetilde{H}_s(e^{j\Omega}, n') = \sum_{l=0}^{L} h_s(l, n') e^{-j l \varphi_a(\Omega)} . \tag{2.71}$$

If the real filter coefficients have the symmetry of Eq. 2.51, it can be shown that the transfer function of Eq. 2.71 can be expressed by

$$\widetilde{H}_s(e^{j\Omega}, n') = e^{-j \frac{L}{2} \varphi_a(\Omega)} \begin{cases} \sum_{l=0}^{\frac{L}{2}} 2\mathcal{A}(\Omega, l, L, n') - h_s\left(\frac{L}{2}, n'\right) & ; \; L \text{ even} \\ \sum_{l=0}^{\frac{L-1}{2}} 2\mathcal{A}(\Omega, l, L, n') & ; \; L \text{ odd} \end{cases} \tag{2.72a}$$

$$\mathcal{A}(\Omega, l, L, n') = h_s(l, n') \cos\left(\left[\frac{L}{2} - l\right] \varphi_a(\Omega)\right) . \tag{2.72b}$$

The non-linear phase term $\varphi_a(\Omega) L/2$ can be compensated by a phase equalizer which has to fulfill Eq. 2.23 with $d_P = L/2$ [48]. The expressions to the right of the curly brace are real and cause only phase shifts of $\pm\pi$. (This can also be shown for the more general case of a complex prototype filter with linear phase response, cf. [44].) Thus, the system has a *generalized linear phase response* despite the time-varying coefficients, if a sufficient phase compensation is performed. A system with a generalized linear phase response has a constant group delay, if the discontinuities that result from the addition of constant phase shifts due to the real function are neglected [59]. For the transposed direct form filter, Eq. 2.72 is approximately fulfilled in case of moderately time-varying coefficients.

For speech and audio processing, a phase equalizer might be omitted for smaller phase distortions ($d_P < 20$) as the human ear is relatively insensitive towards minor phase distortions.

2.3.7 Comparison between FBE and AS FB

A comparison with the uniform and warped AS FB of Sec. 2.2 shows that the concept of the FBE exhibits the following benefits:

- The FBE is not affected by aliasing distortions for the reconstructed signal. Hence, the difficult problem of achieving a sufficient aliasing cancellation, especially for time-varying spectral gain factors, does not occur. One consequence is that the FBE does not require a prototype filter with high degree and/or a low subsampling rate R to reduce aliasing distortions. Moreover, the prototype filter design for the FBE is less complex than for the AS FB.
- The allpass transformation of the FBE does not cause aliasing distortions for the reconstructed signal. Thus, only the phase modifications due to the frequency warping need to be compensated. Therefore, near-perfect reconstruction can be achieved with an arbitrarily small error and much lower efforts than for the AS FB.
- The signal delay d_0 for the FBE is significantly lower than for the corresponding AS FB. The uniform AS FB with linear-phase prototype filter has a signal delay of $d_0 = L$. In contrast, the uniform FBE with linear-phase prototype filter has a signal delay of $d_0 = L/2$, which can be further reduced by a non-linear phase filter. Accordingly, the design objective of Eq. 2.23 for the phase equalizer applies with $d_p = L$ for the warped AS FB (structure II) and with $d_P = L/2$ for the warped FBE. Therefore, a phase equalizer with a significantly lower filter degree can be used for the FBE which results a lower signal delay.
- The warped FBE can achieve an almost linear overall phase characteristic even for time-varying coefficients, which can be beneficial for multi-channel processing.

A drawback of the FBE is its higher algorithmic complexity for some configurations, which is exposed in the following.

2.3.8 Algorithmic Complexity

Tab. 2.1 contrasts the algorithmic complexity of the developed uniform and warped FBE to that of the corresponding uniform and warped AS FB for the same values L, M, R, and N_p. The real DFT of size M can be computed in-place by a radix-2 FFT, e.g., [63]. The FFT of a real sequence of size M can

Table 2.1. Algorithmic complexity for different realizations of a polyphase network (PPN) DFT filter-bank with real valued prototype filter of length $L+1 = l_M M$.

	2 real FFTs	Remaining operations	Additional operations due to allpass transformation
AS FB			
Multiplications	$2\frac{M}{R}\log_2 M$	$\frac{1}{R}(2L+2+M)$	$4L + N_p + 1$
Summations	$3\frac{M}{R}\log_2 M$	$\frac{1}{R}(L-M+1) + L$	$4L + N_p$
Delay elements	$2M$	$2L$	N_p
Direct form FBE			
Multiplications	$2\frac{M}{R}\log_2 M$	$L+1+M$	$2L + N_p + 1$
Summations	$3\frac{M}{R}\log_2 M$	L	$2L + N_p$
Delay elements	$2M$	L	N_p
Transposed direct form FBE			
Multiplications	$2\frac{M}{R}\log_2 M$	$L+1+M+\frac{L}{R}$	$4L + N_p + 1$
Summations	$3\frac{M}{R}\log_2 M$	$L + \frac{1}{R}(L+1-M)$	$4L + N_p$
Delay elements	$2M$	$2L$	N_p

be calculated by a complex FFT of size $M/2$, which requires approximately half the algorithmic complexity as a complex M-point FFT [62]. The GDFT can be computed by the FFT with similar complexity as for the DFT.

The last column contains the additional operations and delay elements due to the allpass transformation. The implementation of an allpass filter according to Fig. 2.4 is considered. This requires two real multiplications, two real summations, and one delay element for a real allpass coefficient $a = \alpha$. A LS phase equalizer of degree N_P is applied to compensate phase distortions.

It should be noted that allpass transformed filter-banks are usually operated with a smaller number of channels M than uniform filter-banks. As reasoned before, a higher subsampling rate R and a lower phase equalizer degree N_p are needed for the warped FBE in comparison to the warped AS FB. Therefore, the warped FBE has a lower algorithmic complexity than the corresponding warped AS FB for most parameter configurations. Contrariwise, the uniform AS FB has a lower complexity than the uniform FBE. A design example for these filter-banks is given later in Sec. 2.5.

2.4 Further Measures for Signal Delay Reduction

Even though the FBE causes only about half the algorithmic signal delay than the corresponding AS FB, a further reduced delay might be required for applications with very demanding system delay constraints. One example are the initially mentioned hearing aid devices. For such cases, a modification of the FBE concept will be discussed, which allows a further lowering of the signal delay and computational complexity with almost no loss for the perceived quality of the enhanced speech.

2.4.1 Concept

One approach to reduce the signal delay of a filter-bank is to reduce the transform size M to allow for a lower prototype filter degree and to adjust the gain calculation to the altered time-frequency resolution (smoothing factors etc.), e.g., [28]. For the FBE, a further reduction of the signal delay can also be accomplished by approximating the original time-domain filter by a filter of lower degree. This approach offers a greater flexibility in the choice of the time-domain filter and requires no adjustment of the spectral gain calculation. The principle is illustrated in Fig. 2.17. In contrast to the FBE of Fig. 2.11, an additional module for the filter approximation is inserted, which evaluates the new $L_D + 1$ filter coefficients $a_l(n')$ from the $L + 1$ original FIR filter coefficients $h_s(l, n')$.

In the following, we investigate an FIR and IIR filter approximation for the uniform FBE first, before the results are extended to the more general case of allpass transformed filters.

2.4.2 Approximation by a Moving-Average Filter

The time-domain filter of the FBE can be approximated by an FIR filter of lower degree $L_D < L$ following a technique very similar to FIR filter design by

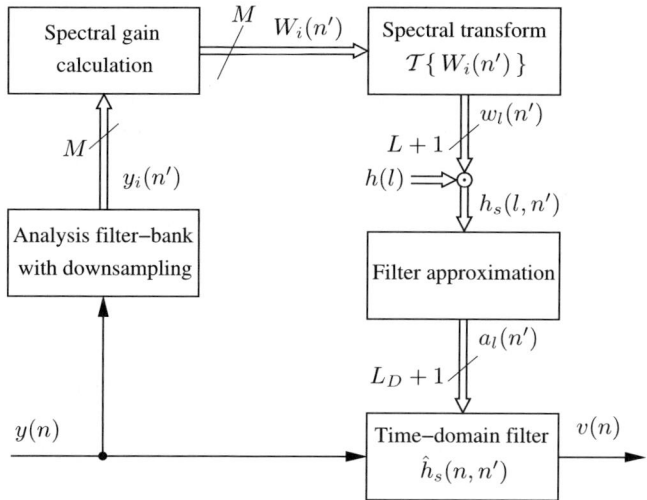

Fig. 2.17. Modification of the FBE to achieve a further reduced signal delay.

windowing, e.g., [63]. The impulse response $h_s(l, n')$ of Eq. 2.38 is truncated by a window sequence of length $L_D + 1$. This yields the FIR filter coefficients

$$a_l(n') = \hat{h}_s(l, n') = h_s(l + l_c, n') \, \text{win}_{L_D}(l) \, ; \quad l \in \{0, 1, \ldots, L_D\} \quad (2.73)$$

where the window sequence is defined by Eq. 2.44. The value for l_c determines the part of the impulse response to be truncated, e.g., to maintain the symmetry of Eq. 2.51 for linear-phase filters. The truncation by a window results in a smoothed frequency response which is influenced by the choice of the window sequence, cf. [59].

This modified FBE based on an FIR filter approximation is named as *moving-average low delay filter* (MA LDF) [46]. The term 'low delay filter' (LDF) refers to the overall system according to Fig. 2.17, while the term 'MA filter' denotes the actual time-domain filter.

2.4.3 Approximation by an Auto-Regressive Filter

Instead of a (linear-phase) FIR filter, a *minimum-phase* IIR filter is now considered for the filter approximation. A filter can always be decomposed into an allpass filter and a minimum-phase filter, e.g., [59]. The group delay of the minimum-phase filter is lower than or equal to the group delay of the original filter for all frequencies. The approximation of a mixed-phase filter by a minimum-phase filter leads to a different phase response. This effect, however, is mostly tolerable for speech and audio processing as the human ear is relatively insensitive towards phase modifications, cf. [80, 84]. An allpole filter or auto-regressive (AR) filter is used here, because the calculation of its coefficients demands a lower computational complexity than for a general minimum-phase IIR filter (with zeros outside the origin), cf. [63]. The filter to be approximated contains no sharp zeros in its spectrum, if the threshold ϵ in Eq. 2.47 is greater than zero, which further supports the idea of an AR filter approximation.

The transfer function of the considered AR filter of degree L_D is given by

$$\hat{H}_s(z, n') = H_{\text{AR}}(z, n') = \frac{a_0(n')}{1 - \sum_{l=1}^{L_D} a_l(n') z^{-l}}. \quad (2.74)$$

The AR filter coefficients can be determined by the Yule-Walker equations, e.g., [63]

$$\begin{bmatrix} \varphi_{\bar{h}\bar{h}}(1) \\ \vdots \\ \varphi_{\bar{h}\bar{h}}(L_D) \end{bmatrix} = \begin{bmatrix} \varphi_{\bar{h}\bar{h}}(0) & \cdots & \varphi_{\bar{h}\bar{h}}(1 - L_D) \\ \vdots & \ddots & \vdots \\ \varphi_{\bar{h}\bar{h}}(L_D - 1) & \cdots & \varphi_{\bar{h}\bar{h}}(0) \end{bmatrix} \cdot \begin{bmatrix} a_1 \\ \vdots \\ a_{L_D} \end{bmatrix} \quad (2.75)$$

where the dependence on n' is omitted for the sake of simplicity. The $L_D + 1$ auto-correlation coefficients $\varphi_{\bar{h}\bar{h}}(\lambda)$ are computed by the rule[12]

$$\varphi_{\bar{h}\bar{h}}(\lambda) = \sum_{l=0}^{L-|\lambda|} \bar{h}(l)\,\bar{h}(l+\lambda)\,; \quad 0 \leq |\lambda| \leq L_D \qquad (2.76\text{a})$$

$$\bar{h}(l) = h_s(l)\,\text{win}_L(l)\,; \quad l \in \{0, 1, \ldots, L\}\,. \qquad (2.76\text{b})$$

The scaling factor a_0 in Eq. 2.74 is given by

$$a_0 = \sqrt{\varphi_{\bar{h}\bar{h}}(0) - \sum_{l=1}^{L_D} a_l\,\varphi_{\bar{h}\bar{h}}(l)} \qquad (2.77)$$

and ensures that the AR filter and the original filter have both the same amplification. The calculation of the auto-correlation coefficients according to Eq. 2.76 guarantees a symmetric Toeplitz structure for the auto-correlation matrix of Eq. 2.75. This allows to solve the Yule-Walker equations efficiently by means of the Levinson-Durbin recursion. The auto-correlation matrix is positive-definite so that the obtained AR filter has minimum-phase property, which implies a stable filter, cf. [63].

The devised modification of the FBE is named as *auto-regressive low delay filter* (AR LDF) in analogy to the terminology of the previous section [46].

2.4.4 Algorithmic Complexity

The algorithmic complexity for the low delay filter concept – in terms of computational complexity and memory consumption – is listed in Tab. 2.2. The complexity for the calculation of the original filter coefficients $h_s(l, n')$ has been discussed in Sec. 2.3.8. The variable $\mathcal{M}_{\text{div}}$ marks the number of multiplications needed for a division operation, and $\mathcal{M}_{\text{sqrt}}$ represents the number of multiplications needed for a square-root operation. Accordingly, the variables $\mathcal{A}_{\text{div}}$ and $\mathcal{A}_{\text{sqrt}}$ denote the additions needed for a division and square-root operation, respectively. Their values depend on the numeric procedure and accuracy used to execute these operations. (An equivalent of 15 operations will be assigned to these variables for the later complexity assessment in Sec. 2.5.). A rectangular window is assumed for Eq. 2.76b.

Most of the computational complexity for the AR filter conversion is required to compute the $L_D + 1$ auto-correlation coefficients according to Eq. 2.76. A lower computational complexity can be achieved by calculating Eq. 2.76 by means of the fast convolution or the Rader algorithm [65] with savings dependent on L_D and L.

---

[12] An alternative to this 'auto-correlation method' is the use of the 'covariance method'. However, this results in a more complex procedure for the calculation of the AR filter coefficients, cf. [81].

Table 2.2. Algorithmic complexity for the MA and AR low delay filter (LDF).

Calculation of $h_s(l,n')$ and MA/AR filtering	
Multiplications	$\frac{1}{R}(2M\log_2 M + 2L + 2) + L_D + 1$
Summations	$\frac{1}{R}(3M\log_2 M + L + 1 - M) + L_D$
Delay elements	$L + 2M + L_D$
Calculation of MA filter coefficients $a_l(n')$	
Multiplications	$\frac{1}{R}(L_D + 1)$
Summations	0
Registers	0
Calculation of AR filter coefficients $a_l(n')$	
Multiplications	$\frac{1}{R}\left((L_D+1)(L+4) + L_D(\mathcal{M}_{\text{div}} + \mathcal{M}_{\text{sqrt}})\right)$
Summations	$\frac{1}{R}\left((L_D+1)(L+2) + L_D(\mathcal{A}_{\text{div}} + \mathcal{A}_{\text{sqrt}})\right)$
Registers	$3L_D$

The MA filter conversion needs no multiplications, if a rectangular window is used for Eq. 2.73. However, the AR filter degree is usually chosen significantly lower than the MA filter degree so that both approaches have a similar overall algorithmic complexity as exemplified later in Sec. 2.5.

2.4.5 Warped Filter Approximation

The discussed filter approximations can also be applied to the allpass transformed FBE [47]. In the process, the delay elements of the analysis filter-bank and the time-domain filter are replaced by allpass filters according to Eq. 2.15. For the obtained warped MA LDF, a phase equalizer can be applied to obtain an approximately linear phase response according to Sec. 2.3.6.2.

The direct realization of the warped AR filter is not possible since the allpass transformation leads to delayless feedback loops. Different approaches have been proposed to solve this problem for a real allpass transformation ($a = \alpha$) [29, 30, 74]. Here, the algorithm of Steiglitz [74] is preferred due to its low computational efforts for time-varying filters. The modified transfer function of the allpass transformed AR filter is given by

$$\widetilde{H}_{\text{AR}}(z) = \frac{a_0 \tilde{a}_0}{1 - \tilde{a}_0 \frac{(1-\alpha^2)z^{-1}}{1-\alpha z^{-1}} \sum_{l=1}^{L_D} \tilde{a}_l H_A(z)^{l-1}} \quad (2.78)$$

with coefficients $\tilde{a}_l$ calculated by the recursion

$$\tilde{a}_{L_D} = a_{L_D} \tag{2.79a}$$
$$\tilde{a}_l = a_l - \alpha\, \tilde{a}_{l+1}\,; \quad l = L_D - 1, \ldots, 1 \tag{2.79b}$$
$$\tilde{a}_0 = \frac{1}{1 + \tilde{a}_1 \alpha}\,. \tag{2.79c}$$

The computation of the new filter coefficients $\tilde{a}_l(n')$ needs only L_D multiplications, L_D summations, and one division at intervals of R sample instants. The warped AR filter according to Eq. 2.78 requires $3L_D + 2$ real multiplications and $3L_D$ real summations per sample instant as well as $L_D + 1$ delay elements.

It can be proven that the minimum-phase property of the AR filter is always maintained for a real allpass transformation but not for a complex allpass transformation in general. This is an important result as it guarantees stability for the warped AR filter. The use of a fixed phase equalizer (as for the warped MA LDF) is neither feasible nor required.

The cross-fading approach of Eq. 2.61 can not be applied to the coefficients of an IIR filter. Instead, we use a second filter with the previous coefficients to achieve a smooth transition by a cross-fading of both output signals. This general approach can be expressed by

$$\bar{H}_{\mathrm{g}}(z, n) = \big(1 - c_f(n)\big) \cdot H_{\mathrm{g}}(z, n' - R) + c_f(n) \cdot H_{\mathrm{g}}(z, n') \tag{2.80a}$$
$$c_f(n) = \frac{n - n'}{R}\,. \tag{2.80b}$$

A second filter is required for this smoothing technique which can be applied to arbitrary filters and does not cause an additional signal delay.

2.5 Application to Noise Reduction

In this section, the treated filter-bank designs are employed for noise reduction to compare the achieved performance with regard to speech quality, computational complexity, and signal delay.

2.5.1 System Configurations

The filtering of the noisy speech is done by the DFT AS FB, the GDFT FBE, and the MA/AR LDF. The uniform and allpass transformed version of these filter-banks are used each.[13] A real allpass coefficient of $a = 0.4$ is considered, which yields a good approximation of the Bark scale for a sampling frequency

---

[13] The low delay filter of Sec. 2.4 can be seen as a filter-bank system as it is derived from the FBE or FBSM, respectively.

of 8 kHz [73]. In all cases, a transform size of $M = 64$ and a linear-phase FIR prototype filter of degree $L = 64$ are used.[14]

A square-root Hann window derived from Eq. 2.45 is employed as common prototype filter for the DFT AS FB. The uniform AS FB uses a subsampling rate of $R = 32$ and the warped AS FB a rate of $R = 8$. A higher subsampling rate R can be used for the warped AS FB as well. However, this increases the signal delay significantly since subband filters with a higher stopband attenuation are needed to achieve a sufficient aliasing cancellation. A LS phase equalizer with a filter degree of $N_p = 141$ is applied to the filter-bank output according to Sec. 2.2.4.2.

The GDFT FBE is implemented in the transposed direct form. The MA LDF possesses a filter degree of $L_D = 48$. The LS FIR phase equalizer with filter degree $N_p = 80$ and $N_p = 56$ is applied to the warped FBE and the warped MA LDF, respectively. The considered AR LDF has a filter degree of $L_D = 16$. The cross-fading technique is used to avoid filter-ringing artifacts. A subsampling rate of $R = 64$ is used for the analysis filter-banks of FBE and LDF.

The spectral gain factors are computed by the super-Gaussian joint MAP estimator [50]. This joint spectral amplitude and phase estimator is derived by the more accurate assumption that the real and imaginary parts of the speech DFT coefficients are rather Laplace distributed (considered here) or Gamma distributed than Gaussian distributed. The needed *a priori* SNR is determined by the decision-directed approach [20] with a fixed smoothing parameter of 0.9. The short-term noise power spectral density is estimated by minimum statistics [54]. Speech presence uncertainty is taken into account by applying soft-gains [52]. Independent of the subsampling rate R of the filter-bank, the spectral gain factors are always adapted at intervals of 64 sample instants and no individual parameter tuning is performed to ease the comparison.

2.5.2 Instrumental Quality Measures

The used audio signals of 8 kHz sampling frequency are taken from the noisy speech corpus NOIZEUS presented in [32]. A total of 20 sentences spoken by male and female speakers is used, each disturbed by five different, instationary noise sequences (airport, babble, car, station, and street noise) with signal-to-noise ratios (SNRs) between 0 dB and 15 dB.

The quality of the enhanced speech is evaluated by informal listening tests and instrumental quality measures. (An overview of this topic is provided by Chap. 9.) A common time-domain measure for the quality of the enhanced speech $v(n) = \hat{s}(n)$ is given by the *segmental SNR*

---

[14] A lower number of frequency channels can be used for warped filter-banks whereas a value of $M = 256$ is often preferred for speech enhancement using the uniform DFT filter-bank (at 8 kHz sampling frequency). However, such different configurations are not considered to ease the comparison of the filter-banks.

$$\mathrm{SNR_{seg}/\,dB} = \frac{1}{\mathcal{C}(\mathbb{F}_s)} \sum_{m \in \mathbb{F}_s} 10 \log_{10} \left(\frac{\sum_{\mu=0}^{K-1} s^2(mK + \mu - \tau_0)}{\sum_{\mu=0}^{K-1} \left(\hat{s}(mK + \mu) - s(mK + \mu - \tau_0)\right)^2} \right). \quad (2.81)$$

The calculation comprises only frames with speech activity ($m \in \mathbb{F}_s$) whose total number is denoted by $\mathcal{C}(\mathbb{F}_s)$.

In a simulation, the clean speech $s(n)$ and the additive background noise $b(n)$ can be filtered separately with coefficients adapted for the disturbed speech $y(n) = s(n) + b(n)$. This provides the filtered speech $\bar{s}(n)$ and filtered noise $\bar{b}(n)$ separately, where

$$v(n) = \hat{s}(n) = \bar{s}(n) + \bar{b}(n) \,. \quad (2.82)$$

The algorithmic *signal delay* of non-linear phase systems is determined here by the maximum of the cross-correlation between the clean speech $s(n)$ and the processed speech $\bar{s}(n)$ (due to their strong correlation) according to

$$\tau_0 = \arg \max_{\lambda \in \mathbb{Z}} \{\varphi_{s\bar{s}}(\lambda)\} \,. \quad (2.83)$$

The achieved *segmental noise attenuation* is calculated by the expression

$$\mathrm{NA_{seg}/\,dB} = \frac{1}{\mathcal{C}(\mathbb{F})} \sum_{m \in \mathbb{F}} 10 \log_{10} \left(\frac{\sum_{\mu=0}^{K-1} b^2(mK + \mu - \tau_0)}{\sum_{\mu=0}^{K-1} \bar{b}^2(mK + \mu)} \right) \quad (2.84)$$

where $\mathbb{F}$ marks the set of all frame indices including speech pauses, and $\mathcal{C}(\mathbb{F})$ denotes the total number of frames.

A frequency-domain measure for the speech quality is provided by the mean *cepstral distance* (CD) between the clean speech $s(n)$ and the processed speech $\bar{s}(n)$, cf. [64]. For all instrumental measures, a frame size of $K = 256$ is used, and 40 cepstral coefficients are considered for the CD measure.

2.5.3 Simulation Results for the Uniform Filter-Banks

The instrumental speech quality obtained with the different uniform filter-banks is plotted in Fig. 2.18. The uniform FBE achieves the same (or even better) objective speech quality as the uniform AS FB. Tab. 2.3 reveals that the FBE possesses a slightly higher algorithmic complexity but achieves a significantly lower signal delay. The MA and AR LDF achieve a further reduction of the signal delay and algorithmic complexity. Contrary to the MA LDF, the enhancement by the AR LDF leads to a significantly decreased *objective* speech quality. The AR filter approximation causes phase modifications

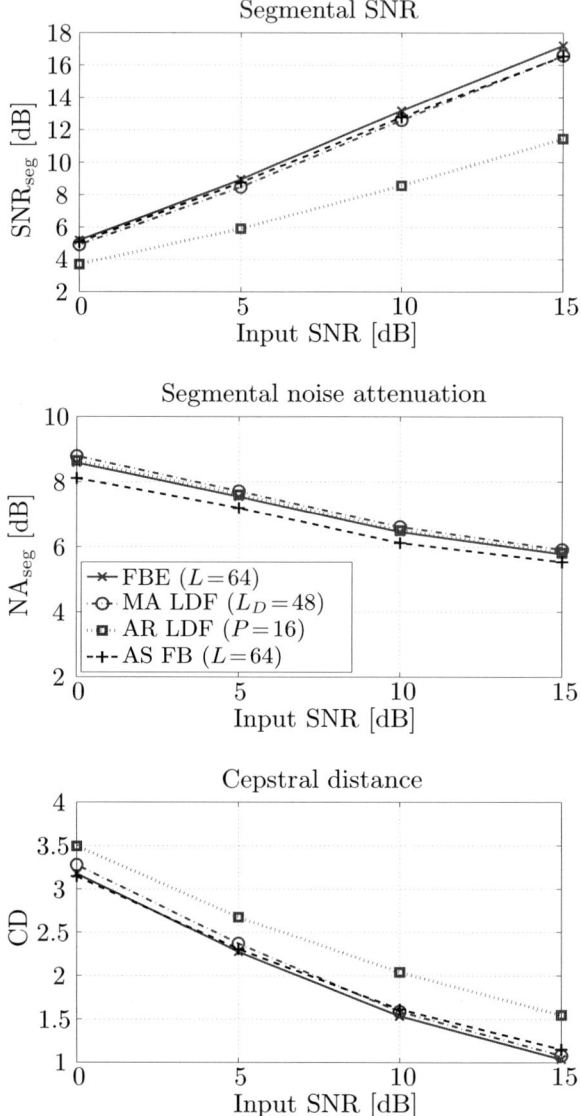

Fig. 2.18. Objective quality measures obtained with the uniform filter-banks.

which have a very detrimental effect on the segmental SNR measure. (Such an effect can also be observed for warped filter-banks with an imperfect phase compensation.) However, informal listening tests have revealed only negligible differences for the perceived *subjective* speech quality. Therefore, a perceptual evaluation of the speech quality (PESQ) according to [33] has been conducted in addition. This PESQ measure ranges from -0.5 (bad quality) to 4.5

Table 2.3. Measured signal delay and average algorithmic complexity per sample for the uniform filter-banks ($M = L = 64$).

Uniform filter-bank	Signal delay [samples]	Summations (real)	Multiplications (real)	Delay elements
AS FB	64	101	31	256
FBE	32	83	142	256
MA LDF	24	67	64	240
AR LDF	0-2	75	74	272

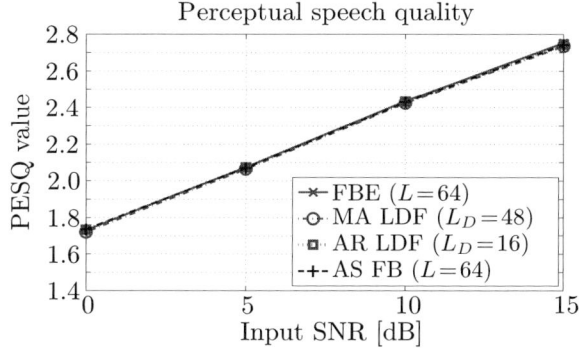

Fig. 2.19. Perceptual evaluation of the speech quality for the enhanced speech $\hat{s}(k)$ achieved with the uniform filter-banks.

(excellent quality). The PESQ measure is mainly used for the assessment of speech codecs, but also employed as a perceptual quality measure for speech enhancement systems, e.g., [4]. The measured PESQ values in Fig. 2.19 show that all four filter-banks achieve an almost identical perceptual speech quality. The PESQ measure is no all-embracing quantity for the subjective speech quality, but it complies well with the results of our informal listening tests. Thus, the low delay filter concept is suitable to achieve a further reduced signal delay in a flexible and simple manner with negligible loss for the perceived (subjective) speech quality.

2.5.4 Simulation Results for the Warped Filter-Banks

The curves for the objective speech quality obtained by means of the different warped filter-banks are plotted in Fig. 2.20. The measured PESQ values are not plotted again since they are as close together as in Fig. 2.19 but all about 0.25 PESQ units higher. Thus, the warped filter-banks achieve an improved instrumental speech quality in comparison to the corresponding uniform filter-banks. These results comply with our informal listening tests where the speech enhanced by the warped filter-banks was rated to be superior.

The measured signal delay and algorithmic complexity of the used allpass transformed filter-banks are listed in Tab. 2.4. It shows the increase of the

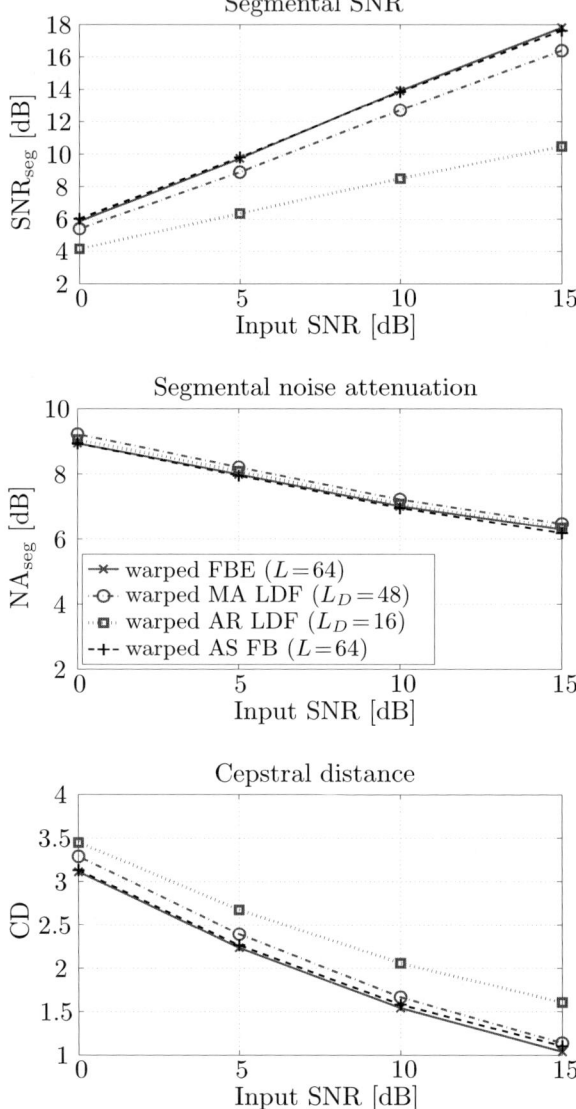

Fig. 2.20. Objective quality measures obtained by the allpass transformed filter-banks.

signal delay and algorithmic complexity due to the allpass transformation if the *same* values for M and L are taken (see also Sec. 2.3.8). The warped FBE causes a significantly lower signal delay and possesses a lower algorithmic complexity than the corresponding warped AS FB, but achieves the same objective and subjective speech quality.

Table 2.4. Measured signal delay and average algorithmic complexity per sample for the allpass transformed filter-banks ($M = L = 64$).

Warped filter-bank	Signal delay [samples]	Summations (real)	Multiplications (real)	Delay elements
AS FB	141	605	518	397
FBE	80	418	478	336
MA LDF	56	347	335	296
AR LDF	0-2	268	268	274

As for the uniform filter-banks, a further reduction of the signal delay and algorithmic complexity can be achieved by the low delay filter approximation with no loss for the subjective speech quality. The AR LDF is a minimum-phase system and causes a very low signal delay of only a few samples.

2.6 Conclusions

Filter-bank systems used for speech and audio processing have to fulfill several, partly conflicting requirements. A low signal delay and low algorithmic complexity are important for many applications such as mobile communication devices or digital hearing aids. A non-uniform, Bark-scaled frequency resolution is desirable to achieve a high speech and audio quality with a small number of frequency bands.

In this chapter, we have investigated different design approaches for such filter-banks. The main focus lies on allpass transformed filter-bank systems. These frequency warped filter-banks are a generalization of the uniform filter-bank. They are attractive for speech and audio processing due to their ability to mimic the Bark frequency bands of human hearing with great accuracy. However, the use of an allpass transformed analysis-synthesis filter-bank (AS FB) leads to a high signal delay as well as a high algorithmic complexity. This is attributed to the fact that synthesis subband filters of high degree are needed to compensate the aliasing and phase distortions caused by the allpass transformation of the analysis filter-bank.

These problems are addressed by the alternative concept of the filter-bank equalizer (FBE). It is derived as an efficient realization of the filter-bank summation method (FBSM) and performs time-domain filtering with coefficients adapted in the frequency-domain. Perfect signal reconstruction is achieved for a broad class of transformations with significantly lower efforts than for the common AS FB. The reconstructed signal is (inherently) aliasing-free so that a prototype filter with a high filter degree to achieve a high stopband attenuation is not essential. It is shown how the FBE can be efficiently implemented by means of a polyphase network (PPN) structure. The explicit consideration of the time-varying coefficients in the derivation has revealed the influence of the filter structure on system delay, computational complexity, and signal quality. This has shown, among others, how the transposed direct form implementation achieves a stronger smoothing effect for time-varying coefficients in comparison to the direct form implementation, which is beneficial to avoid artifacts for the processed signal.

The presented allpass transformed FBE achieves a near-perfect, aliasing-free signal reconstruction with significantly lower efforts than allpass transformed AS FBs. The uniform FBE has a higher algorithmic complexity than the corresponding uniform AS FB for most parameter configurations (L, M, R), while the opposite applies for the allpass transformed FBE in comparison to the allpass transformed AS FB. The uniform and warped FBE achieve a significantly lower algorithmic signal delay than the corresponding AS FBs. A nearly linear phase response can be maintained even for time-varying coefficients, which can be exploited, e.g., for binaural signal processing in hearing aids.

The proposed filter-bank design provides a versatile concept for applications such as low delay speech enhancement. The uniform and warped FBE achieve the same (or even better) objective and subjective quality for the enhanced speech as comparable AS FBs, but with a significantly lower signal delay. The frequency warping can be utilized either to achieve an improved speech quality or to use a lower number of frequency channels.

The concept of the low delay filter (LDF) is a modification of the FBE to achieve a further reduction of signal delay and algorithmic complexity with almost no compromise on the perceived (subjective) speech quality. In this process, the time-domain filter of the FBE is approximated by a moving-average (MA) filter or auto-regressive (AR) filter of lower degree. The use of the uniform and warped MA filter allows to maintain a constant (near-linear) phase characteristic, which is beneficial, e.g., for multi-channel processing. The uniform and warped AR filter are minimum-phase systems and can achieve an algorithmic signal delay of only a few sample instants.

The use for noise reduction has been considered primarily here, but the presented low delay filter-bank concepts are also suitable for other speech and audio processing algorithms.

References

1. J. Agnew, J. M. Thornton: Just noticeable and objectionable group delays in digital hearing aids, *Journal of the American Academy of Audiology*, **11**(6), 330–336, 2000.
2. K. G. Beauchamp: *Walsh Functions and Their Applications,* London, GB: Academic Press, 1975.
3. M. G. Bellanger, G. Bonnerot, M. Coudreuse: Digital filtering by polyphase network: application to sample-rate alteration and filter banks, *IEEE Trans. on Acoustics, Speech, and Signal Processing*, **ASSP-24**(2), 109–114, April 1976.
4. J. Benesty, S. Makino, J. Chen: *Speech Enhancement,* Berlin, Germany: Springer, 2005.

5. S. F. Boll: Suppression of acoustic noise in speech using spectral subtraction, *IEEE Trans. on Acoustics, Speech, and Signal Processing*, **ASSP-27**(2), 113–120, April 1979.
6. C. Braccini, A. V. Oppenheim: Unequal bandwidth spectral analysis using digital frequency warping, *IEEE Trans. on Acoustics, Speech, and Signal Processing*, **ASSP-22**(4), 236–244, August 1974.
7. C. S. Burrus, R. A. Gopinath, H. Guo: *Introduction to Wavelets and Wavelet Transforms: A Primer*, Upper Saddle River, NJ, USA: Prentice-Hall, 1998.
8. O. Cappé: Elimination of the musical noise phenomenon with the Ephraim and Malah noise suppressor, *IEEE Trans. on Speech and Audio Processing*, **2**(2), 345–349, April 1994.
9. I. Cohen: Enhancement of speech using Bark-scaled wavelet packet decomposition, *Proc. EUROSPEECH '01*, 1933–1936, Aalborg, Denmark, September 2001.
10. A. G. Constantinides: Frequency transformation for digital filters, *IEE Electronic Letters*, **3**(11), 487–489, November 1967.
11. R. E. Crochiere: A weighted overlap-add method of short-time Fourier analysis/synthesis, *IEEE Trans. on Acoustics, Speech, and Signal Processing*, **ASSP-28**(10), 99–102, February 1980.
12. R. E. Crochiere, L. R. Rabiner: *Multirate Digital Signal Processing*, Upper Saddle River, NJ, USA: Prentice-Hall, 1983.
13. Z. Cvetković, J. D. Johnston: Nonuniform oversampled filter banks for audio signal processing, *IEEE Trans. on Speech and Audio Processing*, **11**(5), 393–399, September 2003.
14. R. Czarnach: Recursive processing by noncausal digital filters, *IEEE Trans. on Acoustics, Speech, and Signal Processing*, **ASSP-30**(3), 363–370, June 1982.
15. I. Daubechies, W. Sweldens: Factoring Wavelet transforms into lifting steps, *Journal of Fourier Analysis and Applications*, **4**(3), 247–269, May 1998.
16. Y. Deng, V. J. Mathews, B. Farhang-Boroujeny: Low-delay nonuniform pseudo-QMF banks with application to speech enhancement, *IEEE Trans. on Signal Processing*, **55**(5), 2110–2121, May 2007.
17. G. Doblinger: An efficient algorithm for uniform and nonuniform digital filter banks, *Proc. ISCAS '91*, **1**, 646–649, Singapore, June 1991.
18. B. Dumitrescu, R. Bregović, T. Saramäki, R. Niemistö: Low-delay nonuniform oversampled filterbanks for acoustic echo control, *Proc. EUSIPCO '06*, Florence, Italy, September 2006.
19. A. Engelsberg: *Transformation-Based Systems for Single-Channel Noise Reduction in Speech Signals,* PhD thesis, Christian-Albrechts University, Ulrich Heute (ed.), Kiel, Germany: Shaker Verlag, 1998 (in German).
20. Y. Ephraim, D. Malah: Speech enhancement using a minimum mean-square error short-time spectral amplitude estimator, *IEEE Trans. on Acoustics, Speech, and Signal Processing*, **ASSP-32**(6), 1109–1121, December 1984.
21. C. Feldbauer, G. Kubin: Critically sampled frequency-warped perfect reconstruction filterbank, *Proc. ECCTD '03*, Krakow, Poland, September 2003.
22. S. Franz, S. K. Mitra, J. C. Schmidt, G. Doblinger: Warped discrete Fourier transform: a new concept in digital signal processing, *Proc. ICASSP '02*, **2**, 1205–1208, Orlando, FL, USA, May 2002.
23. E. Galijašević: *Allpass-Based Near-Perfect-Reconstruction Filter Banks,* PhD thesis, Christian-Albrechts University, Ulrich Heute (ed.), Kiel, Germany: Shaker Verlag, 2002.

24. E. Galijašević, J. Kliewer: Design of allpass-based non-uniform oversampled DFT filter banks, *Proc. ICASSP '02*, **2**, 1181–1184, Orlando, FL, USA, May 2002.
25. R. C. Gonzalez, P. Wintz: *Digital Image Processing,* London, GB: Addison-Wesley, 1977.
26. T. Gülzow, A. Engelsberg, U. Heute: Comparison of a discrete Wavelet transformation and a nonuniform polyphase filterbank applied to spectral-subtraction speech enhancement, *Signal Processing*, Elsevier, **64**(1), 5–19, January 1998.
27. T. Gülzow, T. Ludwig, U. Heute: Spectral-subtraction speech enhancement in multirate systems with and without non-uniform and adaptive bandwidths, *Signal Processing*, Elsevier, **83**(8), 1613–1631, August 2003.
28. H. Gustafsson, S. E. Nordholm, I. Claesson: Spectral subtraction using reduced delay convolution and adaptive-averaging, *IEEE Trans. on Speech and Audio Processing*, **9**(8), 799–807, November 2001.
29. A. Härmä: Implementation of recursive filters having delay free loops, *Proc. ICASSP '98*, **3**, 1261–1264, Seattle, WA, USA, May 1998.
30. A. Härmä: Implementation of frequency-warped recursive filters, *Signal Processing*, Elsevier, **80**(3), 543–548, March 2000.
31. U. Heute: Noise reduction, in E. Hänsler, G. Schmidt (eds.), *Topics in Acoustic Echo and Noise Control*, 325–384, Berlin, Germany: Springer, 2006.
32. Y. Hu, P. C. Loizou: Subjective comparison of speech enhancement algorithms, *Proc. ICASSP '06*, Tolouse, France, May 2006.
33. ITU-T Rec. P.862: Perceptual evaluation of speech quality (PESQ): an objective method for end-to-end speech quality assessment of narrow-band telephone networks and speech codecs, February 2001.
34. M. Kahrs, K. Brandenburg: *Applications of Digital Signal Processing to Audio and Acoustics*, Boston, MA, USA: Kluwer, 1998.
35. M. Kappelan: *Characteristics of Allpass Chains and their Application for Non-Equispaced Spectral Analysis and Synthesis,* PhD thesis, RWTH Aachen University, Peter Vary (ed.), Aachener Beiträge zu Digitalen Nachrichtensystemen, Aachen, Germany: Mainz Verlag, 1998 (in German).
36. M. Kappelan, B. Strauß, P. Vary: Flexible nonuniform filter banks using allpass transformation of multiple order, *Proc. EUSIPCO '96*, **3**, 1745–1748, Trieste, Italy, 1996.
37. T. Karp, A. Mertins: Lifting schemes for biorthogonal modulated filter banks, *Proc. of Intl. Conf. on Digital Signal Processing (DSP) '97*, **1**, 443–446, Santorini, Greece, July 1997.
38. T. Karp, A. Mertins, G. Schuller: Efficient biorthogonal cosine-modulated filter banks, *Signal Processing*, Elsevier, **81**(5), 997–1016, May 2001.
39. J. M. Kates, K. H. Arehart: Multichannel dynamic-range compression using digital frequency warping, *EURASIP Journal on Applied Signal Processing*, **18**, 3003-3014, 2005.
40. J. Kliewer, A. Mertins: Oversampled cosine-modulated filter banks with arbitrary system delay, *IEEE Trans. on Signal Processing*, **46**(4), 941–955, April 1998.
41. A. M. Kondoz: *Digital Speech - Coding for Low Bit Rate Communication Systems,* Chichester, UK: Wiley, 2004.
42. T.-Y. Leou, J. K. Aggarwal: Recursive implementation of LTV filters – frozen-time transfer function versus generalized transfer function, *Proc. of the IEEE*, **72**(7), 980–981, July 1984.

43. J. Li, T. Q. Nguyen, S. Tantaratana: A simple design for near-perfect-reconstruction nonuniform filter banks, *IEEE Trans. on Signal Processing*, **45**(8), 2105–2109, August 1997.
44. H. W. Löllmann, P. Vary: Efficient non-uniform filter-bank equalizer, *Proc. EUSIPCO '05*, Antalya, Turkey, September 2005.
45. H. W. Löllmann, P. Vary: Generalized filter-bank equalizer for noise reduction with reduced signal delay, *Proc. INTERSPEECH '05*, 2105–2108, Lisbon, Portugal, September 2005.
46. H. W. Löllmann, P. Vary: Low delay filter for adaptive noise reduction, *Proc. IWAENC '05*, 205–208, Eindhoven, The Netherlands, September 2005.
47. H. W. Löllmann, P. Vary: A warped low delay filter for speech enhancement, *Proc. IWAENC '06*, Paris, France, September 2006.
48. H. W. Löllmann, P. Vary: Parametric phase equalizers for warped filter-banks, *Proc. EUSIPCO '06*, Florence, Italy, September 2006.
49. H. W. Löllmann, P. Vary: Improved design of oversampled allpass transformed DFT filter-banks with near-perfect reconstruction, *Proc. EUSIPCO '07*, Poznan, Poland, September 2007.
50. T. Lotter, P. Vary: Speech enhancement by MAP spectral amplitude estimation using a super-Gaussian speech model, *EURASIP Journal on Applied Signal Processing*, **7**, 1110–1126, May 2005.
51. A. Makur, S. K. Mitra: Warped discrete Fourier transform: theory and application, *IEEE Trans. on Circuits and Systems I*, **48**(9), 1086–1093, September 2001.
52. D. Malah, R. V. Cox, A. J. Accardi: Tracking speech-presence uncertainty to improve speech enhancement in non-stationary noise environments, *Proc. ICASSP '99*, 789–792, Phoenix, AR, USA, May 1999.
53. R. Martin, H.-G. Kang, R. V. Cox: Low delay analysis synthesis schemes for joint speech enhancement and low bit rate speech coding, *Proc. EUROSPEECH '99*, **3**, 1463–1466, Budapest, Hungary, 1999.
54. R. Martin: Noise power spectral density estimation based on optimal smoothing and minimum statistics, *IEEE Trans. on Speech and Audio Processing*, **9**(5), 504–512, July 2001.
55. R. Martin: Statistical methods for the enhancement of noisy speech, in J. Benesty, S. Makino, J. Chen (eds.), *Speech Enhancement*, 43–65, Berlin, Germany: Springer, 2005.
56. S. K. Mitra, C. D. Creusere, H. Babic: A novel implementation of perfect reconstruction QMF banks using IIR filters for infinite length signals, *Proc. ISCAS '92*, 2312–2315, San Diego, CA, USA, May 1992.
57. D. R. Morgan, J. C. Thi: A delayless subband adaptive filter architecture, *IEEE Trans. on Signal Processing*, **43**(8), 1819–1830, August 1995.
58. A. V. Oppenheim, D. Johnson, K. Steiglitz: Computation of spectra with unequal resolution using the fast Fourier transform, *Proc. of the IEEE*, **59**(2), 299–301, February 1971.
59. A. V. Oppenheim, R. W. Schafer, J. R. Buck: *Discrete-Time Signal Processing*, 2nd edition, Upper Saddle River, NJ, USA: Prentice-Hall, 1999.
60. T. W. Parks, C. S. Burrus: *Digital Filter Design*, Chichester, GB: Wiley, 1987.
61. A. Petrovsky, M. Parfieniuk, A. Borowicz: Warped DFT based perceptual noise reduction system, *Convention Paper of Audio Engineering Society*, Berlin, Germany, May 2004.

62. W. H. Press, S. A. Teukolsky, W. T. Vetterling, B. P. Flannery: *Numerical Recipes in C*, 2nd edition, Cambridge, GB: Cambridge University Press, 1992.
63. J. G. Proakis, D. G. Manolakis: *Digital Signal Processing: Principles, Algorithms, and Applications*, 3rd edition, Upper Saddle River, NJ, USA: Prentice-Hall, 1996.
64. S. R. Quackenbush and T. P. Barnwell III and M. A. Clements: *Objective Measures of Speech Quality*, Upper Saddle River, NJ, USA: Prentice-Hall, 1988.
65. C. M. Rader: An improved algorithm for high-speed autocorrelation with application to spectral estimation, *IEEE Trans. on Audio and Electroacoustics*, **18**(4), 439–441, December 1970.
66. K. R. Rao and P. Yip: *Discrete Cosine Transform*, New York, NY, USA: Academic Press, 1990.
67. M. Renfors, T. Saramäki: Recursive Nth-band digital filters – part I: design and properties, *IEEE Trans. on Circuits and Systems*, **34**(1), 24–39, January 1987.
68. M. Schönle, C. Beaugeant, K. Steinert, H. W. Löllmann, B. Sauter, P. Vary: Hands-free audio and its application to telecommunication terminals, *Proc. of Intl. Conf. on Audio for Mobile and Handheld Devices (AES)*, Seoul, Korea, September 2006.
69. G. D. T. Schuller and T. Karp: Modulated filter banks with arbitrary system delay: efficient implementation and the time-varying case, *IEEE Trans. on Signal Processing*, **48**(3), 737–748, March 2000.
70. H. W. Schüßler, W. Winkelnkemper: Variable digital filters, *Archiv der Elektrischen Übertragung (AEÜ)*, **24**(11), 524–525, 1970.
71. H. W. Schüßler: Implementation of variable digital filters, *Proc. EUSIPCO '80*, 123–129, Lausanne, Switzerland, September 1980.
72. B. Shankar M. R., A. Makur: Allpass delay chain-based IIR PR filterbank and its application to multiple description subband coding, *IEEE Trans. on Signal Processing*, **50**(4), 814–823, April 2002.
73. J. O. Smith, J. S. Abel: Bark and ERB bilinear transforms, *IEEE Trans. on Speech and Audio Processing*, **7**(6), 697–708, November 1999.
74. K. Steiglitz: A note on variable recursive digital filters, *IEEE Trans. on Acoustics, Speech, and Signal Processing*, **ASSP-28**(1), 111–112, February 1980.
75. M. A. Stone, B. C. J. Moore: Tolerable hearing aid delays II: estimation of limits imposed during speech production, *Ear and Hearing*, **32**(4), 325–338, 2002.
76. W. Sweldens: The lifting scheme: a custom-design construction of biorthogonal wavelets, *Applied and Computational Harmonic Analysis*, **3**(2), 186–200, 1996.
77. P. P. Vaidyanathan: *Multirate Systems and Filter Banks*, Upper Saddle River, NJ, USA: Prentice-Hall, 1993.
78. P. Vary: On the design of digital filter banks based on a modified principle of polyphase, *AEÜ (Archive for Electronics and Communications)*, **33**, 293–300, 1979.
79. P. Vary: Digital filter banks with unequal resolution, *Short Communication Digest of European Signal Processing Conf. (EUSIPCO)*, 41–42, Lausanne, Switzerland, September 1980.
80. P. Vary: Noise suppression by spectral magnitude estimation – mechanism and theoretical limits, *Signal Processing, Elsevier*, **8**(4), 387–400, July 1985.
81. P. Vary, R. Martin: *Digital Speech Transmission: Enhancement, Coding and Error Concealment*, Chichester, GB: Wiley, 2006.
82. P. Vary: An adaptive filter-bank equalizer for speech enhancement, *Signal Processing, Elsevier*, **86**(6), 1206–1214, June 2006.

83. M. Vetterli, J. Kovačević: *Wavelets and Subband Coding,* Upper Saddle River, NJ, USA: Prentice-Hall, 1995.
84. E. Zwicker, H. Fastl: *Psychoacoustics: Facts and Models,* 2nd edition, Berlin, Germany: Springer, 1999.

3

A Pre-Filter for Hands-Free Car Phone Noise Reduction: Suppression of Harmonic Engine Noise Components

Henning Puder

Siemens Audiological Engineering Group, Erlangen, Germany

3.1 Introduction

Single channel noise reduction is a severe challenge, especially when considering the difficulties of estimating the appropriate spectral distribution of interfering noise. Usually, all approaches, e.g. Wiener filters, Ephraim-Malah filters, and Kalman filters, rely on the assumption that the interfering noise components are longer-term stationary when compared to desired signals.

Here, we focus on the application of noise reduction for car phones meaning that the interfering noise is car noise with its specific properties. When analyzing car noise, one observes that the assumption of long-term stationarity is true only as a first approximation. Analyzing car noise in more detail, described later in this chapter in the second section, one discovers that the different components vary in dependence of the car velocity or the engine speed.

Engine noise, in particular, exhibits clear and predicable properties. One can show that engine noise consists mainly of a harmonic structure with strong spectral components at multiples of one half of the engine frequency. This allows the design of methods for specifically reducing these engine noise components since the engine frequency is available in all modern cars on the internal CAN[1] bus. The description and the analysis of different methods for specific engine noise reduction are the topic of this chapter. The target is to design a kind of pre-filter for the classic noise reduction approaches mentioned above and specifically pre-filter the engine noise components. This pre-filter facilitates the work of the classic noise reduction methods by removing the highly non-stationary and strong engine noise components.

In this chapter we present two different types of pre-filter approaches for the reduction of the engine noise components:

---

[1] CAN (Controller Area Network): Serial data network in cars for the data exchange of controller devices.

- The first type of approach, described in the third section, is a frequency selective periodic filtering of the engine noise components. Here, we propose two different types of specific notch filters which are applied at the multiples of the car engine frequency.
- The second approach, described in the fourth section, is based on a compensation filter. The engine noise components are modelled as sinusoidals with the appropriate amplitudes and phases. An adaptive filter is applied for the determination of the amplitudes and phases. The modelled components are then subtracted in the time domain.

Both approaches are described for a full-band signal with the typical sampling frequency of 8 kHz. As described in [9], the approaches can also be applied to sub-band signals which may show advantages for the following noise reduction procedures, when those are applied in sub-bands such as the Kalman filter described in [10].

In the fifth section the results of the notch filter and the compensation approaches are compared before concluding this chapter in the sixth section.

3.2 Analysis of the Different Car Noise Components

When analyzing the spectral distribution of car noise, one mainly observes very strong low frequency components followed by a sharp decrease and then a flatter decrease towards higher frequency components. Concerning the time dependence of typical car noises, one mainly observes slow variances. Both analyses are shown in an example in Fig. 3.1.

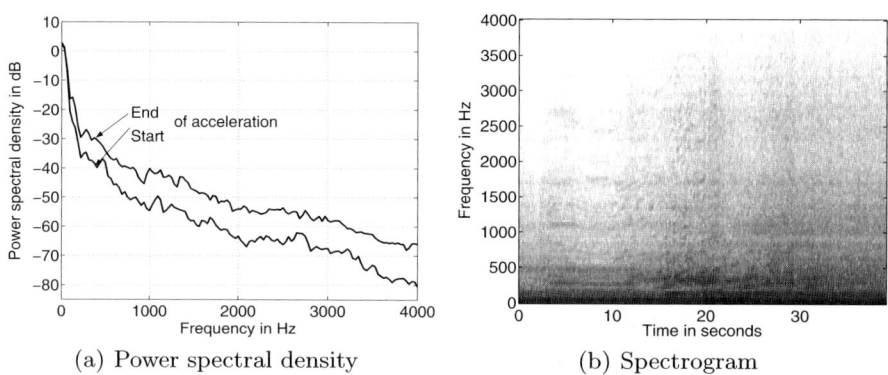

(a) Power spectral density (b) Spectrogram

Fig. 3.1. Power spectral density of a typical car noise at the start and end of an acceleration period (a) and the corresponding spectrogram (b). The acceleration start at about 10 seconds and ends at about 30 seconds.

For the given example, the noise of a car has been analyzed during an acceleration period from 50 km/h to 110 km/h. Here, an average increase in signal power of approximately 10 dB can be observed.

When one increases the spectral resolution as given in Fig. 3.2, prominent spectral lines become obvious. These spectral lines change the characteristic of the spectrum which, according to Fig. 3.1, was expected to have a rather smooth characteristic.

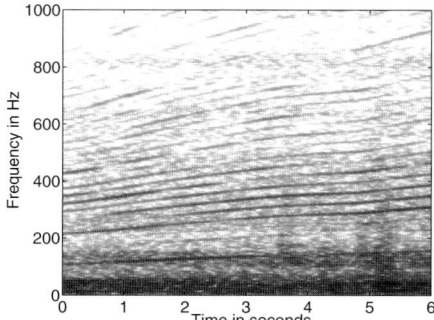

Fig. 3.2. Spectrogram of car noise when accelerating analyzed with a high frequency resolution.

In literature [12], one can find out that the strongest components of car cabin noise are due the engine, the tires and wind.

3.2.1 Wind Noise

Typical wind noise components, as depicted in Fig. 3.3, exhibit a rather smooth spectrum. Optimized car cabin design avoids the occurrence of whistling at certain frequencies. Additionally, the c_{wA} value[2] is optimized. The combination of these provokes considerably reduced wind noise power inside the car.

3.2.2 Tire Noise

The source of tire noises is the rolling of the tires on the street surface. In detail the compression and following decompression of the tire, as well as the compression of the air volume within the profile generate tire noise. The acoustic tire noise components are propagated into the car cabin. Typically, an increase in velocity results in a strong increase in noise level. However, the noise level is also influenced by the types of tires and the street surface.

---

[2] Usual standardized value to indicate the air resistance of a car.

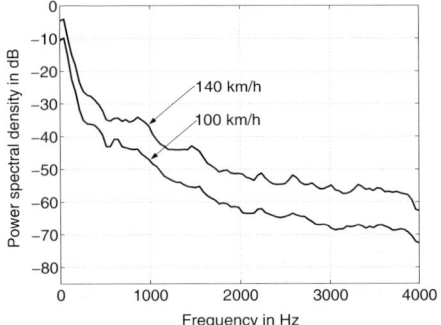

Fig. 3.3. Power spectral density of wind noise components recorded in a wind tunnel at two different wind speeds.

This dependence on the surface is shown in Fig. 3.4. Here, the power spectral density (PSD) (a) and the spectrogram (b) are depicted for two different street surfaces but at the same velocity and with the same car. Remarkable differences can be observed for frequency components below 1 kHz.

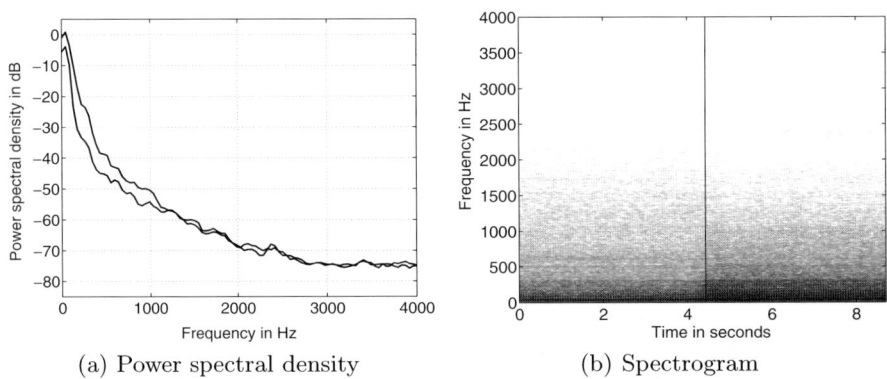

(a) Power spectral density (b) Spectrogram

Fig. 3.4. Power spectral density of a typical tire noise for two different street surfaces (a) and the corresponding spectrogram (b). Here the different surfaces are separated by the vertical line.

3.2.3 Engine Noise

Engine noise components, however, show a considerably higher character of predictability. Analyzing engine noise, which is present in the car cabin, in detail and with a high frequency resolution, one observes strong spectral components at multiples of one half of the engine frequency. For four or six cylinder engines the spectral components at double and triple the engine speed are especially strong, respectively, as shown in Fig. 3.5.

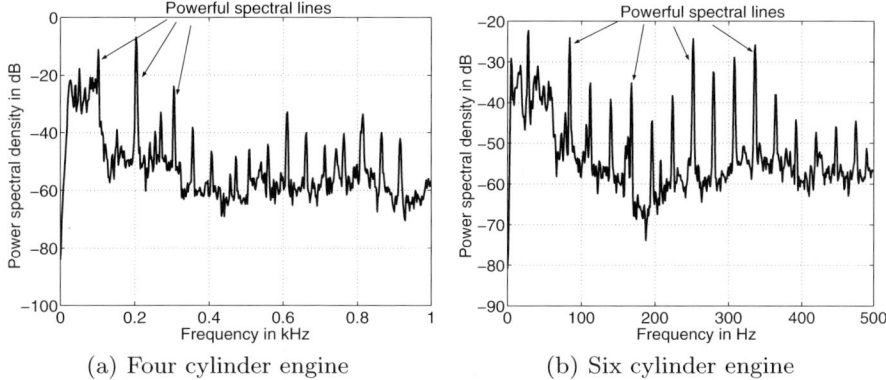

(a) Four cylinder engine (b) Six cylinder engine

Fig. 3.5. Power spectral densities of typical engine noises for a four cylinder engine (a) and six cylinder engine (b).

These specific observations can be explained by the

- gas forces and
- mass forces or moments

present in an combustion engine.

Gas forces occur every fourth cycle when the gas air mixture explodes [1]. This is equivalent to a combustion at every second rotation of the engine corresponding to a frequency which is one half the engine frequency. Unbalanced masses, especially in four cycle engines, provoke periodic forces and moments [2]. They are the reason for the stronger spectral components every second or third line for four and six cylinder engines, respectively. The most important properties of engine noise are the specific harmonic noise components at multiples of one half of the engine frequency. Additionally, based on the engine frequency which is available on the CAN bus for nearly all modern cars, the frequency location can be predicted.

Based on this prediction, selective noise reduction methods can be developed. These noise reduction methods will be the content of the following sections: In Sec. 3.3, a method for a narrow band attenuation with comb filters is described where the focus is to selectively remove the engine noise components while distorting the desired signal as little as possible. In Sec. 3.4, a compensation filter method is presented in combination with an optimal stepsize control which allows a removal of the periodic engine noise components without attenuating or distorting the desired signal components.

3.3 Engine Noise Removal Based on Notch Filters

The basic idea for this notch filter approach [8] is to attenuate the engine noise components selectively in frequency by a narrow-band filter without noticeably disturbing the desired signal. The typical notch filter is characterized by a pair of poles and nulls as described in Fig. 3.6(a). The frequency response is given by [7]

$$H_{\mathrm{N}}(z) = \frac{(1+r_{\mathrm{p}})^2}{4} \frac{1 - 2\cos(\Omega_0)\, z^{-1} + z^{-2}}{1 - 2\, r_{\mathrm{p}} \cos(\Omega_0)\, z^{-1} + r_{\mathrm{p}}^2\, z^{-2}}, \qquad (3.1)$$

and is depicted in Fig. 3.6(b).

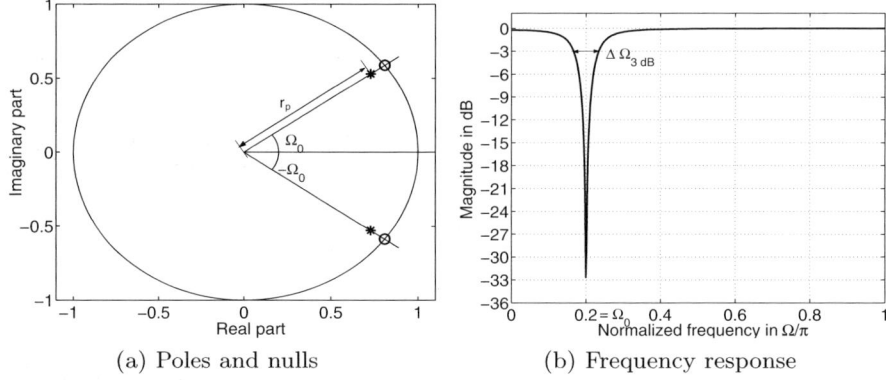

(a) Poles and nulls (b) Frequency response

Fig. 3.6. Poles (o) and nulls (∗) of notch filter $H_{\mathrm{N}}(z)$ resulting in a notch at the normalized frequency $\Omega_0(n) = 0.2\,\pi$ with pole radius $r_{\mathrm{p}} = 0.9$ (a). The right graph (b) shows the frequency response where the 3 dB notch width is depicted.

This filter attenuates signal components at the frequency

$$f_0(n) = \frac{\Omega_0(n)}{2\pi}\, f_{\mathrm{s}}, \qquad (3.2)$$

where f_{s} is the sampling frequency of the signal and

$$\Omega_0(n) = n_0\, \Omega_{\mathrm{M}}(n) \qquad (3.3)$$

is a multiple of one half of the normalized engine frequency, $rpm(n)$:

$$\Omega_{\mathrm{M}}(n) = \frac{2\pi}{f_{\mathrm{s}}} \frac{rpm(n)}{2}. \qquad (3.4)$$

The pole radius r_{p} determines the 3 dB cut-off frequency of the filter, and as such, the selectivity of the filter. The closer r_{p} is to 1, i.e. the closer the pole is to unity circle, the more selective the filter and the smaller the filter width.

The cut-off frequency $\Delta\Omega_{3\,\mathrm{dB}}$ can be denoted as

$$\cos\left(\frac{\Delta\Omega_{3\,\mathrm{dB}}}{2}\right) = \frac{2r_\mathrm{p}}{1+r_\mathrm{p}^2}, \tag{3.5}$$

and the value r_p can be determined dependent on the 3 dB notch width according to the following equation:

$$r_\mathrm{p} = \frac{1}{\cos\left(\frac{\Delta\Omega_{3\,\mathrm{dB}}}{2}\right)} - \sqrt{\frac{1}{\cos^2\left(\frac{\Delta\Omega_{3\,\mathrm{dB}}}{2}\right)} - 1}. \tag{3.6}$$

For the choice of the notch width, the best compromise based on the following criteria has to be found:

- The notch width must be large enough to encompass the engine noise components for optimum attenuation.
- The notch width should be as small as possible in order to cause the least amount of disturbance to the desired signal.

Analyzing the results obtained with these kinds of notch filters, one observes an effective noise reduction for comb filters with a notch width of approximately 1.9 Hz (see Fig. 3.7(c)). The distortion of the desired speech is low but already noticeable. A slight signal distortion can be observed as well as some howling. However, compared to the well-known signal distortions typically provoked by Wiener filtering, the distortion is very moderate.

One problem associated with the type of notch filters, investigated thus far, is that the steepness of the notch filter is limited due to the filter design. Filters of first order with one pole and null at the notch frequency do not allow the desired selectivity with respect to frequency.

In the following, we want to investigate the design of second order notch filters and analyze possibilities to reduce the trade-off between the engine noise attenuation and distortion of the desired signal.

The design of second order notch filters is based on a second order high pass with two poles and two nulls. The focus here is to design notch filters with steeper slopes but with certain cut-off widths to accomplish two goals:

- better suppression of the engine noise components and
- less distortion of the desired signal.

The prototype high-pass filter can be denoted follows:

$$H_{\mathrm{HP}}(z) = \frac{c_0 + c_1\,z^{-1} + c_2\,z^{-2}}{1 + d_1\,z^{-1} + d_2\,z^{-2}}, \tag{3.7}$$

where the filter values are

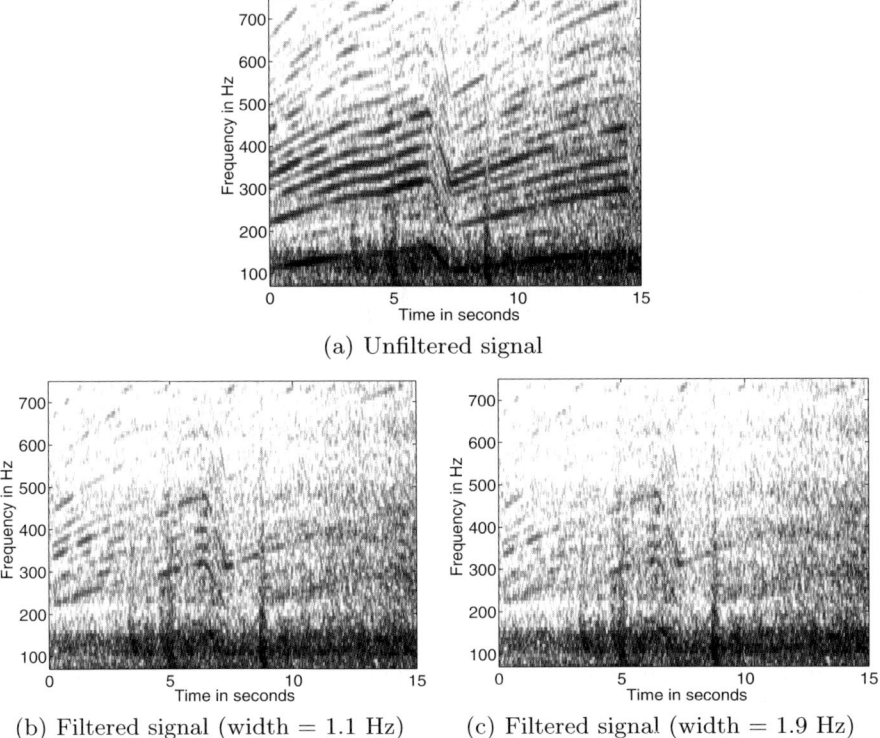

Fig. 3.7. Comparing the results of notch filters with a width of 1.1 Hz (bottom left) and 1.9 Hz (bottom right) with the spectrogram of unfiltered engine noise (top). The noise reduction is increased for the wider notch filter. However, speech distortions can be observed applying the filter with 1.9 Hz width to speech signals.

$$\begin{aligned} c_0 &= 0.7917, \\ c_1 &= -1.5736, \\ c_2 &= 0.7917, \\ d_1 &= -1.5296, \\ d_2 &= 0.6273 \end{aligned} \tag{3.8}$$

and where the 3 dB cut-off frequency is the following:

$$\cos\left(\Omega_{3\,\mathrm{dB,HP}}\right) = -\frac{4c_0c_1 - d_1(1+d_2)}{4(2c_0^2 - d_2)} \\ + \sqrt{\frac{(4c_0c_1 - d_1(1+d_2))^2}{16(2c_0^2 - d_2)^2} - \frac{c_1^2 - \frac{1}{2}(1+d_1^2+d_2^2-2d_2)}{4c_0^2 - 2d_2}}. \tag{3.9}$$

The frequency response as well as the location of poles and nulls are depicted in Fig. 3.8.

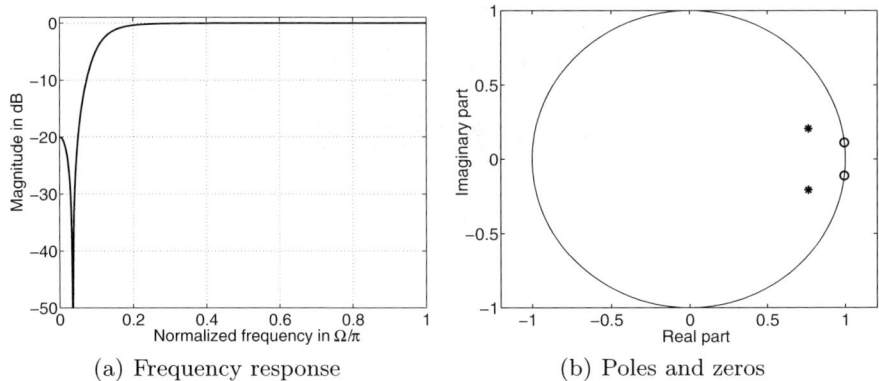

Fig. 3.8. Frequency response (a) and locations of poles and nulls (b) of the prototype high-pass filter.

The poles $z_{p,i}$ and its nulls $z_{n,i}$ with $i \in \{1,2\}$ of the prototype high-pass filter are located at

$$z_{n,1} = e^{j\phi_1} \quad \text{and} \quad z_{n,2} = e^{-j\phi_1} \quad \text{as well as}$$
$$z_{p,1} = r_p\, e^{j\phi_2} \quad \text{and} \quad z_{p,2} = r_p\, e^{-j\phi_2}. \tag{3.10}$$

A modified notch filter can be designed by squeezing the poles and nulls closer together by a factor M, which determines the reduction of the filter width, and by shifting them to the desired notch frequencies $+\Omega_1$ and $-\Omega_1$ in order to obtain a real-value filter.

The 3 dB width of the resulting second order notch filter is given by

$$\Delta\Omega_{3\,\mathrm{dB}} = 2\,\frac{\Omega_{3\,\mathrm{dB,HP}}}{M}. \tag{3.11}$$

The filter $H^+(\Omega)$ corresponding to the frequency shift of $+\Omega_1$ has following poles and nulls

$$\tilde{z}_{n,1} = e^{j\,(\phi_1/M + \Omega_1)} \quad \text{and} \quad \tilde{z}_{n,2} = e^{-j\,(\phi_1/M + \Omega_1)} \quad \text{as well as}$$
$$\tilde{z}_{p,1} = r_p^{1/M}\, e^{j\,(\phi_2/M + \Omega_1)} \quad \text{and} \quad \tilde{z}_{p,2} = r_p^{1/M}\, e^{-j\,(\phi_2/M + \Omega_1)}. \tag{3.12}$$

The frequency response of the modified notch filter can be denoted as the product of this filter $H^+(z)$ and the filter $H^-(z)$ which denotes the filter corresponding to the shift to the frequency $-\Omega_1$:

$$H_{N,\text{mod}}(z) = H^+(z)\, H^-(z)$$
$$= |H^+(z)|^2$$
$$= |k_N|^2 \left| \frac{b_0 + b_1\, z^{-1}\, e^{j\Omega_1} + b_2\, z^{-2}\, e^{j2\Omega_1}}{1 + a_1\, z^{-1}\, e^{j\Omega_1} + a_2\, z^{-2}\, e^{j2\Omega_1}} \right|^2, \qquad (3.13)$$

with

$$k_N = \frac{1 + r_p^{2/M} + 2 r_p^{1/M} \cos\left(\frac{\phi_2}{M}\right)}{2 + 2\cos\left(\frac{\phi_1}{M}\right)},$$
$$b_0 = 1,$$
$$b_1 = -2\cos\left(\frac{\phi_1}{M}\right),$$
$$b_2 = 1,$$
$$a_1 = -2\, r_p^{1/M} \cos\left(\frac{\phi_2}{M}\right),$$
$$a_2 = r_p^{2/M}. \qquad (3.14)$$

For $M = 10$, in Fig. 3.9 the frequency response of such a notch filters is

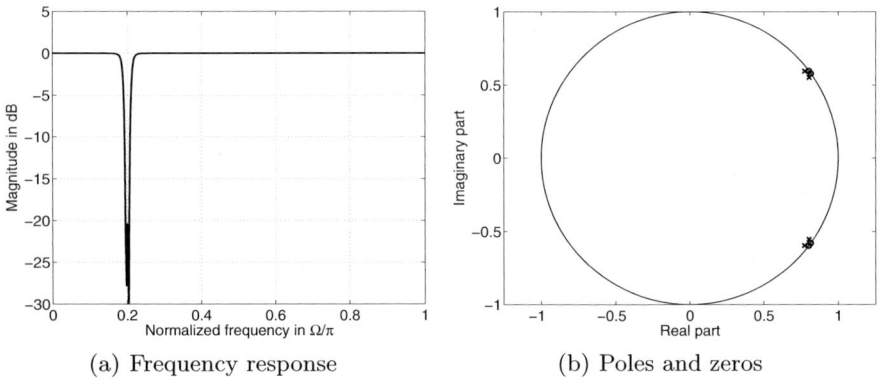

Fig. 3.9. Frequency response (a) and locations of poles and nulls (b) of the modified notch filter for $M = 10$ and a notch frequency of $\Omega = 0.2\pi$.

depicted together with the distribution of zeros and nulls where the notch frequency is $\Omega = 0.2\,\pi$.

In order to suppress N_1 engine harmonics at multiples of the engine frequency Ω_M the following filter has to be utilized:

$$H_{N,\text{mult}}(z) = |k_N|^{2N_1} \prod_{i=1}^{N_1} \left| \frac{b_0 + b_1\, z^{-1}\, e^{j\, i\Omega_M} + b_2\, z^{-2}\, e^{j\, 2i\Omega_M}}{1 + a_1\, z^{-1}\, e^{j\, i\Omega_M} + a_2\, z^{-2}\, e^{j\, 2i\Omega_M}} \right|^2. \qquad (3.15)$$

This type of second order notch filters have been tested with different notch widths of 0.53 Hz and 1.06 Hz. The attenuation of the engine harmonics which can be obtained with these choices is equivalent to first order notch filters with widths of 1.1 or 1.9 Hz, respectively. Results for a second order notch filter are depicted in Fig. 3.10.

Due to the reduced filter width, speech distortion can be reduced. For the filter width of 0.53 Hz, no distortion can be observed. Nevertheless, not all engine noise components can be completely cancelled.

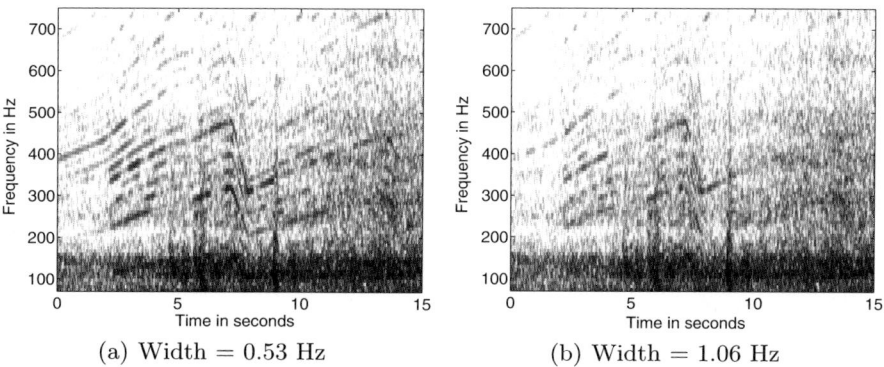

Fig. 3.10. Spectrograms for the modified notch filters with a width of 0.53 Hz (a) and 1.06 Hz (b).

An approach which achieves both goals, no signal distortion and a complete cancelling of the engine harmonics, is still missing. This leads us to investigate another approach which will be described in the following section. The basic idea here is to model the harmonic components with adaptive filters and to subtract them from the noisy signal. Supposing this approach works as desired, it allows the cancellation of engine harmonics without distorting the desired signal.

3.4 Compensation of Engine Harmonics with Adaptive Filters

The compensation method is an alternative approach to cancel engine noise harmonic components. Here, the harmonics are modelled in the time domain with the correct phase and amplitude [11] at the known frequencies which are multiples of one half of the engine frequency. These modelled components are then subtracted from the noisy signal. This approach can be expressed by the following equation:

$$v(n) = x(n) - \sum_{i=-N_1(n)}^{N_1(n)} \hat{c}_i(n)\, e^{j\phi_i(n)}, \quad (3.16)$$

where $x(n)$ denotes the desired signal disturbed by car noise. Furthermore, we have

$$\hat{c}_0(n) = 0. \quad (3.17)$$

The number of modelled harmonics, $N_1(n)$, is chosen dependent on the engine frequency such that the main contributing components can be cancelled. The compensation method is demonstrated in Fig. 3.11.

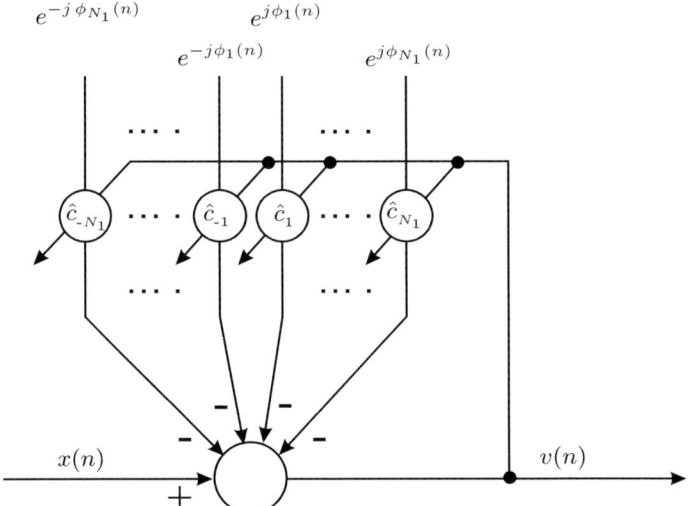

Fig. 3.11. Scheme of the compensation approach.

The time dependent phases of the harmonics are determined dependent on the time varying engine frequency $\Omega_M(n)$:

$$\phi_i(n) = \phi_i(n-1) + i\,\Omega_M(n). \quad (3.18)$$

Utilizing the least mean square (LMS) algorithm [3–5] for adapting the coefficients $\hat{c}_i(n)$, the update equation can be expressed as follows:

$$\hat{\mathbf{c}}(n+1) = \hat{\mathbf{c}}(n) - \frac{\mu}{2} \nabla_{\hat{\mathbf{c}}(n)} |v(n)|^2, \quad (3.19)$$

where $\hat{\mathbf{c}}(n)$ denotes the vector of filter coefficients

$$\hat{\mathbf{c}}(n) = \left[\hat{c}_{-N_1}(n),\, \ldots,\, \hat{c}_{N_1}(n)\right]^{\mathrm{T}}. \quad (3.20)$$

The m-th component of the gradient can be expressed as:

$$\nabla_{\hat{c}_m(n)} |v(n)|^2 = \frac{\partial |v(n)|^2}{\partial \mathrm{Re}\{\hat{c}_m(n)\}} + j \frac{\partial |v(n)|^2}{\partial \mathrm{Im}\{\hat{c}_m(n)\}}, \quad (3.21)$$

where $\mathrm{Re}\{\ldots\}$ and $\mathrm{Im}\{\ldots\}$ denote the real and imaginary component of a complex value. Calculating this gradient based on

$$|v(n)|^2 = \left[x(n) - \sum_i \hat{c}_i(n)\, e^{j\phi_i(n)}\right] \left[x(n) - \sum_i \hat{c}_i^*(n)\, e^{-j\phi_i(n)}\right], \quad (3.22)$$

one obtains

$$\begin{aligned}
\nabla_{\hat{c}_m(n)} |v(n)|^2 &= -e^{j\phi_m(n)}\, v^*(n) - e^{-j\phi_m(n)}\, v(n) \\
&\quad + j\left[-j\, e^{j\phi_m(n)}\, v^*(n) + j\, e^{-j\phi_m(n)}\, v(n)\right] \\
&= -2\, e^{-j\phi_m(n)}\, v(n),
\end{aligned} \quad (3.23)$$

what finally results in the following adaptation rule:

$$\hat{c}_i(n+1) = \hat{c}_i(n) + \mu\, v(n)\, e^{-j\phi_i(n)}. \quad (3.24)$$

Filter and adaptation equations then can be expressed as follows:

$$v(n) = x(n) - \mathbf{e}^{\mathrm{H}}(n)\, \hat{\mathbf{c}}(n), \quad (3.25)$$

$$\hat{\mathbf{c}}(n+1) = \hat{\mathbf{c}}(n) + \mu\, v(n)\, \mathbf{e}(n), \quad (3.26)$$

where the complex exponential functions are denoted with the vector

$$\mathbf{e}(n) = \left[e^{-j\phi_{-N_1}(n)}, \ldots, e^{-j\phi_{N_1}(n)}\right]^{\mathrm{T}}. \quad (3.27)$$

An appropriate initialization of the phases $\phi_i(0)$, e.g. as

$$\phi_i(0) = 0, \quad (3.28)$$

results in symmetric phases and filter coefficients

$$\phi_i(n) = -\phi_{-i}(n), \quad (3.29)$$
$$\hat{c}_i(n) = \hat{c}_{-i}^*(n) \quad (3.30)$$

if the input signal $x(n)$ is assumed to be real. Then also the output signal $v(n)$ stays a real-value signal.

3.4.1 Step-Size Control

Step-size control is an essential issue for the substantial reduction of engine noise harmonics. Since all other signal components besides the harmonics are disturbing the adaptation, a powerful step-size control has to be developed in order to allow a fast, but reliable, adaptation towards the correct filter values. Before deriving the optimum step-size, first, the relation of the step-size and the adaptation of the filter coefficients is examined.

Considering the adaptation of only one single spectral component and combining the filter equation (Eq. 3.24) and adaptation equation (Eq. 3.25), one obtains the following:

$$\hat{c}_i(n+1) = (1-\mu)\,\hat{c}_i(n) + \mu\, x(n)\, e^{-j\phi_i(n)}. \qquad (3.31)$$

This is equivalent to a low-pass filter of the frequency shifted signal $\tilde{x}(n) = x(n)\, e^{-j\phi_i(n)}$. The corresponding frequency response can be denoted as

$$H(e^{j\Omega},\mu) = \frac{\hat{C}_i(e^{j\Omega})}{\tilde{X}(e^{j\Omega})} = \frac{\mu e^{-j\Omega}}{1-(1-\mu)\,e^{-j\Omega}}, \qquad (3.32)$$

which has the following 3 dB cut-off frequency:

$$\cos\left(\Omega_{3\,\mathrm{dB}}(\mu)\right) = 1 - \frac{\mu^2}{2(1-\mu)}. \qquad (3.33)$$

For $\mu \ll 1$ also $\Omega_{3\,\mathrm{dB}}(\mu) \ll 1$ the following approximation

$$\cos\left(\Omega_{3\,\mathrm{dB}}(\mu)\right) \approx 1 - \frac{\Omega_{3\,\mathrm{dB}}(\mu)^2}{2} \qquad (3.34)$$

is valid and the 3 dB cut-off frequency can be approximated by (see [13]):

$$\Omega_{3\,\mathrm{dB}}(\mu) = \sqrt{\frac{\mu^2}{1-\mu}} \approx \mu. \qquad (3.35)$$

Thus, the adaptation equation is equivalent to an IIR low-pass filter for the input signal which has been shifted by the frequency for which the engine noise component has to be cancelled.

The step-size μ determines the width of the IIR filter. The larger the step-size is, the faster the filter can adapt. However, signals other than the engine noise component are disturbing for the adaptation. The lower the step-size is, the less the other signals disturb the adaptation. These other signals are other noise components as well as speech components. In the presence of these components, the step-size has to be lowered in order not to provoke a distortion of the adaptation and, consequently, to avoid a distortion of the filtered output signal. Since these disturbing components are different for the specific engine noise harmonics, a different step-size for each harmonic is necessary.

For the now different and time-dependent step-sizes $\mu_i(n)$ summarized within the matrix

$$\boldsymbol{\mu}(n) = \operatorname{diag}\left\{\left[\mu_{-N_1}(n), \ldots, \mu_{N_1}(n)\right]^{\mathrm{T}}\right\}$$
$$= \begin{bmatrix} \mu_{-N_1}(n) & 0 & \cdots & 0 \\ 0 & \mu_{-N_1+1}(n) & \cdots & 0 \\ \vdots & \vdots & \ddots & \vdots \\ 0 & 0 & \cdots & \mu_{N_1}(n) \end{bmatrix}, \quad (3.36)$$

the adaptation equation is the following:

$$\hat{\boldsymbol{c}}(n+1) = \hat{\boldsymbol{c}}(n) + \boldsymbol{\mu}(n)\, v(n)\, \boldsymbol{e}(n)\,. \qquad (3.37)$$

Denoting the true weighting of the engine harmonics as c_i, supposing that they are time-independent, and putting them together within the vector $\boldsymbol{c}$, the vector containing the weighting error can be denoted as

$$\boldsymbol{w}(n) = \boldsymbol{c} - \hat{\boldsymbol{c}}(n)\,, \qquad (3.38)$$

with the following recursion:

$$\boldsymbol{w}(n+1) = \boldsymbol{w}(n) - \boldsymbol{\mu}(n)\, v(n)\, \boldsymbol{e}(n). \qquad (3.39)$$

Deriving the expectation value of the norm of the error vector

$$\mathrm{E}\{\boldsymbol{w}^{\mathrm{H}}(n+1)\,\boldsymbol{w}(n+1)\}$$
$$= \mathrm{E}\Big\{\boldsymbol{w}^{\mathrm{H}}(n)\,\boldsymbol{w}(n) - \boldsymbol{w}^{\mathrm{H}}(n)\boldsymbol{\mu}(n)\, v(n)\, \boldsymbol{e}(n)$$
$$\qquad - v^*(n)\, \boldsymbol{e}^{\mathrm{H}}(n)\, \boldsymbol{\mu}(n)\, \boldsymbol{w}(n) + |v(n)|^2\, \boldsymbol{e}^{\mathrm{H}}(n)\, \boldsymbol{\mu}^2(n)\, \boldsymbol{e}(n)\Big\}$$
$$= \mathrm{E}\Big\{\boldsymbol{w}^{\mathrm{H}}(n)\,\boldsymbol{w}(n) - 2\mathrm{Re}\{v^*(n)\, \boldsymbol{e}^{\mathrm{H}}(n)\, \boldsymbol{\mu}(n)\, \boldsymbol{w}(n)\}$$
$$\qquad + |v(n)|^2\, \boldsymbol{e}^{\mathrm{H}}(n)\, \boldsymbol{\mu}^2(n)\, \boldsymbol{e}(n)\Big\} \qquad (3.40)$$

with respect to each component of the step-size vector,

$$\frac{\partial}{\partial \mu_i}\mathrm{E}\{\boldsymbol{w}^{\mathrm{H}}(n+1)\,\boldsymbol{w}(n+1)\} \overset{!}{=} 0\,, \qquad (3.41)$$

one obtains the following equation for the optimal step-size:

$$\mu_{i,\mathrm{opt}}(n) = \frac{\mathrm{E}\big\{\mathrm{Re}\{v^*(n)\, e_i^*(n)\, w_i(n)\}\big\}}{\mathrm{E}\big\{v^2(n)\, |e_i(n)|^2\big\}}. \qquad (3.42)$$

With $e_i(n) = e^{-j\phi_i(n)}$ this expression may be noted as follows:

$$\mu_{i,\mathrm{opt}}(n) = \frac{\mathrm{E}\big\{\mathrm{Re}\{v^*(n)\, e^{j\phi_i(n)}\, w_i(n)\}\big\}}{\mathrm{E}\{v^2(n)\}}. \qquad (3.43)$$

In order to simplify this equation, we assume that the error signal $v(n)$ is composed of three components:

$$v(n) = u_i(n) + \left[w_i(n) + p_i(n)\right] e^{j\phi_i(n)}, \tag{3.44}$$

where

- $u_i(n)$ denotes all components separate from the i-th engine harmonic,
- $w_i(n)$ denotes the unknown error of the estimated amplitude of the i-th engine harmonic, and
- $p_i(n)$ denotes the desired signal (speech) component at the i-th engine harmonic.

With these signals the numerator of the optimal step-size can be denoted as follows:

$$\begin{aligned}
&\mathrm{E}\Big\{\mathrm{Re}\big\{v^*(n)\, e^{j\phi_i(n)}\, w_i(n)\big\}\Big\} \\
&= \mathrm{E}\Big\{\mathrm{Re}\big\{\left[u_i^*(n) + [w_i^*(n) + p_i^*(n)]\, e^{-j\phi_i(n)}\right] e^{j\phi_i(n)}\, w_i(n)\big\}\Big\} \\
&= \underbrace{\mathrm{E}\Big\{\mathrm{Re}\big\{u_i^*(n)\, e^{j\phi_i(n)}\, w_i(n)\big\}\Big\}}_{=0} + \mathrm{E}\Big\{\mathrm{Re}\big\{[w_i^*(n) + p_i^*(n)]\, w_i(n)\big\}\Big\} \\
&= \mathrm{E}\Big\{|w_i(n)|^2\Big\}.
\end{aligned} \tag{3.45}$$

Two components are zero due to the orthogonality of $w_i(n)$ with $p_i(n)$ and with $u_i(n)$.

The optimal step-size can then be denoted as

$$\mu_{i,\mathrm{opt}}(n) = \frac{\mathrm{E}\big\{|w_i(n)|^2\big\}}{\mathrm{E}\big\{|v(n)|^2\big\}} \tag{3.46}$$

which is the quotient of the residual error power $\mathrm{E}\{|w_i(n)|^2\}$ and the overall signal power at the filter output $\mathrm{E}\{|v(n)|^2\}$ comparable to the optimum step-size for echo cancellation applications [6].

3.4.2 Calculating the Optimal Step-Size

According to Eq. 3.46, the power of the residual engine noise component (numerator) and the overall power of the signal after the compensation filter (denominator) have to be calculated or estimated.

The calculation of the denominator $\mathrm{E}\{|v(n)|^2\}$ can be easily achieved since the signal $v(n)$ is available. The power can be estimated simply by a recursive averaging:

$$\overline{|v(n)|^2} = \alpha \, \overline{|v(n-1)|^2} + (1-\alpha)\,|v(n)|^2, \qquad (3.47)$$

$$\mathrm{E}\bigl\{|v(n)|^2\bigr\} \approx \overline{|v(n)|^2}. \qquad (3.48)$$

The calculation of the numerator $\mathrm{E}\{|w_i(n)|^2\}$ is more complex since the signal $w_i(n)$ is not available, and the estimation has to be based only on the available signal $v(n)$. The other car noise components and the speech signals interfere with the desired estimation. Therefore, for a reliable estimation two targets have to be fulfilled:

- Signal components $u_i(n)$ separate from the engine noise harmonics have to be removed.
- Speech components overlapping the engine noise harmonics $p_i(n)$ should not be considered for the estimation.

The first task can be obtained by a frequency transposition of the signal with regard to the respective frequency of the engine noise harmonic

$$\tilde{v}_i(n) = v(n)\,e^{-j\phi_i(n)} \qquad (3.49)$$

followed by a low-pass filtering in order to obtain estimates for the sum of speech and engine noise components at the frequency of interest:

$$\tilde{v}_i^{(\mathrm{LP})}(n) = h_{\mathrm{LP}}(n) * \tilde{v}_i(n) \approx w_i(n) + p_i(n). \qquad (3.50)$$

For the low-pass filter, an IIR filter of fifth order is chosen with a 3 dB cut-off frequency at approximately 10 Hz and a stop-band attenuation of about 60 dB. The high frequency selectivity is necessary to select single engine noise harmonics for engine frequencies above 1200 rpm (equivalent to 10 Hz). Beginning at approximately 2000 rpm, it is apparent by engine noise components that the filter selectivity is sufficient.

The second task is to separate engine noise and speech components at the engine's harmonic frequencies. This has to be performed in order to avoid a misadaptation of the filter coefficients in the presence of speech, cancellation of speech components, and speech distortion.

Having investigated several different possibilities, a conditional recursive averaging has been chosen. The motivation for this conditional estimation can be based on Fig. 3.14 (a). Analyzing this spectrogram of speech which is disturbed by car noise with strong engine noise components, one observes that speech components are only present in limited spectral sections and only for short time periods. Therefore, not all engine noise harmonics are superimposed by speech components, and the adaptation has to be stopped during speech activity only for those components. A stop of the estimation during speech activity for all harmonics is not appropriate as this prevents tracking the engine noise harmonics during speech activity. This alternative approach with voice activity detection (VAD) will be analyzed and compared to the results obtained with the preferred method.

The following estimation procedure provides the time and frequency selective adaptation. This procedure does what is intuitively possible for the human observer. It differentiates short time, powerful spectral speech components and engine noise harmonics. Latter have to be considered during the averaging procedure in order to obtain an estimation for $\mathrm{E}\{|w_i(n)|^2\}$.

The formal description of the estimator is the following:

$$\overline{\left|\tilde{v}_i^{(\mathrm{LP})}(n)\right|^2} = \begin{cases} \alpha \overline{\left|\tilde{v}_i^{(\mathrm{LP})}(n-1)\right|^2} + (1-\alpha)\left|\tilde{v}_i^{(\mathrm{LP})}(n)\right|^2 & : \text{cond. true,} \\ \overline{\left|\tilde{v}_i^{(\mathrm{LP})}(n-1)\right|^2} & : \text{else,} \end{cases} \quad (3.51)$$

where the condition (cond. true) is the following

$$\left|\tilde{v}_i^{(\mathrm{LP})}(n)\right|^2 < \beta \overline{\left|\tilde{v}_i^{(\mathrm{LP})}(n-1)\right|^2}. \quad (3.52)$$

Thus, the recursive averaging is stopped when the current input signal power at the frequency of the i-th harmonic exceeds its mean estimated power by the factor of β. This is equivalent to a nearly instantaneous increase of the power. This generally only occurs when speech components are present. The power of engine noise harmonics vary more slowly. A choice of $\beta \in [2, 2.5]$ showed the best results for several test signals.

The required estimate can thus be obtained by

$$\mathrm{E}\{|w_i(n)|^2\} \approx \overline{\left|\tilde{v}_i^{(\mathrm{LP})}(n)\right|^2}. \quad (3.53)$$

The estimation method for the optimal step size is summarized in Fig. 3.12.

In addition to the mentioned procedures, two further measures allow longer-term stable estimation and adaptation results.

1. In the case that the CAN bus provides the engine moment, the compensation of the engine harmonics is only activated when the moment exceeds 15 Nm. Otherwise tire noise is especially dominant and the risk of misadaptation and target signal distortion is higher than a possible slight gain in noise reduction.
2. After a gear change or a reactivation of the cancellation (see item 1.), the condition according Eq. 3.51 is not checked for a short time period in order to allow an adaptation to the new condition. Adhering to the known condition could prevent the estimator from adapting to the true values. The reason is that the conditional estimation only allows for a slight value change which is too small in such completely changed conditions.

3.4.3 Results of the Compensation Approach

The performance of the compensation approach is analyzed for engine noise only as well as for engine noise superimposed by speech signals. In contrast

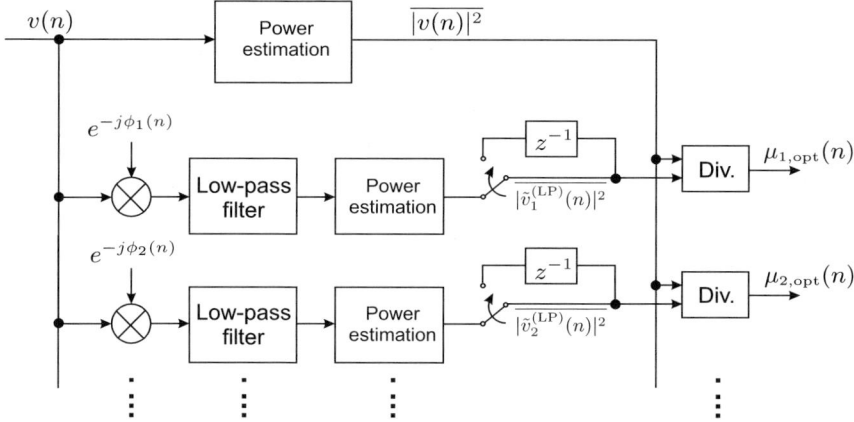

Fig. 3.12. Overview of the step-size calculation of the cancellation of engine noise harmonics with adaptive compensation filters. The switches are turned whenever the condition according to Eqn. 3.51 is not true, which means that the adaptation is stopped.

to the filter approach from Sec. 3.3, here, it is necessary to investigate the behavior of this approach also for the case of speech presence. The reason is that speech signals are a significant disturbance for the adaptation.

The spectrograms depicted in Fig. 3.13 are the results for the noise only case. The desired results are obtained since the engine noise components are sufficiently removed. Only slight shadows remain.

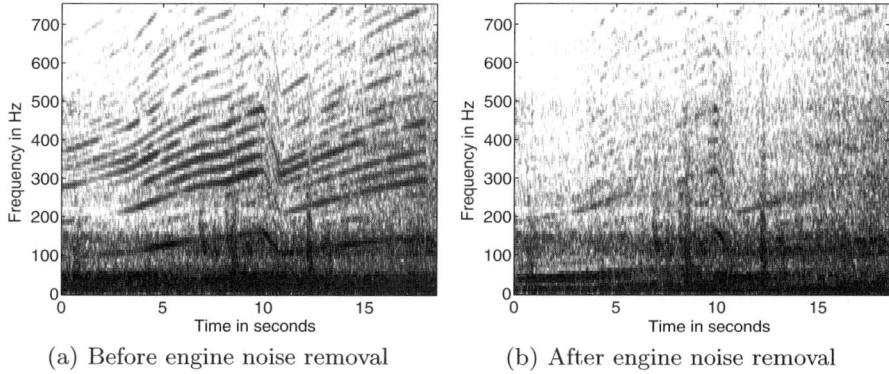

(a) Before engine noise removal (b) After engine noise removal

Fig. 3.13. Spectrograms of car noise before (a) and after (b) removal of the engine noise components.

The results for the second case of investigation, where the car noise is superimposed by speech, are depicted in Fig. 3.14. These results are very satisfying since a considerable reduction of engine noise components – also in the presence of speech components – has been obtained.

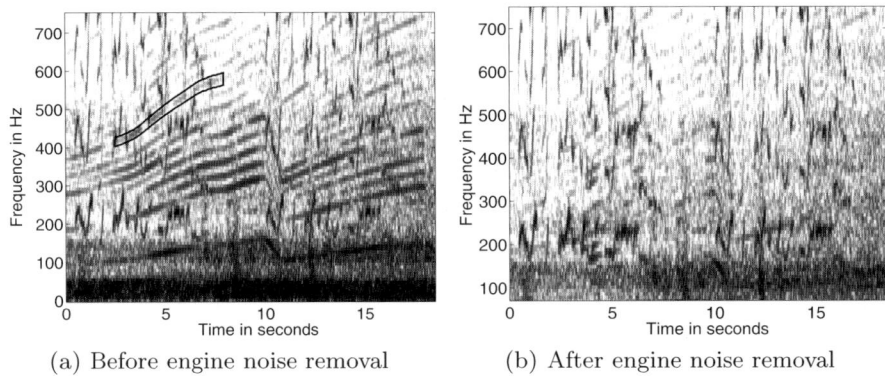

(a) Before engine noise removal   (b) After engine noise removal

Fig. 3.14. Spectrograms of car noise superimposed with speech before (a) and after (b) removal of the engine noise components. Especially for the highlighted component (left) several speech components are present which do not disturb the adaptation.

The estimation of the residual power of the engine noise components $E\{|w_i(n)|^2\}$ according to Eq. 3.51 shows the desired results. The engine noise components are well reduced without distorting the speech components. The fact that speech components are not disturbed has also been subjectively evaluated by informal listening tests.

In order to analyze the results in more detail, the estimated values for $E\{|w_i(n)|^2\}$ and the step-size are also compared to the values obtained without speech presence. In the case where the car noise signal only is available, the conditional recursive smoothing according to Eq. 3.51 can be performed continuously since no speech components are disturbing the adaptation. The values obtained with this procedure are used as reference in the following.

The results are also compared to the alternative with a VAD. This means that the estimation in Eq. 3.51 is stopped when voice activity is detected. Thus, a VAD is used instead of the proposed power based condition.

One specific engine noise component has been chosen for this comparison. This component is highlighted in Fig. 3.14 (a). The adaptation control is specifically challenging for this component since during speech activity the power of this engine noise harmonic varies severely.

Considering first the estimation of the residual power $E\{|w_i(n)|^2\}$, according to Fig. 3.15, above, one observes that the conditional smoothing generally follows the true values rather well. Sometimes an overestimation occurs which is rather a consequence of the different adaptation performance in the presence of speech than a wrong consideration of speech components for the estimate. The VAD alternative, however, leads to severe false estimates which can be observed around 4 sec. Here a following of the changes is not possible.

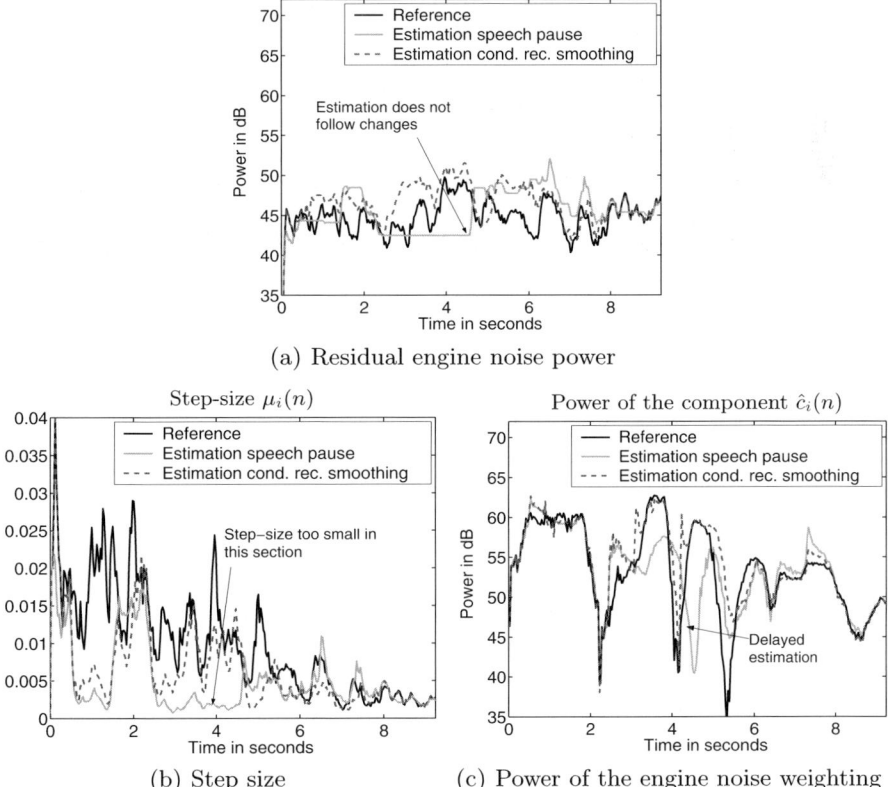

Fig. 3.15. Estimates for the residual power of the engine noise components $E\{|w_i(n)|^2\}$ (a), the step-size $\mu_i(n)$ (b), and the power of the engine noise weighting $\hat{c}_i(n)$ (c). Here, a reference value obtained for car noise only is compared to the proposed step-size control and to an alternative control where a VAD is utilized instead of the conditional recursive estimation in Eqn. 3.51. This is done for the engine noise component which is highlighted in Fig. 3.14.

Analyzing the step-size $\mu_i(n)$ according to Fig. 3.15, below left, and comparing it with the reference, one observes that sometimes the estimated step-size is considerably below the reference. The reason, here, is speech components not present in the reference case. This means that with speech present the reduced step-size is appropriate in order to avoid misadaptation. In comparison, the step-size obtained with the VAD alternative is too small.

The consequences of the step-size control can be observed in Fig. 3.15, below right. Here the true and estimated power of one spectral component are depicted. Analyzing the reference value, one can observe that it shows strong power variations of more than 20 dB which is challenging for the adaptation and the step-size control. The tracking in the presence of speech is sufficiently accomplished with the proposed step-size control, whereas the VAD control exhibits a delayed estimation.

Summarizing, in comparison to the reference values as well as in comparison to the alternative VAD-based control, the proposed step-size control shows the desired properties.

3.5 Evaluation and Comparison of the Results Obtained by the Notch Filter and the Compensation Approach

To compare the performance of the two different notch filter types according to Sec. 3.3 and the compensation filter method of the previous section, the time frame at 9 sec of the investigated signal (see Fig. 3.14) has been chosen. In Fig. 3.16 the power spectral density of this time frame is depicted for the reference signal and the different methods for cancelling the engine harmonics. The frequency range between 250 Hz and 550 Hz which contains the strongest engine components has been chosen.

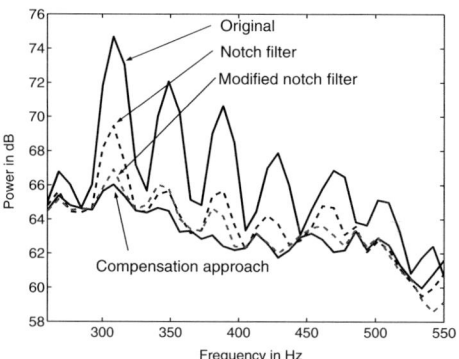

Fig. 3.16. Comparison of the performance of the different methods for attenuating engine noise harmonics for the known signal frame (see Fig. 3.14) at 9 sec. Here the results are depicted for the approaches highlighted in Tab. 3.1 which provoke no or slight speech distortion, only.

For the two notch filter approaches, the filter variants with the narrowest notch, provoking only slight speech distortion, (see bold lines in Tab. 3.1) have been chosen.

Table 3.1. Mean attenuation of car noise with a Diesel engine for the frequency range between 200 Hz and 600 Hz. The highlighted figures show methods which provoke hardly noticeable speech distortion.

Approach	Mean attenuation
Notch filter 0.6 Hz	**2.1 dB**
Notch filter 1.1 Hz	2.5 dB
Notch filter 1.9 Hz	3.0 dB
Mod. notch filter 0.53 Hz	**2.5 dB**
Mod. notch filter 1.06 Hz	3.1 dB
Compensation approach	**2.9 dB**

The reduction of up to 10 dB for the harmonics can be easily observed. In Tab. 3.1 the mean noise reduction for the frequency range between 200 Hz and 600 Hz is given for the complete signal frame registered when accelerating a car with Diesel engine (Fig. 3.14).

One can observe that for the notch filters which provoke only slight signal distortion, the reduction of the engine harmonics is less than that of the compensation filter approach where nearly no signal distortion has been observed. The wider notch filters (1.9 Hz for the normal notch filter and 1.06 Hz for the modified notch filter) offer nearly the same attenuation as the compensation approach, but the speech distortion for these filter widths is noticeable.

3.6 Conclusions and Summary

3.6.1 Conclusion

Both the notch filter and the modified notch filter offer the possibility to guarantee a certain attenuation of the engine noise harmonics. However, the approach presents the risk distorting speech components when filters are chosen that are not very selective in frequency. When the requirement of frequency selectivity is fulfilled, nearly no speech distortion can be observed. However, then the engine noise harmonics usually cannot completely be removed. Here, within the two notch filter versions, the modified notch filter shows the advantage of allowing a better compromise between the obtainable attenuation and the resulting speech distortion.

The compensation filter approach offers the advantage to show no speech distortion when the optimum step-size of the adaptation has been chosen. However, the adaptive filtering requires approximately doubling the computational effort compared to the notch filter approach, especially due to the sophisticated step-size control.

Thus, both approaches exhibit their specific advantages and disadvantages and are worth considering for the described application. For the cancellation of Diesel engine noise harmonics, the compensation filter is advantageous since, here, due to strong gas forces, many harmonics are generated in frequency ranges where speech signals exhibit remarkable components which should not be disturbed. Gasoline engines, however, generate fewer and less powerful harmonics. Distortions of speech in the frequency range where the gasoline engine harmonics are most powerful are normally not perceived as disturbing for the human ear.

3.6.2 Summary

In this chapter a specific pre-filter approach for hands-free car phones has been motivated and described. First, it has been shown that car noise contains strong engine noise components which exhibit a typical harmonics structure in frequency. Exactly at multiples of one half of the engine frequency, remarkable components can be observed. Since these components are usually very powerful, especially for Diesel engines, and are more fluctuating than the other car noise components, their cancellation offers a considerable challenge for traditional noise reduction approaches.

A reduction with a pre-filter, which shows no speech distortion in combination with the classic noise reduction approaches, can show better results with respect to the noise reduction and speech quality than the application of the classic noise reduction approaches alone.

For the design of such a pre-filter, two different approaches have been described and their results have been analyzed.

The first is a notch filter approach which directly applies a frequency selective attenuation at the frequencies where the engine noise components are present. The modification with notch filters of second order allows for a more specific attenuation which exhibits a more selective attenuation with less speech distortion.

The second is a compensation filter approach which models the harmonic components in order to subtract them from the noisy signal. With an optimal step-size control for the adaptation of the correct filter coefficients, it is possible to almost completely reduce the engine noise without disturbing the desired speech signal.

Both approaches have been analyzed in detail and compared and application scenarios for both have been shown.

References

1. Robert Bosch GmbH: *Bosch Kraftfahrtechnisches Taschenbuch*, 22 ed., Berlin, Germany: Springer, 1999 (in German).
2. H. Gahlau: *Fahrzeugakustik: Entwicklung und Einsatz von Systemen zur Lärmreduzierung*, Mod. Industrie, 1998 (in German).
3. E. Hänsler: *Statistische Signale, Grundlagen und Anwendungen*, 3rd ed., Berlin, Germany: Springer, 2001 (in German).
4. M.H. Hayes: *Statistical Digital Signal Processing and Modelling*, New York, NY, USA: Wiley, 1996.
5. S. Haykin: *Adaptive Filter Theory*, 3rd ed., Upper Saddle River, NJ, USA: Prentice Hall, 1996.
6. A. Mader, H. Puder, and G.U. Schmidt: Step-size control for acoustic echo cancellation filters – an overview, *Signal Processing*, **80**(9), 1697-1719, September 2000.
7. S. K. Mitra: *Digital Signal Processing – A Computer-Based Approach,* New York, NY, USA: McGraw-Hill, 1998.
8. A. V. Oppenheim, R. W. Schafer, with J.R. Buck: *Discrete-Time Signal Processing*, Second Edition, Englewood Cliffs, NJ, USA: Prentice Hall, 1999.
9. H. Puder: *Geräuschreduktionsverfahren mit modellbasierten Ansätzen für Freisprecheinrichtungen in Kraftfahrzeugen*, Düsseldorf, Germany, Fortschr.-Ber., VDI-Reihe **10**(721), VDI Verlag, 2003 (in German).
10. H. Puder: Noise reduction with Kalman-filters for hands-free car phones based on parametric spectral speech and noise estimates, in E. Hänsler, G. Schmidt (eds.): *Topics Acoustic Echo and Noise Control*, 599-636, Berlin, Germany: Springer, 2006.
11. M. Tahernezadhi, R.V. Yellapantula: A subband AEC coupled with engine noise cancellation, *Proc. ISCAS '96*, **2**, 241-244, 1996.
12. M. Tandara, V. Tandara: Außenlärm von Personenkraftwagen, *Proc. DAGA '92*, 373-376, 1992 (in German).
13. C. Vaz, X. Kong: An adaptive estimation of periodic signals using a Fourier linear combiner, *IEEE Trans. on Signal Processing*, **42**(1), 1994.

4

Model-Based Speech Enhancement

Mohamed Krini and Gerhard Schmidt

Harman/Becker Automotive Systems, Ulm, Germany

In this chapter partial spectral reconstruction methods for improving noisy speech signals are described. The reconstruction process is performed on the basis of speech models for the short-term spectral envelope and for the so-called excitation signal: the signal that would be recorded directly behind the vocal cords.

Conventional noise suppression methods achieve at low signal-to-noise ratios (SNRs) only a low output quality and, thus, are improvable in these situations. The idea of model-based speech enhancement is first to detect those time-frequency areas that seem to be appropriate for reconstruction. In order to achieve a successful reconstruction it is necessary that at least a few time-frequency areas have a sufficiently high SNR. These signal parts are then used to reconstruct those parts with lower SNR. For reconstruction several speech signal properties such as pitch frequency or the degree of voicing need to be estimated in a reliable manner.

With the reconstruction approach it is possible to generate noise-free signals. But in most cases the resulting signals sound a bit robotic (comparable to low bit rate speech coders). For that reason the reconstructed signal is adaptively combined with a conventionally noise suppressed signal. In those time-frequency parts that exhibit a sufficiently high SNR the output signal of a conventional noise reduction is utilized – in the other parts the reconstructed signal is used.

4.1 Introduction

If a hands-free telephone or a speech-dialog system is used in a car, the speech signals that are recorded by a microphone are superposed with background noise consisting, e.g., of engine, wind, and tyre noise. The amount of noise that is recorded by the microphone depends on the type of the car, on the speed of the car, and also on further boundary conditions such as if one of the windows is open or not. While during stand-still or at slow speed usually a very good

SNR – sometimes even better than in an office – is achieved, a very poor SNR is obtained at medium and high speed. In Fig. 4.1 two time-frequency analyses are depicted. The upper one shows an analysis of a microphone signal that was recorded in a sports car at a speed of 120 km/h. While the typical speech patterns (pitch contour lines) are clearly visible above about 500 Hz, any speech pattern is visible (and audible) below 500 Hz. The lower diagram of Fig. 4.1 shows a time-frequency analysis of the output signal of a conventional noise suppression scheme as it will be described in one of the next sections. Even if the background noise can be reduced considerably it is not possible to recover the highly degraded speech components at low frequencies.

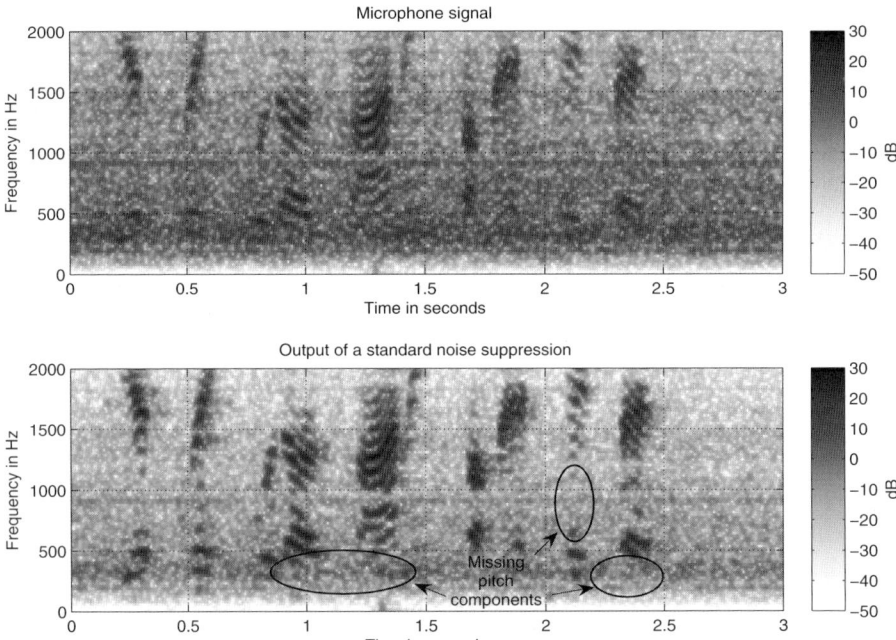

Fig. 4.1. Time-frequency analysis of a microphone signal (top) recorded in a car at a speed of about 120 km/h. The lower diagram shows the same analysis but now applied to the output signal of a conventional noise suppression scheme. Several pitch lines are covered entirely by background noise. These components can not be recovered by conventional noise suppression schemes.

To overcome the drawbacks of conventional noise suppression schemes partial speech reconstruction can be applied. Here the noisy speech components are analyzed and compared with a priori trained speech models. That part of the model that matched best with the parameters extracted out of the current frame determines those speech components that can not be measured due to too much noise. Because of the importance of the underlying speech models we have denoted this kind of speech enhancement as *model-based* approacha.

Model-based speech enhancement can be applied in several manners. Some approaches modify only the spectral estimate of the undisturbed speech components that are used within conventional noise suppression schemes to determine the attenuation factors. Others really try to reconstruct parts of the speech signal. We will describe both approaches but with emphasis on the latter. Before doing so we start with a brief introduction on conventional noise suppression characteristics in the next section. Afterwards speech reconstruction approaches are described in detail and finally it is shown how conventional noise suppression and speech reconstruction can be combined adaptively in order to achieve best results.

4.2 Conventional Speech Enhancement Schemes

To improve the speech quality in hands-free systems or the recognition rate in speech-dialog systems noise suppression is applied to the microphone signal as a preprocessing stage (see Fig. 4.2). Algorithms such as the adaptive Wiener filter [7], spectral subtraction [24], or decision directed approaches [3, 4] are often utilized. Common to all of these approaches is that the disturbed signal components are filtered with an SNR-based filter characteristic.

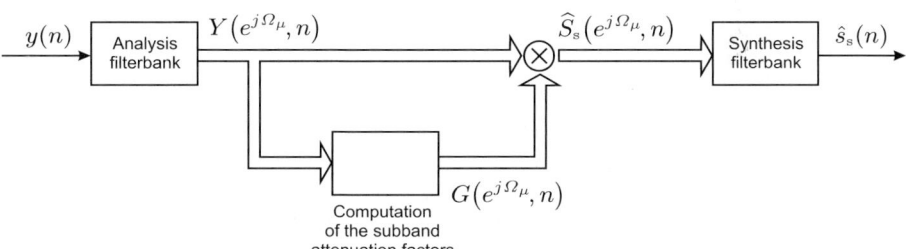

Fig. 4.2. Basic structure of a noise suppression algorithm applied to the microphone signal as a preprocessing stage for hands-free systems or speech dialog systems.

The filter characteristics are often applied in a subband or spectral domain as (time variant) attenuation factors or frequency responses, respectively. For this reason an analysis and a synthesis filterbank are depicted in Fig. 4.2. The term filterbank could mean simple DFT or DCT transformations, but also more complex types of filterbanks such as polyphase, wavelet, or gammatone decompositions [23].

Within this chapter the noisy microphone signal is denoted by $y(n)$. It consists of speech components $s(n)$ and background noise $b(n)$ according to:

$$y(n) = s(n) + b(n). \tag{4.1}$$

By applying an analysis filterbank we obtain subband signals or short-term spectra which we denote with $Y(e^{j\Omega_\mu}, n)$. The quantities Ω_μ with

$$\mu \in \{0, 1, \ldots, M-1\} \tag{4.2}$$

describe the discrete frequency supporting points that are computed with the analysis filterbank. The subband signals are determined every r samples. In all examples in this chapter we have used the following values for the number of subbands M and for the frameshift r:

$$M = 256, \tag{4.3}$$
$$r = 64. \tag{4.4}$$

In case of a filterbank with a uniform frequency resolution – as used in all examples presented in this chapter – we obtain for the frequency supporting points

$$\Omega_\mu = \mu \frac{2\pi}{M}. \tag{4.5}$$

Each frame is windowed with a Hann window h_k [18] of length $N_{\text{ana}} = M = 256$. This means that succeeding frames are overlapping by 75 percent. In most conventional noise suppression schemes the resulting subband signals

$$Y(e^{j\Omega_\mu}, n) = \sum_{k=0}^{M-1} y(n-k) h_k e^{-j\Omega_\mu k} \tag{4.6}$$

are multiplied with attenuation factors $G(e^{j\Omega_\mu}, n)$ in order to obtain a noise reduced output spectrum:

$$\widehat{S}_{\text{s}}(e^{j\Omega_\mu}, n) = Y(e^{j\Omega_\mu}, n) \, G(e^{j\Omega_\mu}, n). \tag{4.7}$$

The resulting subband signals[1] $\widehat{S}_{\text{s}}(e^{j\Omega_\mu}, n)$ are estimations for the undistorted input speech components $S(e^{j\Omega_\mu}, n)$. By using a synthesis filterbank the denoised subband signals are combined in order to obtain the time domain output signal $\hat{s}_{\text{s}}(n)$ (see Fig. 4.2).

If, for example, the subband coefficients $G(e^{j\Omega_\mu}, n)$ are computed according to a modified Wiener rule [9], we obtain:

$$G(e^{j\Omega_\mu}, n) = \max\left\{ G_{\min}(e^{j\Omega_\mu}, n), \, 1 - \beta(e^{j\Omega_\mu}, n) \frac{\widehat{S}_{bb}(\Omega_\mu, n)}{\widehat{S}_{yy}(\Omega_\mu, n)} \right\}. \tag{4.8}$$

---

[1] We have used here (and will use in the following) the hat symbol to indicate estimated quantities. Furthermore, the subscript "s" is meant to show that the quantity results from a noise suppression scheme. For the speech reconstruction part – that is addressed in the following sections – we utilize the subscript "r".

The quantities $\widehat{S}_{bb}(\Omega_\mu, n)$ and $\widehat{S}_{yy}(\Omega_\mu, n)$ denote the estimated short-term power spectral densities of the background noise and of the noisy input signal, respectively. The first named quantity can be estimated, e.g., using recursive smoothing of the current squared input spectrum. The smoothing is performed only in speech pauses, during speech activity the smoothing – and thus the estimation – is stopped. Details about such methods can be found, e.g., in [7] as well as in the references cited there. Another method would be the tracking of the minimum short-term power in each subband. This method – called *minimum statistics* – does not require any speech activity detection. Details can be found in [15–17].

For estimating the power spectral density of the noisy input signal often the squared magnitude of the current input spectrum is used:

$$\widehat{S}_{yy}(\Omega_\mu, n) = \left| Y\left(e^{j\Omega_\mu}, n\right) \right|^2. \qquad (4.9)$$

The quantities $\beta(e^{j\Omega_\mu}, n)$ and $G_{\min}(e^{j\Omega_\mu}, n)$ in Eq. 4.8 are denoting the overestimation parameter of the background noise and the maximum attenuation, respectively. In basic systems these quantities are both time and frequency independent. Enhanced systems adjust these parameters adaptively and independently for each subband. In [12], e.g., a proposal for a time and frequency selective choice for the overestimation factor $\beta(e^{j\Omega_\mu}, n)$ can be found. In [20] an algorithm for adaptive adjustment of the attenuation limits $G_{\min}(e^{j\Omega_\mu}, n)$ is described.

Besides the filter characteristic that is described in Eq. 4.8 a broad variety of other approaches exists (see e.g. [3, 4, 14]). Common to most of these approaches is the basic processing structure (short-term spectral attenuation of the noisy input) that is shown in Fig. 4.2. With such an approach the signal-to-noise ratio and thus also the speech quality and the recognition rate of speech-dialog systems can be improved. However, this is only true for input signals with at least 5 to 10 dB SNR. If the input signal contains very strong distortions – as it appears often at low frequencies in automotive applications – most conventional noise suppression schemes apply the maximum allowed attenuation.

4.3 Speech Enhancement Schemes Based on Nonlinearities

At high noise conditions the harmonics of a speech signal are sometimes distorted to such an extend that common short-term suppression techniques are unable to work properly. One possibility which might help to regenerate missing or degraded harmonics is to use a nonlinearity. In the following method the output signal of a conventional noise suppression is used to generate an artificial harmonic signal by utilizing a nonlinear function. Afterwards, the harmonically extended signal is then used to determine new attenuation factors. Finally, the noisy input short-term spectrum is weighted by the modified suppression factors in order to obtain an enhanced output spectrum with regenerated missing pitch structure.

Fig. 4.3 shows an overview of a speech enhancement method which is based on a nonlinearity. As presented in the previous section, first an analysis

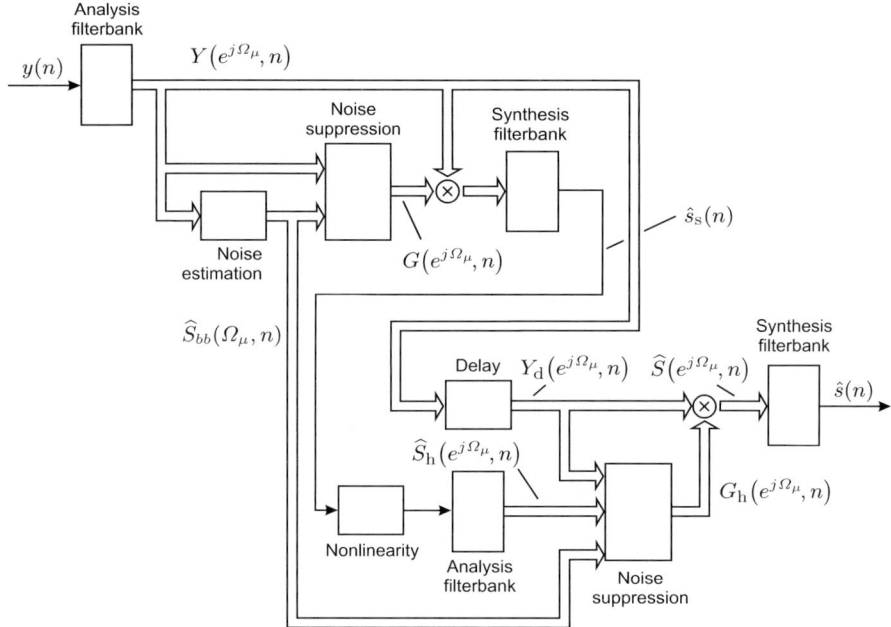

Fig. 4.3. Overview of a speech enhancement method based on nonlinearity.

filterbank is applied to the noisy input signal $y(n)$. Afterwards the power spectral density (PSD) of the background noise $\widehat{S}_{bb}(\Omega_\mu, n)$ as well as the PSD of the input signal $\widehat{S}_{yy}(\Omega_\mu, n)$ are estimated in order to compute attenuation factors $G(e^{j\Omega_\mu}, n)$ utilizing an arbitrary noise suppression characteristic. The noisy input subband signals $Y(e^{j\Omega_\mu}, n)$ are weighted by the attenuation factors $G(e^{j\Omega_\mu}, n)$ to get a noise reduced short-term spectrum $\widehat{S}_{\mathrm{s}}(e^{j\Omega_\mu}, n)$.

At low SNR situations, however, some harmonics of the input signal are considered as noise only components – as a consequence the maximum attenuation of the filter characteristic is applied. The resulting output signal $\hat{s}_{\mathrm{s}}(n)$ – after applying a synthesis filterbank to $\widehat{S}_{\mathrm{s}}(e^{j\Omega_\mu}, n)$ – often suffers from distortions due to the missing harmonics. A possible way to restore signal harmonics, as mentioned before, is to apply a nonlinear function $f_{\mathrm{NL}}(\cdot)$ to the noise reduced signal $\hat{s}_{\mathrm{s}}(n)$:

$$\hat{s}_{\mathrm{h}}(n) = f_{\mathrm{NL}}\left(\hat{s}_{\mathrm{s}}(n)\right), \qquad (4.10)$$

whereas $\hat{s}_\mathrm{h}(n)$ corresponds to an artificially restored harmonic signal. It is well known that applying a nonlinear characteristic to a harmonic signal produces sub- and superharmonics.

A broad variety of different nonlinear functions, such as a quadratic characteristic or a half-way rectification, can be applied. Focusing on the latter one, the harmonic signal from Eq. 4.10 can be rewritten according to:

$$\hat{s}_\mathrm{h}(n) = \hat{s}_\mathrm{s}(n)\, v\bigl(\hat{s}_\mathrm{s}(n)\bigr), \qquad (4.11)$$

whereas $v\bigl(\hat{s}_\mathrm{s}(n)\bigr)$ is defined by:

$$v\bigl(\hat{s}_\mathrm{s}(n)\bigr) = \begin{cases} 1, & \text{if } \hat{s}_\mathrm{s}(n) > 0, \\ 0, & \text{else}. \end{cases} \qquad (4.12)$$

Since a short voiced speech signal block can approximately be considered as a periodic signal with a constant pitch period $\hat{\tau}_\mathrm{p}(n)$, the half-way rectification $v\bigl(\hat{s}_\mathrm{s}(n)\bigr)$ can be interpreted as a harmonic comb filter. For details the interested reader is referred to [19].

The artificial signal $\hat{s}_\mathrm{h}(n)$ can be exploited to compute new attenuation factors $G_\mathrm{h}(e^{j\Omega_\mu}, n)$ which will be able to preserve the harmonics of the speech signal. In order to compute the attenuation factors first an analysis filterbank is applied to $\hat{s}_\mathrm{h}(n)$. Note that due to the additional synthesis and analysis filterbanks a delay is inserted in the signal path. The resulting spectrum $\widehat{S}_\mathrm{h}(e^{j\Omega_\mu}, n)$, however, consists of harmonics at the desired frequencies, but with biased amplitudes due to the used nonlinearity. For this reason the harmonic spectrum $\widehat{S}_\mathrm{h}(e^{j\Omega_\mu}, n)$ has to be combined with the delayed version of the noise reduced subband signals $\widehat{S}_\mathrm{s,d}(e^{j\Omega_\mu}, n)$ according to:[2]

$$\widehat{S}_{s_\mathrm{m} s_\mathrm{m}}(\Omega_\mu, n) = \lambda_\mathrm{m} \bigl|\widehat{S}_\mathrm{s,d}(e^{j\Omega_\mu}, n)\bigr|^2 + (1-\lambda_\mathrm{m})\bigl|\widehat{S}_\mathrm{h}(e^{j\Omega_\mu}, n)\bigr|^2, \qquad (4.13)$$

whereas λ_m is the mixing factor within a range of:

$$0 < \lambda_\mathrm{m} < 1. \qquad (4.14)$$

Using, e.g., the modified Wiener rule from Eq. 4.8 the subband coefficients can be determined according to:

$$G_\mathrm{h}(e^{j\Omega_\mu}, n) = \qquad (4.15)$$
$$\max\left\{ G_\mathrm{min}(e^{j\Omega_\mu}, n),\, \frac{\widehat{S}_{s_\mathrm{m} s_\mathrm{m}}(\Omega_\mu, n) + \bigl(1 - \beta(e^{j\Omega_\mu}, n)\bigr)\widehat{S}_{bb}(\Omega_\mu, n)}{\widehat{S}_{y_\mathrm{d} y_\mathrm{d}}(\Omega_\mu, n)} \right\},$$

whereas the short-term power spectral density $S_{ss}(e^{j\Omega_\mu}, n)$ of the clean speech is replaced by $\widehat{S}_{s_\mathrm{m} s_\mathrm{m}}(\Omega_\mu, n)$. The estimate $\widehat{S}_{y_\mathrm{d} y_\mathrm{d}}(\Omega_\mu, n)$ denotes the delayed power spectral density.

---
[2] A delayed version of the noise reduced spectrum is used due to the additional synthesis and analysis filterbank.

After computing the new attenuation factors the enhanced speech spectrum is finally determined as follows:

$$\widehat{S}(e^{j\Omega_\mu}, n) = G_\mathrm{h}(e^{j\Omega_\mu}, n)\, Y_\mathrm{d}(e^{j\Omega_\mu}, n)\,, \qquad (4.16)$$

whereas $Y_\mathrm{d}(e^{j\Omega_\mu}, n)$ corresponds to a delayed version of the noisy input spectrum.

It should be noted that for practical implementation the time-domain nonlinearity should be approximated by appropriate processing in the subband domain in order to avoid the additional delay in the signal path and to reduce the computational complexity. The corresponding speech enhancement structure based on nonlinear subband processing is depicted in Fig. 4.4.

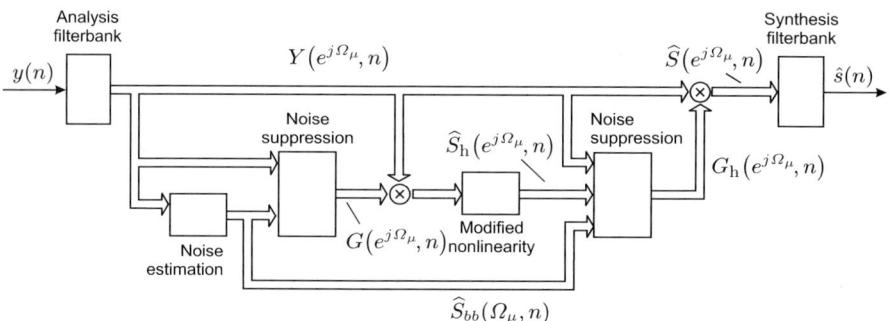

Fig. 4.4. Overview of a speech enhancement method based on nonlinear subband processing.

A time-frequency analysis of a noisy speech signal as well as the corresponding analysis of the outcome using a conventional noise suppression without and with nonlinearity are depicted in Fig. 4.5. For the simulation the modified Wiener rule and the half-way rectification were applied. The mixing parameter λ_m used in Eq. 4.14 was set to $\lambda_\mathrm{m} = 0.2$. Furthermore, a non-frequency selective maximal attenuation of $G_\mathrm{min} = 0.05$ was utilized to make the analysis in Fig. 4.5 clearer. As it can be seen, by employing a nonlinearity some of the missing harmonics can be regenerated.

However, replacing the noisy output signal with the generated artificial signal $s_\mathrm{h}(n)$ will in the majority of cases sound raspy and unnatural. For that reason the artificial signal is simply utilized to compute attenuation factors in order to preserve degraded harmonics from the noisy input signal. It has to be mentioned that in highly disturbed speech signals many harmonics are masked by the noise. In such cases speech enhancement methods based on a nonlinearity are not able to work properly anymore. Sometimes the resulting signal sounds hoarse and unnatural. To overcome this, speech reconstruction – as it is described in the next sections – can be applied.

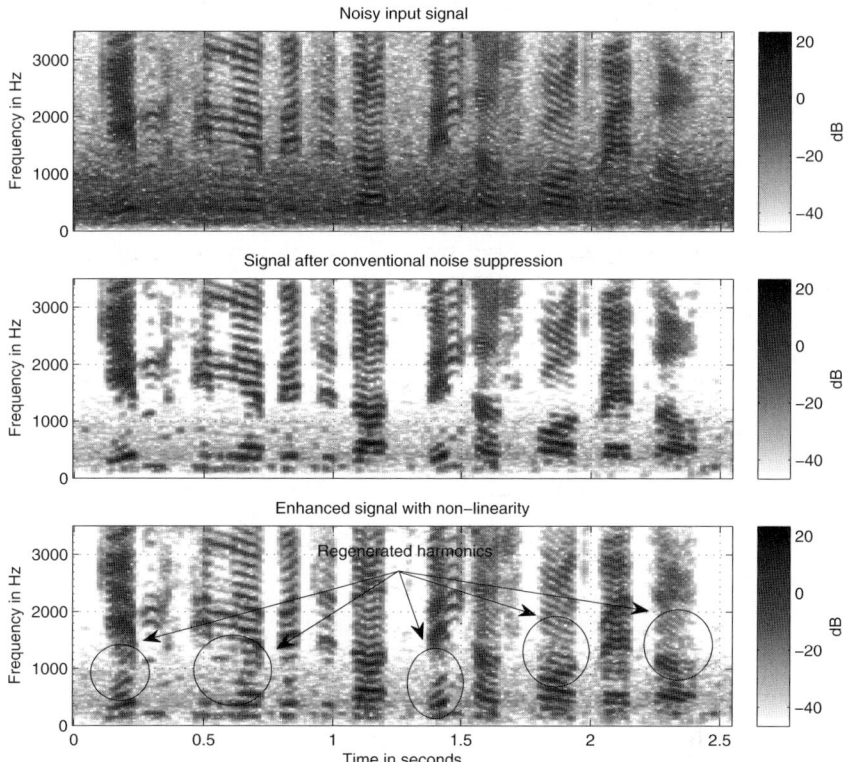

Fig. 4.5. Time-frequency analyses of a noisy speech signal measured in a car at a high speed (upper diagram) and of a noise reduced signal using a conventional noise suppression without (center) and with nonlinearity (lower diagram).

4.4 Speech Enhancement Schemes Based on Speech Reconstruction

The basic speech enhancement schemes that have been described in the previous section can be extended by a signal reconstruction approach (see Fig. 4.6). Here, heavily distorted time-frequency areas of a signal are reconstructed out of those parts of the signal that have a higher SNR.

The reconstruction scheme is based on the so-called *source-filter model*, known, e.g., from speech coding [2, 22].[3] Both, for the source and for the filter part codebooks that contain pitch impulse prototypes and spectral envelope prototypes, respectively, can be trained in advance. Additionally, short-term features of the speech signal such as the excitation type (voiced/unvoiced/mixed) or the pitch frequency need to be extracted. In order to achieve

---

[3] A detailed description of the source-filter model can be found in Chap. 5 of this book.

a successful feature extraction at least a few time-frequency areas of the noisy input signal need to have a sufficiently high SNR. However, as we have seen in Sec. 4.1, some applications, such as hands-free telephony in cars, produce such signals.

The reconstruction of a noisy speech signal does not result – at least in our approach – in a very natural and artefact-free speech signal. Thus, in time-frequency areas with high SNR a conventional noise suppression should be preferred and the reconstructed signal should be utilized only in low SNR regions. For that reason an adaptive frequency-selective mixer needs to be implemented (see right part of Fig. 4.6). Several estimations, such as the short-term SNR, are required both in the reconstruction as well as in the mixing part. Thus, a mutual control unit is also depicted in Fig. 4.6.

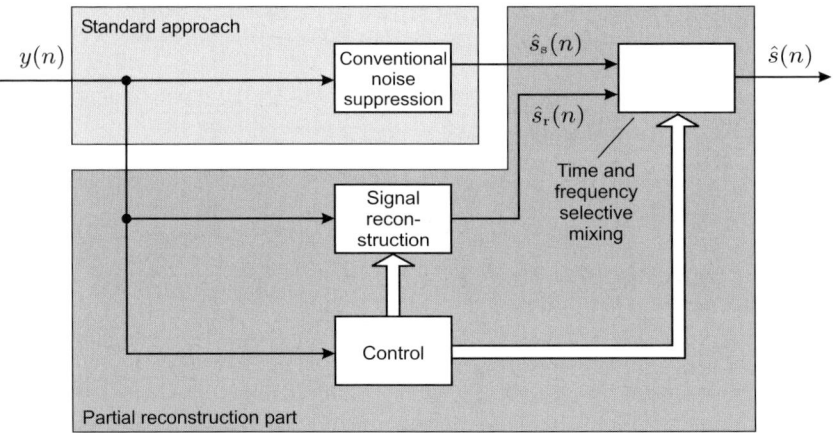

Fig. 4.6. Combined (suppression and reconstruction) approach for speech signal enhancement.

The following sections describe the individual algorithmic parts in details. It is started with the control part of the algorithm (Sec. 4.4.1). Next the signal reconstruction part (Sec. 4.4.2) is described, followed by an overview about the adaptive mixing scheme (Sec. 4.5). Finally, some examples are presented in order to show the advantages, but also the limits, of current speech reconstruction approaches.

While describing all algorithmic parts we assume the basic (time-domain) processing structure depicted in Fig. 4.6. However, most algorithms will be realized in the short-term frequency or subband domain. For that reason a combined implementation of a noise suppression and a speech reconstruction scheme should first decompose the microphone signal into subbands using an appropriate analysis filterbank, apply then all signal processing in this domain, and finally compose the enhanced subband signals using a synthesis filterbank. For better separability of the main algorithmic parts we have not used this option in this chapter.

4.4.1 Feature Extraction and Control

As already mentioned in the previous section we will start with a short description of how several quantities, that are necessary for the final reconstruction and mixing process, are estimated. All of these estimations are summarized in the block entitled *control* in Fig. 4.6. The details of this block are shown in Fig. 4.7.

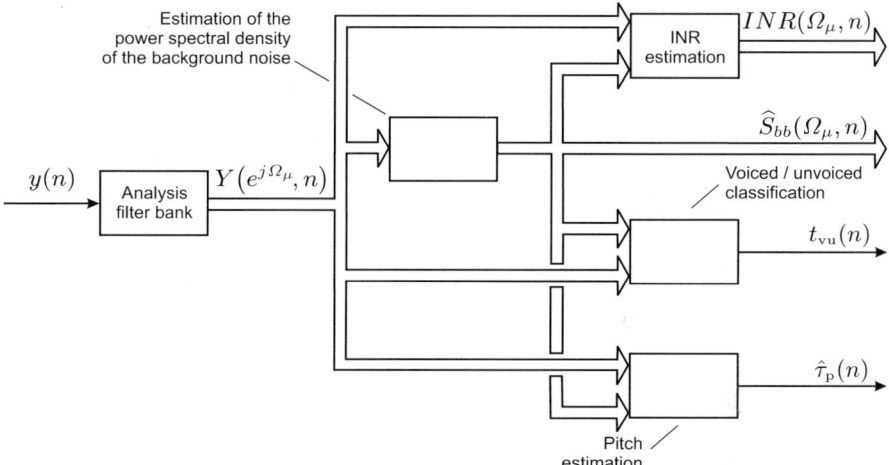

Fig. 4.7. Structure of the feature classification and control unit.

Mainly four estimations are performed:

- The power spectral density of the background noise $\widehat{S}_{bb}(\Omega_\mu, n)$ is estimated.
- Using the estimated power spectral density of the background noise and the current magnitude squared microphone spectrum the input-to-noise ratio $INR(\Omega_\mu, n)$ is computed.
- A voiced/unvoiced classification is determined.
- Finally, a pitch estimation is performed.

Before entering the next sections that show the details of the individual parts of the control unit, it should be highlighted that every description should be regarded only as a basic realization example. Of course other estimation schemes exist and they could replace the approaches presented next.

4.4.1.1 Estimation of the Power Spectral Density of the Background Noise

The main assumptions on which background noise estimation schemes rely is that the noisy speech signal $y(n)$ contains several pauses in which only the noise components $b(n)$ are present and that the noise is not changing significantly in the succeeding speech activity period. Assuming that the speech and the noise components are uncorrelated, a basic approach for background noise estimation is a simple search for the minima of the subband short-term powers.

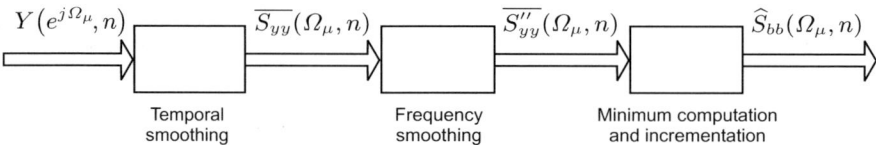

Fig. 4.8. Structure of a basic noise estimation scheme.

The basic structure of the noise estimation scheme that is presented here is depicted in Fig. 4.8. As a first stage an IIR[4] smoothing of first order in temporal direction is applied in order to reduce the variations of the instantaneous squared subband signals:

$$\overline{S_{yy}}(\Omega_\mu, n) = \lambda_t \, \overline{S_{yy}}(\Omega_\mu, n-1) + (1-\lambda_t) \left| Y(e^{j\Omega_\mu}, n) \right|^2. \quad (4.17)$$

The time constant λ_t determines on one hand the maximum possible speed for following changes of the background noise level and on the other hand – as mentioned above – the variance of the estimation. For stable operation λ_t needs to be chosen out of the interval

$$0 \leq \lambda_t < 1. \quad (4.18)$$

The smaller the time constant is chosen the faster the estimation scheme can follow changes of the noise level. To reduce the error variance even more, it is advantageous to smooth the output of the temporal IIR filter also in positive and negative frequency direction. This helps especially to reduce short outliers having very small amplitudes. The smoothing process in positive frequency direction can be described as:

$$\overline{S'_{yy}}(\Omega_\mu, n) = \begin{cases} \overline{S_{yy}}(\Omega_\mu, n), & \text{if } \mu = 0, \\ \lambda_f \overline{S'_{yy}}(\Omega_{\mu-1}, n) + (1-\lambda_f) \overline{S_{yy}}(\Omega_\mu, n), & \text{else}. \end{cases} \quad (4.19)$$

---

[4] The term *IIR* abbreviates *infinite impulse response*.

Afterwards smoothing in negative frequency direction is applied – again by utilizing IIR smoothing of first order:

$$\overline{S_{yy}''}(\Omega_\mu, n) = \begin{cases} \overline{S_{yy}'}(\Omega_\mu, n), & \text{if } \mu = M - 1, \\ \lambda_f \overline{S_{yy}''}(\Omega_{\mu+1}, n) + (1 - \lambda_f) \overline{S_{yy}'}(\Omega_\mu, n), & \text{else}. \end{cases}$$
(4.20)

As for the temporal filtering the constant λ_f for the frequency smoothing needs to be chosen out of the interval

$$0 \leq \lambda_f < 1$$
(4.21)

to guarantee stable operation. Finally, the background noise power spectral density is estimated according to

$$\widehat{S}_{bb}(\Omega_\mu, n) = \max\left\{S_{bb,\min}, \min\left\{\widehat{S}_{bb}(\Omega_\mu, n-1), \overline{S_{yy}''}(\Omega_\mu, n)\right\}(1+\epsilon)\right\}.$$
(4.22)

If the parameter ϵ in Eq. 4.22 would be set to zero, the estimation result could only decrease or stay at its current value. Thus, we would have a global minimum tracker. To allow also for a slow increment the parameter ϵ should be set to a small positive value:

$$0 < \epsilon \ll 1.$$
(4.23)

Now the estimation can recover after short low-power noise periods. Fig. 4.9 depicts a noisy input signal $y(n)$ in the upper diagram and the squared magnitude $|Y(e^{j\Omega_\mu}, n)|^2$ as well as the noise estimate $\widehat{S}_{bb}(\Omega_\mu, n)$ in one of the subbands in the lower diagram.

Assuming that the parameter ϵ would have been chosen larger, the ramps that can be observed in the estimated background noise power would become steeper. Thus, a too large ϵ on the one hand results in heavily overestimated noise levels during speech activity. A too small ϵ on the other hand would lead to a slow increment after an increase of the noise level (e.g. after opening a window or when accelerating the car).

The quantity $S_{bb,\min}$ and the maximum operation have been inserted in Eq. 4.22 to avoid that the value zero can enter the estimation scheme. In this case the entire estimation would never recover without the limitation.

4.4.1.2 Estimation of the Input-to-Noise and the Signal-to-Noise Ratio

For estimating the input-to-noise ratio $INR(\Omega_\mu, n)$ the current squared short-term microphone magnitude spectrum is divided by the estimated power spectral density of the background noise (see previous section):

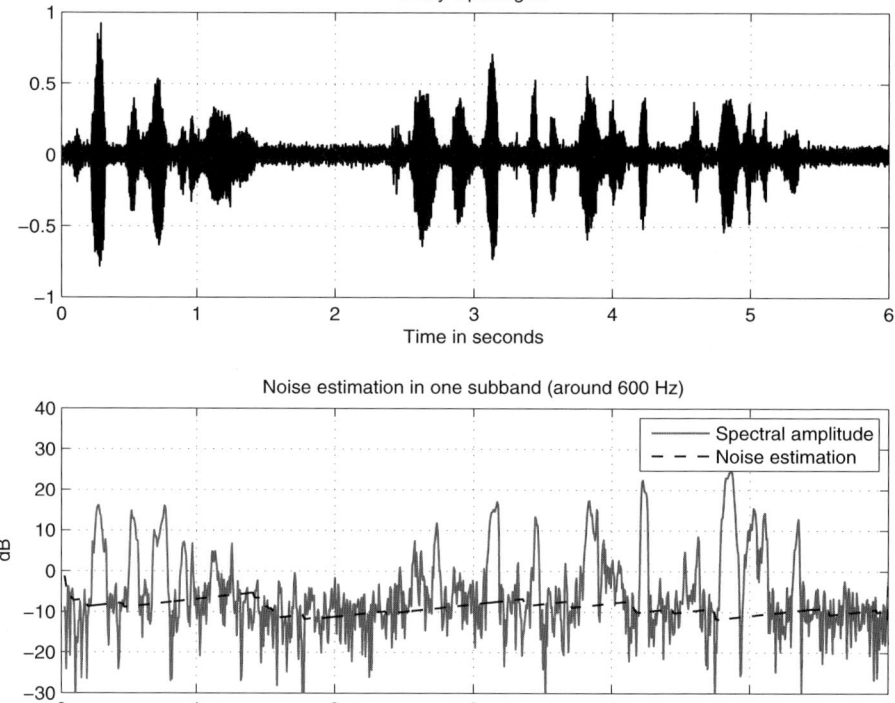

Fig. 4.9. Noise estimation example. In the upper diagram a noisy speech signal is depicted. In the lower diagram the short-term squared magnitude $|Y(e^{j\Omega_\mu}, n)|^2$ as well as the noise estimate $\widehat{S}_{bb}(\Omega_\mu, n)$ in one subband are shown.

$$INR(\Omega_\mu, n) = \frac{\left|Y\left(e^{j\Omega_\mu}, n\right)\right|^2}{\widehat{S}_{bb}(\Omega_\mu, n)}. \quad (4.24)$$

The range of an (optimally estimated) input-to-noise ratio is basically

$$1 \leq INR_{\mathrm{opt}}(\Omega_\mu, n) = \frac{S_{yy}(\Omega_\mu, n)}{S_{bb}(\Omega_\mu, n)} = \frac{S_{ss}(\Omega_\mu, n) + S_{bb}(\Omega_\mu, n)}{S_{bb}(\Omega_\mu, n)} \leq \infty. \quad (4.25)$$

However, due to errors in the estimated power spectral density of the background noise the lower limit can be exceeded. Nevertheless, even with the estimation errors the ratio $INR(\Omega_\mu, n)$ stays within the interval

$$0 \leq INR(\Omega_\mu, n) \leq \infty. \quad (4.26)$$

As we will see in the following sections, the input-to-noise ratio will be utilized for controlling several other algorithmic parts of the speech reconstruction approach. Alternatively, also the more familiar signal-to-noise ratio (SNR) can be computed out of the input-to-noise ratio[5] and be used instead of it:

$$SNR(\Omega_\mu, n) = \max\left\{0,\, INR(\Omega_\mu, n) - 1\right\}. \tag{4.27}$$

The maximum operation prevents that a negative value is used as an SNR estimate.

4.4.1.3 Voiced/Unvoiced Classification

For determining whether the current frame contains either a voiced or an unvoiced speech segment – assuming that the frame has already been classified as not containing a speech pause – it is exploited that unvoiced speech segments have their center of power at higher frequencies than voiced ones. For that reason the average input-to-noise ratio (INR) is computed for a low and a high frequency range:

$$INR_{\text{low}}(n) = \frac{1}{\mu_1 - \mu_0 + 1} \sum_{\mu=\mu_0}^{\mu_1} INR(\Omega_\mu, n), \tag{4.28}$$

$$INR_{\text{high}}(n) = \frac{1}{\mu_3 - \mu_2 + 1} \sum_{\mu=\mu_2}^{\mu_3} INR(\Omega_\mu, n). \tag{4.29}$$

The summation limits Ω_{μ_0} to Ω_{μ_3} are chosen such that they represent the following frequencies:

$$\Omega_{\mu_0} \longleftrightarrow 300\,\text{Hz}, \tag{4.30}$$
$$\Omega_{\mu_1} \longleftrightarrow 1050\,\text{Hz}, \tag{4.31}$$
$$\Omega_{\mu_2} \longleftrightarrow 3800\,\text{Hz}, \tag{4.32}$$
$$\Omega_{\mu_3} \longleftrightarrow 5200\,\text{Hz}. \tag{4.33}$$

Out of the two average INR values the ratio

$$r_{INR}(n) = \frac{INR_{\text{high}}(n)}{INR_{\text{low}}(n) + \Delta_{INR}} \tag{4.34}$$

is computed. By adding the small positive constant Δ_{INR} divisions by zero are avoided.

---

[5] The optimally estimated SNR is defined as $SNR_{\text{opt}}(\Omega_\mu, n) = \frac{S_{ss}(\Omega_\mu, n)}{S_{bb}(\Omega_\mu, n)}$, which can be modified to $SNR_{\text{opt}}(\Omega_\mu, n) = \frac{S_{ss}(\Omega_\mu, n) + S_{bb}(\Omega_\mu, n) - S_{bb}(\Omega_\mu, n)}{S_{bb}(\Omega_\mu, n)} = \frac{S_{yy}(\Omega_\mu, n) - S_{bb}(\Omega_\mu, n)}{S_{bb}(\Omega_\mu, n)} = INR_{\text{opt}}(\Omega_\mu, n) - 1$.

In the upper diagrams of Fig. 4.10 two short-term spectra – one belonging to a voiced (left) and one to an unvoiced speech segment (right) – are depicted together with the estimated power spectral density of the background noise. The lower diagrams show the corresponding input-to-noise ratio as well as the low and the high frequency average of this quantity. If the dotted line that connects both averages has a large positive gradient it is very likely that the current frame contains unvoiced speech. If the gradient is largely negative the current frame is obviously a voiced one.

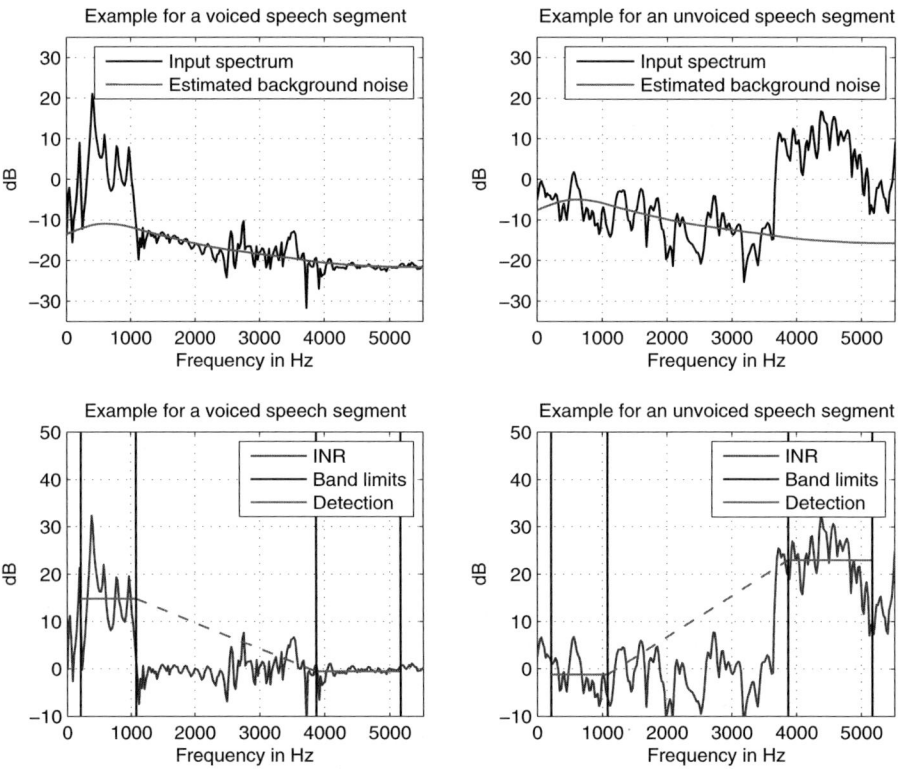

Fig. 4.10. Example for the voiced/unvoiced classification.

To obtain a normalized output range for the either *hard* or *soft* voiced/unvoiced classification the quantity $r_{INR}(n)$ is mapped via a nonlinear characteristic $f_{uv}(...)$ on the output range $[0, 1]$:

$$t_{vu}(n) = f_{uv}\left(r_{INR}(n)\right). \tag{4.35}$$

If a hard classification should be achieved a simple comparison with a fixed threshold r_0 can be applied:

$$f_{\text{vu}}\left(r_{INR}(n)\right) = \begin{cases} 1, & \text{if } r_{INR}(n) < r_0, \\ 0, & \text{else.} \end{cases} \qquad (4.36)$$

For soft classification approaches other characteristics, e.g. the ones utilized as activation functions in neural network applications (see Chap. 5 for examples), can be applied. However, we will assume in the following that such characteristics result in an output range

$$0 \le t_{\text{vu}}(n) \le 1 \qquad (4.37)$$

and generate large values if the frame contains voiced speech and small values for unvoiced speech.

4.4.1.4 Pitch Estimation

In the literature a broad variety of different algorithms for estimating the fundamental frequency (pitch) of a speech signal exists: the analysis in the cepstral domain [8], the harmonic product spectrum [21], or short-term autocorrelation based schemes [1]. In this section a detection method is presented, that is based on the latter approach. We start with an estimate of the short-term power spectral density in terms of the current squared magnitude spectrum $|Y(e^{j\Omega_\mu}, n)|^2$. The determined PSD is then divided by its spectral envelope. Thereby, the short-term envelope of the power spectrum is removed. Afterwards, a linear weighting of the normalized PSD is performed:

$$\widehat{S}_{yy,\text{norm}}(\Omega_\mu, n) = \frac{|Y(e^{j\Omega_\mu}, n)|^2}{S''_{yy,\text{inst}}(\Omega_\mu, n)} W(e^{j\Omega_\mu}). \qquad (4.38)$$

The weighting function $W(e^{j\Omega_\mu})$ has been chosen such that the attenuation rises with the frequency due to the fact that speech has mainly a fundamental frequency structure at low frequencies – which in turn results in an improved estimation. The spectral envelope $S''_{yy,\text{inst}}(\Omega_\mu, n)$ is computed as described in Sec. 4.4.1.1 (Eqs. 4.17 to 4.20), but without temporal smoothing ($\lambda_{\text{t}} = 0$).

After applying an IDFT to Eq. 4.38 an estimated autocorrelation function (ACF) is obtained:

$$\hat{r}_{yy}(m, n) = \frac{1}{M} \sum_{\mu=0}^{M-1} \widehat{S}_{yy,\text{norm}}(\Omega_\mu, n) \, e^{j \frac{2\pi}{M} \mu m}. \qquad (4.39)$$

The ACF $\hat{r}_{yy}(m, n)$ is used in order to estimate the fundamental frequency $f_{\text{p}}(n)$, which can be determined by a search for the maximum in a selected range of indices:

$$\hat{f}_{\text{p}}(n) = \frac{f_{\text{s}}}{\hat{\tau}_{\text{p}}(n)}, \qquad (4.40)$$

with

$$\hat{\tau}_{\text{p}}(n) = \underset{m_{\min} \leq m \leq m_{\max}}{\operatorname{argmax}} \left\{ \hat{r}_{yy}(m,n) \right\} \qquad (4.41)$$

and f_{s} being the sampling frequency. Furthermore, the reliability of $\hat{f}_{\text{p}}(n)$ is determined as the value of the normalized ACF by the power of the input signal vector as:

$$p_{\hat{f}_{\text{p}}}(n) = \frac{\hat{r}_{yy}(\hat{\tau}_{\text{p}}(n), n)}{\hat{r}_{yy}(0, n)}. \qquad (4.42)$$

Large values of Eq. 4.42 indicate a very high confidence of detection, whereas small values indicate a doubtful one. For this reason a detection only takes place for values which exceed a predefined threshold.

Enhancement I – Extending the Autocorrelation Function

Investigations with the method presented above have shown that a reliable estimation can only be determined for fundamental frequencies greater than about 100 Hz (using a sampling frequency $f_{\text{s}} = 11025$ Hz). The idea of the first enhancement is, that not only the present signal frame $\boldsymbol{y}(n)$ is used for fundamental frequency detection but also the signal frame $\boldsymbol{y}(n-r)$, which is delayed by r samples. The fundamental frequency estimation can be improved significantly by utilizing a current frame as well as a frame delayed by r samples. In addition to the first estimation that was presented above the cross power spectral density is estimated by:

$$\widehat{S}_{yy_{\text{d}}}(\Omega_\mu, n) = Y^*\left(e^{j\Omega_\mu}, n\right) Y\left(e^{j\Omega_\mu}, n-1\right). \qquad (4.43)$$

The determined cross PSD is normalized by the smoothed auto PSD and multiplied by the weighting function from Eq. 4.38:

$$\widehat{S}_{yy_{\text{d}},\text{norm}}(\Omega_\mu, n) = \frac{\widehat{S}_{yy_{\text{d}}}(\Omega_\mu, n)}{S''_{yy,\text{inst}}(\Omega_\mu, n)} W\left(e^{j\Omega_\mu}\right). \qquad (4.44)$$

The crosscorrelation function (CCF), $\hat{r}_{yy_{\text{d}}}(m,n)$, is determined by applying the IDFT to $\widehat{S}_{yy_{\text{d}},\text{norm}}(\Omega_\mu, n)$:

$$\hat{r}_{yy_{\text{d}}}(m,n) = \sum_{\mu=0}^{M-1} \widehat{S}_{yy_{\text{d}},\text{norm}}(\Omega_\mu, n)\, e^{j\frac{2\pi}{M}\mu m}. \qquad (4.45)$$

To achieve an enhanced detection, particularly at low fundamental frequencies, the autocorrelation function $\hat{r}_{yy}(m,n)$ and the crosscorrelation function $\hat{r}_{yy_{\text{d}}}(m,n)$ are combined in order to obtain an extended ACF of higher length of indices m:

$$\hat{r}_{yy,\text{ext}}(m,n) = \begin{cases} \hat{r}_{yy}(m,n), & \text{for } 0 \leq m < \frac{M}{2} - r, \\ a(m-r)\hat{r}_{yy}(m,n) \\ \quad + (1 - a(m-r))\hat{r}_{yy_\text{d}}(m-r,n), & \text{for } \frac{M}{2} - r \leq m < \frac{M}{2}, \\ \hat{r}_{yy_\text{d}}(m-r,n), & \text{for } \frac{M}{2} \leq m < \frac{M}{2} + r. \end{cases}$$
(4.46)

The function $a(m)$ can be chosen such that with an increasing index m the weighting decreases in a linear manner from 1 to 0. Now, $\hat{r}_{yy,\text{ext}}(m,n)$ can be used to estimate $\hat{\tau}_\text{p}(n)$ according to Eq. 4.41 but with an extended search range

$$m_{\min} \leq m \leq m_{\max,\text{ext}}, \tag{4.47}$$

with

$$m_{\max,\text{ext}} > m_{\max}. \tag{4.48}$$

To show the benefits of the extended autocorrelation two examples are depicted in Fig. 4.11. In the upper part (a) the standard autocorrelation $\hat{r}_{yy}(m,n)$, the crosscorrelation function $\hat{r}_{yy_\text{d}}(m,n)$, and the combination of both $\hat{r}_{yy,\text{ext}}(m,n)$ are depicted for a female voice (low pitch period, high pitch frequency). In this example it is not really necessary to extend the search range of the standard autocorrelation because the true pitch period is included in the search range of the standard analysis. The extension, however, also does not distort the estimation process (the arguments of the maxima of the standard and the extended autocorrelation functions are equal). In part (b) of Fig. 4.11 the same analyses are depicted – but now for a male voice (large pitch period, low pitch frequency). In this case the search range of the standard autocorrelation is not large enough for detecting the true pitch period. However, the pitch period can be estimated reliably when using the extended autocorrelation function.

Enhancement II – Improving the Selectivity of the Frequency Analysis

Due the windowing of successive signal blocks within the analysis filterbank, a significant frequency overlap arises among neighboring subband channels. Thus, adjacent fundamental frequency trajectories are sometimes hard to separate which is important for fundamental frequency estimation schemes. In order to reduce this overlap the order of the DFT might be increased. However, in this case it should be considered that for hands-free systems several restrictions have to be met. The ITU and the ETSI [5] set up requirements for the maximally tolerable front-end delay. For mobile phones, e.g., the additional delay introduced by hands-free systems should not exceed 39 ms. Thus, increasing the DFT order from $M = 256$ to $M = 512$, e.g., improves the separability of neighboring pitch frequencies but results in an additional delay

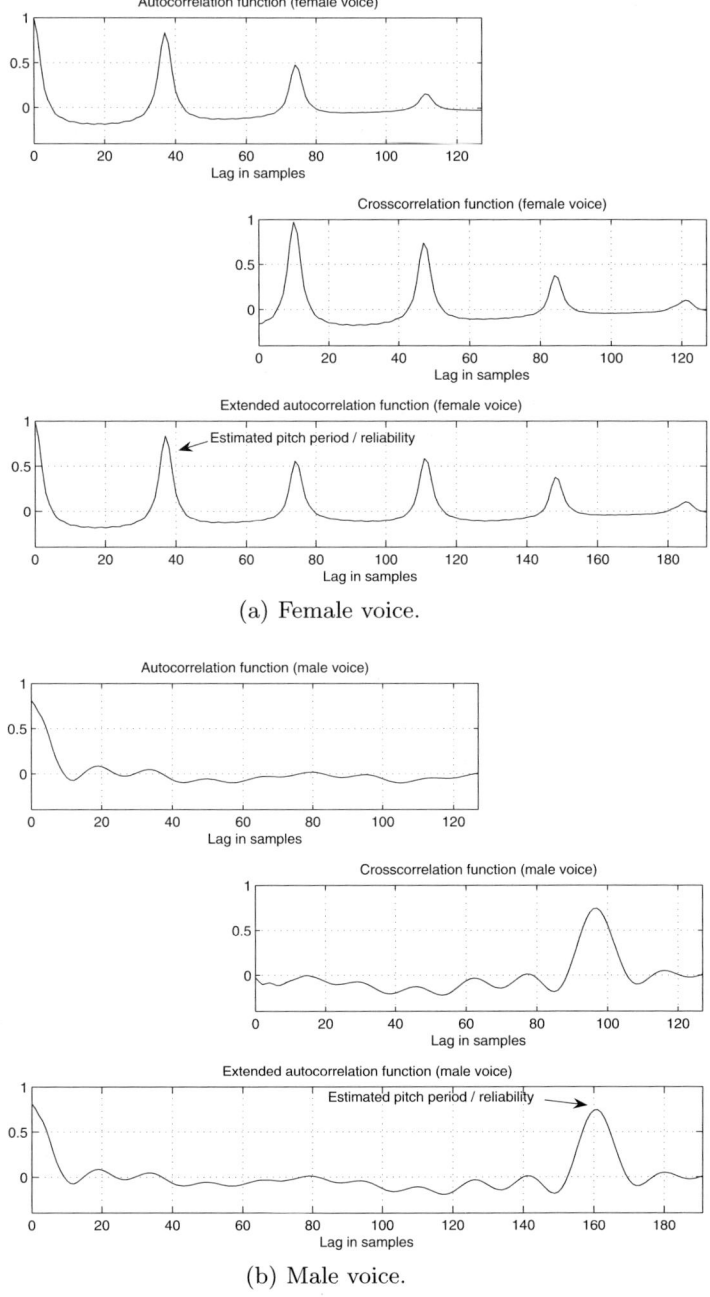

Fig. 4.11. Examples for generating the extended autocorrelation function. Part a) shows the result for a female voice, part b) for a male voice.

of approx. 23 ms at a sampling frequency of $f_s = 11025$ Hz. Now the overall delay of the analysis-synthesis scheme does not fulfill the specifications any more. By applying something that we call *spectral refinement* the separability problem can be solved without increasing the DFT order. A refined version of the input spectrum can be determined by a linear combination of temporal succeeding subband signals:

$$\sum_{i=N_{\text{ref,start}}}^{N_{\text{ref,end}}} K_i\left(e^{j\Omega_\mu}\right) Y\left(e^{j\Omega_\mu}, n-i\right).$$

This could be achieved best if also a few noncausal signals ($N_{\text{ref,start}} > 0$) are utilized. However, even with only causal processing ($N_{\text{ref,start}} = 0$) and using just about $N_{\text{ref,end}} = 1\ldots 4$ coefficients a significant improvement can be achieved. The refinement needs to be applied only for low frequencies (e.g. up to 1 kHz). For the upper frequency range the non-refined input subband signals $Y(e^{j\Omega_\mu}, n)$ can be used:

$$\widetilde{Y}\left(e^{j\Omega_\mu}, n\right) = \begin{cases} \sum_{i=0}^{N_{\text{ref,end}}} K_i\left(e^{j\Omega_\mu}\right) Y\left(e^{j\Omega_\mu}, n-i\right), & \text{for } |\Omega_\mu| < \Omega_{\text{ref}}, \\ Y\left(e^{j\Omega_\mu}, n\right), & \text{else.} \end{cases}$$

(4.49)

The refinement is performed for frequencies below Ω_{ref}. The computation of the weighting coefficients $K_i(e^{j\Omega_\mu})$ is straight forward. Details can be found in [10]. To combine spectral refinement with the correlation based scheme presented above the input subband signals $Y(e^{j\Omega_\mu}, n)$ in Eqs. 4.38 and 4.43 have to be replaced by their refined counterparts $\widetilde{Y}(e^{j\Omega_\mu}, n)$.

To demonstrate the effect of the spectral refinement, first two different sine signals are generated and then added to a small amount of white noise. For the analysis a DFT of order $M = 256$, a Hann window, and a frame shift of $r = 64$ were used. The frequencies of the two sine signals have been increased slowly over time, whereas a distance of 130 Hz has been kept constant – corresponding approximately to the average fundamental frequency of male voices. The simulation results are depicted in Fig. 4.12. In the upper part the analysis without spectral refinement and in the lower part with spectral refinement are shown. Spectral refinement has been applied to all subbands with $N_{\text{ref,end}} = 4$. When comparing the results, the improved separability achieved by spectral refinement is clearly visible.

Simulation Results

To show the performance and the accurateness of the proposed improvements for fundamental frequency estimation a simulation example is introduced. In Fig. 4.13 time-frequency analyses of sinusoidal signals with varying frequency distances from 300 Hz down to 60 Hz at $f_s = 11025$ Hz as well as the results

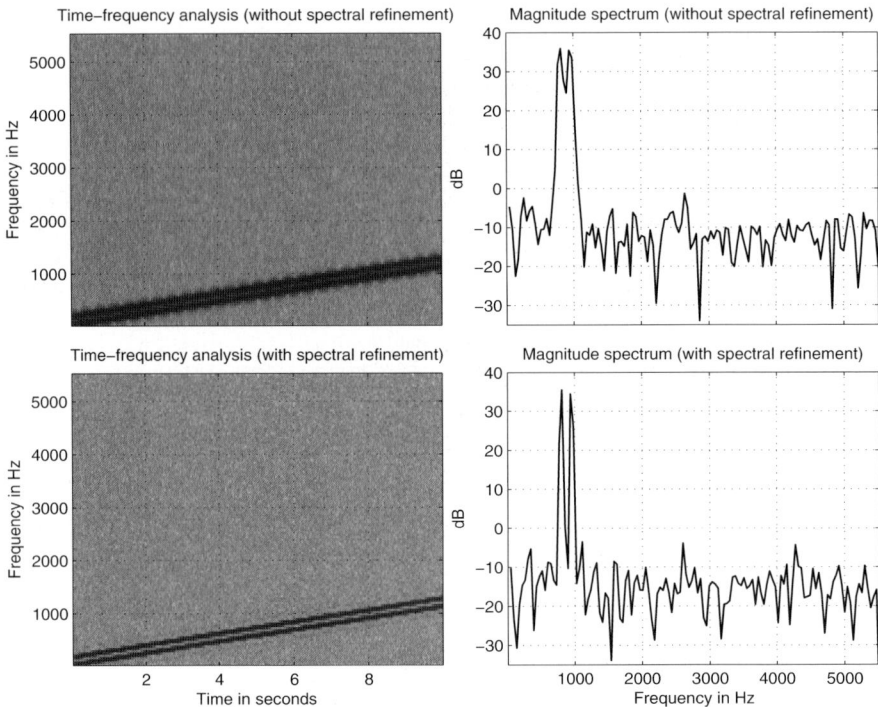

Fig. 4.12. Time-frequency analyses of two sine signals added to white noise: upper graphs without refinement, lower diagrams with refinement.

of the pitch estimation scheme with the different enhancements are shown. The included white curves within the figures demonstrate the estimated fundamental frequencies. In the upper diagram an analysis that uses only the basic fundamental frequency estimation method is depicted. A reliable pitch estimation is possible only for fundamental frequencies greater than about 120 Hz. The diagram in the middle shows the same analysis but now the pitch estimation uses the extended autocorrelation. Due to the extended search range fundamental frequencies that are larger than about 90 Hz can be detected. For lower pitch frequencies the spectral resolution of the Hann window in combination with a DFT of order $M = 256$ is not sufficient. However, if spectral refinement is additionally applied for the low frequency range up to 1000 Hz it is possible to reliably estimate fundamental frequencies down to 60 Hz (depicted in the lowest diagram of Fig. 4.13).

4.4.2 Reconstruction of Speech Signals

The core of this model-based speech enhancement scheme is the entire reconstruction of the signal by using signal parameters, such as the input-to-noise

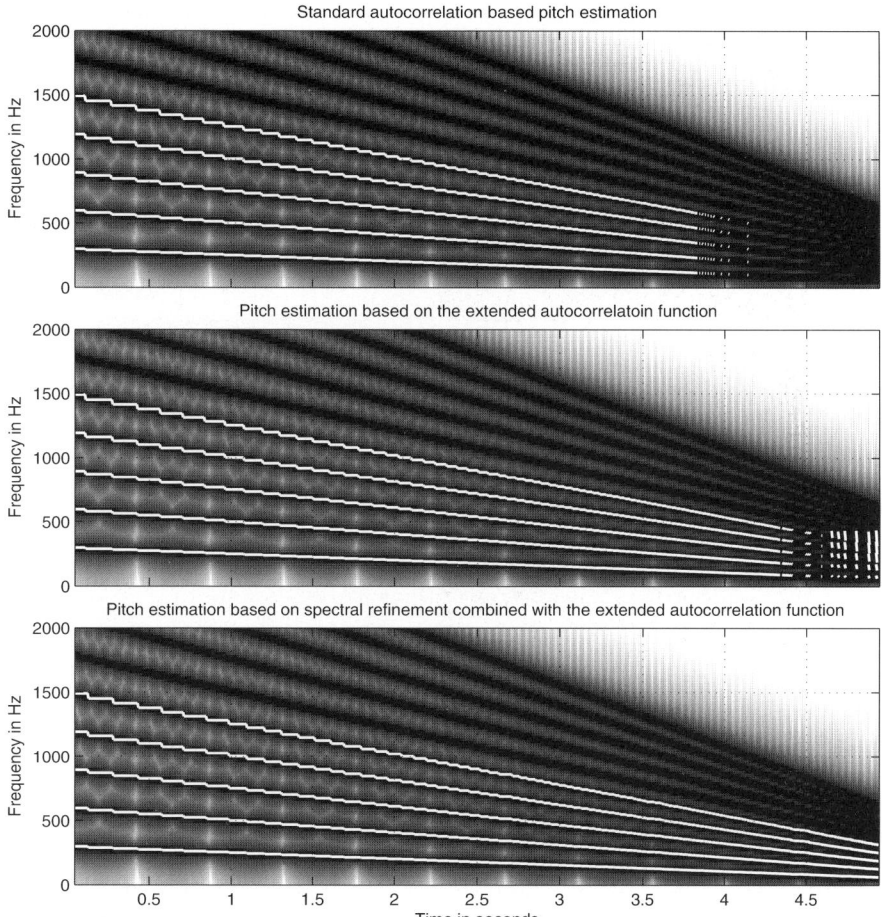

Fig. 4.13. Time-frequency analyses of sinusoidal signals with varying frequency distances. The lowest white line in each graph represents the estimated pitch frequency – the other four lines show multiples of the estimation result.

ratio, the degree of voicing, or the pitch period, that have been extracted out of the noisy speech signal in advance. An overview about the reconstruction part is depicted in Fig. 4.14.

We differentiate between the reconstruction of unvoiced and voiced signal components (uppermost and lowest box in Fig. 4.14). Since the reconstruction of unvoiced signals is simpler compared to voiced ones, we will start with a description of this part in Sec. 4.4.2.2. The reconstruction of voiced signals is described afterwards in Sec. 4.4.2.3.

Both reconstruction methods are based on the current short-term input spectrum $Y(e^{j\Omega_\mu}, n)$. Thus, an analysis filterbank is depicted on the left of

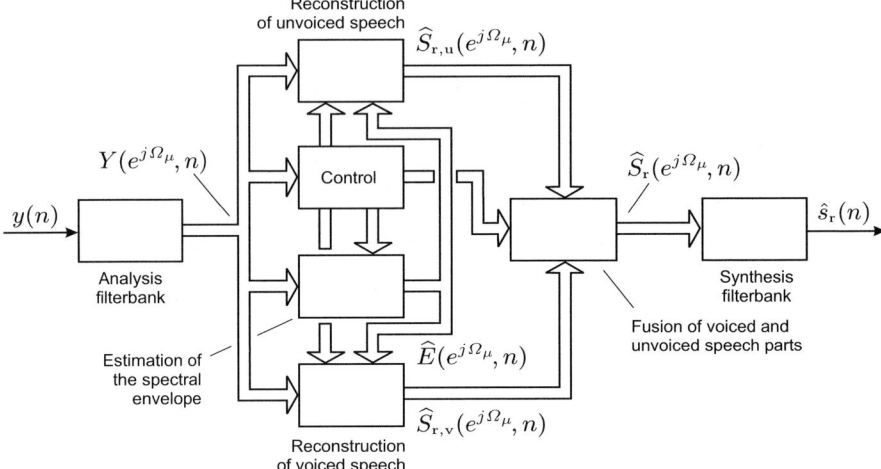

Fig. 4.14. Overview for generating the reconstructed speech signal $\hat{s}_\mathrm{r}(n)$. For the reason of simplicity we have not specified the connections between the control unit and the other algorithmic parts. Details about these connections can be found in the figures that show the individual algorithmic parts.

Fig. 4.14 as a first processing unit. However, as mentioned before, one can omit multiples of this processing stage if the entire scheme – consisting of a noise suppression, a speech reconstruction, and an adaptive mixing device – is realized completely in the subband domain.

Since for both types of reconstruction an estimation of the spectral envelope $\widehat{E}(e^{j\Omega_\mu}, n)$ is required this algorithmic part is performed only once. We will start describing a basic version of this estimation unit in the next section.

4.4.2.1 Estimation of the Spectral Envelope

An overview about a codebook based envelope estimation scheme is depicted in Fig. 4.15. We will describe now a straight forward scheme that is computationally quite expensive. Several extensions can be applied to reduce computational cost. These are described briefly at the end of this section.

As a first step the envelope of the noisy input spectrum $Y(e^{j\Omega_\mu}, n)$ is extracted. Several possibilities such as LPC[6] analysis exist for this task. Since we need something that can be weighted afterwards with a frequency selective cost function we simply smooth the absolute values of the input subband signals in positive and negative direction along the frequency axis:

---

[6] *LPC* abbreviates *linear predictive coding*.

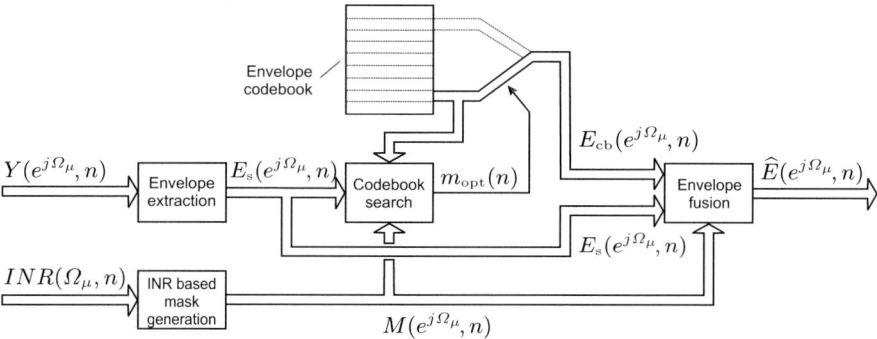

Fig. 4.15. Overview about the estimation of the spectral envelope.

$$E'_\text{s}\left(e^{j\Omega_\mu},n\right) = \begin{cases} \left|Y\left(e^{j\Omega_\mu},n\right)\right|, & \text{if } \mu = 0, \\ \lambda_\text{e}\, E'_\text{s}(e^{j\Omega_{\mu-1}},n) + (1-\lambda_\text{e})\left|Y\left(e^{j\Omega_\mu},n\right)\right|, & \text{else}, \end{cases}$$
(4.50)

$$E_\text{s}\left(e^{j\Omega_\mu},n\right) = \begin{cases} E'_\text{s}(e^{j\Omega_\mu},n), & \text{if } \mu = M-1, \\ \lambda_\text{e}\, E_\text{s}(e^{j\Omega_{\mu+1}},n) + (1-\lambda_\text{e})\, E'_\text{s}(e^{j\Omega_\mu},n), & \text{else}. \end{cases}$$
(4.51)

For stable operation the smoothing constant λ_e must be chosen out of the interval

$$0 \leq \lambda_\text{e} < 1.$$
(4.52)

A value of

$$\lambda_\text{e} = 0.5$$
(4.53)

has been selected in our setup. Fig. 4.16 shows an example of the smoothing process.

The resulting spectral envelope $E_\text{s}(e^{j\Omega_\mu},n)$ represents only at frequencies with high SNR or INR a good estimate for the envelope of the undistorted speech signal. At frequencies with low SNR the background noise is dominating the shape. For achieving a reliable estimation also at frequencies with large noise components, an a priori trained envelope codebook

$$C_\text{env} = \begin{bmatrix} E_{\text{cb,log},0}\left(e^{j\Omega_0}\right) & E_{\text{cb,log},0}\left(e^{j\Omega_1}\right) & \cdots & E_{\text{cb,log},0}\left(e^{j\Omega_{M-1}}\right) \\ E_{\text{cb,log},1}\left(e^{j\Omega_0}\right) & E_{\text{cb,log},1}\left(e^{j\Omega_1}\right) & \cdots & E_{\text{cb,log},1}\left(e^{j\Omega_{M-1}}\right) \\ \vdots & \vdots & \ddots & \vdots \\ E_{\text{cb,log},N_\text{env}-1}\left(e^{j\Omega_0}\right) & E_{\text{cb,log},N_\text{env}-1}\left(e^{j\Omega_1}\right) & \cdots & E_{\text{cb,log},N_\text{env}-1}\left(e^{j\Omega_{M-1}}\right) \end{bmatrix}$$
(4.54)

can be used. The codebook is organized as a matrix, where each row represents a prototype envelope in the logarithmic domain.

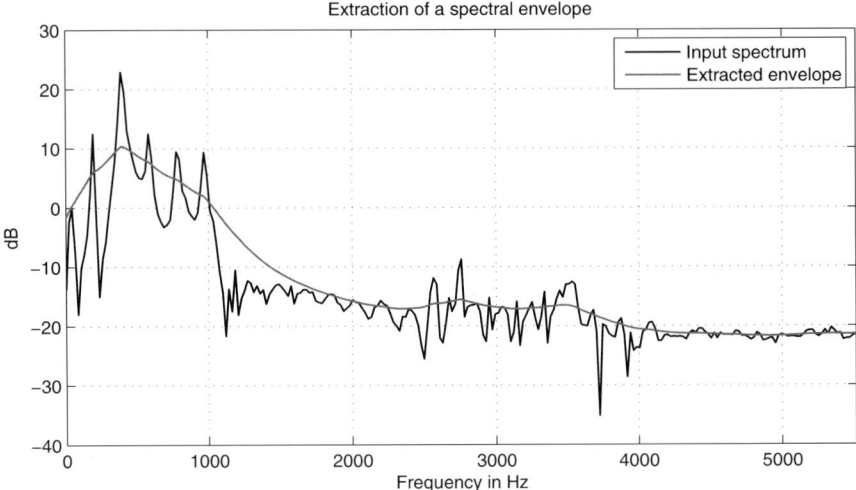

Fig. 4.16. Example for the estimation of the spectral envelope.

As mentioned above such a codebook needs to be training in advance on the basis of a large data base. In our case the *Linde-Buzo-Gray* (LBG) algorithm [11] has been used to extract $N_{\text{env}} = 256$ prototype envelopes out of a very large set of clean speech envelopes. Each entry was computed as defined in Eqs. 4.50 and 4.51. Afterwards a 20 $\log_{10}\{...\}$ operation was applied to each element of the resulting vector and the mean was subtracted. Fig. 4.17 shows such a codebook with $N_{\text{env}} = 8$ entries.

In the codebook the vector that matches best with the logarithm of the spectral envelope of the current input signal is searched. When computing a distance between the codebook entries and the envelope of the noisy input signal only those frequency areas are taken into account that have a sufficiently large input-to-noise ratio. To achieve this first an INR-based masking function is computed:

$$M(\Omega_\mu, n) = f_{\text{mask}}\Big(INR(\Omega_\mu, n)\Big). \quad (4.55)$$

The function $f_{\text{mask}}(x)$ should be designed such that the input INR range is mapped onto the interval [0, 1]. Output values close to one should be generated from large INR values while values close to zero should result from small input-to-noise ratios. A simple realization is a binary masking function according to

$$f_{\text{mask}}\Big(INR(\Omega_\mu, n)\Big) = \begin{cases} 1, & \text{if } INR(\Omega_\mu, n) > INR_0 \,, \\ 0.001, & \text{else} \,. \end{cases} \quad (4.56)$$

With regard to the search procedure a small positive value (instead of zero) was selected for small input-to-noise ratios. For the threshold INR_0 values

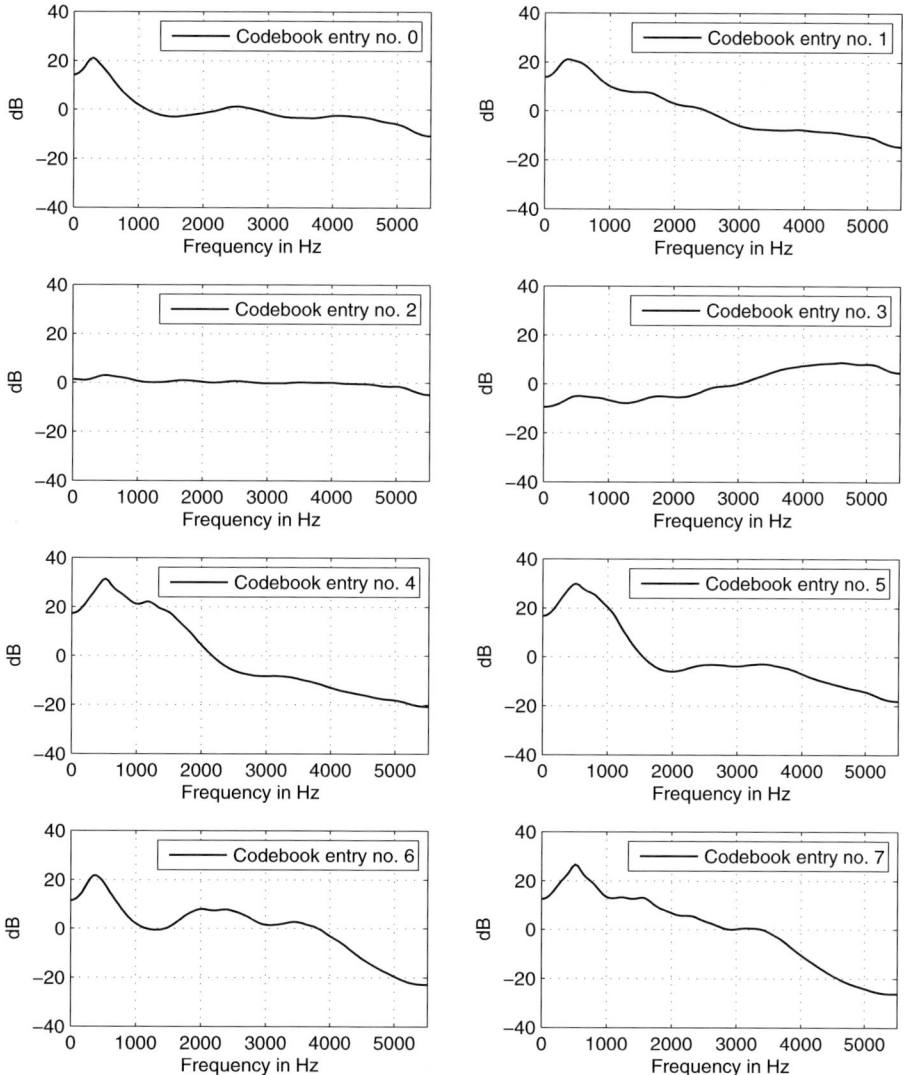

Fig. 4.17. Example for a codebook of size $N_{\text{env}} = 8$ containing prototypes of spectral envelopes.

out of the interval $[2, 4]$ were chosen. Beside this binary masking function also others – especially continuous functions – can be designed.

To perform the codebook search the spectral envelope of the noisy input signal is transformed in the logarithmic domain:

$$E_{\text{s,log}}\bigl(e^{j\Omega_\mu}, n\bigr) = 20 \log_{10}\Bigl\{ E_{\text{s}}\bigl(e^{j\Omega_\mu}, n\bigr) + \epsilon \Bigr\}. \qquad (4.57)$$

Adding the small positive constant ϵ has been done to avoid large negative values resulting from envelope components close to zero. In order to compensate for different speaker loudness or amplification a normalization value according to

$$E_{\text{s,log,norm}}(n) = \frac{\sum\limits_{\mu=0}^{M-1} M(\Omega_\mu, n)\, E_{\text{s,log}}(e^{j\Omega_\mu}, n)}{\sum\limits_{\mu=0}^{M-1} M(\Omega_\mu, n)} \qquad (4.58)$$

is computed. Due to the small positive constant (instead of zero) in Eq. 4.56 a division by zero in periods without speech is avoided. The normalization value is subtracted from the logarithmic spectral envelope to obtain a normalized spectral envelope:

$$\widetilde{E}_{\text{s,log}}(e^{j\Omega_\mu}, n) = E_{\text{s,log}}(e^{j\Omega_\mu}, n) - E_{\text{s,log,norm}}(n). \qquad (4.59)$$

All codebook entries are normalized in the same manner:

$$\widetilde{E}_{\text{cb,log},m}(e^{j\Omega_\mu}, n) = E_{\text{cb,log},m}(e^{j\Omega_\mu}) - E_{\text{cb,log,norm},m}(n). \qquad (4.60)$$

with

$$E_{\text{cb,log,norm},m}(n) = \frac{\sum\limits_{\mu=0}^{M-1} M(\Omega_\mu, n)\, E_{\text{cb,log},m}(e^{j\Omega_\mu})}{\sum\limits_{\mu=0}^{M-1} M(\Omega_\mu, n)} \qquad (4.61)$$

The frame individual normalization of the entire codebook according to Eqs. 4.60 and 4.61 is computationally rather expensive. Tests have shown that a normalization with $M(\Omega_\mu, n) = 1$ performed once before processing achieves nearly the same results as the frame individual adjustment.

For the selection of the codebook entry that is closest to the current envelope spectrum the weighted absolute logarithmic distance is utilized:

$$m_{\text{opt}}(n) = \underset{m}{\arg\min} \sum_{\mu=0}^{M-1} M(\Omega_\mu, n) \left| \widetilde{E}_{\text{s,log}}(e^{j\Omega_\mu}, n) - \widetilde{E}_{\text{cb,log},m}(e^{j\Omega_\mu}, n) \right|.$$

$$(4.62)$$

The resulting normalized logarithmic prototype envelope is first denormalized and then transformed back into the linear domain. However, this is only performed if the masking weights indicate that a sufficiently large frequency range has been used while searching the codebook. If this is not the case the previously selected entry – attenuated by a few decibels (adjustable via the parameter K_{dec}) – is utilized:

$$E_{\text{cb}}\left(e^{j\Omega_\mu},n\right) = \begin{cases} 10^{\left(\widetilde{E}_{\text{cb,log},m_{\text{opt}}(n)}\left(e^{j\Omega_\mu},n\right)+E_{\text{s,log,norm}}(n)\right)/20}, \\ \qquad\qquad\qquad\qquad \text{if } \frac{1}{M}\sum_{\mu=0}^{M-1} M(\Omega_\mu,n) > M_0\,, \\ E_{\text{cb}}\left(e^{j\Omega_\mu},n-1\right)K_{\text{dec}}\,, \\ \qquad\qquad\qquad\qquad \text{else}\,. \end{cases} \quad (4.63)$$

The parameter M_0 can be chosen as $M_0 = 0.12\ldots0.2$ meaning that about 12 to 20 percent of the frequency range has a sufficiently large SNR. Finally, the envelope of the noisy input signal, $E_\text{s}(e^{j\Omega_\mu},n)$, and the codebook based one, $E_\text{cb}(e^{j\Omega_\mu},n)$, are combined. At frequencies with large SNR or INR the original envelope is utilized, at the remaining frequencies the codebook based estimation is used. To avoid abrupt switching first order IIR smoothing in both frequency directions is applied again. First a hard switched envelope is computed according to

$$\check{E}\left(e^{j\Omega_\mu},n\right) = M(\Omega_\mu,n)\, E_\text{s}\left(e^{j\Omega_\mu},n\right) + \left(1 - M(\Omega_\mu,n)\right) E_\text{cb}\left(e^{j\Omega_\mu},n\right). \quad (4.64)$$

Afterward the IIR smoothing is performed in positive frequency direction

$$\widetilde{E}\left(e^{j\Omega_\mu},n\right) = \begin{cases} \check{E}\left(e^{j\Omega_\mu},n\right), & \text{if } \mu = 0\,, \\ \lambda_{\text{mix}}\,\widetilde{E}\left(e^{j\Omega_{\mu-1}},n\right) + (1-\lambda_{\text{mix}})\,\check{E}\left(e^{j\Omega_\mu},n\right), & \text{else}\,, \end{cases} \quad (4.65)$$

and finally in negative frequency direction

$$E\left(e^{j\Omega_\mu},n\right) = \begin{cases} \widetilde{E}\left(e^{j\Omega_\mu},n\right), & \text{if } \mu = M-1\,, \\ \lambda_{\text{mix}}\,E\left(e^{j\Omega_{\mu+1}},n\right) + (1-\lambda_{\text{mix}})\,\widetilde{E}\left(e^{j\Omega_\mu},n\right), & \text{else}\,. \end{cases} \quad (4.66)$$

For the smoothing constant a value of about

$$\lambda_{\text{mix}} = 0.3 \quad (4.67)$$

was selected.

As mentioned at the beginning of this section – the proposed scheme is computationally quite expensive. However, the objective was to explain the basic principle and not the cheapest realization. If, for example, the codebook search is performed on Bark or Mel [25] based groups of subbands the processing comes at much lower cost (e.g. 15 groups instead of 256 subbands). Several other improvements – as they are known from speech coding – can be applied too [24].

4.4.2.2 Reconstruction of Unvoiced Signals

Once we have a reliable estimate of the spectral envelope it is rather straight forward to reconstruct the unvoiced part of a speech signal. In accordance with the source-filter model of human speech generation [2] we can simply multiply the spectral envelope $E(e^{j\Omega_\mu}, n)$ with an unvoiced excitation spectrum $A_\mathrm{u}(e^{j\Omega_\mu}, n)$ as depicted in Fig. 4.18.

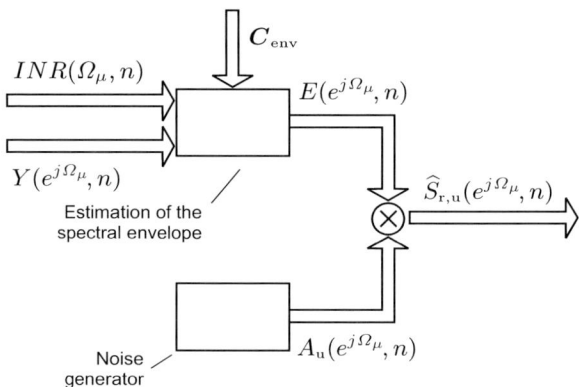

Fig. 4.18. Overview about the reconstruction of the unvoiced speech spectrum.

The unvoiced excitation subband signals $A_\mathrm{u}(e^{j\Omega_\mu}, n)$ can be generated with a noise generator that produces uncorrelated white Gaussian noise for the real and the imaginary part of the complex signals. The statistical properties of the noise generator should be

$$\mathrm{E}\Big\{\mathrm{Re}\big\{A_\mathrm{u}(e^{j\Omega_\mu}, n)\big\}\Big\} = \mathrm{E}\Big\{\mathrm{Im}\big\{A_\mathrm{u}(e^{j\Omega_\mu}, n)\big\}\Big\} = 0 \qquad (4.68)$$

and

$$\mathrm{E}\Big\{\mathrm{Re}^2\big\{A_\mathrm{u}(e^{j\Omega_\mu}, n)\big\}\Big\} = \mathrm{E}\Big\{\mathrm{Im}^2\big\{A_\mathrm{u}(e^{j\Omega_\mu}, n)\big\}\Big\} = \frac{1}{2}. \qquad (4.69)$$

This results in zero-mean complex subband signals with unit variance. The resulting unvoiced speech spectrum

$$\widehat{S}_\mathrm{u,r}\left(e^{j\Omega_\mu}, n\right) = A_\mathrm{u}\left(e^{j\Omega_\mu}, n\right) E\left(e^{j\Omega_\mu}, n\right) \qquad (4.70)$$

is only a good estimate for unvoiced speech periods. In voiced parts the spectrum is much too loud. This problem is addressed in Sec. 4.4.2.4 where it is described how the voiced and the unvoiced spectra are combined.

4.4.2.3 Reconstruction of Voiced Signals

In the previous section a description was given of how unvoiced speech signal parts are synthesized – this section deals with reconstruction of voiced speech components. Short voiced speech signal blocks can approximately be considered as a periodic signal with a constant pitch frequency. In our approach the voiced synthetic speech is generated based on the well-known source-filter model for speech generation [2]. Individual pitch impulses are generated first in the subband domain. Afterwards the individual pitch impulses are combined in the time domain to form a periodic signal. Special care is taken when placing the pitch impulses. An overview of synthesizing voiced speech signals is shown in Fig. 4.19.

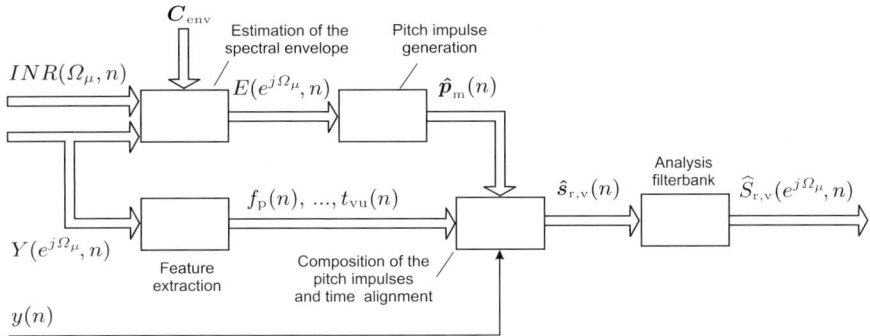

Fig. 4.19. Overview about the reconstruction of the voiced speech spectrum.

Once an estimate for the spectral envelope $E\left(e^{j\Omega_\mu}, n\right)$ is available, it is utilized to generate a so-called pitch impulse spectrum. The principle idea of computing a pitch impulse is depicted in Fig. 4.20.

According to the source-filter model first the extracted spectral envelope $E\left(e^{j\Omega_\mu}, n\right)$ is multiplied by a spectrum of a short generic pitch impulse $P\left(e^{j\Omega_\mu}\right)$ according to:

$$\widehat{P}_\mathrm{m}(e^{j\Omega_\mu}, n) = E(e^{j\Omega_\mu}, n)\, P(e^{j\Omega_\mu}). \tag{4.71}$$

The so-called *generic pitch impulse* corresponds to a short predefined impulse that was measured during a voiced section of a real speech signal, whereas its spectral envelope had been removed. We refer to this impulse as the pitch impulse prototype. The spectral envelope, as already mentioned, approximates the behavior of the vocal tract. Due to the element-wise weighting in Eq. 4.71

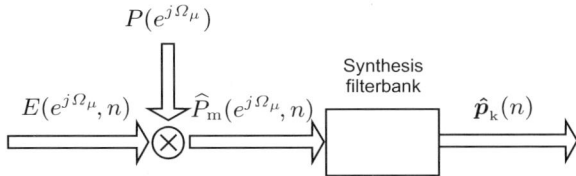

Fig. 4.20. Overview of generating pitch impulses.

a modified pitch impulse prototype spectrum $\widehat{P}_\mathrm{m}(e^{j\Omega_\mu}, n)$ results. Thereby, the power is adjusted to that of the input signal. In order to generate the pitch impulse

$$\widehat{\boldsymbol{p}}_\mathrm{m}(n) = \Big[\hat{p}_\mathrm{m}(0, n), \ldots, \hat{p}_\mathrm{m}(M-1, n)\Big]^\mathrm{T} \qquad (4.72)$$

a synthesis filterbank or an IFFT is applied according to:

$$\hat{p}_\mathrm{m}(k, n) = \frac{1}{M} \sum_{\mu=0}^{M-1} \widehat{P}_\mathrm{m}(e^{j\Omega_\mu}, n) \, e^{j\frac{2\pi}{M}\mu k}. \qquad (4.73)$$

It is important to note that instead of using an excitation signal – as it is commonly done for speech coding [24] – a predefined prototype pitch impulse is employed. However, instead of using only a generic impulse one could also use speaker- or pitch-specific impulses for further improvements. For certain pitch-frequencies (e.g., 80 Hz, 90 Hz, ..., 300 Hz) a priori pitch impulse prototypes may be specified. The desired prototype pitch impulse is selected according to the current pitch estimate from the predefined pitch-specific database.

The next step of the process is the generation of a synthetic voiced speech signal segment. In order to constitute a clean speech signal the resulting modified pitch impulse from Eq. 4.73 is added to those obtained from the preceding ones considering an appropriate time-shift. The time-shift is chosen according to the current pitch period

$$\hat{\tau}_\mathrm{p}(n) = \frac{f_\mathrm{s}}{\hat{f}_\mathrm{p}(n)}, \qquad (4.74)$$

whereas the quantity f_s corresponds to the sampling frequency and $\hat{f}_\mathrm{p}(n)$ denotes the estimated pitch frequency.

For the sake of clarity, an example is introduced in Fig. 4.21 that shows a section of a synthetic speech signal over three sub-frames. A train of modified pitch impulses is depicted which, as mentioned before, vary depending on the current spectral envelope.

It has to be mentioned that for a successful speech synthesis an accurate estimation of the pitch frequency even at highly disturbed speech signals is needed. A method for pitch estimation has been proposed in Sec. 4.4.1.4 that allows a reliable operation at low SNR scenarios even for very low fundamental frequencies.

Furthermore, it should be ensured that no strong variations of the extracted spectral envelope and the estimated pitch frequency exist. In addition, more features have to be extracted from the noisy input signal to achieve a high-quality speech synthesis. Reasonable features are voiced/unvoiced-classification $t_\mathrm{vu}(n)$ and detection of speech presence by exploiting, e.g., the

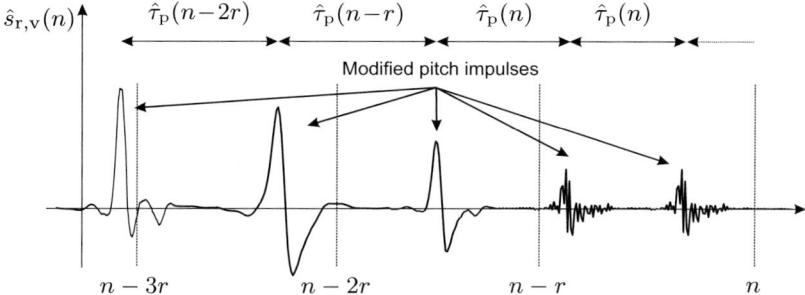

Fig. 4.21. Example of a synthesized speech signal over three sub-frames. Note that in the last sub-frame two pitch impulses are inserted and therefore $\hat{\tau}_p(n)$ is used twice.

$INR(\Omega_\mu, n)$ and the reliability information $p_{\hat{f}_p}(n)$ of the estimated pitch frequency. At transitions from voiced to unvoiced parts the amplitudes of the modified pitch impulse $\hat{\boldsymbol{p}}_m(n)$ should be decreased slowly in order to avoid artifacts. For the same reason the pitch period should be smoothed according to:

$$\hat{\tau}_p(n) = \begin{cases} \dfrac{f_s}{\hat{f}_p(n)}, & \text{if a reliable pitch estimation is possible}, \\ \hat{\tau}_p(n-1) + \Delta_p(n), & \text{else}, \end{cases} \quad (4.75)$$

whereas $\Delta_p(n)$ is updated during voiced speech segments only. It corresponds to the difference between the current and the previous pitch period:

$$\Delta_p(n) = \begin{cases} (1-\lambda)\Delta_p(n-1) + \lambda\big(\hat{\tau}_p(n) - \hat{\tau}_p(n-1)\big), \\ \qquad\qquad\qquad\qquad \text{during voiced speech segments}, \\ \Delta_p(n-1), \\ \qquad\qquad\qquad\qquad \text{else}. \end{cases} \quad (4.76)$$

For the smoothing factor a value of about $\lambda = 0.3$ is chosen at a sampling frequency of $f_s = 11025$ Hz.

Most often the reconstructed phase of the synthetic speech signal is different from the phase of the input signal. For this reason a time alignment is applied before composition of the individual pitch impulses. It is performed in such a way that at pitch onsets first a maximum search over the current input signal $y(n)$ and the pitch impulse $\hat{p}_m(k,n)$ is accomplished:

$$\rho_y(n) = \underset{k}{\operatorname{argmax}}\big\{y(n-k)\big\}, \quad (4.77)$$

$$\rho_r(n) = \underset{k}{\operatorname{argmax}}\big\{\hat{p}_m(k,n)\big\}, \quad (4.78)$$

with

$$k \in \{0, ..., M-1\}. \tag{4.79}$$

The corresponding phase difference: $\rho_y(n) - \rho_r(n)$, is then used to align the pitch impulse correctly in time. Whereas during voiced blocks a maximum search is performed over $y(n)$ in a selected range around the current pitch period. The pitch impulse is then adjusted in time considering the current pitch period and the difference between the maximal lags.

The synthesized signal vector $\tilde{s}_{r,v}(n)$ is subsequently multiplied by a window function, e.g. Hann or Hamming window [18],

$$\hat{s}_{r,v}(k,n) = \tilde{s}_{r,v}(k,n)\, h_k, \tag{4.80}$$

for smoothing of signal parts at the edges of the current frame. In Fig. 4.22 a simulation example of a windowed input signal segment as well as the corresponding synthesized speech signal block before and after time alignment are depicted. As can be seen within the graph after time alignment the phase difference is almost compensated.

In oder to combine the synthetic voiced signal parts with the unvoiced signal components in the frequency domain an analysis filterbank or an FFT as a last processing unit is applied according to:

$$\widehat{S}_{r,v}(e^{j\Omega_\mu}, n) = \sum_{k=0}^{M-1} \hat{s}_{r,v}(k,n)\, e^{-j\frac{2\pi}{M}\mu k}. \tag{4.81}$$

4.4.2.4 Combining the Voiced and the Unvoiced Reconstructed Signals

Combining the voiced and the unvoiced reconstructed signals can be performed by adding weighted versions of both signals in the subband domain:

$$\widehat{S}_r(e^{j\Omega_\mu}, n) = H_v(e^{j\Omega_\mu}, n)\, \widehat{S}_{r,v}(e^{j\Omega_\mu}, n) + H_u(e^{j\Omega_\mu}, n)\, \widehat{S}_{r,u}(e^{j\Omega_\mu}, n). \tag{4.82}$$

The weights for the voiced part are denoted by $H_v(e^{j\Omega_\mu}, n)$, consequently $H_u(e^{j\Omega_\mu}, n)$ are the unvoiced weighting factors. To obtain these factors first preliminary desired weights are computed as:

$$\widetilde{H}_v(e^{j\Omega_\mu}, n) = \begin{cases} \widetilde{H}_{vu,min}, \\ \qquad \text{if } (INR_{low}(n) < INR_0) \wedge (INR_{high}(n) < INR_0), \\ \max\{t_{vu}(n), \widetilde{H}_{vu,min}\}, \\ \qquad \text{else}, \end{cases} \tag{4.83}$$

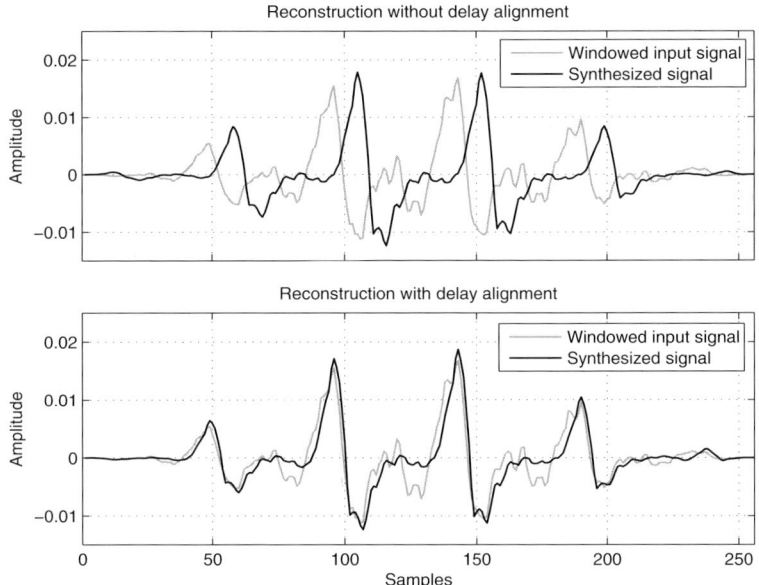

Fig. 4.22. Reconstruction without and with delay alignment.

and

$$\widetilde{H}_{\mathrm{u}}\big(e^{j\Omega_{\mu}}, n\big) = \begin{cases} \widetilde{H}_{\mathrm{vu,min}}, \\ \quad \text{if } \big(INR_{\mathrm{low}}(n) < INR_0\big) \wedge \big(INR_{\mathrm{high}}(n) < INR_0\big), \\ \max\big\{1 - t_{\mathrm{vu}}(n), \widetilde{H}_{\mathrm{vu,min}}\big\}, \\ \quad \text{else}, \end{cases}$$

(4.84)

respectively. In case of speech pauses a lower bound for both weights of about $\widetilde{H}_{\mathrm{vu,min}} = 0.1 \ldots 0.5$ is applied. Those signal periods are determined if neither the input-to-noise ratio at lower nor at higher frequencies is above the threshold INR_0. This value can be chosen out of the interval

$$INR_0 \in [2, 10].$$

(4.85)

If a large INR is achieved either in one of both frequency ranges or in both the voiced/unvoiced classification $t_{\mathrm{vu}}(n)$ – described in Sec. 4.4.1.3 – is utilized to determine the voiced weights. Consequently its counterpart, $1 - t_{\mathrm{vu}}(n)$, is used for the unvoiced weights. To prevent the mixing factors from getting close to zero, the lower bound $\widetilde{H}_{\mathrm{vu,min}}$ is applied again. This helps to improve the robustness against false decisions of the voiced/unvoiced classification. To achieve a smooth transition from opening to closing periods IIR smoothing of first order is employed to the preliminary weights:

$$H_\text{v}\bigl(e^{j\Omega_\mu},n\bigr) = \lambda_\text{vu}\,H_\text{v}\bigl(e^{j\Omega_\mu},n-1\bigr) + \bigl(1-\lambda_\text{vu}\bigr)\widetilde{H}_\text{v}\bigl(e^{j\Omega_\mu},n\bigr),\quad (4.86)$$
$$H_\text{u}\bigl(e^{j\Omega_\mu},n\bigr) = \lambda_\text{vu}\,H_\text{u}\bigl(e^{j\Omega_\mu},n-1\bigr) + \bigl(1-\lambda_\text{vu}\bigr)\widetilde{H}_\text{u}\bigl(e^{j\Omega_\mu},n\bigr). \quad (4.87)$$

The smoothing constant λ_vu should be chosen out of the interval

$$\lambda_\text{vu} \in [0.5,\ 0.7]. \qquad (4.88)$$

If the reconstructed signal is required in its time-domain version $\hat{s}_\text{r}(n)$, a synthesis filterbank – as shown in Fig. 4.23 – should follow the mixing stage.

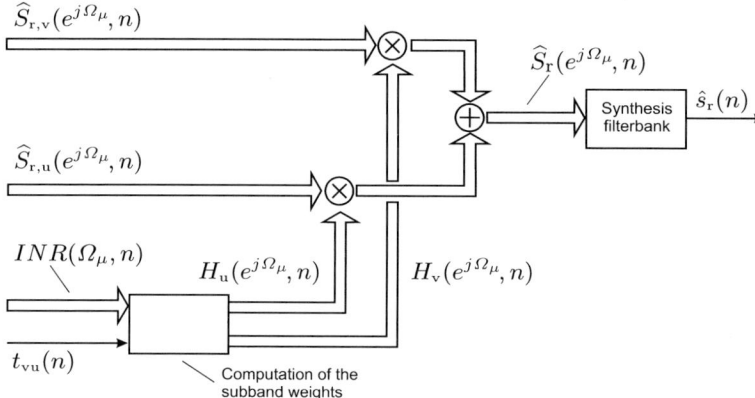

Fig. 4.23. Mixing the voiced and the unvoiced components of the speech reconstruction.

4.5 Combining the Reconstructed and the Noise Suppressed Signal

In a last algorithmic part the output signals of the conventional noise suppression and the speech reconstruction are combined in an adaptive manner. We differentiate between two versions of this mixing process:

- If the entire reconstruction – consisting of a voiced and an unvoiced part, as well as a combination of both as described in the last section – is computed a combination with the noise suppressed signal can be performed for all frequencies. Details about this type of mixing scheme are described in the next section.
- For some applications, e.g. for speech enhancement in automotive environments, it is sufficient to reconstruct only the low frequency range. Since low frequency speech consists mainly of voiced segments it is enough to perform only the voiced part of the signal reconstruction. In this case a different type of mixing weight computation should be applied. Details about this kind of mixing are described in Sec. 4.5.2.

4.5.1 Adding the Fully Reconstructed Signal

The basic structure of this mixing process is depicted in Fig. 4.24. Combining both signals is performed in the subband domain – for that reason analysis filterbanks for the input signals $\hat{s}_\mathrm{s}(n)$ and $\hat{s}_\mathrm{r}(n)$ and a synthesis filterbank for the mixed signal $\hat{s}(n)$ are depicted. Again, it should be mentioned that the analysis filterbanks can be omitted if the input signals are available in the subband domain. In this case vectors containing the subband signals $\widehat{S}_\mathrm{s}(e^{j\Omega_\mu}, n)$ and $\widehat{S}_\mathrm{r}(e^{j\Omega_\mu}, n)$, respectively, can be passed to the adaptive mixing scheme directly.

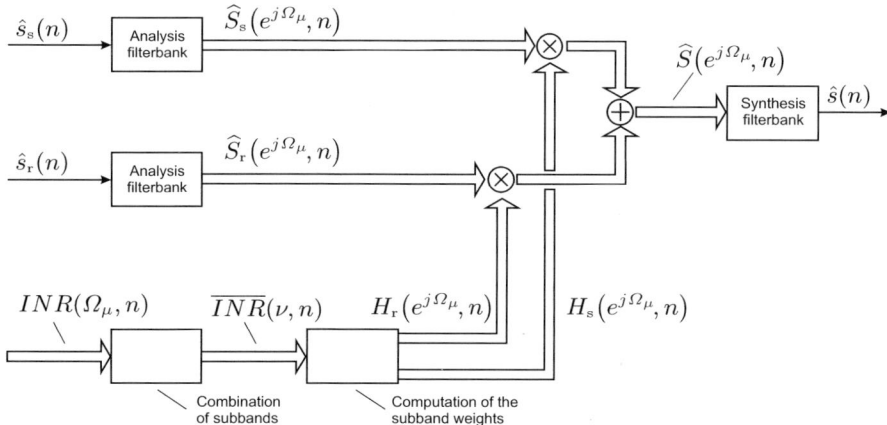

Fig. 4.24. Overview about the adaptive combination of the output signal of a conventional noise suppression $\hat{s}_\mathrm{s}(n)$ and the reconstructed speech signal $\hat{s}_\mathrm{r}(n)$.

For the computation of the mixing weights $H_\mathrm{s}(e^{j\Omega_\mu}, n)$ and $H_\mathrm{r}(e^{j\Omega_\mu}, n)$, respectively, the input-to-noise ratios of neighboring subbands are grouped together on the basis of a Mel [25] scale:

$$\overline{INR}(\nu, n) = \frac{\sum_{\mu=0}^{M-1} F_\nu\left(e^{j\Omega_\mu}\right) INR(\Omega_\mu, n)}{\sum_{\mu=0}^{M-1} F_\nu\left(e^{j\Omega_\mu}\right)}, \qquad (4.89)$$

with

$$\nu \in \{0, 1, \ldots, M_\mathrm{mel} - 1\}. \qquad (4.90)$$

The quantity $F_\nu(e^{j\Omega_\mu})$ denote the frequency response of the ν-th Mel filter. Using a Mel based frequency resolution reduces the computational complexity of the mixer weights, improves the robustness against outliers within the INR estimation, and achieves a frequency resolution that is close to that of the human auditory system. In contrast to standard Mel resolutions we have combined the first two Mel bands to a common one. Further details about a Mel based frequency scale can be found in [25]. For a sample rate $f_\mathrm{s} = 11025$ Hz we have utilized

$$M_\mathrm{mel} = 16 \tag{4.91}$$

Mel bands. The frequency responses of the individual filters are depicted in Fig. 4.25 (center diagram). For better visibility the Mel filters are depicted alternatingly in black and gray.

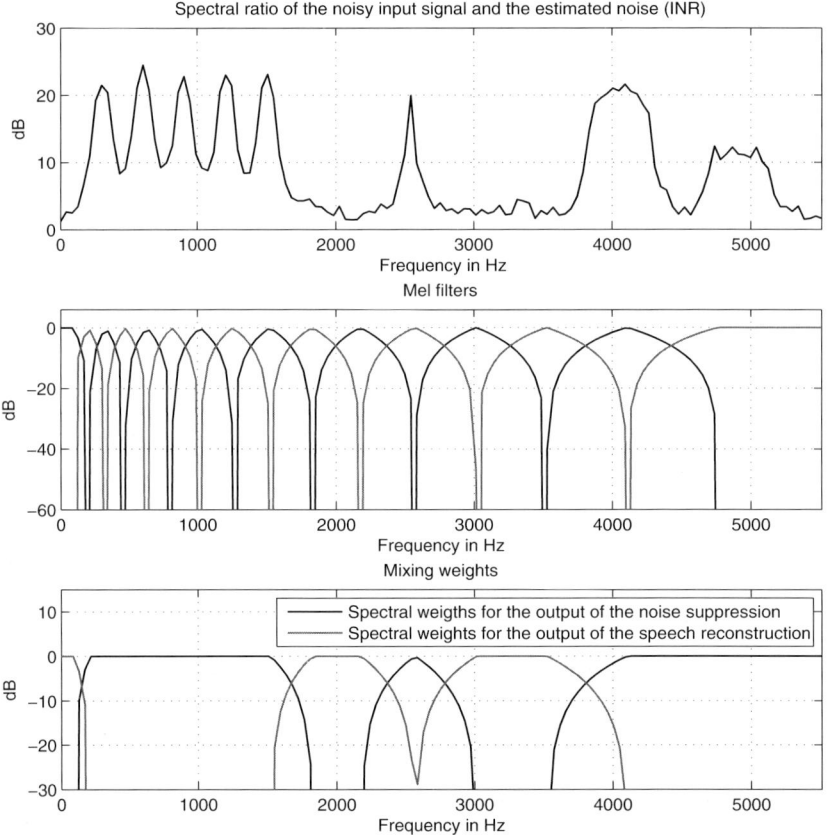

Fig. 4.25. Example for a frequency selective input-to-noise ratio $INR(\Omega_\mu, n)$ (top), magnitude frequency responses $F_\nu(e^{j\Omega_\mu})$ of the utilized Mel filters (center), and resulting mixer weights for the conventional noise suppression output spectrum $\widehat{S}_\mathrm{s}(e^{j\Omega_\mu}, n)$ without limitation ($H_\mathrm{s,min}(e^{j\Omega_\mu}, n) = 0$) and for the reconstructed speech spectrum $\widehat{S}_\mathrm{r}(e^{j\Omega_\mu}, n)$ (bottom).

By applying a binary characteristic to the Mel averaged input-to-noise ratios the mixing weights are computed first in the Mel domain:

$$f_{\mathrm{mix}}\left(\overline{INR}(\nu,n)\right) = \begin{cases} 1, \text{ falls } \overline{INR}(\nu,n) > \overline{INR}_0, \\ 0, \text{ else}. \end{cases} \quad (4.92)$$

The threshold $\overline{INR}_0$ should be chosen out of the interval

$$4 < \overline{INR}_0 < 10. \quad (4.93)$$

Besides the binary characteristic described before also more advanced continuous characteristics can be applied. With the usage of the Mel based mixer weights and the frequency responses of the Mel filters preliminary subband weights for the output of the noise suppression can be computed according to

$$\widetilde{H}_{\mathrm{s}}\left(e^{j\Omega_\mu},n\right) = \sum_{\nu=0}^{M_{\mathrm{mel}}-1} f_{\mathrm{mix}}\left(\overline{INR}(\nu,n)\right) F_\nu\left(e^{j\Omega_\mu}\right). \quad (4.94)$$

If a residual noise floor is desired the final mixing weights should be bounded by a lower limit:

$$H_{\mathrm{s}}\left(e^{j\Omega_\mu},n\right) = \max\left\{\widetilde{H}_{\mathrm{s}}\left(e^{j\Omega_\mu},n\right), H_{\mathrm{s,min}}\left(e^{j\Omega_\mu},n\right)\right\}. \quad (4.95)$$

The floor parameter $H_{\mathrm{s,min}}(e^{j\Omega_\mu},n)$ should be chosen somewhere between $-4\,\mathrm{dB}$ and $0\,\mathrm{dB}$ in order to allow for slightly more noise suppression as adjusted via the maximum attenuation factors $G_{\min}(e^{j\Omega_\mu},n)$ (see Eq. 4.8) in the reconstructed subbands. The weights for the reconstructed spectrum are calculated as:

$$H_{\mathrm{r}}\left(e^{j\Omega_\mu},n\right) = 1 - \widetilde{H}_{\mathrm{s}}\left(e^{j\Omega_\mu},n\right). \quad (4.96)$$

Finally, the output spectrum is computed according to

$$\widehat{S}\left(e^{j\Omega_\mu},n\right) = H_{\mathrm{s}}\left(e^{j\Omega_\mu},n\right) \widehat{S}_{\mathrm{s}}\left(e^{j\Omega_\mu},n\right) + H_{\mathrm{r}}\left(e^{j\Omega_\mu},n\right) \widehat{S}_{\mathrm{r}}\left(e^{j\Omega_\mu},n\right). \quad (4.97)$$

To clarify the entire mixing process an input-to-noise ratio $INR(\Omega_\mu,n)$ is depicted exemplarily in the upper diagram of Fig. 4.25. By utilizing the scheme for computing the mixing weights described above the coefficients $H_{\mathrm{r}}(e^{j\Omega_\mu},n)$ and $H_{\mathrm{s}}(e^{j\Omega_\mu},n)$, which are depicted in the lowest diagram of Fig. 4.25, result. Note that in this example no limitation has been applied ($H_{\mathrm{s,min}}(e^{j\Omega_\mu},n) = 0$) for the reason of better visibility.

Finally, an example for the proposed algorithm is presented in Fig. 4.26. A microphone signal measured in a car at a speed of about 90 km/h is used as an input signal $y(n)$. In the uppermost diagram of Fig. 4.26 a time-frequency analysis of the input signal is depicted. The average SNR of this recording is about 15 dB.

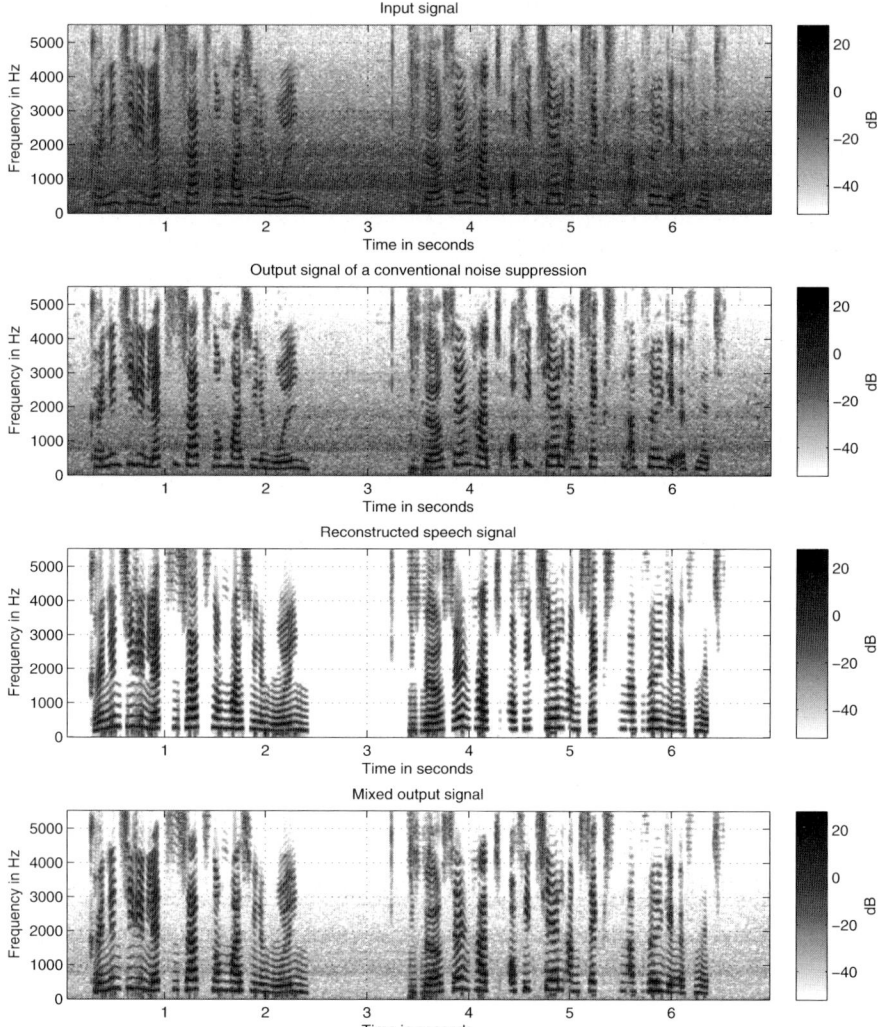

Fig. 4.26. Time-frequency analyses of the noisy input signal $y(n)$ (top), of the output signal of a conventional noise suppression $\hat{s}_\mathrm{s}(n)$ (second diagram), of the reconstructed signal $\hat{s}_\mathrm{r}(n)$ (third diagram), and of the adaptively mixed signal $\hat{s}(n)$ (bottom).

The output signal $\hat{s}_\mathrm{s}(n)$ of a conventional noise suppression scheme (respectively its time-frequency analysis) is shown in the second diagram of Fig. 4.26. A modified Wiener rule according to Eq. 4.8 with an adaptive overestimation

$$\beta\left(e^{j\Omega_\mu}, n\right) = \max\left\{2, \frac{1}{G\left(e^{j\Omega_\mu}, n-1\right)}\right\} \quad (4.98)$$

according to [12] has been utilized. The maximum attenuation was adjusted as

$$G_{\min}\left(e^{j\Omega_\mu}, n\right) = 0.13 \tag{4.99}$$

for all subbands which corresponds to $-18\,\mathrm{dB}$. In time-frequency areas with low SNR the speech signal disappears within the noise components. With the help of the speech reconstruction these signal parts can be recovered. The output signal of the reconstruction $\hat{s}_\mathrm{r}(n)$ is depicted in the third diagram. By mixing of the reconstructed and the conventionally noise suppressed signals the output signal $\hat{s}(n)$ is generated. Its time-frequency analysis is depicted in the lowest diagram of Fig. 4.26.

4.5.2 Adding only the Voiced Part of the Reconstructed Signal

In some applications – such as hands-free telephony for cars – it is sufficient to perform a signal reconstruction only for the low frequency range. Even if the background noise in a car can get very strong at high speed in most cases a sufficiently large SNR can be observed at frequencies above 500 Hz to 1000 Hz due to the Lombard effect [6,13]. However, the increase of the speaking level is not large enough to compensate for the large noise power of engine noise at low frequencies.

For these applications we suggest a slightly different approach compared to the full range mixing scheme described in the last section. As a first step we define a maximum reconstruction frequency $\Omega_{\mu_\mathrm{r,max}}$. Typically this value is chosen to represent a frequency out of the interval

$$\Omega_{\mu_\mathrm{r,max}} \in \frac{2\pi}{f_\mathrm{s}}\left[500\,\mathrm{Hz},\ 1000\,\mathrm{Hz}\right]. \tag{4.100}$$

For frequencies above that threshold only a conventional noise suppression is performed, below an adaptive mixing of the noise suppressed and the reconstructed signal is performed:

$$\widehat{S}(e^{j\Omega_\mu}, n) = \begin{cases} \widehat{S}_\mathrm{s}(e^{j\Omega_\mu}, n) + H_\mathrm{r}(e^{j\Omega_\mu}, n)\, \widehat{S}_\mathrm{r,v}(e^{j\Omega_\mu}, n), \\ \qquad \text{if } \mu \leq \mu_\mathrm{r,max}, \\ \widehat{S}_\mathrm{s}(e^{j\Omega_\mu}, n), \\ \qquad \text{else}. \end{cases} \tag{4.101}$$

As mentioned at the beginning of Sec. 4.5 it is sufficient to utilize only the voiced part of the reconstruction since most of the speech utterances are completely voiced below $\Omega_{\mu_\mathrm{r,max}}$.

A car can be driven at different speeds which leads also to a very wide range of signal-to-noise ratios. While the reconstruction should be switched off completely during stand-still or at low speed it can improve the signal quality significantly in high noise conditions. To differentiate these scenarios we propose to analyze the weights $G(e^{j\Omega_\mu}, n)$ of the conventional noise suppression filter. In our setup this leads to slightly better results compared to the scheme presented in the previous section. In Fig. 4.27 an overview about the second mixing scheme is presented.

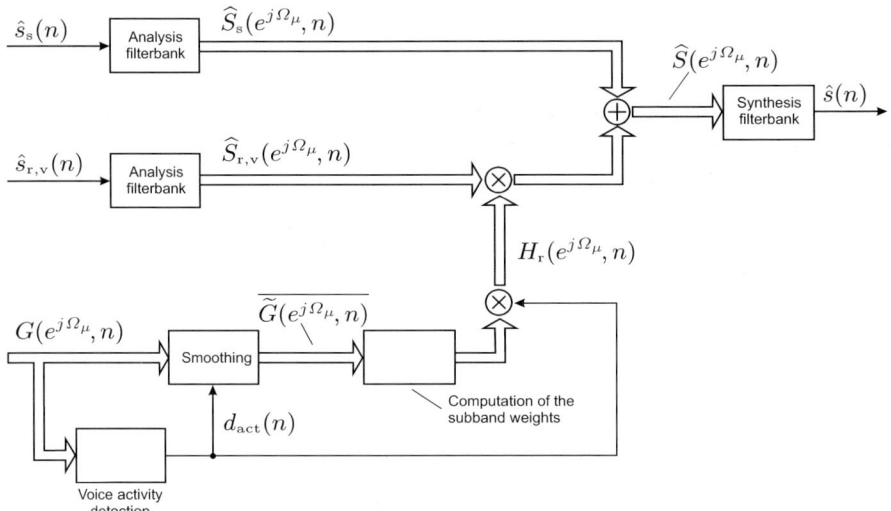

Fig. 4.27. Overview about the adaptive combination of the output signal of a conventional noise suppression $\hat{s}_s(n)$ and the reconstructed speech signal $\hat{s}_r(n)$.

In order to reduce the dependance of the noise floor parameter of the noise suppression rule we first subtract the maximum attenuations $G_{\min}(e^{j\Omega_\mu}, n)$ from the noise reduction weights and second scale the resulting attenuation range such that values between 0 and 1 are obtained:

$$\widetilde{G}\left(e^{j\Omega_\mu}, n\right) = \frac{G\left(e^{j\Omega_\mu}, n\right) - G_{\min}\left(e^{j\Omega_\mu}, n\right)}{1 - G_{\min}\left(e^{j\Omega_\mu}, n\right)}. \qquad (4.102)$$

In the next step speech activity is detected by computing the average normalized weight of the noise suppression characteristic within a limited frequency range $\Omega_{\mu_4} \leq \Omega_\mu \leq \Omega_{\mu_5}$:

$$\widetilde{G}_{\text{avg}}(n) = \frac{1}{\mu_5 - \mu_4 + 1} \sum_{\mu=\mu_4}^{\mu_5} \widetilde{G}\left(e^{j\Omega_\mu}, n\right). \qquad (4.103)$$

If the average normalized weight is larger than a threshold G_0 voice activity is detected:

$$d_{\text{act}}(n) = \begin{cases} 1, & \text{if } \widetilde{G}_{\text{avg}}(n) > G_0, \\ 0, & \text{else}. \end{cases} \qquad (4.104)$$

Threshold values of about $G_0 = 0.4 \ldots 0.6$ have proven to be reasonable. The frequency range of this detection should not start at very low frequencies, since in very noisy scenarios the very low frequency weights mostly stay at the maximum attenuation. In our setup we have chosen the lower and upper frequency limit as

$$\Omega_{\mu_4} \longleftrightarrow 700\,\text{Hz}, \qquad (4.105)$$
$$\Omega_{\mu_5} \longleftrightarrow 3000\,\text{Hz}. \qquad (4.106)$$

During periods of detected voice activity the normalized filter weights are smoothed slowly over time. An IIR filter of first order, e.g., with a smoothing constant of about $\lambda_G = 0.8 \ldots 0.99$ can be utilized:

$$\overline{\widetilde{G}\left(e^{j\Omega_\mu},n\right)} = \begin{cases} \lambda_G \, \overline{\widetilde{G}\left(e^{j\Omega_\mu},n-1\right)} + (1-\lambda_G)\, \widetilde{G}\left(e^{j\Omega_\mu},n\right), & \text{if } d_{\text{act}}(n) = 1, \\ \overline{\widetilde{G}\left(e^{j\Omega_\mu},n-1\right)}, & \text{else}. \end{cases} \qquad (4.107)$$

For those subbands that show permanently a large attenuation the reconstruction is (preliminarily) activated. This can be accomplished by simply comparing the smoothed weights with a threshold $\widetilde{G}_0$:

$$\widetilde{H}_{\text{r}}\left(e^{j\Omega_\mu},n\right) = \begin{cases} 1, & \text{if } \overline{\widetilde{G}\left(e^{j\Omega_\mu},n\right)} < \widetilde{G}_0, \\ 0, & \text{else}. \end{cases}$$

Due to the normalization of the noise suppression factors a good choice for the threshold is $\widetilde{G}_0 = 0.5$. Since the reconstruction should be applied in periods of voice activity the final filter weights are computed by multiplying the preliminary ones with the voice activity detection flag:

$$H_{\text{r}}\left(e^{j\Omega_\mu},n\right) = \widetilde{H}_{\text{r}}\left(e^{j\Omega_\mu},n\right) d_{\text{act}}(n). \qquad (4.108)$$

As in the description of the mixing scheme for the entire frequency range we will now show an example in terms of time-frequency analyses. Fig. 4.28 shows in its upper diagram a spectrogram of a microphone signal recorded in a car at a speed of about 160 km/h. The low SNR at low frequencies is clearly visible – most of the time the lowest two pitch trajectories are not visible. Consequently, a conventional noise suppression only inserts its maximum attenuation – which was set to -10 dB in this example – in these subbands (see center diagram). By applying the voiced speech reconstruction and the mixing scheme presented in this section an output signal that is depicted in the lowest part of Fig. 4.28 is generated.

For the speech reconstruction an upper frequency limit $\Omega_{\mu_{\text{r,max}}}$ of about 1000 Hz was selected. Monitoring the weights of the conventional noise suppression, however, adaptively reduces this threshold to about 800 Hz. When

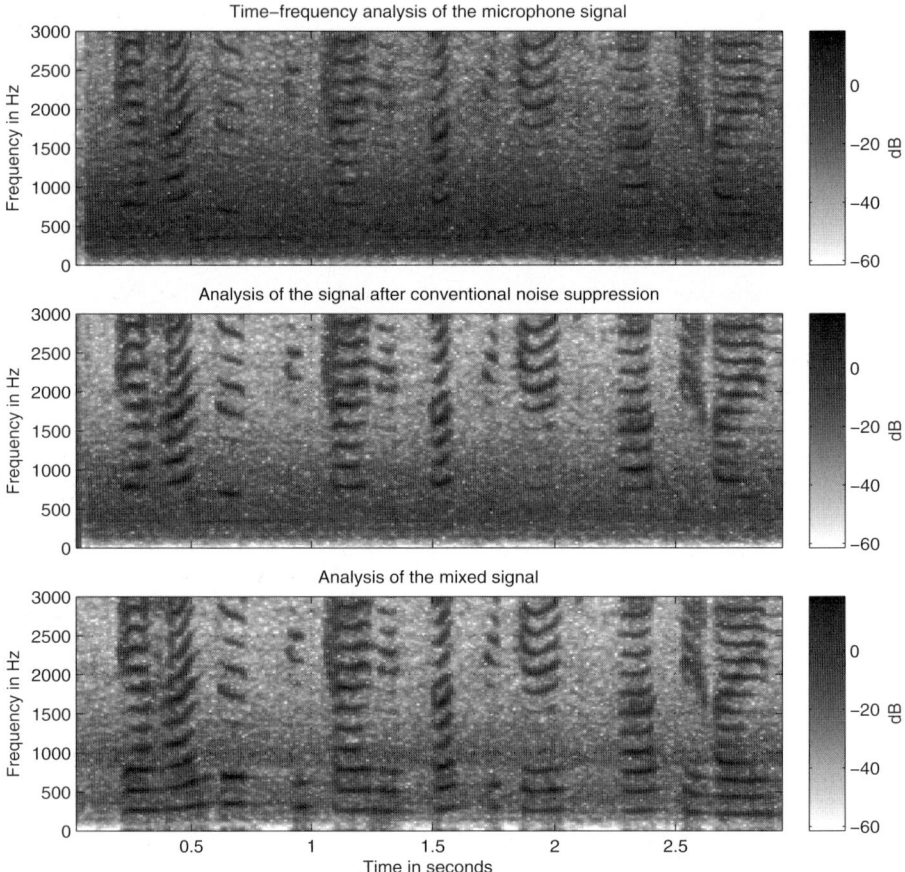

Fig. 4.28. Time-frequency analyses of the noisy input signal $y(n)$ (top), of the output signal of a conventional noise suppression $\hat{s}_\mathrm{s}(n)$ (center), and of the adaptively mixed signal $\hat{s}(n)$ (bottom).

activating the reconstruction part we have increased the maximum attenuation of the noise suppression by 3 dB to −13 dB, since artifacts that appear in standard noise suppression schemes are masked now mostly by the reconstructed parts.

Comparing the time-frequency analyses of the conventionally noise suppressed signal and the reconstructed signal shows clearly the recovery of the lowest two pitch trajectories.

4.6 Summary and Outlook

In this chapter an overview about a special type of model-based speech enhancement schemes – speech reconstruction – has been presented. Special emphasis was put on partial speech reconstruction approaches, applied either only in the computation of the frequency selective attenuation factors of conventional noise suppression schemes or directly to the output signal. For selected algorithms basic realizations as well as *measured results* were presented.

Partial speech reconstruction starts where conventional noise suppression approaches such as spectral subtraction, Wiener filtering, or decision-directed approaches fail – in time-frequency areas of noisy speech signals with signal-to-noise ratios lower than 0 dB. Thus, all *ingredients* of these new schemes, such as voice-activity detection, pitch and spectral envelope estimation, need to work reliably in these high noise conditions. Each detector or estimator improvement directly improves also the quality of the reconstructed signal.

A fully reconstructed signal is free of noise but sounds with the current approaches still some kind of *robotic* or *metallic*. However, if mixed properly with a conventionally noise suppressed signal significant quality improvements can be achieved – especially when evaluating the speech quality after coding/decoding schemes. The authors are convinced that this kind of model-based speech enhancement will be a valuable add-on for standard speech enhancement approaches for high noise conditions.

References

1. A. de Cheveigne, H. Kawahara: Yin, a fundamental frequency estimator for speech and music, *JASA*, **111**(4), 1917–1930, 2002.
2. J. Deller, J. Hansen, J. Proakis: *Discrete-Time Processing of Speech Signals*, New York, NY, USA: IEEE Press, 1993.
3. Y. Ephraim, D. Malah: Speech enhancement using a minimum mean-square error short-time spectral amplitude estimator, *IEEE Trans. Acoust. Speech Signal Process.*, **32**(6), 1109–1121, 1984.
4. Y. Ephraim, D. Malah: Speech enhancement using a minimum mean-square error log-spectral amplitude estimator, *IEEE Trans. Acoust. Speech Signal Process.*, **33**(2), 443–445, 1985.
5. ETS 300 903 (GSM 03.50): Transmission planning aspects of the speech service in the GSM public land mobile network (PLMS) system, *ETSI*, France, 1999.
6. J. H. L. Hanson: Morphological constrained feature enhancement with adaptive cepstral compensation (MCE-ACC) for speech recognition in noise and Lombard effect, *EEE Trans. Speech Audio Process.*, **2**(4), 598–614, 1994.
7. E. Hänsler, G. Schmidt: *Acoustic Echo and Noise Control*, Hoboken, NJ, USA: Wiley, 2004.
8. W. Hess: *Pitch Determination of Speech Signals*, Berlin, Germany: Springer, 1983.

9. U. Heute: Noise reduction, in E. Hänsler, G. Schmidt (eds.), *Topics in Acoustic Echo and Noise Control,* Berlin, Germany: Springer, 325–384, 2006.
10. M. Krini, G. Schmidt: Spectral refinement and its application to fundamental frequency estimation, *Proc. WASPAA '07,* New Paltz, NY, USA, 2007.
11. Y. Linde, A. Buzo, R. M. Gray: An algorithm for vector quantizer design, *IEEE Trans. Comm.,* **COM-28**(1), 84–95, Jan. 1980.
12. K. Linhard, T. Haulick: Spectral noise subtraction with recursive gain curves, *Proc. ICSLP '98,* **4**, 1479–1482, Sydney, Australia, 1998.
13. E. Lombard: Le signe de l'elevation de la voix, *Ann. Maladies Oreille, Larynx, Nez. Pharynx,* **37**, 101–119, 1911 (in French).
14. T. Lotter, P. Vary: Noise reduction by joint maximum a posteriori spectral amplitude and phase estimation with super-Gaussian speech modelling, *Proc. EUSIPCO '04,* **2**, 1457–1460, Wien, Austria, 2004.
15. R. Martin: An efficient algorithm to estimate the instantaneous SNR of speech signals, *Proc. EUROSPEECH '93,* 1093–1096, 1994.
16. R. Martin: Spectral subtraction based on minimum statistics, *Proc. EURASIP '94,* 1182–1185, Elsevier, Amsterdam, Netherlands, 1994.
17. R. Martin: Noise power spectral density estimation based on optimal smoothing and minimum statistics, *IEEE Trans. Speech Audio Process.,* **T-SA-9**(5), 504–512, 2001.
18. A. V. Oppenheim, R. W. Schafer, J. R. Buck: *Discrete-Time Signal Processing,* 2nd ed., Englewood Cliffs, NJ, USA: Prentice Hall, 1998.
19. C. Plapous, C. Marro, P. Scalart: Speech enhancement using harmonic regeneration, *Proc. ICASSP '05,* 157–160, Philadelphia, Pennsylvania, USA, 2005.
20. H. Puder, O. Soffke: An approach for an optimized voice-activity detector for noisy speech signals, *Proc. EUSIPCO '02,* **1**, 243–246, Toulouse, France, 2002.
21. M. R. Schroeder: Period histogram and product spectrum: New methods for fundamental frequency measurements, *JASA,* **43**(4), 829–834, 1968.
22. A. Spanias: Speech coding – a tutorial review, *Proc. IEEE,* **82**(10), 1541–1582, 1994.
23. P. P. Vaidyanathan: *Mulitrate Systems and Filter Banks,* Englewood Cliffs, NJ, USA: Prentice Hall, 1992.
24. P. Vary, R. Martin: *Digital Speech Transmission,* Hoboken, NJ, USA: Wiley, 2006.
25. E. Zwicker, H. Fastl: *Psychoacoustics – Facts and Models,* 2nd ed., Berlin, Germany: Springer, 1999.

5

Bandwidth Extension of Telephony Speech

Bernd Iser and Gerhard Schmidt

Harman/Becker Automotive Systems, Ulm, Germany

In this chapter an introduction on bandwidth extension of telephony speech is given. It is presented why current telephone networks apply a limiting bandpass, what kind of bandpass is used, and what can be done to (re)increase the bandwidth on the receiver side without changing the transmission system. Therefore, several approaches – most of them based on the source-filter model for speech generation – are discussed. The task of bandwidth extension algorithms that make use of this model can be divided into two subtasks: excitation signal extension and wideband envelope estimation. Different methods like non-linear processing, the use of signal and noise generators, or modulation approaches on the one hand and codebook approaches, linear mapping schemes or neural networks on the other hand, are presented.

5.1 Introduction

Speech is the most natural and convenient way of human communication. This is the reason for the big success of the telephone system since its invention in the nineteenth century [2]. At that time costumers didn't request high quality speech. Nowadays, however, costumer are aften not satisfied with the quality of service provided by the telephone system especially when compared to other audio sources, such as radio or compact disc. The degradation of speech quality using analog telephone systems is caused by the introduction of band limiting filters within amplifiers that are required to maintain a certain signal level in long local loops [26]. These filters exhibit a passband from approx. 300 Hz up to 3400 Hz (see Fig. 5.1 a) and are applied to reduce crosstalk between different channels.

As we can see in Fig. 5.3 b the application of such a bandpass attenuates large speech portions. Digital networks, such as *Integrated Service Digital Network* (ISDN) and *Global System for Mobile Communication* (GSM), are able to transmit speech in higher quality since additionally signal components below 300 Hz as well as components between 3.4 kHz and 4 kHz can be transmitted

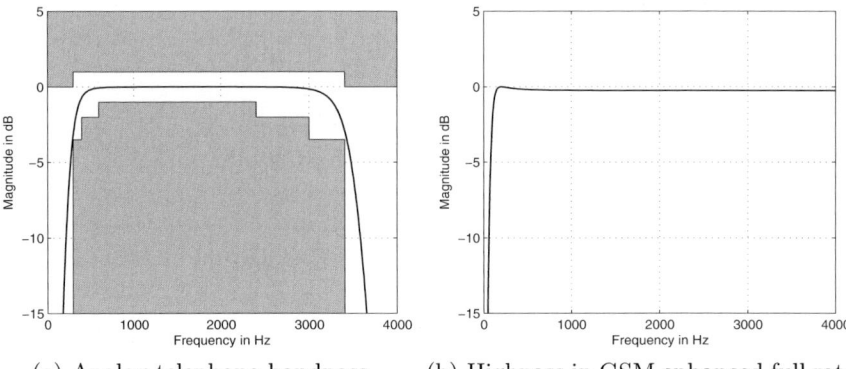

(a) Analog telephone bandpass. (b) Highpass in GSM enhanced full rate codec.

Fig. 5.1. (a): Telephone bandpass according to ITU recommendation G.151 [20] (used in analog phone connections) and (b): highpass according to [11] (used in the GSM cell phone network).

(see Fig. 5.1 b). However, this is only true if the entire call (in terms of its routing) remains in those networks – when leaving into an analog telephone network the speech signal is once again bandlimited. Furthermore, we still have a band limitation of 4 kHz.

Thus, great efforts have been made to increase the quality of telephone speech signals in recent years. Wideband codecs are able to increase the bandwidth up to 7 kHz or even higher at only moderate complexity increase [21]. Nevertheless, applying these codecs would mean to modify current networks. Another possibility is to increase the bandwidth after transmission by means of bandwidth extension. The basic idea of these enhancements is to estimate the speech signal components above 3400 Hz and below 300 Hz and to complement the signal in the new frequency bands with this estimate. In this case the telephone networks are left untouched. Fig. 5.2 shows the basic structure of a telephone connection with a bandwidth extension (BWE) system inserted in the receiving path of a telephone connection. The incoming signal $\tilde{s}_{\text{tel}}(n)$ at the local terminal is first upsampled from 8 kHz to the desired sampling rate, e.g. 11 or 16 kHz. The resulting signal $s_{\text{tel}}(n)$ has now the desired rate but contains no signal components above 4 kHz (assuming anti-imaging filters of sufficient quality within the upsampling process). In the bandwidth extension unit the signal is analyzed within the telephone band and the missing frequency components are generated and added to the original input signal. The resulting signal is denoted as $s_{\text{ext}}(n)$.

Additionally, three time-frequency analyses are presented in Fig. 5.3. The first analysis depicts a wideband speech signal $s(n)$ as it would be recorded close to the mouth of the communication partner on the remote side. If we assume not to have any kind of errors or distortions on the transmission a

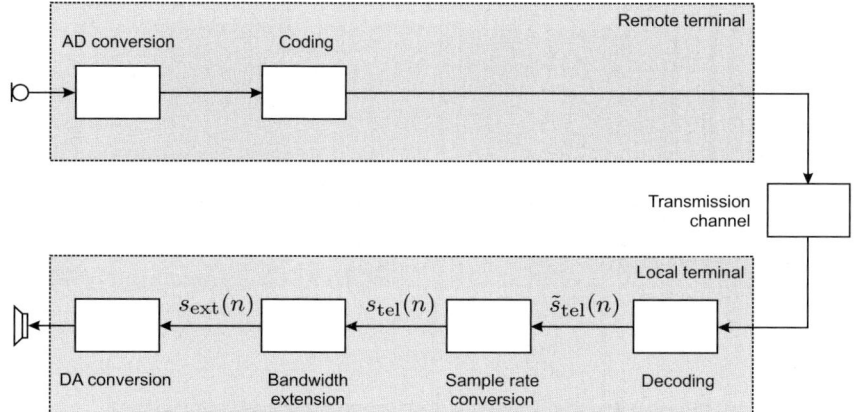

Fig. 5.2. Overall system for bandwidth extension.

bandlimited signal $s_{\text{tel}}(n)$ as depicted in the center diagram would be received at the local side. The truncation of the frequency range is clearly visible. Without any additional processing the local communication partner would be listening to this signal. If bandwidth extension is applied a signal $s_{\text{ext}}(n)$ as depicted in part (c) of Fig. 5.3 would be reconstructed. Even if the signal is not exactly the same as the original one, it sounds more natural and – as a variety of listening tests indicate – the speech quality in general is increased as well [17].

Additionally, we have to note that the application of bandwidth extension is not limited to the bandwidth of current telephone networks. If wideband codecs such as [21] will be used BWE schemes can extend the frequency range above 7 kHz. Early investigations show that in these scenarios the improvement of speech quality is even better. However, in this chapter we will focus on extending the bandwidth of 4 kHz telephone speech.

5.2 Organization of the Chapter

Since most of the recent schemes for bandwidth extension utilize the so-called source-filter model of human speech generation [8] we will first introduce this model and a few basics in Sec. 5.3. Early attempts of bandwidth extension, however, do not rely on this model assumption. Thus, we will give a short overview about non-model based schemes in Sec. 5.4. The main part of this contribution is about model-based extension schemes, which are discussed in Sec. 5.5. Finally, an outlook on the evaluation of bandwidth extension systems for telephony speech is presented in Sec. 5.6.

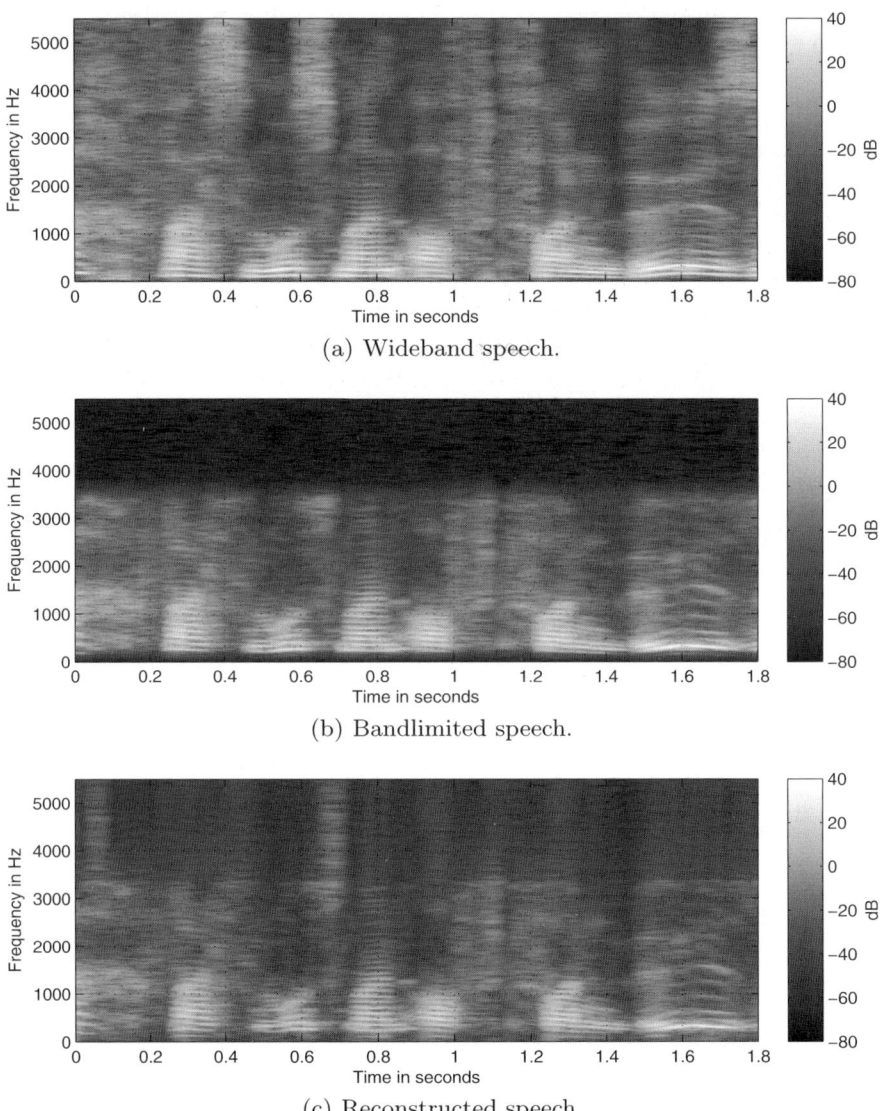

Fig. 5.3. Spectrograms of (a): wideband speech (b): bandlimited speech (c): reconstructed speech.

5.3 Basics

The basic idea of bandwidth extension algorithms is to extract information on the missing components out of the available narrowband signal $s_{\text{tel}}(n)$.

For finding information that is suitable for this task most of the algorithms employ the so-called *source-filter model* of speech generation [8, 13, 38].

5.3.1 Human Speech Generation

This model is motivated by the anatomical analysis of the human speech apparatus (see Fig. 5.4). By breathing, the lungs get filled with air. This air being expelled through the trachea causes the tensed vocal cords within the larynx tube to vibrate. The resulting air flow through the opening of the vocal cords, the so-called *glottis*, gets chopped into periodic pulses by the periodically opening and closing vocal cords. This is the case when voiced sounds like [ə], [æ], [ɑː], [ɛ], [ɪ], or [iː] are being produced.[1] The inverse of the corresponding time period is called fundamental frequency or pitch frequency and ranges from about 80 Hz for male speakers up to 300 Hz for female speakers or children. In the case of unvoiced sounds like the fricatives [ʃ], [s], [θ], or [f] the vocal cords are loose causing a turbulent, noise-like air flow [38].

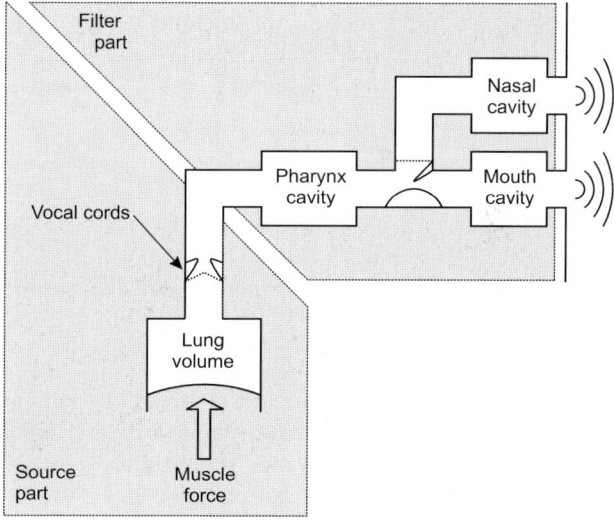

Fig. 5.4. Human speech apparatus.

In Fig. 5.5 the glottal sound pressure difference that could be observed directly behind the vocal cords is depicted for a voiced and an unvoiced sound. In the case of the voiced utterance the periodicity is clearly visible. The period of 10 ms is corresponding to a pitch frequency of 100 Hz. The recording was done with a male participant.

---

[1] The notation of the phonetic symbols was made according to [19].

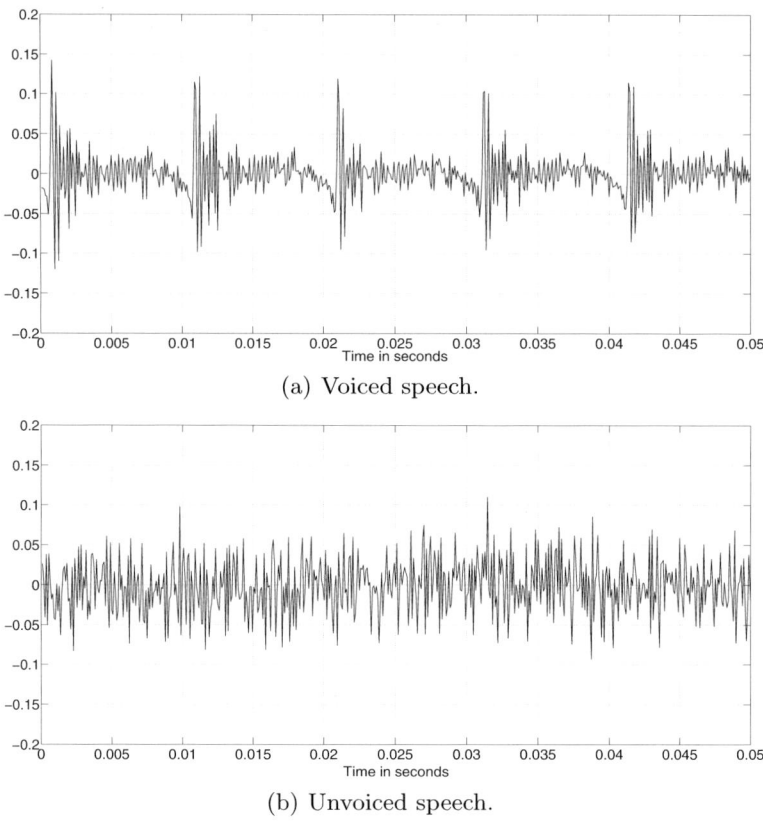

Fig. 5.5. Glottal air flow respectively the sound pressure difference for (a) a voiced sound and (b) an unvoiced sound.

After passing the vocal cords the air flow reaches several cavities namely the pharynx cavity, the nasal cavity, and the mouth cavity which all together form the so-called *vocal tract*. The cavities within the vocal tract can additionally be separated from each other and changed in size and shape by the tongue, the velum, the jaw, the mouth, and the lips. In these cavities with their special characteristics concerning resonance the final sounds are produced. The characteristic of distinct human voices is build up by the mixture of a pitch frequency which varies around a working point and the characteristics of the vocal tract [8, 37, 38]. In Fig. 5.6 the short-term spectra of a voiced and unvoiced utterance are depicted. Since speech is a quasi-stationary process the approach using these short-term spectra is feasible (in the following these short-term spectra will mostly be called spectra for reasons of brevity). One can see clearly the pitch structure in Fig. 5.6 (a) with its maxima every 100 Hz. The local maxima of the envelope spectrum are called formants and are denoted with F_1 to F_4. The terms pitch and envelope spectrum will become clearer in the next section when a model for the speech production process is introduced.

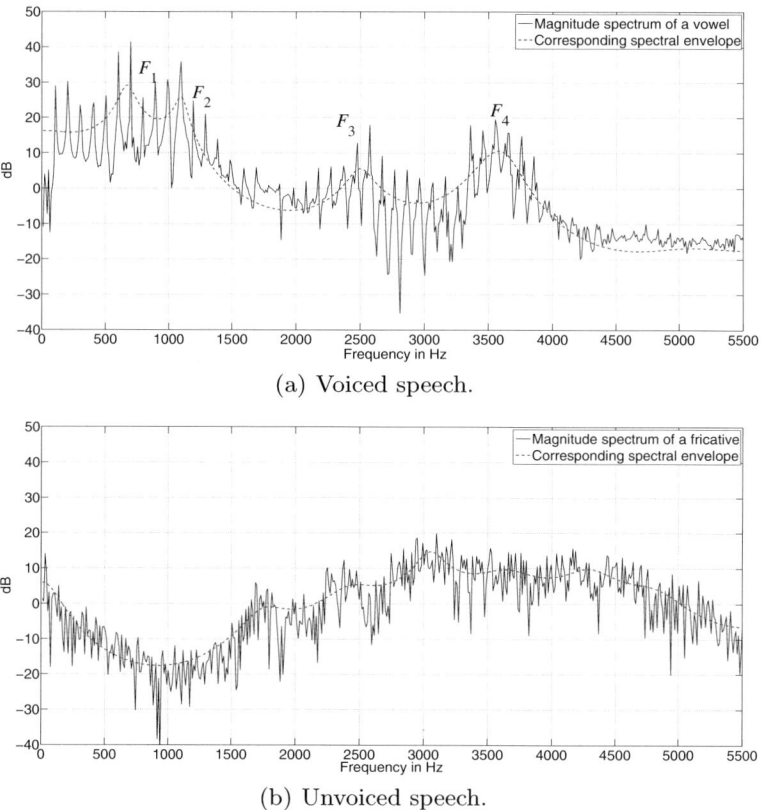

Fig. 5.6. Short-term spectrum of voiced (a) and unvoiced (b) utterances.

5.3.2 Source-Filter Model

As explained in the previous section a flow of air coming from the lungs is pressed through the vocal cords. At this point two scenarios can be distinguished:

- In the first scenario the vocal cords are loose causing a turbulent (noise-like) air flow.
- In the second scenario the vocal cords are tense and closed. The pressure of the air coming from the lungs increases until it causes the vocal cords to open. Now the pressure decreases rapidly and the vocal cords close once again. This scenario results in a periodic signal.

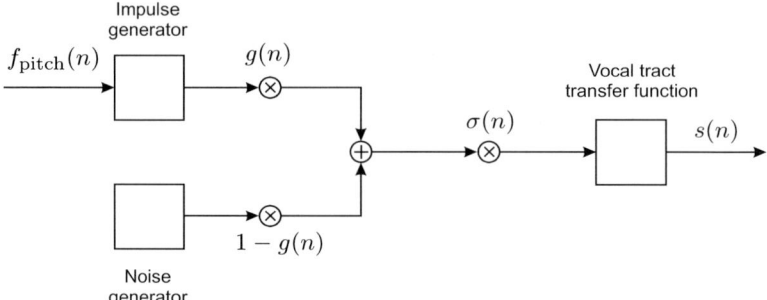

Fig. 5.7. Source-filter model for the process of human speech generation.

The signal that could be observed directly behind the vocal cords is called *excitation signal*. This excitation signal has the property of being spectrally flat. After passing the vocal cords the air flow goes through the cavities described before.

The source-filter model tries to rebuild the two scenarios that are responsible for the generation of the excitation signal by using two different signal generators (see Fig. 5.7):

- a noise generator for rebuilding unvoiced (noise-like) utterances and
- a pulse train generator for rebuilding voiced (periodic) utterances.

The inverse of the pulse duration is called pitch frequency – in Fig. 5.7 denoted as $f_{\text{pitch}}(n)$. These signal generators can be accessed either in a binary manner,

$$g(n) \in \{0, 1\}, \tag{5.1}$$

or continuously,

$$0 \leq g(n) \leq 1. \tag{5.2}$$

For rebuilding the influence of the cavities a low-order all-pole filter with the frequency response

$$H(e^{j\Omega}, n) = \frac{\sigma(n)}{A(e^{j\Omega}, n)} = \frac{\sigma(n)}{1 - \sum_{i=1}^{N_{\text{pre}}} \tilde{a}_i(n) e^{-j\Omega i}} \tag{5.3}$$

is employed.[2] The order N_{pre} of the all-pole model is chosen usually in the range of 10 to 20. Since the excitation signal is a spectrally flat signal, the

---

[2] We have used the "tilde notation" for the coefficients $\tilde{a}_i(n)$ in order to avoid a conflict with the definitions of standard transformations such as Fourier or z-transform. The Fourier transform is defined as $A(e^{j\Omega}, n) = \sum_{i=-\infty}^{\infty} a_i(n) e^{-j\Omega i}$. By comparing the coefficients we get $a_0(n) = 1$, $a_i(n) = -\tilde{a}_i(n)$ for $1 \leq i \leq N_{\text{pre}}$, and $a_i(n) = 0$ else.

transfer function of this all-pole model represents the spectral envelope of the speech signal. The parameters $\tilde{a}_i(n)$ of the all-pole model can be computed by solving the so-called *Yule–Walker equation system*

$$\underbrace{\begin{bmatrix} r_{ss,0}(n) & r_{ss,1}(n) & \cdots & r_{ss,N_{\text{pre}}-1}(n) \\ r_{ss,1}(n) & r_{ss,0}(n) & \cdots & r_{ss,N_{\text{pre}}-2}(n) \\ \vdots & \vdots & \ddots & \vdots \\ r_{ss,N_{\text{pre}}-1}(n) & r_{ss,N_{\text{pre}}-2}(n) & \cdots & r_{ss,0}(0) \end{bmatrix}}_{\boldsymbol{R}_{ss}(n)} \underbrace{\begin{bmatrix} \tilde{a}_1(n) \\ \tilde{a}_2(n) \\ \vdots \\ \tilde{a}_{N_{\text{pre}}}(n) \end{bmatrix}}_{\tilde{\boldsymbol{a}}(n)} = \underbrace{\begin{bmatrix} r_{ss,1}(n) \\ r_{ss,2}(n) \\ \vdots \\ r_{ss,N_{\text{pre}}}(n) \end{bmatrix}}_{\boldsymbol{r}_{ss}(n)}.$$

(5.4)

The coefficients $r_{ss,i}(n)$ represent the short-term autocorrelation at lag i estimated around the time index n. Finally, the gain parameter $\sigma(n)$ in Eq. 5.3 is computed as the square root of the output power of a predictor error filter with coefficients $\tilde{a}_i(n)$:

$$\begin{aligned} \sigma(n) &= \sqrt{r_{ss,0}(n) - \sum_{i=1}^{N_{\text{pre}}} \tilde{a}_i(n)\, r_{ss,i}(n)} \\ &= \sqrt{r_{ss,0}(n) - \tilde{\boldsymbol{a}}^{\mathrm{T}}(n)\, \boldsymbol{r}_{ss}(n)}\,. \end{aligned} \quad (5.5)$$

Due to the special character of the matrix $\boldsymbol{R}_{ss}(n)$ the equations 5.4 and 5.5 can be solved in an order-recursive manner by using, e.g., the Levinson-Durbin recursion [10, 30]. Since speech can be assumed to have a stationary character only for short periods of time, the parameters of the model need to be estimated periodically every 5 to 10 ms. By utilizing about 10 coefficients $a_i(n)$ and the gain $\sigma(n)$ one is able to estimate the spectral envelope of a speech signal in a reliable manner.

5.3.3 Parametric Representations of the Spectral Envelope

Due to their compact representation of short-term spectral envelopes the prediction parameters $a_i(n)$ (or $\tilde{a}_i(n)$) play a major role in speech coding and bandwidth extension. However, if cost functions that take the human auditory perception into account are applied the prediction parameters are often transformed into so-called *cepstral* coefficients [33]:

$$c_i(n) = \frac{1}{2\pi} \int_{-\pi}^{\pi} \ln\left\{ H\left(e^{j\Omega}, n\right) \right\} e^{j\Omega i}\, d\Omega\,. \quad (5.6)$$

By applying the natural logarithm on the complex filter spectrum we have to use its complex type which is defined as [3]

$$\ln\{z\} = \ln|z| + j\arg\{z\}\,, \quad (5.7)$$

where arg$\{z\}$ denotes the angle of z within the complex plain. As we will see later on, the resulting cepstral coefficients $c_i(n)$ are real due to the special symmetry properties of $H\left(e^{j\Omega}, n\right)$ as well as due to the fact that $H\left(e^{j\Omega}, n\right)$ represents a stable all-pole filter. Applying a Fourier transform on both sides of Eq. 5.6 leads us to

$$\sum_{i=-\infty}^{\infty} c_i(n)\, e^{-j\Omega i} = \ln\left\{H\left(e^{j\Omega}, n\right)\right\}. \tag{5.8}$$

The cepstral coefficients used in this chapter are so-called *linear predictive cepstral coefficients*. As indicated by the name these coefficients are computed on the basis of linear predictive coefficients. By exchanging the filter spectrum $H(e^{j\Omega}, n)$ by its representation using an all-pole model we obtain

$$\sum_{i=-\infty}^{\infty} c_i(n)\, z^{-i}\bigg|_{z=e^{j\Omega}} = \ln\left\{\frac{\sigma(n)}{A(z,n)}\right\}\bigg|_{z=e^{j\Omega}}$$

$$= \ln\{\sigma(n)\} - \ln\{A(z,n)\}\bigg|_{z=e^{j\Omega}}. \tag{5.9}$$

Let us now have a closer look at the last term $\ln\{A(z,n)\}$ in Eq. 5.9:

$$\ln\{A(z,n)\} = \ln\left\{1 - \sum_{i=1}^{N_{\text{pre}}} \tilde{a}_i(n)\, z^{-i}\right\}$$

$$= \ln\left\{\sum_{i=0}^{N_{\text{pre}}} a_i(n)\, z^{-i}\right\}. \tag{5.10}$$

By representing this expression as a product of zeros with modified coefficients $b_i(n)$ we can write

$$\ln\left\{\sum_{i=0}^{N_{\text{pre}}} a_i(n)\, z^{-i}\right\} = \ln\left\{\prod_{i=0}^{N_{\text{pre}}} \left[1 - b_i(n)\, z^{-1}\right]\right\}$$

$$= \sum_{i=0}^{N_{\text{pre}}} \ln\left\{1 - b_i(n)\, z^{-1}\right\}. \tag{5.11}$$

By exploiting the following series expansion [3]

$$\ln\left\{1 - b\, z^{-1}\right\} = -\sum_{k=1}^{\infty} \frac{b^k}{k}\, z^{-k}, \qquad \text{for } |z| > |b|, \tag{5.12}$$

that holds for factors that converge within the unit circle, which is the case here since $A(z,n)$ is analytic inside the unit circle [33], we can further rewrite Eq. 5.11 (and thus also Eq. 5.10)

$$\ln\{(A(z,n)\} = -\sum_{i=0}^{N_{\text{pre}}} \sum_{k=1}^{\infty} \frac{b_i^k(n)}{k} z^{-k}$$
$$= -\sum_{k=1}^{\infty} \sum_{i=0}^{N_{\text{pre}}} \frac{b_i^k(n)}{k} z^{-k}. \quad (5.13)$$

If we now compare Eq. 5.9 with Eq. 5.13 we observe that the two sums do not have equal limits. This means that the $c_i(n)$ are equal to zero for $i < 0$. For $i > 0$ we can set the $c_i(n)$ equal to the inner sum in Eq. 5.13. For $i = 0$ we have $c_0(n) = \ln\{\sigma(n)\}$ from Eq. 5.9. In conclusion, we can state

$$c_i(n) = \begin{cases} \sum_{m=0}^{N_{\text{pre}}} \frac{b_m^i(n)}{i}, & \text{for } i > 0, \\ \ln\{\sigma(n)\}, & \text{for } i = 0, \\ 0, & \text{for } i < 0. \end{cases} \quad (5.14)$$

This leads to the assertion

$$\ln\left\{\frac{\sigma(n)}{A(z,n)}\right\} = \ln\{\sigma(n)\} + \sum_{i=1}^{\infty} c_i(n) z^{-i}. \quad (5.15)$$

If the N_{pre} predictor coefficients $a_i(n)$ (or $\tilde{a}_i(n)$) are known we can derive a simple recursive computation of the linear predictive cepstral coefficients $c_i(n)$ by differentiating both sides of Eq. 5.15 with respect to z and equating the coefficients of alike powers of z:

$$-\frac{d}{dz}\left[\ln\left\{1 - \sum_{i=1}^{N_{\text{pre}}} a_i(n) z^{-i}\right\}\right] = \frac{d}{dz}\left[\sum_{i=1}^{\infty} c_i(n) z^{-i}\right]$$
$$\sum_{i=1}^{N_{\text{pre}}} i\, a_i(n) z^{-i-1} \left[-1 + \sum_{i=1}^{N_{\text{pre}}} a_i(n) z^{-i}\right]^{-1} = -\sum_{i=1}^{\infty} i\, c_i(n) z^{-i-1}. \quad (5.16)$$

Multiplying both sides of Eq. 5.16 with $\left[-1 + \sum_{i=1}^{N_{\text{pre}}} a_i(n) z^{-i}\right]$ leads to

$$\sum_{i=1}^{N_{\text{pre}}} i\, a_i(n) z^{-i-1} = \sum_{i=1}^{\infty} i\, c_i(n) z^{-i-1} - \sum_{k=1}^{\infty} \sum_{i=1}^{N_{\text{pre}}} k\, c_k(n) a_i(n) z^{-k-i-1}. \quad (5.17)$$

If we now consider the equation above for equal powers of z, we find that starting from left, the first two terms only contribute a single term each up to $z^{-N_{\text{pre}}-1}$. We will label the order with i. The last term in contrast produces a number of terms that depend on i. Equating all terms that belong to the same power of z results in

$$i\,a_i(n) = i\,c_i(n) - \sum_{k=1}^{i-1} k\,c_k(n)\,a_{i-k}(n)\,, \quad \text{for } i \in \{1, ..., N_{\text{pre}}\}\,. \quad (5.18)$$

Solving this equation for $c_i(n)$ results in

$$c_i(n) = a_i(n) + \frac{1}{i}\sum_{k=1}^{i-1} k\,c_k(n)\,a_{i-k}(n)\,, \quad \text{for } i \in \{1, ..., N_{\text{pre}}\}\,. \quad (5.19)$$

For $i > N_{\text{pre}}$ the first term on the right hand side still needs to be considered in Eq. 5.17 whereas the term on the left hand side does not contribute to powers of z larger than $N_{\text{pre}} - 1$ and can therefore be omitted. Therefore, we can solve Eq. 5.17 for $i > N_{\text{pre}}$ as

$$c_i(n) = \frac{1}{i}\sum_{k=1}^{i-1} k\,c_k(n)\,a_{i-k}(n)\,, \quad \text{for } i > N_{\text{pre}}\,. \quad (5.20)$$

By summarizing all results we can formulate a recursive computation of cepstral coefficients from linear predictive coefficients as

$$c_i(n) = \begin{cases} 0\,, & \text{for } i < 0\,, \\ \ln\{\sigma(n)\}\,, & \text{for } i = 0\,, \\ a_i(n) + \frac{1}{i}\sum_{k=1}^{i-1} k\,c_k(n)\,a_{i-k}(n)\,, & \text{for } 1 \leq i \leq N_{\text{pre}}\,, \\ \frac{1}{i}\sum_{k=1}^{i-1} k\,c_k(n)\,a_{i-k}(n)\,, & \text{for } i > N_{\text{pre}}\,. \end{cases} \quad (5.21)$$

This means that the average gain of the filter is represented by the coefficient $c_0(n)$ while its shape is described by the coefficients $c_1(n)$, $c_2(n)$, etc. In most cases the series $c_i(n)$ fades towards zero rather quickly. Thus, it is possible to approximate all coefficients above a certain index by zero. We will compute in the following only the coefficients from $i = 0$ to $i = N_{\text{cep}} - 1$, with

$$N_{\text{cep}} = \frac{3}{2} N_{\text{pre}}\,, \quad (5.22)$$

and assume all other coefficients to be zero. We have elaborated the relation between cepstral and predictor coeffients since we will use a cost function based on cepstral coeffients in most of the succeeding bandwidth extension schemes – either directly or within a parameter training stage.

Furthermore, the model according to Eq. 5.3 represents an IIR filter. Thus, after any modification of the coefficients $a_i(n)$ (or $\tilde{a}_i(n)$, respectively) it needs to be checked whether stability is still ensured. These tests can be avoided if a transformation into so-called *line spectral frequencies* (LSFs) [23] is applied. These parameters are the angles of the zeros of the polynomials $P(z)$ and $Q(z)$ in the z-domain

$$P(z,n) = A(z,n) + z^{-(N_{\text{pre}}+1)}A(z^{-1},n),\quad (5.23)$$

$$Q(z,n) = A(z,n) - z^{-(N_{\text{pre}}+1)}A(z^{-1},n).\quad (5.24)$$

where P(z) is a mirror polynomial and Q(z) an anti-mirror polynomial. The mirror property is characterized by

$$P(z,n) = z^{-(N_{\text{pre}}+1)}P(z^{-1},n),\quad (5.25)$$

$$Q(z,n) = z^{-(N_{\text{pre}}+1)}Q(z^{-1},n).\quad (5.26)$$

As shown in [39] $A(z,n)$ is guaranteed to be a minimum phase filter (and therefore the synthesis filter $H(z,n) = \sigma(n)/A(z,n)$ is guaranteed to be a stable filter) if the zeros of $P(z,n) = 0$ and $Q(z,n) = 0$ lie on the unit circle and if they are increasing monotonously and alternating. In fact $P(z,n)$ has a real zero at $z = -1$ and $Q(z,n)$ at $z = 1$.

In Fig. 5.8 two versions of an original spectral envelope (also depicted) quantized with 6 bits are depicted. Representations using LPC coefficients as well as LSF coefficients have been used. It is obvious that LSF coefficients are more robust against quantization than LPC coefficients [40]. However, the benefit concerning robustness grained by the usage of line spectral frequencies does not always justify the increased computational complexity that results form the search for the zeroes.

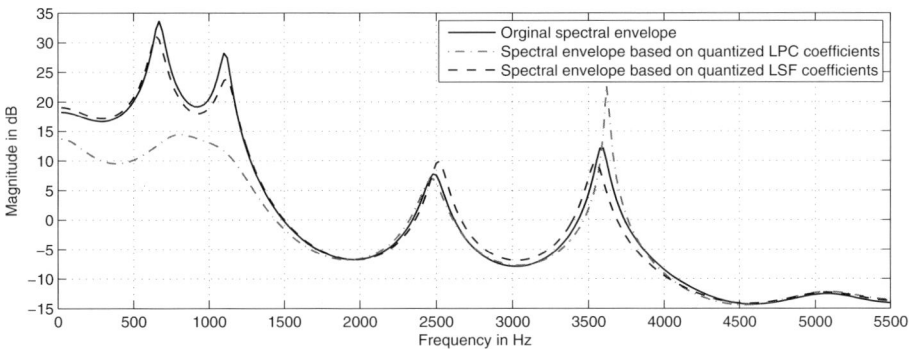

Fig. 5.8. Original spectral envelope and two quantized versions (6 bits).

5.3.4 Distance Measures

Distance measures play an important role in speech quality assessment, speech coding, and – as we will see in the next sections – also in training and search procedures in bandwidth extension schemes. Plenty of different distance measures for all kinds of applications exist [41]. Since we are dealing with bandwidth extension and one major task is the extension of the spectral envelope, we will focus here on distance measures that are appropriate to evaluate distances between parametric representations of spectral envelopes. Most of the spectral distance measures are L_p-norm based measures

$$d_p\Big(H\left(e^{j\Omega},n\right),\widehat{H}\left(e^{j\Omega},n\right)\Big) = \left[\frac{1}{2\pi}\int_{-\pi}^{\pi}\left|H\left(e^{j\Omega},n\right) - \widehat{H}\left(e^{j\Omega},n\right)\right|^p d\Omega\right]^{\frac{1}{p}}. \tag{5.27}$$

The most common choices for p are $p = 1$, $p = 2$, and $p = \infty$ resulting in the so-called *city block* distance, *Euclidean* distance and *Minkowski* distance. Since the representation of the magnitude spectrum in a logarithmic manner is very popular, another well known distance measure has emerged from this kind of representation. Furthermore, this distance measure can be computed directly on the parametric representation of the spectral envelope. It is the so-called *cepstral* distance, which is defined as

$$d_{\text{cep}}\big(c_i(n),\hat{c}_i(n)\big) = \sum_{i=-\infty}^{\infty}\Big(c_i(n) - \hat{c}_i(n)\Big)^2, \tag{5.28}$$

where the $c_i(n)$ and $\hat{c}_i(n)$ denote linear predictive cepstral coefficients as described before. An interesting property of this definition can be shown using Parsevals theorem [38] (compare Eq. 5.6)

$$d_{\text{cep}}\big(c_i(n),\hat{c}_i(n)\big) = \frac{1}{2\pi}\int_{-\pi}^{\pi}\left|\ln\left\{H\left(e^{j\Omega},n\right)\right\} - \ln\left\{\widehat{H}\left(e^{j\Omega},n\right)\right\}\right|^2 d\Omega. \tag{5.29}$$

For the tasks required in bandwidth extension schemes it is necessary to compare only the shape of two spectral envelopes but not the gain. For this reason, we start the summing index at $i = 1$ in Eq. 5.28. Furthermore, we assume – as explained before – that the cepstral series does not show significant values for large indices. Thus, we modify the original definition (Eq. 5.28) for our purposes to

$$\tilde{d}_{\text{cep}}\big(c_i(n),\hat{c}_i(n)\big) = \sum_{i=1}^{N_{\text{cep}}}\Big(c_i(n) - \hat{c}_i(n)\Big)^2. \tag{5.30}$$

This represents an estimate for the average squared logarithmic difference in the frequency domain of two gain normalized envelopes:

$$\tilde{d}_{\text{cep}}\big(c_i(n),\hat{c}_i(n)\big) \approx \frac{1}{2\pi}\int_{-\pi}^{\pi}\left|\ln\left\{\frac{H\left(e^{j\Omega},n\right)}{\sigma(n)}\right\} - \ln\left\{\frac{\widehat{H}\left(e^{j\Omega},n\right)}{\hat{\sigma}(n)}\right\}\right|^2 d\Omega$$

$$= \frac{1}{2\pi}\int_{-\pi}^{\pi}\left|\ln\left\{A\left(e^{j\Omega},n\right)\right\} - \ln\left\{\widehat{A}\left(e^{j\Omega},n\right)\right\}\right|^2 d\Omega. \tag{5.31}$$

For the last modification in Eq. 5.31 it has been used that $\ln\{1/A(e^{j\Omega})\} = -\ln\{A(e^{j\Omega})\}$. This distance measure will be utilized for most of the schemes presented in the following sections of this chapter.

5.4 Non-Model-Based Algorithms for Bandwidth Extension

The first approaches for bandwidth extension that have been documented do neither make use of any particular model for the speech generation process nor do they make use of any a priori knowledge concerning speech properties. These so-called *non-model-based* algorithms are the most simple methods for increasing the bandwidth. As reported in [24] the first experiments have been conducted in 1933 using a non-linear (quadratic) characteristic. Later on the British Broadcasting Corporation (BBC) was interested in increasing the bandwidth during telephone contributions [7]. Mainly three different non-model based algorithms exist. In the following we will discuss these (historical) methods in a succinct manner as some of these ideas will show up in later sections once again.

5.4.1 Oversampling with Imaging

This method makes use of the spectral components that occur when upsampling (inserting zeros) a signal either without any anti-imaging filter at all or with a filter $H_{\text{AI}}(e^{j\Omega})$ showing only some slow decay above half of the desired sampling frequency (see Fig. 5.9). The upsampling process results in a spec-

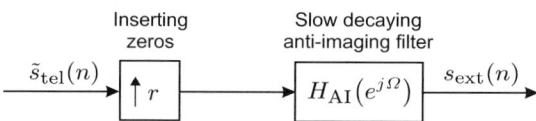

Fig. 5.9. Bandwidth extension based on oversampling and imaging.

trum that is mirrored at half of the original sampling frequency. This method profits from the noise-like nature of the excitation signal concerning unvoiced utterances. Unvoiced utterances result in a spectrum which has most of its energy in the higher frequency regions and therefore exactly this portions are mirrored. The temporal behavior of unvoiced components below the cut-off frequency correlates strongly with them above the cut-off frequency which is preserved applying this mirroring.

A drawback of this method concerning telephony speech is that the lower frequency part of the spectrum is not extended at all. Furthermore if a signal that is bandlimited according to the frequency response of an analog telephone system with 8 kHz sampling rate is upsampled a spectral gap is produced around 4 kHz. A rather costly alternative would be to first downsample the signal to a sampling rate that is equal to twice of the maximum signal bandwidth (e.g. 2 · 3600 Hz) before upsampling once again. An example is depicted in Fig. 5.10. The upper plot shows a time-frequency analysis of the incoming bandlimited telephone signal $\tilde{s}_{\text{tel}}(n)$. After inserting zeros in between neighboring samples of the input signal (upsampling with $r = 2$) and filtering with an anti-imaging filter that has its 3 dB cut-off frequency at about 7 kHz (instead of 4 kHz) a wideband signal $s_{\text{ext}}(n)$ as depicted in the lower diagram of Fig. 5.10 is generated.

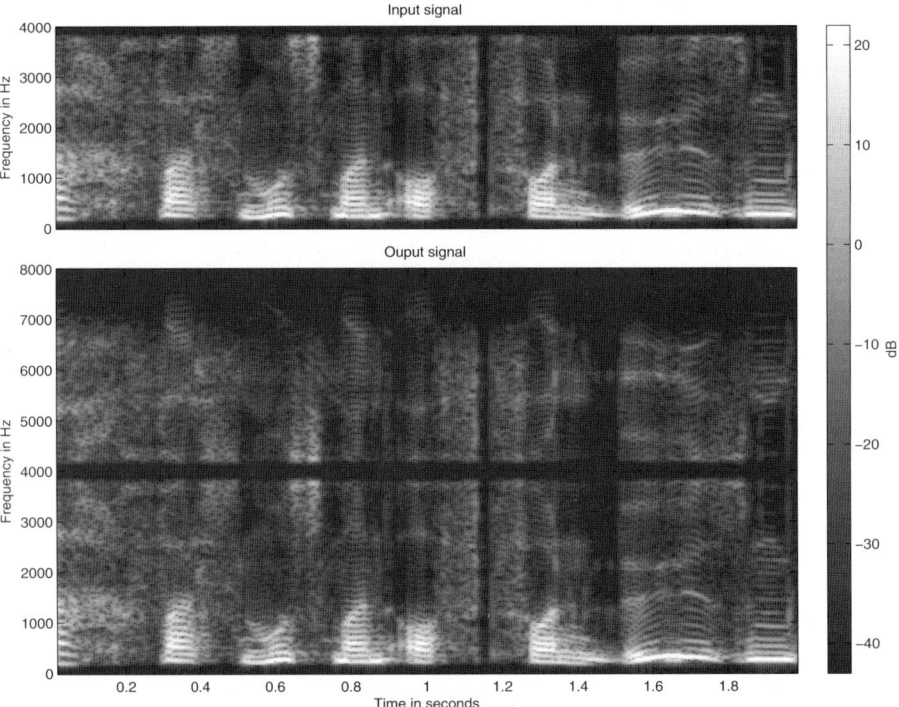

Fig. 5.10. Bandwidth extension using upsampling and low-quality anti-imaging filters. A time-frequency analysis of the input signal $\tilde{s}_{\text{tel}}(n)$ is depicted in the upper diagram. After upsamping with an upsampling ratio of $r = 2$ and filtering with an anti-imaging filter having a cut-off frequency of about 7 kHz an output signal $s_{\text{tel}}(n)$ as depicted in the lower diagram is generated.

The results crucially depend on the effective bandwidth of the original signal and the upsampling ratio. For increasing the bandwidth of signals that are bandlimited to 8 kHz up to 12 kHz, for example, this method works surprisingly well, but in the case of telephony speech this method produces poor results [6, 9].

5.4.2 Spectral Shifting

The basic idea behind spectral shifting approaches is to exploit the presence of an existing harmonic (or noise-like) structure in the telephone band assuming voiced (unvoiced) utterances and shifting a weighted copy of that part of the short-term spectrum that belongs to the telephone band in different manners into the extension regions. This is usually performed using the short-term Fourier transform. If the desired output sampling rate is an integer multiple of the input sampling rate, spectral shifting can easily be combined with the upsampling process. After block extraction and appropriate windowing an FFT of lower order (compared to the synthesis stage, e.g. $\tilde{N}_{\text{FFT}} = 256$) is performed. The resulting vector is denoted as

$$\tilde{\boldsymbol{S}}_{\text{tel}}(n) = \left[\tilde{S}_{\text{tel}}\left(e^{j\frac{2\pi}{\tilde{N}_{\text{FFT}}}0}, n\right), \ldots, \tilde{S}_{\text{tel}}\left(e^{j\frac{2\pi}{\tilde{N}_{\text{FFT}}}(\tilde{N}_{\text{FFT}}-1)}, n\right)\right]^T. \quad (5.32)$$

The FFT bin around π can be set to zero and the resulting spectral vector can be extended by inserting zeros next to that bin (e.g. also 256 zeros for doubling the sampling rate):

$$\boldsymbol{S}_{\text{tel}}(n) = \left[\tilde{S}_{\text{tel}}\left(e^{j\frac{2\pi}{\tilde{N}_{\text{FFT}}}0}, n\right), \ldots, \tilde{S}_{\text{tel}}\left(e^{j\frac{2\pi}{\tilde{N}_{\text{FFT}}}(\frac{\tilde{N}_{\text{FFT}}}{2}-1)}, n\right), \underbrace{0, 0, 0, \ldots 0}_{N_{\text{FFT}} - \tilde{N}_{\text{FFT}} + 1 \text{ zeros}},\right.$$

$$\left.\tilde{S}_{\text{tel}}\left(e^{j\frac{2\pi}{\tilde{N}_{\text{FFT}}}(\frac{\tilde{N}_{\text{FFT}}}{2}+1)}, n\right), \ldots, \tilde{S}_{\text{tel}}\left(e^{j\frac{2\pi}{\tilde{N}_{\text{FFT}}}(\tilde{N}_{\text{FFT}}-1)}, n\right)\right]^T.$$

$$(5.33)$$

Next the extended spectral vector $\boldsymbol{S}_{\text{tel}}(n)$ can be modified according to one of the methods that will be presented below. Afterwards an inverse FFT of appropriate size (e.g. $N_{\text{FFT}} = 512$) and again a window function (e.g. a Hann window of order 512) is applied. The basic structure of this approach is depicted in Fig. 5.11.

5.4.2.1 Fixed Spectral Shifting

Fixed spectral shifting indicates a shift with a fixed frequency offset. The starting index of the copy for the upper and lower extension region can be chosen independently. However, when choosing the starting indices (Ω_{low}, Ω_{high}) for the copy of the upper extension region one has to keep in mind not to exceed the upper band limit given by the telephone bandpass, nor the Nyquist limit. The so derived signal is then weighted in order to take care of the average

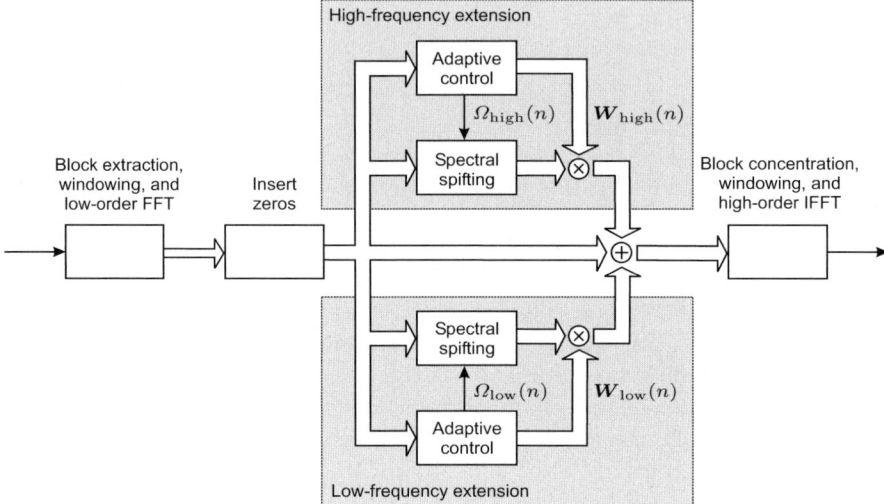

Fig. 5.11. Structure of spectral shifting approaches that incorporate the upsampling process.

spectral decay towards higher frequencies of most utterances and added to the spectral vector that was extended with zeros before. The reason for performing such a shift can easily be understood when we consider the processing of voiced utterances. The periodic signal that is typical for voiced utterances can be extended by these modulation techniques. As seen in Sec. 5.3.1 the excitation signal of unvoiced utterances is noise-like and therefore no need to take care of a specific structure exists. This means that we can also apply this excitation extension method for unvoiced utterances but without having to worry about discontinuing a structure. The possibility of continuing a structure in an erroneous manner using a fixed shifting scheme in the case of voiced utterances and the small noticeable artifacts affiliated, lead to a pitch-adaptive implementation presented in the following. A disadvantage that has to be mentioned is the fact that the phase information that is also copied into the extension regions might not be correct and could therefore lead to audible artifacts. These artifacts are much stronger for the low frequency extension than for high frequency signal reconstruction.

5.4.2.2 Adaptive Spectral Shifting

Incorporating a pitch detection and estimation [16] one could perform an adaptive shift and thereby maintain the right pitch structure even at the transition regions from the telephone bandpass to the extension regions, [12, 15]. This can be done choosing the starting index for the copy such that the gap between the last harmonic component in the telephone band and the first one in the copy just equals the fundamental frequency that is known through the pitch determination algorithm.

Experiments using a pitch adaptive spectral shifting showed that the resulting speech quality critically depends on the quality of the pitch determination algorithm. Jitter in the pitch detection results in audible artifacts – however, only for the excitation of the lower frequency range. The improvement that can be achieved by an adaptive shift concerning the upper extension region is slightly smaller compared to the lower extension. Nonetheless, the extension of the high frequency range is not very sensitive towards pitch estimation errors (even in voiced periods).

Additionally, one can add an adaptive weighting of the extended frequency ranges ($\boldsymbol{W}_{\text{low}}(n)$, $\boldsymbol{W}_{\text{high}}(n)$). If on the one hand much lower short-term power in the frequency range around 3 kHz compared to the range around 600 Hz is detected (this is the case for several vowels), the high frequency range should be extended only softly. If on the other hand a higher short-term power in the upper frequency region compared to the lower one is detected (as it is the case for fricatives), the lower extension should be attenuated and the higher extension should be boosted.

5.4.3 Application of Non-Linear Characteristics

The application of non-linear characteristics to periodic signals produces harmonics as we will see in more detail in section 5.5.1.2. This can be exploited for increasing the bandwidth similar to the above described method. The components generated out of the telephone band are usually attenuated by an empirically determined filter. An advantage of this method compared to the above presented one is that concerning telephony speech the lower frequency part is extended as well. Also no spectral gap occurs within the higher frequency part. A drawback might be the aliasing that can occur depending on the effective bandwidth, the sampling rate, and the non-linear characteristic that has been applied. In Sec. 5.5.1.2 a small selection of non-linear characteristics and their properties is presented.

The results produced by this approach are similar to the results of the method described in Sec. 5.4.1. Concerning telephony speech the results achieved are not satisfying [7, 35].

5.5 Model-Based Algorithms for Bandwidth Extension

In contrast to the methods described before the following algorithms make use of the already discussed source-filter model and thereby of a priori knowledge. As already stated in Sec. 5.3 the task of bandwidth extension following the source-filter model can be divided into two subtasks,

- namely the generation of a broadband excitation signal and

- the estimation of the broadband spectral envelope.

In the next two sections we will discuss possibilities to accomplish these tasks. In Fig. 5.12 a basic structure of a model-based bandwidth extension scheme is depicted.

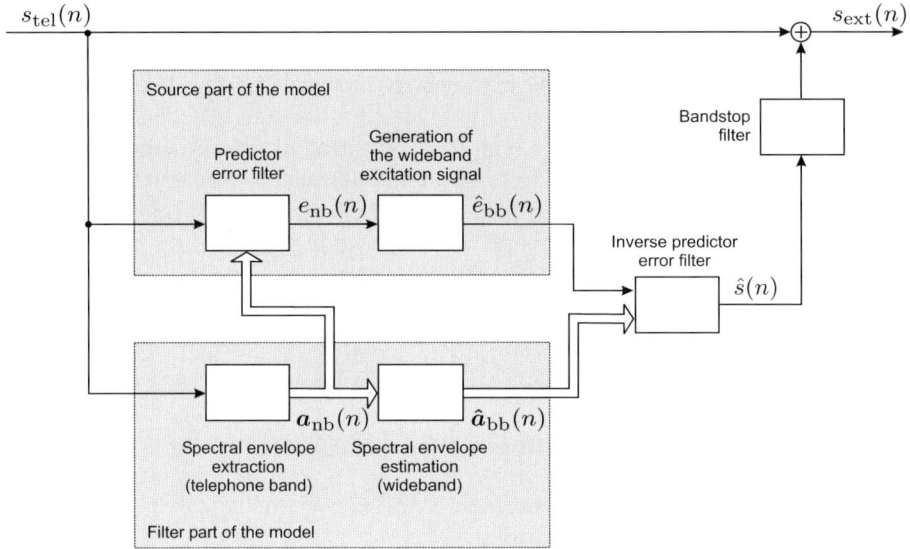

Fig. 5.12. Basic structure of a model-based scheme for bandwidth extension.

We assume an input signal $s_{\text{tel}}(n)$ that has already been upsampled to the desired sample rate but with a conventional sampling rate conversion (meaning that we do not have significant signal energy above 4 kHz). To generate the estimated broadband excitation signal $\hat{e}_{\text{bb}}(n)$ often the spectral envelope of the input signal is first removed by means of a predictor error filter:

$$e_{\text{nb}}(n) = s_{\text{tel}}(n) + \sum_{i=1}^{N_{\text{pre, nb}}} a_{\text{nb},i}(n)\, s_{\text{tel}}(n-i). \qquad (5.34)$$

The coefficients $a_{\text{nb},i}(n)$ are computed on a block basis every 5 to 10 ms using the methods described in Sec. 5.3.2. The index "nb" indicates that telephone or *narrow band* quantities are addressed, while "bb" denotes extended or *broad band* signals and parameters. The spectrally whitened input signal $e_{\text{nb}}(n)$ is used for the estimation of the broadband excitation signal $\hat{e}_{\text{bb}}(n)$ while the extracted spectral envelope, described by its predictor coefficient vector

$$\boldsymbol{a}_{\text{nb}}(n) = \Big[a_{\text{nb},1}(n),\, a_{\text{nb},2}(n),\, \ldots a_{\text{nb},N_{\text{pre,nb}}}(n)\Big]^{\text{T}}, \qquad (5.35)$$

is utilized for estimating the broadband spectral envelope. We will describe the latter term also by a vector containing the coefficients of a prediction filter

$$\hat{\boldsymbol{a}}_{\text{bb}}(n) = \left[\hat{a}_{\text{bb},1}(n),\, \hat{a}_{\text{bb},2}(n),\, \ldots \hat{a}_{\text{bb},N_{\text{pre,bb}}}(n)\right]^{\text{T}}. \tag{5.36}$$

Usually, this vector consists of more coefficients than its narrowband equivalent

$$N_{\text{pre,bb}} \geq N_{\text{pre,nb}}. \tag{5.37}$$

Finally, both model parts are combined using an inverse predictor error filter with coefficients $\hat{a}_{\text{bb},i}(n)$ that is excited with the estimated broadband excitation signal $\hat{e}_{\text{bb}}(n)$:

$$\hat{s}(n) = \hat{e}_{\text{bb}}(n) - \sum_{i=1}^{N_{\text{pre,bb}}} \hat{s}(n-i)\,\hat{a}_{\text{bb},i}(n). \tag{5.38}$$

In some implementations a power adjustment of this signal is necessary. Since we want to reconstruct the signal only in those frequency ranges that are not transmitted over the telephone line a bandstop filter is applied. Only those frequencies should pass the filter that are not transmitted (e.g. frequencies below 200 Hz and above 3.8 kHz). Finally the reconstructed and the upsampled telephone signals are added (see Fig. 5.12).

5.5.1 Generation of the Excitation Signal

For the generation of the broadband excitation signal mainly three classes of approaches exist. These classes include modulation techniques, non-linear processing, and the application of function generators. We will describe them briefly in the next sections.

5.5.1.1 Modulation Techniques

Modulation techniques is a term that implies the processing of the excitation signal in the time domain by performing a multiplication with a modulation function

$$\hat{e}_{\text{bb}}(n) = e_{\text{nb}}(n) \cdot 2\cos(\Omega_0 n). \tag{5.39}$$

This multiplication with a cosine function in the time domain corresponds to the convolution with two dirac impulses in the frequency domain:

$$\begin{aligned}\widehat{E}_{\text{bb}}\left(e^{j\Omega}\right) &= E_{\text{nb}}\left(e^{j\Omega}\right) * \left[\delta(\Omega - \Omega_0) + \delta(\Omega + \Omega_0)\right] \\ &= E_{\text{nb}}\left(e^{j(\Omega - \Omega_0)}\right) + E_{\text{nb}}\left(e^{j(\Omega + \Omega_0)}\right).\end{aligned} \tag{5.40}$$

In Fig. 5.13 an example for a modulation approach is depicted. Part (a) shows a time-frequency analysis of an input signal consisting of several harmonic sine terms with a basic fundamental frequency that slowly increases over time. If we multiply this signal with a 3800 Hz modulation function a signal as depicted in part (b) of Fig. 5.13 is generated.

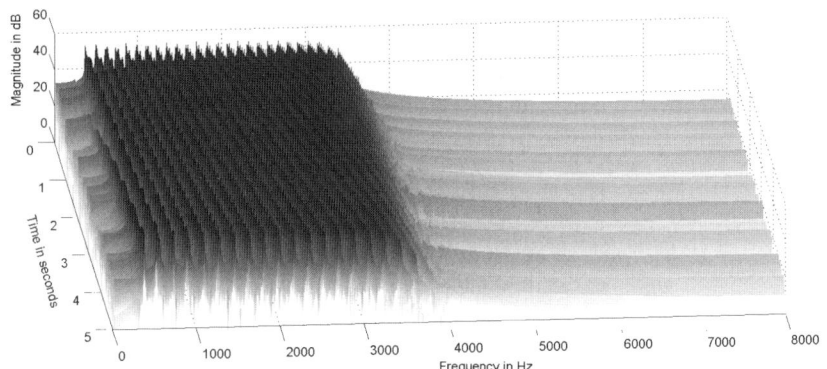

(a) Harmonic (bandlimited) signal $e_{\text{nb}}(n)$.

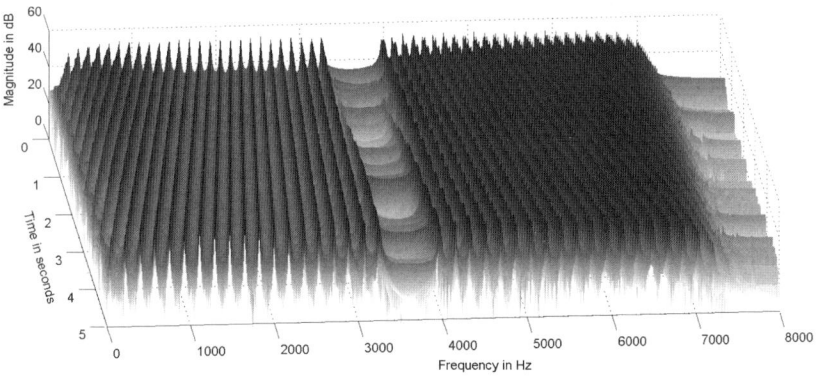

(b) Harmonic (extended) signal $\hat{e}_{\text{bb}}(n)$.

Fig. 5.13. The effect of a cosine modulation. The spectral gap around 3800 Hz in the extended signal $\hat{e}_{\text{bb}}(n)$ appears due to missing low frequency components within the input signal $e_{\text{nb}}(n)$.

Depending on the modulation frequency Ω_0 sometimes the two shifted spectra overlap. For that reason bandpass filtering before applying the modulation function might be necessary. Additionally the output signal might contain signal components within the telephone band. This is the case in the example that we have depicted in Fig. 5.13. In those cases also a filter can be applied – this time the filter is applied to the output $\hat{e}_{\text{bb}}(n)$ – that lets only those frequencies pass that are desired. Additionally the original signal, respectively its whitened version $e_{\text{nb}}(n)$, can be used within the telephone band.

The reason for performing such a frequency shift can easily be understood when we consider the processing of voiced utterances. The periodic signal that is typical for voiced utterances can be extended by modulation techniques. By incorporating a pitch detection one could even perform an adaptive shift and thereby keep the pitch structure even at the transition regions from the telephone band to the extension area. As seen in Sec. 5.3 the excitation signal of unvoiced utterances is noise-like and therefore we do not have a structure we have to take care of. This means that we also can apply this excitation extension method for unvoiced utterances without having to worry about continuing the signal structure along the frequency axis.

5.5.1.2 Non-Linear Processing

One major problem of the above discussed modulation techniques is the pitch detection if the algorithm is designed (pitch-) adaptive. Especially in the low frequency part bothersome artifacts occur if, concerning voiced utterances, the harmonics of the pitch frequency are misplaced. This means the performance of the algorithm crucially depends on the performance of the pitch detection.

Another possibility to extend the excitation signal is the application of non-linear characteristics. Non-linearities have the property that they produce harmonics when applied to a periodic signal. This once again takes the case of voiced utterances into account. There exist a variety of non-linear characteristics which all have different properties. A quadratic characteristic on one hand produces only even harmonics. A cubic characteristic on the other hand produces only odd harmonics. The effect of the application of a non-linear characteristic can be explained best for the quadratic characteristic. The application of a quadratic characteristic in the time domain corresponds to the convolution of the signal with itself in the frequency domain

$$\hat{e}_{\text{bb}}(n) = e_{\text{nb}}^2(n) \quad \circ\!\!-\!\!\bullet \quad E_{\text{nb}}\left(e^{j\Omega}\right) * E_{\text{nb}}\left(e^{j\Omega}\right) = \widehat{E}_{\text{bb}}\left(e^{j\Omega}\right). \quad (5.41)$$

If we assume a line spectrum in the case of voiced sounds the effect becomes clear. Every time the lines match during the shift within the convolution the resulting signal will have a strong component. In contrast to the above presented method where we had the convolution with a dirac impulse at the arbitrary frequency Ω_0 (or a frequency determined by pitch estimation algorithm) we convolve the signal with dirac impulses at proper harmonics of Ω_0. The effect of other non-linear characteristics can be explained by their representation as a power series. Some possible characteristics are:

- Half-way rectification: $f(x) = \begin{cases} x, & \text{if } x > 0, \\ 0 & \text{else.} \end{cases}$

- Full-way rectification: $f(x) = |x|$.

- Saturation characteristic: $f(x) = \begin{cases} a + (x-a)b, & \text{if } x > a, \\ a - (x+a)b, & \text{if } x < a, \\ x, & \text{else.} \end{cases}$

- Quadratic characteristic: $f(x) = x^2$ or $f(x) = x \cdot |x|$.

- Cubic characteristic: $f(x) = x^3$.

- Tanh characteristic: $f(x) = \tanh(\mu x)$.

Here a, b and μ are arbitrary positive parameters. Another property is that they (depending on the effective bandwidth, the sampling rate and the kind of characteristic) produce components out of the Nyquist border. Therefore, the signal has to be upsampled before applying the non-linear characteristic and filtered by a lowpass with the cut-off frequency corresponding to half of the desired sampling rate to avoid aliasing completely before downsampling. A second property is that these non-linear characteristics produce strong components around 0 Hz, which have to be removed. After the application of a non-linear characteristic the extended excitation signal might have undergone an additional coloration. This can be taken into account by applying a prediction error filter (whitening filter). A disadvantage of non-linear characteristics is that in the case of harmonic noise (e.g. engine noise in the car environment) the harmonics of this noise signal are complemented as well. Furthermore, the processed signals of quadratic and cubic characteristics show a wide dynamic range and a power normalization is required. Fig. 5.14 shows the effect of a half way rectification without any post processing (no predictor error filter).

5.5.1.3 Function Generators

The last class of algorithms for excitation signal extension are so called function generators. The simplest form of a function generator is a sine generator. Similar to the use of adaptive modulation techniques this method needs a pitch estimation. The sine generators are working in the time domain. The parameters (amplitude, frequency) of the sine generators are obtained by employing the estimated broadband spectral envelope and the estimate for the fundamental frequency and its harmonics respectively. The advantage of the use of sine generators lies in the discrimination between actual and desired values for amplitude and frequency. The sine generators are designed to change their actual parameters within a maximum allowed range of the parameters towards the desired value. This prevents artifacts due to a step of amplitude or frequency from one frame to another. Another advantage of these sine generators is that artifacts due to a step in phase of the low frequency components

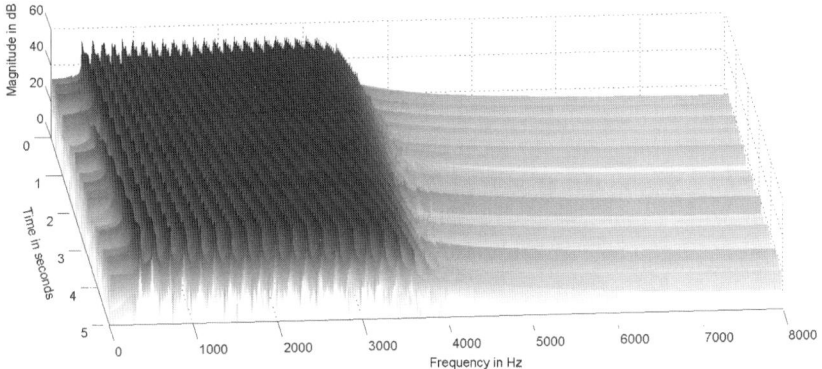

(a) Harmonic (bandlimited) signal $e_{\mathrm{nb}}(n)$.

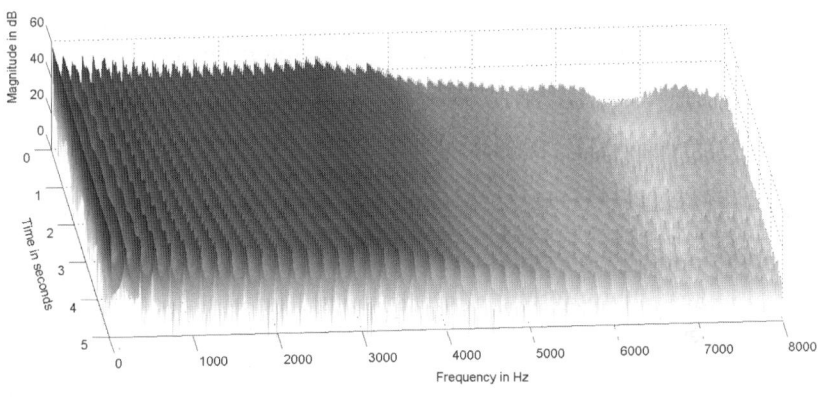

(b) Harmonic (extended) signal $\hat{e}_{\mathrm{bb}}(n)$.

Fig. 5.14. The effect of non-linear characteristics (half way rectification).

do not appear due to the time domain processing. Furthermore, the sine generators do not need an estimated value for the fundamental frequency with every sample or frame but for example only whenever such an estimate is very reliable.

5.5.2 Vocal Tract Transfer Function Estimation

Besides the generation of the excitation signal $\hat{e}_{\mathrm{bb}}(n)$ the wideband spectral envelope needs to be estimated. Several classes of estimation techniques have been suggested:

- One of the simplest approaches is linear mapping. In this case the feature vector containing the parameters of the bandlimited envelope is multiplied with a matrix in order to estimate the wideband feature vector. The advantage of the linear mapping approach is its simplicity in terms of computational complexity and memory requirements. However, its performance is – due to the unrealistic assumption of a linear relation between bandlimited and wideband parameters – only very limited.
- Originating in the field of speech coding codebook approaches have been proposed. In codebook based methods the spectral envelope in terms of low order all-pole models of the current (bandlimited) frame is first mapped on one of N_{cb} codebook entries according to a predefined distance measure [27]. The codebook is usually trained with a large corpus of pairs of bandlimited and wideband speech sequences. Each entry of the codebook for bandlimited envelopes has its wideband equivalent. This counterpart is utilized as an estimate for the wideband spectral envelope of the current frame.
- The third class of estimation techniques is based on the application of neural networks. In this case the assumption of a linear relationship between the bandlimited parameter set and its wideband counterpart is dropped and also a nonlinear relation can be modeled.

Furthermore, combinations of these estimation techniques can be realized. Details about the above introduced approaches are described within the next sections. Common to all approaches is that the parameters of each method are trained using a large database of pairs of bandlimited and wideband speech sequences. As known from speech and pattern recognition best results are achieved when the training database is as close as possible to the final application. Thus, boundary conditions such as recording devices and involved coding schemes should be chosen carefully. Better results are achieved if the training is performed speaker dependent. However, for most applications this is not possible.

5.5.2.1 Generation and Preparation of Training Data

Before being able to train a codebook, neural network, or linear mapping matrices, one important precondition is the availability of a sufficient amount of training data that has to be processed carefully to extract the required features. Starting point is a broadband speech corpus with at least the desired target sampling rate and bandwidth, respectively. Such a speech corpus should meet the requirements of the later application as closely as possible concerning environmental noise level, the kind of utterances (read text or natural speech) and so on. Hence for the training of the approaches presented here an additional speech corpus has been recorded. This speech corpus consisted of recordings of 31 male and 16 female speakers including spontaneous speech as well as read texts. Another requirement from the car environment was the recording of Lombard speech. Lombard speech determines the effect that humans try to speak louder when the surrounding noise level is higher. Simultaneously the formants are shifted towards higher frequencies by speaking at a higher volume. The effect of exiting car noise has been simulated by using headphones playing a binaural recording of a car interior noise environment when driving with a speed of 100 km/h at the appropriate sound pressure level [4]. The recordings have been equalized for the microphone frequency response of the respective headset. The method used for doing so is presented in more detail later on.

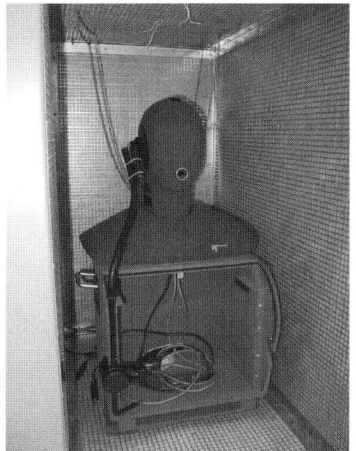 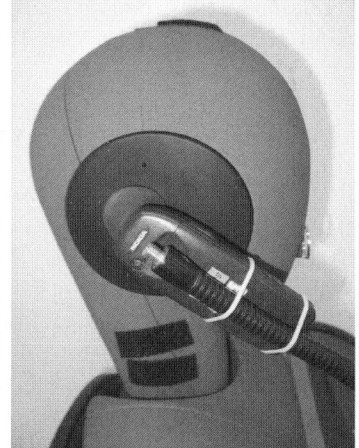

(a) HATS in anechoic chamber. (b) Mobile phone attached to HATS in side view.

Fig. 5.15. Setup using a head and torso simulator (HATS) and an anechoic chamber for playing back the speech corpus and succeeding transmission via GSM.

This speech data can then be downsampled if needed. After removing the speech pauses the speech corpus then has to be processed in an appropriate manner to simulate a realistic band limiting transmission line. Another possibility is not to simulate such a behavior but to record the real transmitted data. The speech corpus in this chapter has been played back using a head and torso simulator (HATS) within an anechoic chamber (see Fig. 5.15). The corpus has been equalized concerning the frequency response as well as the sound pressure level so that the signal at the mouth reference point (MRP) of the HATS equals the signal as it would have been recorded at the MRP of each person. For this task a test sequence $x(n)$ of white noise of sufficient length has been generated and played back over the HATS. This signal has been recorded by a calibrated measurement microphone in the MRP. If we denote the recorded signal with $y(n)$ we can estimate the frequency response of the HATS as

$$H_{\text{HATS}}\left(e^{j\Omega}\right) = \sqrt{\frac{\widehat{S}_{yy}(\Omega)}{\widehat{S}_{xx}(\Omega)}}, \qquad (5.42)$$

with $\widehat{S}_{yy}(\Omega)$ and $\widehat{S}_{xx}(\Omega)$ being estimations for the auto power spectral density functions of the processes $y(n)$ and $x(n)$, respectively. The resulting frequency response has been smoothed by applying an IIR-filter in frequency direction in forward as well as in backward operation. After inverting the frequency response we obtain the frequency response of an appropriate preemphasis filter for equalizing the frequency response of the HATS.

A Nokia 6310i mobile phone has been attached at the HATS (see Fig. 5.15) and the far-end signal has been recorded after transmission over the German D1-network (GSM 900 MHz). As device on the far-end side a Siemens MC35i GSM modem has been used which was accessible through a serial port and an appropriate user interface.

After the band limitation the most crucial task concerning later quality is to compensate for the delay introduced by the band limiting components. The two data sets (narrowband and broadband) need to be synchronous. This synchronization has been performed in this approach by filtering both, the broadband signal as well as the narrowband signal, by an additional bandpass with a passband that has to have the cut off frequencies within the passband of the bandpass, the narrowband signal has been processed with. The filtering has been performed in a forward and backward manner so that no additional delay has been introduced. After the filtering, a blockwise crosscorrelation has been computed between the narrowband and the former broadband signal. The offset of the position of the maximum of the crosscorrelation to the index 0 then indicates the delay which has been introduced into the narrowband signal by the processing done within the transmission. Additional variable delay introduced by so-called *clock drift* between the sound card used for playback and the one used for recording occurred but was negligible and has therefore not been compensated.

After ensuring that the signals are synchronous and they correlate sufficiently the features can be extracted. This has been done in a blockwise manner. These features can comprise several parametric representations of the spectral envelope or any other interesting features. If the transmission was not simulated but real time, the correlation between the narrowband and the broadband features may get destroyed by drop outs or comfort noise injection during the transmission. Therefore signal blocks that did not correlate sufficiently concerning a quadratic distance measure comparing the spectral envelope representation within the passband, have been deleted. The remaining speech corpus consisted of $N = 4.7 \cdot 10^6$ features which equals an overall length of the speech corpus of 116 minutes (as this does not look like a sufficient amount of speech data we should keep in mind, that all speech pauses have been removed). The whole process of preparing the training data is depicted in Fig. 5.16.

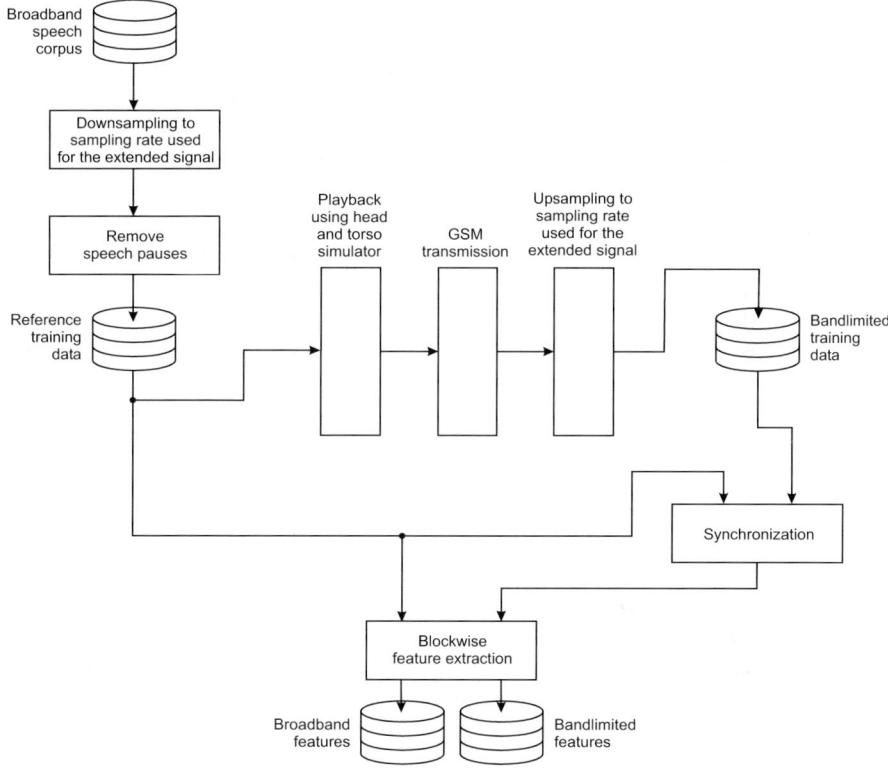

Fig. 5.16. Work flow of the appropriate preparation of training data.

5.5.2.2 Linear Mapping

If the parameter set containing the bandlimited envelope information is described by a vector

$$\boldsymbol{x}(n) = \begin{bmatrix} x_0(n),\, x_1(n),\, ...,\, x_{N_x}(n) \end{bmatrix}^\mathrm{T} \tag{5.43}$$

and its wideband counterpart by a vector

$$\boldsymbol{y}(n) = \begin{bmatrix} y_0(n),\, y_1(n),\, ...,\, y_{N_y}(n) \end{bmatrix}^\mathrm{T}, \tag{5.44}$$

then a linear estimation scheme can be realized by a simple linear operation

$$\hat{\boldsymbol{y}}(n) = \boldsymbol{W}\left(\boldsymbol{x}(n) - \boldsymbol{m}_x\right) + \boldsymbol{m}_y\,. \tag{5.45}$$

In general, the entries of the vector $\boldsymbol{x}(n)$ and $\boldsymbol{y}(n)$ could be predictor coefficients, cepstral coefficients, line spectral frequencies, or any other set of features that describes the spectral envelope. However, since we will use a quadratic cost function cepstral coefficients are a good choice. The multiplication of the bandlimited feature vector $\boldsymbol{x}(n)$ with the $N_y \times N_x$ matrix $\boldsymbol{W}$ can be interpreted as a set of N_y FIR filter operations. Each row of $\boldsymbol{W}$ corresponds to an impulse response which is convolved with the signal vector $\boldsymbol{x}(n)$ resulting in one element of the wideband feature vector $y_i(n)$. As common in linear estimation theory the mean values of the feature vectors $\boldsymbol{m}_x$ and $\boldsymbol{m}_y$ are estimated within a preprocessing stage. For obtaining the matrix $\boldsymbol{W}$ a cost function has to be specified. A very simple approach would be the minimization of the sum of the squared errors over a large database:

$$F(\boldsymbol{W}) = \sum_{n=0}^{N-1} \left\| \boldsymbol{y}(n) - \hat{\boldsymbol{y}}(n) \right\|^2 \to \min . \tag{5.46}$$

In case of cepstral coefficients this results in the distance measure described in Sec. 5.3.4 (see Eqs. 5.30 and 5.31). If we define the entire data base consisting of N zero-mean feature vectors by two matrices

$$\boldsymbol{X} = \Big[\boldsymbol{x}(0) - \boldsymbol{m}_x, \, \boldsymbol{x}(1) - \boldsymbol{m}_x, \, ..., \, \boldsymbol{x}(N-1) - \boldsymbol{m}_x \Big], \tag{5.47}$$

$$\boldsymbol{Y} = \Big[\boldsymbol{y}(0) - \boldsymbol{m}_y, \, \boldsymbol{y}(1) - \boldsymbol{m}_y, \, ..., \, \boldsymbol{y}(N-1) - \boldsymbol{m}_y \Big], \tag{5.48}$$

the optimal solution [32] is given by

$$\boldsymbol{W}_{\text{opt}} = \boldsymbol{Y}\,\boldsymbol{X}^T \left(\boldsymbol{X}\,\boldsymbol{X}^T \right)^{-1} . \tag{5.49}$$

Since the sum of the squared differences of cepstral coefficients is a well-accepted distance measure in speech processing often cepstral coefficients are utilized as feature vectors. Even if the assumption of the existence of a single matrix $\boldsymbol{W}$ which transforms all kinds of bandlimited spectral envelopes into their broadband counterparts is quite unrealistic, this simple approach results in astonishing good results. However, the basic single matrix scheme can be enhanced by using several matrices, where each matrix was optimized for a certain type of feature class. In a two matrices scenario one matrix $\boldsymbol{W}_\text{v}$ can be optimized for voiced sounds and the other matrix $\boldsymbol{W}_\text{u}$ for non-voiced sounds. In this case it is first checked to which class the current feature vector $\boldsymbol{x}(n)$ belongs. In a second stage the corresponding matrix is applied to generate the estimated wideband feature vector[3]

$$\hat{\boldsymbol{y}}(n) = \begin{cases} \boldsymbol{W}_\text{v}\,\big(\boldsymbol{x}(n) - \boldsymbol{m}_{x,\text{v}}\big) + \boldsymbol{m}_{y,\text{v}}, & \text{if the classification indicates a voiced frame,} \\ \boldsymbol{W}_\text{u}\,\big(\boldsymbol{x}(n) - \boldsymbol{m}_{x,\text{u}}\big) + \boldsymbol{m}_{y,\text{u}}, & \text{else.} \end{cases} \tag{5.50}$$

---

[3] Besides two different matrices $\boldsymbol{W}_\text{v}$ and $\boldsymbol{W}_\text{u}$ also different mean vectors $\boldsymbol{m}_{x,\text{v}}$ and $\boldsymbol{m}_{x,\text{u}}$, respectively $\boldsymbol{m}_{y,\text{v}}$ and $\boldsymbol{m}_{y,\text{u}}$, are applied for voiced and unvoiced frames.

The classification of the type of sound can be performed by analyzes such as the zero-crossing rate [8] or the gradient index [24, 36]. Fig. 5.17 shows the structure of the linear mapping approach using cepstral coefficients.

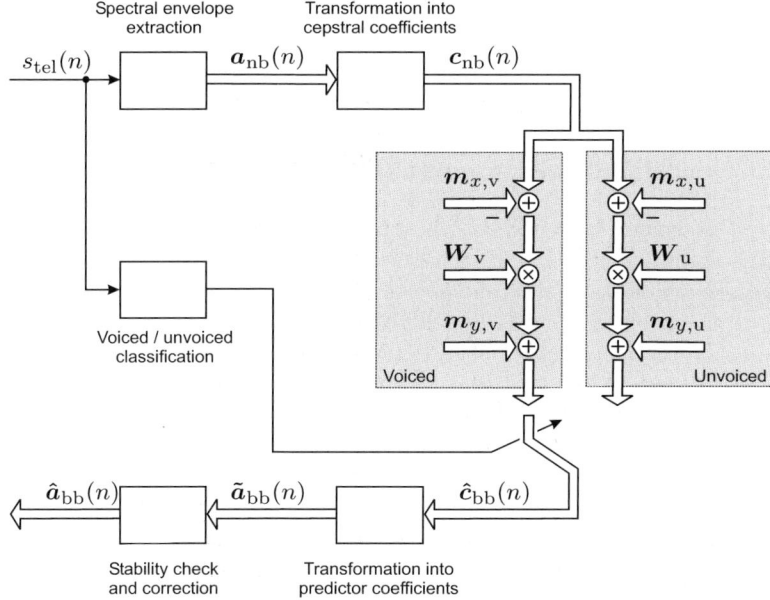

Fig. 5.17. Structure of the linear mapping approach for estimating the broadband spectral envelope (using different parameter sets for voiced and unvoiced utterances).

Since the first analysis stage results in predictor coefficients a feature vector transformation according to Eq. 5.21 is applied, resulting in a vector $c_{\mathrm{nb}}(n)$ containing the narrowband linear predictive cepstral coefficients. After narrowband mean value subtraction, application of the mapping matrices, and broadband mean value addition the resulting broadband cepstral vectors are transformed back into predictor coefficients. This can be achieved also in a recursive manner according to

$$\tilde{a}_{\mathrm{bb},i}(n) = \hat{c}_{\mathrm{bb},i}(n) - \frac{1}{i}\sum_{k=1}^{i-1} k\,\hat{c}_{\mathrm{bb},k}(n)\,\tilde{a}_{\mathrm{bb},i-k}(n)\,, \qquad (5.51)$$

$$\text{for } i \in \{1, ..., N_{\mathrm{bb,pre}}\}\,.$$

$N_{\mathrm{bb,pre}}$ denotes the length of the broadband predictor coefficients vector. Since stability of the resulting IIR filter cannot be guaranteed any more, a stability check needs to be performed. If poles outside the unit circle are detected, one can use, e.g., the last valid coefficient vector or use one of the broadband mean vectors $m_{y,\mathrm{v}}$ or $m_{y,\mathrm{u}}$ instead. Additionally, one can compute the location of each pole and mirror the ones outside the unit circle into the unit circle. This approach, however, is quite costly since the computation of the pole locations is computationally much more demanding than just the detection of instability.

Linear mapping can be applied as a postprocessing stage of codebook approaches. In this case the nonlinear mapping between bandlimited and wideband feature vectors is modeled as a piecewise linear mapping. We will describe this approach in more detail in Sec. 5.5.2.5.

5.5.2.3 Neural Network Approaches

The type of neural network that is most often applied for the estimation of the vocal tract transfer function is the multi-layer perceptron (MLP) in feed forward operation with three layers. As in the case of linear mapping neural networks are often excited with cepstral coefficients, since the sum of squared differences between two sets of cepstral coefficients is a common distance measure in speech processing (see Sec. 5.3.4). Furthermore, such a cost function is a simple least square approach and standard training algorithms such as the well-known back propagation algorithm can be applied to obtain the network parameters.

Before the cepstral coefficients are utilized to excite a neural network a normalization is applied. Usually the input features are normalized to the range $[-1, 1]$. To achieve this – at least approximately – the mean value of each cepstral coefficients is subtracted and the resulting value is divided by three times its standard derivation. If we assume a Gaussian distribution more than 99.7 percent of the values are within the desired interval. In Fig. 5.18 the structure of a vocal tract transfer function estimation with a feed forward multi-layer perceptron is depicted.

Within the input layer the nodes do not perform any processing. The output $a_{\lambda,0}(n)$ of the input neurons equals the input. Processing is only done within the nodes of the hidden layer and the output layer. The amount of nodes $\lambda \in \{0, ... N_\nu\}$ within one layer ν is independent of the amount of nodes within the preceding or succeeding layer. The amount of inputs and outputs equals the amount of coefficients used for the narrowband spectral envelope representation and broadband one, respectively. The function that is implemented in each node λ within the hidden layer or within the output layer ν producing the output is

$$a_{\lambda,\nu}(n) = f_{\text{act}} \left(\theta_{\lambda,\nu} + \sum_{\mu=0}^{N_{\nu-1}-1} a_{\mu,\nu-1}(n)\, w_{\mu,\nu-1,\lambda} \right), \tag{5.52}$$

for

$$\nu \in \{1,2\},\ \lambda \in \{0, ..., N_\nu - 1\},$$

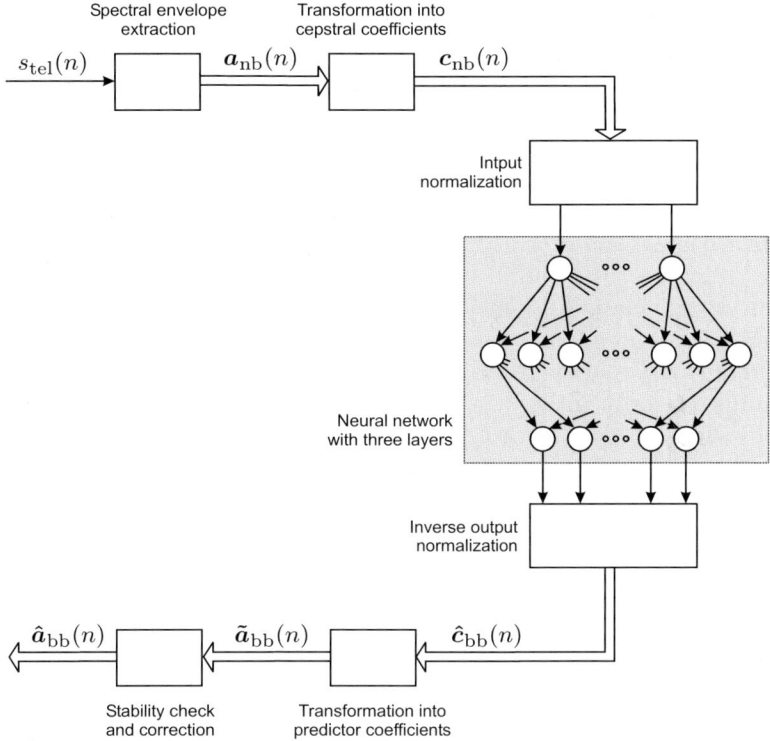

Fig. 5.18. Structure of the vocal tract transfer function estimation using a neural network.

where $a_{\lambda,\nu}(n)$ denotes the output of the node with index λ in layer ν. N_ν is representing the amount of nodes in layer ν. The quantities $w_{\mu,\nu-1,\lambda}$ denote the weights that are applied to the output of node μ of the preceding layer $\nu - 1$, which serves then after weighting as input of node λ of the current layer in process ν. The parameter $\theta_{\lambda,\nu}$ denotes the bias of each node. As an activation function a sigmoid function has been used

$$f_{\text{act}}(x) = \frac{1}{1 + e^{-x}}. \tag{5.53}$$

Since the feature vectors of the desired output (the cepstral coefficients of the broadband envelope) that have been used for training the neural network are normalized too, an inverse normalization has to be applied to the network output (as depicted in Fig. 5.18). As in the linear mapping approach it is not guaranteed that the resulting predictor coefficients belong to a stable all-pole filter. For this reason a stability check has to be applied, too.

In its very basic version the estimation of the vocal tract transfer function does not take any memory into account. This means that any correlation among succeeding frames is not exploited yet. This can be achieved if not only the normalized feature vector of the current frame is fed into the network but also the feature vectors of a few preceding and a few subsequent frames. For the latter case the network output can be computed only with a certain delay. That makes the usage of non-causal features not appropriate for several applications. However, if more than the current feature vector is fed into the network, the network parameters do not only model a direct mapping from the bandlimited to the broadband envelope. Also the temporal transition model between adjacent frames is learned.

The training of the neural networks is accomplished similar to the computation of the mapping matrices of the last section by providing pairs of bandlimited and broadband data. The training itself then has to be performed very carefully to avoid overtraining. Overtraining denotes the optimization on the training set only without further generalization. This can be observed by using a validation data set to control if the network still is generalizing its task or beginning to learn the training data set by heart. Fig. 5.19 shows such a typical behavior.

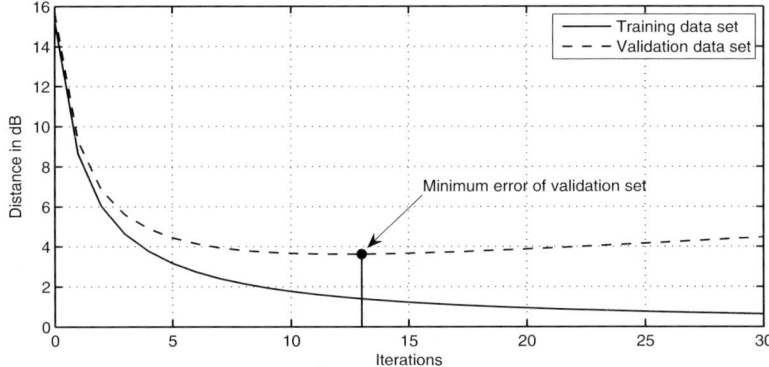

Fig. 5.19. Typical progression of the distance between actual output and desired output of the artificial neural network during the training phase.

The optimum iteration to stop the training is marked in Fig. 5.19. This optimum is characterized by the minimum overall distance between the actual and the desired output of the artificial neural network produced using the validation data set as input.

When it comes to the operation mode of the artificial neural network, after the training phase has been completed, there are two major characteristics of such a network that have to be mentioned. The first one is the low computational complexity needed by such an artificial neural network if not too many layers or nodes are used respectively. Such networks are able to learn complex tasks using comparatively few layers and neurons. This is an advantage over codebook approaches as described in the next section since the computational effort used to evaluate a distance measure for each codebook entry is omitted using artificial neural networks. On the other hand artificial neural networks do not offer possibilities to interfere manually the way codebook approaches do. For example by observing the distance between the narrowband spectral envelope of the input signal and the narrowband codebook entry producing minimum distance one is able to predict if the bandwidth extension runs out of the rudder completely and therefore switch off such a system.

5.5.2.4 Codebook Approaches

A third approach for estimating the vocal tract transfer function is based on the application of codebooks [5]. A codebook contains a representative set of either only broadband or both bandlimited and broadband vocal tract transfer functions. Typical codebook sizes range from $N_{\mathrm{cb}} = 32$ up to $N_{\mathrm{cb}} = 1024$. The spectral envelope of the current frame is computed, e.g. in terms of $N_{\mathrm{pre,nb}} = 10$ predictor coefficients, and compared to all entries of the codebook. In case of codebook pairs the narrowband entry that is closest according to a distance measure to the current envelope is determined and its broadband counterpart is selected as the estimated broadband spectral envelope. If only one codebook is utilized the search is performed directly on the broadband entries. In this case the distance measure should weight the non-excited frequencies much smaller than the excited frequency range from, e.g., 200 Hz to 3600 Hz. The basic structure of a codebook approach utilizing a bandlimited and a broadband codebook is depicted in Fig. 5.20. In the depicted structure a mixed codebook approach, consisting of a narrowband cepstral codebook and a broadband predictor codebook, is used.

As in the neural network approach the predictor coefficients can be transformed into another feature space, such as line spectral frequencies or cepstral coefficients (as depicted in Fig. 5.20), which might be more suitable for applying a cost function. However, also for predictor coefficients well-suited cost functions exist. The likelihood ratio distance measure, that is defined as

$$d_{\mathrm{lhr}}(n,i) = \frac{1}{2\pi} \int_{-\pi}^{\pi} \left(\frac{|\hat{A}_{i,\mathrm{nb}}(e^{j\Omega})|^2}{|A_{\mathrm{nb}}(e^{j\Omega},n)|^2} - 1 \right) d\Omega, \quad (5.54)$$

is sometimes applied for this application. The quantities $A_{\mathrm{nb}}(e^{j\Omega}, n)$ and $\hat{A}_{i,\mathrm{nb}}(e^{j\Omega})$ denote the narrowband spectral envelopes of the current frame and of the ith codebook entry, respectively. The integration of the difference between the ratio of the two spectral envelopes and the optimal ratio

$$\left. \frac{|\hat{A}_{i,\mathrm{nb}}(e^{j\Omega})|^2}{|A_{\mathrm{nb}}(e^{j\Omega},n)|^2} \right|_{\mathrm{opt}} = 1 \quad (5.55)$$

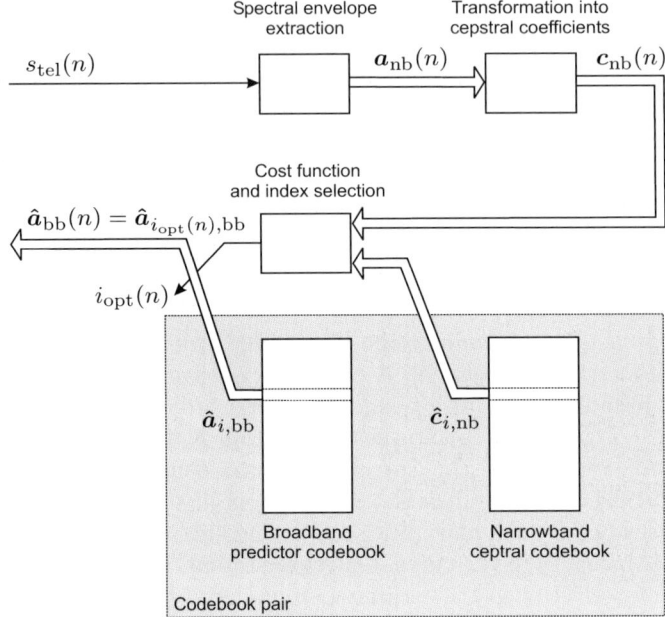

Fig. 5.20. Basic structure of the vocal tract transfer function estimation based on codebook approaches.

can be computed very efficient by using only the predictor coefficients and the autocorrelations matrix of the current frame:

$$d_{\text{lhr}}(n, i) = \frac{\hat{\boldsymbol{a}}_{i,\text{nb}}^{\text{T}} \boldsymbol{R}_{ss}(n) \, \hat{\boldsymbol{a}}_{i,\text{nb}}}{\boldsymbol{a}_{\text{nb}}^{\text{T}}(n) \, \boldsymbol{R}_{ss}(n) \, \boldsymbol{a}_{\text{nb}}(n)} - 1. \tag{5.56}$$

Since only the index $i = i_{\text{opt}}(n)$ corresponding to the minimum of all distances (and not the minimum distance itself) is needed, it is sufficient to evaluate

$$\begin{aligned} i_{\text{opt}}(n) &= \operatorname*{argmin}_{i} \left\{ \tilde{d}_{\text{lhr}}(n, i) \right\} \\ &= \operatorname*{argmin}_{i} \left\{ \hat{\boldsymbol{a}}_{i,\text{nb}}^{\text{T}} \boldsymbol{R}_{ss}(n) \, \hat{\boldsymbol{a}}_{i,\text{nb}} \right\}. \end{aligned} \tag{5.57}$$

Note that Eq. 5.57 can be computed very efficiently since the autocorrelation matrix $\boldsymbol{R}_{ss}(n)$ has Toeplitz structure. Beside the cost function according to Eq. 5.54, which weights the difference between the squared transfer function in a linear manner, a variety of others can be applied [14]. Most of them apply a spectral weighting function within the integration over the normalized frequency Ω. Furthermore, the difference between the spectral envelopes is often weighted in a logarithmic manner (instead of a linear or quadratic). The logarithmic approach takes the human loudness perception in a better way into account. In the approaches discussed in this chapter, we have again used a cepstral distance measure according to Eq. 5.30.

For obtaining the codebook entries iterative procedures such as the method of Linde, Buzo, and Gray (LBG algorithm [31]) can be applied. The LBG-algorithm is an efficient and intuitive algorithm for vector quantizer design based on a long training sequence of data. Various modifications exit (see [29] and [34] for example). In our approach the LBG algorithm is used for the generation of a codebook containing the spectral envelopes that are most representative in the sense of a cepstral distortion measure for a given set of training data. For the generation of this codebook the following iterative procedure is applied to the training data:

1. Initializing:

   Compute the centroid for the whole training data. The centroid is defined as the vector with minimum distance in the sense of a distortion measure to the complete training data.

2. Splitting:

   Each centroid is split into two near vectors by the application of a perturbance.

3. Quantization:

   The whole training data is assigned to the centroids by the application of a certain distance measure and afterwards the centroids are calculated again. Step 3 is executed again and again until the result does not show any significant changes. Is the desired codebook size reached $\Rightarrow$ abort. Otherwise continue with step 2.

Fig. 5.21 shows the functional principle of the LBG algorithm by applying the algorithm to a training data consisting of two clustering points (see Fig. 5.21 (a)). Only one iteration is depicted. Starting with step 1 the centroid over the entire training data is calculated which is depicted in Fig. 5.21 (b). In step 2 this centroid is split into two near initial centroids by the application of a perturbance as can be seen in Fig. 5.21(c). Afterwards in step 3 the training data is quantized to the new centroids and after quantization the new centroids are calculated. This procedure is repeated until a predefined overall distortion is reached or a maximum amount of iterations or no significant changes occur. This final state of the first iteration is depicted in Fig. 5.21(d). Using the LBG algorithm this way, it is only possible to obtain codebooks with an amount of entries equal to a power of two. This could be circumvented by either splitting in more than just two new initial centroids or by starting with an initial guess of the desired n centroids or codebook entries, respectively.

Within these methods the summed distances between the optimum codebook entry and the feature vector of the current frame are minimized for a

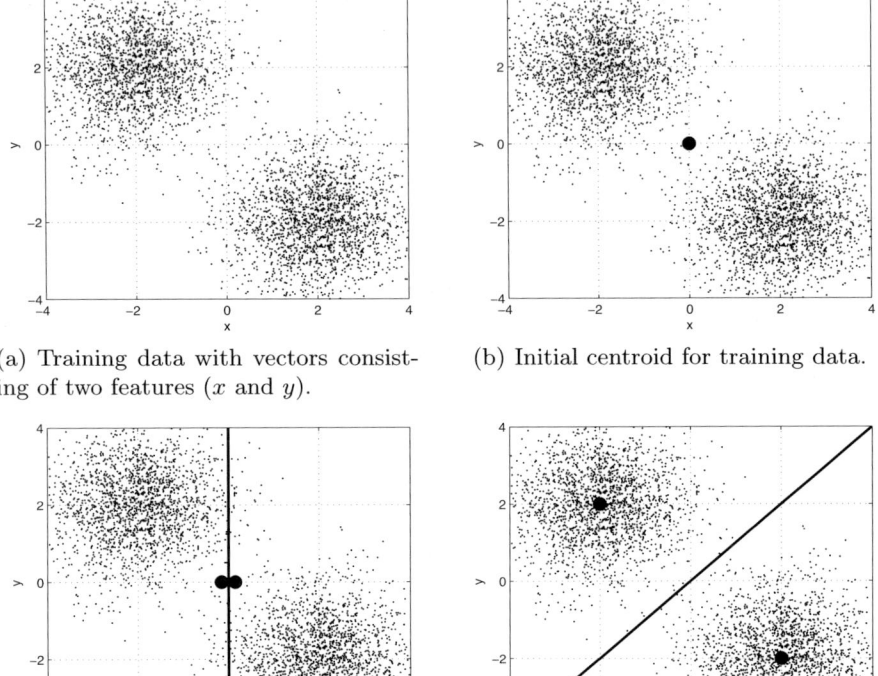

(a) Training data with vectors consisting of two features (x and y).

(b) Initial centroid for training data.

(c) Initial splitting of previous centroid during first iteration and initial cell border (step 2).

(d) Final centroids at end of first iteration and final cell border (end of step 3).

Fig. 5.21. Example for the vector quantization of a training data with two clustering points following the LBG-algorithm.

large training data base with N_{feat} feature sets:[4]

$$D = \sum_{n=0}^{N_{\text{feat}}-1} \min_i \left\{ d(n,i) \right\} \longrightarrow \min . \qquad (5.58)$$

Since the codebook entries should be determined by minimizing Eq. 5.58 on one hand but are also required to find the minimum distance for each data set on the other hand, iterative solutions, such as the LBG algorithm, suggest themselves.

---

[4] The term $d(n,i)$ denotes an arbitrary (non-negative) cost function between the codebook entry with codebook index i and the current input feature vector with frame index n.

During the generation of the codebook it can be assured that all codebook entries are valid feature vectors. As in the examples before, the term *valid* means that the corresponding all-pole filter consists only of poles within the unit circle in the z-domain. For this reason computational expensive stability checks – as required in neural network and linear mapping approaches – can be omitted. Furthermore, the transformation from, e.g. cepstral coefficients into predictor coefficients, can be performed at the end of the training process. For that reason also the feature transformation that was necessary in linear mapping and neural network approaches can be omitted.

Another advantage of codebook based estimation schemes is the inherent discretization of the spectral envelopes. Due to this discretization (broadband spectral envelope is one of N_{cb} prototypes postprocessing schemes that take also the temporal properties of human speech generation into account can be applied. Examples for such postprocessing approaches are

- the application of a hidden Markov model [25] for modeling the temporal transition between the broadband prototype envelopes or
- the consideration of the codebook entry that has been selected during the codebook search of the previous frame for obtaining the current codebook entry [27]. In this case transitions between two rather different broadband envelopes are punished even if each of the succeeding codebook entries are optimum according to a distance measure without memory.

5.5.2.5 Combined Approaches

Prosecuting consequently the idea presented at the end of Sec. 5.5.2.2 of classifying the current frame of speech signal first and then providing specific matrices leads to the awareness that linear mapping can be applied as a postprocessing stage of codebook approaches. In this case the nonlinear mapping between bandlimited and wideband feature vectors is modeled as a piecewise linear mapping. Fig. 5.22 shows such a design.

First the narrowband codebook is searched for the narrowband entry or its index $i_{\text{opt}}(n)$, respectively, producing minimum distance to the actual input feature vector considering a specific distance measure

$$i_{\text{opt}}(n) = \operatorname*{argmin}_{i}\{d(n,i)\}. \tag{5.59}$$

Here the cepstral distance measure as defined in Eq. 5.30

$$d(n,i) = d_{\text{ceps}}(n,i) = \left\| \boldsymbol{c}_{\text{nb}}(n) - \hat{\boldsymbol{c}}_{i,\text{nb}} \right\|^2 \tag{5.60}$$

has been used. An initial estimated broadband spectral envelope is generated by multiplying the matrix $\boldsymbol{W}_{i_{\text{opt}}(n)}$ with the cepstral representation of the mean value compensated input narrowband spectral envelope $\boldsymbol{c}_{\text{nb}}(n) - \boldsymbol{m}_{i_{\text{opt}}(n),\text{nb}}$. Afterwards the broadband mean value $\boldsymbol{m}_{i_{\text{opt}}(n),\text{bb}}$ is added, resulting in

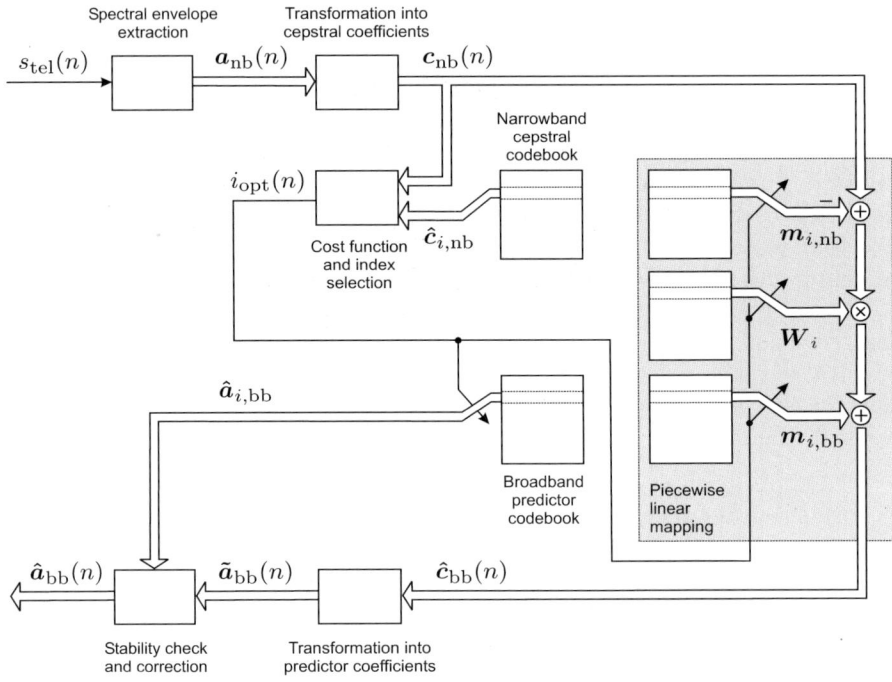

Fig. 5.22. Combined approach using codebook preclassification and linear mapping.

$$\hat{c}_{\mathrm{bb}}(n) = \boldsymbol{W}_{i_{\mathrm{opt}}(n)} \left(c_{\mathrm{nb}}(n) - \boldsymbol{m}_{i_{\mathrm{opt}}(n),\mathrm{nb}} \right) + \boldsymbol{m}_{i_{\mathrm{opt}}(n),\mathrm{bb}} \,. \quad (5.61)$$

After transformation of the vector containing the cepstral coefficients into a corresponding all-pole filter coefficients vector $\tilde{\boldsymbol{a}}_{\mathrm{bb}}(n)$ stability needs to be checked. Depending on the result either the result of the linear mapping operation or the corresponding entry of the broadband predictor codebook $\hat{\boldsymbol{a}}_{i_{\mathrm{opt}}(n),\mathrm{bb}}$ serves as output $\hat{\boldsymbol{a}}_{\mathrm{bb}}(n)$:

$$\hat{\boldsymbol{a}}_{\mathrm{bb}}(n) = \begin{cases} \hat{\boldsymbol{a}}_{i_{\mathrm{opt}}(n),\mathrm{bb}}\,, & \text{if instability was detected,} \\ \tilde{\boldsymbol{a}}_{\mathrm{bb}}(n)\,, & \text{else.} \end{cases} \quad (5.62)$$

Note that stability of the broadband predictor codebook entries can be ensured during the training stage of the codebook.

The training of this combined approach can be split into two separate training stages. The training of the codebook is independent of the succeeding linear mapping vectors and matrices and can be conducted as described in the previous section. Then the entire training data is grouped into $N_{i,\mathrm{cb}}$ sets containing all feature vectors classified to the specific codebook entries $\hat{\boldsymbol{c}}_{i,\mathrm{nb}}$. Now for each subset of the entire training material a single mapping matrix and two mean vectors (narrowband and wideband) are trained according to the method described in Sec. 5.5.2.2. In Fig. 5.23 the function principle of the

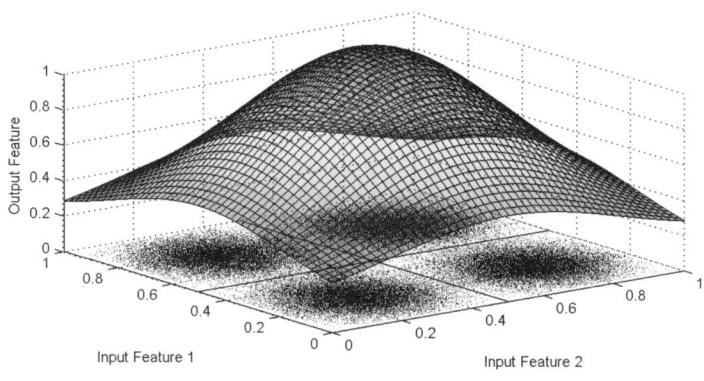

(a) True mapping function.

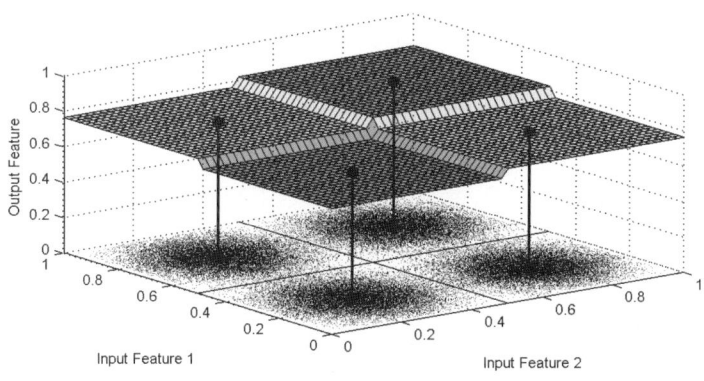

(b) Codebook only approach.

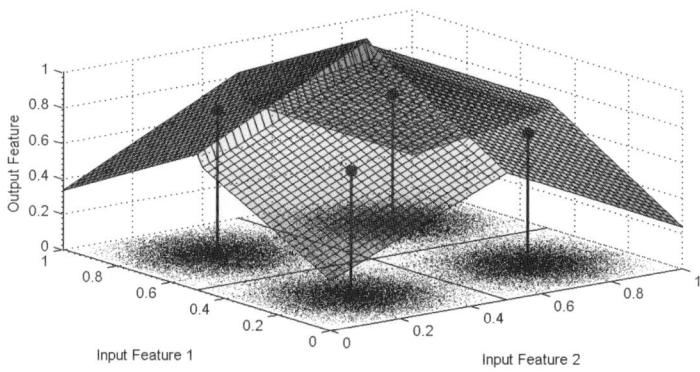

(c) Combined approach using codebooks and linear mapping.

Fig. 5.23. Illustration of the function of an approach using codebook classification and succeeding linear mapping.

combined approach using a preclassification by a codebook and afterwards doing an individual linear mapping corresponding to each codebook entry is illustrated as an example. The little dots at the floor of each diagram represent data points in the two-dimensional input feature space (for this example we limit ourselves to a two-dimensional space). The big dot represents the centroid which is the euclidean mean over all input vectors. The surface in Fig. 5.23 (a) represents the mapping of the input vectors to one feature of the output vectors – this would be the desired function for the extrapolation task. In pure codebook approaches we map the output feature of all vectors that fall into one cell (see part (b) of Fig. 5.23). If we now combine a codebook classification with linear mapping as illustrated in part (c) a plane is placed according to the data points within each cell with minimum overall distance to the original surface resulting in a smaller error when processing an input vector which is close to a cell border. When comparing the approximations of part (b) and (c) with the true mapping as depicted in part (a) the improvement due to the linear postprocessing stage is clearly visible.

Beside a postprocessing with linear mapping also individually trained neural networks can be utilized. Again, an improved performance can be achieved. Even if the computational complexity does not increase that much the memory requirements of combined approaches are significantly larger compared to single approaches.

5.5.2.6 Concluding Remarks

We have mentioned only the very basic approaches in this section and we do not claim for completeness. Several important approaches, such as Gaussian mixture models (GMMs) or mapping matrixes that operate directly on DFT-based input vectors, have not been treated. For these approaches we refer to books and PhD theses such as [18] or [24, 28].

5.6 Evaluation of Bandwidth Extension Algorithms

In many situations and particularly in the context of developing and improving bandwidth extension algorithms, performance measures are needed to evaluate whether one algorithm or one algorithmic version is in some sense superior to another algorithm or to its preceding version. Bandwidth extension systems are often evaluated utilizing objective measures, such as distance measures [14, 17] between the original broadband and the extended signal. These measures can be quantified in straightforward objective terms. We will describe a few of them in the next section. However, due to the complexity of current algorithms these methods deliver only a general sense of the speech quality.

A better – but also more expensive – way to evaluate the speech quality are subjective listening tests. These tests are the most reliable tool available for the evaluation of bandwidth extension algorithms. The challenge of these tests is to design them in such a way that the quality of the extension system can be measured in a reliable and repeatable manner. A variety of listening test have been published, each of them optimized for a special purpose. For further details about listening tests the interested reader is referred to Chaps. 9 and 10 of this book. We will focus here only on comparison mean opinion scores (CMOS). For this kind of subjective test standards have been published by the International Telecommunication Union (ITU) [22].[5]

5.6.1 Objective Distance Measures

Some well known objective distance measures are L_2-norm based logarithmic spectral distortion measures [14], such as

$$d_{\text{Eu,log}}(n) = \frac{1}{2\pi} \int_{-\pi}^{\pi} \left| 10 \log_{10} \left\{ \frac{|S_{\text{ext}}(e^{j\Omega}, n)|^2}{|S_{\text{bb}}(e^{j\Omega}, n)|^2 + \epsilon} \right\} \right|^2 d\Omega. \quad (5.63)$$

Here $S_{\text{ext}}(e^{j\Omega}, n)$ denotes the short-term spectrum of the bandwidth extended signal, $S_{\text{bb}}(e^{j\Omega}, n)$ is the original broadband spectrum and ϵ is a small positive constant to avoid divisions by zero. In order to achieve a reliable measurement, the distances $d_{\text{Eu,log}}(n)$ between the original and the estimated broadband envelope are computed for all frames of the large database and averaged finally:

$$D_{\text{Eu,log}} = \frac{1}{N} \sum_{n=0}^{N-1} d_{\text{Eu,log}}(n). \quad (5.64)$$

The distance measure presented above is applied usually only on the spectral envelopes of the original and the estimated signal, meaning that we can set

$$S_{\text{ext}}\left(e^{j\Omega}, n\right) = \widehat{A}_{\text{bb}}\left(e^{j\Omega}, n\right), \quad (5.65)$$
$$S_{\text{bb}}\left(e^{j\Omega}, n\right) = A_{\text{bb}}\left(e^{j\Omega}, n\right). \quad (5.66)$$

For this reason they are an adequate measure for evaluating the vocal tract transfer function estimation. If the entire bandwidth extension should be evaluated distance measures which take the spectral fine structure as well as the characteristic of the human auditory system into account need to be applied.[6]

---

[5] Note, that the reason why we focus here only on the speech quality and not on the speech intelligibility using, e.g., diagnostic rhyme tests [1,42] is because of the nature of bandwidth extension algorithms. Even if a bandwidth extension system increases the bandwidth and makes the speech sound better it does not add any new information to the signal.

[6] If the logarithmic Euclidean distance is applied directly on the short-term spectra no reliable estimate is possible. For example, a zero at one frequency supporting point in one of the short-term spectra and a small but nonzero value in the other would result in a large distance even if no or nearly no difference would be audible.

For deriving such a distance measure we first define the difference between the squared absolute values of the estimated and the original broadband spectra of the current frame as

$$\Delta\left(e^{j\Omega}, n\right) = 20\log_{10}\left\{\frac{\left|S_{\text{ext}}\left(e^{j\Omega}, n\right)\right|}{\left|S_{\text{bb}}\left(e^{j\Omega}, n\right)\right| + \epsilon}\right\}, \qquad (5.67)$$

One basic characteristic of human perception is that with increasing frequency the perception resolution decreases. We can take this fact into account by adding an exponentially decaying weighting factor for increasing frequency. Another basic characteristic is, that if the magnitude of the estimated spectrum is above the magnitude of the original one, there will occur bothersome artifacts. In the other case the estimated spectrum has less magnitude than the original one. This leads to artifacts that are less bothersome. This characteristic implies the use of a non-symmetric distortion measure which we simply call *spectral distortion measure* (SDM)

$$d_{\text{SDM}}(n) = \frac{1}{2\pi}\int_{-\pi}^{\pi}\xi\left(e^{j\Omega}, n\right)d\Omega. \qquad (5.68)$$

Where $\xi\left(e^{j\Omega}, n\right)$ is defined as

$$\xi\left(e^{j\Omega}, n\right) = \begin{cases} \Delta\left(e^{j\Omega}, n\right)\cdot e^{\alpha\Delta\left(e^{j\Omega}, n\right) - \beta\Omega}, & \text{if } \Delta\left(e^{j\Omega}, n\right) \leq 0, \\ \ln\left(-\Delta\left(e^{j\Omega}, n\right) + 1\right)\cdot e^{-\beta\Omega}, & \text{else.} \end{cases} \qquad (5.69)$$

Fig. 5.24 shows the behavior of this distortion measure for different normalized frequencies Ω. It is clearly visible that positive errors (indicating that the enhanced signal is louder than the original one) leading to a larger distortion than negative ones. Furthermore, errors at lower frequencies are leading also to a larger distortion than errors at high frequencies. For computing the spectral distortion measure the integral in Eq. 5.68 is replaced by a sum over a significantly large number of FFT bins (e.g., $N_{\text{FFT}} = 256$).

To evaluate the quality of some of the approaches presented in Sec. 5.5.2 for the estimation of the broadband spectral envelope, the distance between the original broadband spectral envelope and the spectral envelope of the extended output signal has been measured. This has been done using an evaluation set that has not been part of the training data set. As distance measures the spectral distortion measure described above averaged over the evaluation data set has been employed:

$$\overline{d}_{\text{SDM}} = \frac{1}{N_{\text{eval}}}\sum_{n=0}^{N_{\text{eval}}-1} d_{\text{SDM}}(n). \qquad (5.70)$$

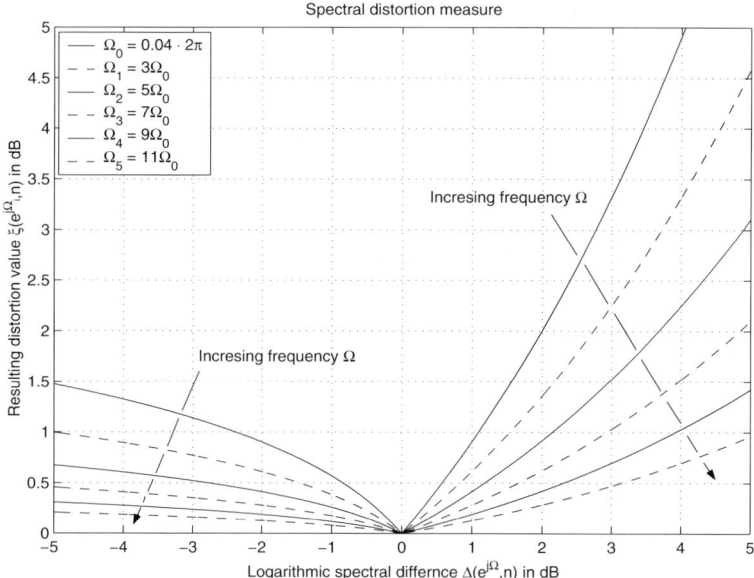

Fig. 5.24. Branches of the spectral distortion measure for $\alpha = 0.1$ and $\beta \approx 0.0005$.

The evaluation has been performed for the pure codebook approach as well as for the combined approach consisting of a preclassification using a codebook approach and a postprocessing stage using a linear mapping model. For both schemes the same method for generating the excitation signal was used. Tab. 5.1 shows the results for different codebook sizes.

Table 5.1. Average spectral distortions ($\bar{d}_{\mathrm{SDM}}$).

Codebook size	Codebook only	Codebook and linear mapping
2	38.47	15.36
4	23.66	11.54
8	17.12	9.21
16	14.64	8.71
32	13.30	8.10
64	12.44	7.64
128	11.89	7.38
256	11.41	7.23

An interesting aspect of the results in Tab. 5.1 is the fact that using a codebook with four entries and subsequent linear mapping produces the same quality as using a codebook consisting of 256 entries. This means that we can either increase quality and maintain the computational complexity or maintain the quality and decrease the computational complexity drastically.

Subjective tests as described in the next section showed that this measure exhibits a better correlation with human quality evaluations than standard distance measures. However, a variety of other distance measures that take not only the spectral envelope into account might be applied for objective evaluations, too.

5.6.2 Subjective Measures

In order to evaluate the subjective quality of the extended signals subjective tests such as mean-opinion-score (MOS) tests should be executed. As indicated at the beginning of this section we will focus here on the evaluation of the progress during the design of a bandwidth extension algorithm. Thus, we will compare not only the enhanced and the original (bandlimited) signals but also signals that have been enhanced by two different algorithmic versions.

If untrained listeners perform the test it is most reliable to perform comparison ratings (CMOS tests). Usually about 10 to 30 people of different age and gender participate in a CMOS test. In a seven-level CMOS test the subjects are asked to compare the quality of two signals (pairs of bandlimited and extended signals) by choosing one of the statements listed in Tab. 5.2.

Table 5.2. Conditions of a 7-level CMOS test.

Score	Statement
-3	A is much worse than B
-2	A is worse than B
-1	A is slightly worse than B
0	A and B are about the same
1	A is slightly better than B
2	A is better than B
3	A is much better than B

This is done for both versions of the bandwidth extension system. Furthermore, all participants were asked whether they prefer version A or version B of the bandwidth extension. For the latter question no "equal-quality" statement should be offered to the subjects since in this case the statistical significance would be reduced drastically if the subjective differences between the two versions are only small.

As an example for such a subjective evaluation we will present a comparison between two bandwidth extension schemes which differ only in the method utilized for estimating the vocal tract transfer function. For the first version a neural network as described in Sec. 5.5.2.3 has been implemented, the second method is based on a codebook approach as presented in Sec. 5.5.2.4. Two questions were of main interest:

- Do the extension schemes enhance the speech quality?
- Which of both methods produces the better results?

Before we present the results of the evaluation some general comments are made. An interesting fact is that both the codebook as well as the neural network are not representing the higher frequencies appropriately where the behavior of the neural network is even worse. At least the power of the higher frequencies produced by the neural network and the codebook is mostly less than the power of the original signal so that bothersome artifacts do not attract attention to a certain degree. Further results are presented in [17].

The participants rated the signals of the network approach with an average mark of 0.53 (between equal and slightly better than the bandlimited signals) and the signals resulting from the codebook scheme with 1.51 (between slightly better and better than the bandlimited signals). So the first question can be answered positively for both approaches. When choosing which approach produces better results around 80 percent voted for the codebook based scheme. Fig. 5.25 shows the results of the CMOS test. Note, that these results do not indicate that codebook approaches are generally better than neural networks. The results depend strongly on the specific implementation.

5.7 Conclusions

In this contribution a basic introduction into bandwidth extension algorithms for the enhancement of telephony speech is presented. Since nearly all bandwidth extension approaches can be split into one part that is generating the required excitation signal and another part that estimates the wideband vocal tract transfer function, we have described both tasks in detail in the main body of this chapter.

Even if the resulting quality is not as good as wideband coded speech, significant improvement (compared to narrowband transmission over the network) is possible. However, wideband coding and bandwidth extension are not competitors, since the latter one is also able to extend even wideband coded speech (e.g. from 7 kHz bandwidth in case of a transmission according to the ITU standard G.722 [21] to 12 or 16 kHz). Due to the ability of bandwidth extension schemes to enhance the speech quality of the incoming signal without modifying the network, research as well as product development in this topic will continue with increasing expense.

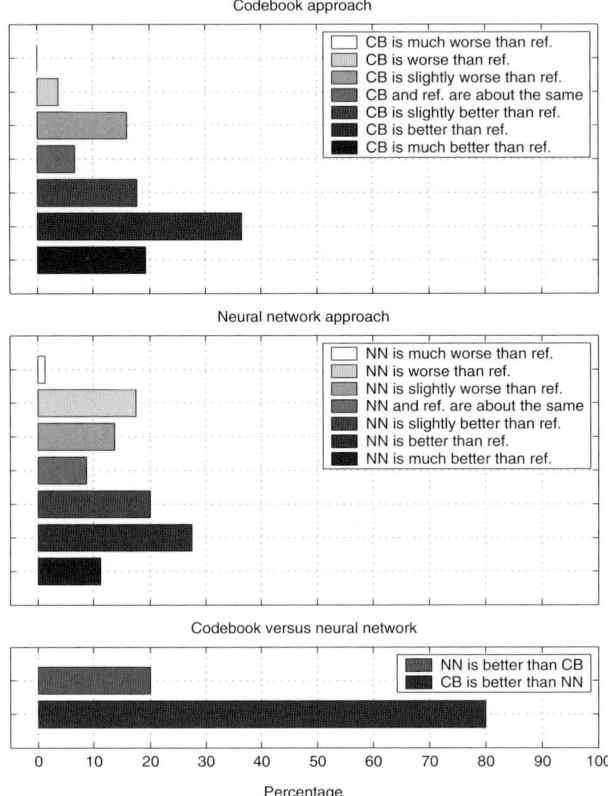

Fig. 5.25. Results of the CMOS test (NN abbreviates neural network, CB stands for codebook, and ref. is denoting the bandlimited signal).

References

1. ANSI S3.2-1989: Method for measuring the intelligibility of speech over communication systems, *American National Standard*, 1989.
2. The New Bell Telephone, *Sci. Am.*, **37**(1), 1, 1877.
3. I. N. Bronstein, K. A. Semendjajew: *Handbook of Mathematics,* Berlin, Germany: Springer, 2004.
4. M. Buck, H.-J. Köpf, T. Haulick: Lombard-Sprache für Kfz-Anwendungen: eine Analyse verschiedener Aufnahmekonzepte, *Proc. DAGA '06*, Braunschweig, Germany, 2006 (in German).
5. H. Carl, U. Heute: Bandwidth enhancement of narrow-band speech signals, *Proc. EUSIPCO '94*, **2**, 1178–1181, Edinburgh, Scotland, U.K., 1994.
6. H. Carl: *Untersuchung verschiedener Methoden der Sprachcodierung und eine Anwendung zur Bandbreitenvergrößerung von Schmalband-Sprachsignalen*, PhD thesis, Ruhr-Universität Bochum, 1994 (in German).
7. M. G. Croll: Sound quality improvement of broadcast telephone calls, *BBC Research Report*, RD1972/26, British Broadcasting Corporation, 1972.

8. J. Deller, J. Hansen, J. Proakis: *Discrete-Time Processing of Speech Signals*, Piscataway, NJ, USA: IEEE Press, 2000.
9. M. Dietrich: Performance and implementation of a robust ADPCM algorithm for wideband speech coding with 64 kbit/s, *Proc. Int. Zürich Seminar Digital Communications*, 1984.
10. J. Durbin: The fitting of time series models, *Rev. Int. Stat. Inst.*, **28**, 233–244, 1960.
11. ETSI: Digital cellular telecommunications system (phase 2+); Enhanced full rate (EFR) speech transcoding, ETSI EN 300 726 V8.0.1, Nov. 2000.
12. J. A. Fuemmeler, R. C. Hardie, W. R. Gardner. Techniques for the regeneration of wideband speech from narrowband speech. *EURASIP J. Appl. Signal. Process.*, **2001**(4), 266–274, 2001.
13. B. Gold, N. Morgan: *Speech and Audio Signal Processing*, New York, NY, USA: Wiley, 2000.
14. R. Gray, A. Buzo, A. Gray Jr., Y. Matsuyama: Distortion measures for speech processing, *IEEE Trans. Acoust., Speech, Signal Processing*, **ASSP-28**(4), 367–376, Aug. 1980.
15. H. Gustafsson, I. Claesson, U. Lindgren: Speech bandwidth extension *Proc. ICME '01*, 206–209, Tokyo, Japan, 2001.
16. W. Hess: *Pitch Determination of SPeech Signals*, Berlin, Germany: Springer, 1983.
17. B. Iser, G. Schmidt: Neural networks versus codebooks in an application for bandwidth extension of speech signals, *Proc. EUROSPEECH '03*, **1**, 565–568, Geneva, Switzerland, 2003.
18. B. Iser, W. Minker, G. Schmidt: *Bandwidth Extension of Speech Signals*, Berlin, Germany: Springer, 2008.
19. IPA: The principles of the International Phonetic Association (IPA), being a description of the International Phonetic Alphabet and the manner of using it, illustrated by texts in 51 languages, University College, Department of Phonetics, London, 1949.
20. ITU: General performance objectives applicable to all modern international circuits and national extension circuits, ITU-T recommendation G.151, 1988.
21. ITU: 7 kHz audio coding within 64 kbit/s, ITU-T recommendation G.722, 1988.
22. ITU: Methods for subjective determination of transmission quality, ITU-T recommendation P.800, August 1996.
23. F. Itakura: Line spectral representation of linear prediction coefficients of speech signals, *JASA*, **57**(1), 1975.
24. P. Jax: *Enhancement of Bandlimited Speech Signals: Algorithms and Theoretical Bounds*, PhD thesis, RWTH Aachen, 2002.
25. P. Jax, P. Vary: On artificial bandwidth extension of telephone speech, *Signal Processing*, **83**(8), 1707–1719, August 2003.
26. K. D. Kammeyer: *Nachrichtenübertragung,* Stuttgart, Germany: Teubner, 1992 (in German).
27. U. Kornagel: Spectral widening of telephone speech using an extended classification approach, *Proc. EUSIPCO '02*, **2**, 339–342, Toulouse, France, 2002.
28. U. Kornagel: *Synthetische Spektralerweiterung von Telefonsprache,* PhD thesis, Forschritt-Berichte VDI, **10**(736), Düsseldorf, Germany, 2004 (in German).
29. W. P. LeBlanc, B. Bhattacharya, S. A. Mahmoud, V. Cuperman: Efficient search and design procedures for robust multi-stage VQ of LPC parameters for 4 kb/s

speech coding, *IEEE Transactions on Speech and Audio Processing*, **1**(4), 373-385, October 1993.
30. N. Levinson: The Wiener RMS error criterion in filter design and prediction, *J. Math. Phys.*, **25**, 261–268, 1947.
31. Y. Linde, A. Buzo, R. M. Gray: An algorithm for vector quantizer design, *IEEE Trans. Comm.*, **COM-28**(1), 84–95, Jan. 1980.
32. D. G. Luenberger: *Opitmization by Vector Space Methods,* New Yory, NY, USA: Wiley, 1969.
33. A. V. Oppenheim, R. W. Schafer: *Discrete-Time Signal Processing,* Englewood Cliffs, NJ, USA: Prentice Hall, 1989.
34. K. K. Paliwal, B. S. Atal: Efficient vector quantization of LPC parameters at 24 bits/frame, *IEEE Transactions on Speech and Audio Processing*, **1**(1), 3-14, January 1993.
35. P. J. Patrick: *Enhancement of Bandlimited Speech Signals,* PhD thesis, Loughborough University of Techology, 1983.
36. J. Paulus: Variable rate wideband speech coding using perceptually motivated thresholds, *Proc. of IEEE Workshop on Speech Coding*, 35–36, Annapolis, MD, USA, 1995.
37. L. Rabiner, R. W. Schafer: *Digital Processing of Speech Signals,* Englewood Cliffs, NJ, USA: Prentice Hall, 1978.
38. L. Rabiner, B. H. Juang: *Fundamentals of Speech Recognition,* Englewood Cliffs, NJ, USA: Prentice Hall, 1993.
39. H. W. Schüsssler: *Digitale Signalverarbeitung 1,* 4th edition, Berlin, Germany: Springer, 1994 (in German).
40. F. K. Soong, B.-H. Juang: Optimal quantization of LSP parameters, *IEEE Transactions on Speech and Audio Processing*, **1**(1), 15-24, January 1993.
41. P. Vary, R. Martin: *Digital Speech Transmission: Enhancement, Coding and Error Concealment,* New York, NY, USA: Wiley, 2006.
42. W. Voiers: Evaluating processed speech using the diagnostic rhyme test, *Speech Technology*, 30–39, Jan./Feb. 1983.

6

Dereverberation and Residual Echo Suppression in Noisy Environments[*]

Emanuël Anco Peter Habets[1,2], Sharon Gannot[1], and Israel Cohen[2]

[1] School of Engineering, Bar-Ilan University, Ramat-Gan, Israel
[2] Dept. of Electrical Engineering, Technion - Israel Institute of Technology, Haifa, Israel

Hands-free devices, such as mobile phones, are often used in noisy and reverberant environments. Therefore, the received microphone signal contains not only the desired speech (commonly called *near-end speech*) signal, but also interferences such as reverberations of the desired source, background noise, and a far-end echo signal that results from a sound that is produced by the loudspeaker. These interferences degrade the fidelity and intelligibility of the near-end speech and decrease the performance of automatic speech recognition systems.

Acoustic echo cancellers are widely used to cancel the far-end echo. Post-processors, employed in conjunction with acoustic echo cancellers, further enhance the near-end speech. Most post-processors that are described in the literature only suppress background noise and residual echo, i.e., echo which is not suppressed by the acoustic echo canceller. The intelligibility of the near-end speech also depends on the amount of reverberation. Dereverberation techniques have been developed to cancel or suppress reverberation. Recently, practically feasible spectral enhancement techniques to suppress reverberation have emerged that can be incorporated into the post-processor.

After a short introduction, the problems encountered in a hands-free device are formulated. A general purpose post-filter is developed, which can be used to suppress non-stationary, as well as stationary, interferences. The problem of dereverberation of noisy speech signals is addressed by using the general purpose post-filter employed to suppress reverberation and background noise. Next, suppression of residual echo is discussed. Finally, a post-processor is developed for the joint suppression of reverberation, residual echo, and background noise. An experimental study demonstrates the beneficial use of the proposed post-processor for jointly reducing reverberation, residual echo, and background noise.

---

[*] This research was partly supported by the Israel Science Foundation (grant no. 1085/05) and by the Technology Foundation STW, applied science division of NWO and the technology programme of the Dutch Ministry of Economic Affairs.

6.1 Introduction

Conventional and mobile telephones are often used in noisy and reverberant environments. When such a device is used in hands-free mode, a loudspeaker is used to reproduce the sound of the far-end speaker. In addition, the distance between the desired speaker and the microphone is usually larger than the distance encountered in handset mode. Therefore, the microphone signal contains not only the desired speech signal (commonly called *near-end speech signal*), but also interfering signals. These interferences, that distort the near-end speech signal, result from the

i) echoes from sounds that are produced by the loudspeaker,
ii) background noise, and
iii) reverberation caused by the desired source.

These distortions degrade the fidelity and intelligibility of the near-end speech that is perceived by the listener at the far-end side.

Acoustic echo cancellation is the most important and well-known technique to cancel the acoustic echo. This technique enables one to conveniently use a hands-free device while maintaining a high user satisfaction in terms of low speech distortion, high speech intelligibility, and acoustic echo attenuation. Furthermore, in driving cars the hands-free functionality is often required by law. The acoustic echo cancellation problem is usually solved by using an adaptive filter in parallel to the acoustic echo path [8,32,34,54]. The adaptive filter is used to generate a signal that is a replica of the acoustic echo signal. An estimate of the near-end speech signal is then obtained by subtracting the estimated acoustic echo signal, i.e., the output of the adaptive filter, from the microphone signal. Sophisticated control mechanisms have been proposed for fast and robust adaptation of the adaptive filter coefficients in realistic acoustic environments [34,45]. In practice, residual echo, i.e., echo that is not suppressed by the echo canceller, always exists. The residual echo is caused by the

i) deficient length of the adaptive filter,
ii) mismatch between the true and the estimated echo path, and
iii) non-linear system components.

It is widely accepted that echo cancellers alone do not provide sufficient echo attenuation [32,34,45]. Turbin et al. compared three post-filtering techniques to reduce the residual echo and concluded that the spectral subtraction technique, which is commonly used for noise suppression, was the most efficient [60]. In a reverberant environment there can be a large amount of so-called late residual echo due the deficient length of the adaptive filter. In [18] Enzner proposed a recursive estimator for the short-term power spectral density (PSD) of the late residual echo signal. The late residual echo was suppressed by a spectral enhancement technique using the estimated short-term PSD of the late residual echo signal.

In some applications, like hands-free terminal devices, noise reduction becomes necessary due to the relative large distance between the microphone and the speaker. The first attempts to develop a combined echo and noise reduction system can be attributed to Grenier et al. [24,63] and to Yasukawa [64]. Both employ more than one microphone. A survey of these systems can be found in [7,34]. Gustafsson et al. [28] proposed two post-filters for residual echo and noise reduction. The first post-filter was based on the *log spectral amplitude* (LSA) estimator [19] and was extended to attenuate multiple interferences. The second post-filter was psychoacoustically motivated.

In case the hands-free device is used in a noisy reverberant environment, the acoustic path becomes longer, and the microphone signal contains reflections of the near-end speech signal as well as noise. Martin and Vary proposed a system for joint acoustic echo cancellation, dereverberation and noise reduction using two microphones [42]. A similar system was developed by Dörbecker and Ernst in [17]. In both papers, dereverberation was performed by exploiting the coherence between the two microphones as proposed by Allen et al. in [3]. Bloom [6] found that this dereverberation approach had no statistically significant effect on intelligibility, even though the measured average reverberation time, which is defined as the time for the reverberation level to decay to 60 dB below the initial level, and the perceived reverberation time were considerably reduced by the processing. A major drawback of these approaches is that they require two microphones, whereas current hands-free devices are commonly equipped with a single microphone.

Recently, practically feasible single microphone speech dereverberation techniques have emerged. Lebart proposed a single microphone dereverberation method based on spectral subtraction of late reverberant energy [41]. The late reverberant energy is estimated using a statistical model of the acoustic impulse response (AIR). This method was extended to multiple microphones by Habets [29]. Recently, Wen et al. [62] presented experimental results from listening tests using the algorithm developed by Habets. These results showed that the algorithm in [29] can significantly increase the subjective speech quality. The methods in [41] and [29] do not require an estimate of the AIR. However, they do require an estimate of the reverberation time of the room which might be difficult to estimate blindly. Furthermore, both methods do not consider any interferences. Since this is the case in many hands-free applications the latter problem needs to be addressed.

In Sec. 6.2 of this chapter the general problem of suppressing residual echo, reverberation, and background noise is formulated. An *optimally-modified log spectral amplitude* (OM-LSA) estimator that can be used to estimate a desired signal in the presence of one non-stationary interference (e.g., residual echo or reverberation) and one stationary interference (e.g., background noise) is developed in Sec. 6.3. In the following sections it is shown that this estimator can be used for different applications. In Sec. 6.4 a short introduction to speech dereverberation is presented, and a recently developed spectral enhancement technique for dereverberation of noisy speech is described. In Sec. 6.5 a

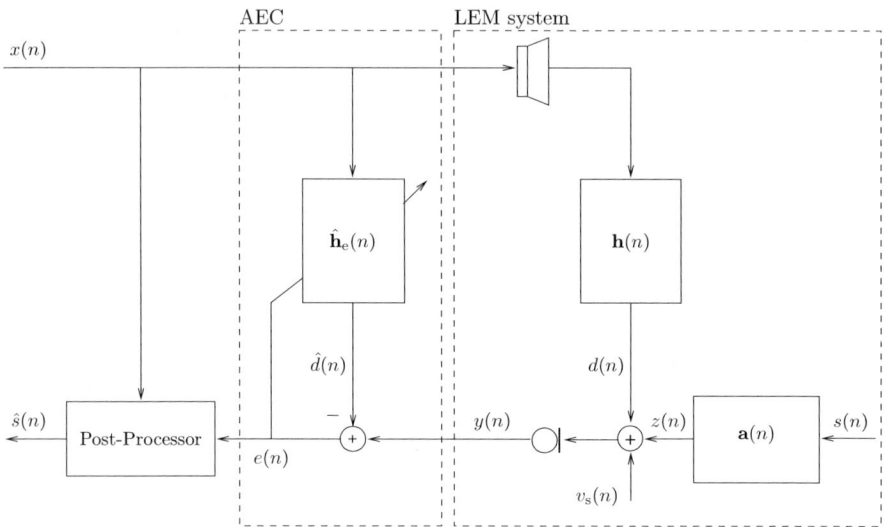

Fig. 6.1. Conventional acoustic echo canceller with post-processor.

residual echo suppression technique which focusses on the suppression of late residual echo is developed. Finally, in Sec. 6.6, a post-processor is developed that is able to jointly suppress reverberation, residual echo, and background noise. In Sec. 6.7, experimental results are presented that demonstrate the beneficial use of post-processors in Secs. 6.5-6.6. This chapter is concluded with a summary of the developed post-processor and future developments in Sec. 6.8.

6.2 Problem Formulation

A loudspeaker enclosure microphone (LEM) system, a conventional acoustic echo canceller (AEC), and a post-processor are depicted in Fig. 6.1. The microphone signal is denoted by $y(n)$ and consists of a reverberant speech component $z(n)$, an acoustic echo $d(n)$, and a noise component $v_\mathrm{s}(n)$, i.e.,

$$y(n) = z(n) + d(n) + v_\mathrm{s}(n), \qquad (6.1)$$

where n denotes the discrete time index.

The reverberant near-end speech component $z(n)$ is given by

$$z(n) = \sum_{j=0}^{N_\mathrm{a}-1} a(n,j) s(n-j), \qquad (6.2)$$

where $a(n,j)$ denotes the jth coefficient of the AIR between the near-end source and the microphone at time n, N_a is the length of the AIR, and $s(n)$ denotes the anechoic speech signal. The reverberant speech signal $z(n)$ can be divided into two components:

i) the early speech component $z_\text{e}(n)$, which consists of a direct sound and early reverberation that is caused by early reflections, and
ii) the late reverberant speech component $z_\text{r}(n)$, which consists of late reverberation that is caused by the reflections that arrive after the early reflections.

Independent research [2, 5, 62] has shown that the speech quality and intelligibly are most affected by late reverberation. In addition it has been shown that the first reflections that arrive shortly after the direct path usually contribute to the intelligibility of speech. In Sec. 6.4 we focus on the estimation of the early speech component in the presence of a stationary interference.

The acoustic echo given by

$$d(n) = \sum_{j=0}^{N_\text{h}-1} h(n,j)\, x(n-j), \qquad (6.3)$$

where $h(n,j)$ denotes the jth coefficient of the acoustic echo path at time n, N_h is the length of the acoustic echo path, and $x(n)$ denotes the far-end speech signal. The acoustic echo path

$$\boldsymbol{h}(n) = [h(n,0),\, h(n,1),\, \ldots,\, h(n, N_\text{h}-1)]^\text{T} \qquad (6.4)$$

can approximately be modelled using $T_{60}\, f_\text{s}$ coefficients, where f_s denotes the sample frequency. However, an adaptive filter of length N_e ($N_\text{e} < N_\text{h}$) is commonly used to cancel the acoustic echo signal $d(n)$ [8, 32, 34, 54]. In this chapter a standard normalized least mean square (NLMS) algorithm is used to estimate part of the acoustic echo path $\boldsymbol{h}(n)$. The update equation for the NLMS algorithm is given by

$$\widehat{\boldsymbol{h}}_\text{e}(n+1) = \widehat{\boldsymbol{h}}_\text{e}(n) + \mu(n) \frac{\boldsymbol{x}(n)\, e(n)}{\boldsymbol{x}^\text{T}(n)\, \boldsymbol{x}(n) + \rho_\text{NLMS}}, \qquad (6.5)$$

where

$$\widehat{\boldsymbol{h}}_\text{e}(n) = \left[\widehat{h}_\text{e}(n,0),\, \widehat{h}_\text{e}(n,1),\, \ldots,\, \widehat{h}_\text{e}(n, N_\text{e}-1)\right]^\text{T}, \qquad (6.6)$$

is the estimated impulse response vector, $\mu(n)$ ($0 < \mu(n) < 2$) denotes the step-size, ρ_NLMS ($\rho_\text{NLMS} > 0$) a regularization factor, and

$$\boldsymbol{x}(n) = \left[x(n),\, \ldots,\, x(n - N_\text{e} + 1)\right]^\text{T} \qquad (6.7)$$

denotes the far-end speech signal state-vector. It should be noted that other, more advanced, algorithms can be used, e.g., recursive least squares (RLS) or affine projection (AP), see for example [34] and the references therein. These advanced adaptive algorithms are beyond the scope of this chapter. The estimated echo signal $\hat{d}(n)$ is obtained by

$$\hat{d}(n) = \sum_{j=0}^{N_\text{e}-1} \widehat{h}_\text{e}(n,j)\, x(n-j). \qquad (6.8)$$

The error signal $e(n)$ of the AEC is given by

$$\begin{aligned} e(n) &= y(n) - \hat{d}(n) \\ &= z(n) + d(n) + v_{\mathrm{s}}(n) - \hat{d}(n) \\ &= z(n) + e_{\mathrm{r}}(n) + v_{\mathrm{s}}(n), \end{aligned} \quad (6.9)$$

where $e_r(n) = d(n) - \hat{d}(n)$ denotes the residual echo signal. In general the residual echo signal $e_{\mathrm{r}}(n)$ will not be zero [34, 45] since:

1) Due to practical reasons, e.g., complexity, slow convergence of long adaptive filters, and robustness, the length of the estimate $\widehat{\boldsymbol{h}}_{\mathrm{e}}(n)$ has to be limited to a certain length, N_{e}. This length is shorter than the length of the acoustic echo path. Thus, the amount of echo resulting from the unmodelled part of the echo path cannot be removed by the AEC.
2) The adaptation is not ideal, i.e., the system mismatch vector,

$$\boldsymbol{h}_{\Delta}(n) = \boldsymbol{h}'(n) - \widehat{\boldsymbol{h}}_{\mathrm{e}}(n), \quad (6.10)$$

where $\boldsymbol{h}'(n) = [h(n, 0), \ldots, h(n, N_{\mathrm{e}} - 1)]^{\mathrm{T}}$ is a truncated version of $\boldsymbol{h}(n)$, is not zero.
3) The echo may contain some components, that are non-linear with respect to $x(n)$. Since the model is linear, these components cannot be handled correctly. This non-linearity has a twofold influence. First, it contributes to the residual echo. Second, it influences the convergence of the adaptive filter, which may lead to even larger residual echo components.

In Sec. 6.5 we focus on the residual echo that is caused by the deficient length of the adaptive filter $\widehat{\boldsymbol{h}}_{\mathrm{e}}(n)$.

Double-talk occurs during periods when the far-end speaker and the near-end speaker are talking simultaneously and can seriously affect the convergence and tracking ability of the adaptive filter. Double-talk detectors and optimal step-size control methods have been presented to alleviate this problem [22,33,34,45]. These methods are beyond the scope of this chapter. In this chapter it is assumed that the filter is adapted in those periods where only the far-end speech signal is active, which have been chosen manually. Therefore, the step-size $\mu(n)$ is chosen as follows,

$$\mu(n) = \begin{cases} 0, & \text{during double-talk}, \\ \mu, & \text{otherwise}. \end{cases} \quad (6.11)$$

While the AEC estimates and subtracts the far-end echo signal a post-processor is used to enhance the near-end speech signal. Most post-processors are designed to estimate the reverberant speech signal $z(n)$ or the noisy reverberant speech signal $z(n) + v_{\mathrm{s}}(n)$. Since the speech intelligibility of the reverberant speech signal $z(n)$ is mainly determined by the amount of late reverberation, we focus on the estimation of the early speech component $z_{\mathrm{e}}(n)$.

6.3 OM-LSA Estimator for Multiple Interferences

In many practical situations the desired signal is degraded by stationary and non-stationary interferences. For example, in the previous section it was shown that the near-end speech signal is degraded by two non-stationary interferences, i.e., reverberation and acoustic echo, and a stationary interference, i.e., background noise. While part of the acoustic echo can be cancelled by the AEC, the remaining residual echo needs to be estimated and suppressed by a post-processor. While the spectral properties of stationary interferences can be estimated using so-called minimum statistics approaches [9,11,43], the estimation of the spectral properties of non-stationary interferences is often very difficult. However, in the application that is considered in this chapter it is possible to obtain an estimate of the spectral properties of the non-stationary interferences (see Sec. 6.4 and Sec. 6.5). In this section a spectral amplitude estimator is developed that can be used to estimate the spectral component in the presence of multiple (stationary and non-stationary) interferences.

Let us define a signal $y(n)$ which contains a desired signal $z(n)$, a non-stationary interference $v_{\text{ns}}(n)$, and a stationary interference $v_{\text{s}}(n)$, i.e.,

$$y(n) = z(n) + v_{\text{ns}}(n) + v_{\text{s}}(n). \tag{6.12}$$

In the short-time Fourier transform (STFT) domain we have

$$Y(l,k) = Z(l,k) + V_{\text{ns}}(l,k) + V_{\text{s}}(l,k), \tag{6.13}$$

where l denotes the time frame index, and k denotes the frequency index.

6.3.1 OM-LSA Estimator

In this section a modified version of the OM-LSA estimator is developed to obtain an estimate of the desired spectral component $Z(l,k)$ in the presence of a non-stationary and stationary interference.

The Log Spectral Amplitude (LSA) estimator proposed by Ephraim and Malah [19] minimizes

$$\mathrm{E}\left\{\left(\log_e\{A(l,k)\} - \log_e\{\widehat{A}(l,k)\}\right)^2\right\}, \tag{6.14}$$

where $A(l,k) = |Z(l,k)|$ denotes the spectral speech amplitude, and $\widehat{A}(l,k)$ is its optimal estimator. Assuming conditionally independent spectral coefficients, given their variances [12], the optimal estimator is defined as

$$\widehat{A}(l,k) = \exp\left\{\mathrm{E}\left\{\log_e\{A(l,k)\}|Y(l,k)\right\}\right\}, \tag{6.15}$$

with $\exp\{x\} = e^x$. The LSA gain function depends on the *a posteriori* and *a priori* signal-to-interference ratio (SIR), which are given by

$$\gamma(l,k) = \frac{|Y(l,k)|^2}{\lambda_{v_{\mathrm{ns}}}(l,k) + \lambda_{v_{\mathrm{s}}}(l,k)}, \tag{6.16}$$

and

$$\xi(l,k) = \frac{\lambda_z(l,k)}{\lambda_{v_{\mathrm{ns}}}(l,k) + \lambda_{v_{\mathrm{s}}}(l,k)}, \tag{6.17}$$

respectively. Here $\lambda_z(l,k)$, $\lambda_{v_{\mathrm{ns}}}(l,k)$ and $\lambda_{v_{\mathrm{s}}}(l,k)$ denote the spectral variances of the desired signal component, the non-stationary interference, and the stationary interference, respectively. It should be noted that the spectral variance $\lambda_z(l,k)$ of the desired signal in (6.17) is unobservable. The estimation of the *a priori* SIR is treated in Sec. 6.3.2.

The LSA gain function is given by [19]

$$G_{\mathrm{LSA}}(l,k) = \frac{\xi(l,k)}{1+\xi(l,k)} \exp\left\{ \frac{1}{2} \int_{\zeta(l,k)}^{\infty} \frac{e^{-t}}{t} \, dt \right\}, \tag{6.18}$$

where

$$\zeta(l,k) = \frac{\xi(l,k)}{1+\xi(l,k)} \gamma(l,k). \tag{6.19}$$

The OM-LSA spectral gain function, which minimizes the mean-square error of the log-spectra, is obtained as a weighted geometric mean of the hypothetical gains associated with the speech presence uncertainty [10]. Given two hypotheses, $H_0(l,k)$ and $H_1(l,k)$, which indicate speech absence and speech presence, respectively, we have

$$H_0(l,k): \quad Y(l,k) = V_{\mathrm{ns}}(l,k) + V_{\mathrm{s}}(l,k),$$
$$H_1(l,k): \quad Y(l,k) = Z(l,k) + V_{\mathrm{ns}}(l,k) + V_{\mathrm{s}}(l,k).$$

Based on a Gaussian statistical model, the *a posteriori* speech presence probability is given by

$$p(l,k) = \left[1 + \frac{q(l,k)}{1-q(l,k)} (1+\xi(l,k)) \exp\left\{ -\zeta(l,k) \right\} \right]^{-1}, \tag{6.20}$$

where $q(l,k)$ is the *a priori* signal absence probability [10].

The OM-LSA gain function is given by,

$$G_{\mathrm{OM\text{-}LSA}}(l,k) = \left[G_{H_1}(l,k) \right]^{p(l,k)} \left[G_{H_0}(l,k) \right]^{1-p(l,k)}, \tag{6.21}$$

with $G_{H_1}(l,k) = G_{\mathrm{LSA}}(l,k)$ and $G_{H_0}(l,k) = G_{\min}$. The lower-bound for the gain when the signal is absent is denoted by $G_{\min}$, and specifies the maximum amount of suppression in those frames.

An estimate of the desired spectral speech component $Z(l,k)$ can now be obtained using

$$\widehat{Z}(l,k) = G_{\text{OM-LSA}}(l,k)\, Y(l,k). \tag{6.22}$$

To avoid speech distortions $G_{\min}$ is usually set between -12 and -18 dB. However, in practice the non-stationary interference needs to be reduced more than 12-18 dB. Due to the constant lower-bound the non-stationary interference will still be audible in some time-frequency frames [30]. To be able to suppress the non-stationary interference down to the residual level of the stationary interference we have to modify $G_{H_0}(l,k)$. In case $G_{H_0}(l,k)$ is applied to those time-frequency frames where the desired signal is assumed to be absent, i.e., the hypothesis $H_0(l,k)$ is assumed to be true, we obtain

$$\widehat{Z}(l,k) = G_{H_0}(l,k)\bigl(V_{\text{ns}}(l,k) + V_{\text{s}}(l,k)\bigr). \tag{6.23}$$

The desired solution for $\widehat{Z}(l,k)$ is

$$\widehat{Z}(l,k) = G_{\min}\, V_{\text{s}}(l,k). \tag{6.24}$$

Hence, the desired gain function $G_{H_0}(l,k)$ can be obtained by minimizing

$$\mathrm{E}\left\{\left|G_{H_0}(l,k)\bigl(V_{\text{ns}}(l,k) + V_{\text{s}}(l,k)\bigr) - G_{\min}\, V_{\text{s}}(l,k)\right|^2\right\}. \tag{6.25}$$

Under the assumption that the interferences are uncorrelated we obtain

$$G_{H_0}(l,k) = G_{\min}\, \frac{\widehat{\lambda}_{v_{\text{s}}}(l,k)}{\widehat{\lambda}_{v_{\text{ns}}}(l,k) + \widehat{\lambda}_{v_{\text{s}}}(l,k)}. \tag{6.26}$$

In case $\widehat{\lambda}_{v_{\text{ns}}}(l,k) = 0$ the standard gain function $G_{H_0}(l,k) = G_{\min}$ is obtained. However, when $\widehat{\lambda}_{v_{\text{ns}}}(l,k) > 0$ the time and frequency dependent lower bound $G_{H_0}(l,k)$ minimizes Eq. 6.25 in such a that a stationary residual noise level is obtained. The results of an informal listening test showed that the residual interference was more pleasant than the residual interference that was obtained using $G_{H_0}(l,k) = G_{\min}$.

6.3.2 A priori SIR Estimator

The *a priori* SIR in (6.17) can be written as

$$\frac{1}{\xi(l,k)} = \frac{1}{\xi_{v_{\text{ns}}}(l,k)} + \frac{1}{\xi_{v_{\text{s}}}(l,k)}, \tag{6.27}$$

with

$$\xi_{\vartheta}(l,k) = \frac{\lambda_z(l,k)}{\lambda_{\vartheta}(l,k)}, \tag{6.28}$$

where $\vartheta \in \{v_{\text{ns}}, v_{\text{s}}\}$. Hence, the total *a priori* SIR can be calculated using the *a priori* SIRs of each interference separately [27, 28, 30]. By doing this, one gains control over

i) the trade-off between the interference reduction and the distortion of the desired signal, and
ii) the *a priori* SIR estimation approach of each interference.

Note that in some cases it might be desirable to reduce one of the interferences at the cost of larger speech distortion, while other interferences are reduced less to avoid distortion.

In case the desired signal and the non-stationary interference are very small, the *a priori* SIRs $\xi_{v_{\mathrm{ns}}}(l,k)$ may be unreliable since $\lambda_z(l,k)$ and $\lambda_{v_{\mathrm{ns}}}(l,k)$ are close to zero. In the following we assume that there is always a certain amount of stationary interference, i.e., $\lambda_{v_{\mathrm{s}}}(l,k) > 0$. We propose to calculate $\xi(l,k)$ using only the most important and reliable *a priori* SIRs as follows

$$\xi(l,k) = \begin{cases} \xi_{v_{\mathrm{s}}}(l,k), & \text{if } 10\log_{10}\left\{\frac{\lambda_{v_{\mathrm{s}}}(l,k)}{\lambda_{v_{\mathrm{ns}}}(l,k)}\right\} > \beta_{\mathrm{dB}}, \\ \dfrac{\xi_{v_{\mathrm{ns}}}(l,k)\xi_{v_{\mathrm{s}}}(l,k)}{\xi_{v_{\mathrm{ns}}}(l,k) + \xi_{v_{\mathrm{s}}}(l,k)}, & \text{otherwise}, \end{cases} \qquad (6.29)$$

where the threshold β_{dB} specifies the level difference between $\lambda_{v_{\mathrm{s}}}(l,k)$ and $\lambda_{v_{\mathrm{ns}}}(l,k)$ in dB. In case the noise level is β_{dB} higher than the level of the non-stationary interference, the total *a priori* SIR, $\xi(l,k)$, will be equal to $\xi_{v_{\mathrm{s}}}(l,k)$. Otherwise $\xi(l,k)$ will depend on both $\xi_{v_{\mathrm{s}}}(l,k)$ and $\xi_{v_{\mathrm{ns}}}(l,k)$.

A well-known *a priori* SIR estimator is the decision-directed estimator, proposed by Ephraim and Malah [19]. We now show how this estimator can be used to estimate the individual *a priori* SIRs. It should be noted that each *a priori* SIR could be estimated using a different approach, e.g., the non-causal *a priori* SIR estimator presented by Cohen in [12].

The decision-directed *a priori* SIR estimator is given by

$$\widehat{\xi}(l,k) = \max\left\{\eta\,\frac{\widetilde{A}^2(l-1,k)}{\lambda_{v_{\mathrm{ns}}}(l,k) + \lambda_{v_{\mathrm{s}}}(l,k)} + (1-\eta)\,\psi(l,k),\,\xi_{\min}\right\}, \qquad (6.30)$$

where $\widetilde{A}(l,k) = |G_{H_1}(l,k)\,Y(l,k)|$, $\psi(l,k) = \gamma(l,k) - 1$ is the *instantaneous SIR*, $\gamma(l,k)$ is the *a posteriori* SIR as defined in Eq. 6.16, and $\xi_{\min}$ is a lower-bound on the *a priori* SIR. The weighting factor η ($0 \le \eta \le 1$) controls the tradeoff between the amount of noise reduction and distortion. To estimate $\xi_\vartheta(l,k)$ we use the following expression

$$\widehat{\xi}_\vartheta(l,k) = \max\left\{\eta_\vartheta\,\frac{\widetilde{A}^2(l-1,k)}{\lambda_\vartheta(l-1,k)} + (1-\eta_\vartheta)\,\psi_\vartheta(l,k),\,\xi_{\min,\vartheta}\right\}, \qquad (6.31)$$

where

$$\begin{aligned}\psi_\vartheta(l,k) &= \frac{\lambda_{v_{\mathrm{ns}}}(l,k) + \lambda_{v_{\mathrm{s}}}(l,k)}{\lambda_\vartheta(l,k)}\,\psi(l,k) \\ &= \frac{|Y(l,k)|^2 - \left(\lambda_{v_{\mathrm{ns}}}(l,k) + \lambda_{v_{\mathrm{s}}}(l,k)\right)}{\lambda_\vartheta(l,k)},\end{aligned} \qquad (6.32)$$

and $\xi_{\min,\vartheta}$ is the lower-bound on the *a priori* SIR $\xi_\vartheta(l,k)$.

6.4 Dereverberation of Noisy Speech Signals

In this section the dereverberation of noisy speech signals is addressed. Furthermore, a post-processor that suppresses late reverberation and background noise is developed.

In Sec. 6.4.1 a short introduction to speech dereverberation is presented. The speech dereverberation problem is formulated in Sec. 6.4.2. In Sec. 6.4.3 a statistical reverberation model is described. This model is used in Sec. 6.4.4 in the development of a spectral variance estimator for the late reverberant signal component. The post-processing algorithm is summarized in Sec. 6.4.5.

6.4.1 Short Introduction to Speech Dereverberation

Speech signals that are received by a microphone at a distance from the speech source usually contain reverberation. Reverberation is the process of multipath propagation of an acoustic signal from its source to a microphone. The received signal generally consists of a direct sound, reflections that arrive shortly after the direct sound (commonly called *early reverberation*), and reflections that arrive after the early reverberation (commonly called *late reverberation*). The combination of the direct sound and early reverberation is sometimes referred to as the *early speech component*. Early reverberation changes the short-term temporal characteristics of the anechoic signal and contributes to spectral coloration, while late reverberation changes the waveform's temporal envelope as exponentially decaying tails are added at sound offsets [47]. The coloration can be characterized by the spectral deviation σ which is defined as the standard deviation of the log-amplitude frequency response of the AIR [38]. Reverberation can degrade the fidelity and intelligibility of speech and the recognition performance of automatic speech recognition systems.

The detrimental effects of reverberation on speech intelligibility have been attributed to two types of masking. Nábělek et al. [46] found evidence of *overlap-masking*, whereby a preceding phoneme and its reflections mask a subsequent phoneme, and of *self-masking*, which refers to the time and frequency alterations of an individual phoneme. In 1982 Allen [5] reported a formula to predict the *subjective preference* of reverberant speech. His main result is given by the equation

$$P = P_{\max} - \sigma T_{60}, \qquad (6.33)$$

where P is the subjective preference in some arbitrary units, $P_{\max}$ is the maximum possible preference, σ is the spectral deviation in decibels, and T_{60} is the reverberation time in seconds. According to this formula, decreasing either the spectral deviation σ or the reverberation time T_{60} results in an increased reverberant speech quality.

Reverberation reduction processes can be divided into many categories. They may, for example, be divided into single- or multi-microphone methods and into those primarily affecting coloration or those affecting late reverberation. We categorized the reverberation reduction processes depending on whether or not the AIR needs to be estimated. We then obtain two main categories, i.e., *reverberation cancellation* and *reverberation suppression*.

The first category, i.e., reverberation cancellation, consists of methods known as blind deconvolution. Much research has been undertaken on the topic of blind deconvolution, cf. [35] and the references therein. Multi-channel methods appear particularly interesting because theoretically perfect inverse-filtering can be achieved if the AIRs could be obtained *a priori*, and they do not have any common-zeros in the z-plane [44]. To achieve dereverberation without *a priori* knowledge of the room acoustics, i.e., blind dereverberation, many traditional methods assume that the target signal is independent and identically-distributed (i.i.d.). However, the i.i.d. hypothesis does not hold for speech-like signals. When applying such traditional deconvolution methods to speech, the speech generating process is somehow deconvolved and the target speech signal is excessively whitened. A method which explores the null-space of the spatial correlation matrix, calculated from the received signals, was developed by Gürelli and Nikias [26]. It was shown that the null space of the correlation matrix of the received signals contains information on the acoustic transfer functions. This method, although originally aimed at solving communication problems, has also potential in the speech processing framework and was extended by Gannot and Moonen [21]. Hopgood used a realistic source signal model and Baysian parameter estimation techniques to estimate the unknown parameters [36]. The speech signal is modelled using a block stationary auto-regressive process while the room acoustics are modelled using an all-pole model. Other methods in this category try to estimate the equalization filter directly [15,59]. The methods in this category suffer from several limitations

i) they have been shown to be little robust to small changes in the AIR [51],
ii) channels can not be identified uniquely when they contain common zeros,
iii) observation noise causes severe problems, and
iv) some methods require knowledge of the order of the unknown system [37].

Methods in the second category, i.e., reverberation suppression, do not require an estimate of the AIR and explicitly exploit the characteristics of speech, the effect of reverberation on speech, or the characteristics of the AIR. Methods based on processing of the linear prediction (LP) residual belong to this category [23,25,65]. The residual signal following the LP analysis has been observed to contain the effects of reverberation. The peaks in the residual signal correspond to excitation events in voiced speech together with additional peaks due to the reverberation. Other, so-called, spatial processing methods, use multiple microphones placed at different locations. They often use a limited amount of *a priori* knowledge of the AIR, such as the direction of arrival of the desired speech source. The multiple input signals can be manipulated to enhance or attenuate signals emanating from particular directions.

Some spatial processing methods are inspired by the mechanisms of audition in the hearing system of animals and humans. The well-known delay and sum beamformer is a good example of such a method and belongs to the reverberation suppression category. In [41] a single-channel speech dereverberation method based on spectral subtraction was introduced to reduce the effect of overlap-masking. The described method estimates the short-term PSD of late reverberation based on a statistical reverberation model. This model exploits the exponential decay of the AIR, and depends on a single parameter which is related to the reverberation time of the room. In [29] Habets showed that the estimated short-term PSD of late reverberation can be improved using multiple microphones. Additionally, the fine-structure of the speech signal is partially restored by means of spatial averaging of the received power spectra.

6.4.2 Problem Formulation

The reverberant signal results from the convolution of the anechoic speech signal $s(n)$ and the causal AIR $\boldsymbol{a}(n)$. In this section we assume that the AIR is time-invariant, i.e., $a(n,j) = a(j) \ \forall \ n$, and that the length of the AIR is infinite. To simplify the notation we assume (without loss of generality) that the direct sound arrives at $n = 0$. We can now write Eq. 6.2 as

$$z(n) = \sum_{j=0}^{\infty} a(j)\, s(n-j)$$
$$= \sum_{j=-\infty}^{n} s(j)\, a(n-j). \qquad (6.34)$$

Since our goal is to suppress late reverberation we split the AIR into two components (see Fig. 6.2) such that

$$a(j) = \begin{cases} a_{\mathrm{e}}(j), & \text{for } 0 \leq j < N_{\mathrm{r}}, \\ a_{\mathrm{r}}(j), & \text{for } N_{\mathrm{r}} \leq j \leq \infty, \\ 0, & \text{otherwise}, \end{cases} \qquad (6.35)$$

where N_{r} is chosen such that $\boldsymbol{a}_{\mathrm{e}}$ consists of the direct path and a few early reflections, and $\boldsymbol{a}_{\mathrm{r}}$ consists of all later reflections. The fraction $N_{\mathrm{r}}/f_{\mathrm{s}}$ usually ranges from 40 to 80 ms.

Using (6.35) we can write the microphone signal $y(n)$ as

$$y(n) = \underbrace{\sum_{j=n-N_{\mathrm{r}}+1}^{n} s(j) a_{\mathrm{e}}(n-j)}_{z_{\mathrm{e}}(n)} + \underbrace{\sum_{j=-\infty}^{n-N_{\mathrm{r}}} s(j)\, a_{\mathrm{r}}(n-j)}_{z_{\mathrm{r}}(n)} + v_{\mathrm{s}}(n), \qquad (6.36)$$

where $z_{\mathrm{e}}(n)$ is the early speech component, and $z_{\mathrm{r}}(n)$ denotes the late reverberant speech component. Note that $z_{\mathrm{e}}(n)$ is affected only by the early

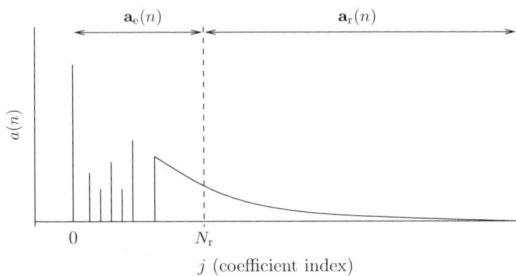

Fig. 6.2. Schematic representation of the acoustic impulse response $\boldsymbol{a}(n)$.

reflections. Suppression of the late reverberant speech component $z_r(n)$ will partially dereverberate the speech signal, and reduce the effective reverberation time. According to Eq. 6.33 this treatment will increase the subjective preference of the processed speech. Since the response of the first part of the AIR, i.e., $z_e(n)$, remains unaltered we do not reduce the colorations caused by the early reflections.

The post-processor is depicted in Fig. 6.3. The noisy and reverberant speech signal is denoted by $y(n)$, and is first transformed to the STFT domain. The resulting spectral component $Y(l,k)$ is used to estimate the spectral variance $\lambda_{v_s}(l,k)$ of the background noise, and to estimate the spectral variance $\lambda_{v_{ns}}(l,k)$ of the non-stationary interference which is equal to the spectral variance $\lambda_{z_r}(l,k)$ of the late reverberant signal component $z_r(n)$. The spectral variance $\lambda_{v_s}(l,k)$ is estimated used the *improved minima controlled recursive averaging* (IMCRA) approach [11]. An estimator for the spectral variance of $z_r(n)$ is derived in Sec. 6.4.4. The post-filter introduced in Sec. 6.3 can then be used to estimate the early speech component $Z_e(l,k)$. The partially

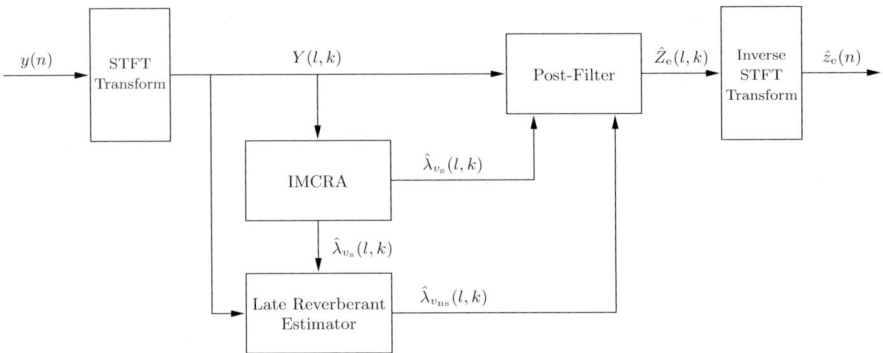

Fig. 6.3. Post-processor for dereverberation of a noisy speech signal.

dereverberated speech signal $\hat{z}_\mathrm{e}(n)$ can then be obtained using the inverse STFT and the weighted overlap-add method [14].

6.4.3 Statistical Reverberation Model

Since the acoustic behaviour in real rooms is too complex to model explicitly we make use of statistical room acoustics (SRA). SRA provides a statistical description of the transfer function of the system between the source and the receiver in terms of a few key quantities, e.g., source-receiver distance, room volume, and reverberation time. The crucial assumption of SRA is that the distribution of amplitudes and phases of individual plane waves, which sum up to produce sound pressure at some point in a room, is so close to random that the sound field is fairly uniformly distributed throughout the room volume. The validity of this description is subjected to a set of conditions that must be satisfied to ensure the accuracy of calculations. Our analysis therefore implicitly assumes that the following conditions hold [40, 51, 58]:

1. The dimensions of the room are relatively large compared to the longest wavelength of the sound of interest.
2. The average spacing of the resonance frequencies of the room must be smaller than one third of their bandwidth. In a room with volume V (measured in m^3), and reverberation time T_{60} (in seconds), this condition is fulfilled for frequencies that exceed the Schroeder frequency:

$$f_\mathrm{g} = 2000\,\mathrm{Hz}\,\sqrt{\frac{T_{60}}{V}\frac{\mathrm{m}^3}{\mathrm{s}}}. \tag{6.37}$$

3. The source and the microphone are located in the interior of the room, at least a half-wavelength away from the walls.

Sabine's [53] major contribution was the introduction of statistical methods to calculate the reverberation time of an enclosed space without considering the details of the space geometry. Schroeder has extended Sabine's fundamental work [55, 56] and derived a frequency domain model, and a set of statistical properties about the frequency response of the random impulse response. Polack [49] developed a time-domain model complementing Schroeder's frequency domain model. In this model, an AIR is described as a realization of a non-stationary stochastic process. This model is defined as

$$a(j) = \begin{cases} b(j)\,\mathrm{e}^{-\rho j}, & \text{for } j \geq 0; \\ 0, & \text{otherwise,} \end{cases} \tag{6.38}$$

where $b(j)$ is a white zero-mean Gaussian stationary noise and ρ is linked to the reverberation time T_{60} through

$$\rho \triangleq \frac{3\log_\mathrm{e}(10)}{T_{60}f_\mathrm{s}}. \tag{6.39}$$

In contrast to the model described in Eq. 6.38 the reverberation time is frequency dependent due to frequency dependent reflection coefficients of walls and other objects and the frequency dependent absorption coefficient of air [40]. This dependency can be taken into account by using a different model in each sub-band. This extension is rather straightforward because processing is performed in the time-frequency domain.

In the early nineties, Polack [50] proved that the most interesting properties of room acoustics are statistical when the number of "simultaneous" arriving reflections reaches a limit of about 10. In this case the so-called echo density (as defined in [40]) is high enough, such that the space can be considered to be in a fully diffused or mixed state. The essential requirement is ergodicity, which requires that any given reflection trajectory in the space will eventually reach all points. The ergodicity assumption is determined by the shape of the enclosure and the surface reflection properties. It should be noted that non-ergodic shapes will exhibit much longer mixing times and may not even have an exponential decay. Nevertheless, while it may not be true that all acoustic environments can be modelled using this stochastic model, it is sufficiently accurate for most spaces. Kuttruff [40] concluded that the model is valid in case the distance between the source and the measurement point is greater than the critical distance. The critical distance indicates the distance at which the steady-state reverberant energy equals the direct sound energy. For an omnidirectional source it can be shown [40] that the critical distance is defined as

$$r_c = 0.1\text{m} \sqrt{\frac{V}{\pi T_{60}} \frac{\text{s}}{\text{m}^3}}. \tag{6.40}$$

This implies that Polack's statistical model is valid whenever the source-microphone distance is larger than the critical distance.

The energy envelope of the AIR can be expressed as

$$\mathrm{E}_a\{a^2(j)\} = \sigma^2 \, e^{-2\rho j}, \tag{6.41}$$

where σ^2 denotes the variance of $b(j)$ and $\mathrm{E}_a\{\cdot\}$ denotes ensemble averaging over a, i.e., over different realizations of the stochastic process in Eq. 6.38. Under the assumption that our space is ergodic we may evaluate the ensemble average in Eq. 6.41 by spatial averaging so that different realizations of this stochastic process are obtained by varying either the position of the receiver or the source [39]. Note that the same stochastic process will be observed, for all allowable positions (in terms of the third SRA condition), provided that the time origin be defined with respect to the signal emitted by the source, and not with respect to the arrival time of the direct sound at the receiver.

6.4.4 Late Reverberant Spectral Variance Estimator

Before the spectral variance $\lambda_{z_r}(l,k)$ can be estimated we need to obtain an estimate of the spectral variance of the reverberant spectral component $Z(l,k)$, i.e., $\lambda_z(l,k)$. The spectral variance $\lambda_z(l,k)$ is estimated by minimizing

$$\mathrm{E}\left\{\left(|Z(l,k)|^2 - |\widehat{Z}(l,k)|^2\right)^2\right\}, \qquad (6.42)$$

where $\widehat{Z}(l,k) = G_{\mathrm{SP}}(l,k)Y(l,k)$.

As shown in [1] this leads to the following spectral gain function

$$G_{\mathrm{SP}}(l,k) = \sqrt{\frac{\xi_{\mathrm{SP}}(l,k)}{1+\xi_{\mathrm{SP}}(l,k)}\left(\frac{1}{\gamma_{\mathrm{SP}}(l,k)} + \frac{\xi_{\mathrm{SP}}(l,k)}{1+\xi_{\mathrm{SP}}(l,k)}\right)} \qquad (6.43)$$

where

$$\xi_{\mathrm{SP}}(l,k) = \frac{\lambda_z(l,k)}{\lambda_{v_{\mathrm{s}}}(l,k)}, \qquad (6.44)$$

and

$$\gamma_{\mathrm{SP}}(l,k) = \frac{|Y(l,k)|^2}{\lambda_{v_{\mathrm{s}}}(l,k)}, \qquad (6.45)$$

denote the *a priori* and *a posteriori* SIRs, respectively. The *a priori* SIR is estimated using the decision-directed estimator [19]. Estimates of the spectral variance $\lambda_{v_{\mathrm{s}}}(l,k)$ of the noise in the received signal $y(n)$ are obtained using the IMCRA approach [11]. An estimate of the spectral variance of the reverberant signal $z(n)$ is then obtained by:

$$\widehat{\lambda}_z(l,k) = \left(G_{\mathrm{SP}}(l,k)\right)^2 |Y(l,k)|^2. \qquad (6.46)$$

In order to derive an estimator for the spectral variance of the late reverberant signal component $z_{\mathrm{r}}(n)$ we start by analysing the auto-correlation of the reverberant signal $z(n)$. The auto-correlation of the reverberant signal $z(n)$ at discrete time n and lag τ for a fixed source-receiver configuration is defined as

$$r_{zz}(n, n+\tau) = \mathrm{E}_z\{z(n)z(n+\tau)\}, \qquad (6.47)$$

where $\mathrm{E}_z\{\cdot\}$ denotes ensemble averaging with respect to z. Using Eq. 6.34 we have for one realization of $a(j)$

$$r_{zz}(n, n+\tau, a) = \sum_{j=-\infty}^{n}\sum_{j'=-\infty}^{n+\tau} \mathrm{E}_s\{s(j)\,s(j')\}\,a(n-j)\,a(n+\tau-j'), \qquad (6.48)$$

where $\mathrm{E}_s\{\cdot\}$ denotes ensemble averaging with respect to $s(n)$. The spatially averaged auto-correlation results in

$$r_{zz}(n, n+\tau) = \mathrm{E}_a\{r_{zz}(n, n+\tau, a)\}$$
$$= \sum_{j=-\infty}^{n}\sum_{j'=-\infty}^{n+\tau} \mathrm{E}_s\{s(j)\,s(j')\}\,\mathrm{E}_a\{a(n-j)\,a(n+\tau-j')\}.$$
$$(6.49)$$

Note that the stochastic processes $a(j)$ and $s(n)$ are independent, and therefore can be separated. Using Eq. 6.38 and the fact that $b(j)$ consists of a zero-mean white Gaussian noise sequence, it follows that

$$E_a\{a(n-j)\,a(n+\tau-j')\} = \sigma^2\,e^{-2\rho n}\,e^{\rho(j+j'-\tau)}\,\delta(j-j'+\tau), \qquad (6.50)$$

where $\delta(\cdot)$ denotes the Kronecker delta function. Eq. 6.49 leads to

$$r_{zz}(n, n+\tau) = e^{-2\rho n} \sum_{j=-\infty}^{n} E_s\{s(j)\,s(j+\tau)\}\,\sigma^2\,e^{2\rho j}. \qquad (6.51)$$

The auto-correlation at time n can be divided into two terms. The first term depends on the early speech component between time $n - N_r + 1$ and n, whereas the second depends on the late reverberant speech component and is responsible for overlap-masking, i.e.,

$$r_{zz}(n, n+\tau) = e^{-2\rho n} \sum_{j=n-N_r+1}^{n} E_s\{s(j)\,s(j+\tau)\}\,\sigma^2\,e^{2\rho j}$$

$$+ e^{-2\rho n} \sum_{j=-\infty}^{n-N_r} E_s\{s(j)\,s(j+\tau)\}\,\sigma^2\,e^{2\rho j}. \qquad (6.52)$$

Let us now consider the spatially averaged auto-correlation at time $n - N_r$

$$r_{zz}(n-N_r, n-N_r+\tau) = e^{-2\rho(n-N_r)} \sum_{j=-\infty}^{n-N_r} E_s\{s(j)\,s(j+\tau)\}\,\sigma^2\,e^{2\rho j}. \qquad (6.53)$$

We can now see that the auto-correlation at time n can be expressed as

$$r_{zz}(n, n+\tau) = r_{z_e z_e}(n, n+\tau) + r_{z_r z_r}(n, n+\tau), \qquad (6.54)$$

with

$$r_{z_e z_e}(n, n+\tau) = e^{-2\rho n} \sum_{j=n-N_r+1}^{n} E_s\{s(j)\,s(j+\tau)\}\,\sigma^2\,e^{2\rho j}, \qquad (6.55)$$

$$r_{z_r z_r}(n, n+\tau) = e^{-2\rho N_r}\,r_{zz}(n-N_r, n-N_r+\tau). \qquad (6.56)$$

In practice the signals can be considered as stationary over periods of time that are short compared to the reverberation time T_{60}. This is justified by the fact that the exponential decay is very slow, and that speech is quasi-stationary. We consider that $N_r \ll T_{60} f_s$ and that N_r/f_s is larger than the time span over which the speech signal can be considered stationary, which is usually around 20-40 ms [16]. Under these assumptions, the counterparts of Eqs. 6.54 and 6.56 in terms of the spectral variances are:

$$\lambda_z(l,k) = \lambda_{z_e}(l,k) + \lambda_{z_r}(l,k), \qquad (6.57)$$

$$\lambda_{z_r}(l,k) = e^{-2\rho(k)N_r}\lambda_z(l - N_r/R, k), \qquad (6.58)$$

where R denotes the time shift (in samples) between two subsequent STFT frames. Note that the value N_r should be chosen such that N_r/R is an integer value, and that the decay-rate ρ is now frequency dependent. We can now estimate the spectral variance of the early speech signal $\lambda_{z_e}(l,k)$ by using the late reverberant spectral variance $\lambda_{z_r}(l,k)$.

6.4.5 Summary and Discussion

In the previous subsection we have developed a spectral variance estimator for the late reverberant signal component. The spectral variance estimator depends on the spectral variance of the reverberant speech signal $z(n)$ and the reverberation time T_{60} of the room.

Partially blind methods to estimate the reverberation time have been developed in which the characteristics of the room are estimated using neural network approaches [13]. Another method uses a segmentation procedure for detecting gaps in the signal, and tracks the sound decay curve [41]. Recently, a blind method was proposed by Ratnam et al. based on a maximum-likelihood estimation procedure [52]. Most of these methods can also be applied to sub-band signals in order to estimate the frequency dependent reverberation time. For some applications, e.g, audio or video-conferencing where a fix setup is used the reverberation time could also be obtained using a calibration process. Currently, the performance of blind estimation methods for speech that is degraded by background noise is still limited, and therefore its application can be problematic in some situations. However, in some applications the reverberation time can be estimated in a non-blind way, as will be shown in Sec. 6.5.

The complete algorithm for the post-processor depicted in Fig. 6.3 is summarized in Tab. 6.1.

6.5 Residual Echo Suppression

As explained in Sec. 6.2 the residual echo signal $e_r(n)$ is usually not zero. In this section a post-processor is developed that can be used in conjunction with an AEC. The post-processor suppresses the residual echo and background noise, and therefore enhances the near-end speech signal.

In Sec. 6.5.1 the problem of residual echo and noise suppression is formulated. In Sec. 6.5.2 an estimator for the spectral variance of the so-called late residual echo signal is derived. The developed spectral variance estimator requires two parameters. The estimators for these parameters are developed in Sec. 6.5.3. Finally, the algorithm for the developed post-processer is summarized in Sec. 6.5.4.

Table 6.1. Summary of the post-processor that suppresses late reverberation and background noise.

Number	Algorithmic part
1)	**STFT:** Calculate the STFT of the noisy and reverberant input signal $y(n)$.
2)	**Estimate background noise:** Estimate $\lambda_{v_\mathrm{s}}(l,k)$ using [11].
3)	**Estimate the reverberation time:** The reverberation time needs to be estimated online, e.g., using [13,41,52].
4)	**Estimate late reverberant energy:** Calculate $G_\mathrm{SP}(l,k)$ using Eqs. 6.43-6.45. Estimate $\lambda_z(l,k)$ using Eq. 6.46, and calculate $\widehat{\lambda}_{v_\mathrm{ns}}(l,k) = \widehat{\lambda}_{z_\mathrm{r}}(l,k)$ using Eq. 6.58.
5)	**Post-filter:** • Calculate the *a-posteriori* SIR using Eq. 6.16 and the individual *a priori* SIRs using Eqs. 6.31-6.32 with $\vartheta \in \{v_\mathrm{ns}, v_\mathrm{s}\}$, the total *a priori* SIR can then be calculated using Eq. 6.29. • Calculate the speech presence probability using (6.20). • Calculate the gain function $G_\mathrm{OM\text{-}LSA}(l,k)$ using Eqs. 6.18, 6.26, 6.20, and 6.21. • Calculate $\widehat{Z}_\mathrm{e}(l,k)$ using Eq. 6.22.
6)	**Inverse STFT:** Calculate the output $\hat{z}_\mathrm{e}(n)$ by applying the inverse STFT to $\widehat{Z}_\mathrm{e}(l,k)$.

6.5.1 Problem Formulation

In a reverberant room the length of the acoustic echo path N_h is approximately given by $f_\mathrm{s} T_{60}$ [8]. At a sampling frequency of 8 kHz, the length of the acoustic echo path in an office with a reverberation time of 0.5 seconds would be approximately 4000 coefficients. Due to practical reasons, e.g., complexity and required convergence rate, the length of the adaptive filter is smaller than the length of the acoustic echo path, i.e., $N_\mathrm{e} < N_\mathrm{h}$. Furthermore, movements in the room can have a large influence on the acoustic echo path. These changes will result in an increase in residual echo. The tail part of the acoustic echo path has a very specific structure. In this section we show that this structure can be exploited to estimate the spectral variance of the late residual echo

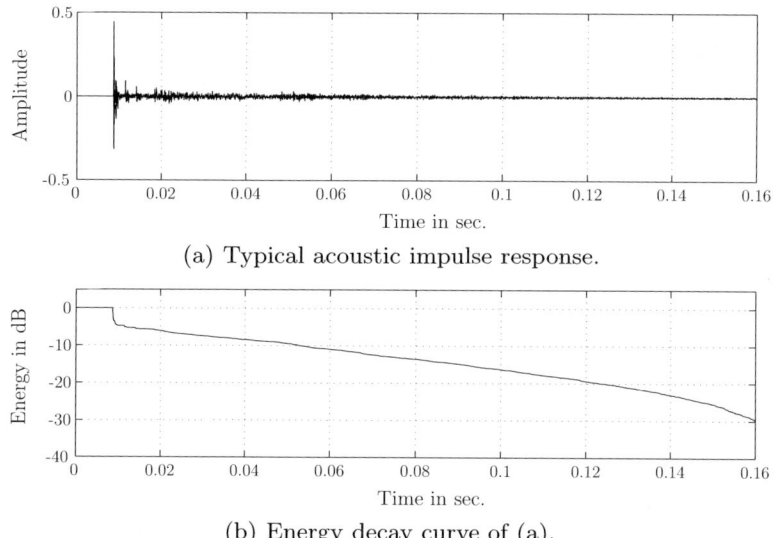

Fig. 6.4. Typical acoustic impulse response and related energy decay curve.

which is related to the part of the acoustic echo path that is not modelled by the adaptive filter.

In Fig. 6.4 a typical AIR and its energy decay curve (EDC) are depicted. The EDC is obtained by backward integration of the squared AIR [57], and is normalized with respect to the total energy of the AIR. In Fig. 6.4 we can see that the tail of the AIR exhibits an exponential decay, and the tail of the EDC exhibits a linear decay.

In the sequel we assume that the acoustic echo path is time-invariant and that it exhibits an infinite length, i.e., $N_h = \infty$. The late residual echo $e_r(n)$ can then be expressed as

$$e_r(n) = \sum_{j=0}^{\infty} h_r(j) \, x_r(n-j), \tag{6.59}$$

where $x_r(n) = x(n - N_e)$.

Enzner [18] proposed a recursive estimator for the short-term PSD of the late residual echo which is related to $\boldsymbol{h}_r = \begin{bmatrix} h(N_e), h(N_e+1), \ldots, h(N_h-1) \end{bmatrix}^T$. The recursive estimator exploits the fact that the exponential decay rate of the AIR is directly related to the reverberation time of the room, which can be estimated using the estimated echo path $\widehat{\boldsymbol{h}}_e$. Additionally, the recursive estimator requires a second parameter that specifies the initial power of the late residual echo. In Sec. 6.5.2 an essentially equivalent recursive estimator is derived, starting in the time-domain rather than directly in the frequency domain as in [18].

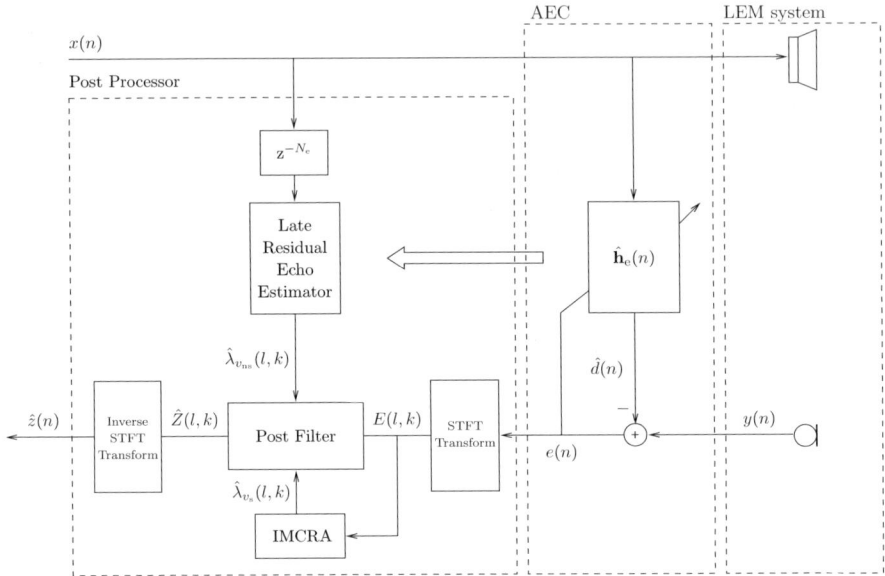

Fig. 6.5. Acoustic echo canceller with late residual echo and background noise suppression.

In Fig. 6.5 it is shown how the developed spectral variance estimator for the late residual echo can be used in the post-processor. First the error signal $e(n)$ is calculated by subtracting the estimated echo signal $\hat{d}(n)$ from the microphone signal $y(n)$. The error signal is first transformed to the STFT domain. The spectral component $E(l,k)$ is then used to estimate the spectral variance $\lambda_{v_s}(l,k)$ of the stationary interference using the IMCRA approach [11]. The spectral variance of the non-stationary interference $\lambda_{v_{ns}}(l,k)$ is equal to the spectral variance of the late residual echo signal $\lambda_{e_r}(l,k)$. The OM-LSA estimator presented in Sec. 6.3 can then be used to estimate the near-end spectral speech component $Z(l,k)$. The estimated near-end speech signal $\hat{z}(n)$ is then obtained using the inverse STFT and the weighted overlap-add method [14].

6.5.2 Late Residual Echo Spectral Variance Estimator

Using the statistical reverberation model introduced in Sec. 6.4.3 the spectral variance of the late residual echo can be estimated. The spectral variance of $e_r(n)$ is defined as

$$\lambda_{e_r}(l,k) \triangleq \mathrm{E}\Big\{\big|E_r(l,k)\big|^2\Big\}. \tag{6.60}$$

In the STFT domain we can approximate $E_r(l,k)$ by

$$E_r(l, k) \approx \sum_{i=0}^{\infty} H_r(i, k) \, X\left(l - i - N_e/R, k\right), \qquad (6.61)$$

where $H_r(i, k)$ is related to the STFT of $\boldsymbol{h}_r$, and i denotes the frame coefficient index. Note that N_e should be chosen such that N_e/R is an integer value.

Using Eqs. 6.60, 6.61, and the assumption that

$$\mathrm{E}\{H_r(i, k) \, H_r(i + \tau, k)\} = 0 \quad \forall \tau \neq 0, \qquad (6.62)$$

we can express $\lambda_{e_r}(l, k)$ as

$$\begin{aligned}\lambda_{e_r}(l, k) &\approx \mathrm{E}\left\{\sum_{i=0}^{\infty} |H_r(i, k)|^2 \, |X(l - i - N_e/R, k)|^2\right\} \\ &\approx \sum_{i=0}^{\infty} \mathrm{E}\{|H_r(i, k)|^2\} \, \mathrm{E}\{|X(l - i - N_e/R, k)|^2\}. \end{aligned} \qquad (6.63)$$

Using Polack's statistical reverberation model [49] the energy envelope of $H_r(i, k)$ can be expressed as

$$\mathrm{E}\{|H_r(i, k)|^2\} = c(k) \, \alpha^i(k), \qquad (6.64)$$

where R denotes the time shift (in samples) between two subsequent STFT frames, $c(k)$ denotes the initial power of $H_r(0, k)$ in the kth sub-band[3],

$$\alpha(k) \triangleq e^{-2\rho(k)R} \quad (0 \leq \alpha(k) < 1), \qquad (6.65)$$

where $\rho(k)$ denotes the exponential decay rate. The decay rate $\rho(k)$ is related to the frequency dependent reverberation time $T_{60}(k)$ through Eq. 6.39.

Using Eq. 6.64, and the fact that $\lambda_x(l, k) = \mathrm{E}\{|X(l, k)|^2\}$, Eq. 6.63 can be rewritten as

$$\lambda_{e_r}(l, k) \approx \sum_{i=0}^{\infty} c(k) \, \alpha^i(k) \, \lambda_x\left(l - i - N_e/R, k\right). \qquad (6.66)$$

In practice $\lambda_x(l, k)$ can be estimated using

$$\widehat{\lambda}_x(l, k) = \eta_x \, \widehat{\lambda}_x(l - 1, k) + (1 - \eta_x) \, |X(l, k)|^2, \qquad (6.67)$$

where η_x ($0 \leq \eta_x < 1$) denotes the smoothing parameter. We can now derive a recursive expression for $\lambda_{e_r}(l, k)$ for which only the spectral variance $\lambda_x(l - N_e/R, k)$ is required, i.e.,

$$\begin{aligned}\lambda_{e_r}(l, k) &\approx \sum_{i'=-\infty}^{l} c(k) \, \alpha^{l-i'}(k) \, \lambda_x\left(i' - N_e/R, k\right) \\ &\approx \alpha(k) \, \lambda_{e_r}(l - 1, k) + c(k) \, \lambda_x\left(l - N_e/R, k\right). \end{aligned} \qquad (6.68)$$

---

[3] Since we will estimate the quantity $c(k)$ later on by using information from the echo cancellation filter, it will be replaced by its time-variant smoothed estimate soon.

6.5.3 Parameter Estimation

In order to estimate the spectral variance $\lambda_{e_r}(l,k)$ we need to estimate $\alpha(k)$, which depends on the reverberation time T_{60} and the number of samples R between two subsequent STFT frames, and the initial power $c(k)$. In this section it is shown how both parameters can be estimated.

Reverberation Time

A well known technique to estimate the reverberation time was proposed by Schroeder in [57]. First the EDC of $\widehat{\boldsymbol{h}}_e(n)$ is calculated. Secondly, a straight line is fitted to a selected part of the EDC values to obtain the slope of the EDC. In our case, the last EDC values are not useful due to the finite length of $\widehat{\boldsymbol{h}}_e(n)$ and due to the final misalignment of the adaptive filter coefficients. Therefore, we use only a dynamic range of 20 dB[4] to determine the slope of the EDC. Finally, the reverberation time is updated using an adaptive scheme. A detailed description of this algorithm can be found in Tab. 6.2.

In general, the reverberation time T_{60} is frequency dependent due to frequency dependent reflection coefficients of walls and other objects and the frequency dependent absorption coefficient of air [40]. Instead of applying the above procedure to $\widehat{\boldsymbol{h}}_e(n)$, we can apply the above procedure to a band-pass filtered version of $\widehat{\boldsymbol{h}}_e(n)$. We used 1-octave band filters to acquire a higher frequency resolution. First, the reverberation time is estimated for each band-pass filtered version of $\widehat{\boldsymbol{h}}_e(n)$. Secondly, the subband reverberation time values are interpolated and extrapolated to obtain an estimate of $\widehat{T}_{60}(k)$ for each frequency bin k.

To reduce the complexity of the estimator we can estimate the reverberation time at regular intervals, i.e., for $n = uR_{\text{EDC}}$, where $u = 1, 2, \ldots$.

Using the estimated reverberation time $\widehat{T}_{60}(k)$ we can compute the decay-rate $\rho(k)$ using Eq. 6.39. The parameter $\alpha(k)$ can then be computed directly using its definition given in Eq. 6.65.

Initial Power

The estimated initial power $\hat{c}(l,k)$ can be calculated using the following expression

$$\hat{c}(l,k) = \left| \sum_{j=0}^{N_w-1} \widehat{h}_r(lR,j) \, e^{-\iota \frac{2\pi k}{N_{\text{DFT}}} j} \right|^2 \quad \text{for } k = \{0, \ldots, N_{\text{DFT}} - 1\}, \quad (6.69)$$

---

[4] It might be necessary to decrease the dynamic range when N_e is small or the reverberation time is short.

Table 6.2. Estimation of the reverberation time using $\widehat{\boldsymbol{h}}_e(n)$.

Number	Algorithmic part
1)	**Energy decay curve estimation:** Calculate the energy decay curve of $\widehat{\boldsymbol{h}}_e(n)$, where n equals $u\, R_{\text{EDC}}$ ($u = 1, 2, \ldots$) and R_{EDC} denotes the estimation rate, using $$EDC(u,m) = 20 \log_{10} \left\{ \sum_{j=m}^{N_e-1} \widehat{h}_e^2(u R_{\text{EDC}}, j) \right\} \quad \text{for } 0 \leq m \leq N_e - 1.$$
2)	**Straight line fit:** A straight line is fitted to part of the EDC data points, using a least squares approach. The line at time u is described by $p(u) + q(u)\, m$, where $p(u)$ and $q(u)$ denotes the offset and the regression coefficient of the line, respectively. The regression coefficient $q(u)$, of the line is obtained by minimizing the following cost function: $$J\big(p(u), q(u)\big) = \sum_{m=m_s}^{m_e} \Big(EDC(u,m) - \big(p(u) + q(u)\, m\big) \Big)^2,$$ where m_s ($0 \leq m_s < N_e - 1$) and m_e ($m_s < m_e \leq N_e - 1$) denote the start-time and end-time of EDC values that are used, respectively. A good choice for m_s and m_e is given by $$m_s = \operatorname*{argmin}_{m} \left| \frac{EDC(u,m)}{EDC(u,0)} + 5 \right|$$ and $$m_e = \operatorname*{argmin}_{m} \left| \frac{EDC(u,m)}{EDC(u,0)} + 25 \right|,$$ respectively.
3)	**Reverberation time:** The reverberation time $\widehat{T}_{60}(u)$ can now be calculated using $$\widehat{T}_{60}(u) = \widehat{T}_{60}(u-1) + \mu_{T_{60}} \left(\frac{60}{q(u) f_s} - \widehat{T}_{60}(u-1) \right),$$ where $\mu_{T_{60}}$ denotes the adaptation step-size.

where $\iota = \sqrt{-1}$, N_{DFT} denotes the length of the discrete Fourier transform (DFT), and N_w is the length of the analysis window. Since $\widehat{\boldsymbol{h}}_r(n)$ is not available, we use the last N_w coefficients of $\widehat{\boldsymbol{h}}_e(n)$ and extrapolate the energy using the estimated decay. We then obtain an estimate of $c(l,k)$ by

$$\hat{c}(l,k) = \alpha^{\frac{N_{\mathrm{w}}}{R}}(k) \left| \sum_{j=0}^{N_{\mathrm{w}}-1} \widehat{h}_{\mathrm{e}}(lR, N_{\mathrm{e}} - N_{\mathrm{w}} + j)\, e^{-\iota \frac{2\pi k}{N_{\mathrm{DFT}}} j} \right|^2$$

$$\text{for } k = \{0, \ldots, N_{\mathrm{DFT}} - 1\}. \quad (6.70)$$

The estimated initial power might contain some spectral zeros which will result in a zero spectral variance for these frequencies. The spectral zeros can easily be removed by smoothing $\hat{c}(l,k)$ along the frequency axis using

$$\tilde{c}(l,k) = \sum_{i=-w}^{w} b_i\, \hat{c}(l, k+i), \quad (6.71)$$

where b is a normalized window function with

$$\sum_{i=-w}^{w} b_i = 1, \quad (6.72)$$

that determines the frequency smoothing. To reduce the complexity of the late residual echo estimator, the initial power $\tilde{c}(l,k)$ can be updated at a lower rate.

6.5.4 Summary

In the previous subsections a spectral variance estimator for the late residual echo signal has been developed, together with two estimators for the required parameters. The estimated spectral variance can be used by the post-filter described in Sec. 6.3 to suppress the residual echo. The complete algorithm for the post-processor is summarized in Tab. 6.3.

6.6 Joint Suppression of Reverberation, Residual Echo, and Noise

In Sec. 6.4 and Sec. 6.5 it has been shown how the spectral variance of the late residual echo and late reverberation can be obtained by exploiting a statistical reverberation model. Although the dereverberation technique described in Sec. 6.4 is very appealing it requires an estimate of the reverberation time of the room. A major advantage of the hands-free application is that due to the existence of the acoustic echo the reverberation time can be estimated in a non-blind way (as described in Sec. 6.5.3). Since, it is commonly expected that the reverberation time in the enclosure is spatially invariant we propose to use the estimate T_{60} for the estimation of the late reverberant spectral variance.

In the joint system (see Fig. 6.6), the echo signal $d(n)$, the spectral variance $\lambda_{e_r}(l,k)$ of the late residual echo, and the spectral variance $\lambda_{v_s}(l,k)$ of

Table 6.3. Summary of the post-processor that suppresses late residual echo and background noise.

Number	Algorithmic part
1)	**Acoustic Echo Cancellation:** Update the adaptive filter $\hat{\boldsymbol{h}}_\mathrm{e}(n)$ and calculate $\hat{d}(n)$.
2)	**STFT:** Calculate the STFT of $e(n) = y(n) - \hat{d}(n)$, and $x(n)$.
3)	**Estimate Background Noise:** Estimate $\lambda_{v_\mathrm{s}}(l,k)$ using [11].
4)	**Estimate Reverberation Time:** Estimate T_{60} using the scheme presented in Tab. 6.2.
5)	**Estimate Late Residual Echo:** Calculate $\alpha(k)$ using Eq. 6.65, $\tilde{c}(l,k)$ using Eq. 6.71, and $\hat{\lambda}_{v_\mathrm{ns}}(l,k) = \hat{\lambda}_{e_r}(l,k)$ using Eq. 6.68.
6)	**Post-Filter:** Since the post-filter is applied to the output of the AEC we should replace $Y(l,k)$ by $E(l,k)$ in the hereafter mentioned equations. • Calculate the *a-posteriori* SIR using Eq. 6.16 and the individual *a priori* SIRs using Eqs. 6.31-6.32 with $\vartheta \in \{v_\mathrm{ns}, v_\mathrm{s}\}$, the total *a priori* SIR can then be calculated using Eq. 6.29. • Calculate the speech presence probability using Eq. 6.20. • Calculate the gain function $G_\text{OM-LSA}(l,k)$ using Eqs. 6.18, 6.26, 6.20, and 6.21. • Calculate $\hat{Z}(l,k)$ using Eq. 6.22.
7)	**Inverse STFT:** Calculate the output $\hat{z}(n)$ by applying the inverse STFT to $\hat{Z}(l,k)$.

the background noise are estimated first. In case the late reverberant spectral variance estimator is used in the hands-free application, it should be noted that the received microphone signal also contains the acoustic echo signal. Therefore, the estimated echo signal $\hat{d}(n)$ is first subtracted from the microphone signal $y(n)$. In case the error signal $e(n)$ would have been used directly to estimate the spectral variance $\lambda_{z_r}(l,k)$, the spectral variance will be biased due to the residual echo and the background noise. In Sec. 6.4 it has been shown how to cope with the background noise. To reduce the bias in the late reverberant spectral variance estimator, the spectral variance $\lambda_{e_r}(l,k)$ of the

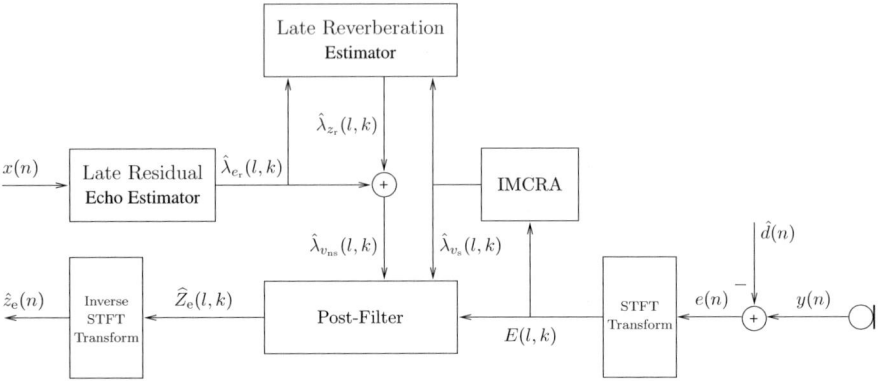

Fig. 6.6. Post-processor for the suppression of late reverberation, late residual echo and background noise.

late residual echo is included in Eqs. 6.44 and 6.45, such that

$$\xi_{\mathrm{SP}}(l,k) = \frac{\lambda_z(l,k)}{\lambda_{v_s}(l,k) + \lambda_{e_r}(l,k)}, \tag{6.73}$$

and

$$\gamma_{\mathrm{SP}}(l,k) = \frac{|E(l,k)|^2}{\lambda_{v_s}(l,k) + \lambda_{e_r}(l,k)}. \tag{6.74}$$

When $\lambda_{e_r}(l,k)$ and $\lambda_{z_r}(l,k)$ are estimated, the estimate of the spectral variance $\lambda_{v_{\mathrm{ns}}}(l,k)$ of the non-stationary interference can be obtained by

$$\lambda_{v_{\mathrm{ns}}}(l,k) = \lambda_{e_r}(l,k) + \lambda_{z_r}(l,k) \tag{6.75}$$

Now the post-filter described in Sec. 6.3 can be used to obtain an estimate of the partially dereverberated near-end speech signal $\widehat{Z}_e(l,k)$. The near-end speech signal $\hat{z}_e(n)$ can then be obtained using the inverse STFT and the weighted overlap-add method [14]. The complete algorithm for the joint suppression of reverberation, residual echo, and noise is summarized in Tab. 6.4.

6.7 Experimental Results

In this section, experimental results that demonstrate the beneficial use of the proposed spectral variance estimators and post-filter are presented. The experimental setup is described in Sec. 6.7.1. In Sec. 6.7.2 the dereverberation and noise suppression performance of the post-processor developed in Sec. 6.4 are evaluated. In Sec. 6.7.3 the late residual echo and noise suppression performance of the post-processor developed in Sec. 6.5 are analyzed. In

Table 6.4. Summary of the acoustic echo canceller and the developed post-processor for the suppression of late reverberation, late residual echo, and background noise.

Number	Algorithmic part
1)	**Acoustic Echo Cancellation:** Update the adaptive filter $\widehat{\boldsymbol{h}}_\mathrm{e}(n)$ and calculate $\hat{d}(n)$.
2)	**STFT:** Calculate the STFT of $e(n) = y(n) - \hat{d}(n)$, and $x(n)$.
3)	**Estimate Background Noise:** Estimate $\lambda_{v_\mathrm{s}}(l,k)$ using [11].
4)	**Estimate Reverberation Time:** Estimate T_{60} using the scheme presented in Tab. 6.2.
5)	**Estimate Late Residual Echo:** Calculate $\alpha(k)$ using Eq. 6.65, $\tilde{c}(l,k)$ using Eq. 6.71, and $\widehat{\lambda}_{e_r}(l,k)$ using Eq. 6.68.
6)	**Estimate Late Reverberant Energy:** Calculate $G_\mathrm{SP}(l,k)$ using Eq. 6.43 and Eqs. 6.73-6.74. Calculate $\widehat{\lambda}_z(l,k) = \left(G_\mathrm{SP}(l,k)\right)^2 \left\|E(l,k)\right\|^2$, and calculate $\widehat{\lambda}_{z_\mathrm{r}}(l,k)$ using Eq. 6.58.
7)	**Post-Filter:** Since the post-filter is applied to the output of the AEC we should replace $Y(l,k)$ by $E(l,k)$ in the hereafter mentioned equations. • Calculate the spectral variance of the non-stationary interference $v_\mathrm{ns}(n) = e_r(n) + z_r(n)$ using Eq. 6.75. • Calculate the *a-posteriori* SIR using Eq. 6.16 and the individual *a priori* SIRs using Eqs. 6.31 – 6.32 with $\vartheta \in \{v_\mathrm{ns}, v_\mathrm{s}\}$, the total *a priori* SIR can then be calculated using Eq. 6.29. • Calculate the speech presence probability using Eq. 6.20. • Calculate the gain function $G_\mathrm{OM\text{-}LSA}(l,k)$ using Eqs. 6.18, 6.26, 6.20, and 6.21. • Calculate $\widehat{Z}_\mathrm{e}(l,k)$ using Eq. 6.22.
8)	**Inverse STFT:** Calculate the output $\hat{z}_\mathrm{e}(n)$ by applying the inverse STFT to $\widehat{Z}_\mathrm{e}(l,k)$.

addition, the robustness with respect to changes in the tail of the acoustic echo path is evaluated. In Sec. 6.7.4, the performance of the entire system,

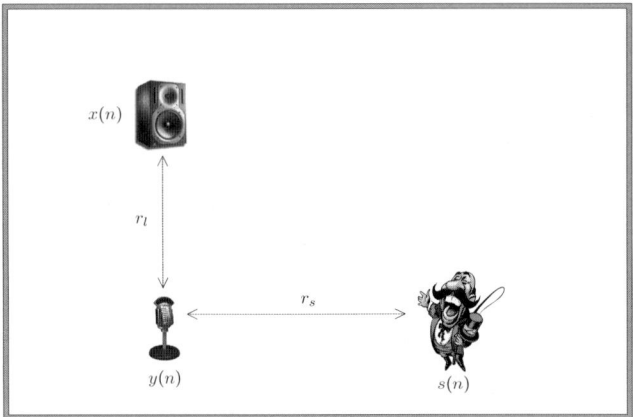

Fig. 6.7. Experimental setup.

as described in Sec. 6.6, is evaluated in case all interferences are present, i.e., during double-talk.

6.7.1 Experimental Setup

The experimental setup is depicted in Fig. 6.7. The room dimensions are 5 m × 4 m × 3 m (length × width × height). The distance between the near-end speaker and the microphone (r_s) is 1.5 m, the distance between the loudspeaker and microphone (r_l) is 0.25 m. All AIRs were generated using Allen and Berkley's image method [4,48]. The wall absorption coefficients were chosen such that the reverberation time is approximately 500 milliseconds. The microphone signal $y(n)$ was generated using Eq. 6.1. The analysis window $w(n)$ of the STFT is a 256 point Hamming window, i.e., $N_w = 256$, and the overlap between two successive frames is set to 75%, i.e., $R = 0.25\,N_w$. The remaining parameter settings are shown in Tab. 6.5. The additive noise $v(n)$ is speech-like noise, taken from the NOISEX-92 database [61].

6.7.2 Joint Suppression of Reverberation and Noise

The dereverberation performance is evaluated using the segmental SIR and the log spectral distance (LSD). The reverberation time is obtained using the AIR [57].

The instantaneous SIR of the lth frame is defined as

$$SIR(l) = 10\log_{10}\left\{\frac{\sum_{n=lR'}^{lR'+L'-1} z_e^2(n)}{\sum_{n=lR'}^{lR'+L'-1} \left(z_e(n) - v(n)\right)^2}\right\} \text{ dB}, \qquad (6.76)$$

where $v(n) \in \{y(n), \hat{z}_\mathrm{e}(n)\}$, $L' = 256$, and $R' = L'/4$. The segmental SIR is defined as the average instantaneous SIR over the set of frames where the desired speech is active.

The LSD between $z_\mathrm{e}(n)$ and the dereverberated signal is used as a measure of distortion. The distance in the lth frame is calculated using

$$LSD(l) = \frac{1}{K} \sum_{k=0}^{K-1} \left| 10 \log_{10} \left\{ \frac{\mathcal{C}\{|Z_\mathrm{e}(l,k)|^2\}}{\mathcal{C}\{|\Upsilon(l,k)|^2\}} \right\} \right| \text{ dB}, \qquad (6.77)$$

where $\Upsilon(l,k) \in \{Y(l,k), \widehat{Z}_\mathrm{e}(l,k)\}$, K denotes the number of frequency bins, and $\mathcal{C}\{|A(l,k)|^2\} \triangleq \max\{|A(l,k)|^2, \delta\}$ denotes a clipping operator which confines the log-spectrum dynamic range to about 50 dB, i.e., $\delta = 10^{-50/10} \cdot \max_{l,k}\{|A(l,k)|^2\}$. Finally, the LSD is defined as the average distance over all frames.

The dereverberation performance is tested under various segmental signal to noise ratios (SNRs) conditions. The segmental SNR value is determined by averaging the instantaneous SNR of those frames where the speech is active. Since the late reverberation is suppressed down to the residual background noise level the post-filter will always include the noise suppression. To show the improvement related to the dereverberation process the segmental SIR and LSD measures are evaluated for the unprocessed signal, the processed signal (noise suppression (NS) only), and the processed signal (noise and reverberation suppression (NS+RS)). The results, presented in Tab. 6.6, show that compared to the unprocessed signal the segmental SIR and LSD are improved in all cases.

The instantaneous SIR and LSD results obtained with a segmental SNR of 25 dB together with the anechoic, reverberant and processed signals are presented in Fig. 6.8. Since the SNR is relatively high, the instantaneous SIR mainly relates to the amount of reverberation, such that the SIR improvement is related to the reverberation suppression. The instantaneous SIR and LSD

Table 6.5. Parameters used for experiments.

Parameter	Value
f_s	8000 Hz
N_e	$0.128\, f_\mathrm{s}$
N_r	$0.032\, f_\mathrm{s}$
$G_{\min}^{\mathrm{dB}}$	18 dB
β^{dB}	3 dB
w	3

Table 6.6. Segmental SIR and LSD for different segmental signal to noise ratios.

Type of processing	Measured quantity		
	SNR$_{\text{seg}}$	SIR$_{\text{seg}}$	LSD
Unprocessed	5 dB	-4.89 dB	10.54 dB
	10 dB	-1.97 dB	7.90 dB
	25 dB	0.72 dB	4.88 dB
Processed (NS)	5 dB	0.20 dB	5.34 dB
	10 dB	0.68 dB	4.75 dB
	25 dB	1.00 dB	4.40 dB
Processed (RS+NS)	5 dB	2.01 dB	5.12 dB
	10 dB	3.36 dB	4.33 dB
	25 dB	4.45 dB	3.46 dB

are, respectively, increased and decreased, especially in those areas where the SIR of the unprocessed signal is low.

The spectrograms and waveforms of the speech signals $y(n)$, $z_e(n)$ and the processed signal $\hat{z}_e(n)$ are shown in Fig. 6.9. From these plots it can be seen that the smearing in time due to the reverberation has been reduced significantly.

6.7.3 Suppression of Residual Echo

To be able to suppress late residual echo an estimate of the spectral variance of the interference is required. The estimation of the spectral variance of the late residual echo signal $e_r(n)$ is therefore very important. In Sec. 6.5.2 such estimator has been derived. In Fig. 6.10 the spectrogram of the late residual echo $e_r(n)$ (as defined in Eq. 6.59) and the estimated spectral variance $\widehat{\lambda}_{e_r}(l,k)$ are depicted. The estimator is computational efficient and only requires two parameters, i.e., the initial power $\tilde{c}(l,k)$ and the, possibly frequency dependent, reverberation time T_{60}. Here, the two parameters were obtained using the method described in Sec. 6.5.3. The close resemblance between the two spectrograms demonstrates the effectiveness of the developed estimator.

The echo cancellation performance, and more specifically the improvement due to the post-filter, was evaluated using the ERLE. To be able to measure the ERLE this experiment was conducted without noise. For this experiment the gain function $G_{H0}(l,k)$ was equal to $G_{\min} = 10^{-60/20}$, i.e., the maximum suppression was set to 60 dB. The ERLE achieved by the adaptive filter was calculated using

Dereverberation and Residual Echo Suppression in Noisy Environments

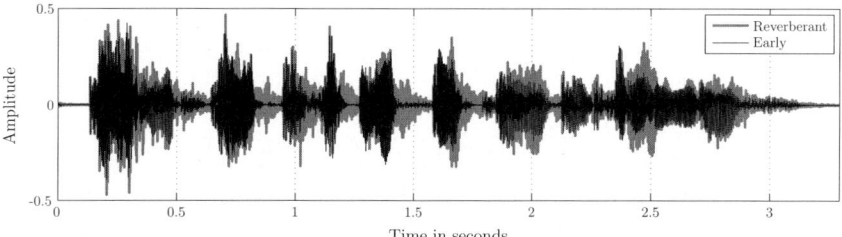

(a) Reverberant speech signal $z(n)$, and the early speech component $z_e(n)$.

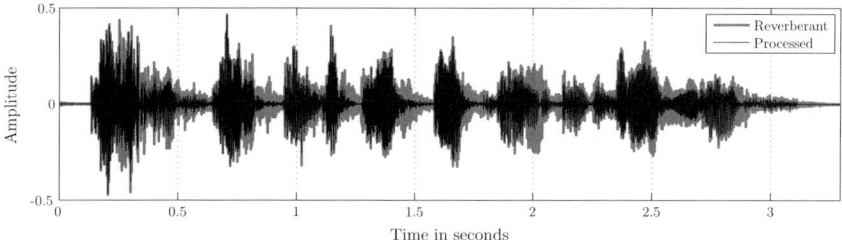

(b) Reverberant and processed speech signal.

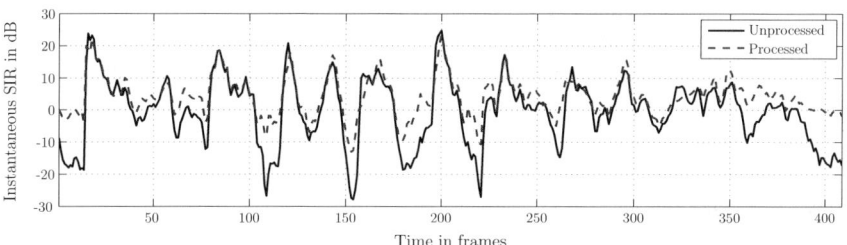

(c) SIR of the unprocessed and processed speech signal.

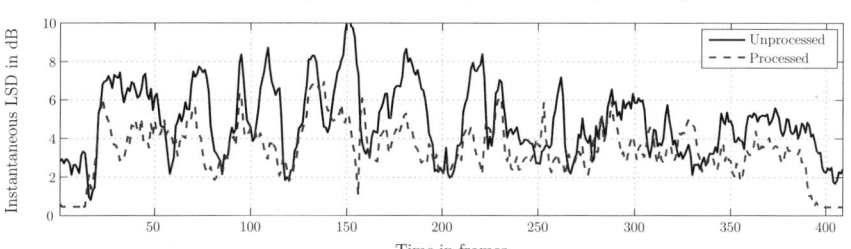

(d) LSD of the unprocessed and processed speech signal.

Fig. 6.8. Reverberation suppression performance of the post-processor described in Tab. 6.1 on page 204 ($T_{60} \approx 0.5$ s).

$$ERLE(l) = 10\log_{10}\left\{\frac{\sum_{n=lR'}^{lR'+L'-1} d^2(n)}{\sum_{n=lR'}^{lR'+L'-1} \left(d(n) - \hat{d}(n)\right)^2}\right\} \text{ dB}, \qquad (6.78)$$

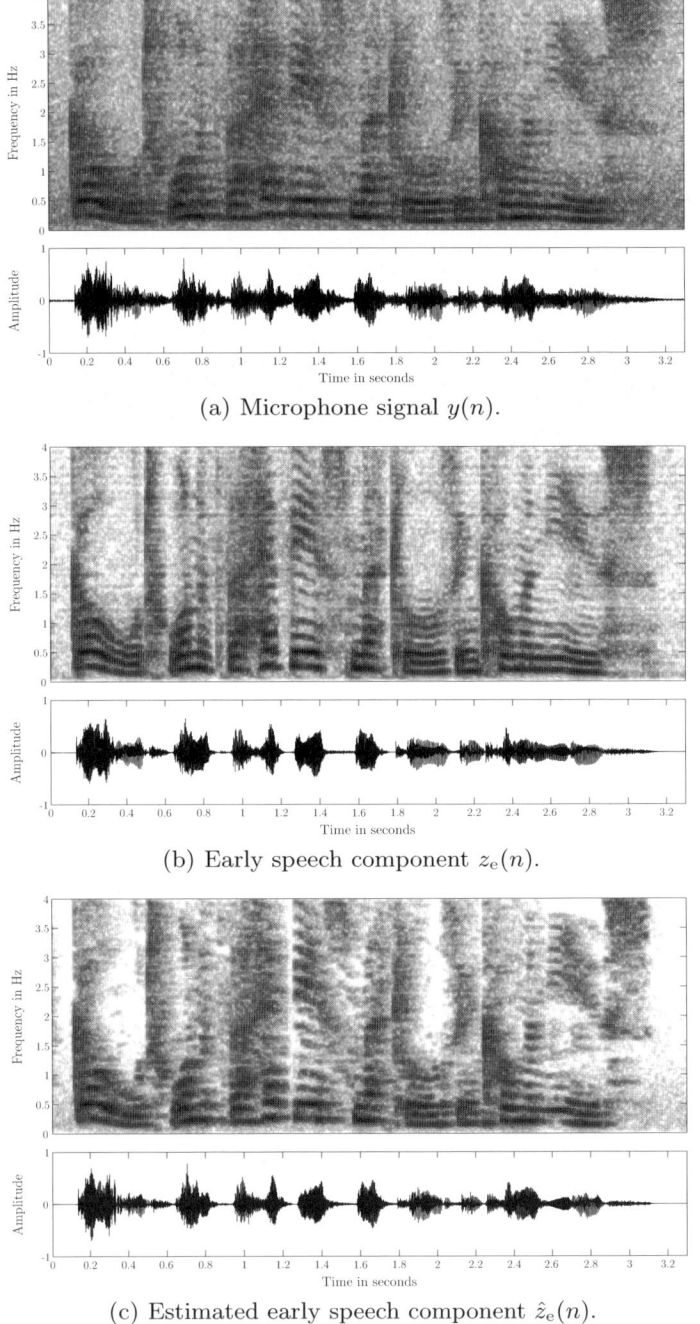

(a) Microphone signal $y(n)$.

(b) Early speech component $z_e(n)$.

(c) Estimated early speech component $\hat{z}_e(n)$.

Fig. 6.9. Spectrogram and waveform of (a) the microphone signal $y(n)$, (b) the speech signal $z_e(n)$, and (c) the dereverberated speech signal $\hat{z}_e(n)$ (segmental SNR = 25 dB, $T_{60} \approx 0.5$ s).

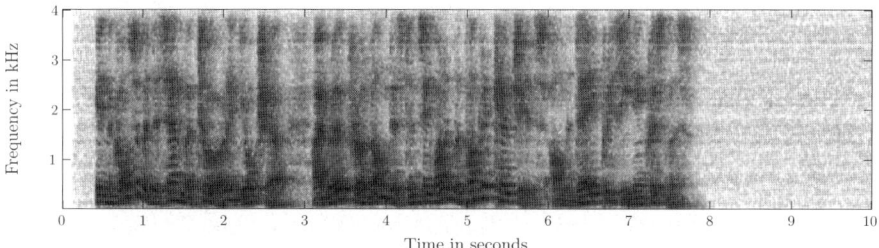

(a) The spectrogram of the late residual echo signal $e_\mathrm{r}(n)$ as defined in Eq. 6.59.

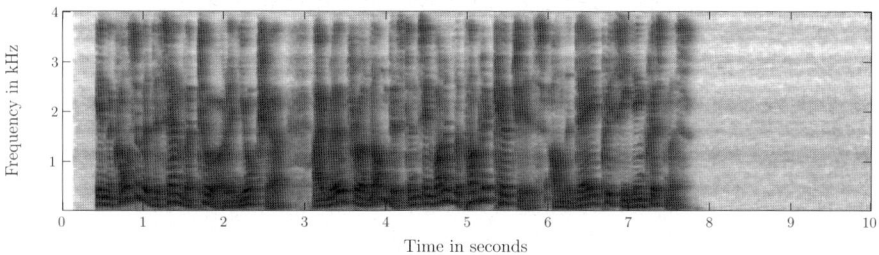

(b) The estimated spectral variance $\widehat{\lambda}_{e_\mathrm{r}}(l,k)$.

Fig. 6.10. Spectrograms of the true and estimated late residual echo signals. The spectral variance was estimated using the estimator developed in Sec. 6.5.2.

where $L' = 0.032 f_\mathrm{s}$ is the frame length and $R' = L'/4$ denotes the number of samples between two subsequent frames. To evaluate the total echo suppression, i.e., with post-filter, we calculated the ERLE using (6.78) and replaced $(d(n) - \hat{d}(n))$ by the residual echo at the output of the post-filter which is given by $(\hat{z}(n) - z(n))$. Note that by subtracting $z(n)$ from the output of the post-filter $\hat{z}(n)$, we avoid any bias in the ERLE due to the near-end speech signal. The final misalignment of the NLMS adaptive filter was -24 dB. It should be noted that the residual echo which is caused by the system mismatch of the adaptive filter cannot be compensated by the post-filter. The microphone signal $y(n)$, the error signal $e(n)$, and the ERLE with and without post-filter are shown in Fig. 6.11. We can see that the ERLE is significantly increased when the post-filter is used.

We evaluate the robustness of the developed late residual echo suppressor with respect to changes in the tail of the acoustic echo path when the far-end speech signal was active. Let us assume that the AEC is working perfectly at all times, i.e., the $\widehat{h}_c(n) = h_e(n)\ \forall\ n$. We have compared three systems:

i) the (perfect) AEC
ii) the AEC followed by an NLMS adaptive filter of length 1024 which compensates for the late residual echo, and
iii) the AEC followed by the developed post-filter.

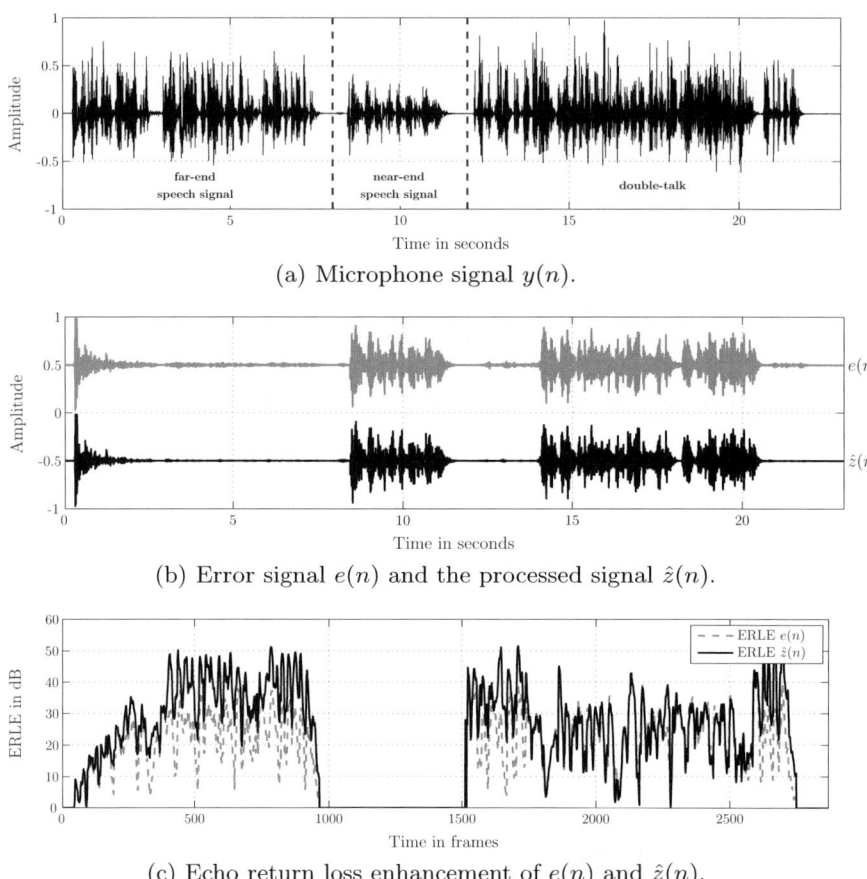

Fig. 6.11. Echo suppression performance of the AEC and the post-processor described in Sec. 6.5.

It should be noted that the total length of the linear time invariant filter that is used to cancel the echo in system ii is still shorter than the acoustic echo path. The output of system ii is denoted by $e'(n)$. At 4 seconds the acoustic echo path was changed by changing the position of the loudspeaker. Since our main objective is to study the robustness with respect to changes in the tail of the acoustic echo path we shifted the loudspeaker by 30 degrees (clockwise) with respect to the position of the microphone such that the direct path remains constant. The microphone signal $y(n)$, the error signal $e(n)$ of the standard AEC, the signals $e'(n)$ and $\hat{z}(n)$, and the ERLEs are shown in Fig. 6.12. The time at which the position of the loudspeaker changes is marked with a dash-dotted line. From the results we can see that the ERLEs of $e'(n)$ and $\hat{z}(n)$ are improved compared to the ERLE of $e(n)$. The late residual echo estimator is mainly based on the exponential decaying envelope of the AIR, which does

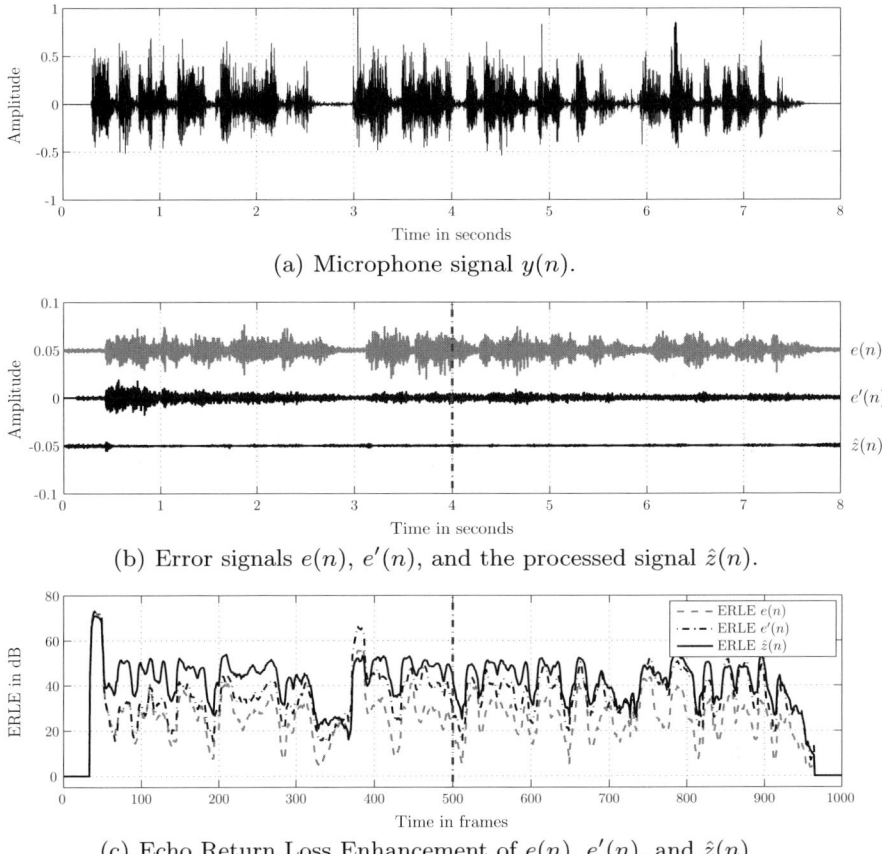

Fig. 6.12. Echo suppression performance with respect to echo path changes.

not change over time. Hence, the estimator does not suffer from the change in the tail of the acoustic echo path. In addition, the estimator does not require any convergence time. Furthermore, in practice the adaptive filter might not even be able to converge due to the echo to near-end speech plus noise ratio of the microphone signal $y(n)$.

6.7.4 Joint Suppression of Reverberation, Residual Echo, and Noise

We now evaluate the performance of the post-processor described in Tab. 6.4 during double-talk. The reverberation time $\widehat{T}_{60}$ was obtained using the procedure described in Sec. 6.5.3. After convergence of the adaptive filters T_{60} was 493 ms. The performance is evaluated at three different segmental SNR values

Table 6.7. Segmental SIR and LSD for different segmental signal to noise ratios during double-talk.

Type of processing	Measured quantity		
	SNR_{seg}	SIR_{seg}	LSD
Unprocessed	5 dB	-10.65 dB	13.53 dB
	10 dB	-9.73 dB	11.93 dB
	25 dB	-9.17 dB	10.38 dB
AEC	5 dB	-5.96 dB	11.14 dB
	10 dB	-3.12 dB	8.21 dB
	25 dB	-0.52 dB	4.27 dB
AEC + Post-Filter (NS)	5 dB	-1.56 dB	5.00 dB
	10 dB	-0.37 dB	3.80 dB
	25 dB	0.20 dB	3.17 dB
AEC + Post-Filter (NS+RES)	5 dB	-1.34 dB	4.71 dB
	10 dB	-0.20 dB	3.57 dB
	25 dB	0.57 dB	2.87 dB
AEC + Post-Filter (NS+RES+RS)	5 dB	0.11 dB	4.59 dB
	10 dB	2.08 dB	2.36 dB
	25 dB	3.68 dB	2.27 dB

using the segmental SIR and the LSD. Since all non-stationary interferences are reduced down to the residual background noise level, the background noise is suppressed first. The intermediate results that were obtained using

i) the AEC,
ii) the AEC and post-filter (noise suppression (NS)),
iii) the AEC and post-filter (noise and residual echo suppression(NS+RES)), and
iv) the AEC and post-filter (noise, residual echo, and reverberation suppression (NS+RES+RS))

are presented in Tab. 6.7. These results show a consistent and significant improvement in terms of segmental SIR and LSD.

The spectrograms of the microphone signal, near-end speech signal and the estimated signal $\hat{z}_e(n)$ for a segmental SNR of 25 dB, are shown in Fig. 6.13. The spectrograms demonstrate how well the interferences are suppressed during double-talk.

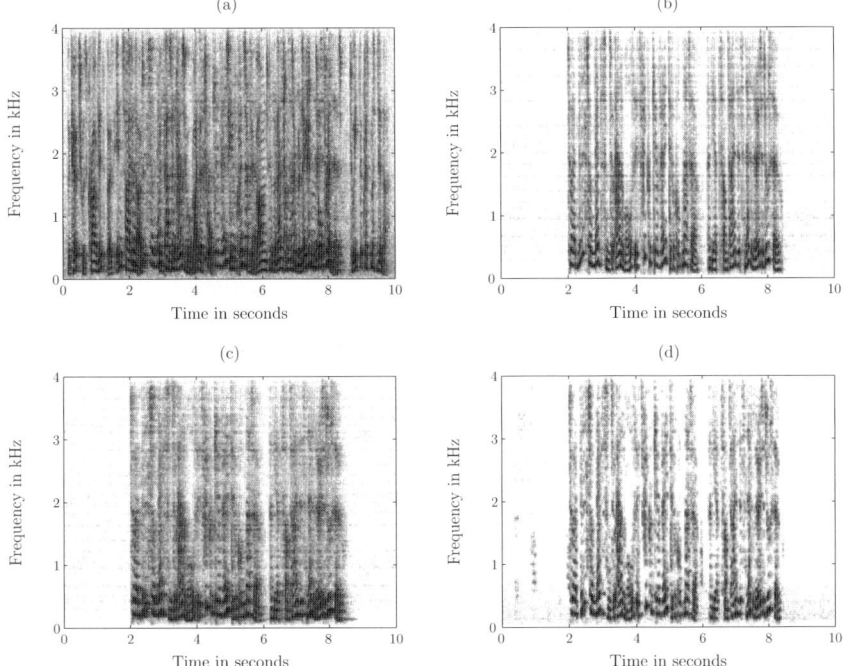

Fig. 6.13. Spectrograms of (a) the microphone signal $y(n)$, (b) the near-end speech signal $z_e(n)$, (c) the reverberant near-end speech signal $z(n)$, and (d) the estimated signal $\hat{z}_e(n)$, during double-talk (segmental SNR = 25 dB, $T_{60} \approx 0.5$ s).

6.8 Summary and Outlook

We have described a general purpose post-filter that is suitable for suppression of stationary and non-stationary interferences. The post-filter requires an estimate of the spectral variance of these interferences. Spectral variance estimators for two important non-stationary interferences, i.e., late reverberation and late residual echo, have been developed. Both estimators exploit the exponentially decaying structure of the AIR by using a statistical reverberation model. The decay rate of the AIR is directly related to the reverberation time of the room, and is the most important model parameter. A major advantage of the acoustic echo cancellation application is that due to the existence of the echo an estimate of the reverberation time can be obtained in a non-blind way. A post-processor has been developed which can effectively and efficiently suppress late reverberation, late residual echo, and background noise. Experimental results demonstrate the performance of the developed post-processor, and its robustness to small position changes in the room.

Polack's statistical reverberation model that is used in this chapter is only valid when the source-receiver distance is larger than the critical distance. When the source-receiver distance is smaller than the critical distance the energy of the direct path is larger than the energy of all reflections. In this case the described spectral variance estimator will be biased. In many hands-free applications the source-receiver distance can be smaller than the critical distance. Recently, a generalized statistical model was developed [31], which includes the contribution of the direct path. The estimator which is based on this model can be used in case the source-receiver distance is smaller than the critical distance. The generalized model requires an additional parameter which is related to the direct to reverberation ratio. Blind estimation of this parameter is a topic for future research.

Extending the developed post-processor to the case where a microphone array is available, rather than a single microphone, is another topic for future research. The ability to combine temporal-spatial and spectral techniques may further enhance the ability to reduce coherent and diffuse background noise and reverberation.

References

1. A. J. Accardi, R. V. Cox: A modular approach to speech enhancement with an application to speech coding, *Proc. ICASSP '99*, **1**, 201–204, 1999.
2. F. Aigner, M. J. O. Strutt: On a physiological effect of several sources of sound on the ear and its consequences in architectural acoustics, *JASA*, **6**(3), 155–159, 1935.
3. J. B. Allen, L. Radiner: A unified approach to short-time fourier analysis and synthesis, *Proc. of the IEEE*, **65**(11), 1558–1564, 1977.
4. J.B. Allen, D.A. Berkley: Image method for efficiently simulating small room acoustics, *JASA*, **65**(4), 943–950, 1979.
5. J. B. Allen: Effects of small room reverberation on subjective preference, *JASA*, **71**, 1–5, 1982.
6. P. Bloom, G. Cain: Evaluation of two input speech dereverberation techniques, *Proc. ICASSP '82*, **1**, 164–167, 1982.
7. R. Le Bouquin Jeannès, P. Scalart, G. Faucon, C. Beaugeant: Combined noise and echo reduction in hands-free systems: a survey, *IEEE Trans. Speech Audio Processing*, **9**(8), 808–820, 2001.
8. C. Breining, P. Dreiseitel, E. Hänsler, A. Mader, B. Nitsch, H. Puder, T. Schertler, G. Schmidt, J. Tilp: Acoustic echo control – an application of very-high-order adaptive filters, *IEEE Signal Processing Mag.*, **16**(4), 42–69, 1999.
9. I. Cohen, B. Berdugo: Noise estimation by minima controlled recursive averaging for robust speech enhancement, *IEEE Signal Processing Lett.*, **9**(1), 12–15, Jan. 2002.
10. I. Cohen: Optimal speech enhancement under signal presence uncertainty using log-spectral amplitude estimator, *IEEE Signal Processing Lett.*, **9**(4), 113–116, Apr. 2002.

11. I. Cohen: Noise spectrum estimation in adverse environments: Improved minima controlled recursive averaging, *IEEE Trans. Speech Audio Processing*, **11**(5), 466–475, Sept. 2003.
12. I. Cohen: Relaxed statistical model for speech enhancement and a priori SNR estimation, *IEEE Trans. Speech Audio Processing*, **13**(5), 870–881, Sept. 2005.
13. T. J. Cox, F. Li, P. Darlington: Extracting room reverberation time from speech using artificial neural networks, *Journal of the Audio Engineering Society*, **49**(4), 219–230, 2001.
14. R. E. Crochiere, L. R. Rabiner: *Multirate Digital Signal Processing*, Englewood Cliffs, NJ, USA: Prentice-Hall, 1983.
15. M. Delcroix, T. Hikichi, M. Miyoshi: Precise dereverberation using multichannel linear prediction, *IEEE Trans. Audio, Speech, Language Processing*, **15**(2), 430–440, 2006.
16. J. R. Deller, J. G. Proakis, J. H. L. Hansen: *Discrete-Time Processing of Speech Signals*, New York, NY, USA: MacMillan, 1993.
17. M. Dörbecker, S. Ernst: Combination of two-channel spectral subtraction and adaptive Wiener post-filtering for noise reduction and dereverberation, *Proc. EUSIPCO '96*, Triest, Italy, 1996.
18. G. Enzner: *A Model-Based Optimum Filtering Approach to Acoustic Echo Control: Theory and Practice*, Ph.D. thesis, RWTH Aachen University, Aachen, Germany: Wissenschaftsverlag Mainz, 2006.
19. Y. Ephraim, D. Malah: Speech enhancement using a minimum mean square error short-time spectral amplitude estimator, *IEEE Trans. Acoust., Speech, Signal Processing*, **32**(6), 1109–1121, Dec. 1984.
20. Y. Ephraim, D. Malah: Speech enhancement using a minimum mean square error log-spectral amplitude estimator, *IEEE Trans. Acoust., Speech, Signal Processing*, **33**(2), 443–445, Apr. 1985.
21. S. Gannot, M. Moonen: Subspace methods for multimicrophone speech dereverberation, *EURASIP Journal on Applied Signal Processing*, **11**, 1074–1090, 2003.
22. T. Gänsler, J. Benesty: The fast normalized cross-correlation double-talk detector, *Signal Processing*, **86**, 1124–1139, June 2006.
23. N. D. Gaubitch, P. A. Naylor, D. Ward: On the use of linear prediction for dereverberation of speech, *Proc. IWAENC '03*, 99–102, Kyoto, Japan, 2003.
24. Y. Grenier, M. Xu, J. Prado, D. Liebenguth: Real-time implementation of an acoustic antenna for audio-conference, *Proc. IWAENC '89*, Berlin, Germany, Sept. 1989.
25. S. Griebel, M. Brandstein: Wavelet transform extrema clustering for multichannel speech deverberation, *Proc. WASPAA '99*, 1999.
26. M. Gürelli, C. Nikias: EVAM: an eigenvector-based algorithm for multichannel blind deconvolution of input colored signals, *IEEE Trans. Signal Processing*, **43**(1), 134–149, 1995.
27. S. Gustafsson, R. Martin, P. Vary: Combined acoustic echo control and noise reduction for hands-free telephony, *Signal Processing*, **64**, 21–32, 1998.
28. S. Gustafsson, R. Martin, P. Jax, P. Vary: A psychoacoustic approach to combined acoustic echo cancellation and noise reduction, *IEEE Trans. Speech Audio Processing*, **10**(5), 245–256, 2002.
29. E. A. P. Habets: Multi-channel speech dereverberation based on a statistical model of late reverberation, *Proc. ICASSP '05*, 173–176, Philadelphia, USA, Mar. 2005.

30. E. A. P. Habets, I. Cohen, S. Gannot: MMSE log spectral amplitude estimator for multiple interferences, *Proc. IWAENC '06*, 1–4, Paris, France, Sept. 2006.
31. E.A.P. Habets: Single- and Multi-Microphone Speech Dereverberation using Spectral Enhancement, *Ph.D. Thesis*, Technische Universiteit Eindhoven, June 2007.
32. E. Hänsler: The hands-free telephone probleman annotated bibliography, *Signal Processing*, **27**(3), 259–271, 1992.
33. E. Hänsler, G. Schmidt: Hands-free telephones – joint control of echo cancellation and postfiltering, *Signal Processing*, **80**, 2295–2305, 2000.
34. E. Hänsler, G. Schmidt: *Acoustic Echo and Noise Control: A Practical Approach*, Hoboken, NJ, USA: Wiley-IEEE Press, 2004.
35. S. Haykin: *Blind Deconvolution,* fourth ed., Englewood Cliffs, NJ, USA: Prentice-Hall, 1994.
36. J. Hopgood: *Nonstationary Signal Processing with Application to Reverberation Cancellation in Acoustic Environments*, Ph.D. thesis, Cambridge University, 2001.
37. Y. Huang, J. Benesty: A class of frequency-domain adaptive approaches to blind multichannel identification, *IEEE Trans. Signal Processing*, **51**(1), 11–24, Jan. 2003.
38. J. J. Jetzt: Critical distance measurement of rooms from the sound energy spectral response, *JASA*,**65**(5), 1204–1211, 1979.
39. J.-M. Jot, L. Cerveau, O. Warusfel: Analysis and synthesis of room reverberation based on a statistical time-frequency model, *Audio Engineering Society, 103th Convention*, Aug. 1997.
40. H. Kuttruff: *Room Acoustics,* fourth ed., London, GB: Spon Press, 2000.
41. K. Lebart, J. M. Boucher, P. N. Denbigh: A new method based on spectral subtraction for speech dereverberation, *Acta Acoustica*, **87**(3), 359–366, 2001.
42. R. Martin, P. Vary: Combined acoustic echo cancellation, dereverberation and noise reduction: a two microphone approach, *Annales des Telecommunications*, **49**(7-8), 429–438, 1994.
43. R. Martin: Noise power spectral density estimation based on optimal smoothing and minimum statistics, *IEEE Trans. Speech Audio Processing*, **9**(5), 504–512, 2001.
44. M. Miyoshi, Y. Kaneda: Inverse filtering of room acoustics, *IEEE Trans. Speech Audio Processing*, **36**(2), 145–152, 1988.
45. V. Myllylä: Residual echo filter for enhanced acoustic echo control, *Signal Processing*, **86**(6), 1193–1205, June 2006.
46. A. K. Nábělek, T. R. Letowski, F. M. Tucker: Reverberant overlap- and self-masking in consonant identification, *JASA*, **86**(4), 1259–1265, 1989.
47. P. A. Naylor, N.D. Gaubitch: Speech dereverberation, *Proc. IWAENC '05*, 2005.
48. P. M. Peterson: Simulating the response of multiple microphones to a single acoustic source in a reverberant room, *JASA*, **80**(5), 1527–1529, Nov. 1986.
49. J. D. Polack: *La transmission de l'énergie sonore dans les salles,* Thèse de doctorat d'etat, Université du Maine, La mans, 1988.
50. J. D. Polack: Playing billiards in the concert hall: the mathematical foundations of geometrical room acoustics, *Applied Acoustics*, **38**(2), 235–244, 1993.
51. B. D. Radlović, R. Williamson, R. Kennedy: Equalization in an acoustic reverberant environment: robustness results, *IEEE Trans. Speech Audio Processing*, **8**(3), 311–319, 2000.

52. R. Ratnam, D. L. Jones, B. C. Wheeler, W. D. O'Brien Jr., C. R. Lansing, A. S. Feng: Blind estimation of reverberation time, *JASA*, **114**(5), 2877–2892, Nov. 2003.
53. W. C. Sabine: *Collected Papers on Acoustics (Originally 1921)*, Los Altos, CA, USA: Peninsula Publishing, 1993.
54. G. Schmidt: Applications of acoustic echo control - an overview, *Proc. EUSIPCO '04*, Vienna, Austria, 2004.
55. M. R. Schroeder: Statistical parameters of the frequency response curves of large rooms, *Journal of the Audio Engineering Society*, **35**, 299–306, 1954.
56. M. R. Schroeder: Frequency correlation functions of frequency responses in rooms, *JASA*, **34**(12), 1819–1823, 1962.
57. M. R. Schroeder: Integrated-impulse method measuring sound decay without using impulses, *JASA*, **66**(2), 497–500, 1979.
58. F. Talantzis, D. B. Ward: Robustness of multichannel equalization in an acoustic reverberant environment, *JASA*, **114**(2), 833–841, 2003.
59. M. Triki, D. T. M. Slock: Delay and predict equalization for blind speech dereverberation, *Proc. ICASSP '06*, **5**, 97–100, Toulouse, France, May 2006.
60. V. Turbin, A. Gilloire, P. Scalart: Comparison of three post-filtering algorithms for residual acoustic echo reduction, *Proc. ICASSP '97*, **1**, 307–310, 1997.
61. A. Varga, H. J. M. Steeneken: Assessment for automatic speech recognition: II. NOISEX-92: a database and an experiment to study the effect of additive noise on speech recognition systems, *Speech Communication*, **12**, 247–251, July 1993.
62. J. Y. C. Wen, N. D. Gaubitch, E. A. P. Habets, T. Myatt, P. A. Naylor: Evaluation of speech dereverberation algorithms using the MARDY database, *Proc. IWAENC '06*, 1–4, Paris, France, Sept. 2006.
63. M. Xu, Y.Grenier: Acoustic echo cancellation by adaptive antenna, *Proc. IWAENC '89*, Berlin, Germany, Sept. 1989.
64. H. Yasukawa: An acoustic echo canceller with sub-band noise cancelling, *IEICE Trans. on Fundamentals of Electronics, Communications and Computer Sciences*, **E75-A**(11), 1516–1523, 1992.
65. B. Yegnanarayana, P. S. Murthy: Enhancement of reverberant speech using LP residual signal, *IEEE Trans. Speech Audio Processing*, **8**(3), 267–281, 2000.

7

Low Distortion Noise Cancellers – Revival of a Classical Technique

Akihiko Sugiyama

NEC Corporation, Japan

This chapter presents low-distortion noise cancellers with their applications to communications and speech recognition. This classical technique, originally proposed by Widrow et al. in mid 70's, is first reviewed from a view point of output-signal distortion to show that interference and crosstalk are the primary reasons. As a solution to the interference problem, a paired filter (PF) structure introduces an auxiliary adaptive filter for estimating a signal-to-noise ratio (SNR) that is used to control the coefficient-adaptation stepsize in the main adaptive filter. A small stepsize for high SNRs, when the desired signal seriously interferes the misadjustment, provides steady and accurate change of coefficients, leading to low-distortion. This PF structure is extended to more general cases in which crosstalk from the desired-signal source to the auxiliary microphone is not negligible. A cross-coupled paired filter (CCPF) structure and its generalized version are solutions that employ another set of paired filters. The generalized CCPF (GCCPF) is applied to speech recognition in a human-robot communication scenario where improvement in distortion is successfully demonstrated by evaluations in the real environment. This robot had been demonstrated for six months at 2005 World Exposition in Aichi, Japan.

7.1 Introduction

Adaptive noise cancellers (ANCs) were first proposed by Widrow et al. for two-microphone speech enhancement [22]. Widrow's ANC has two microphones. The primary microphone captures a mixture of a desired signal and noise. A secondary (or reference) microphone is placed sufficiently close to the noise source to pick up a reference noise. The reference noise drives an adaptive filter to generate a noise replica at the primary microphone. By subtracting the noise replica from the primary microphone signal, an enhanced speech is obtained as the output. The output contains residual noise and is used for coefficient adaptation. Due to the auxiliary information by the reference microphone, Widrow's ANC is a more effective technique, especially in low SNR environments, than single-microphone noise suppressors based on, for example, spectral subtraction (SS) [2] and minimum mean squared error short time spectral amplitude (MMSE STSA) estimation [6]. However, the quality of the output speech may be degraded for two reasons, namely, interference and crosstalk.

Interference to coefficient adaptation is caused by the desired signal. Coefficients should be updated by the residual noise that is the difference between the noise at the primary microphone and the noise replica. However, the residual noise cannot be obtained separately from the desired signal. It serves as an interference when the error, which is composed of the residual noise and the desired signal, is used for coefficient adaptation of the ANC. As a result, the performance of the ANC is limited [22] with insufficient noise cancellation and distortion in the desired signal.

Crosstalk happens at the reference microphone. In Widrow's ANC, it is assumed that the reference microphone is placed sufficiently close to the noise source. It is necessary for the reference microphone to pick up only the noise. However, this assumption is often violated in reality. There are desired-signal components that leak into the reference microphone. Such a leak-in signal is called crosstalk. Crosstalk contaminates the reference noise that is the adaptive-filter input in Widrow's ANC. The adaptive-filter output is no longer a good replica of the noise and the ANC output may be distorted.

This chapter presents low-distortion noise cancellers as solutions to the interference and the crosstalk problems. In Sec. 7.2, interference and crosstalk are investigated from a viewpoint of distortion in the output signal. A solution to the interference problem is presented in Sec. 7.3. Some ANCs are described in Sec. 7.4 in search of a good structure for the crosstalk problem. Finally, Secs. 7.5 and 7.6 are devoted to more advanced ANC structures for successful applications.

7.2 Distortions in Widrow's Adaptive Noise Canceller

7.2.1 Distortion by Interference

Fig. 7.1 shows a block diagram of Widrow's ANC. $s(n)$, $n(n)$ and $n_1(n)$ are the signal, the noise, and the noise component in the primary-microphone signal, all with a time index n. $\boldsymbol{h}(n)$ represents the impulse response of the noise path from the noise source to the primary microphone. The primary signal $x_P(n)$ and the reference signal $x_R(n)$ can be written as

$$x_P(n) = s(n) + n_1(n), \tag{7.1}$$
$$x_R(n) = n(n). \tag{7.2}$$

The output $e_1(n)$ of the ANC is given by

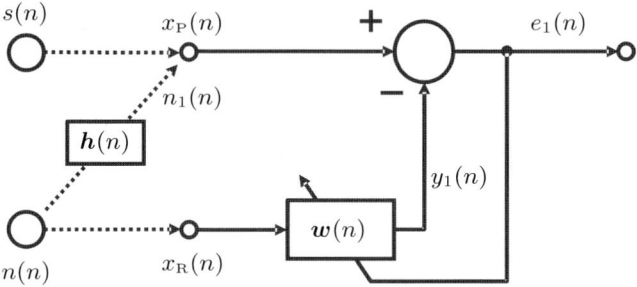

Fig. 7.1. Widrow's adaptive noise canceller.

$$e_1(n) = s(n) + n_1(n) - y_1(n), \qquad (7.3)$$
$$n_1(n) = \boldsymbol{h}^{\mathrm{T}}(n)\,\boldsymbol{n}(n), \qquad (7.4)$$
$$y_1(n) = \boldsymbol{w}^{\mathrm{T}}(n)\,\boldsymbol{x}_{\mathrm{R}}(n) = \boldsymbol{w}^{\mathrm{T}}(n)\,\boldsymbol{n}(n), \qquad (7.5)$$

where $n_1(n)$ and $y_1(n)$ are the noise component in $x_\mathrm{P}(n)$ and the output of the adaptive filter. $\boldsymbol{w}(n)$ is the coefficient vector of the adaptive filter, and $\boldsymbol{h}(n)$ is the impulse-response vector of the noise path. $\boldsymbol{x}_\mathrm{R}(n)$ and $\boldsymbol{n}(n)$ are the reference signal and the noise vectors with a size of N. These vectors are defined by

$$\boldsymbol{h}^{\mathrm{T}}(n) = \big[h_0(n),\, h_1(n),\, \cdots,\, h_{N-1}(n)\big], \qquad (7.6)$$
$$\boldsymbol{w}^{\mathrm{T}}(n) = \big[w_0(n),\, w_1(n),\, \cdots,\, w_{N-1}(n)\big], \qquad (7.7)$$
$$\boldsymbol{n}^{\mathrm{T}}(n) = \big[n(n),\, n(n-1),\, \cdots,\, n(n-N+1)\big], \qquad (7.8)$$
$$\boldsymbol{x}_{\mathrm{R}}^{\mathrm{T}}(n) = \big[x_\mathrm{R}(n),\, x_\mathrm{R}(n-1),\, \cdots,\, x_\mathrm{R}(n-N+1)\big] = \boldsymbol{n}^{\mathrm{T}}(n). \qquad (7.9)$$

Eqs. 7.3 – 7.5 reduce to

$$e_1(n) = s(n) + \big[\boldsymbol{h}(n) - \boldsymbol{w}(n)\big]^{\mathrm{T}} \boldsymbol{n}(n). \qquad (7.10)$$

Assuming that the noise path $\boldsymbol{h}(n)$ is estimated with the NLMS algorithm [9], the update of $\boldsymbol{w}(n)$ is performed by

$$\boldsymbol{w}(n+1) = \boldsymbol{w}(n) + \frac{\mu\, e_1(n)\, \boldsymbol{x}_\mathrm{R}(n)}{\|\boldsymbol{x}_\mathrm{R}(n)\|^2} = \boldsymbol{w}(n) + \frac{\mu\, e_1(n)\, \boldsymbol{n}(n)}{\|\boldsymbol{n}(n)\|^2}, \qquad (7.11)$$

where μ is a stepsize. From Eqs. 7.3 and 7.10, it can be seen that $e_1(n)$ is close to $s(n)$ when $y_1(n) \approx n_1(n)$ or equivalently, $\boldsymbol{h}(n) \approx \boldsymbol{w}(n)$ near convergence. Because $e_1(n)$ is the output, the desired signal $s(n)$ is obtained at the output after convergence. However, the misadjustment $y_1(n) - n_1(n)$ which is needed for coefficient adaptation is severely contaminated by the desired signal $s(n)$, resulting in signal-distortion such as reverberation [3]. This distortion can be removed if the coefficient update is performed only in the absence of the desired signal $s(n)$. However, it requires an accurate speech detector.

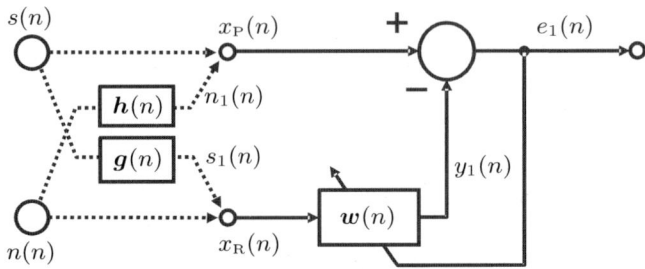

Fig. 7.2. Widrow's adaptive noise canceller with crosstalk.

7.2.2 Distortion by Crosstalk

When there is crosstalk, Eq. 7.2 does not hold anymore because a crosstalk term $s_1(n)$ should be considered as depicted in Fig. 7.2. In the presence of crosstalk, $x_R(n)$ is expressed by

$$x_R(n) = n(n) + s_1(n), \tag{7.12}$$

where

$$s_1(n) = \bm{g}^T(n)\,\bm{s}(n), \tag{7.13}$$
$$\bm{g}^T(n) = \bigl[g_0(n),\, g_1(n),\, \cdots,\, g_{N-1}(n)\bigr]. \tag{7.14}$$

In reference to Eqs. 7.2, 7.12, and 7.5, the output $y_1(n)$ with crosstalk is given by

$$\begin{aligned} y_1(n) &= \bm{w}^T(n)\,\bm{x}_R(n) + \bm{w}^T(n)\,\bm{s}_1(n), \\ &= \bm{w}^T(n)\,[\bm{n}(n) + \bm{s}_1(n)], \end{aligned} \tag{7.15}$$
$$\bm{s}_1^T(n) = \bigl[s_1(n),\, s_1(n-1),\, \cdots,\, s_1(n-N+1)\bigr]. \tag{7.16}$$

Eqs. 7.3, 7.4, and 7.15 result in

$$e_1(n) = s(n) + \bigl[\bm{h}(n) - \bm{w}(n)\bigr]^T \bm{n}(n) - \bm{w}^T(n)\,\bm{s}_1(n). \tag{7.17}$$

When $\bm{w}(n)$ perfectly identifies $\bm{h}(n)$, i.e. $\bm{w}(n) = \bm{h}(n)$, Eq. 7.17 reduces to

$$e_1(n) = s(n) - \bm{h}^T(n)\,\bm{s}_1(n), \tag{7.18}$$

which is not equal to the desired signal. In addition, Eqs. 7.16 and 7.18 suggest that past activities of $s(n)$ affect the output $e_1(n)$. Therefore, the output $e_1(n)$ is not equal to the desired signal $s(n)$, resulting in distortion, unless $\bm{h}(n)$ or $\bm{s}_1(n)$ is a zero vector.

For the environment with crosstalk, multi-stage ANCs [5, 8, 13] have been developed as extensions of Widrow's ANC. They all try to perform coefficient

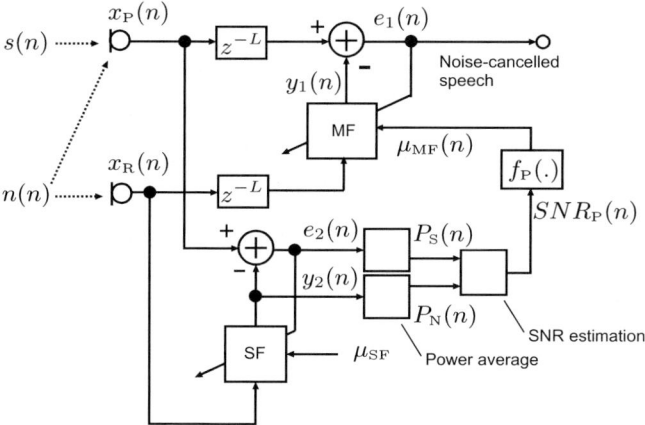

Fig. 7.3. ANC with a paired filter (PF) structure.

adaptation during absence of the desired signal so that the undesirable effect of the crosstalk does not appear neither in the input signal nor in the error that is used for coefficient adaptation. Eq. 7.2 holds in this case. It implies that a sufficiently accurate signal detector for $s(n)$ is essential, which is not available in reality. In addition, this strategy does not work in the presence of $s(n)$. The output is still expressed by Eq. 7.18. Therefore, multi-stage ANCs [5,8,13] are not effective for the crosstalk problem.

7.3 Paired Filter (PF) Structure

A paired filter (PF) structure [11] has been developed as a solution to the interference problem. Fig. 7.3 shows an ANC with a paired filter structure, which consists of two adaptive filters, namely, a main filter (MF) and a subfilter (SF). They operate in parallel to generate noise replicas. The SF is used for estimating the signal-to-noise ratio (SNR) of the primary input. The stepsize for the MF is controlled based on the estimated SNR.

7.3.1 Algorithm

7.3.1.1 SNR Estimation by Subfilters

The SF works in the same way as Widrow's ANC. Filter coefficients are updated by the NLMS algorithm [9]. The stepsize, μ_{SF}, is set large and fixed for fast convergence and rapid tracking of the noise-path change. A small value of μ_{SF} results in more precise estimation. For the estimation of SNR, an average power of the noise replica, $P_{\text{N}}(n)$, and the power of the error signal, $P_{\text{S}}(n)$, are calculated by

$$P_N(n) = \sum_{j=0}^{M-1} y_2^2(n-j), \qquad (7.19)$$

$$P_S(n) = \sum_{j=0}^{M-1} \left(x_P(n-j) - y_2(n-j)\right)^2,$$

$$= \sum_{j=0}^{M-1} e_2^2(n-j), \qquad (7.20)$$

where $y_2(n)$ is the SF output. M is the number of samples used for calculating $P_N(n)$ and $P_S(n)$. From $P_S(n)$ and $P_N(n)$, an estimated signal-to-noise ratio, $SNR_P(n)$, of the primary signal is calculated by

$$SNR_P(n) = 10 \, \log_{10} \left\{ \frac{P_S(n)}{P_N(n)} \right\} \, \text{dB}. \qquad (7.21)$$

7.3.1.2 Stepsize Control for the Main Filter

The stepsize for the MF is controlled by the estimated signal-to-noise ratio, $SNR_P(n)$. If $SNR_P(n)$ is low, the stepsize is set large for fast convergence because low $SNR_P(n)$ means small interference for the coefficient adaptation. Otherwise, the stepsize is set small for smaller signal-distortion in the ANC output. The following equation shows a function which determines the stepsize, $\mu_{\text{MF}}(n)$, based on $SNR_P(n)$:

$$\mu_{\text{MF}}(n) = \begin{cases} \mu_{\text{M}_{\min}}, & \text{if } SNR_P(n) > SNR_{P_{\max}}, \\ \mu_{\text{M}_{\max}}, & \text{if } SNR_P(n) < SNR_{P_{\min}}, \\ f_P(SNR_P(n)), & \text{otherwise,} \end{cases} \qquad (7.22)$$

where $\mu_{\text{M}_{\max}}$, $\mu_{\text{M}_{\min}}$ and $f_P(\cdot)$ are the maximum and the minimum stepsizes and a function of $SNR_P(n)$, respectively. It is natural that $f_P(\cdot)$ is a decreasing function since a small stepsize is suitable for a large SNR. For simplicity, let us assume that $f_P(\cdot)$ is a first-order function of $SNR_P(n)$. Then, it may be given by

$$f_P(SNR_P(n)) = A \cdot SNR_P(n) + B, \qquad (7.23)$$

where A ($A < 0$) and B are constants. $\mu_{\text{M}_{\min}}$ determines the signal distortion in the utterance. If $\mu_{\text{M}_{\min}}$ is set to zero, the adaptation is skipped when $SNR_P(n)$ is higher than $SNR_{P_{\max}}$. In this case, this algorithm works as the adaptation-stop method [10] with a speech detector.

7.3.1.3 Delay Compensation for the Main Filter

The estimated SNR, $SNR_P(n)$, is given with a time delay, which depends on the number of samples M used for calculation of $P_N(n)$ and $P_S(n)$. This time delay directly raises the signal distortion in the processed speech because the stepsize remains large in the beginning of the utterance. To compensate for this delay, the delay unit z^{-L} is incorporated only for the MF. L is set to $M/2$ since the time delay is $M/2$.

7.3.2 Evaluations

The performance of the ANC with a paired filter structure was evaluated in comparison with that of a variable stepsize algorithm [18] assuming a communication scenario in a military tank. The tank operators wear headsets with a reference microphone on the earpiece. A diesel-engine noise recorded in a tank was used as a noise source. Shown in Fig. 7.4 at the top is a noise-path impulse response measured in a room with a dimension of 3.05 m (width) × 2.85 m (depth) × 1.80 m (height). In order to evaluate the tracking capability, the polarity of the noise path was inverted at 12.5 sec.[1] The noise component, which was generated by convolution of the noise source with the noise-path, was added to the speech source to obtain the noise-corrupted signal. This signal contains the uncorrelated noise component which should exist in the recording environment. The sampling frequency was 8 kHz and other parameter values are shown in Tab. 7.1. Parameters for the variable stepsize algorithm were adjusted such that fast convergence and the final misadjustment equivalent to that of the ANC with the PF structure are obtained.

Table 7.1. Parameters and corresponding values.

Parameter	Value
N	64
M	128
L	64
μ_{SF}	0.4
$\mu_{\text{M}_{\max}}$	0.4
$\mu_{\text{M}_{\min}}$	2^{-6}
$SNR_{\text{P}_{\max}}$	-10 dB
$SNR_{\text{P}_{\min}}$	-50 dB
A	-0.01
B	0.4

---

[1] This is nothing more than an example. An abrupt polarity change was imposed as an extreme example. Path changes in the real environment are slower and less significant, thus, easier to track.

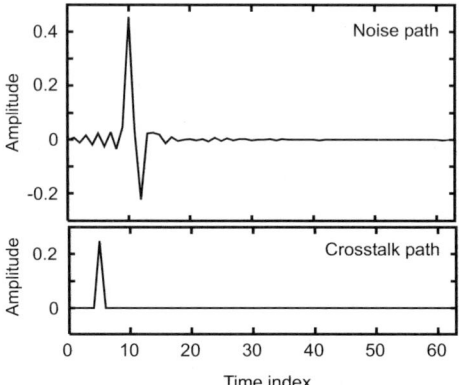

Fig. 7.4. Impulse responses of noise and crosstalk paths.

7.3.2.1 Objective Evaluations

Fig. 7.5 illustrates the desired signal, the primary signal and the output signal. The SNR in the primary signal was around 0 dB in the utterance. The ANC with the PF structure successfully cancels the noise and tracks the noise-path change at 12.5 sec.

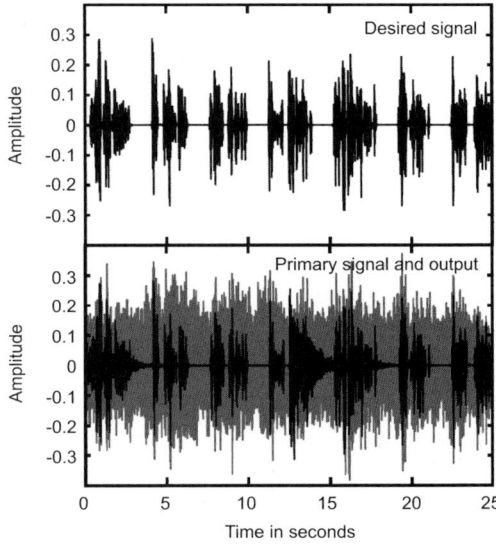

Fig. 7.5. Desired signal (upper diagram), primary signal (gray, lower diagram), and output speech signal (black, lower diagram).

The original SNR and the SNR estimated by the SF are compared in part (a) and (b) of Fig. 7.6. Since the peaks of the estimated SNR approximate those of the original SNR in a good manner, it is considered reliable. Part (c) of Fig. 7.6 exhibits the stepsize behaviors of the ANC with the PF structure and that of the variable stepsize algorithm. The stepsize of the ANC with the PF structure remains small in the utterance, while the other becomes larger in the beginning of the utterance.

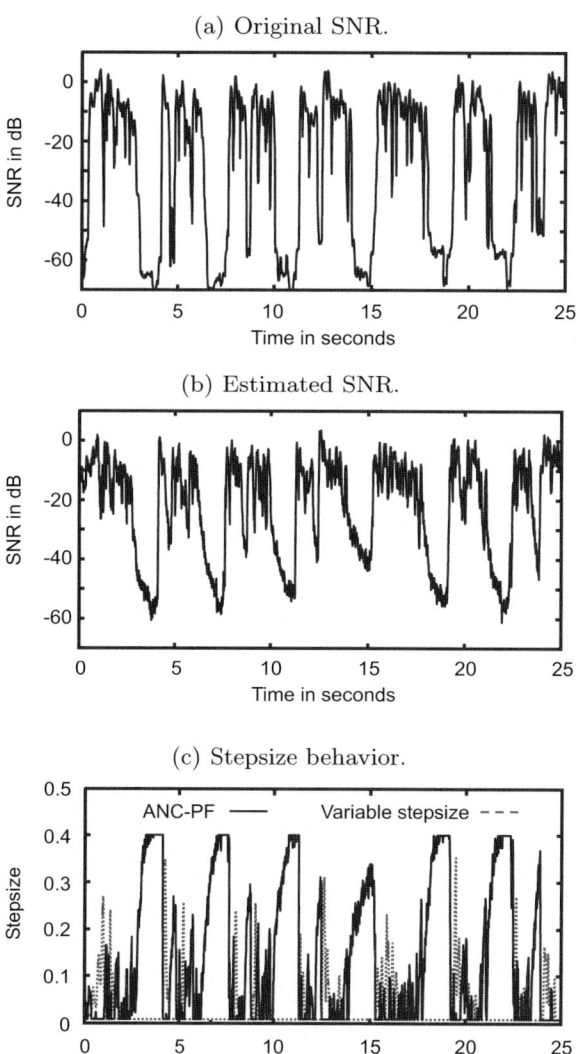

Fig. 7.6. Original (a) and estimated SNR (b), as well as the stepsize behavior (c).

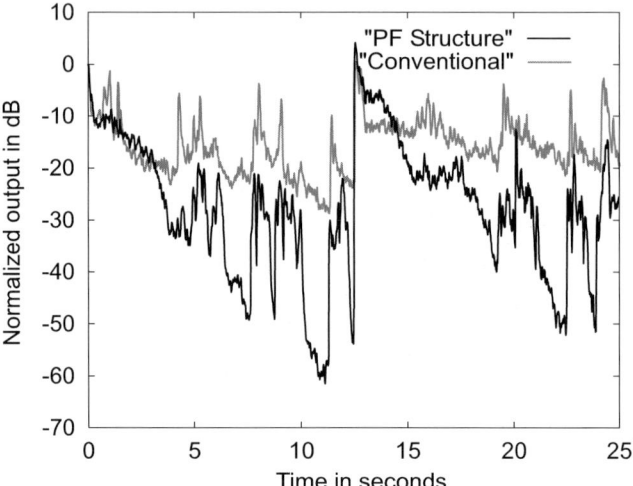

Fig. 7.7. Normalized output.

Fig. 7.7 shows a normalized output $\epsilon(n)$ defined by

$$\epsilon(n) = 10\ \log_{10} \left\{ \frac{\sum_{j=0}^{M_\epsilon -1} e_1^2(n-j)}{\sum_{j=0}^{M_\epsilon -1} x_\mathrm{P}^2(n-j)} \right\}\ \mathrm{dB}, \qquad (7.24)$$

where M_ϵ is the number of samples for average and was set to 512. In case of good noise cancellation, $\epsilon(n)$ should take a large negative value in nonspeech sections. The normalized output of the ANC with the PF structure is approximately 10 dB smaller in the utterance compared with the variable stepsize algorithm.

Fig. 7.8 depicts signal distortion $\delta(n)$ in the output defined by

$$\delta(n) = 10\ \log_{10} \left\{ \frac{\sum_{j=0}^{M_\delta -1} \left(e_1(n-j) - s(n-j)\right)^2}{\sum_{j=0}^{M_\delta -1} s^2(n-j)} \right\}\ \mathrm{dB}, \qquad (7.25)$$

where M_δ is the number of samples for average and was set to 512. The ANC with the PF structure reduces the signal distortion by up to 15 dB compared with the variable stepsize algorithm in the utterance.

Similar performance of the ANC with the PF structure has been confirmed for ±6 dB and 0 dB SNR with respect to the estimated SNR, the normalized

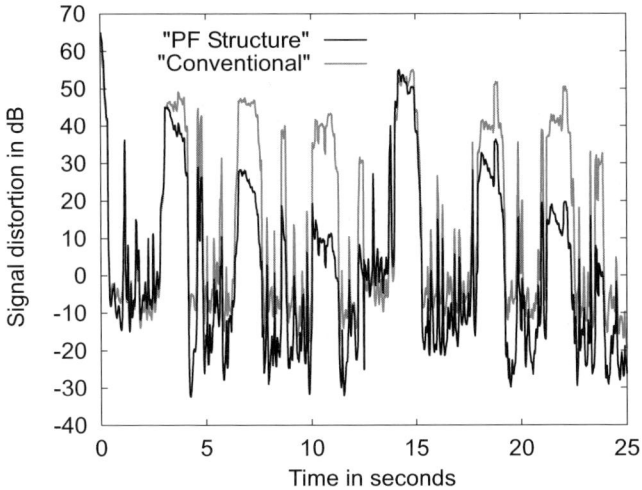

Fig. 7.8. Signal distortion.

output $\epsilon(n)$, and the signal distortion $\delta(n)$ [11]. The performance does not change for different SNRs.[2]

7.3.2.2 Subjective Evaluation

To evaluate the subjective performance of the ANC with a paired filter structure, a listening test was carried out. Widrow's ANC with the NLMS algorithm [9] and the ANC with the PF structure were compared for the case where the SNR in the primary signal is 0 dB. The 5-point mean opinion scores (MOS's) were given by 20 listeners. A noise-free (*i.e.* clean) speech sample and a noisy speech sample before noise-cancellation were included as the highest and the lowest anchors. The same speech source as in Sec. 7.3.2.1 was employed for the subjective test. Fig. 7.9 shows the subjective evaluation results. The vertical line centered in the shaded area and the numeral represent the mean value of the MOS. The width of the shaded area corresponds to the standard deviation. The mean values of the MOS for the ANC with the PF structure are higher than Widrow's ANC by about 1 point.

7.4 Crosstalk Resistant ANC and Cross-Coupled Structure

Crosstalk resistant ANC (CTRANC) [14, 15, 23] and an ANC with a cross-coupled structure [1] have been developed independently for crosstalk resistance.

---

[2] For hardware implementation and evaluation, please refer to [11].

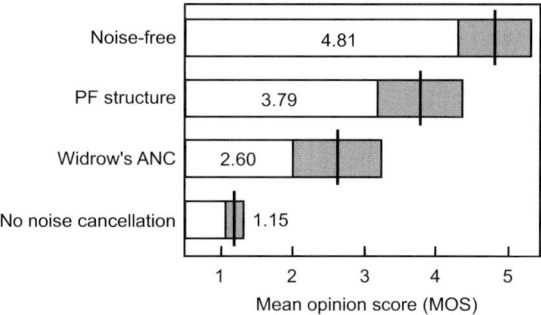

Fig. 7.9. Results of the subjective evaluation.

They both employ an auxiliary filter for crosstalk, however, its usage is different from that in the multi-stage ANCs [5, 8, 13].

7.4.1 Crosstalk Resistant ANC

Fig. 7.10 depicts a block diagram of CTRANC. The secondary adaptive filter, F2, is driven by the ANC output $e_1(n)$ to generate a crosstalk replica. When the ANC operation is ideal, its output should be equal to the desired signal. It suggests that the ANC output could be used as a replica of the desired signal. The output $y_3(n)$ of F2, approximating the crosstalk, is subtracted from the reference input $x_R(n)$. The result $e_3(n)$ is used as the input of the primary adaptive filter, F1. Because the crosstalk components that were originally contained in the reference input are cancelled by F2, F1 has little crosstalk contamination in its input. Therefore, F1 works as if there were no crosstalk.

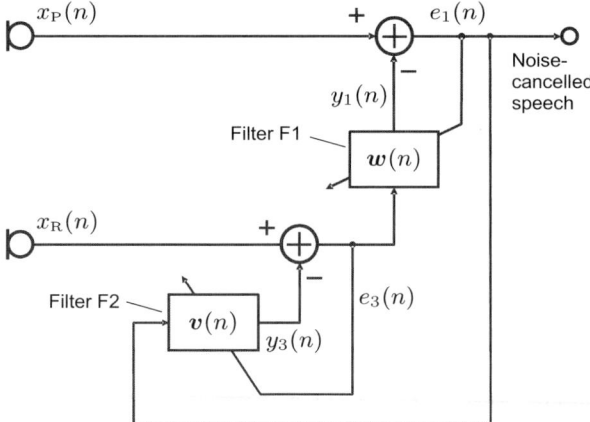

Fig. 7.10. Crosstalk resistant adaptive noise canceller (CTRANC).

Referring to Fig. 7.10, the input of F1 represented by $e_3(n)$ is given by

$$e_3(n) = x_R(n) - \boldsymbol{v}^T(n)\,\boldsymbol{e}_1(n). \tag{7.26}$$
$$e_1(n) = x_P(n) - \boldsymbol{w}^T(n)\,\boldsymbol{e}_3(n), \tag{7.27}$$

From Eqs. 7.1, 7.4, 7.12, 7.13, 7.26, and 7.27, the following equations are obtained:

$$e_1(n) = s(n) + \boldsymbol{h}^T(n)\,\boldsymbol{n}(n) - \boldsymbol{w}^T(n)\,\boldsymbol{e}_3(n), \tag{7.28}$$
$$e_3(n) = \boldsymbol{g}^T(n)\,\boldsymbol{s}(n) + n(n) - \boldsymbol{v}^T(n)\,\boldsymbol{e}_1(n), \tag{7.29}$$

where

$$\boldsymbol{e}_1^T(n) = [e_1(n-1),\ \cdots,\ e_1(n-N+1),\ e_1(n-N)], \tag{7.30}$$
$$\boldsymbol{e}_3^T(n) = [e_3(n),\ e_3(n-1),\ \cdots,\ e_3(n-N+1)], \tag{7.31}$$
$$\boldsymbol{v}^T(n) = [v_0(n),\ v_1(n),\ \cdots,\ v_{N-1}(n)]. \tag{7.32}$$

$\boldsymbol{v}(n)$ is the coefficient vector of F2. It should be noted that the elements of $\boldsymbol{e}_1(n)$ are one-sample shifted to the left. This is because $e_1(n)$ is not available when $y_3(n)$ is calculated.[3]

For perfect cancellation of the crosstalk $s_1(n)$, $e_3(n) = n(n)$ should be satisfied. Applying this condition to Eqs. 7.28 and 7.29 leads to

$$e_1(n) = s(n), \tag{7.33}$$
$$\boldsymbol{w}(n) = \boldsymbol{h}(n), \tag{7.34}$$
$$\boldsymbol{v}(n) = \boldsymbol{g}(n). \tag{7.35}$$

Eq. 7.33 implies that the output $e_1(n)$ theoretically has no distortion.

7.4.2 Cross-Coupled Structure

An ANC with a cross-coupled structure [1] is an equivalent form to CTRANC. The cross-coupled structure is illustrated in Fig. 7.11. It has a paired structure with dedicated adaptive filters for noise and crosstalk paths. Combining Eqs. 7.1 and 7.2, with Fig. 7.11, it is straightforward to derive Eqs. 7.28 and 7.29. It is clearly seen in Fig. 7.11 that the filters F1 and F2 are cooperating to make the input of its counterpart less contaminated by crosstalk or noise. Therefore, if one works better with a clean input, the other also works better with its own input that is cleaner. Finally, both F1 and F2 operate with crosstalk-free and noise-free inputs, respectively, which is the ideal situation.

---

[3] This fact implies that $v_0(n)$, the first element of $\boldsymbol{v}(n)$, approximates $g(1)$ in Fig. 7.2. Although $g(0)$ cannot be modeled by $\boldsymbol{v}(n)$, it does not cause a problem as far as there is a one-sample delay in the crosstalk path. This is usually the case in practice.

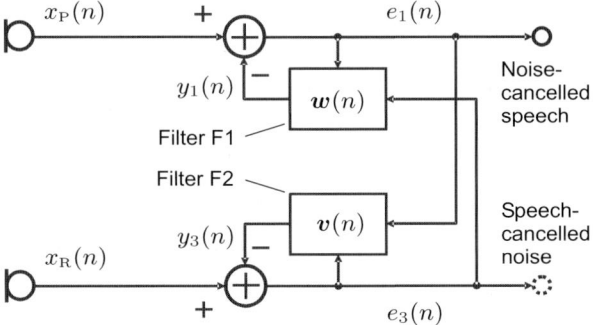

Fig. 7.11. ANC with a cross-coupled structure.

However, [1] points out that the output of the cross-coupled ANC may not be sufficiently good. This is because the desired signal in the output interferes the coefficient adaptation in F1. Similarly, the noise in $e_3(n)$ interferes the coefficient adaptation in F2. The ANC structure by itself does not make a good solution to the crosstalk problem. From this viewpoint, integration of a good coefficient adaptation algorithm and a good ANC structure is essential. Such integrations are discussed in the following sections.

7.5 Cross-Coupled Paired Filter (CCPF) Structure

To cancel crosstalk caused by the desired signal in the reference input, the cross-coupled paired filter (CCPF) structure [12] employs a cross-coupled structure [1] in which cross-coupled adaptive filters should cancel the noise component in the primary signal and the crosstalk in the reference signal simultaneously. For the interference problem the paired filter structure is extended to the cross-coupled structure.

7.5.1 Algorithm

Fig. 7.12 depicts a block diagram of an ANC with a CCPF structure. The filters MF1 and MF2 form the cross-coupled structure. The filters SF1 and SF2 make a pair with MF1 and MF2 for the interference resistance.

MF1 and SF1 take the roles of the main filter (MF) and the subfilter (SF) in the PF structure. Another set of paired filters, namely MF2 and SF2, operate in the same manner as MF1 and SF1 except that they try to cancel the crosstalk instead of the noise. The stepsizes of MF1 and MF2 are controlled with the help of SF1 and SF2 in a similar way to that in MF in the PF structure. When no crosstalk is present, this structure in Fig. 7.12 works as that in Fig. 7.3. Because there are no correlated components to $s(n)$ in either $e_3(n)$ or $e_4(n)$, the coefficients of SF2 and MF2 do not grow from the initial values (*i.e.* zero).

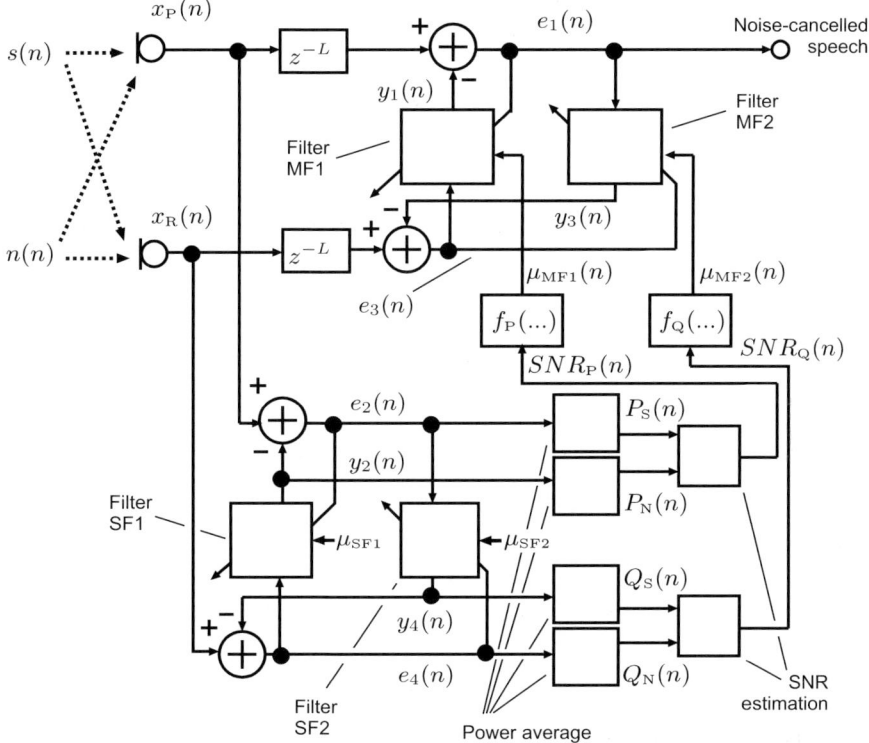

Fig. 7.12. ANC with a cross-coupled paired filter (CCPF) structure.

7.5.1.1 SNR Estimation by Subfilters

SF1 carries out SNR estimation for the primary signal in the same way as in the PF structure by Eqs. 7.19 – 7.21. SF2, on the other hand, estimates the SNR for the reference signal $x_\mathrm{R}(n)$ by generating a replica of the desired signal component (crosstalk) in the reference signal. $SNR_\mathrm{R}(n)$ of the reference signal can be estimated using a replica of the desired signal at the subtractor output and the estimated noise. The crosstalk estimated by SF2 is generally a speech signal, which naturally contains silent sections. When the crosstalk is small, the signal component to be estimated may be much smaller than the noise component. To perform a stable SNR estimation, the stepsize for the coefficient adaptation μ_SF2 is therefore set smaller than the stepsize for SF1.

The average power $Q_\mathrm{S}(n)$ of the desired-signal replica and the average power $Q_\mathrm{N}(n)$ of the error signal of SF2 can be calculated using the following equations:

$$Q_{\text{S}}(n) = \sum_{j=0}^{M-1} y_4^2(n-j), \qquad (7.36)$$

$$Q_{\text{N}}(n) = \sum_{j=0}^{M-1} \Big(x_{\text{R}}(n-j) - y_4(n-j)\Big)^2,$$

$$= \sum_{j=0}^{M-1} e_4^2(n-j), \qquad (7.37)$$

where $y_4(n)$, $e_4(n)$ are respectively the desired-signal replica which is the output of SF2 and the reference signal. M is the number of samples averaged for calculation of $Q_{\text{S}}(n)$ and $Q_{\text{N}}(n)$. From $Q_{\text{S}}(n)$ and $Q_{\text{N}}(n)$, the signal-to-noise ratio $SNR_{\text{Q}}(n)$ can be obtained by

$$SNR_{\text{Q}}(n) = 10 \log_{10} \left\{ \frac{Q_{\text{S}}(n)}{Q_{\text{N}}(n)} \right\} \text{ dB}. \qquad (7.38)$$

7.5.1.2 Stepsize Control in the Main Filters

The adjustment of the stepsize $\mu_{\text{MF1}}(n)$ in MF1 is controlled based on the estimated $SNR_{\text{P}}(n)$ in the same manner as in the PF structure. $\mu_{\text{MF1}}(n)$ is determined by Eqs. 7.21 and 7.22 where $\mu_{\text{MF}}(n)$, $\mu_{\text{M}_{\max}}$, and $\mu_{\text{M}_{\min}}$ should be replaced with $\mu_{\text{MF1}}(n)$, $\mu_{\text{M1}_{\max}}$, and $\mu_{\text{M1}_{\min}}$. The stepsize $\mu_{\text{MF2}}(n)$ is controlled by the estimated $SNR_{\text{Q}}(n)$. When $SNR_{\text{Q}}(n)$ is low, the stepsize is set small since there is a large noise component which interferes with MF2 in the crosstalk estimation. On the other hand, when $SNR_{\text{Q}}(n)$ is high with a large crosstalk component, the stepsize is set large. $\mu_{\text{MF2}}(n)$ is determined by Eq. 7.39 based on $SNR_{\text{Q}}(n)$:

$$\mu_{\text{MF2}}(n) = \begin{cases} \mu_{\text{M2}_{\min}}, & \text{if } SNR_{\text{Q}}(n) < SNR_{\text{Q}_{\min}}, \\ \mu_{\text{M2}_{\max}}, & \text{if } SNR_{\text{Q}}(n) > SNR_{\text{Q}_{\max}}, \\ f_{\text{Q}}\big(SNR_{\text{Q}}(n)\big), & \text{otherwise,} \end{cases} \qquad (7.39)$$

where $\mu_{\text{M2}_{\max}}$ and $\mu_{\text{M2}_{\min}}$ are the maximum and the minimum values of the stepsize, respectively. It is desirable that $f_{\text{Q}}(\cdot)$ is a monotonically increasing function. For simplicity, let us assume that $f_{\text{Q}}(\cdot)$ is a first-order function. Then, it may be given by

$$f_{\text{Q}}\big(SNR_{\text{Q}}(n)\big) = C \cdot SNR_{\text{Q}}(n) + D, \qquad (7.40)$$

where C ($C > 0$) and D are constants.

7.5.1.3 Delay Compensation for Main Filters

As in the PF structure, the estimated $SNR_{\text{P}}(n)$ and $SNR_{\text{Q}}(n)$ generate time delays depending on M. To compensate for these delays, L-sample delay units

z^{-L} are incorporated into the input paths of the primary and the reference signals of MF1 and MF2. L is set to $M/2$ since the delay in the sense of the moving-average for M samples is a half of M.

7.5.2 Evaluations

The performance of the ANC with the CCPF structure was evaluated in comparison with that of the PF structure. The same recorded speech and the noise as in Sec. 7.3.2 as well as the impulse response of the noise path in Fig. 7.4 were used. Since the crosstalk path can be approximated by a delay when the primary and the reference microphones are located close to each other, like using a headset, a unit impulse response with an amplitude of 0.25 and a time delay of 5 samples was used.

The noise component, which was generated by convolution of the noise source with the impulse response of the noise path, was added to the desired signal to create the primary signal. The reference signal was generated by adding the noise to the crosstalk generated by convolution of the desired signal with the impulse response of the crosstalk path. SNRs of the primary and the reference signals in the utterance were 6 dB and -12 dB, respectively. To evaluate the influence by an SNR change, these SNRs were increased by 10 dB by decreasing the noise power by 10 dB after 15 seconds[4]. The sampling frequency was 8 kHz. Other parameter values are shown in Tab. 7.2.

Table 7.2. Parameters and corresponding values.

Parameter	Value	Parameter	Value
N	64	$SNR_{P_{min}}$	-30 dB
M	128	A	-0.01
L	64	B	0.1
μ_{SF}	0.1	$\mu_{M2_{max}}$	0.02
μ_{SF1}	0.1	$\mu_{M2_{min}}$	0.0
μ_{SF2}	0.002	$SNR_{Q_{max}}$	0 dB
$\mu_{M1_{max}}$	0.2	$SNR_{Q_{min}}$	-10 dB
$\mu_{M1_{min}}$	0.0	C	0.002
$SNR_{P_{max}}$	-10 dB	D	0.02

---

[4] A severer condition was selected on purpose. A smaller noise power means a stronger interference (*i.e.* stronger desired signal) for adaptation of SF1 and MF1.

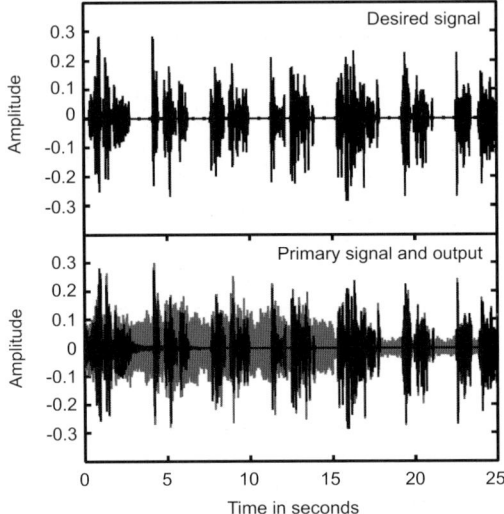

Fig. 7.13. Desired signal (upper diagram), primary signal (gray, lower diagram) and output (black, lower diagram) speech.

Fig. 7.13 shows the desired signal (upper diagram), the primary signal (gray, lower diagram), and the output (black, lower diagram) of the ANC with the CCPF structure. They indicate that the CCPF structure cancels the noise to a satisfactory level.

Fig. 7.14 shows the SNR of the primary signal, its estimate using SF1, the SNR of the reference signal, and its estimate utilizing SF2. Where the SNR is high, the estimated SNR agrees well with the actual one.

Fig. 7.15 shows the stepsizes of MF1 and MF2. The stepsize of MF1 is large when the speech signal is absent. The stepsize is generally small after 15 seconds since the SNR was increased by 10 dB. On the other hand, the stepsize of MF2 is large in those sections where the speech is present. After 15 seconds, the stepsize is generally large due to the increased SNR.

Part (a) of Fig. 7.16 depicts the normalized output, $\epsilon(n)$, in Eq. 7.24. A large negative value of $\epsilon(n)$ in nonspeech sections represents good noise cancellation. The ANC with the CCPF structure achieves as much as 10 dB lower output level than the ANC with the PF structure.

Part (b) of Fig. 7.16 exhibits the signal distortion $\delta(n)$ in Eq. 7.25 at the ANC output. The ANC with a CCPF structure reduces the distortion by as much as 15 dB in utterances compared to that with the PF structure. Although there are sections where the CCPF structure creates larger distortion, it does not result in the degradation of the subjective voice quality since these sections are limited to silent sections with small signal power. In addition, the CCPF structure creates no increase in distortion even when the SNR is increased by 10 dB.

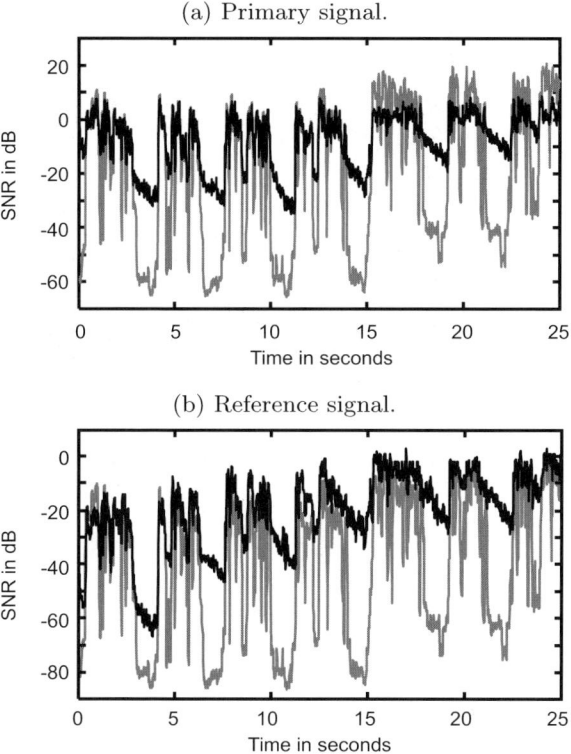

Fig. 7.14. True (gray) and estimated (black) SNRs of the primary (upper diagram) and the reference signals (lower diagram).

7.6 Generalized Cross-Coupled Paired Filter (GCCPF) Structure

The CCPF structure has two pairs of cross-coupled adaptive filters, each of which consists of a main filter and a subfilter. The subfilters serve as pilot filters whose output is used to estimate the signal-to-noise ratios (SNRs) of the primary and the reference signals. The stepsizes for the adaptation of the main filters are controlled according to the estimated SNRs. Since the stepsize is controlled by the subfilters, good noise cancellation and low signal-distortion in the output are simultaneously achieved. However, the CCPF structure was developed for communication headsets in noisy environment. The fixed stepsize for each subfilter may not provide satisfactory performance for a wide range of SNRs that are encountered in other applications. Actually, application to human-robot communication is attracting more interests from a viewpoint of noise and interference cancellation [19]. It is possible to follow the path from the PF structure to the CCPF structure by introducing yet

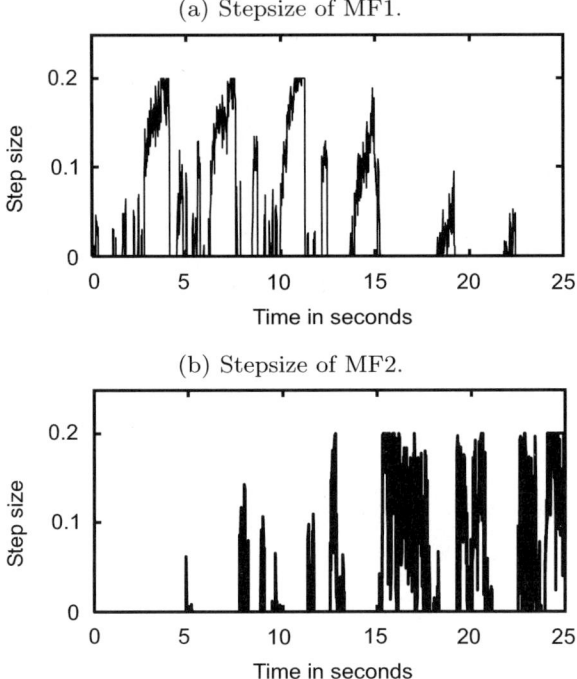

Fig. 7.15. Stepsizes of MF1 and MF2.

another set of subfilters for stepsize control of SF1 and SF2. However, this path may lead to an infinite chain of such extensions.

A generalized cross-coupled paired filter (GCCPF) structure [16] utilizes the primary and the reference signals for approximating their SNRs instead of another pair of pilot filters for SF1 and SF2. Fig. 7.17 shows a block diagram of an ANC with a GCCPF structure. Four adaptive filters, namely, the main adaptive filters (MF1, MF2) and the sub adaptive filter (SF1, SF2) generate noise and crosstalk replicas as in the CCPF structure. Adaptive control of the stepsizes for SF1 and SF2 forms the most significant difference from the CCPF structure. Average powers $R_S(n)$ and $R_N(n)$ of the primary signal $x_P(n)$ and the reference signal $x_R(n)$ are first calculated. A ratio of $R_S(n)$ to $R_N(n)$, representing a rough estimate of the SNR at the primary input, is used for controlling the stepsizes of SF1 and SF2.

The SF1 output $y_2(n)$ and the subtraction result $e_2(n)$ are used to estimate a more precise SNR at the primary input. $e_2(n)$ serves as an approximation to the desired signal, and $y_2(n)$ is used as that to the noise. The stepsize for MF1 is controlled based on the estimated SNR calculated from the output signal of SF1. SF2 works for crosstalk instead of noise in a similar way to

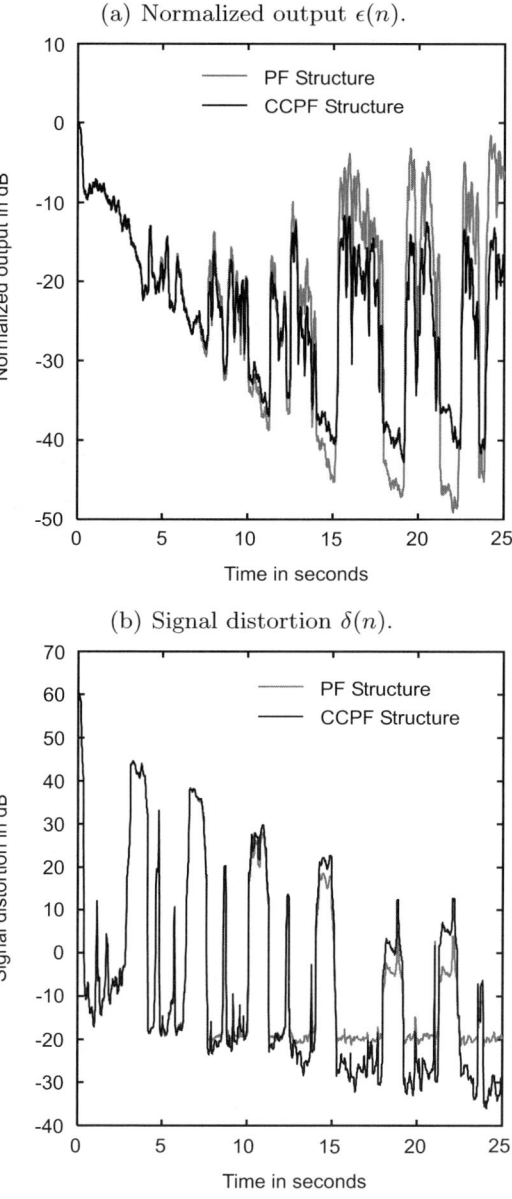

Fig. 7.16. Normalized output (a) and signal distortion (b).

that of SF1. The resulting SNR estimate from SF2 output signals is used to control the MF2 stepsize.

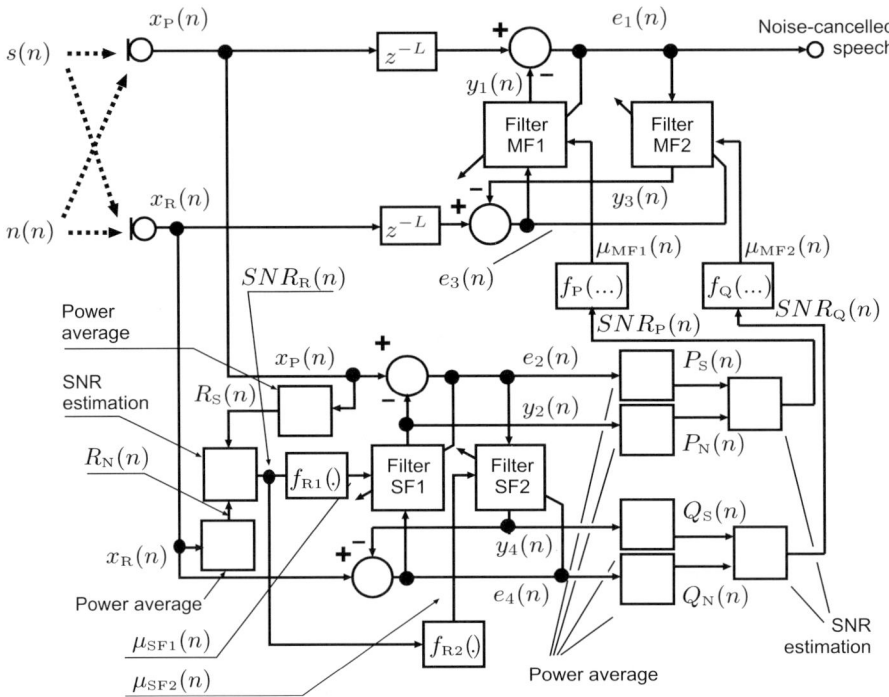

Fig. 7.17. ANC with a generalized cross-coupled paired filter (GCCPF) structure.

7.6.1 Algorithm

The stepsize for the coefficient adaptation should be kept smaller when there is more interference in the error. In the structure in Fig. 7.17, the stepsize, $\mu_{\text{SF1}}(n)$, for SF1 and the stepsize, $\mu_{\text{MF1}}(n)$, for MF1 should be set to a small value when the SNR at the primary input is high to avoid distortion at the ANC output. On the other hand, $\mu_{\text{SF1}}(n)$ and $\mu_{\text{MF1}}(n)$ can be set large when this SNR is low for fast convergence and rapid tracking of noise-path changes. A similar rule applies to $\mu_{\text{SF2}}(n)$ and $\mu_{\text{MF2}}(n)$ for coefficient adaptation in SF2 and MF2 with respect to the SNR at the reference input. All these stepsizes can be controlled appropriately once the SNRs for the adaptive filters become available.

The SNR for the primary signal, $SNR_{\text{R}}(n)$, is approximated by

$$SNR_{\text{R}}(n) = 10 \log_{10} \left\{ \frac{R_{\text{S}}(n)}{R_{\text{N}}(n)} \right\} \text{ dB}, \tag{7.41}$$

$$R_{\text{S}}(n) = \sum_{j=0}^{M-1} x_{\text{P}}^2(n-j), \tag{7.42}$$

$$R_{\text{N}}(n) = \sum_{j=0}^{M-1} x_{\text{R}}^2(n-j). \tag{7.43}$$

$\mu_{\text{SF1}}(n)$ and $\mu_{\text{SF2}}(n)$ are controlled by the estimated SNR, $SNR_{\text{R}}(n)$, as in the following equations:

$$\mu_{\text{SF1}}(n) = \begin{cases} \mu_{\text{S}_{\min}}, & \text{if } SNR_{\text{R}}(n) > SNR_{\text{R}_{\max}}, \\ \mu_{\text{S}_{\max}}, & \text{if } SNR_{\text{R}}(n) < SNR_{\text{R}_{\min}}, \\ f_{\text{R1}}(SNR_{\text{R}}(n)), & \text{otherwise}, \end{cases} \tag{7.44}$$

$$\mu_{\text{SF2}}(n) = \begin{cases} \mu_{\text{S}_{\min}}, & \text{if } SNR_{\text{R}}(n) < SNR_{\text{R}_{\min}}, \\ \mu_{\text{S}_{\max}}, & \text{if } SNR_{\text{R}}(n) > SNR_{\text{R}_{\max}}, \\ f_{\text{R2}}(SNR_{\text{R}}(n)), & \text{otherwise}. \end{cases} \tag{7.45}$$

$\mu_{\text{S}_{\max}}$ and $\mu_{\text{S}_{\min}}$ are the maximum and the minimum stepsizes for $\mu_{\text{SF1}}(n)$ and $\mu_{\text{SF2}}(n)$. $f_{\text{R1}}(\cdot)$ and $f_{\text{R2}}(\cdot)$ are functions of $SNR_{\text{R}}(n)$. $f_{\text{R1}}(\cdot)$ should be a decreasing function because a small stepsize is suitable for a large SNR. On the other hand, it is desirable that $f_{\text{R2}}(\cdot)$ is an increasing function. Eqs. 7.44 and 7.45 enable the ideal stepsize control described earlier, leading to small residual error and distortion in the noise-cancelled signal. $\mu_{\text{MF1}}(n)$ and $\mu_{\text{MF2}}(n)$ are controlled by the estimated SNRs, $SNR_{\text{P}}(n)$ and $SNR_{\text{Q}}(n)$, in the same way as in the CCPF structure based on Eqs. 7.19 – 7.22 and Eqs. 7.36 – 7.39. MF1 and MF2 are equipped with L-sample delay units z^{-L} for time-delay compensation.

7.6.2 Evaluation by Recorded Signals

7.6.2.1 Noise Reduction and Distortion

The performance of the ANC with the GCCPF structure was evaluated by computer simulations from the viewpoints of noise reduction and distortion in comparison with the ANC with the CCPF structure [12]. TV sound and a male voice were recorded in a carpeted room with a dimension of 5.5 m (width) × 5.0 m (depth) × 2.4 m (height) in a human-robot communication scenario. The primary microphone was mounted on the forehead and the reference microphone was attached to upper back of a robot, whose height is approximately 0.4 m. The impulse responses of the noise path and the crosstalk path were measured for a direction of noise arrival of 180 degrees with this set-up. An example with a speaker distance of 0.5 m is depicted in

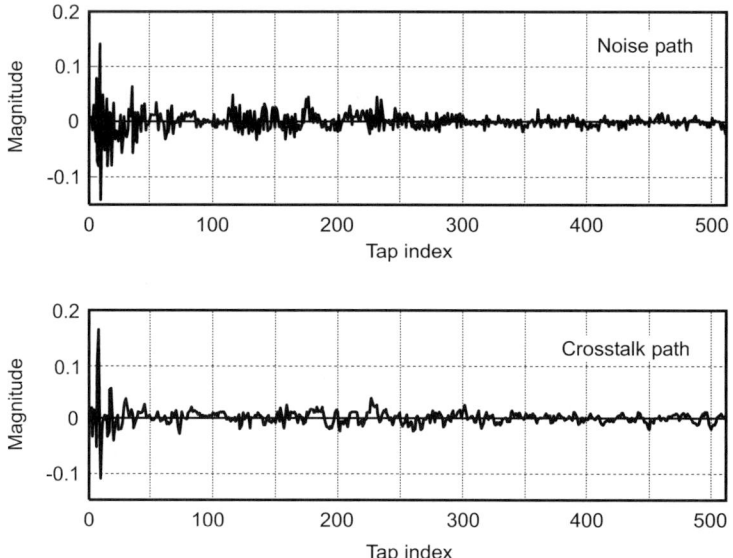

Fig. 7.18. Impulse responses of the noise and the crosstalk path.

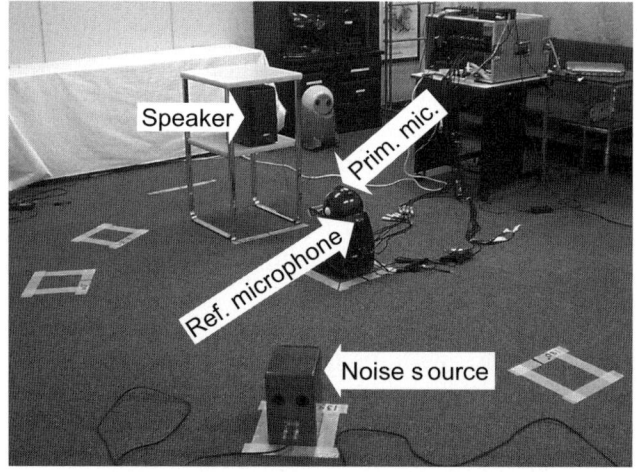

Fig. 7.19. Evaluation environment.

Fig. 7.18. The robot was placed on the straight line between the speaker and the noise source, facing the speaker. This environment is shown in Fig. 7.19.[5]

---

[5] This figure shows an example where the direction of noise arrival is set to 135 degrees.

Low Distortion Noise Cancellers – Revival of a Classical Technique 253

Table 7.3. Speaker and noise-source layout.

SNR [primary, reference]	Speaker distance	Noise level
[35, 15] dB	0.5 m	low
[25, 5] dB	1.0 m	low
[20, 0] dB	0.5 m	high
[10, -10] dB	1.0 m	high

The noise component, which is obtained by convolution of the noise source with the impulse response of the noise path, was added to the speech signal to create the primary signal. The reference signal was generated by adding the noise to the crosstalk generated by convolution of the speech signal with the impulse response of the crosstalk path. SNRs of the primary and the reference signals in the utterance were set to [35, 15], [25, 5] dB, [20, 0] dB, and [10, −10] dB, which correspond to four different speaker and noise-source layouts summarized in Tab. 7.3. The sound level for "low" and "high" respectively correspond to 55 − 60 dB and 65 − 70 dB. The sampling frequency was 11.025 kHz. Other parameters are show in Tab. 7.4. These specific layouts have been selected for evaluations because they represent typical scenarios in human-robot communication at home. Moreover, they include some difficult situations such as 35 dB primary SNR where the interference is significant.

Figs. 7.20 and 7.21 show the stepsize for MF1 (upper diagram) and that for MF2 (lower diagram) in the cases of [35, 15] dB, [25, 5] dB, [20, 0] dB, and [10, -10] dB SNRs. Dips in the upper figure and peaks in the lower figure both correspond to speech sections. To highlight speech and nonspeech sections, a rectangular waveform is added to the top of each figure. The waveform has two levels: SP and NSP. SP represents speech sections and NSP corresponds to nonspeech sections.

The stepsizes of the ANC with the GCCPF structure represented by the black solid line show better match with speech sections than those of the conventional ANC expressed in a gray dotted line. Such characteristics, which are closer to the ideal behavior already described in Sec. 7.6.1, are achieved by newly introduced stepsize control for SF1 and SF2 based on the estimated primary and the reference signal powers.

The normalized output (upper diagram) and distortion (lower diagram) at the output are illustrated in Figs. 7.22 and 7.23 for SNR settings of [35, 15] dB, [25, 5] dB, [20, 0] dB, and [10, -10] dB. Speech and nonspeech sections are specified by the same rectangular waveform to that in Figs. 7.20 and 7.21. The results by the ANCs with the GCCPF and the CCPF structures are represented by a black solid and a gray dotted lines. The normalized output, $\epsilon(n)$, and the distortion, $\delta(n)$, were calculated by Eqs. 7.24 and 7.25. In case of

Table 7.4. Parameters and corresponding values.

ANC structure	Parameter	Selected value
Common	N	512
	M	128
	L	64
CCPF	μ_{SF1}	0.02
	μ_{SF2}	0.002
	$SNR_{\mathrm{P_{min}}}$	-7 dB
	$SNR_{\mathrm{P_{max}}}$	5 dB
GCCPF	$SNR_{\mathrm{P_{min}}}$	0 dB
	$SNR_{\mathrm{P_{max}}}$	10 dB
	$\mu_{\mathrm{S_{min}}}$	0.002
	$\mu_{\mathrm{S_{max}}}$	0.02
	$SNR_{\mathrm{Q_{min}}}$	-10 dB
	$SNR_{\mathrm{Q_{max}}}$	0 dB
	$\mu_{\mathrm{M1_{min}}}, \mu_{\mathrm{M2_{min}}}$	0.002
	$\mu_{\mathrm{M1_{max}}}, \mu_{\mathrm{M2_{max}}}$	0.02

good noise cancellation, $\epsilon(n)$ should take a small value in nonspeech sections. When the SNR is low, it takes a lower value than 0 dB even in speech sections. It goes without saying that a smaller distortion, represented by a smaller value of $\delta(n)$, is desirable. Peaks in the normalized output and dips in the distortion correspond to speech sections.

Both noise reduction and distortion are improved by as much as 20 dB in part (a) of Fig. 7.22. Part (b) of Figs. 7.22 and part (a) of Fig. 7.23 also exhibit as much as 15 and 10 dB improvement in both measures. In the case of part (b) of Fig. 7.23, noise reduction is improved by as much as 8 dB. The improvement in distortion in part (b) of Fig. 7.23 is not as evident as that in part (a) of Fig. 7.22. This is because the parameters for the ANC with the CCPF structure are optimal for $[10, -10]$ dB SNR. However, an improvement close to 10 dB between the dotted and the solid lines can be observed in circled areas.

These lower residual-noise levels and smaller distortions for a wide range of SNRs are both due to the adaptive control of the stepsizes for SF1 and SF2. The SF1 stepsize takes relatively small values in speech sections and large

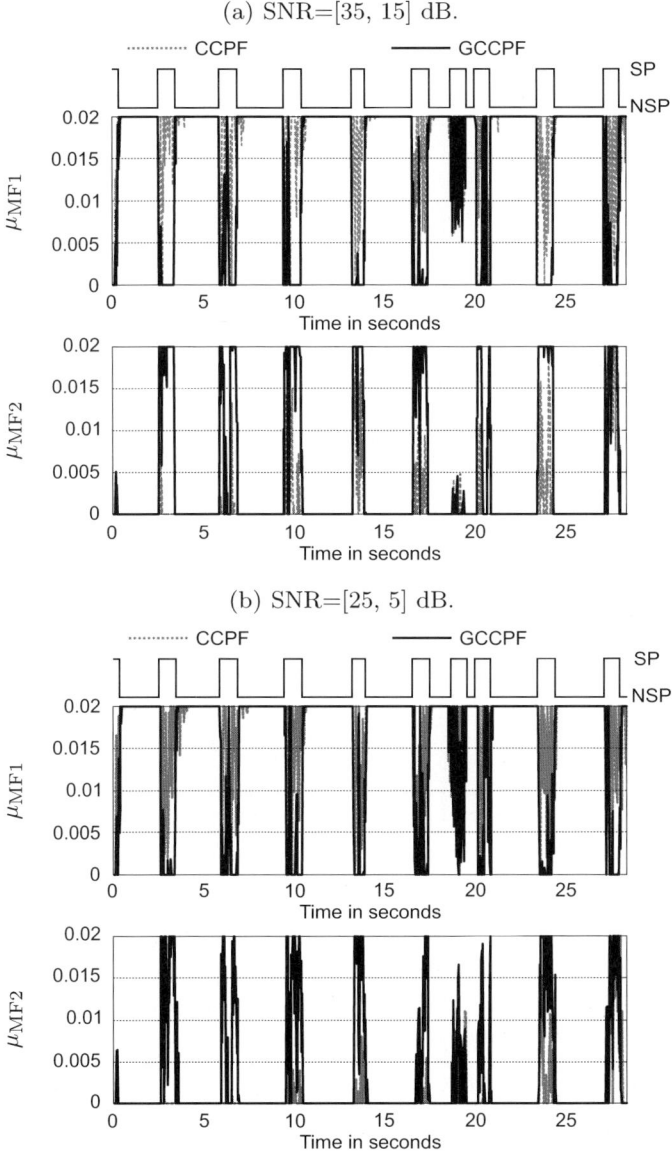

Fig. 7.20. Stepsize of MF1 and MF2 ((a) SNR=[35, 15] dB, (b) SNR=[25, 5] dB).

ones in nonspeech sections to implement the ideal behavior as was described earlier in Sec. 7.6.1. The SF2 stepsize takes the opposite pattern.

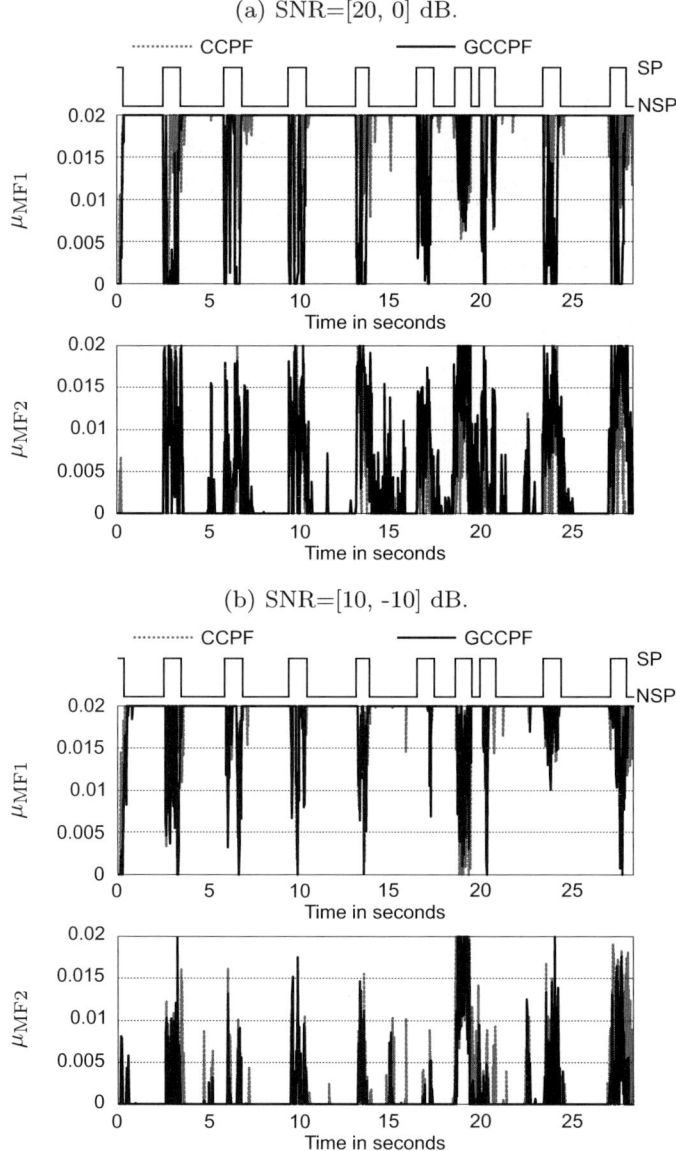

Fig. 7.21. Stepsize of MF1 and MF2 ((a) SNR=[20, 0] dB, (b) SNR=[10, -10] dB).

7.6.2.2 Speech Recognition

Speech recognition was performed with noise-cancelled speech by the ANC with the GCCPF structure. This is because the conventional ANC does not

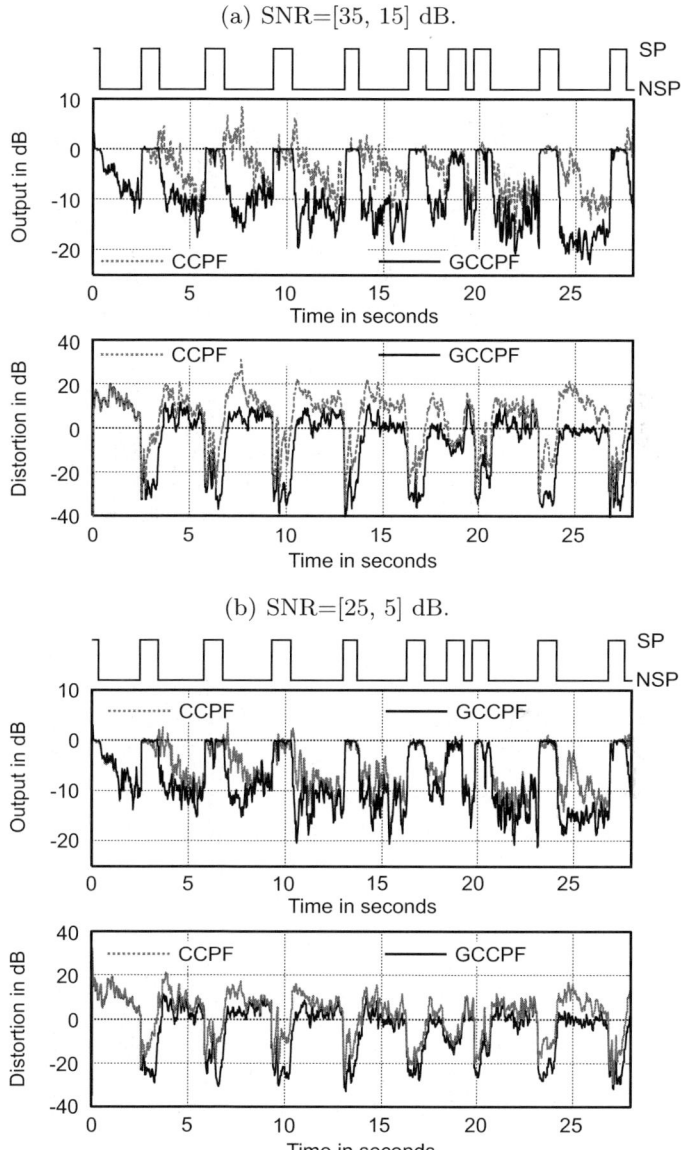

Fig. 7.22. Normalized output and distortion ((a) SNR=[35, 15] dB, (b) SNR=[25, 5] dB).

achieve sufficiently low residual noise nor low distortion for a wide range of SNRs. Distortion in the noise-cancelled speech degrades the speech recognition rate because its acoustic characteristics are less likely to match the HMM

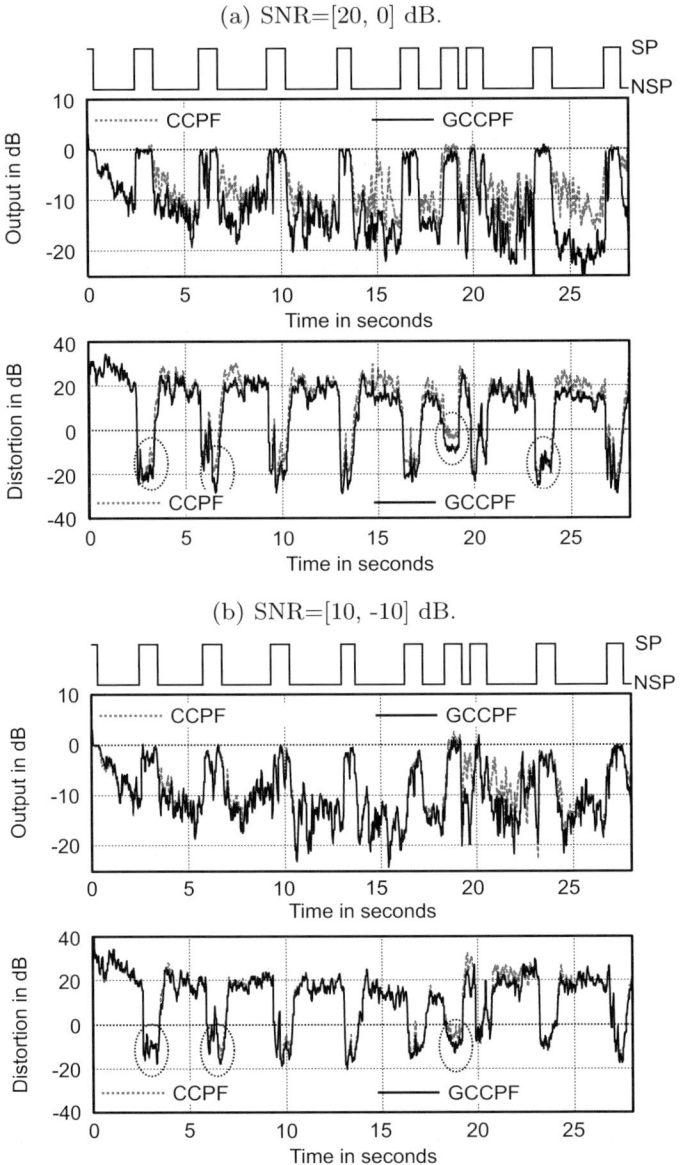

Fig. 7.23. Normalized output and distortion ((a) SNR=[20, 0] dB, (b) SNR=[10, -10] dB).

(hidden Markov model) in the recognition system. The residual noise, on the other hand, leads to wrong detection of speech sections. Because speech segmentation is important in speech recognition, it also results in low recognition

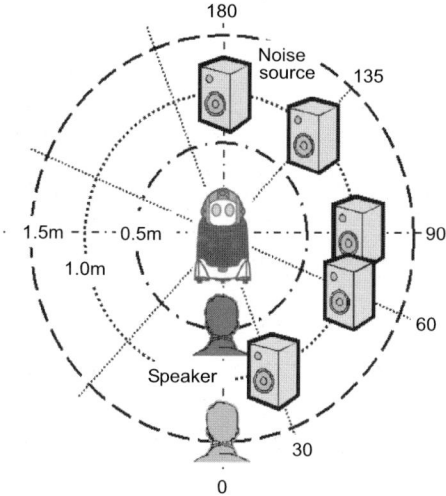

Fig. 7.24. Experimental set-up for speech recognition.

rates. 10 − 20 dB decrease by the ANC with the CCPF structure compared to the ANC with the GCCPF structure apparently predicts its inferior recognition rates.

The experimental set-up is illustrated in Fig. 7.24. 150 utterances by 30 different male, female, and child speakers were presented at a distance of 0.5 and 1.5 m. The noise source was placed at a distance of 1.0 m in a direction of 30, 60, 90, 135, or 180 degrees. Speaker independent speech recognition based on semi-syllable hidden Markov models [21] was used with a dictionary of 600 robot commands.

Fig. 7.25 depicts the speech recognition rate for a commercial and a news TV-programs as the noise source. Shaded columns represent improvements, *i.e.* the difference in the recognition rate with and without the ANC. For the commercial program, the recognition rate is equivalent to that in the noise-free condition when the speaker distance is 0.5 m with a 57 dB noise in directions of 90 to 180 degrees. The maximum improvement reaches 65%. The recognition rate is degraded accordingly for off-direction noise placement, longer distance of the speech source, and/or a higher noise level of 67 dB.

For the news program, the recognition rate is slightly degraded compared to that in part (a) of Fig. 7.25 for noise directions of 135 and 180 degrees. However, with a noise directions of 30, 60, and 90 degrees, the recognition rate is significantly lower. The improvement is also degraded accordingly. This degradation is caused by similar spectral components in the the speech to be recognized and the news program. It should be noted that such a difference have some variance and its maximum and the minimum are shown. A transition from significant effect to moderate effect is observed in the direction of

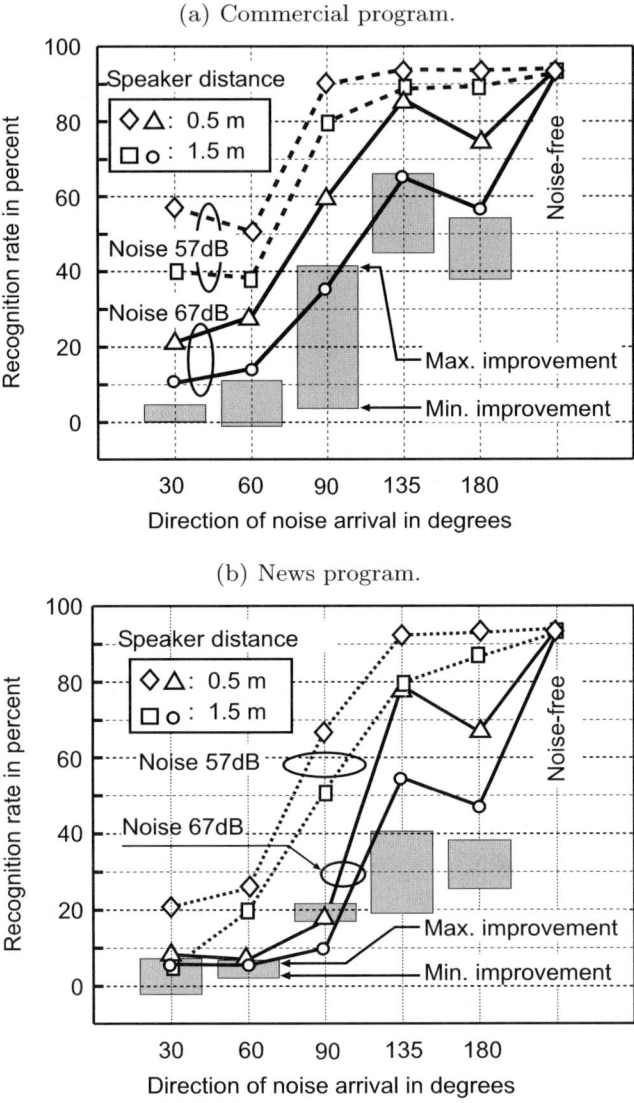

Fig. 7.25. Speech recognition results (top: commercial program, bottom: news program).

arrival of 90 degrees in part (a) and 135 degrees in part (b) of Fig. 7.25. In the case of a distant speaker, the recognition rates in the direction of arrival of 180 degrees are degraded. It is caused by significant contribution by increased reverberation and decreased SNR due to the attenuated speech.

Fig. 7.26. Child-care robot, PaPeRo.

7.7 Demonstration in a Personal Robot

In speech recognition for robots, microphone arrays [4] have become popular [20]. However, they are not as effective for diffuse noise as for directional interference. On the contrary, ANCs produce no directivity and are useful for diffuse noise.

The ANC with the GCCPF structure is equipped with in a personal robot, PaPeRo [7], whose child-care model is depicted in Fig. 7.26. PaPeRo has a laptop PC inside to perform all necessary computations and controls.[6] The primary microphone and the reference microphone are mounted on the forehead and the back of the neck. PaPeRo had been exhibited at the 2005 World Exposition (EXPO 2005), Aichi, Japan, as a childcare robot. Several children at a time with a dedicated instructor played a variety of games with the robot as shown in Fig. 7.27. 27000 children aged 3 to 12 enjoyed playing with PaPeRo. The total number of visitors reached 780000. In such a noisy environment, the speech recognition was successful due to the ANC with the GCCPF structure. This success demonstrates a revival of a classical technique originally proposed by Widrow et al.

7.8 Conclusions

Low-distortion noise cancellers and their applications have been presented. Distortion in Widrow's adaptive noise canceller (ANC) has been investigated

---

[6] A more compact implementation based on an embedded processor accomodating three ARM9 and a DSP cores is also available [17].

Fig. 7.27. PaPeRo demonstration with children and instructors.

to show that interference in coefficient adaptation and crosstalk are problems. As a solution to the interference problem, a paired filter (PF) structure has been described. For the crosstalk problem, it has been pointed out that CTRANC and a cross-coupled structure without a reliable adaptation control are not sufficiently good. As a good solution to both interference and crosstalk, an ANC with a cross-coupled paired filter (CCPF) structure has been presented. For more adverse environment, the CCPF structure has been extended to a generalized cross-coupled paired filter (GCCPF) structure. Evaluation results of the GCCPF structure have demonstrated its superior performance with respect to residual noise and distortion in a human-robot interaction scenario.

Although Widrow's adaptive noise canceller is a classical technique and has found a few applications, its descendants have found their ways in robotics where nondirectional interference plays a significant role. A successful demonstration of a partner-type robot PaPeRo at 2005 World Exposition in Aichi, Japan, for six months tells us that it is a revival of a classical technique.

References

1. M. J. Al-Kindi, J. Dunlop: A low distortion adaptive noise cancellation structure for real time applications, *Proc. ICASSP '87*, 2153–2156, Apr. 1987.
2. S. F. Boll: Suppression of acoustic noise in speech using spectral subtraction, *IEEE Trans. Acoust., Speech, Signal Processing*, **ASSP-27**(2), 113–120, Apr. 1979.
3. S. F. Boll, D. C. Pulsipher: Suppression of acoustic noise in speech using two microphone adaptive noise cancellation, *IEEE Trans. Acoust., Speech, and Signal Processing*, **ASSP-28**, 752–753, 1980.
4. M. Brandstein, D. Ward (eds.): *Microphone Arrays*, Berlin, Germany: Springer, 2001.
5. J. Dunlop, M. J. Al-Kindi, L. E. Virr: Application of adaptive noise cancelling to diver voice communications, *Proc. ICASSP '87*, 1708–1711, Apr. 1987.
6. Y. Ephraim, D. Malah: Speech enhancement using a minimum mean-square error short-time spectral amplitude estimator, *IEEE Trans. Acoust., Speech, Signal Processing*, **ASSP-32**(6), 1109–1121, Dec. 1984.
7. Y. Fujita: Personal robot PaPeRo, *J. of Robotics and Mechatronics*, **14**(1), Jan. 2002.
8. W. A. Gardner, B. G. Agee: Two-stage adaptive noise cancellation for intermittent-signal applications, *IEEE Trans. IT*, **IT-26**(6), 746–750, Nov. 1980.
9. G. C. Goodwin, K. S. Sin: *Adaptive Filtering, Prediction and Control*, Englewood Cliffs, NJ, USA: Prentice-Hall, 1985.
10. W. A. Harrison, J. S. Lim, E. Singer: A new application of adaptive noise cancellation, *IEEE Trans. Acoust., Speech, and Signal Processing*, **ASSP-34**, 21–27, 1986.
11. S. Ikeda, A. Sugiyama: An adaptive noise canceller with low signal-distortion for speech codecs, *IEEE Trans. Sig. Proc.*, 665–674, Mar. 1999.
12. S. Ikeda, A. Sugiyama: An adaptive noise canceller with low signal-distortion in the presence of crosstalk, *IEICE Trans. Fund*, 1517–1525, Aug. 1999.
13. H. Kubota, T. Furukawa, H. Itakura: Pre-processed noise canceller design and its performance, *IEICE Trans. Fund.*, **J69-A**(5), 584–591, May 1986 (in Japanese).
14. G. Mirchandani, R. L. Zinser, J. B. Evans: A new adaptive noise cancellation scheme in the presence of crosstalk, *IEEE Trans. CAS-II*, 681–694, Oct. 1992.
15. V. Parsa, P. A. Parker, R. N. Scott: Performance analysis of a crosstalk resistant adaptive noise canceller, *IEEE Trans. Circuits and Systems*, **43**, 473–482, 1996.
16. M. Sato, A. Sugiyama, S. Ohnaka: An adaptive noise canceler with low signal-distortion based on variable stepsize subfilters for human-robot communication, *IEICE Trans. Fund.*, **E88-A**(8), 2055–2061, Aug. 2005.
17. M. Sato, T. Iwasawa, A. Sugiyama: A noise-robust speech recognition on a compact speech dialogue module, *Proc. SIG AI-Challenge 2007*, Nov. 2007 (in Japanese).
18. A. Sugiyama, M. N. S. Swamy, E. I. Plotkin: A fast convergence algorithm for adaptive FIR filters, *Proc. ICASSP '89*, 892–895, 1989.
19. A. Sugiyama, M. Sato: Robust speech recognition in noisy environment for robot applications, *J. of Acoust. Soc. Japan*, **63**(1), 47–53, Jan. 2007 (in Japanese).
20. J.-M. Valin, J. Rouat, F. Michaud: Enhanced robot audition based on microphone array source separation with post-filter, *Proc. ICRSJ 2004*, **3**(28), 2123–2128, Oct. 2004.

21. T. Watanabe: Problems in the design of a speech recognition system and their solution, *Trans.*, **J.79-D-II**(12), 2022–2031, Dec. 1996 (in Japanese).
22. B. Widrow, J. R. Glover, Jr., J. M. McCool, J. Kaunitz, C. S. Williams, R. H. Hearn, J. R. Zeidler, E. Dong, Jr., R. C. Goodlin: Adaptive noise cancelling: principles and applications, *Proc. IEEE*, **63**(12), 1692–1716, 1975.
23. R. L. Zinser, G. Mirchandani, J. B. Evans: Some experimental and theoretical results using a new adaptive filter structure for noise cancellation in the presence of crosstalk, *Proc. ICASSP '85*, 1253–1256, Mar. 1985.

Part II

Echo Cancellation

8

Nonlinear Echo Cancellation Based on Spectral Shaping

Osamu Hoshuyama and Akihiko Sugiyama

Common Platform Software Research Laboratories,
NEC Corporation, Kawasaki, Japan

8.1 Introduction

Acoustic echo cancellation or suppression for hands-free cellphones is a challenging topic. One of the biggest problems is nonlinearity of the echo-path such as loudspeaker distortion and vibrations of the cellphone shell [4, 18, 19, 28]. When a speech signal with large power is injected into a small loudspeaker, the loudspeaker itself and many mechanical contacts in the shell generate distorted echo. Among the sources of nonlinearity (distortion), mechanical nonlinearity with shell vibrations is said to be the dominant factor [4].

An ordinary echo canceller with a linear adaptive filter is used to eliminate linear echo. However, it can not suppress nonlinear echo components that may be mixed with the linear echo. A linear adaptive filter models only the linear echo. The remaining nonlinear echo is one tenth of its linear counterpart or even larger in amplitude. It is audible and degrades the quality of communication. A common solution to such a significant nonlinear echo is to mute the whole residual signal obtained at the output of the echo canceller. However, it often causes discontinuous speech during double talk periods.

Using a nonlinear adaptive filter is also a popular solution to the nonlinear echo problem. Volterra adaptive filters fit the nonlinear echo-path model and can theoretically cancel nonlinear echoes [16, 25, 31]. Their drawbacks are heavy computational load and slow convergence. These problems become more distinct when a high-order Volterra filter is used. In the case of a cellphone handset, it is not possible for Volterra filters to track fast and frequent changes of the echo path. Simplified nonlinear adaptive filters that have fast tracking capability with reasonable computations have also been proposed [7, 20, 21, 27, 30, 32]. Unfortunately, the improvement by these simplified filters is limited.

Single-input postfilter is another approach to suppressing uncancelled nonlinear echo as well as ambient noise [2, 6, 10–12, 15]. When the uncancelled nonlinear echo is sufficiently small, or has different statistics from those of the near-end speech, a single-input postfilter is applicable to suppressing the

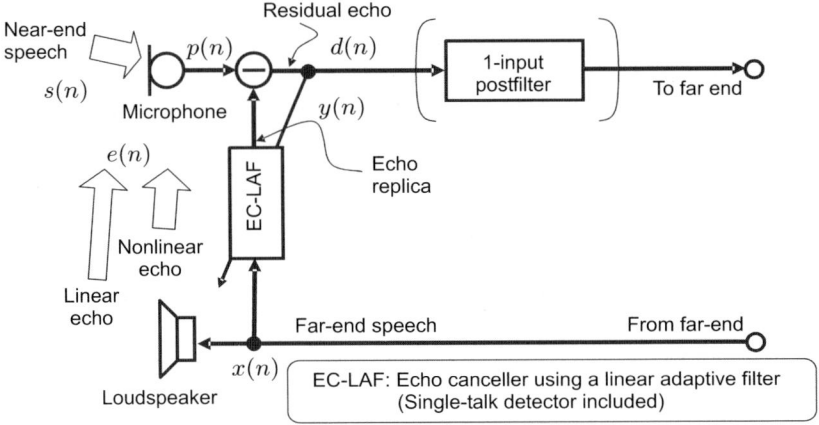

Fig. 8.1. Echo canceller using a linear adaptive filter.

uncancelled nonlinear echo. However, they are not useful for high nonlinearity often encountered in hands-free cellphone handsets equipped with inexpensive loudspeakers. The uncancelled nonlinear echo violates the condition that it is relatively small compared to the near-end speech.

Two-input postfilter based on spectral shaping with intentional oversubtraction is a more robust approach [9, 23, 24, 33]. In the nonlinear echo cancellers using two-input postfilters, the nonlinear residual echo is suppressed as well as the linear echo by spectral shaping. Thanks to intentional oversubtraction, these methods are robust against the nonlinear echo. However, the design procedures for the oversubtraction factors are not disclosed in the literature.

This chapter presents a nonlinear echo canceller using spectral shaping based on a frequency-domain model of highly nonlinear residual echo [14]. In the next section, the residual-echo model based on the spectral correlation between the residual echo and the echo replica is developed from experimental results. An echo suppressor structure based on the new model is shown in Sec. 8.3. Evaluation results using a hands-free cellphone handset are demonstrated in Sec. 8.4.

8.2 Frequency-Domain Model of Highly Nonlinear Residual Echo

Fig. 8.1 shows a basic structure of an echo canceller using linear adaptive filter (EC-LAF). The signal at the microphone, $p(n)$, consists of the near-end signal $s(n)$, and the echo signal $e(n)$:

$$p(n) = s(n) + e(n), \tag{8.1}$$

where n is the time index. $e(n)$ contains both the linear and nonlinear components of the echo. In the echo canceller with a linear adaptive filter (EC-LAF), the residual signal $d(n)$ is calculated by subtracting the echo replica $y(n)$, which is generated by LAF from the far-end signal $x(n)$, from $p(n)$. The residual signal $d(n)$ after the EC-LAF is expressed as a sum of the near-end signal $s(n)$ and the residual echo $q(n)$ as

$$d(n) = s(n) + q(n). \tag{8.2}$$

When the EC-LAF cancels the linear echo almost completely, $q(n)$ mainly consists of the nonlinear components of the echo.

8.2.1 Spectral Correlation Between Residual Echo and Echo Replica

The highly nonlinear residual echo and the echo replica of an EC-LAF under single-talk conditions were investigated in the frequency domain. The experimental setup to obtain the data for analysis is shown in Fig. 8.2. A loudspeaker mounted on the backside of the cellphone shell was used. The diaphragm diameter of the loudspeaker was 2.5 cm, and the distance between the loudspeaker and the microphone was 6 cm. The far-end signal was a male-and-female dialogue with a signal bandwidth of 4 kHz.

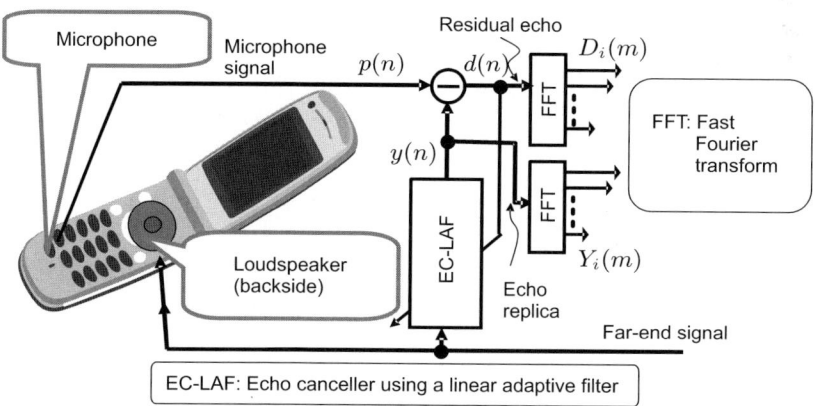

Fig. 8.2. Experimental setup.

A frequency-domain representation of Eq. 8.2 is obtained as

$$\boldsymbol{D}(m) = \boldsymbol{S}(m) + \boldsymbol{Q}(m), \tag{8.3}$$

$$\boldsymbol{D}(m) = \left[D_0(m), D_1(m), \ldots, D_{L-1}(m)\right]^{\mathrm{T}} \triangleq \mathrm{FFT}\{\boldsymbol{d}(mM)\} \tag{8.4}$$

$$\boldsymbol{S}(m) = \left[S_0(m), S_1(m), \ldots, S_{L-1}(m)\right]^{\mathrm{T}} \triangleq \mathrm{FFT}\{\boldsymbol{s}(mM)\} \tag{8.5}$$

$$\boldsymbol{Q}(m) = \left[Q_0(m), Q_1(m), \ldots, Q_{L-1}(m)\right]^{\mathrm{T}} \triangleq \mathrm{FFT}\{\boldsymbol{q}(mM)\} \tag{8.6}$$

where $\mathrm{FFT}[\,\cdot\,]$ is a windowed fast Fourier transform with overlap, and m, M, and L represent the frame index, the frame size, and the window size, respectively. The residual-signal vector $\boldsymbol{d}(n)$, the near-end speech vector $\boldsymbol{s}(n)$, and the residual-echo vector $\boldsymbol{q}(n)$ are defined by

$$\boldsymbol{d}(n) \triangleq \left[d(n), d(n-1), \ldots, d(n-L+1)\right]^{\mathrm{T}}, \tag{8.7}$$

$$\boldsymbol{s}(n) \triangleq \left[s(n), s(n-1), \ldots, s(n-L+1)\right]^{\mathrm{T}}, \tag{8.8}$$

$$\boldsymbol{q}(n) \triangleq \left[q(n), q(n-1), \ldots, q(n-L+1)\right]^{\mathrm{T}}. \tag{8.9}$$

For the i-th frequency bin, Eq. 8.3 becomes

$$D_i(m) = S_i(m) + Q_i(m), \tag{8.10}$$
$$(i = 0, 1, \ldots, L-1).$$

A frequency-domain representation $\boldsymbol{Y}(m)$ of the echo replica $y(n)$ is obtained by

$$\boldsymbol{Y}(m) = \left[Y_0(m), Y_1(m), \ldots, Y_{L-1}(m)\right]^{\mathrm{T}} \triangleq \mathrm{FFT}\{\boldsymbol{y}(mM)\}, \tag{8.11}$$

where

$$\boldsymbol{y}(n) \triangleq \left[y(n), y(n-1), \ldots, y(n-L+1)\right]^{\mathrm{T}}. \tag{8.12}$$

Fig. 8.3 plots the distribution of the residual echo $|D_i(m)|$ and echo replica $|Y_i(m)|$, for the same frame-index after convergence of the EC-LAF at various frequencies. Dots in the figure exhibit linear regression that is a sign of significant correlation between the residual echo and the echo replica at all the frequencies. Of course, distribution of the dots varies with the spectrum and level of the far-end signal. However, it stays in a limited range for various types of speech. At some frequencies (e.g. 313 Hz), the correlation is not strong, however, at those frequencies, the residual echo is relatively smaller than those at other frequencies with strong correlation, therefore, it is less important. Similar correlation was observed with other sets of dynamic loudspeakers and piezo loudspeakers. This correlation means that when the linear echo is large, the nonlinear echo is also large, which sounds natural. The nonlinear residual echo can be modeled by the dotted line as a good 1st order approximation, although harmonics generated by the nonlinearity are not

Nonlinear Echo Cancellation Based on Spectral Shaping 271

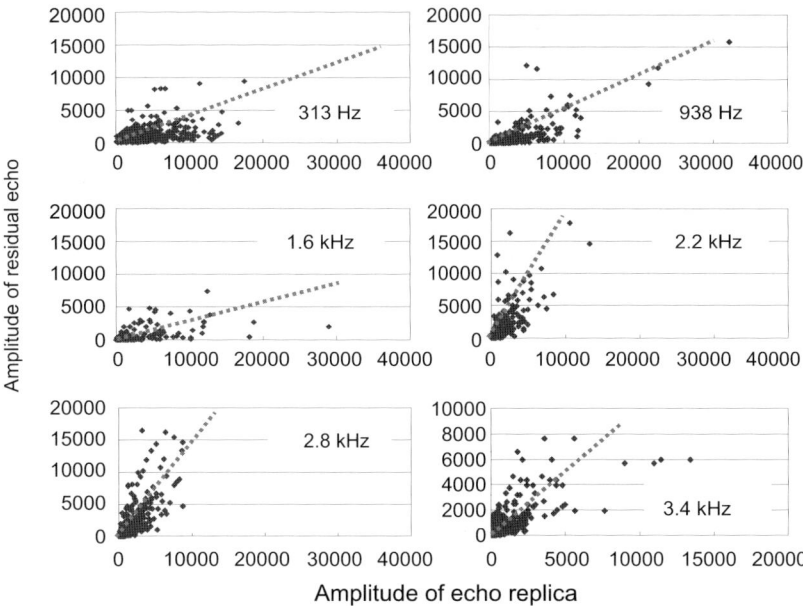

Fig. 8.3. Spectral correlation between residual echo and echo replica at various frequencies.

taken into account. The dotted line can be obtained by a regression analysis or a simple averaging.

In order to investigate the dependency of the correlation on conditions, a regression analysis was performed for different positions of the loudspeaker and various far-end signals with different characteristics. Experimental setup is shown in Fig. 8.4. An independent sealed enclosure with a loudspeaker was

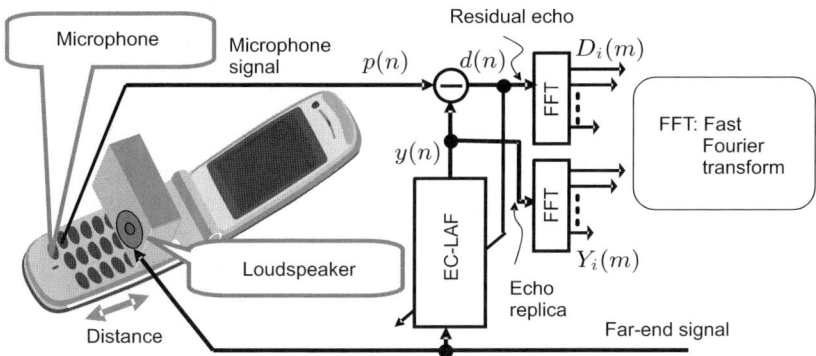

Fig. 8.4. Experimental setup.

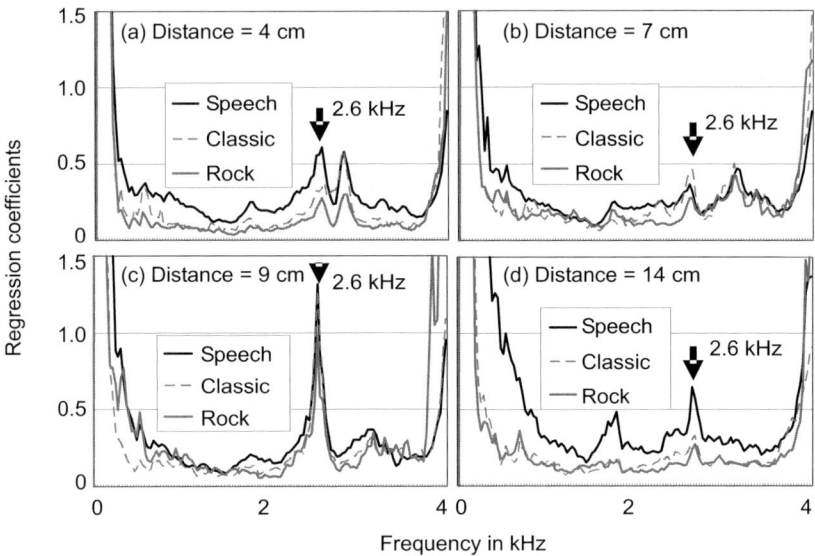

Fig. 8.5. Regression coefficients as a function of frequency for different distances between loudspeaker and microphone, and various far-end signal categories.

attached to the surface of the cellphone shell by scotch tape to change the distance between the loudspeaker and the microphone. Diaphragm diameter of the loudspeaker was 1.2 cm, and the distances were 4, 7, 9, and 14 cm. The enclosure volume was 110 cm^3, which is much bigger than ordinary loudspeaker enclosures in cellphones for smaller enclosure nonlinearity. The far-end signals were male-and-female dialogue, classical music (brass), and rock music. They were bandlimited to 0.3–3.4 kHz, and preprocessed by the AMR codec [1] at 12.2 kbps. The lengths of the signals were from 20 to 40 seconds. Then, the regression coefficients between the spectral amplitudes of the residual echo and the echo replica were calculated.

Regression coefficients as a function of frequency are shown in Fig. 8.5 (a), (b), (c), and (d) for different loudspeaker-microphone distances. A larger regression coefficient means higher nonlinearity. The regression coefficients at low-end and high-end frequencies are large for all the curves in Fig. 8.5. It is because the input signals were bandlimited, however, the loudspeaker has large distortion.

The curves of the regression coefficients depend on the distance between the loudspeaker and the microphone, and the characteristics of the far-end signal. By comparing all the curves, it can be seen that the influence of the loudspeaker-microphone distance is larger than that of the far-end signal. All the curves have a peak at 2.6 kHz due to resonance of the cellphone shell.

However, the height of the resonance peak basically depends on the distance, but not on the characteristics of the far-end signal.

The speech and music signals have significantly different spectra. However, the shapes of the corresponding curves are similar in each graph. The variation in the regression coefficients is smaller than 0.3 in the frequency range from 0.5 to 3.5 kHz. It is sufficiently small for the nonlinear-echo canceller using spectral shaping which is introduced in Sec. 8.3. A common set of regression coefficients can be used for suppressing nonlinear echo generated from various far-end signals.

8.2.2 Model of Residual Echo Based on Spectral Correlation

Based on Fig. 8.3, when there is no near-end signal, a ratio of $|Q_i(m)|$ to $|Y_i(m)|$ is defined as $a_i(m)$. Using $a_i(m)$, a residual echo component $|Q_i(m)|$ is given by

$$|Q_i(m)| = a_i(m) \cdot |Y_i(m)|. \tag{8.13}$$

Eq. 8.13 suggests that $|Q_i(m)|$ can be obtained from the echo replica $|Y_i(m)|$ if $a_i(m)$ is available. However, in reality, it is not the case. The ratio $a_i(m)$ should somehow be estimated.

Let us consider approximating $a_i(m)$ by $\widehat{a}_i$, the slope ratio of the dotted line in Fig. 8.3, using averaged absolute values of the residual echo and the echo replica. Then we have

$$\widehat{a}_i = \frac{\overline{|Q_i(m)|}}{\overline{|Y_i(m)|}}, \tag{8.14}$$

where overline $\overline{\cdots}$ means an averaging operation. When there is no near-end speech, i.e. $S_i(m) = 0$, Eq. 8.10 reduces to

$$D_i(m)_{\text{Single talk}} = Q_i(m). \tag{8.15}$$

From Eqs. 8.14 and 8.15, $\widehat{a}_i$ is given by

$$\widehat{a}_i = \frac{\overline{|D_i(m)|_{\text{Single talk}}}}{\overline{|Y_i(m)|}}. \tag{8.16}$$

The estimated ratio $\widehat{a}_i$ can be obtained by advance experimental measurements in quiet environments, or can be estimated during a single talk period. One set of $\widehat{a}_i$ is needed for each set up of the loudspeaker and the microphone.

By approximating $a_i(m)$ in Eq. 8.13 with $\widehat{a}_i$, the residual echo $|Q_i(m)|$ is modeled as the product of $\widehat{a}_i$ and $|Y_i(m)|$.

$$|Q_i(m)| \simeq |\widehat{Q}_i(m)| \tag{8.17}$$

$$\stackrel{\triangle}{=} \widehat{a}_i \cdot |Y_i(m)|. \tag{8.18}$$

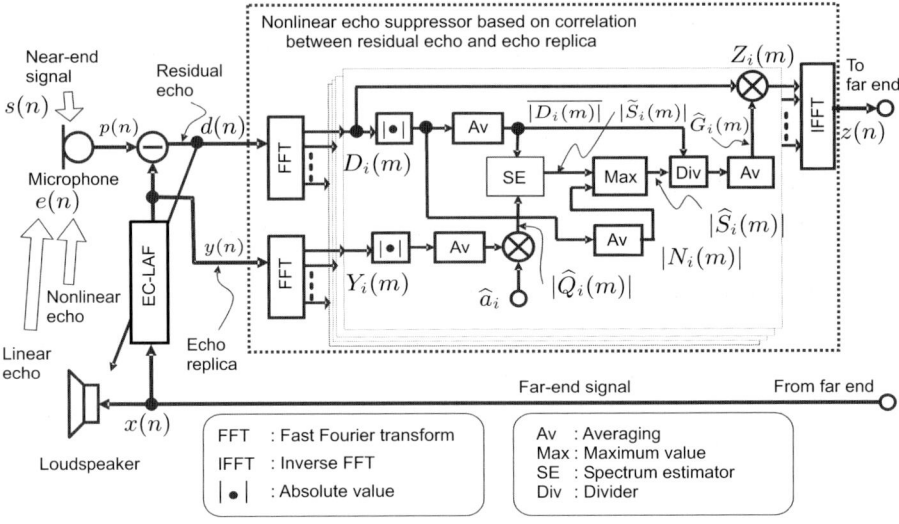

Fig. 8.6. Hands-free system with a nonlinear echo canceller based on the correlation between the nonlinear residual echo and the echo replica.

8.3 Echo Canceller Based on the New Residual Echo Model

8.3.1 Overall Structure

Eq. 8.10 can be viewed as an additive model of the residual signal, which is widely known in noise suppression. The near-end signal corresponds to the speech to be enhanced and the nonlinear echo, to the additive noise. Spectral subtraction [5] is a popular technique in noise suppressors and can be directly applied to the nonlinear-echo model in Eq. 8.18. However, minimum mean-square error short-time spectral amplitude estimation (MMSE-STSA) [8] is adopted to reduce subjectively annoying musical noise as in [17]. Fig. 8.6 shows the structure of the nonlinear-echo canceller including an EC-LAF.

In the framework of MMSE-STSA, the output signal, $|Z_i(m)|$, is obtained as a product of a spectral gain $\widehat{G}_i(m)$ and the residual signal $|D_i(m)|$ as

$$|Z_i(m)| = \widehat{G}_i(m) \cdot |D_i(m)|. \tag{8.19}$$

$|Z_i(m)|$ is combined with $\angle D_i(m)$ to reconstruct $Z_i(m)$ as

$$Z_i(m) = |Z_i(m)| \cdot \exp(j\angle D_i(m)) \tag{8.20}$$
$$(i = 0, 1, \ldots, L-1).$$

In Eq. 8.20, $\angle D_i(m)$ is not modified at all because the phase is not important for speech intelligibility [22].

The output signals $z(n)$ in the time domain are obtained as elements of the segmented frame which is reconstructed by the inverse Fourier transform as follows.

$$\boldsymbol{z}(n) \triangleq \big[z(n),\, z(n-1),\, \ldots,\, z(n-L+1)\big]^{\mathrm{T}} = \mathrm{IFFT}\{\boldsymbol{Z}(m)\} \quad (8.21)$$

$$\boldsymbol{Z}(m) \triangleq \big[Z_0(m),\, Z_1(m),\, \ldots,\, Z_{L-1}(m)\big]^{\mathrm{T}}, \quad (8.22)$$

where IFFT{ ... } is a windowed inverse FFT with overlap-add operations.

8.3.2 Estimation of Near-End Speech

The spectral gain $\widehat{G}_i(m)$ in Eq. 8.19 is ideally equal to the ratio of $|S_i(m)|$ to $|D_i(m)|$. However, it is not available because $|S_i(m)|$ is the ideal output. To obtain $\widehat{G}_i(m)$, the spectral amplitude of the near-end signal $|S_i(m)|$ is estimated by substituting $Q_i(m)$ in Eq. 8.10 with Eq. 8.18.

$$|S_i(m)| = |D_i(m) - Q_i(m)| \quad (8.23)$$

$$\simeq |D_i(m) - \widehat{Q}_i(m)| \quad (8.24)$$

$$\simeq |D_i(m) - \widehat{a}_i \cdot Y_i(m)|. \quad (8.25)$$

$|D_i(m)|$ and $|Y_i(m)|$ have almost no cross correlation because they are decorrelated by the EC-LAF. Therefore, by taking averaged power of both sides of Eq. 8.25, we obtain

$$\overline{|S_i(m)|}^2 \simeq \overline{|D_i(m)|}^2 - \widehat{a}_i^2 \cdot \overline{|Y_i(m)|}^2. \quad (8.26)$$

By taking the square root of Eq. 8.26, $|\widetilde{S}_i(m)|$, an approximation to $|S_i(m)|$, is obtained as follows.

$$|S_i(m)| \simeq \overline{|S_i(m)|} \quad (8.27)$$

$$\simeq |\widetilde{S}_i(m)| \triangleq \sqrt{\overline{|D_i(m)|}^2 - \widehat{a}_i^2 \cdot \overline{|Y_i(m)|}^2}. \quad (8.28)$$

$\overline{|D_i(m)|}$ and $\overline{|Y_i(m)|}$ are recursively calculated by two averaging operations as follows.

$$\overline{|D_i(m)|} = \begin{cases} \beta_{DA}\,|D_i(m)| + (1-\beta_{DA})\,\overline{|D_i(m-1)|}, \\ \quad \text{if } |D_i(m)| > \overline{|D_i(m-1)|}, \\ \beta_{DD}\,|D_i(m)| + (1-\beta_{DD})\,\overline{|D_i(m-1)|}, \\ \quad \text{otherwise,} \end{cases} \quad (8.29)$$

$$\overline{|Y_i(m)|} = \begin{cases} \beta_{YA}\,|Y_i(m)| + (1-\beta_{YA})\,\overline{|Y_i(m-1)|}, \\ \quad \text{if } |Y_i(m)| > \overline{|Y_i(m-1)|}, \\ \beta_{YD}\,|Y_i(m)| + (1-\beta_{YD})\,\overline{|Y_i(m-1)|}, \\ \quad \text{otherwise,} \end{cases} \quad (8.30)$$

where β_{DA}, β_{DD}, β_{YA}, and β_{YD} are constants to determine the time-constant of the averages ($0 < \beta_{DD} < \beta_{DA} \leq 1$ and $0 < \beta_{YD} < \beta_{YA} \leq 1$). Each averaging operation has two averaging constants for superior tracking capability typically represented by "fast attack and slow decay" [13].

The estimated spectral amplitude $|\tilde{S}_i(m)|$ usually contains some error, because the model of the residual echo is not precise. When the error is large, oversubtraction may occur, which brings modulations by the far-end signal to the near-end signal. When the near-end signal is nonstationary like speech, most of the modulation effects are masked by the nonstationarity. However, when the near-end signal is stationary like air conditioner noise, the modulations are perceived as fluctuations.

To reduce the modulations, a spectral flooring [3] is introduced in the nonlinear-echo canceller. The floor value is proportional to the amount of the stationary component of the near-end signal. The stationary component $|N_i(m)|$ is calculated from $\overline{|D_i(m)|}$ by an averaging operation with a very-slow-attack-and-fast-decay response.

$$|N_i(m)| = \begin{cases} \beta_{NA}\,\overline{|D_i(m)|} + (1-\beta_{NA})\,|N_i(m-1)|, \\ \quad \text{if } \overline{|D_i(m)|} > |N_i(m-1)|, \\ \beta_{ND}\,\overline{|D_i(m)|} + (1-\beta_{ND})\,|N_i(m-1)|, \\ \quad \text{otherwise,} \end{cases} \quad (8.31)$$

where β_{NA} and β_{ND} are averaging constants satisfying $0 < \beta_{NA} < \beta_{ND} \leq 1$. The two averaging constants enables slow-attack-and-fast-decay response which is needed for minimum tracking to estimate the stationary component of the near-end signal. An improved spectral amplitude $|\widehat{S}_i(m)|$ is obtained by the flooring operation as follows.

$$|\widehat{S}_i(m)| \triangleq \max\left\{\gamma_D\,|N_i(m)|,\,|\tilde{S}_i(m)|\right\}, \quad (8.32)$$

where γ_D is a gain parameter for the flooring operation, and an operator $\max\{...,...\}$ stands for the maximum of the two arguments.

8.3.3 Spectral Gain Control

The spectral gain $\widehat{G}_i(m)$ is calculated by smoothing a ratio of $|\widehat{S}_i(m)|$ to $\overline{|D_i(m)|}$ as follows:

$$\widehat{G}_i(m) = \begin{cases} \beta_{GA}\,\widetilde{G}_i(m) + (1-\beta_{GA})\,\widehat{G}_i(m-1), \\ \quad \text{if } \widetilde{G}_i(m) > \widehat{G}_i(m-1), \\ \beta_{GD}\,\widetilde{G}_i(m) + (1-\beta_{GD})\,\widehat{G}_i(m-1), \\ \quad \text{otherwise,} \end{cases} \quad (8.33)$$

with

$$\widetilde{G}_i(m) = \frac{|\widehat{S}_i(m)|}{|D_i(m)| + \delta_G}, \tag{8.34}$$

where $\widetilde{G}_i(m)$ is a temporary variable for calculating $\widehat{G}_i(m)$, and δ_G is a positive constant for avoiding zero division. The averaging constants β_{GA} and β_{GD} correspond to β_{DA} and β_{DD} in Eq. 8.29 satisfying $0 < \beta_{GD} < \beta_{GA} \leq 1$.

8.4 Evaluations

8.4.1 Objective Evaluations

Simulations of quiet and noisy environments were performed with recorded data using a folding hands-free cellphone with a 1-inch loudspeaker mounted at a lower backside. The distance between the loudspeaker and the microphone was approximately 6 cm. Loudness of the loudspeaker was set so that the echo level at the microphone is comparable to the near-end speech.

For the EC-LAF, the number of filter coefficients was 128 and the coefficients were updated by the normalized least-mean-square (NLMS) algorithm with a double-talk control detector based on two echo-path models [26]. The nonlinear-echo canceller employed the window size L of 256, and the frame size M of 160 following the AMR codec standard [1]. A Hanning window was used as the windowing function. Constants for averaging operations were set to $\beta_{DA}=\beta_{YA}=1.0$, $\beta_{DD}=\beta_{YD}=0.8$, $\beta_{NA}=0.001$, $\beta_{ND}=0.2$, $\beta_{GA}=0.8$, and $\beta_{GD}=0.3$. The gain parameter γ_D for the flooring operation was 1.0. As a set of $\widehat{a}_i$, a curve for a female was used, though the far-end talker in the evaluations was a male to demonstrate the robustness of $\widehat{a}_i$.

8.4.1.1 Quiet Environment

Fig. 8.7 depicts simulation results under the quiet environment, where only near-end speech and far-end speech exist with no noise. The curve in (a) represents the echo and the near-end speech signal. (b) is the near-end signal, which is the ideal output signal. (c) and (d) are the output signals of the EC-LAF and the nonlinear-echo canceller (NL-EC).

In the left half of Fig. 8.7, a single talk condition is implemented. The amplitude in (c) is one tenth or more of that in (a), which indicates that the EC-LAF canceled the echo only by 15 to 20 dB due to the distortion of the echo path. Referring to section A in (c), where distortion of the echo path is significant, the residual echo of the EC-LAF is audible. On the other hand, the residual echo of the NL-EC is inaudible as shown in (d).

In the right half of Fig. 8.7, a near-end speech signal was added to the echo to implement a double-talk condition. Ideally, the signal to the far end should be the near-end speech shown in Fig. 8.7(b). Referring to section B of (b), the ideal output should be zero, that is achieved in (d) by the NL-EC.

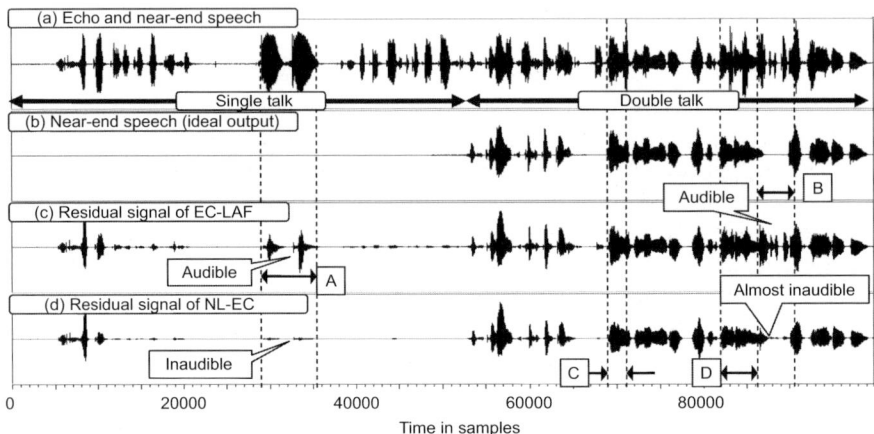

Fig. 8.7. Input and output signals in quiet environment.

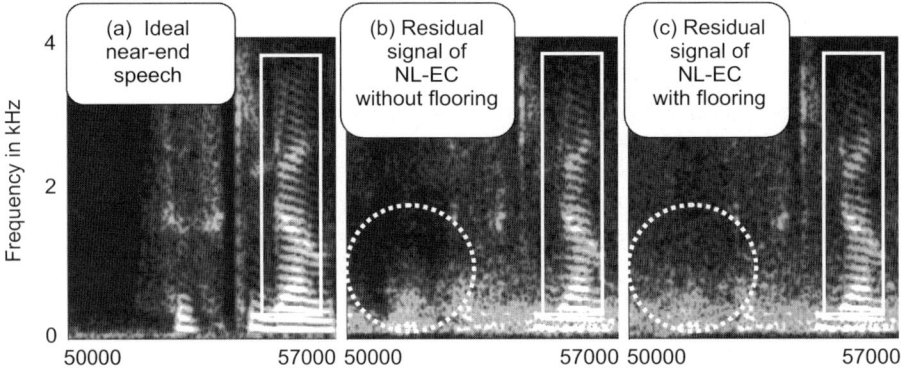

Fig. 8.8. Spectrograms in noisy environment.

The residual echo of the NL-EC is almost inaudible as shown in the figure. However, the EC-LAF fails to cancel the nonlinear echo, whose residual echo is depicted in section B in (c). The residual echo after the EC-LAF is clearly audible.

It is also seen that the output signal of the NL-EC, (d), exhibits some difference from the ideal near-end speech (b), in double-talk periods as sections C and D. Such a difference is mainly caused by attenuation at high frequencies. Despite the degradation, the quality of the output (d) in Fig. 8.7 was sufficiently good for speech communication.

8.4.1.2 Noisy Environment

To demonstrate the effectiveness of the flooring operation, simulations of a noisy environment were also performed. A station noise was added to the signal in Fig. 8.7 (a). Fig. 8.8 shows spectrograms of the ideal near-end speech and the output signals of the NL-ECs with and without the spectral flooring. In (b) and (c), the NL-ECs caused attenuation in the middle of the figures at high frequencies. However, as shown in the rectangular boxes, they keep the harmonic structure of the near-end speech. The NL-EC without the spectral flooring caused oversubtraction as depicted in Fig. 8.8 (b) as a dark spot in the dotted circle. On the contrary, the NL-EC with the spectral flooring preserves uniformity of the background noise as shown in (c), in which the dark spot in (b) disappeared. The difference is perceived as less modulated background noise.

8.4.2 Subjective Evaluation

A subjective evaluation were performed with 5 sets of recorded data obtained in quiet and noisy environments using the same hands-free cellphone as in Sec. 8.4.1. Loudness of the loudspeaker was set so that the echo level at the microphone is comparable to the near-end speech. All the parameters are the same as those in the previous section.

The output signals were evaluated by mean opinion score (MOS) with headphone listening by 10 nonprofessional subjects. As anchors, the near-end signal with nonlinear echo was used for grade 1, and the original near-end signal without echo for grade 5. The subjects were instructed with examples showing that there may be attenuation of high frequency components or near-end signal modulation with the oversubtraction during double-talk periods.

Evaluation results are shown in Fig. 8.9. The number beside each bar represents the score obtained by the corresponding method. The vertical stroke

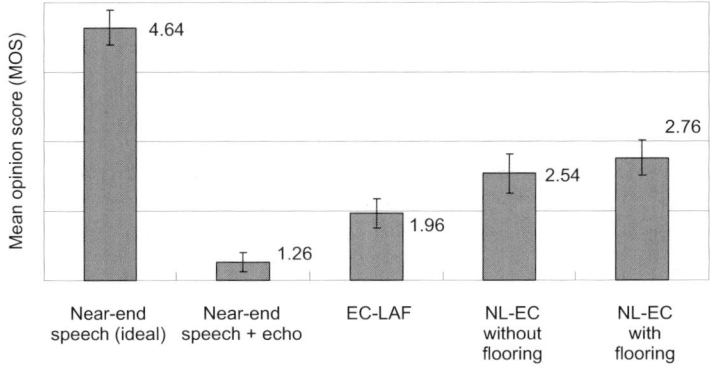

Fig. 8.9. Subjective test results.

on each bar indicates the 95 % confidence interval. The EC-LAF obtained 1.96 points because the residual echo is still audible. The nonlinear-echo canceller without flooring (NL-EC without flooring) can suppress the nonlinear residual echo almost completely, however, the near-end signal modulation is serious, thus, it scored 2.54 points. Its confidence interval has no overlap with that of the EC-LAF, which means that the nonlinear-echo canceller has statistically significant improvement at the output signal quality. The nonlinear-echo canceller with flooring (NL-EC with flooring) obtained 2.76 points, which is the highest of all the methods and 0.8 point higher than that of the EC-LAF. Though its confidence interval is overlapped with that of NL-EC without flooring, further subjective evaluations by more subjects will verify the difference.

8.5 DSP Implementation and Real-Time Evaluation

In order to evaluate the resource requirements, the nonlinear-echo canceller was implemented on a DSP starter kit (DSK) of TMS320C6416T running at 1 GHz [29]. The programming was carried out in C language with a compiler provided by Texas Instruments. In the implementation, three typical memory allocation methods were compared to evaluate the computational load. Fig. 8.10 shows computational loads for different memory allocations. When both program codes and data are allocated to internal memory (SRAM: synchronized RAM), the computations including EC-LAF are minimized to 6.1 MIPS (million instructions per second), although total usage of internal memory is 88 kBytes. When all the codes and data are allocated to external memory (SDRAM: synchronized dynamic RAM), which is the worst case, the total computations are 16.8 MIPS. In case 32 kBytes of fast cash memory (L2 cash) is available on the internal memory, even when only the external memory is used, the total computations are as small as 9.2 MIPS.

Real-time hands-free communication tests were performed using the DSK with the set of loudspeaker and microphone on the real cellphone mockup in Sec. 8.2. The users' comments were positive and agree with the subjective evaluation results shown in Sec. 8.4.2. They said that the echo was sufficiently small for conversation and the degradation of the near-end signal was acceptable even in double-talk periods.

Even when the users changed the echo path by touching the cellphone shell, the echo-path change caused small echo and acceptable attenuation of the near-end signals, thanks to the robustness of the nonlinear-echo canceller. Most of the users said that the quality of the hands-free communication was almost the same as that of hand-set communication.

8.6 Conclusions

This chapter has presented a nonlinear-echo canceller based on the correlation between spectral amplitudes of residual echo and echo replica in the

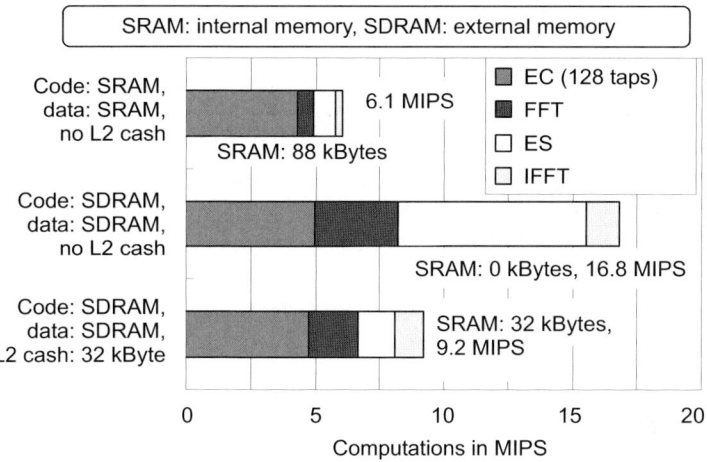

Fig. 8.10. Computational loads for different memory allocations.

frequency domain. The nonlinear-echo canceller structure controls the gain in each frequency bin based on the new model where the highly nonlinear echo is approximated as a product of a regression coefficient and the linear echo replica. To reduce annoying modulation by the error of the model, a flooring operation of the estimated near-end signal level is introduced to the gain control. Simulation results with speech data have demonstrated that the nonlinear-echo canceller reduces the highly nonlinear residual echo to an almost inaudible level. The subjective evaluation has shown that the MOS of the nonlinear-echo suppressor with flooring is superior to that of an echo canceller with linear adaptive filter by 0.8 points on a 5-point scale.

Acknowledgement

The authors would like to thank Dr. Masahiro Serizawa, Senior Manager of Media Processing Technology Group, Common Platform Software Research Laboratories, NEC Corporation, for his guidance and valuable comments.

References

1. 3GPP TS26.90: Adaptive multi-rate speech codec; transcode functions, March 2001.
2. C. Beaugeant, P. Scalart: Combined systems for noise reduction and echo cancellation, *Proc. EUSIPCO '98*, 957–960, Island of Rhodes, Greece, Sep. 1998.
3. M. Berouti, R. Schwartz, J. Makhoul: Enhancement of speech corrupted by acoustic noise, *Proc. ICASSP '79*, 208–211, Apr. 1979.

4. A. N. Birkett, R. A. Goubran: Limitations of handsfree acoustic echo cancellers due to nonlinear loudspeaker distortion and enclosure vibration effects, *Proc. WASPAA '95*, 13–16, 1995.
5. S. F. Boll: Suppression of acoustic noise in speech using spectral subtraction, *IEEE Trans. on Acoustics, Speech, and Signal Processing*, **ASSP-27**(2), 113–120, April 1979.
6. A. S. Chhetri, A. C. Surendran, J. W. Stokes, J. C. Platt: Regression-based residual acoustic echo suppression, *Proc. IWAENC '05*, Eindhoven, The Netherlands, Sep. 2005.
7. G. Enzner, P. Vary: Robust and elegant, purely statistical adaptation of acoustic echo canceller and postfilter, *Proc. IWAENC '03*, 43–46, Kyoto, Japan, Sep. 2003.
8. Y. Ephraim, D. Malah: Speech enhancement using a minimum mean-square error short-time spectral amplitude estimator, *IEEE Trans. on Acoustics, Speech, and Signal Processing*, **ASSP-32**(6), 1109–1121, 1984.
9. C. Faller, J. Chen: Suppressing acoustic echo in a spectral envelope space, *IEEE Trans. on Speech and Audio Processing*, **13**(5), 1048–1062, Sep. 2005.
10. S. Gustafsson, P. Jax: Combined residual echo and noise reduction: a novel psychoacoustically motivated algorithm, *Proc. EUSIPCO '98*, 961–964, Island of Rhodes, Greece, Sep. 1998.
11. S. Gustafsson, P. Jax, A. Kamphausen, P. Vary: A postfilter for echo and noise reduction avoiding the problem of musical tones, *Proc. ICASSP '99*, 873–876, Phoenix, AZ, USA, March 1999.
12. S. Gustafsson, R. Martin, P. Jax, P. Vary: A psychoacoustic approach to combined acoustic echo cancellation and noise reduction, *IEEE Trans. on Speech and Audio Processing*, **10**(5), 245–256, July 2002.
13. E. Hänsler, G. Schmidt: *Acoustic Echo and Noise Control*, New York, NY, USA: Wiley, 2004.
14. O. Hoshuyama, A. Sugiyama: An acoustic echo suppressor based on a frequency-domain model of highly nonlinear residual echo, *Proc. ICASSP '06*, **5**, 269–272, Toulouse, France, May 2006.
15. W. L. B. Jeannes, P. Scalart, G. Faucon, C. Beaugeant: Combined noise and echo reduction in hands-free systems: a survey, *IEEE Trans. on Speech and Audio Processing*, **9**(8), 808–820, Nov. 2001.
16. A. J. M. Kaizer: Modeling the nonlinear response of an electrodynamic loudspeaker by a Volterra series expansion, *J. AES*, **35**(6), June 1997.
17. M. Kato, A. Sugiyama, M. Serizawa: A family of 3GPP-standard noise suppressors for the AMR codec and the evaluation results, *Proc. ICASSP '03*, **1**, 916–919, Hong Kong, Apr. 2003.
18. W. Klippel: Nonlinear large-signal behaviour of electrodynamic loudspeakers at low frequencies, *J. AES*, **40**(6), June 1992.
19. M. E. Knappe, R. A. Goubran: Steady-state performance limitations of full-band acoustic echo cancellers, *Proc. ICASSP '94*, **4**, 81–84, 1994.
20. F. Kuech, A. Mitnacht, W. Kellerman: Nonlinear acoustic echo cancellation using adaptive orthogonalized power filters, *Proc. ICASSP '05*, **3**, 105–108, Philadelphia, PY, USA, Mar. 2005.
21. F. Kuech, W. Kellerman: Nonlinear residual echo suppression using a power filter model of the acoustic echo path, *Proc. ICASSP '07*, **1**, 73–76, Honolulu, HI, USA, April 2007.

22. J. S. Lim, A. V. Oppenheim: Enhancement and bandwidth compression of noisy speech, *Proc. IEEE*, **67**(12), 1586–1604, Dec. 1979.
23. X. Lu, B. Champagne: A centralized acoutic echo canceller exploiting masking properties of the human ear, *Proc. ICASSP '03*, **5**, 377–380, Hong Kong, Apr. 2003.
24. X. Lu, B. Champagne: Acoustic echo cancellation with post-filtering in subband, *Proc. WASPAA '03*, 29–32, Mohonk, New Paltz, NY, USA, Oct. 2003.
25. V. J. Mathews: Adaptive polynomial filters, *IEEE Sig. Pro. Mag.*, **8**(3), 10–26, July 1991.
26. K. Ochiai, T. Araseki, T. Ogihara: Echo canceller with two echo path models, *IEEE Trans. Com.*, **COM-25**(6), 589–595, June 1977.
27. G. B. Sentoni: Nonlinear echo cancellation: a real implementation, *Proc. IS-SPIT '03*, 656–659, Darmstadt, Germany, Dec. 2003.
28. M. Soria-Rodoriguez, M. Gabbouj, N. Zacharov, M. S. Hamalainen, K. Koivuniemi: Modeling and real-time auralization of electrodynamic loudspeaker nonlinearity, *Proc. ICASSP '04*, **4**, 81–84, Montreal, Canada, 2004.
29. Spectrum Digital Inc.: TMS320C6416T DSK technical reference, (http://c6000.spectrumdigital.com/dsk6416/V3/docs/dsk6416_TechRef.pdf), Nov. 2004.
30. A. Stenger, R. Rabenstein: An acoustic echo canceller with compensation of nonlinearities, *Proc. EUSIPCO '98*, 969–972, Island of Rhodes, Greece, Sep. 1998.
31. A. Stenger, L. Trautmann, R. Rabenstein: Nonlinear acoustic echo cancellation with 2nd order adaptive Volterra filters, *Proc. ICASSP '99*, 877–880, Phoenix, AZ, USA, March 1999.
32. A. Stenger, W. Kellermann: Nonlinear acoustic echo cancellation with fast convergence memoryless preprocessor, *Proc. ICASSP '00*, 805–808, Istanbul, Turkey, June 2000.
33. F. Wallin, C. Faller: Perceptual quality of hybrid echo canceller/suppressor, *Proc. ICASSP '04*, **4**, 157–160, Montreal, Canada, May 2004.

Part III

Signal and System Quality Evaluation

9
Telephone-Speech Quality

Ulrich Heute

Christian-Albrechts University, Kiel, Germany

9.1 Telephone-Speech Signals

9.1.1 Telephone Scenario

A telephone user speaks and produces an acoustical sound pressure $s_0(t)$. This signal, perhaps together with some surrounding disturbance, enters the microphone of the user's handset, headset, or hands-free equipment. The electrical microphone output $\check{x}_0(t)$ is then generally filtered and band-limited, before it is transmitted – nowadays digitally, of course. With some further possible distortions, it reaches the receiver. There it may be post-processed digitally and/or filtered again as an electrical signal, before an acoustical output $s_1(t)$ is produced. The *quality* of this signal is our concern.

9.1.2 Telephone-Scenario Model

Fig. 9.1 models this telephone-communication scenario such as it is of interest in the following: A speaker produces $s_0(t)$ and thereby the microphone signal $\check{x}_0(t)$. A band limitation by an "anti-aliasing lowpass" (AA-LP) with a cutoff

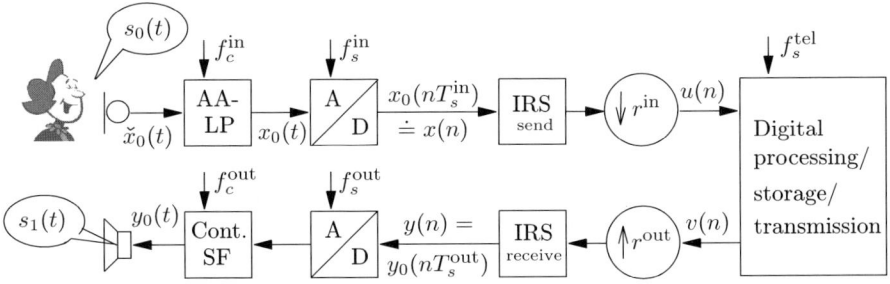

Fig. 9.1. Schematic model of speech processing in a transmission or storage application, defining acoustical signals and their electrical and digital equivalents.

frequency f_c^{in} avoids spectral overlap by the succeeding sampling of the filter output $x_0(t)$, which happens in the analog-to-digital converter (ADC) together with a linear, high-wordlength, i.e., negligible, quantization. The sampling frequency f_s^{in} has to be chosen according to the sampling theorem, i.e., with

$$f_s^{\text{in}} \doteq \frac{1}{T_s^{\text{in}}} \geq 2 \cdot f_c^{\text{in}}. \tag{9.1}$$

The digitized sequence

$$x(n) = x_0\left(n \cdot T_s^{\text{in}}\right) \tag{9.2}$$

is then digitally processed. The processing system may include a further band limitation actually due to the telephone apparatus in use, indicated by the block "IRS$_{\text{send}}$" in Fig. 9.1. This "intermediate reference system", standardized by the International Telecommunication Union (ITU) in a recommendation [44], is a bandpass filter imitating the usual narrow handset-microphone transmission band by an averaged frequency response; a modified version ($[IRS_{\text{send}}]_{\text{mod}}$) takes care of the less sharp upper-band limit in more recent handsets (see Fig. 9.2 (a), [50]).

The processing then continues possibly with a downsampling of $x(n)$ to the telephone-system sampling rate

$$f_s^{\text{tel}} = \frac{f_s^{\text{in}}}{r^{\text{in}}} = 8\,\text{kHz}. \tag{9.3}$$

The decimated sequence $u(n)$ is further filtered, e.g., for equalization purposes; non-linear operations follow, e.g., by adaptive compression or low-wordlength quantization; error insertions, concealments, or corrections happen during and after transmission or storage, and some final filtering may try to remove unwanted or to add missing components. Also, the receiving telephone apparatus' band limitation may be covered in a block "IRS$_{\text{receive}}$" or a modified version (see Fig. 9.2 (b), [44, 50]). The output signal $v(n)$ may be upsampled and digitally interpolated, yielding the sequence

$$y(n) = y_0\left(nT_s^{\text{out}}\right) \tag{9.4}$$

with

$$\frac{1}{T_s^{\text{out}}} = f_s^{\text{out}} = r^{\text{out}} \cdot f_s^{\text{tel}}. \tag{9.5}$$

From $y(n)$, the final acoustic speech signal $s_1(t)$ is created by a digital-to-analog converter (DAC) and a continuous-time smoothing filter (SF) with cutoff frequency f_c^{out}.

Due to the various influences, the sequences $u(n)$ and $v(n)$ as well as $x(n)$ and $y(n)$, the corresponding electrical signals $x_0(t)$ and $y_0(t)$, and the analog acoustical signals $s_0(t)$ and $s_1(t)$ will differ. The processing influences may be unwanted, as in the case of coarse quantization or transmission errors, or they may be partly desired, if deficiencies of a given signal are alleviated. Anyhow, the output signals do not have the original signal's quality.

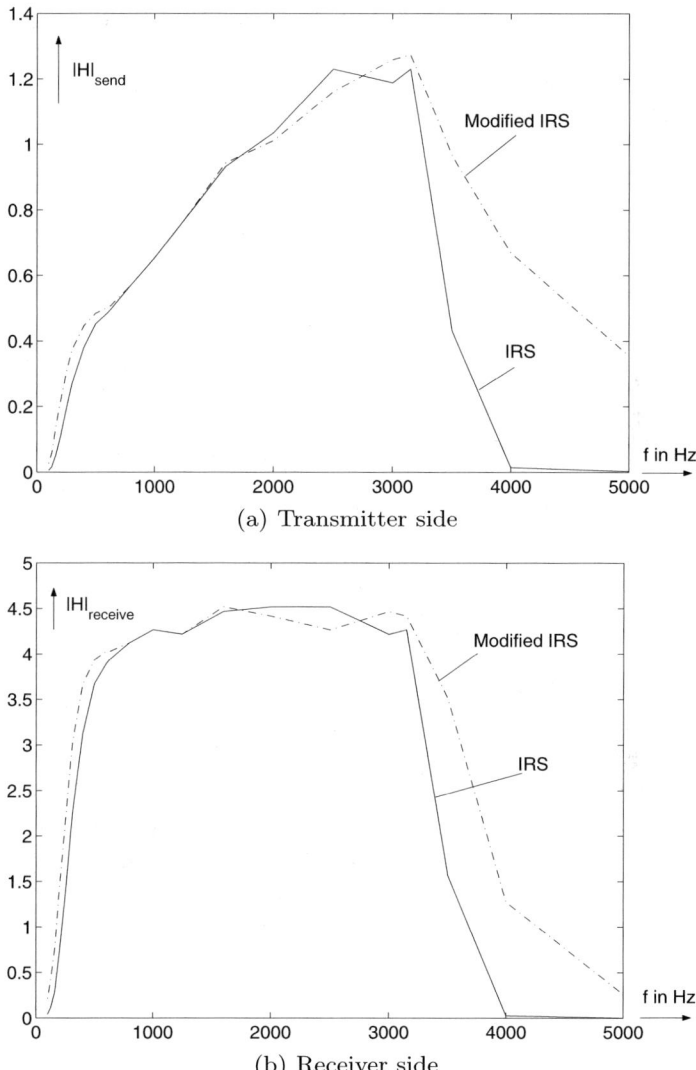

Fig. 9.2. Intermediate-reference system (IRS) filters modeling telephone band limitations, on the a) transmitter, b) receiver side.

9.2 Speech-Signal Quality

9.2.1 Intelligibility

Speech carries information. Thus, the basic requirement concerning $s_1(t)$ is its intelligibility. The correct understanding of single phonemes, phoneme groups, syllables, words, or full sentences can be assessed in appropriate tests. For instance, listeners may be asked to indicate which sound or group of phonemes they heard [54] or which one of a pair or larger set of similar sounds they believed to hear, in a so-called *rhyme test* [20, 33, 91], or they may have to write down whole sentences, note essential information from sentences [45], or insert perceived words into the gaps of given, incomplete sentences [85]. It is worth noting that rhyme tests do not just give a percentage of correct understanding, but also diagnostic insight due to the analysis of confusions between initial, central, or final phonemes. For a list and discussion of such possibilities, the reader is referred to [54]. In the scenario of interest here, intelligibility will, however, not be of concern: Single-sound understanding of 97 % has been given [15] and deemed sufficient for decades of telephone use.

9.2.2 Speech-Sound Quality

Rather than comprehensibility, "sound quality" in a more general sense is of interest, perhaps intuitively described by terms like "naturalness", "pleasantness", "fullness", or "clearness" of the signal $s_1(t)$.

In this chapter, the case of a telephone transmission is considered with the restriction of the presently still wide-spread system: The *integrated-services digital network* (ISDN) operates digitally, thus avoids the noise accumulation of the analog *plain old telephone system* (POTS), it applies a logarithmic pulse-code modulation (log-PCM) with 8 bits/sample quantization [41], it uses better handsets or even hands-free terminals and adds many new service functions, but it is still closely linked to POTS because of its channel width of less than 4 kHz as needed in POTS for its frequency-multiplex transmission with a 4 kHz channel spacing. For Fig. 9.1, this means

$$f_c^{in} = f_c^{out} = 4\,\text{kHz}, \qquad (9.6)$$
$$f_s^{in} = f_s^{out} = 8\,\text{kHz}. \qquad (9.7)$$

In fact, the band limitation primarily determines the general quality impression for a telephone user: The speech is felt to be "muffled" rather than "bright", and, due to the strong low-frequency attenuation (see Fig. 9.2 (a)), also "thin" rather than "full". Special quality aspects of a transmission or storage of wideband speech with a doubled bandwidth are addressed in [30].

For narrow-band speech, many new digital transcoding methods, aiming at lower data rates than used in ISDN, have been investigated over several decades. They consist of both frequency-domain approaches, like in sub-band or adaptive transform coding (SBC, ATC), and time-domain, namely, predictive techniques, as in adaptive differential pulse-code modulation (AD-PCM), adaptive predictive coding (APC), or code-excited linear predictive (CELP) coders. For an overview, the reader is referred to [28], for more details to [88, 89]. Especially the predictive methods have entered the real telephone world – ADPCM in the cordless-telephone DECT[1] standard [43], APC with a reduced "residual signal" in the GSM[2] mobile-phone standard [11], CELP in the enhanced GSM system [12] and in numerous new ITU standards for medium-to-low bit-rate transmission. In all cases, it was required that the speech quality would be close to that of log-PCM with its bit rate

$$f_B^{\text{log-PCM}} = 8\,\frac{\text{bits}}{\text{sample}} \cdot 8000\,\frac{\text{samples}}{\text{s}} = 64\,\text{kbit/s}. \tag{9.8}$$

The quality of this system was actually derived from the situation found in an analogue link over not too long a distance within the international, so-called *trunk* network; it was termed *trunk* or *toll* quality and described by the above-named band limitation defined by the tolerance scheme in Fig. 9.3 [42] plus the signal-to-noise ratio

$$SNR = 10\,\log_{10}\left\{\frac{\sigma_s^2}{\sigma_e^2}\right\}\,\text{dB} \stackrel{!}{\approx} 40\,\text{dB}. \tag{9.9}$$

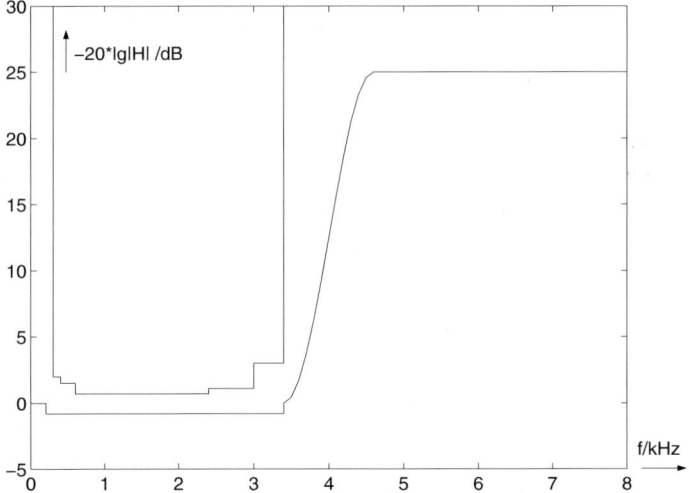

Fig. 9.3. Tolerance scheme for a PCM-transmission input filter.

Here, σ_s^2 is the average-signal, σ_e^2 the distortion or noise power. The log-PCM system itself has an $SNR \approx 38$ dB, the DECT version of ADPCM reaches $SNR \approx 35$ dB. More refined approaches like those mentioned above, however, have much lower SNR values – but still produce good sound quality. This perceived goodness was the basis for their selection as transmission

---

[1] *DECT* abbreviates *digital enhanced cordless telecommunications*.
[2] *GSM* stands for *global system for mobile communications*.

standards: Among various proposals, the solution with the best, but at least perceptually close-to-toll quality for a given bit-rate at limited algorithmic delay and limited hardware cost would be the choice. So, *quality assessment* procedures are of vital interest.

9.3 Speech-Quality Assessment

9.3.1 Auditory Quality Assessment

Quality is the result of a judgment after a perception [54]. Both steps require a human being or "subject" to be involved. Therefore, the term *subjective quality test* is often used for the determination of a communication-system's quality by asking test persons to "use" the telephone in some sense under various conditions and then "grade" its performance. Such tests are strictly formalized and normalized, however, to make them reproducible, comparable, and objective; thus, the term "auditory" is preferred to "subjective", in the following. Still, it has to be admitted that *some subjectivity* remains – due to users' peculiarities in terms of their voices and hearing abilities, or because of the speech material chosen by the test organizer. Voice quality is an interesting field [61], but outside the scope of this chapter, and both for the speech production and the hearing side, "normal" test persons are assumed. This means the exclusion of "extraordinary" speakers and "handicapped" listeners. The former classification is not easy; the latter one is easier if, e.g., an audiometric analysis reveals a too strong increase of the hearing threshold, say, by more than 20 dB at some frequency. Also for the material, the usual rules for a selection of "phonologically balanced" words and sentences are assumed to be kept: Sounds should appear with the same statistics as in "everyday speech".

Auditory quality is analyzed in a test laboratory. This means that, when test users are asked to "use the telephone and check its quality", a model for the actual telephony situation is created.

9.3.2 Aims

Two distinct aims can be followed in such a test:

- The "integral" or "total" quality can be determined, as needed for the decision on a best candidate for a new standard. "Acceptance" or "effort of use" are similar overall ratings.
- Single quality-related features may be asked for, like the above-mentioned "brightness" or "fullness", or other attributes like "dullness", "noisiness", or "sharpness".

Like rhyme tests in intelligibility assessments, the latter measures allow a deeper insight into a system's behavior: A developing engineer, knowing the technical details of a realization and understanding the attributes, may use these diagnostic results to optimize corresponding parameters. This is not possible from just one integral term: Two systems may be equally bad because of quite different reasons. Beyond, it must be possible to derive the total quality from diagnostic results, since a human subject will base a judgment on various single impressions. So, an attribute-based assessment is advantageous.

9.3.3 Instrumental Quality Assessment

Auditory tests are reproducible only within limits, and they are costly and time-consuming. During a system-development phase, where many parameters are changed frequently for quality optimization, they are prohibitive, at least in the above formal version. Since "everything that can be heard must also be measurable" [7], it should be possible to measure all signal features which are relevant for a subject's judgment and finally "read the quality" from an instrument. As, then, no subjects are involved anymore (with the exception of the speech-signal production), such tests are often called "objective"; for the reasons discussed in Sec. 9.3.1, the term "instrumental" is preferred in the following.

Instrumental evaluations happen in a test laboratory again, where now the model of the telephone situation in the auditory test is itself modeled by a measurement device.

Also instrumental measures may concern the integral quality or single attributes. From instrumentally well estimated, suitably chosen attributes, an integral quality should then also be predictable.

9.4 Compound-System Quality Prediction

9.4.1 The System-Planning Task

From instrumental measurements on existing signal-processing equipment, the perceived quality may be estimated, as explained above. A completely different task is the prediction of the quality of a planned compound network of subsystems from knowledge on the components' effects, but before realization. For this network-planning phase, the so-called *E-Model* is an appropriate tool. Transmission planning is actually outside the scope of this chapter; parts of the philosophy behind the E-Model, however, turn out to be valuable also for the assessment aspects addressed here, especially the combination of attributes into an integral quality estimate.

9.4.2 ETSI Network-Planning Model (E-Model)

First within ETSI, the European Telecommunications Standards Institute, then in the ITU, the proposal [55] was discussed and finally formalized as a standard [40]. The model allows to calculate the expected total quality of a speech signal after transmission. It is not based directly on the usual *mean-opinion score* (MOS) of typical auditory quality tests (see Sec. 9.5.1 – 9.5.4). Its basic terms are, instead, the "transmission rating" $R \in [0, 100]$ and several "impairments" $I \in [0, 100]$ caused by network elements.

These terms are defined such that they act additively on a ratio scale. A "perfect" transmission would have a rate $R = 100$; a "clean" ISDN link would be rated by $R = 93.2$, being "impaired" by the typical band-limitation and quantization (see Sec. 9.2.2). Any further impairment would further reduce R according to the formula

$$R = R_\mathrm{o} - I_\mathrm{s} - I_\mathrm{d} - I_{\mathrm{e,eff}} + A. \tag{9.10}$$

Here, I_s denotes signal-synchronous, I_d delayed quality reductions, while $I_{\mathrm{e,eff}}$ sums (also literally) all impairments of the equipment responsible for any signal-processing, coding, and decoding. By A, the (subjective, circumstance-depending) "advantage" of some special system ability, like, e.g., mobility, is expressed, which may lead to a better quality judgment. R_o is the basic rating of a system without further distortions. For the above-mentioned clean ISDN link, this means $R_\mathrm{o} = 93.2$, $I_\mathrm{s} = I_\mathrm{d} = I_{\mathrm{e,eff}} = 0$, $A = 0$, and thus $R = R_\mathrm{o}$.

9.5 Auditory Total-Quality Assessment

9.5.1 Conversation Tests

As said before, a valid assessment can be carried out by asking human subjects to "use a telephone". Although happening in a laboratory, the connection should be made as natural as possible – which means that two partners should talk and listen: A conversation with natural beginning, flow, and ending should be arranged.

The last word indicates the difficulty: An arranged communication is not a natural one. However, the less arrangement is applied, i.e., the closer the conversation comes to a "chat" on some everyday items, by participants knowing each other, the smaller is the quality information gained from the test. The users will not take much care of the transmission conditions in detail. So, a certain structure has to be enforced more or less artificially, also in order to keep the conversation going in cases with two test persons not acquainted before. Various scenarios have been developed for this reason [70].

An example is a discussion about pictures in different ways. This idea suffices to keep a balanced communication (in the sense of both partners being equally involved, a broad enough vocabulary and varying speaking style being used, and neither contents nor signal quality being in the focus alone). A long test duration of about 5 to 6 minutes is, however, necessary to arrive at valid judgments, and the topic is not really felt to be a natural case for a telephone use.

Enhancements in terms of naturalness as well as length are possible by means of so-called *short-conversation tests*, dealing with information requests or ordering items, as often practiced via a telephone. The conversation structure (e.g., for an ordering task) is depicted in Fig. 9.4, derived from [70], where a halved duration is shown to be achieved.

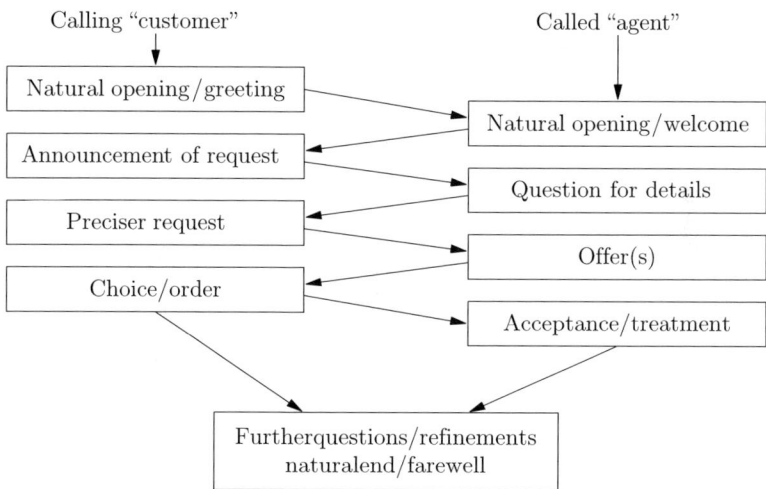

Fig. 9.4. Dialog structure in a telephone-ordering scenario.

In standardized tests following [49], verbal quality statements from a five-point list are given by the test persons (see Tab. 9.1) together with binary decisions whether the conversation was found to be difficult or not. The latter are evaluated in terms of a "difficulty percentage" of all subjects, the former opinions, as expressed by the according numbers in Tab. 9.1, are averaged and thus yield a numerical mean-opinion score (MOS) between 1.0 and 5.0.

Table 9.1. Numerical and verbal descriptions for an absolute-category conversation-quality rating.

Numerical grade	5	4	3	2	1
Verbal grade	Excellent	Good	Fair	Poor	Bad

A different type of conversation becomes more frequent nowadays, namely, a dialog between a human customer and a computer. Many additional questions arise in such cases, not to be covered here. Interested readers are referred to, e.g., [9, 73, 74] for recent advances, to [71] for an overview, and to [72] for a thorough general treatment.

9.5.2 Listening Tests

Conversation tests model a real telephone communication quite closely, but they are very expensive. Whenever possible, they are therefore replaced by assessments where test persons only listen to (pre-fabricated or online-produced) speech samples of relatively short length like, e.g., two to five sentences of two seconds each with short breaks in-between. A variety of partly phonologically balanced, partly specialized texts is in use, e.g.,

- examples like "you will have to be very quiet", as mentioned in [49],
- the (German) sentences of the so-called *Marburg* and *Berlin collections* [75, 87] contained in a large corpus found in [35] and described in [86],
- very popular sentences like "these days, a chicken leg is a rare dish", described in [1], or "she had your dark suit in greasy wash water all year", which are found in the TIMIT[3] data base [14] or its narrowband version [53],
- semantically unpredictable sentences, like "the great car met the milk", as prepared at several institutions in various languages and discussed in [54], where also more information on data bases is to be found.

Listening to a few such utterances has the consequence that the subjects focus their attention much more on the signal characteristics than on the contents; this is good in terms of the quality grading, but it reduces the naturalness of the model, too. Beyond, such a *listening-only test* (LOT) is unnatural also because the test user does not speak, and another draw-back consists in the exclusion of certain transmission effects: Delays cannot be evaluated, echoes would possibly result only from two-way connections, double-talk will not occur. The strongest positive argument, however, is the short duration allowing for a highly increased number of systems and conditions to be tested within a certain time span. Still, quite a variety of LOTs exist, with different expenditures on one side, but also different amounts of insight to be gained on the other side.

9.5.3 LOTs with Pair Comparisons

Two signals are played to the test subjects who have to compare them. This is another violation of the naturalness requirements: In a "normal" telephone situation, a user expects a certain quality from her or his experience and compares the heard sound with this "internal reference" only, not with a second signal. Due to definite advantages, this approach is very useful anyway. It is mainly practiced in two different ways; details and variants are to be found in [70]. One advantage, in both cases, comes from the use of the same speaker for the compared signals: Voice peculiarities are not so important anymore. The other advantage is a fine resolution of small differences.

---

[3] The abbreviation *TIMIT* results from *Texas Instruments (TI)* and *Massachusetts Institute of Technology (MIT)*.

9.5.3.1 Comparison-Category Rating (CCR)

The comparison may concern two differently processed signals $s_{1,1}(t)$ and $s_{1,2}(t)$ in Fig. 9.1. The perceived difference is denoted according to the categories of Tab. 9.2. Usually, one of the signals is a "clean" version, though not necessarily the original input $s_0(t)$. The two signals follow each other randomly.

Table 9.2. Numerical and verbal descriptions of a second stimulus compared to a first one for a comparison-category total-quality LOT rating [49].

Numerical grade	3	2	1	0	-1	-2	-3
Verbal grade	Much better	Better	Slightly better	About the same	Slightly worse	Worse	Much worse

9.5.3.2 Degradation-Category Rating (DCR)

Here, the clean signal comes first, and the second one is checked for its quality loss according to Tab. 9.3.

Table 9.3. Numerical and verbal descriptions for a degradation-category total-quality LOT rating [49].

Numerical grade	5	4	3	2	1
Verbal grade	inaudible	The degradation is ... audible but not annoying	slightly annoying	annoying	very annoying

9.5.4 Absolute-Category Rating (ACR) LOTs

The fine resolution mentioned above would of course be especially helpful in a system-optimization phase, where some parameter may have to be fixed within a narrow interval. But comparison tests, being of course less expensive than conversation experiments, are still quite costly. In a cheaper LOT, listeners hear the output signal $s_1(t)$ of one system only, compare it to their expectation, and grade it absolutely, according to the absolute-category rating (ACR) scale given already in Tab. 9.1 for conversational opinion tests.

The resulting MOS ∈ [1.0, 5.0] after averaging the ratings of the test persons in such an ACR LOT is the most frequently applied performance description when speech-signal processing systems are compared in terms of their qualities. The reason is that ACR LOTs are the least expensive version of a formal auditory quality-assessment. They are reasonably reproducible within ±0.5 MOS, they are natural in the sense of an internal-reference basis instead of a signal comparison, but they are unnatural due to the missing interaction between two telephone users as in all LOTs, and they are far less sensitive than paired comparisons to small system differences, while being more susceptible to voice or speech-material peculiarities.

9.6 Auditory Quality-Attribute Analysis

9.6.1 Quality Attributes

As mentioned in Sec. 9.3.2, a user's total quality judgment is internally created from a set of single impressions. Here, the listening-only case will be considered alone. Numerous single-aspect quality terms called attributes have been proposed in the literature. A well accepted list is that proposed by [90], used also in the thorough investigations of [77]. It separates signal and background effects as presented in Tab. 9.4.

Table 9.4. Quality attributes for signal and background evaluation as applied to the Diagnostic Acceptability Measure (DAM).

Signal	Fluttering bubbling	Distant thin	Rasping crackling	Muffled smothered	Irregular interrupted	Nasal whining
Background	Hissing rushing	Buzzing humming	Chirping bubbling	Rumbling thumping		

Other attributes were mentioned earlier, namely, fullness (= opposite of "thin"?), or brightness (= opposite of "muffled"?); clearness, sharpness, roughness, or loudness are alternatives, which have no direct counterparts in Tab. 9.4 or are mixtures of several terms. Others may be searched for, if reasons and suitable criteria are given (see Sec. 9.6.3).

9.6.2 Attribute-Oriented LOTs

9.6.2.1 Loudness

Despite its obvious meaning, the measurement of loudness needs a careful definition [94]. As a loudness level L_N, given in the unit phon, the sound-pressure level L in dB of an equally loud 1-kHz-tone is defined. Constant-loudness curves, say, with x phon, for instance, show the necessary sound pressure level of a tone at some frequency f to be perceived as equally loud as an x-dB-tone with $f = 1$ kHz. The 3-phon-curve corresponds to the hearing threshold, measured usually in audiometry for evaluation of an individual's hearing abilities. Smaller refinements are needed if, instead of a tone, narrowband noise is used, or if diffuse rather than plane-wave sound fields are of interest. The comparison with a 1-kHz-tone, however, is the basis. Besides the use of the logarithmic phon scale, also an absolute loudness N can be given: A 1-kHz-tone of 40 dB pressure level ($\hat{=}$ 40 phon) is defined to have a loudness $N = 1$ sone. A doubled loudness needs a level increase of 10 dB; this behavior does not hold below 1 sone: $L_N = 3$ phon corresponds naturally to zero loudness, i.e., $N = 0$ sone (see Fig. 9.5).

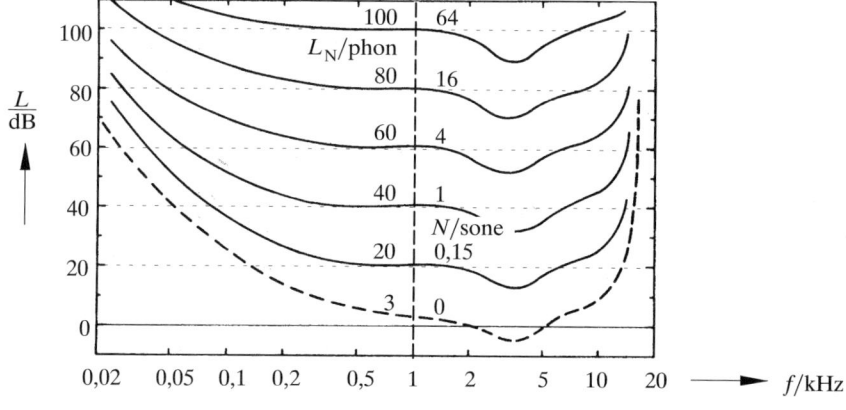

Fig. 9.5. Curves of equal loudness (see [88]).

An output signal $s_1(t)$ of some processing or transmission system in Fig. 9.1 which is too faint or too loud will certainly cause a negative impression, whatever other quality impacts may be included. In auditory tests, however, loudness is often not used as a direct feature. Instead, a suitable amplification or attenuation is chosen such that $s_1(t)$ has a (listener-) preferred or an optimum (in terms of the final quality score) loudness.

9.6.2.2 Sharpness

Like loudness, sharpness [94] can be measured on a ratio scale: A pure noise signal bandlimited to $f \in [920, 1080]$ Hz, i.e., centered at $f_{\text{center}} = 1$ kHz, is defined to have a sharpness $S = 1$ acum. If the center frequency and, proportionally, the bandwidth is increased, the perceived sharpness grows. For $f_{\text{center}} = 3$ kHz and a bandwidth of about 500 Hz, $S = 2$ acum indicates a doubled sharpness feeling, while at $f_{\text{center}} = 350$ Hz and $f \in [300, 400]$ Hz, the sharpness $S = 0.5$ acum is halved (see Fig. 9.6).

A sharpness perception is also evoked if the lower band-edge is kept constant and the upper one is enlarged; a fixed upper edge frequency with falling lower cut-off frequency reduces sharpness. (In all these experiments, the loudness is kept constant.) These observations may be generalized: Unpleasant, sharp sounds are caused by, relatively, too strong high-frequency components. So, an auditory determination of S is valuable in terms of diagnostic insight. A slightly more critical point-of-view will be seen in Sec. 9.8.3.

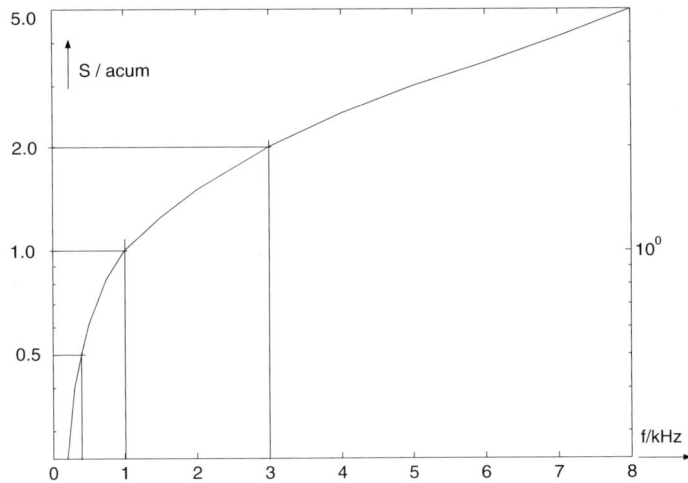

Fig. 9.6. Sharpness growth as a function of the center frequency (and proportional bandwidth) of narrowband noise at constant loudness (after [94]).

9.6.2.3 Roughness

The impression of a rough signal is caused by relatively fast, (quasi-) periodic variations of a signal's (short-time-mean) amplitude.

Very fast amplitude modulations are not heard as such – the time-domain post-masking of our ear hides a fast decay and recovery of the signal size. Slow variations are perceived as fluctuations, known, e.g., as the unpleasant "pumping" effect of certain non-optimum adaptive signal-processing methods. This is of course also a signal impairment, but it is to be separated from this paragraph's item.

Modulation frequencies $f_{\text{mod}} \in [15, 300]$ Hz, together with sufficiently large level changes ΔL, are the reason for a perceived roughness R, given in the unit asper. A value $R = 1$ asper is defined as the roughness of a tone with frequency $f_1 = 1$ kHz modulated with modulation degree $m = 1$ according to

$$s_{\text{mod}}(t) = \left[1 + m \cdot \sin(2\pi \cdot f_{\text{mod}} \cdot t)\right] \cdot \sin(2\pi f_1 \cdot t). \tag{9.11}$$

The strongest roughness impact is found at $f_{\text{mod}} = 70$ Hz, in this case, and R depends on m exponentially according to $R \approx m^{1.6}$ (see Fig. 9.7 (a)). For lower and higher values of f_{mod}, R decays, as said above (see Fig. 9.7 (b)).

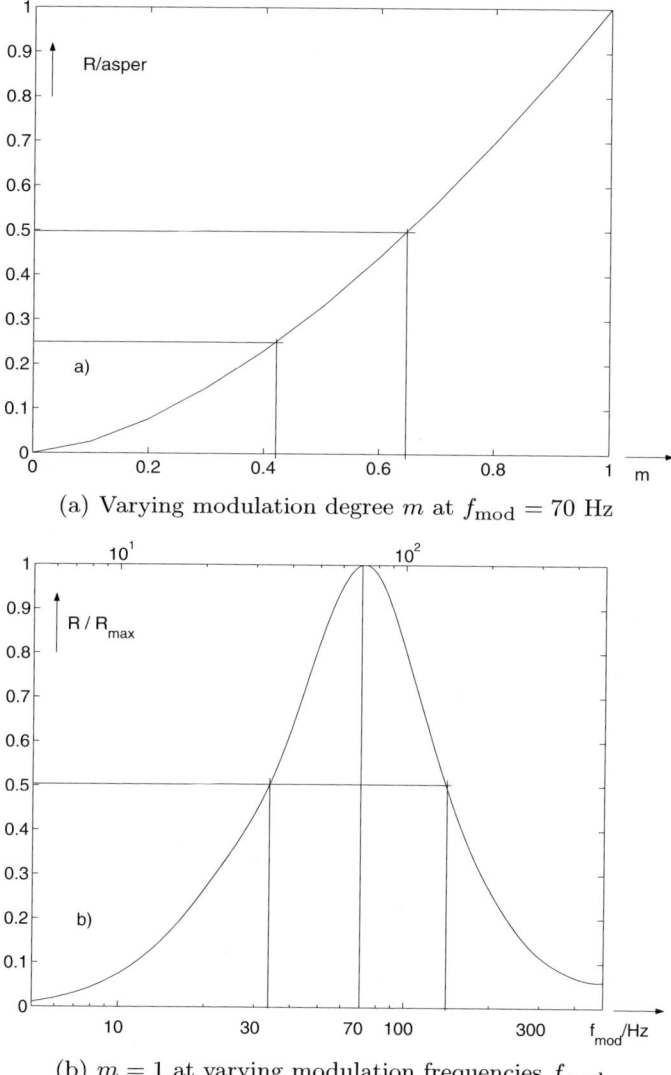

(a) Varying modulation degree m at $f_{\text{mod}} = 70$ Hz

(b) $m = 1$ at varying modulation frequencies f_{mod}

Fig. 9.7. Roughness R of a 1-kHz-tone modulated a) with varying modulation degree m at $f_{\text{mod}} = 70$ Hz, b) with $m = 1$ at varying modulation frequencies f_{mod} (simplified approximation of measurements in [94]).

9.6.2.4 Other Attributes

While the above-named attributes can be checked in a LOT by comparing stimuli with tones of 1 kHz under appropriate conditions, this is not possible for other attributes like clearness, fullness, brightness, or their counterparts like noisiness, distortion, thinness, or dullness. For some of the characterizations in Tab. 9.4, links may be found to the above three attributes: Hissing $\approx$ sharp, rasping $\approx$ rough, distant $\approx$ reduced loudness – but this is not well defined, and for the other terms, no simple relation is obvious.

In any case, however, test listeners can be asked to indicate "how much" of a certain attribute is contained in a speech sample they heard. This may be done with a not too large group of attributes at the same time or consecutively, using appropriate scales, which may be bipolar and each named at both ends with a pair of antonyms.

9.6.3 Search for Suitable Attributes

A fundamental question arises with the aim of a diagnostic-attribute analysis: How many attributes are indeed needed? As mentioned in the above discussion, many given terms are mixtures of impressions themselves – clearness or fullness being almost as broad as quality itself. Of course, a system diagnosis can be based on a multitude of evaluations, but this is certainly difficult, even for an expert. Also numerically, the total quality may be estimated via a regression including many, also strongly overlapping features (see, e.g., [77]) – the regression coefficients will take care of their correct weightings. This, however, requires that a really large number of systems and conditions have to be used in order not to just "interpolate" K measurements by using (nearly) K parameters.

Orthogonal attributes, each one indicating just one specific perceptual impact, would be better, for the sake of both their diagnostic interpretation and their application within an integral-quality estimation formula. They can be developed in two ways.

9.6.3.1 Multi-Dimensional Scaling (MDS)

This approach stems originally from general psychometrics [57, 58] and was applied to telephone-circuit quality analysis in [65].

A variety of K stimuli are prepared. All $K \cdot (K-1)$ possible stimuli pairs (i, j) (both j following i and vice-versa) are presented to a listener group. The test persons do not grade the speech quality but, rather, rate the (dis-) similarity of the two sounds, on a continuous scale between "very similar" and "not similar at all". From the test, after a subject-related normalization and a monotonic mapping, $K \cdot (K-1)$ "disparity" values result. From these, all stimuli can be represented ideally in a $(K-1)$-dimensional space, if identical distance measurements for (i, j) and (j, i) are assumed. As the number of test signals is chosen as large as possible, while the perceptual effects searched for are hoped to be few, a "map" with a lower dimensionality $P < (K - 1)$ is constructed, with unavoidable inaccuracies. These are expressed by two common figures of merit (see, e.g., [77]):

- The "squeezing" of K points onto P dimensions causes a so-called "stress" term S_d, which should be made small:

$$S_d = \sqrt{\frac{\sum_{i=1}^{K}\sum_{j=1}^{K}\left(d_{i,j} - \hat{d}_{i,j}\right)^2}{\sum_{i=1}^{K}\sum_{j=1}^{K}\hat{d}_{i,j}^2}}. \qquad (9.12)$$

S_d describes the normalized root-mean-square error of the distances $d_{i,j}$ in the reduced-dimension map in comparison to the original distances $\hat{d}_{i,j}$ between two stimuli.

- Dropping dimensions has the consequence that not all variability can be covered in the n-dimensional space – the covered normalized variance R^2 should be made close to 1:

$$R^2 = \frac{\sum_{i=1}^{K}\sum_{j=1}^{K}\left(d_{i,j} - \bar{d}\right)^2}{\sum_{i=1}^{K}\sum_{j=1}^{K}\left(\hat{d}_{i,j} - \bar{\hat{d}}\right)^2}. \qquad (9.13)$$

By $\bar{d}$ and $\bar{\hat{d}}$ the mean distances after averaging over i and j are denoted. Since, due to the "squeezing" to a lower dimension, distances never grow, $R \leq 1$ holds.

The necessary dimensionality n is found stepwise, observing the decrease of S_d or/and the growth of R^2 when incrementing P.

An MDS within a thorough investigation of attribute-oriented quality determination is described in [92]. Dimensionalities of $P = 3\ldots4$ appear to be appropriate (see Tab. 9.5).

Table 9.5. Stress S_d and covered variance R^2 for an MDS-based dimensionality reduction.

P	S_d	R^2
3	0.232	0.74
4	0.195	0.79

The dimensions are, however, abstract in the sense that they can not yet be interpreted as "named attributes". Such names can now be found with the help of (system) experts knowing how some processing creates a certain effect on the signal sound. In a succeeding attribute-oriented LOT (see Sec. 9.6.2), the chosen dimensionality and attributes have to be verified. Such a test can, however, be also directly applied for the definition of the reduced perceptual space, as outlined in the next paragraph.

9.6.3.2 Semantic Differential (SD)

Listeners are asked to grade the single stimuli on scales of a highly redundant, large set of predefined descriptors. The list may also be found in a pre-experiment, asking listeners for their own intuitive descriptions. In [92], as many as 217 candidate names were found. They were reduced, by expert inspection, to 13 "antonym pairs", with which the actual LOTs were carried out. Finally, a principal-component analysis was performed with the reduced group, leading to P final attributes. It was found that $P = 3$ still nameless "factors" $F_{1,2,3}$ carry a variance $R^2 = 0.935$.

Names are then found by rotating the P axes such that high correlations with P of the pre-defined antonyms appear, and/or by observing the expert interpretations of the reduced MDS-space results. The names are – naturally – not unique, due to, e.g., the redundancy in the predetermined set.

After the above study, the following choices were made (see Tab. 9.6): As the first of three dimensions was linked, by the listeners, with frequency-content descriptions both in terms of pairs like "dark/bright" and "distant/close", the ambiguous term "directness/frequency content" was selected. The second factor was clearly related to short-time effects in the signals, appearing either as interruptions or as instantaneous sound insertions; it was termed "continuity". The third attribute turned out to describe "hissing" distortions and "noisy" components; so the term "noisiness" is appropriate. Tab. 9.6 also indicates how much variance ΔR_i^2 is explained by taking a factor F_i into account.

Table 9.6. Attributes for telephone-band quality analysis from [92]).

P	F_1	ΔR_1^2	F_2	ΔR_2^2	F_3	ΔR_3^2
3	Directness/ frequency content	0.427	Continuity	0.342	Noisiness	0.166

A subsequent analysis shall reveal whether further terms or sub-dimensions are helpful: "Frequency content" has to do, on one hand, with the linearly transmitted input spectrum, thereby, with the system's average frequency response and especially the band-limitations. On the other hand, it also describes further components, which may be artificially added on purpose or by an incorrectness. "Noisiness" may hint to a hissing speech reproduction, e.g., due to an (over-) emphasis of higher frequencies in fricatives, as well as to added noise. So, either brightness may be positively affected or sharpness in a negative sense. Interruptions due to, e.g., frame erasure in mobile telephony or packet loss in VoIP[4] transmission, are different from short-time effects like the well-known "musical tones" in some less refined noise-reduction methods. Such a deeper study requires again, however, the availability of a very large variety of systems and conditions.

9.6.4 Integral-Quality Estimation from Attributes

After an auditory assessment of P features F_i, the overall listening quality LQ of an investigated system may be estimated from a superposition of the single results. The simplest way is a linear combination:

$$LQ \doteq \sum_{i=1}^{P} b_i \cdot F_i \qquad (9.14)$$

with

$b_1 = 0.46$ for $F_1 \triangleq$ frequency content / directness,
$b_2 = 0.70$ for $F_2 \triangleq$ continuity,
$b_3 = -0.47$ for $F_3 \triangleq$ noisiness (see Tab. 9.6).

This was proposed in [92], and 90 % of the LQ variance could be covered. It has to be noted that, here, LQ is not calculated as a MOS' estimate directly, but as a bipolar, zero-mean version thereof.

Since b_2 is largest, a dominant influence of (dis-) continuities on the perceived quality can be concluded – quite plausibly, as real interruptions create a feeling of an unreliable connection. Still, the doubts about differently weighted sub-dimensions (like "true interruptions" vs. "other short-time effects") remain, as mentioned in Sec. 9.6.3.2, especially as the above test included only a set of 14 conditions.

In [77], the "total acceptability" A is predicted from the auditorily evaluated ten attributes of Tab. 9.4 [90] by a non-linear formula: The six signal-describing terms yield a "total signal quality" TSQ, the remaining six features lead to a "total background quality" TBQ, both found from arithmetic and geometric means of the single results. From both TSQ and TBQ and a linear superposition of single averaged attributes $\bar{F}_i$ with $i \in \{1, ..., 10\}$, the integral value A is computed:

---

[4] The term *VoIP* abbreviates *voice over internet protocol*.

$$A = b_0 + \sum_{i=1}^{10} b_i \cdot \bar{F}_i + b_{11} \cdot TSQ \cdot TBQ. \tag{9.15}$$

In an experiment with more than 200 speech samples, 90 % of the auditorily observed variance is covered. Thus, also a large number of overlapping attributes can be successfully applied, though with a quite complex relation and with a sufficiently broad experimental basis.

9.7 Instrumental Total-Quality Measurement

9.7.1 Signal Comparisons

An instrument has, a priori, no "expectation" or "internal reference from experience" as to how a telephone-output signal should behave. So, a transmission system's impact on quality can only be measured by an analysis of, in Fig. 9.1, the acoustic signals $s_1(t)$ and $s_0(t)$ or, digitally, the sequences $y(n)$ and $x(n)$. This can be done in two distinct ways:

a) Comparison of $y(n)$ and $x(n)$ directly with respect to certain (e.g., spectral) characteristics.
b) Comparison of $x(n)$ or $y(n)$ with a distortion signal

$$e(n) = y(n) - x(n). \tag{9.16}$$

When $e(n)$ is determined, of course, a suitable time and amplitude alignment of $y(n)$ and $x(n)$ has to be carried out in a preprocessing phase before subtraction. This is, however, also necessary in case a) in order to really compare "corresponding" signals. There is a technique replacing $x(n)$ indeed by an "internal reference", namely, by estimating a clean signal from $y(n)$ [48]. This "reference-free" approach is, however, still not generally applicable; it is, therefore, not dealt with here. Both comparisons a) and b) can, again, be carried out in two distinct ways:

α) Description of the compared signals in their full lengths first, then comparison of their long-time, i.e., averaged behaviors.
β) Description of short signal segments (of, e.g., a "frame length" $T_F = 20$ ms), comparison of the short-time behaviors, and averaging at the end.

All variants are depicted in Fig. 9.8.

9.7.2 Evaluation Approaches

The integral-quality term, searched for finally by a signal comparison, may be based on very simple signal features, like their powers, on more refined parameters derived from the signals, on spectral properties, or, somehow most naturally, on a representation modeling the human hearing process. In the following, the most commonly used measures are listed and commented on.

Telephone-Speech Quality 307

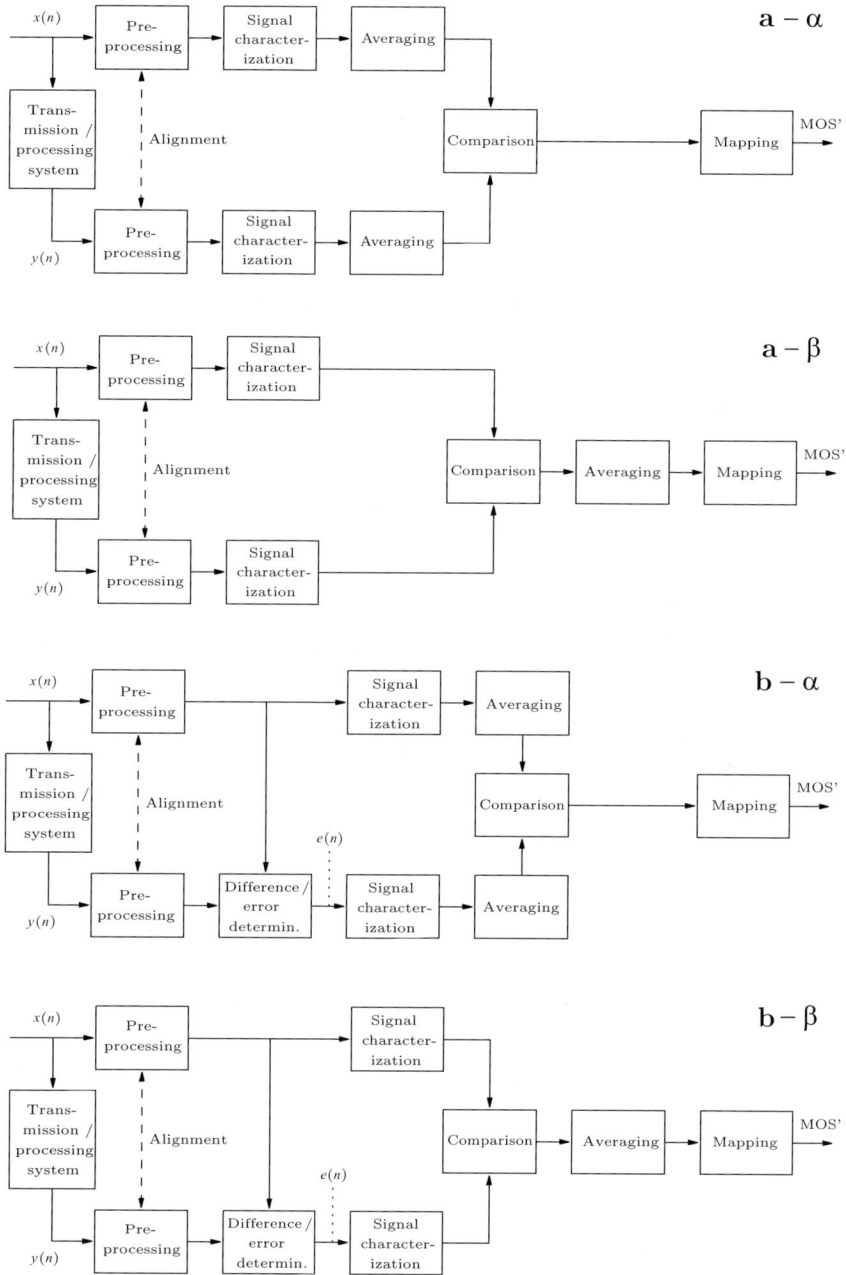

Fig. 9.8. Possible variants of signal comparisons for quality estimation.

9.7.2.1 Signal-to-Noise Ratios

The popular logarithmic distortion measure defined in Eq. 9.9 can be measured digitally from $x(n)$ and $e(n)$ in Fig. 9.1 and Eq. 9.16, for a signal of (possibly very large) length M:

$$\frac{SNR}{\text{dB}} = 10 \cdot \log_{10} \left\{ \frac{\sum_{n=0}^{M-1} x^2(n)}{\underbrace{\sum_{n=0}^{M-1} (y(n) - x(n))^2}_{=e^2(n)}} \right\}. \tag{9.17}$$

This so-called *global SNR* is of type b - α, in the nomenclature of Sec. 9.7.1. It is known to reliably reflect perceived quality as long as it is large enough, i.e., for "high-quality" speech signals with, say, $SNR > 25$ dB. For stronger distortions, the *segmental SNR* [76] works better:

$$\frac{SNR_{\text{seg}}}{\text{dB}} = \frac{10}{N_{\text{seg}}} \cdot \sum_{i=1}^{N_{\text{seg}}} \log_{10} \left\{ \frac{\sum_{n=(i-1)L_{\text{seg}}}^{iL_{\text{seg}}-1} x^2(n)}{\sum_{n=(i-1)L_{\text{seg}}}^{iL_{\text{seg}}-1} e^2(n)} \right\}, \tag{9.18}$$

where L_{seg} now indicates the length of the N_{seg} segments into which the signal has been split, with $M = L_{\text{seg}} \cdot N_{\text{seg}}$. This is of type b - β now, and it gives a reliable quality indication as long as, say, $SNR_{\text{seg}} > 12$ dB holds, assuming already that additional measures are taken to avoid inclusion of signal overdrive or of signal pauses, i.e., extremely negative logarithmic values falsifying the average measure. The latter case can also be avoided by a slight, simple modification of Eq. 9.18 due to [67]:

$$\frac{SNR_{\text{seg}}}{\text{dB}} = \frac{10}{N_{\text{seg}}} \cdot \sum_{i=1}^{N_{\text{seg}}} \log_{10} \left\{ 1 + \frac{\sum_{n=(i-1)L_{\text{seg}}}^{iL_{\text{seg}}-1} x^2(n)}{\sum_{n=(i-1)L_{\text{seg}}}^{iL_{\text{seg}}-1} e^2(n)} \right\}. \tag{9.19}$$

Refinements try to make SNR measurements valid also for even stronger distortions. Enhancements are possible with, e.g., the following steps [22, 77]:

- Eqs. 9.18 or 9.19 are evaluated in narrow frequency bands, and the final value is the average of all bands ("frequency-depending segmental SNR").
- The log-term in Eqs. 9.18 and 9.19 is evaluated per frequency band and averaged over all bands, before the segmental averaging takes place.
- The single-band results in both variants may be weighted by factors proportional to (a power of) the signal energy within each band ("frequency-weighted segmental SNR").

An implicit weighting can also be achieved by the choice of non-constant bandwidths. For example, a set of 16 bands in the telephone-frequency range may be used, with proportionally growing widths and center frequencies as defined in [6] and similar to the so-called *critical bands* in [94]. Then, as a special case of an implicitly frequency-weighted global SNR, the so-called *articulation index* results [16,59]. It could, of course, also be evaluated segment-wise. Another non-linear mapping of the segmental SNR leads to the so-called *information index* as defined in [52].

9.7.2.2 Spectral Distances

The frequency-depending SNR variants are closely related to the class of spectral distances between the signals $x(n)$ and $y(n)$. Here, no difference signal is used, and segmental results are averaged at the end – so, a type a-β analysis takes place (see Fig. 9.8).

A general formulation is given by

$$\Delta_{\text{spec}} \doteq \frac{1}{N_{\text{seg}}} \sum_{i=1}^{N_{\text{seg}}} \left\{ \frac{\sum_{\mu=0}^{M-1} \left|X_{\text{LP}}\left(e^{j\Omega_\mu}, i\right)\right|^\gamma \cdot f^p\left(\left|X_{\text{LP}}\left(e^{j\Omega_\mu}, i\right)\right|, \left|Y_{\text{LP}}\left(e^{j\Omega_\mu}, i\right)\right|\right)}{\sum_{\mu=0}^{M-1} \left|X_{\text{LP}}(e^{j\Omega_\mu}, i)\right|^\gamma} \right\}^{1/p}. \tag{9.20}$$

By $X_{\text{LP}}(e^{j\Omega_\mu}, i)$ and $Y_{\text{LP}}(e^{j\Omega_\mu}, i)$, short-time spectral values of the input and output signal segment number i are denoted as found from a discrete Fourier transformation; so, μ is a frequency index. An L_p-norm of the function $f(*)$ is evaluated. A weighting with the γ−th power of the input spectrum and a normalization are included.

The choice of $f(*)$ determines the exact name of the distance measure:

$f(X, Y) \doteq \bigl|\,|X| - |Y|\,\bigr|$: "linear spectral distance",

$f(X, Y) \doteq \bigl|\,|X|^\delta - |Y|^\delta\,\bigr|$: "non-linear spectral distance",

$f(X, Y) \doteq 20 \cdot \bigl|\log_{10}|X| - \log_{10}|Y|\bigr|$: "log-spectral distance".

As a good choice for the free parameter p in Eq. 9.20, the investigations of [77] showed $p = 6$ especially for the log-spectral distance. Such a high exponent says that larger deviations have to be emphasized in the measure – they have an increased influence on the total quality perception.

In all cases, the DFT is not applied to the signals themselves: The index "LP" is to indicate the use of linear-prediction coefficients. A short paragraph is helpful for an explanation, especially also for links to other measures, to be pointed out later.

9.7.2.3 Linear Prediction – a Brief Excursion

A linear predictor computes an estimation

$$\hat{x}(n) \doteq \sum_{\nu=1}^{\hat{p}} a_\nu \cdot x(n-\nu) \qquad (9.21)$$

for the present value $x(n)$ from $\hat{p}$ past samples. The residual signal

$$d(n) \doteq x(n) - \hat{x}(n) = x(n) - \sum_{\nu=1}^{\hat{p}} a_\nu \cdot x(n-\nu) \qquad (9.22)$$

describes the prediction error. The coefficient vector

$$\boldsymbol{a} \doteq \begin{bmatrix} a_1, a_2, ..., a_{\hat{p}} \end{bmatrix}^{\mathrm{T}} \qquad (9.23)$$

is optimized in the minimum-mean-square prediction-error sense and found as

$$\boldsymbol{a}_{\mathrm{opt}} = \boldsymbol{R}_{xx}^{-1} \cdot \boldsymbol{r}_x. \qquad (9.24)$$

$\boldsymbol{R}_{xx}$ denotes the correlation matrix, $\boldsymbol{r}_x$ the correlation vector of the signal $x(n)$ in total or, better and used in adaptive schemes, of a segment of length L_{seg}, chosen as appropriate for the "short-time speech stationarity" within ca. 20 ms. Due to the strong correlation of speech signals, the variance of the residual signal is much smaller than that of the signal, and

$$\sigma_d^2 = \sigma_x^2 - \boldsymbol{a}^{\mathrm{T}} \cdot \boldsymbol{r}_x \ll \sigma_x^2 \qquad (9.25)$$

means that ("front") bits can be saved when $d(n)$ is transmitted rather than $x(n)$. This allows the well-known data-rate reduction in predictive coding schemes. Corresponding to the z-transform of Eq. 9.22, the transfer function of the non-recursive filter with output $d(n)$ is given by

$$H(z) = 1 - \sum_{\nu=1}^{\hat{p}} a_\nu \cdot z^{-\nu}. \qquad (9.26)$$

Since the LP principle exploits correlation according to Eq. 9.24, $d(n)$ is necessarily decorrelated (up to the prediction order) – ideally: white. From $d(n)$, the receiver recovers $x(n)$ by filtering the residual with the inverse of this transfer function,

$$\frac{1}{H(z)} = \frac{1}{1 - \sum_{\nu=1}^{\hat{p}} a_\nu \cdot z^{-\nu}}. \qquad (9.27)$$

As the excitation $d(n)$ is white while the speech signal $x(n)$ is not, the spectral shape is obviously given by the magnitude frequency response $|1/H(e^{j\Omega})|$ of the above all-pole filter. It can be computed in equi-spaced frequency points by a DFT according to

$$\left|\frac{1}{H(e^{j\Omega_\mu}, i)}\right| = \left|X_{\text{LP}}(e^{j\Omega_\mu}, i)\right| = \frac{1}{\left|\text{DFT}\left\{[1, -(\boldsymbol{a}^{x_i})^{\text{T}}, 0, 0, ..., 0]\right\}\right|}. \tag{9.28}$$

The coefficient vector calculated for the $i-th$ segment of $x(n)$ (indicated by the superscript x_i) according to Eq. 9.24 has to be inserted here and appended by $(M-\hat{p})$ zeros in order to produce a large enough number of spectral samples. Inserting the output segment y_i and its coefficient vector correspondingly yields

$$\left|Y_{\text{LP}}(e^{j\Omega_\mu}, i)\right| = \frac{1}{\left|\text{DFT}\left\{[1, -(\boldsymbol{a}^{y_i})^{\text{T}}, 0, 0, ..., 0]\right\}\right|}. \tag{9.29}$$

9.7.2.4 Other Measures Based on Predictor Coefficients

As prediction works best in an adaptive, i.e., frame-by-frame, evaluation, it is clear that the following measures are evaluated segmentwise; also, all of them rely on the two signals $x(n)$ and $y(n)$ without pre-calculating an error signal $e(n)$. So, they belong to class a-β, in Fig. 9.8.

Variants of the so-called *log-likelihood ratio* [37] have been proposed and analyzed:

$$d_{\text{ll}}(x,y) \doteq \ln\left\{\frac{(\boldsymbol{a}^y)^{\text{T}} \cdot \boldsymbol{R}_{xx} \cdot \boldsymbol{a}^y}{(\boldsymbol{a}^x)^{\text{T}} \cdot \boldsymbol{R}_{xx} \cdot \boldsymbol{a}^x}\right\}. \tag{9.30}$$

By Eq. 9.30, the power ratios of the residual signals are compared which would result from filtering $x(n)$ by predictors optimized for $y(n)$ and $x(n)$ – see Eqs. 9.24 and 9.25; so, a mismatch of the respective frequency responses is described in a logarithmic way. Obviously, this is closely related to a log-spectral distance as discussed in Sec. 9.7.2.2.

The *cepstral distance* compares the cepstra, i.e., the (inverse) Fourier transforms of the logarithmic magnitude spectra of $x(n)$ and $y(n)$. The L_{cep} cepstral values $c_x(k), c_y(k), k \in \{0, 1, ..., L_{\text{cep}}\}$ to be used describe the LP spectrum again, and they can be calculated recursively from the coefficient vectors $\boldsymbol{a}^x, \boldsymbol{a}^y$. The measure defined by

$$CD(x,y) \doteq [c_x(0) - c_y(0)]^2 + 2 \cdot \sum_{k=1}^{L_{\text{cep}}} [c_x(k) - c_y(k)]^2 \tag{9.31}$$

is obviously again closely related to a logarithmic spectral distance described in Sec. 9.7.2.2.

The so-called *log-area (ratio) distance* (LAR distance) has to do with a simplified model of the vocal tract during the utterance of a given speech segment. From the predictor coefficients, the staircase-type cross-sectional area function of $(\hat{p}+1)$ short vocal-tract pieces can be derived (see, e.g., [88, 89]). A logarithmic comparison of two adjacent cross-sections A_ν and $A_{\nu+1}$, recursively computed from $\boldsymbol{a}$, leads to the so-called *log-area ratio*

$$LAR_\nu \doteq \log_{10}\left\{\frac{A_\nu}{A_{\nu+1}}\right\}, \quad \nu \in \{0, 1, ..., \hat{p}-1\}. \quad (9.32)$$

These terms can again be determined both for $x(n)$ and $y(n)$. The difference of corresponding LARs can be averaged over ν in terms of their magnitudes and finally averaged over the speech segments, indexed by i. This average log-area distance

$$\begin{aligned}d_{\mathrm{LAR}} &\doteq \frac{20}{\hat{p}\cdot N_{\mathrm{seg}}} \cdot \sum_{i=1}^{N_{\mathrm{seg}}} \sum_{\nu=0}^{\hat{p}} \left|LAR_\nu^y - LAR_\nu^x\right|_i \\ &= \frac{20}{\hat{p}\cdot N_{\mathrm{seg}}} \cdot \sum_{i=1}^{N_{\mathrm{seg}}} \sum_{\nu=0}^{\hat{p}} \left|\log_{10}\left\{\frac{A_\nu^y \cdot A_{\nu+1}^x}{A_\nu^x \cdot A_{\nu+1}^y}\right\}\right|_i\end{aligned} \quad (9.33)$$

describes the mismatch of the vocal-tract models calculated from $x(n)$ and $y(n)$ and, thereby, a spectral-shape mismatch in a logarithmic way again. In [77], this measure was analyzed also with the modification of an L_p norm of the magnitude term; $p = 1$ was, however, found to be the best choice.

9.7.3 Psychoacoustically Motivated Measures

9.7.3.1 Basis

The thorough investigation of [77], referenced already frequently, showed that the above ideas were partially successful, but only for a limited class of distorting systems and conditions. The correlation between "true" auditory quality judgments and instrumentally estimated ones was around 0.9 for simple distortions, but only about 0.7 or less for more complex cases. Among the best measures, there were the LAR, cepstral, and log-spectral distances – all of them describing spectral changes.

This basis was kept by more recent approaches, but refined in the sense of the corresponding human-ear processing: The first step is a spectral analysis indeed, but along a *warped frequency* axis. The spectral components are then evaluated, but only after a non-linear transformation to *loudness*, possibly including *masking* effects in time and frequency directions. Beyond, highly non-linear *cognitive* effects take place, like asymmetric reactions to added vs. subtracted components or threshold behaviors towards dominant or minor influences. These ideas have produced a group of quality measures with close relationships and internal differences. Always, short-time comparisons are carried out first, and the results are averaged at the end. While an early proposal [82] used a signal-to-distortion comparison, i.e, a type b-β procedure again (see Fig. 9.8), the more successful versions all rely on the type a-β.

9.7.3.2 Critical Bands and Bark Scale

By Θ, the so-called *Bark-frequency scale* is denoted; it describes the non-linearly growing bandwidth and center frequencies of the inner-ear critical-bands filter bank (see Fig. 9.9) by the approximation formula [95]

$$\frac{\Theta}{\text{Bark}} \doteq 13 \cdot \arctan\left(0.76 \cdot \frac{f}{\text{kHz}}\right) + 3.5 \cdot \arctan\left(\left(\frac{f}{7.5\,\text{kHz}}\right)^2\right). \quad (9.34)$$

The bandwidth growth reflected in Fig. 9.9 means that the basilar membrane of our ear realizes a 1/3-octave filter bank, approximately.

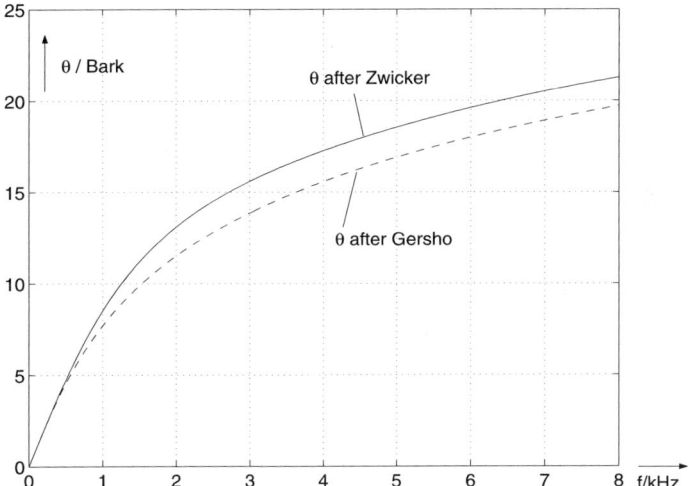

Fig. 9.9. Non-linear mapping of the frequency f to the inner-ear Bark-scale critical-band index θ, as described by Zwicker [95] and in the BSD measure of Gersho [93].

9.7.3.3 Loudness Transformation

As a specific loudness $N'(\Theta)$, the non-linearly mapped excitation E within a critical band Θ is calculated [95]. The mentioned mapping acts as a compression by, roughly, a fourth-root law:

$$N'(\Theta) \sim E^{0.23}. \quad (9.35)$$

The specific loudness is given in sone/Bark. Eq. 9.35 is valid for large values of the excitation. For smaller ones, a more complicated expression is found in the literature:

$$N'(\Theta) = c \cdot \left((\alpha + \beta \cdot E)^{0.23} - 1\right) \frac{\text{sone}}{\text{Bark}}. \tag{9.36}$$

The total loudness N, given in sone, is found by integration over $N'(\Theta)$:

$$\frac{N}{\text{sone}} = \int_{\Theta=0}^{\Theta_{max}} N'(\Theta) \cdot d\Theta. \tag{9.37}$$

As observed in Sec. 9.6.2.1, the total loudness N grows by a factor of two if the loudness level L_N increases by 10 dB. This corresponds to a relation between the total sound energy E_{tot} and the total loudness according to

$$\frac{N}{\text{sone}} \sim E_{\text{tot}}^{0.3}, \tag{9.38}$$

i.e., a replacement of the exponent 0.23 in Eq. 9.35 by 0.3 or a step from the fourth to the third root.

9.7.3.4 Bark-Spectral Distance (BSD)

A first well-known procedure following the above path was published in 1991 already by the speech-processing group of A. Gersho in [93], reporting a success figure of 85 % and 98 % correlation between real and estimated MOS values, for mixed and male-only voices, respectively. It was the basis of a thorough investigation [26]. The claimed performance could not quite be reproduced; still, the underlying principles were reckoned so reasonable that they were used as a prototype for several newly developed variants (see Secs. 9.7.3.7 and 9.7.3.8). These principles are also the foundation of other, well-known systems developed in the same period of time (see Secs. 9.7.3.5 and 9.7.3.6).

The BSD measurement begins with a pre-processing by a simple first-order filter, taking care of the frequency-dependent sensitivity of the ear, also due to the outer and middle-ear transfer functions and visible in Fig. 9.5. A DFT of speech frames with $T_F = 10$ ms creates a spectral representation. The energies in $M = 15$ non-uniform bands are computed; the band separation obeys the expression

$$\frac{\Theta}{\text{Bark}} = 6 \cdot \text{asinh}\left(1.67 \cdot \frac{f}{\text{kHz}}\right), \tag{9.39}$$

which approximates Eq. 9.34 more or less well, according to Fig. 9.9. The energies are expressed by equivalent loudness levels L_N and then mapped to specific loudnesses by a law

$$\frac{N}{\text{sone}} = \begin{cases} \left(\frac{L_N}{40\,\text{phon}}\right)^{2.642}, & L_N \leq 40\,\text{phon}, \\ 2^{((L_N/\text{phon})-40)}, & L_N > 40\,\text{phon}, \end{cases} \qquad (9.40)$$

whose graphical representation in Fig. 9.10 also perfectly matches Eq. 9.38. The squared differences between the specific loudnesses of the signals $x(n)$ and $y(n)$ are averaged over the Θ and time axes, yielding the value

$$BSD = \frac{1}{N_\text{seg}} \cdot \sum_{i=1}^{N_\text{seg}} \sum_{\mu=1}^{M} \left[N'_y(\Theta_\mu, i) - N'_x(\Theta_\mu, i)\right]^2. \qquad (9.41)$$

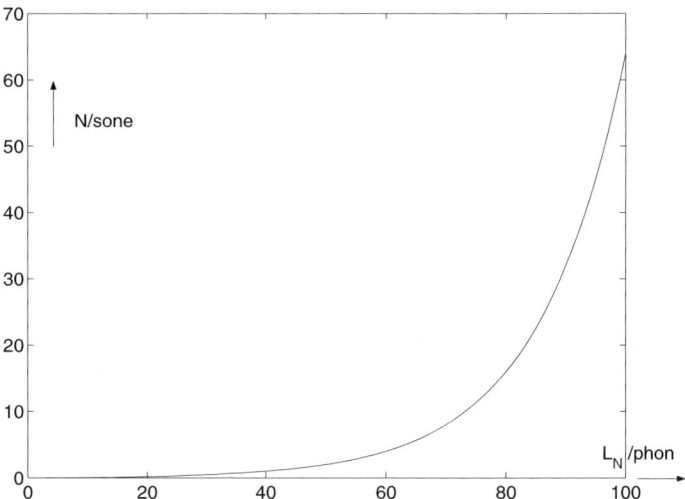

Fig. 9.10. Transformation from loudness level to loudness.

A second-order parabola with optimized coefficients is finally applied to estimate a MOS' value from this perceptually motivated signal distance.

9.7.3.5 Perceptual Evaluation of Speech Quality (PESQ)

The most widely accepted measure [4, 80, 81] has been standardized internationally [51]. It emerged from, essentially, a procedure which was developed in the research laboratories of the Royal PTT, The Netherlands, and named PSQM (Perceptual Speech-Quality Measure, [3]), plus amendments contributed by the British-Telecom developers of PAMS (Perceptual Analysis Measurement System, [79]).

The PSQM calculations follow similar steps as those of the BSD: Outer and middle-ear transfer functions are taken into account, a DFT spectrum is computed (via a fast, i.e., FFT algorithm), spectral power distributions are determined along a Bark scale, and loudnesses are calculated and compared for $x(n)$ and $y(n)$. Differences between PSQM and BSD concern the following points: In PSQM,

- also IRS filtering is included,
- the usually unavoidable background disturbance during a telephone use is modeled by adding "Hoth noise", a lowpass-type room-noise model [32],
- longer segments with $T_\mathrm{F} = 32$ ms (i.e., 256 samples) are used with Hann windowing and an overlap of 50 %,
- the loudness mapping follows Eq. 9.36 but with a heuristically modified exponent of 0.001 (instead of 0.23),
- the total loudnesses of $x(n)$ and $y(n)$ are adaptively equalized within limits,
- an asymmetric weighting within the loudness-difference averaging emphasizes positive errors more than negative ones, since the former are assumed to be more annoying, and
- another unequal weighting is applied to speech-activity and pause segments.

A non-linear mapping of the resulting weighted-mean loudness distance finally yields an estimate MOS'.

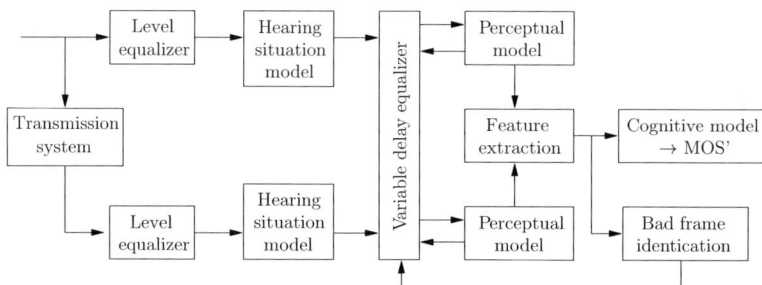

Fig. 9.11. Gross block diagram of PESQ.

In PESQ, the main structure remains (see Fig. 9.11). The equalizations are refined considerably, however, and especially a careful, recursive time alignment with plausibility checks is implemented. The warped frequency axis is modified and non-equally sampled in $M = 42$ points. The loudness transformation returns to Eq. 9.36 with an exponent 0.23, except for the first four subbands, were it is enlarged. Masking of very small, partial masking of small loudness differences is introduced. The asymmetry considerations contained already in PSQM are further refined in a separate averaging of unweighted and asymmetry-weighted distances (partially with, interestingly, an L_p norm using $p = 6$, see Sec. 9.7.2.2). A linear superposition of these separately determined distance measures yields the final estimate MOS'.

9.7.3.6 Telecommunication Objective Speech-Quality Assessment (TOSQA)

The original TOSQA procedure was first developed in the research laboratories of German Telekom as depicted in Fig. 9.12 [7]. The close relations to PESQ in Fig. 9.11 as well as to the principles of BSD in Sec. 9.7.3.4 are obvious.

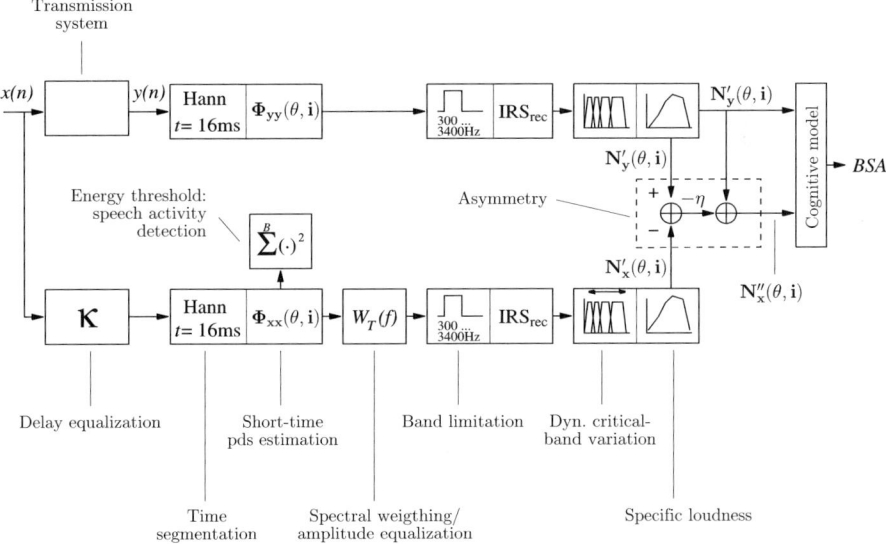

Fig. 9.12. Gross block diagram of TOSQA.

Again, equalizations are carried out for both delays (simpler than in PESQ) and amplitude changes (by a weighting factor); IRS filtering and telephone-band limitations take place, though now in the frequency domain, which again is reached via a DFT/FFT; Hann-windowed speech segments are used, but with a shorter frame length ($T_F = 16$ ms/128 samples). Loudnesses are found after critical-band energy aggregation; the transformation uses Eq. 9.36 but with factors α and β different from those in PESQ and [95].

A very particular feature of TOSQA is a dynamical variation of the critical-band widths in the energy summations: The edges are adaptively steered by observed frequency shifts, especially of formants, in the output-signal's spectral envelope. These shifts are assumed to be speaker, but not speech-quality relevant, and they are therefore equalized.

After asymmetric weighting of added/subtracted loudness errors (here by an adaptive factor η), instead of a distance, a Bark-spectral approximation (BSA) term is computed via cross-correlation of the loudnesses N_x'' and N_y', and a non-linear mapping finally estimates a value MOS'.

A commercially available version of TOSQA contains later enhancements: The voice-activity decision and the delay equalization are refined and adapted. Room-background noise is also inserted, although as a white signal. Packet loss, non-linearity, or noise components are separately detected and taken into account [38, 39]. Beyond, it contains an extension to wideband-speech assessment; for aspects of speech quality at higher bandwidth, the reader is referred to [30].

9.7.3.7 Speech-Quality Evaluation Tool (SQET)

The thorough study mentioned above [26] was carried out in the DSP group of Kiel University. Here, numerous variants of the BSD approach were systematically developed and analyzed. A final, optimized version follows the same principles as PESQ and TOSQA (see Fig. 9.13). The spectral information is, however, gained with an allpass-transformed DFT/FFT-based polyphase filterbank with 18 non-uniform channels, whose frequency responses closely approximate the critical-band structure according to Eq. 9.34. Also frequency and time-domain masking are included.

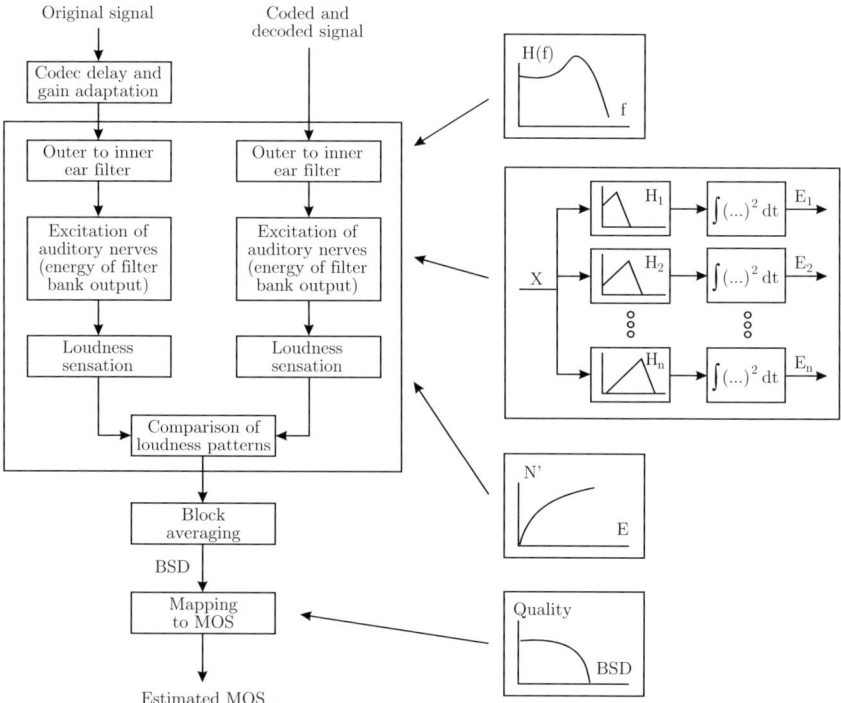

Fig. 9.13. Gross block diagram of SQET (by courtesy of M. Hauenstein).

9.7.3.8 Hair-Cell Model

Within the same study, the model of the human perception process was extended one step further: The spectral information is now the output of a "gamma-tone filterbank" [27]. This is a model of the biologically measured inner-ear spectral analysis system of a cat (see Fig. 9.14). After all the above steps, finally, nerve-fiber impulses are generated as a model of the hair-cell reaction on Corti's organ [31, 66]. The comparison between $x(n)$ and $y(n)$ is then based on the respective nervous activities (see Fig. 9.15).

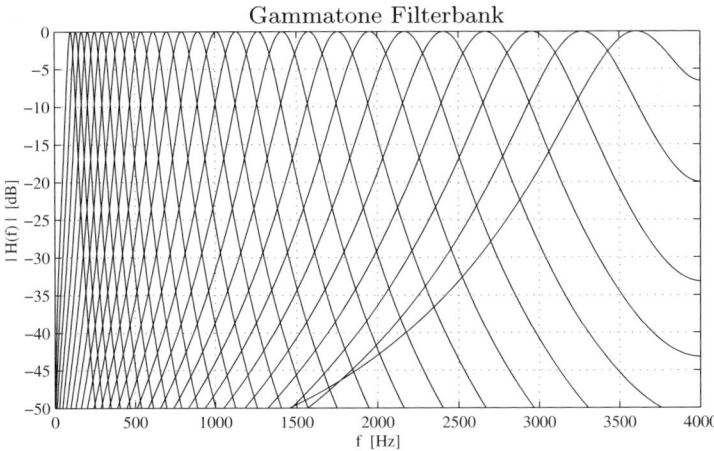

Fig. 9.14. Gammatone filterbank with 17 channels within the telephone bandwidth.

9.7.3.9 Performance and Comments

Beside the above discussed four BSD variants, several other proposals have been published and analyzed in depth [23–25, 64].

Some of them prefer a quite detailed psychoacoustic model; others prefer a sophisticated "cognitive" part. Their performances are quite similar, as depicted in Fig. 9.16: Ideally, estimates MOS' and true MOS values should be identical, i.e., they should all be on the diagonal of a MOS'/MOS plane. In reality, they vary along that line, with different variances and correlations between estimates and auditory results. All figures and all correlation factors, here $\rho = 0.92 \ldots 0.97$, indicate the usability of an instrumental prediction; care must, however, be taken in an evaluation: In [51], the applicability of PESQ to certain classes of systems or distorting conditions is carefully defined. Beyond, all measures have a common difficulty, named "ranking problem" in [29]: Two different systems transmitting speech under varying conditions (as to background noise, bandwidth, frame losses, etc.) may be graded subjectively with

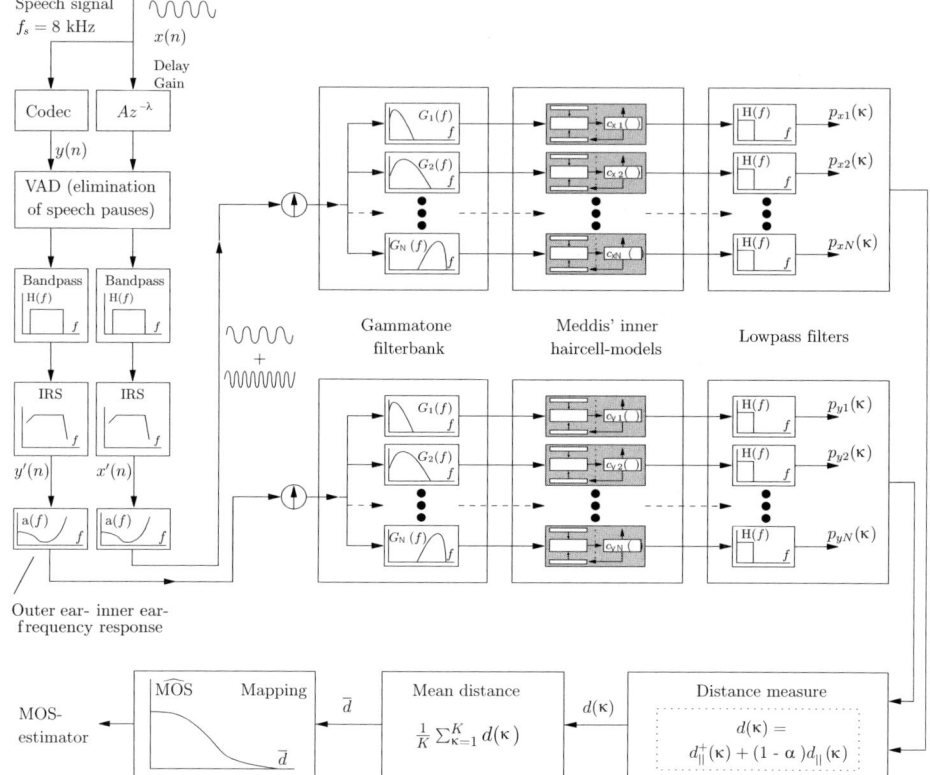

Fig. 9.15. SQET variant with a hair-cell model after a gammatone filterbank.

(almost) the same MOS, while the estimates cover a range $\Delta MOS' \approx 1.0$. Vice-versa, in other cases, the "true" MOS may vary over $\Delta MOS \approx 1.0$, while the estimates are (almost) identical.

Anyhow, as said in Sec. 9.3.2, diagnostic abilities of a quality assessment would be desirable. They are not provided by any integral-quality measure.

9.8 Instrumental Attribute-Based Quality Measurements

9.8.1 Basic Ideas

The application of attributes has been discussed in the context of auditory evaluation in Sec. 9.6. Now, logically, instrumental estimations of attributes are of interest. A decomposition of a quality assessment into a number of attribute assessments is advantageous: It gives diagnostic insight into system strengths and weaknesses, it models the idea closely that a human listener first has separate impressions during a perception time, and it allows, as in a

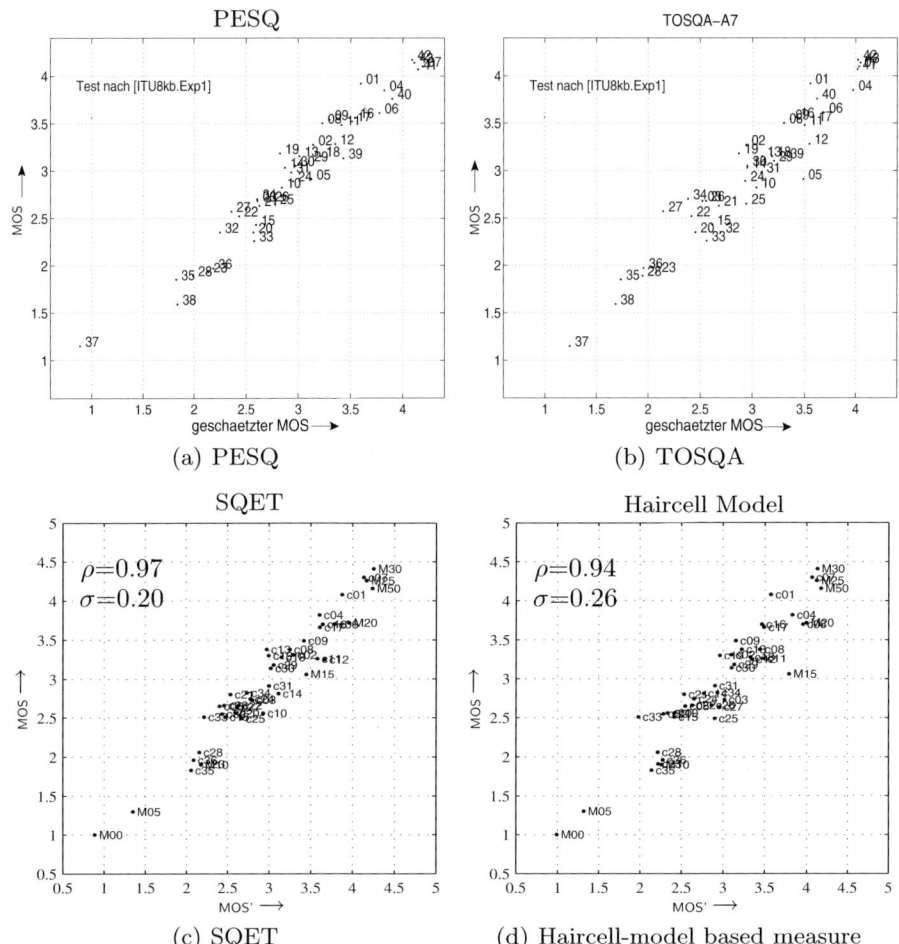

Fig. 9.16. Predicted vs. true MOS values for a set of codecs under various conditions with a) PESQ, b) TOSQA, c) SQET, d) a haircell-model based measure.

human brain, to construct an integral judgment from such separate, hopefully even orthogonal-scale grades. Possible formalisms were discussed in Sec. 9.6.4. Now,

- instrumental attribute estimation is the aim, from which, finally, with similar formulas, a total-quality may be predicted. All measurements should be built such that as few parameters as possible appear, in order to allow a parameter optimization from an always limited set of stimuli and auditory evaluations. For a validation of the measures and the parameter meanings,

- idealized model systems should be developed, which, with suitably chosen parameter settings, would
  - create the "same" perception (within error bounds),
  - yield the "same" parameter values again in an instrumental measurement (within error bounds).

In the following, corresponding techniques are presented, partially well known for the classical attributes of Sec. 9.6.4, partially developed newly for the more recent three dimensions presented in Sec. 9.6.3.

9.8.2 Loudness

In Sec. 9.6.2.1, the loudness of a sound was defined via its comparison with a 1 kHz tone. For a technical measurement, standardized methods exist [36]. Especially for a signal like speech which is only short-time stationary within some $10\ldots 20$ ms, the basis is the determination of the time-frequency behavior. Frames of length $T_\text{F} = 10\ldots 20$ ms undergo a spectral analysis, e.g., a (windowed) DFT. The steps toward a hearing-oriented representation include a frequency warping according to Eq. 9.34, the aggregation of energies in critical bands, the computation of the excitation pattern based on the absolute hearing threshold and smearing/masking along the Basilar membrane, the specific-loudness transformation of Eq. 9.36, and the total-loudness integration of Eq. 9.37. Beyond, time- (post-) masking has to be taken into account, with, due to the short stationarity times, a relatively fast but not negligible decay. The procedure is sketched in Fig. 9.17 in a simplified form; for details, the reader is referred to [10, 13].

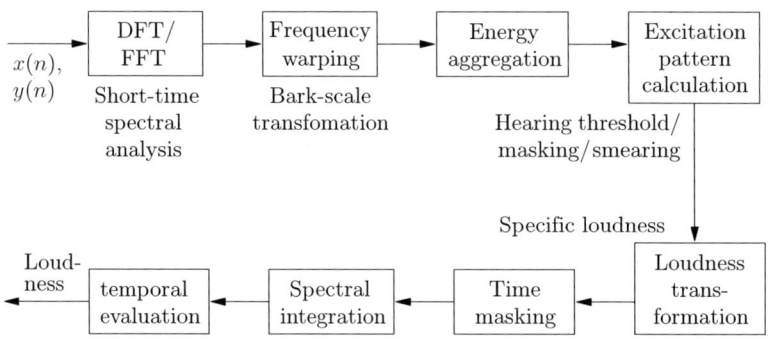

Fig. 9.17. Dynamic loudness evaluation (after [13]).

In the context of a quality assessment, the loudnesses of the two signals $x(n)$ and $y(n)$ in Fig. 9.1 have to be compared, or that of $y(n)$ with a reference loudness. This leads to the concept of loudness rating [70]: The difference between the loudness of a reference system (including the IRS filters of Fig. 9.2) and that of a system under investigation is calculated; it describes the loss of loudness due to the processing system and indeed describes an influence on the perceived quality: As said before, too low or too high a loudness will be annoying for a user, leading to a negative judgment. There is, however, no direct diagnosis contained: The loss is given by just one numerical value, hiding the reasons for an increased or decreased loudness integral, namely, possible frequency-depending attenuations or amplifications as well as possibly compensating non-linear or additive disturbances. For this reason, also no idealized model system can be given here: The separate effects hidden in loudness, i.e., perceptive sub-dimensions, are needed first and may then be modeled.

9.8.3 Sharpness

Like loudness, also sharpness was explained, in Sec. 9.6.2.2, by a comparison of a sound with a 1 kHz-centered signal. The technical measurement, too, is closely related to that of loudness [2, 95]: Sharpness is measured from a loudness integral as in Eq. 9.37, but with a growing weight $g(\Theta)$ for higher frequencies:

$$S \doteq c \cdot \frac{\int_{\Theta=0}^{\Theta_{\max}} N'(\Theta) \cdot g(\Theta) \cdot \Theta \cdot d\Theta}{\int_{\Theta=0}^{\Theta_{\max}} N'(\Theta) \cdot d\Theta} \qquad (9.42)$$

Variants are known from literature for the weighting function $g(\Theta)$.

Sharpness describes a possible emphasis of higher frequencies within a measured loudness. This is indeed a diagnostic result, saying that an unpleasant impression may stem from an unnaturally strong upper part of the spectrum. This can be especially used in a comparison between values measured for the signals $x(n)$ and $y(n)$ in Fig. 9.1. It remains, however, unclear, whether the annoying components emerged from a partly differentiating, i.e., growing frequency-response or from artificially created components. Again, no directly applicable model is therefore given.

9.8.4 Roughness

In Sec. 9.6.2.3, a signal's roughness R was attributed to fast signal-amplitude variations. In [95], $R \sim \Delta_L \cdot f_{\text{mod}}$ is shown to hold, with Δ_L describing the modulation depth after taking time-masking into account, and f_{mod} being the modulation frequency. For an instrumental analysis, modulation detection and estimation are therefore needed.

Amplitude variations can be measured by averaging a suitable number of short-time signal energies within suitable-length frames – or within fundamental periods of voiced-speech sections. The latter, with a normalization, is termed "shimmer" and used, in the context of voice analyses, for a roughness description. If the necessary "pitch detection" is available, detecting the usually slowly time-varying fundamental period $T_0(t)$, also faster variations of $T_0(t)$ can be quantified. A normalized average measure is termed "jitter"; it describes frequency rather than amplitude modulations, but it also contributes to a voice's roughness. The estimation can be further augmented by exploiting the cross-correlation of succeeding pitch periods [68, 69].

For an analysis of the perceived speech-signal roughness R as a diagnostic attribute for a transmitting or processing system, the above approaches seem to be quite appropriate, too. Also, an idealized-model system creating roughness of the same degree R can be imagined easily by means of amplitude and frequency-modulation algorithms, given the shimmer and jitter values – plus estimated modulation frequencies.

In the attribute analysis discussed further in the following, roughness has not been included. It should, however, deserve deeper consideration: Somewhat "hoarse" codecs, for instance, are quite well-known in the speech-coding community, and fast amplitude fluctuations happen in some adaptive noise-reduction systems. So, a roughness detector would indeed be valuable.

9.8.5 Directness/Frequency Content (DFC)

Plausibly, a good system passes much of the input spectrum to the output without change. A real system's ability to do this is described by its average frequency response. The left part of Fig. 9.18 displays a corresponding measurement result, as an attenuation given in dB over the Bark-frequency scale Θ (see Eq. 9.34), with several quite typical effects to be observed:

- There is a band limitation, to be expressed by a bandwidth BW.
- The band has a certain position on the Θ-axis, to be described by the band edges or, together with some shape parameters, by the center of gravity

$$\Theta_G = \frac{\int_{\Theta_1}^{\Theta_2} G(\Theta) \cdot \Theta \cdot d\Theta}{\int_{\Theta_1}^{\Theta_2} G(\Theta) \cdot d\Theta}. \qquad (9.43)$$

By $\Theta_{1,2}$, the band edges are described within which the Bark-scale-transformed magnitude frequency response $G(\Theta)$ is above a certain threshold.
- The passband shows a certain average tilt, with an angle β.
- The passband behavior is not smooth; beyond the tilt, there are oscillations, to be characterized by their rate γ_R and their depth γ_D, under the simplifying assumption of a cosine-type "ripple".

This may be insufficient in some cases. In the left part of Fig. 9.18, e.g., there is no unique tilt; in fact, two tilts add to a "hat"-shape. A better description is possible, but requires at least another parameter. Still, a "dominant"

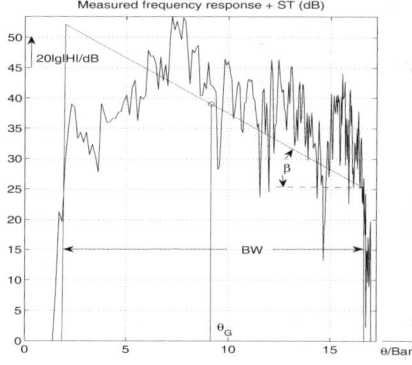
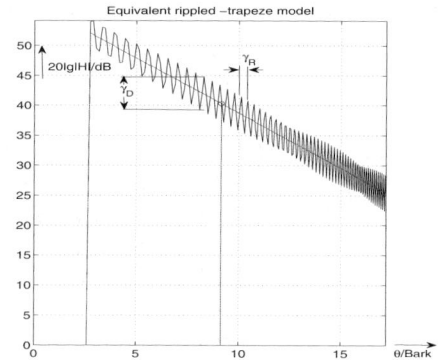

Fig. 9.18. Measured (left) and modelled (right) average frequency response of an example system.

angle can be found from averaging. Also, the other features can be calculated from the measured curve. Especially, BW can be expressed by an equivalent rectangular bandwidth

$$ERB = \frac{\int_{\Theta_1}^{\Theta_2} G(\Theta) \cdot d\Theta}{\max_{\Theta \in [\Theta_1, \Theta_2]} \{G(\Theta)\}}. \tag{9.44}$$

With these parameters, the basis is also found for a model system, realized by, e.g., a non-recursive digital filter such that $ERB, \Theta_G, \beta, \gamma_R,$ and γ_D appear in a stylized, i.e., idealized manner (see right part of Fig. 9.18). The necessary signal-based measurements follow classical methods [84], based on (cross-) periodogram computations. The derivation of the parameters and the model is found in [34, 83].

In [83], it was shown that width and position of the averaged passband of the system alone are able to quite well predict DFC-LOT results by a simple linear formula:

$$DFC' = -20.5865 + 0.2466 \frac{ERB}{\text{Bark}} + 1.873 \frac{\Theta_G}{\text{Bark}}. \tag{9.45}$$

The correlation between DFC' and subjective DFC values is $\rho \approx 0.96$. Although this seems to say that the estimation 9.45 suffices, it is admitted in [83] that further parameters have to be included, the high correlation being due to the too small number of systems in the test. Also some psychoacoustic considerations confirm this necessity:

- ERB and Θ_G may be able to model "correct frequency content" as parts of "fullness" and "brightness". Still, an average-slope parameter β inside

the passband, i.e., a pre- or de-emphasis behavior, should be explicitly introduced, since, above some limit, it has certainly less to do with the above attributes than with, e.g., sharpness (see Sec. 9.8.3).
- "Directness" needs quantities reflecting "distance effects" – but not only loudness reductions, but especially also echoes, reflected by a comb-filter behavior, i.e., a "rippled" frequency response; so, the "ripple rate" γ_R and a "ripple depth" γ_D should take care of this, at least partly.

9.8.6 Continuity

Signal discontinuities have a strong impact on the perceived quality: Clipping of speech onsets (e.g. by a misadjusted voice-activity driven switch) or gaps inside the signal flow (due to lost data blocks, in a frame or packet-based transmission) create the feeling of a randomly "loose connection", even causing intelligibility problems – and heavy dissatisfaction anyhow.

For such "drop-outs" of longer signal sequences, techniques can be used which were developed for online monitoring of services by so-called "in-service non-intrusive measurement devices" (INMD) [46, 47].

Lost frames in a block-coding system, with a duration of $T_F \approx 10\ldots 20$ msec, may be technically "hidden"; the simplest such "concealment" is a repetition of the last completely received frame. Then, obviously, the signal samples at frame distance must be identical or, with unknown frame boundaries, at least highly correlated. This can be exploited for detection. Erroneous alarms may happen if the signal itself contains a periodicity of a similar duration; this happens if the fundamental ("pitch") frequency f_0 fulfills the condition $f_0 = \lambda \cdot 1/T_F$, $\lambda \in Z$. False alarms can the be avoided by a pitch detection and a check of the above condition [62, 63]. Similarly, other simple packet-loss concealments may be detected.

Unconcealed losses of data blocks create simply a longer zero sequence in the signal. This can be found from an observation of an "energy gradient" $\Delta E(i)$ [62]. It is defined as the change of the energy between two succeeding blocks indexed by (i-1) and i:

$$\Delta E(i) \doteq \frac{E(i)}{E(i-1)} - 1. \tag{9.46}$$

The block energies are calculated from the squared signal samples after appropriate band limitation to the telephone band or, due to Parseval's theorem, by summation/integration over squared spectral values within this band. Occurrence of $E(i) = 0$ indicates a lost block. Then, $\Delta E(i) = -1.0$ appears. If the following block is transmitted again, $E(i-1) = 0$ and $E(i) > 0$ lead to a devision by zero; a limitation to $E_{\max} = +1$ is useful. Then, a sequence $\Delta E(i) = -1$ followed by $\Delta E(i+1) = +1$ indicates a single block loss. If more losses happen in a sequel, $E(i) = E(i+1) = 0$ are found, where $\Delta E(i)$ is set to zero. The first value $\Delta E(i) = +1$ thereafter will indicate the recovered transmission. Fig. 9.19 shows an example of a signal with several single frame losses visible both in the corresponding spectrogram and in the sequence of energy gradients. The detection potential was also evaluated in [62], and a reliability of 95 % correctly identified losses was stated. A further refinement is proposed for cases where lost frames are first replaced by the last correct one, but then a slow muting takes place if no recovery is detected, and zero sequences appear thereafter.

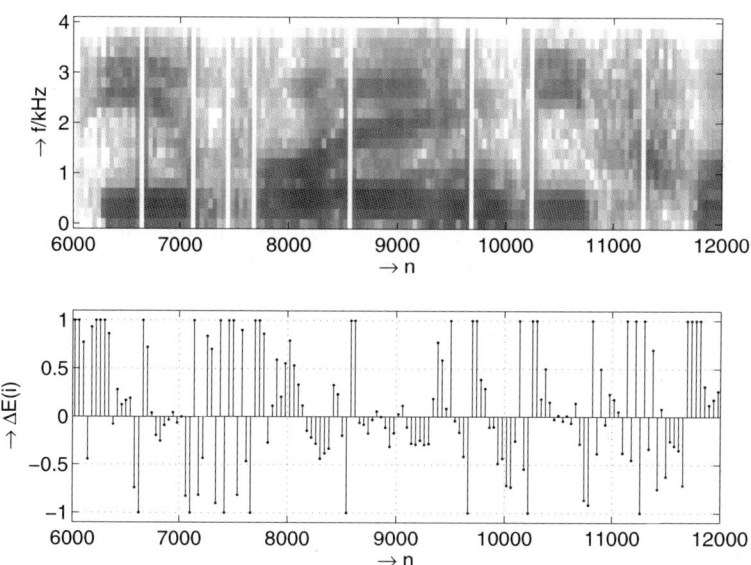

Fig. 9.19. Spectrogram and corresponding energy-gradient sequence for a signal with single frame losses without concealment.

More carefully concealed losses are less easily identified [62], but also less important: The quality loss will be correspondingly smaller.

For a quality-impact estimation, loss parameters have to be derived now from the detection results. They should consist of a loss-probability estimate, first, and, if detectable, an indication of a (simple) concealment. Beyond, the loss type must be characterized: Erasures may occur as independent, single random events as well as in longer "bursts" of drop-outs, with a certain burst-length probability to be estimated. Beyond, these descriptors may be constant or time-varying during a connection. Once suitable parameters are found, an idealized model for this type of disturbance can be realized, dropping or possibly replacing blocks with appropriate probabilities.

Other short-time effects include pulse-like disturbances, e.g., due to bit errors, which are instantaneous and therefore occupy the whole frequency band. Short-time blocks of narrowband distortions may also appear, like the so-called "musical noise" created by simple noise-reduction techniques. Fig. 9.20 depicts a spectrogram of a corresponding case. The typical small, rectangular spots indicate the insertion of artificial narrowband, "almost tonal" sounds. These spots may be found obviously by a technical observation of the spectrogram. The so-called "relative approach" of [17] is a good candidate. It finds "unexpected" short-time components by a comparison with an averaged Bark-scale and loudness-transformed spectrogram, the latter describing the "expected" behavior. This method was also applied to packet-loss concealment in [56]; it could therefore be devised for a combined lost-data and tonal-noise detection.

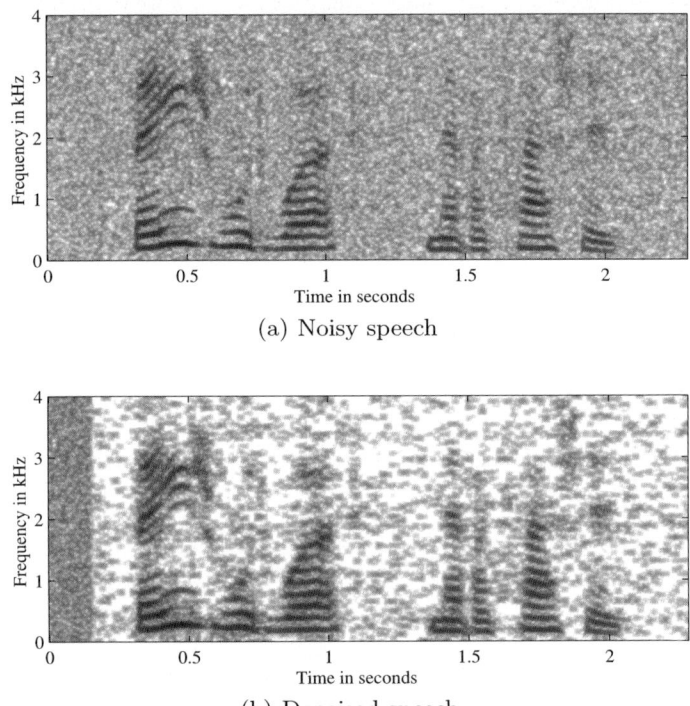

Fig. 9.20. Spectrogram of a speech signal with additive white noise before a) and after application of a simple de-noising technique b) showing "musical-noise" artifacts [19].

A proposal due to [18] for post-processing of spectral-subtraction results is also of interest. It aims at a removal of the tonal residual by monitoring the variation of the values along a set of straight lines with different angles in a time-frequency plane. In pure-noise or unvoiced-speech regions, all variances will be relatively low; in regions of voiced speech, a small variance will occur at least in one of the directions, namely, along the pitch contour; in musical-noise regions, however, relatively large variances will be found in all directions. In our context, this can be exploited for the desired detection and characterization of short-time narrow-band signal insertions.

An idealized model creating data loss with a prescribed character is conceptually simple, and the same may hold for insertion of impulse disturbances. The generation of musical-noise-type effects is, however, not simple at all, since they may vary considerably in reality, concerning the shape of the spectrogram spots, i.e., their time and frequency structure.

Discontinuity is a dominant quality factor, as seen in Sec. 9.6.4. The definition of a continuity measure C, however, remains difficult, even if it is assumed that all single "discontinuous" effects can be reliably found, described, and modeled. The above discussion shows that quite different influences are "somehow summarized" up to now. An estimator C' for a true continuity C may therefore not be derived until C becomes clear. Especially, possible subdimensions may be searched for, and their relations to the perceived quality. As to the drop-out part, hints may be found from the thorough investigations of quality vs. loss statistics in [78]. Especially the use of impairment terms as intermediate quantities, as applied there, could be also transferred to the other short-time effects.

9.8.7 Noisiness

While, once more, this attribute seems to be intuitively clear at a first glance, it turns out to be quite complex. This may be explained by a simple example: The German language has two translations for noise, namely, "Rauschen" and "Geräusch". The first term is what in electrical and information engineering is usually linked to noise, namely, a more or less wideband, random signal, though with possibly various spectral shapes. The second term may be just *any* acoustical signal – that of a motor with perhaps a few harmonic components only, that of screeching car brakes, or that of a group of background speakers ("babble").

Also from inspection of the examples included in the MDS and SD analysis addressed in Secs. 9.6.3.1 and 9.6.3.2, it becomes clear that systems with considerable "noisiness" values on the negative part of the related scale do not simply suffer from additive white noise. Colored background noise, broadband decoding and output-terminal noise, and "signal-correlated" noise appear as well as a noise-like hoarseness. A systematic investigation is reported in [60]. In the telephone-transmission case dealt with, it is known that background noise must have passed both the send and receive handset filters (see Fig. 9.2). The system noise is filtered much less heavily by the receiver only, where the signal-correlated part is limited to $f_c = 4$ kHz according to Eq. 9.6.

A possible approach is then to determine background noise N_{BG} inside the telephone band up to ~ 3.4 kHz during speech pauses, evaluate the band between 3.4 kHz and 4 kHz in speech periods for correlated noise N_{corr}, and check for higher-frequency system noise N_{HF} above 4 kHz. Interestingly, "hoarseness" was interpreted as "very short interruptions" SI, not influencing "continuity", but still measurable by a suitably modified energy-gradient observation (see Eq. 9.46 in Sec. 9.8.6). A measurement of higher-frequency parts requires, of course, that the signal between the actual transmission/processing and the acoustical output. i.e., $y(n)$ before becoming $y_0(t)$ in Fig. 9.1, is up-sampled to a rate $f_s > 8$ kHz. From the measured powers within the above bands plus the term SI, a linear superposition can be made up [60] to predict the noisiness NS by an estimate NS'. This estimation is found to be well correlated with auditory noisiness ratings and also orthogonal with respect to other dimensions (see Tab. 9.7). However, this is still a field of investigation.

The corresponding idealized distortion system is defined easily, on the other hand, inserting correspondingly filtered noise types and "short interruptions" – which, by their explanation, point to a link with "roughness" (see Sec. 9.8.4).

Table 9.7. Correlations between auditory dimension evaluations and the linear formula for noisiness [60].

Dimension	ρ
Noisiness	0.9109
Directness/frequency content	0.0362
Continuity	-0.1213

9.8.7.1 Total-Quality Calculation

As said in Secs. 9.3.2 and 9.6.4, an integral-quality prediction can be constructed from the single attribute predictions. If sufficiently verified estimations C' for the continuity C, DFC' for the directness/frequency content DFC, and NS' for the noisiness NS are given, it is an obvious step to go to the MDS and SD analysis in Sec. 9.6.4: The auditively identified three factors $F_1 = DFC$, $F_2 = C$, and $F_3 = NS$ are able to describe the difference between the systems and conditions and, thereby, their perceived qualities, in the simplest approach via a linear combination as given in Eq. 9.14 [92]. For a MOS' estimation, all dimensions in Eq. 9.14 have to be replaced by their estimates; the numerical values given for the necessary weighting coefficients are, however, preliminary. Also the linear combination as such is, of course, questionable. There have been non-linear approaches with similar success (see Sec. 9.6.4). An alternative may be an intermediate step to impairments as used in the E-Model (see Sec. 9.4), since they are indeed devised to really add linearly.

9.8.8 Combined Direct and Attribute-Based Total Quality Determination

For years, the direct integral-quality estimation on the basis of a Bark-spectral loudness distance by PESQ and similar systems, on one hand, and, on the other hand, the search for attributes, their instrumental estimation and their combination into an integral-quality prediction have been dealt with separately. Now, within ITU-T, since 2005, the enhancement of total-quality estimators by dimension analysis in parallel is investigated. Hopefully orthogonal, "decomposed" perceptual effects are to be addressed, hopefully leading to improved predictions, but, more important, also delivering a quality diagnosis [8]. On the other side, in the present overall-algorithms, some quantities are measured which are used to emphasize or diminish certain perceptually more or less important features; also, knowledge about specific effects from certain systems was exploited to broaden the validity: These algorithms may be used in a dimension-based analysis to develop alternative attribute descriptors.

9.9 Conclusions, Outlook, and Final Remarks

For telephone-band speech quality, PESQ is claimed to be a "world standard" [5] for an instrumental quality measurement, predicting the MOS results of well-defined ACR-LOTs reliably under numerous, carefully explained conditions [51]. In the same context, however, amendments are stated to be desirable, reducing the sensitivity of PESQ, TOSQA, SQET, and similar techniques towards unknown types of distorting systems. The way via decomposed, generic attributes first and a derivation of an overall quality estimation thereafter is natural and helpful, and it adds diagnostic insight. A cross-exploitation of internal features from both approaches is promising. Modeling both attribute-specific disturbance generation and total-quality perception needs more integration of psychoacoustical and system-theoretical knowledge to be taken into account. The present investigations of dimensionality as such and "named" attributes or especially (perhaps: numerous) sub-dimensions requires a rigid continuation, with a really large variety of stimuli included. This is not only a question of sufficiently many speech data, but a problem of voluminous auditory tests as a basis.

In such continued research efforts, some topics should be covered which were not even touched in this chapter:

One point concerns the test signals used in the instrumental measurements. Short speech samples, like those of auditory tests, seem to be a natural choice. But they have drawbacks, in terms of speaker peculiarities or reproducibility of measurement results. In an early investigation of attribute-based instrumental measures [21, 22], stochastic signals were applied instead, digitally generated such that they have speech-like characteristics concerning, *e.g.*, the PDF, the short-time PDS, time-varying harmonic components, etc. The above-mentioned approach lead to high correlations with real and estimated MOS values, though not outside the limited world of the included processing systems. The reason for this failure is, however, found in the facts that

- a type b - α measure, in terms of Fig. 9.8, was devised, which is the poorest choice, based on a comparison of *averaged* signal and *error* behaviors,
- a frequency-dependent SNR with some additional parameters was adapted to pre-selected attributes numerically, i.e., without modeling the attribute interpretation as it is done in the recent techniques discussed above.

The reason is thus not assumed in the use of stochastic speech-model signals, which deserve to be re-visited. These signals have also advantages in the sense that, on one hand, really average signals without peculiarities are available, and, on the other hand, special features may easily be inserted, like special features of female/male speech or of specific languages.

Another point not discussed here is the fact that all presented measures aim at a system's quality as such, that is: A user is asked *afterwards* about the quality impression. The dynamic evolvement of the final judgment is out of scope – but it would be very interesting, since modern transmission systems (like VoIP or mobile phones) have a time-varying performance. Beyond, research results would help to refine other measures by an inclusion of dynamic features, as this is partly already included in PESQ. Efforts in this direction, discussed in [24] or in [78], deserve a continuation.

An apology and an acknowledgment are due, at this point. The former concerns the fact that this chapter has mainly the work of the author's team and of those groups in its focus, who have cooperated with him over years. Other work has been given credit, but of course, regrettably but necessarily, in an incomplete way. The work of many cooperating friends, colleagues, and students, however, who have contributed to the background of this text, is deeply acknowledged. This concerns especially the team of S. Möller in Berlin and, above all, the research students in Kiel.

References

1. J. Allen, M. S. Hunnicut, D. Klatt: *From Text to Speech: The MITalk System,* Cambridge, USA: Cambridge Univ. Press, 1987.
2. W. Aures: *Computational Methods for the Sound of Arbitrary Acoustic Signals – a Contribution to Hearing-Related Sound Analysis,* Diss., Techn. Univ. Munich, Germany, 1984 (in German).
3. J. G. Beerends, J. A. Stemerdink:. A perceptual speech-quality measure based on a psychoacoustic sound representation, *J. Audio Eng. Soc.*, **42**, 115–123, 1994.
4. J. G. Beerends, A. P. Hekstra, A. W. Rix, M. P. Hollier: Perceptual evaluation of speech Qquality (PESQ), the new ITU standard for end-to-end speech-quality assessment, part II – psychoacoustic model, *J. Audio Eng. Soc.*, **50**, 765–778, 2002.

5. J. G. Beerends, J. M. van Vugt: *Speech-Quality Degradation Decomposition*, 4-th IEEE Benelux Sig. Process. Symp., Hilvarenbeek, The Netherlands, 2004.
6. L. L. Beranek: The design of speech-communication systems, *Proc. IRE*, **35**, 880–890, 1947.
7. J. Berger: *Instrumental Methods for Speech-Quality Estimation*. Diss. Univ. Kiel, Arb. Dig. Sig. Process., no. 13, ed.: U. Heute, Aachen, Germany: Shaker, 1998 (in German).
8. J. Berger: Requirements of a New Model for Objective Speech-Quality Assessment P.OLQA, Delayed Contribution 75, Qu. 9, Study Group 12, ITU-T, Geneva, CH, 2005.
9. F. Burkhardt, F. Metze, J. Stegmann: Speaker classification for next-generation voice-dialog systems, in R. Martin, U. Heute, C. Antweiler (eds.), *Advanced Digital Speech Transmission*, Hoboken, USA: Wiley, 2008
10. J. Chalupper, H. Fastl: Dynamic loudness model, *Acustica/Acta Acustica*, **88**, 378–386, 2002.
11. ETSI GSM 06.10: GSM full-rate transcoding, *ETSI Recommendation*, Sophia-Antipolis, France, 1988.
12. ETSI GSM 06.60: Digital cellular telecommunications system: enhanced full-rate (EFR) speech transcoding, *ETSI Recommendation*, Sophia-Antipolis, France, 1996.
13. H. Fastl: Psycho-acoustics and sound quality, in J. Blauert (ed.), *Communication Acoustics*, Berlin, Germany: Springer, 2005.
14. W. M. Fisher, G. R. Doddington, K. M. Goudie-Marshall: The DARPA speech recognition research database: specifications and status, *Proc. DARPA Workshop*, 93–99, California, USA, 1986.
15. H. Fletcher, R. H. Galt: The perception of speech and its relation to telephony, *JASA*, **22**, 89–151, 1950.
16. N. R. French, J. C. Steinberg: Factors governing the intelligibility of speech sounds, *JASA*, **19**, 90–119, 1947.
17. K. Genuit: Objective evaluation of acoustic quality based on a relative approach, *Proc. Internoise '96*, 3233–3238, Liverpool, UK, 1996.
18. Z. Goh, K. Tan, B. T. G. Tan: Postprocessing method for suppressing musical noise generated by spectral subtraction, *IEEE Trans. on Speech and Audio Process.*, **6**(3), 287–292, 1998.
19. T. Gülzow: *Quality Enhancement for Heavily Disturbed Speech Signals - Detection of a Carrier Mismatch and Suppression of Additive Noise,* Diss. Univ. Kiel, Arb. Dig. Sig. Process., **20**, U. Heute (ed.), Aachen, Germany: Shaker, 2001 (in German).
20. J. D. Griffith: Rhyming minimal contrasts: a simplified diagnostic articulation test, *JASA*, **42**, 236–241, 1967.
21. U. Halka, U. Heute: A new approach to objective quality measures based on attribute-matching, *Speech Communication*, **11**, 15–30, 1992.
22. U. Halka: *Objective Quality Assessment of Speech-Coding Methods Applying Speech-Model Processes,* Diss. Ruhr-Univ. Bochum, Arb. Dig. Sig. Process., **3**, U. Heute (ed.), Aachen, Germany: Shaker, 1993 (in German).
23. M. Hansen: *Assessment and Prediction of Speech-Transmission Quality with an Auditory Processing Model,* Diss., Univ. Oldenburg, Germany, 1998.
24. M. Hansen, B. Kollmeier: Continuous assessment of time-varying speech quality, *JASA*, **106**, 2888–2899, 1999.

25. M. Hansen, B. Kollmeier: Objective modeling of speech quality with a psychoacoustically validated auditory model, *J. Audio Eng. Soc.*, **48**, 395–409, 2000.
26. M. Hauenstein: *Psychoacoustically Motivated Measures for Instrumental Speech Quality Assessment,* Diss. Univ. Kiel, Arb. Dig. Sig. Process., **10**, U. Heute (ed.), Aachen, Germany: Shaker, 1997 (in German).
27. M. Hauenstein: Application of Meddis' Inner-Hair-Cell Model to the Prediction of Subjective Speech Quality, *Proc. ICASSP '98*, 545–548, Seattle, USA, 1998.
28. U. Heute: Speech and audio coding – aiming at high quality and low data rates, in J. Blauert (ed.), *Communication Acoustics,* Berlin, Germany: Springer, 2005.
29. U. Heute, S. Möller, A. Raake, K. Scholz, M. Wältermann: Integral and diagnostic speech-quality measurement: state of the art, problems, and new approaches, Proc. Forum Acust. '05, 1695–1700, Budapest, Hungary, 2005.
30. U. Heute: Quality aspects of wideband vs. narrowband speech signals, in R. Martin, U. Heute, C. Antweiler (eds.), *Advanced Digital Speech Transmission,* Hoboken, USA: Wiley, 2008.
31. M. J. Hewitt, R. Meddis: An evaluation of eight computer models of mammalian inner hair-cell function, *JASA*, **90**, 904–917, 1991.
32. D. F. Hoth: Room-noise spectra at subscriber's telephone location, *JASA*, **12**, 499–504, 1941.
33. A. S. House, C. E. Williams, M. H. L. Hecker, K. D. Kryter: Articulation-testing methods: consonantal differentiation with a closed-response set, *JASA*, **37**, 158–166, 1965.
34. L. Huo, K. Scholz, U. Heute: Idealized system for studying the speech-quality dimension "directness/frequency content", Proceed. 2-nd ISCA-DEGA Workshop Percept. Qual. Syst. '06, 109–114, Berlin, Germany, 2006.
35. IPDS: *Kiel Corpus,* www.ipds.uni-kiel.de/publikationen/kcrsp.de.html, 1997.
36. ISO R 532 B: *Acoustics-Method for Calculating Loudness Level,* Int. Standardiz. Org., Geneva, Switzerland, 1975.
37. F. Itakura: Minimum prediction-residual principle applied to speech recognition, *IEEE Trans. on Acoustics, Speech, and Signal Proc.*, **23**(1), 67–72, 1975.
38. ITU COM 12-34-E: *TOSQA – Telecommunication Objective Speech-Quality Assessment,* Contrib. 34, Study Group 12, ITU-T, Geneva, Switzerland, 1997.
39. ITU COM 12-19-E: *Results of objective speech-quality assessment of wideband speech using the advanced TOSQA2001,* Contrib. 19, Study Group 12, ITU-T, Switzerland. CH, 2000.
40. ITU G.107: The E-model, a computational Model for Use in Transmission Planning, *Recommendation ITU-T,* Geneva, Switzerland, 2005.
41. ITU G.711: Pulse-code modulation (PCM) of voice rrequencies, *Recommendation ITU-T,* Geneva, Switzerland, 1993.
42. ITU G.712: Transmission performance characteristics of pulse code modulation channels, *Recommendation ITU-T,* Geneva, Switzerland, 2001.
43. ITU G.726: 40, 32, 24, 16 kbit/s adaptive differential pulse-code modulation (ADPCM), *Recommendation ITU-T,* Geneva, Switzerland, 1990.
44. ITU P.48: Specification of an intermediate reference system, *Recommendation ITU-T,* Geneva, Switzerland, 1993.
45. ITU P.85: A method for subjective performance assessment of the quality of speech voice output eevices, *Recommendation ITU-T,* Geneva, Switzerland, 1995.

46. ITU P.561: In service, non-intrusive measurement device – voice-service measurements, *Recommendation ITU-T,* Geneva, Switzerland, 1996.
47. ITU P.562: Analysis and enterpretation of INMD voice-service measurements, *Recommendation ITU-T,* Geneva, Switzerland, 2000.
48. ITU P.563: Single-ended method for objetive speech-quality assessment in narrowband telephony applications, *Recommendation ITU-T,* Geneva, Switzerland, 2004.
49. ITU P.800: Methods for subjective determination of transmission quality, *Recommendation ITU-T,* Geneva, Switzerland, 1996.
50. ITU P.830: Subjective performance assessment of telephone-band and wideband digital codecs, *Recommendation ITU-T,* Geneva, Switzerland, 1996.
51. ITU P.862: Perceptual evaluation of speech quality (PESQ): An objective method for end-to-end speech quality assessment of narrowband telephone networks and speech codecs, *Recommendation ITU-T,* Geneva. Switzerland, 2001.
52. ITU P-S.3: Models for predicting transmission quality from objective measurements, *Supplement 3, Series P-Recommendtions ITU-T,* Geneva, Switzerland, 1993.
53. C. Jankowski, A. Kalyanswamy, S. Basson, J. Spitz: NTIMIT: a phonetically balanced, continuous speech, telephone-bandwidth speech database, *Proc. ICASSP '90,* **1**, 109–112, Albuquerque, USA.
54. U. Jekosch: *Voice and Speech Quality Perception,* Berlin, Germany: Springer, 2005.
55. N. O. Johannesson: The ETSI computation model: a tool for transmission planning of telephone networks, *IEEE Commun. Mag.,* **35**, 70–79, 1997.
56. F. Kettler, H. W. Gierlich, F. Rosenberger: Application of the Relative Approach to Optimize Packet-Loss Concealment Implementations, *Proc. DAGA '03,* 662–663, Aachen, Germany, 2003.
57. J. Kruskal: Multidimensional scaling by optimizing goodness of fit to a nonmetric hypothesis, *Psychometrika,* **29**, 1–27, 1964.
58. J. Kruskal, M. Wish: Multidimensional scaling, in E. M. Uslaner (ed.), *Quantitative Applications in the Social Sciences,* Newbury Park, USA: Sage, 1978.
59. K. D. Kryter: Methods for the calculation and use of the articulation index, *JASA,* **34**, 1689–1697, 1962.
60. C. Kühnel: *Investigation of Instrumental Measurement and Systematic Variation of the Speech-Quality Dimension "Noisiness",* Dipl. Thesis, Univ. Kiel, Germany, 2007 (in German).
61. J. Laver: *The Phonetic Description of Voice Quality,* Cambridge, UK: Cambr. Univ. Press, 1980.
62. T. Ludwig: *Measurement of Speech Characteristics for Reference-Free Quality Evaluation of Telefone-Band Speech,* Diss., Univ. Kiel, Arb. Dig. Sig. Process., **23**, U. Heute (ed.), Aachen, Germany: Shaker, 2003 (in German).
63. T. Ludwig, K. Scholz, U. Heute: Speech-quality evaluation in telephone networks, *Proc. DAGA '03,* 718–719, Aachen, Germany, 2003.
64. V. V. Mattila: *Perceptual Analysis of Speech Quality in Mobile Communications,* Diss., Tampere Univ. Techn., **340**, Tampere, Finnland, 2001.
65. B. J. McDermott: Multidimensional analyses of circuit quality judgments, *JASA,* **45**, 774–781, 1969.
66. R. Meddis: Implementation details of a computation model of the inner hair-cell auditory-nerve synapse, *JASA,* **87**, 1813-1816, 1990.

67. P. Mermelstein: Evaluation of a segmental SNR measure as an indicator of the quality of ADPCM-coded speech, *JASA*, **66**, 1664–1667, 1979.
68. D. Michaelis, H. W. Strube, E. Kruse: Multidimensional analysis of acoustic voice-quality parameters, in M. Gross (ed.), *Aktuelle phoniatrisch-pädaudiologische Aspekte 1995*, **3**, Berlin, Germany: R. Gross, 1996 (in German).
69. D. Michaelis, M. Fröhlich, H. W. Strube, E. Kruse, B. Story, I. R. Titze: *Some simulations concerning jitter and shimmer measurements*, Proc. Int. Workshop Voice & Speech Res. '98, 71–80, Aachen, Germany, 1998.
70. S. Möller: *Assessment and Prediction of Speech Quality in Telecommunications*, Boston, USA: Kluwer Acad. Press, 2000.
71. S. Möller: Quality of transmitted speech for humans and machines, in J. Blauert (ed.), *Communication Acoustics*, Berlin, Germany: Springer, 2005.
72. S. Möller: *Quality of Telephone-Based Spoken-Dialog Systems*, New York, USA: Springer, 2005.
73. S. Möller, K. P. Engelbrecht, M. Pucher, P. Fröhlich, L. Huo, U. Heute, F. Oberle: TIDE: a testbed for interactive spoken-dialogue system evaluation, *Proc. SPECOM '07*, Moscow, GUS, 2007.
74. S. Möller, K. P. Engelbrecht, M. Pucher, P. Fröhlich, L. Huo, F. Oberle, U. Heute: Testbed for dialogue system evaluation and optimization, in Th. Hempel (ed.), *Usability of Speech-Dialog Systems - Listening to the Target Audience*, Berlin, Germany: Springer, 2008.
75. W. Niemeyer, G. Beckmann: A speech-audiometric sentence test (in German: Ein sprachaudiometrischer Satztest), *Arch. Ohren-, Nasen-, Kopfheilk.*, **180**, 742–749, 1962 (in German).
76. P. Noll: Adaptive quantizing in speech coding systems, *Proc. Int. Zürich Sem. '74*, B3(1)–B3(6), 1974.
77. S. R. Quackenbush, T. P. Barnwell, M. A. Clemens: *Objective Measures of Speech Quality*, Englewood Cliffs, USA: Prentice Hall, 1988.
78. A. Raake: *Speech Quality of VOIP – Assessment and Prediction*, Chichester, UK: Wiley, 2006.
79. A. W. Rix, M. P. Hollier: The perceptual analysis measurement system for robust end-to-end speech quality assessment, *Proc. ICASSP '00*, 1515–1518, Istanbul, Turkey, 2000.
80. A. W. Rix, J. G. Beerends, M. P. Hollier, A. P. Hekstra: Perceptual evaluation of speech quality (PESQ) – a new method for speech-quality assessment of telephone networks and codecs, *Proc. ICASSP '01*, 749–752, Salt Lake City, USA, 2001.
81. A. W. Rix, M. P. Hollier, A. P. Hekstra, J. G. Beerends: Perceptual evaluation of speech quality (PESQ), the new ITU standard for end-to-end speech-quality assessment, part I – time-delay compensation, *J. Audio Eng. Soc.*, **50**, 755–764, 2002.
82. M. R. Schroeder, B. S. Atal, J. L. Hall: Optimizing digital speech coders by exploiting masking properties of the human ear, *JASA*, **66**, 1647–1651.
83. K. Scholz, M. Wältermann, L. Huo, A. Raake, S. Möller, U. Heute: Estimation of the quality dimension "directness/frequency content" for the instrumental assessment of speech quality, *Proc. ICSLP '06*, 1523–1526, Pittsburgh, USA, 2006.
84. H. W. Schüssler, Y. Dong: A new method for measuring the performance of weakly non-linear systems, *Frequenz*, **44**, 82–87, 1990.

85. K. Seget: *Investigations on Auditive Quality of Speech-Synthesis Methods,* Dipl. thesis, LNS/EIT/TF/CAU, Kiel. Germany, 2007 (in German).
86. A. P. Simpson, K. J. Kohler, T. Rettstadt: The Kiel corpus of read/spontaneous speech – acoustic data base, processing tool, and analysis results, *Report AIPUK,* **32**, IPDS, Univ. Kiel, Germany, 1997.
87. J. Sotscheck: Sentences for speech-quality measurements and their phonological adaptation to German language (in German: Sätze für Sprachgütemessungen und ihre phonologische Anpassung an die deutsche Sprache), *Proc. DAGA '84,* 873–876, Darmstadt, Germany, 1984 (in German).
88. P. Vary, U. Heute, W. Hess: *Digital Speech Signal Processing (in German: Digitale Sprachsignalverarbeitung),* Stuttgart, Germany: Teubner, 1998.
89. P. Vary, R. Martin: *Digital Speech Transmission,* Chichester, UK: Wiley, 2006.
90. W. D. Voiers: Evaluating processed speech using the diagnostic rhyme test, *Speech Technology,* **1**, 30–39, 1977.
91. W. D. Voiers: Evaluating processed speech using the diagnostic rhyme test, *Speech Technology,* **1**, 30–39, 1983.
92. M. Wältermann, K. Scholz, A. Raake, U. Heute, S. Möller: Underlying quality dimensions of modern telephone connections, *Proc. ICSLP '06,* 2170–2173, Pittsburgh, USA, 2006.
93. S. Wang, A. Sekey, A. Gersho: Auditory distortion measure for speech coding, *Proc. ICASSP '91,* 493–496, Toronto, Canada, 1991.
94. E. Zwicker: *Psychoakustik,* Berlin, Germany: Springer, 1982 (in German).
95. E. Zwicker, H. Fastl: *Psychoacoustics,* 2-nd Ed., Berlin, Germany: Springer, 1999.

10

Evaluation of Hands-free Terminals

Frank Kettler and Hans-Wilhelm Gierlich

HEAD acoustics GmbH
Aachen, Germany

The "hands-free problem" describes the high acoustical coupling between the hands-free loudspeaker and microphone and the resulting acoustical echo for the subscriber on the far end of this connection. It is basically caused by the high distance between the technical interface, i.e. hands-free loudspeaker and microphone, and the human interface, i.e. mouth and ear. The high playback volume – necessary to provide a sufficient playback level at the users ear – and the high sensitivity of the microphone – necessary to amplify the users voice from far distance – leads to a strong coupling – the acoustic echo. Echo cancellers instead of level switching devices are standard today. Moreover, noise reduction and other algorithms further improve the speech quality in noisy environment. On the other hand the use of mobile hands-free telephones in a wide and important application field, i.e. in vehicles, was further enforced by legislation. Hands-free telephones are standard in the automotive industry – at least in middle and upper class vehicles. As a consequence the test procedures and results described in this section mainly focus on the test of hands-free terminals installed in vehicles.

10.1 Introduction

This chapter gives an overview about current evaluation procedures for hands-free terminals, both subjective and objective methods. Sec. 10.2 outlines the principles that need to be considered when testing quality aspects of hands-free terminals. The relevant speech quality parameters are briefly introduced. Sec. 10.3 describes subjective test methods as they have been developed during the last years – from well-known listening-only tests to specific double talk performance tests. The test environment, test signals and analysis methods are introduced in Secs. 10.4 and 10.5. Practical examples of measurement results on different hands-free implementations are used to show the significance of objective laboratory tests. This directly leads to another important aspect, the appropriate summary and representation of the multitude of results, necessary for an in-depth analysis of hands-free implementations. A graphical representation – best described as a "quality pie" – bridges the gap between the complexity and multitude of tests on one side and the need for a quick and comprehensive summary on the other side. It is introduced and discussed in Sec. 10.6. The last section ends up with a discussion of ideas and related aspects in speech communication over hands-free phones and quality testing.

This chapter and especially the practical examples focus on mobile hands-free implementations – simply due to three facts:

- They are standard in the automotive industry today and probably the most rapidly growing hands-free market over the last and coming years.
- Furthermore, the ambient conditions in a driving car, like vehicle noise as one of the crucial parameters for echo cancellation algorithms in identifying the impulse responses, are very critical.
- Last but not least the costs of these systems increase the user's expectation on quality – but do not always satisfy it.

10.2 Quality Assessment of Hands-free Terminals

Hands-free implementations with their typical components like microphone arrays, echo cancellation, noise reduction and speech coders are highly non linear, time variant, speech controlled devices. The development of both subjective and objective quality assessment methods requires a deep understanding of the complexity of each signal processing component and especially the interaction between them. A principal block diagram can be found in Fig. 10.1.

The sending direction (uplink transmission path) typically comprises the microphone or microphone array with its associated algorithms for beamforming. In addition, acoustic echo cancellers (AEC) combined with additional post processing – often also designated as echo suppression or non linear processor – or automatic gain control (AGC) provide the main functionality for reducing the acoustically coupled echo. The block diagram in Fig. 10.1 represents a general example.

In principal different combinations of beamforming and AEC can be realized ("AEC first", "beamforming first", see [39]). Hands-free algorithms and microphone solutions are typically not provided by the same manufacturer, microphone solutions might even change between a single microphone solution and an array during the life cycle of a vehicle type. Consequently there is a high demand for flexible implementations.

Noise reduction algorithms shall further improve the near end signal by algorithmically reducing the added noise from the near end speech signal. Speech coders then provide the signal conditioning for RF transmission.[1]

---

[1] *RF* abbreviates *radio frequency*.

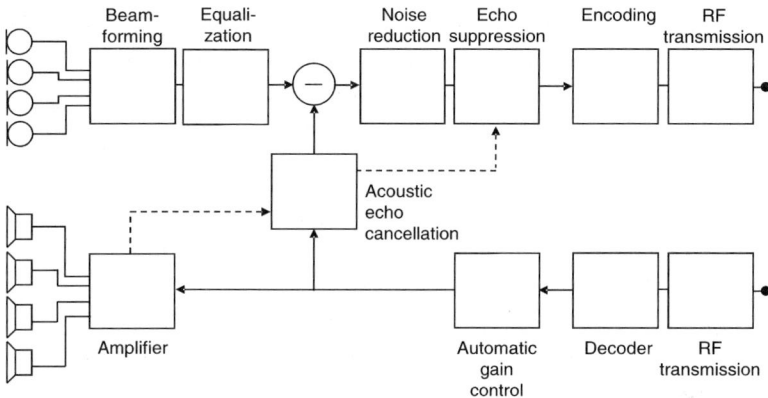

Fig. 10.1. Block diagram of a hands-free implementation with typical components like microphone arrays, echo cancellation, noise reduction and speech coders.

The receive direction (downlink transmission path) typically consists of the speech decoder, potential AGC, e. g. to adapt the volume automatically to the speed and background noise level in the driving car, the playback system and the loudspeakers itself.

The listening speech quality in receiving direction can typically be assessed by subjective listening tests (see Sec. 10.3.5). Objective methods typically reproduce the same situation, the analysis of recorded test signals or real speech at the driver's position using artificial head recording systems on the drivers' seat [25, 48]. The quality is influenced by the loudness of the transmitted speech, the frequency content, the signal to noise ratio in the driving car, the intelligibility and the absence of additional non linear disturbances. The built-in loudspeakers are typically used for playback. In contrary to the sound system used for CD or radio playback, only the front speakers – installed in the door in the driver's and co-driver's footwell – sometimes combined with center speakers are used.

In a similar way the speech quality in sending direction can also be assessed subjectively by listening tests. This transmission path is especially important because all relevant signal processing like microphone characteristics, potential beamforming algorithms and noise reduction are implemented in sending direction. This transmission path is extremely critical in terms of signal to noise ratio caused by the high distance between microphone and driver's mouth. In a technical sense the relevant parameters are the microphone position, the microphone frequency response, the signal to noise ratio, the frequency content of speech and noise, the intelligibility, artifacts like musical tones or other non linear distortions. Objective measures are therefore again based on analyses of transmitted speech or test signals in sending direction with and without background noise.

The two transmission paths can not be regarded as independent from each other. It is obvious that hands-free implementations do not only influence the one-way transmission quality. A comprehensive quality assessment either subjectively or objectively therefore needs to consider all conversational aspects including echo performance, double talk capability – both subscribers act at the same time – and the quality of background noise transmission.

The echo performance is typically assessed by so-called "talking and listening tests" (see Sec. 10.3.4). Test persons assess the echo while they are using a common telephone handset providing standard characteristics at the far end. It is also possible to judge the echo disturbance in listening tests if the self masking effect is considered. Up to a certain extent, this can be reproduced using artificial head technology at the far end [31]. The echo performance is technically influenced by the echo attenuation expressed in dB values, the delay, echo fluctuations vs. time (temporal echo aspects), the spectral content of echo and the "intelligibility" or "clearness" of echoes.

The double talk performance requires conversational tests with two participating subjects at the same time (see detailed description in Sec. 10.3.3). However even this conversational situation can – up to a certain extent – be reproduced by a listening test on simulated conversations using two artificial head testing systems. The double talk capability is mainly determined by three parameters:

- Audible level variations and modulations in sending direction typically introduced by AEC post processing or AGC,
- audible modulation in receiving direction and
- the echo disturbance during double talk.

Last but not least the transmission quality of background noise plays a very important role for mobile hands-free implementations. These devices are typically used in driving cars, thus the background noise situation is critical, the level is rather high. Consequently, this transmission aspect cannot be disregarded any longer. It is important that the background noise itself is not only regarded as a disturbing factor. It carries important information for the conversational partner at the far end side. Consequently, a pleasant and smooth transmission quality needs to be ensured.

10.3 Subjective Methods for Determining the Communicational Quality

The main purpose of telecommunication systems is the bi-directional exchange of information. This has to be considered in the design of terminals and networks. In the first step, quality assessment is always subjective – taking into account the quality as perceived by the user of a communication system including all aspects of realistic conversational situations. When assessing hands-free systems the special conditions where the system is used have to be identified.

For mobile hands-free terminals typical acouctical environmental condtions such as telephoning on a street at the airport or other typical situations have to be considered. For car hands-free systems the acoustical conditions in a car have to be taken into account. Since all quality parameters are based on sensations perceived by the subjects using the communication system, the basis of all objective assessment methods are subjective tests. They have to be defined carefully in order to reflect the real use situation as closely as possible and, on the other hand, to provide reproducible and reliable results under laboratory conditions.

As for other telecommunication systems and services the basic description of the subjective assessment of speech quality in telecommunication is ITU-T Recommendation P.800 [29]. The methods described here are intended to be generally applicable whatever type of degradation factors are present. The ITU-T Recommendation P.832 [32] is of special importance since it addresses the special requirements and test scenarios for hands-free systems. ITU-T Recommendation P.835 [33] is the most relevant recommendation when assessing the speech quality in the presence of background noise which is of major importance in car hands-free systems. It is especially useful when optimizing the design and parameterization of noise canceling techniques based on subjective judgments.

10.3.1 General Setup and Opinion Scales Used for Subjective Performance Evaluation

Subjective testing requires an exactly defined test setup, a well defined selection procedure of the test subjects participating in the tests as well as exact and unambiguous scaling of the scores derived from the subjects participating in subjective experiments.

The setup of the test depends on the type of test to be conducted. Independent of the type of test, as a general rule the test setup should be as realistic as possible. For car hands-free testing the environment chosen for the tests should be "car-like" when evaluating parameters relevant for the user in the car. Other applications require the simulation of their typical use conditions.

Furthermore, the results of a subjective test highly depend on the type of test subjects participating in the test. According to ITU-T Recommendation P.832 the following types of test subjects can be identified:

- **Untrained subjects**

    Untrained subjects are accustomed to daily use of a telephone. However, they are neither experienced in subjective testing nor are they experts in technical implementations of hands-free terminals. Ideally, they have no specific knowledge about the device that they will be evaluating.

- **Experienced subjects**

Experienced subjects (for the purpose of hands-free terminal evaluation) are experienced in subjective testing, but do not include individuals who routinely conduct subjective evaluations. Experienced subjects are able to describe an auditory event in detail and are able to separate different events based on specific impairments. They are able to describe their subjective impressions in detail. However, experienced subjects neither have a background in technical implementations of hands-free terminals nor do they have detailed knowledge of the influence of particular hands-free terminal implementations on subjective quality.

- **Experts**

    Experts (for the purpose of hands-free terminal evaluation) are experienced in subjective testing. Experts are able to describe an auditory event in detail and are able to separate different events based on specific impairments. They are able to describe their subjective impressions in detail. They have a background in technical implementations of hands-free implementations and do have detailed knowledge of the influence of particular hands-free implementations on subjective quality.

For the identification and general evaluation of parameters influencing the communicational quality in hands-free terminals typically untrained subjects are used. Experts and experienced subjects typically are used in order to optimize the performance of a hands-free terminal in a very efficient way. Experts may be used for all types of tests. Care should be taken in case only experts are used in a test since they may focus on parameters not of significance for the average user while missing other parameters average users may find significant. Typically the expert's judgement is validated by untrained subjects representing the average user group the set is intended to be used for.

Subjective testing requires scales easily understandable by the subjects, unambiguous and widely accepted. The design and wording of opinion scales, as seen by subjects in experiments, is very important. Different ways of scaling and different types of scales may be used. Here those most often used in telecommunications are described. More information can be found in the ITU-T P. 800 series Recommendations and [44] and [37].

One of the most frequently used rating is a category rating obtained from each subject at the end of each subjective experiment (see [29, 30]) which is typically based on the following question: Your opinion of (the overall quality, the listening speech quality, etc.) of the connection you have just been using:

   1 – excellent
   2 – good
   3 – fair
   4 – poor
   5 – bad

The averaged result of a this so-called ACR (absolute category rating) test is a mean opinion score MOS. If applied in a conversational test, the result is MOSc (Mean opinion score, conversational). If the scale is used for speech quality rating in listening tests, the result is called MOS (mean listening-quality opinion score).

DCR (degradation category rating) tests are used if degradation occurs in a transmission system. The scale used is given as follows (see [29]):

5 – degradation is inaudible
4 – degradation is audible but not annoying
3 – degradation is slightly annoying
2 – degradation is annoying
1 – degradation is very annoying

The quantity derived from the scores is termed DMOS (degradation mean opinion score).

Sometimes only small differences in quality need to be evaluated. This is relevant e.g. for system optimization or when evaluating higher quality systems. In such experiments a comparison between systems is made and comparison rating is used. The scale used in CCR (comparison category rating) tests is given as follows (see [29]):

The quality of the second system compared to the quality of the first one is

3 – much better
2 – better
1 – slightly better
0 – about the same
−1 – slightly worse
−2 – worse
−3 – much worse

Especially in conversation tests sometimes a binary response is obtained from each subject at the end of a conversation, asking about talking or listening difficulties over the connection used. In such conditions the score is simply "Yes" or "No". Further information can be acquired if the experimenter carefully tries to identify the type of difficulties experienced by the subject in cases where the subject indicates difficulties.

More scales are known, additional information about scales and testing can be found in [28–33].

10.3.2 Conversation Tests

The most realistic test known for communication systems is the conversation test. Both conversational partners exchange information, both act as talker and as listener. Ideally, the test is set up in a way that subjects behave very similar to a real conversation. Therefore the tasks chosen for a conversation test should be mostly natural with respect to the system evaluated. In car hands-free evaluations a situation should be chosen where at least one of the conversational partners is immersed in a (simulated) driving situation. The task chosen for the experiment should be easy to perform for each subject and avoid emotional involvement of the test subjects. Furthermore, the test should be mostly symmetrical (the contribution of each test subject to the conversation should be similar) and it should be independent of the individual personal temperament (the task must stimulate people with low interest in talking and must reduce the engagement of people who always like to talk). Examples for tests used in telecommunication are the so-called "Kandinski tests" (see [31, 32]) or the so-called "short conversational tests" (see [32, 44]).

The "Kandinsky test" is based on pictures with geometrical figures including numbers at different positions in the picture. Each subject has the same picture in front of him but with the numbers at different positions in the picture. The subjects are asked to describe to their partner the position of a set of numbers on a picture. For the subjective evaluation of car hands-free systems this test could be used only in situations where the simulation of the driving task is of minor importance and the focus of the test is mainly on the conversational quality without taking into account additional tasks.

In the so-called "short conversational tests", the test subjects are given a task to be conducted at the telephone similar to a daily-life situation. Finding a specific flight or train connection, ordering a pizza at a pizza service are examples of typical tasks. These tasks can also be performed in the driving situation.

It is advisable to include a sufficient number of talkers in the conversational tests to minimize talker/speaker-dependent effects. The test persons used should be representative with respect to gender, age etc. for the user group of the system evaluated.

Due to the complexity of the task itself, subjects mostly rate their opinion about the overall quality of a connection based on the ACR scale. Often they are asked about difficulties in talking or listening during the conversation. Careful investigation of the nature of these difficulties may require more specialized tests than described below. A more detailed parameter investigation can only be made if experienced subjects or experts are used in the conversation test.

10.3.3 Double Talk Tests

The ability to interact in all conversational situations and especially to interact during double talk with no audible impairments is of critical importance in car hands-free communication. Due to the difficult acoustical situation especially in a car a variety of measures are implemented in a hands-free terminal which may impair the speech quality during double talk. Double talk tests may help to evaluate the system performance under such conditions. The double talk testing method (see [14, 15, 31]) is designed especially for the quality assessment during double talk periods. The test duration is very short, they are very efficient in subjectively evaluating this very important quality aspect in detail.

In general the test setup is the same as for conversational tests. Double talk tests involve two parties. In double talk tests untrained subjects are used when it is important to get an indication of how the general telephone using population would rate the double talk performance of a car hands-free telephone. The test procedure is sensitive enough to let untrained subjects assess the relevant parameters even during sophisticated double talk situations. Experienced subjects are used in situations where it is necessary to obtain information about the subjective effects of individual degradations.

During double talk tests, two subjects read a text. The texts differ slightly. Subject 1 (talking continuously) starts reading the text. It consists of simple, short and meaningful sentences. Subject 2 (double talk) has the text of subject 1 in front of him, follows the text and starts reading his text simultaneously at a clearly defined point. Clearly this situation is less realistic than in a conversation test. Even if the text is very simple, the subjects have to concentrate in a different way compared to a free conversation.

Parameters which are assessed typically using double talk tests are: the dialog capability, the completeness of the speech transmission during double talk, echo and clipping during double talk. In most of the tests ACR or DCR scales are used.

10.3.4 Talking and Listening Tests

Nowadays many hands-free terminals are often used in mobile or IP based transmission systems. As a consequence the delay introduced in a transmission link increases. Complex signal processing in the terminals may add additional significant delay. Therefore the investigation of talking-related disturbances like echo or background noise modulation is of critical importance. In order to investigate such types of impairments in more detail, talking and listening tests can be used. Such tests are mainly used to investigate the performance of speech echo cancellers (EC) and the noise canceller (NC) integrated in the car hands-free terminal. All aspects of EC and NC functions that influence the transmission quality for subscribers while they are either talking-and-listening are covered by this procedure.

The EC and NC implementations of a hands-free terminal are the focus of the setup – it is found on one side of a (simulated) connection. The performance of this implementation is judged by a test subject placed at the opposite end of the (simulated) connection. From the subjects point of view, this is the far-end echo and noise canceller.

A potential far-end subscriber can be simulated by an artificial head if double-talk sequences are required in the test. In this case the artificial mouth is used to produce exactly defined double talk sequences. The environmental conditions used at the far end side (e.g. car-hands-free) should correspond to the typical environmental conditions found in a car especially with respect to background noise.

The test procedure may focus on the examination of the initial performance of a hands-free terminal, e.g., the convergence of an echo canceller or in a second part on the evaluation of the performance under steady-state conditions.

When testing the initial performance, subjects answer an incoming telephone call with the same greeting: e.g. '[name], [greeting]'. After the greeting, the call is terminated and subjects give their rating.

When testing "steady-state conditions", the algorithms of the car hands-free system should be fully converged. Subjects are asked to perform a task, such as to describe the position of given numbers in a picture similar to the "Kandinsky" test procedure described for the conversational tests. An artificial head can be used to generate double talk at defined points in time in order to introduce interfering signal components for the speech echo canceller and test the canceller's ability to handle double talk. After the termination of the call, the subjects are asked to give a rating. The scales used are typically ACR or DCR scales. More information can be found e.g. in [31].

10.3.5 Listening-only Tests (LOT) and Third Party Listening Tests

The main purpose of listening-only tests and third party listening tests is the evaluation of impairments under well-defined and reproducible conditions in the listening situation. Their application for the evaluation of hands-free systems is most useful when evaluating the sending direction of the hands-free system. It should be noted that listening tests are very artificial. Listening-only tests are strongly influenced by the selection of the speech material used in the tests; the influence of the test stimuli is much stronger than e.g. in conversation tests. The tests must be designed carefully including the appropriate selection of test sequences (phoneme distribution), talkers (male, female, age, target groups) and others. A sufficient number of presentations must be integrated into a test, ranging from the best to the worst-case condition of the impairment investigated in the test. Reference conditions (simulated, defined impairments with known subjective rating results) may be included in order to check the validity of the test. More detailed descriptions of the requirements and rules how to conduct listening-only tests for various purposes are found in [29–33].

Pre-recorded, processed speech is presented to the subjects. For car hands-free applications these speech sequences are either recorded in the car (when assessing the listening speech quality in the car) or they are recorded at the output of the hands-free terminal in sending direction. In general two possibilities exist for the presentation of prerecorded speech material. Either reference handset terminals are used in case the sending direction of the car hands-free terminal is judged simulating a handset connection at the far end side. Alternatively third party listening tests can be used. In third party listening tests [31, 32] the speech material is recorded by an artificial head which is used to record the complete acoustical situation including the background.

This procedure can be applied to assess the listening speech quality in the car as well as assessing the speech quality in sending direction.

With this procedure, all types of handset, headset, and hands-free configurations can be evaluated in a listening-only test including the environmental conditions at the terminal location. For playback, equalized headphones are used. The equalization must guarantee that during playback the same ear signals are reproduced which were measured during recording. Thus a binaural reproduction (for details see [9]) is possible which leads to a close-to-original presentation of the acoustical situation during recording.

Furthermore the third party listening setup allows to use this type of test for investigating conversational situations by third parties. Therefore, a complete conversation is recorded and presented to the listeners. Although the listeners are not talking themselves but listening to other persons' voices, these tests have proven their usefulness in investigating conversational impairments in a listening test.

In listening tests all scales are used, mostly ACR or DCR scales. Loudness preference scales can be used as well. More information can be found e.g. in [31] and [32]. Instead of ACR or DCR tests, also CCR (comparison category rating) is used which offers a higher sensitivity and may be used for the quality evaluation of high-quality systems or for the optimization of systems. CCR tests are based on paired comparisons of samples.

10.3.6 Experts Tests for Assessing Real Life Situations

Sometimes besides objective testing of hands-free telephones complementary subjective performance evaluation may be useful. Especially for car hands-free systems a lot of experience has been gained with these types of complementary tests. The general considerations when conducting additional subjective tests are given here with the example of car hands-free systems. Supplementary subjective tests are targeted mainly to "in situ" hands-free tests for optimizing hands-free systems in a target car and under conditions which are not covered by objective test specifications. The main purpose is to investigate the hands-free performance in real live conditions including networks and car to car communication. They are of diagnostic nature and not suitable for parameter identification and value selection. Generally, such tests are based on tests as described above and are found in the ITU-T P.800 series Recommendations but not intended to replace tests as described in the ITU-T P.800 series Recommendations.

For conducting the tests the hands-free system under test is installed in the target car, which is referenced as near-end. The far-end is either a landline phone or an observing car also equipped with the hands-free system under test (car-to-car test). It is recommended to not only test the hands-free system in a landline connection but also in a car-to-car connection because the latter case can be regarded as a worst case scenario resulting in worse hands-free quality compared to landline connections.

The evaluation of the hands-free performance should be done in different driving conditions including different background noise scenarios, different driving speeds, different fan/defrost settings, etc.

Since conversational tests are rather time consuming most of the hands-free tests are conducted as single-talk and double-talk tests as described above. Evaluations are done at the far-end and/or the near-end, depending on the type of impairment to be evaluated.

The performance evaluation of the hands-free system typically covers categories like

- echo cancellation (echo intensity, speed of convergence, etc.),
- double talk performance (echo during double talk, speech level variation, etc.),
- speech and background noise quality in sending direction (level, level variation, speech distortion, etc.),
- speech quality in receiving direction (level, level variation, speech distortion),
- stability of the echo canceller for "closed loop" connection during car-to-car hands-free communication.

The evaluation has to be done by experts who are experienced with subjective testing of hands-free systems. During the tests the signals on near-end and far-end may be recorded to be used for third-party listening evaluation later on. More detailed information can be found in [36].

10.4 Test Environment

In general the evaluation of hands-free terminals is made in a lab-type environment which is simulating the acoustical conditions close to the real use conditions. The test environment described focusses on car hands-free system since car hands-free systems are dedicated to be used in cars only and the car is a quite a special environment from the acoustic point of view the test environment has to be selected carefully. Different approaches can be taken starting from a digital simulation of the transmission paths in the car (mouth to microphone, loudspeaker to the drivers ear and loudspeaker to the microphone) up to the use of a car cabin for installing and testing the hands-free device in a car which is the approach taken in [36] and [47]. The relevant transmission paths in a car are shown in Fig. 10.2.

It is common to most test setups to use a car type environment under lab conditions in order to control the influence of the network as well as the background noise conditions as exactly as possible. Certainly the hands-free evaluation can also be done in real driving situations. However, due to the highly uncontrolled environment (time-variant driving noise, pass-by traffic, unpredictable environmental conditions, unknown network conditions), this type of evaluation is not recommended except for validation tests and design

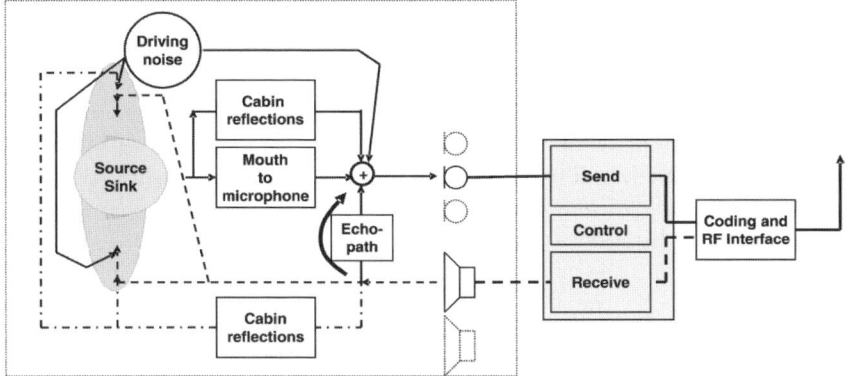

Fig. 10.2. Transmission paths in a car cabin.

optimization taking into account additional influencing factors which could not be simulated in the lab environment.

10.4.1 The Acoustical Environment

The acoustical environment of a car cabin is rather complex: different materials ranging from hard reflecting surfaces such as windows or glass roofs to highly absorbing surfaces such as seats lead to the fact that the simulation of a car type environment is rather difficult. Furthermore different car types have to be considered. Compact cars show completely different properties than luxury cars, trucks, vans or sports cars. Therefore, the coupling between the car hands-free microphone(s) and the hands-free loudspeaker(s) also highly depends on the individual design of the car and the positioning of the microphones and loudspeakers inside the car. The positioning of the hands-free microphone is of special importance. The microphone should be positioned as close as possible to the talker but also in such a way that the coupling between the car hands-free loudspeakers and the car hands-free microphone(s) is minimized. Consequently, it is the easiest solution for most test setups if the actual target car is used when testing complete hands-free systems. This is the best representation of all transmission paths relevant to the hands-free implementation which will finally give the performance of the hands-free system in the target car.

10.4.2 Background Noise Simulation Techniques

Background noise is one of the most influencing factors in hands-free systems but especially in car hands-free systems. Consequently in order to simulate a realistic driving situation, background noise has to be simulated as realistically as possible even in a lab type environment. Background noise simulation techniques are described in [11, 36, 48, 49]. Typically a 4-loudspeaker arrangement with subwoofer is used. This background noise setup is available for simulations under laboratory conditions as well as for car cabins. In Fig. 10.3 the simulation arrangement for car hands-free systems is shown. In order to use this arrangement prior to the tests the background noise produced by a car has to be recorded. This is done under real driving conditions typically using a high quality measurement microphone positioned close to the hands-free microphone. In general all different driving conditions can be recorded. For built-in systems the background noise of the target car is recorded, for after-market systems the background noise of one or more cars considered to be typical target cars is used. If possible the output signal of the hands-free microphone can be used directly. In such a case structure borne noise which might be picked up by the microphone can also be considered in the simulation.

The loudspeaker arrangement used for playback of the recorded background noise signals is equalized and calibrated so that the power density spectrum measured at the microphone position is equal to the recorded one. For equalization either the measurement microphone or the hands-free microphone used for recording is used. The maximum deviation of the A-weighted sound pressure level is required to be less than 1 dB. The third octave power density spectrum between 100 Hz and 10 kHz should not deviate by more than 3 dB from the original spectrum. A detailed description of the equalization procedure as well as a database with background noises can be found e.g. in [11] and [48].

10.4.3 Positioning of the Hands-Free Terminal

The hands-free terminal is installed either as described in the relevant standards (see e.g. [25, 27]) or according to the requirements of the manufacturers. In cars the positioning of the microphone/microphone array and loudspeaker are given by the manufacturer. If no position requirements are given, the test lab has to choose the arrangement. Typically, the microphone is placed close to the in-door mirror, the loudspeaker is typically positioned in the footwell of the driver or the co-driver. In any case the exact location has to be noted. Hands-free terminals installed by the car manufacturer are measured in the original arrangement.

Headset hands-free terminals are positioned according to the requirements of the manufacturer. If no position requirements are given, the test lab has to choose the arrangement. Further information is found in [25, 36, 48].

10.4.4 Positioning of the Artificial Head

The artificial head (HATS Head and Torso Simulator according to ITU-T Recommendation P.58 [21]) is placed as described in the relevant standards

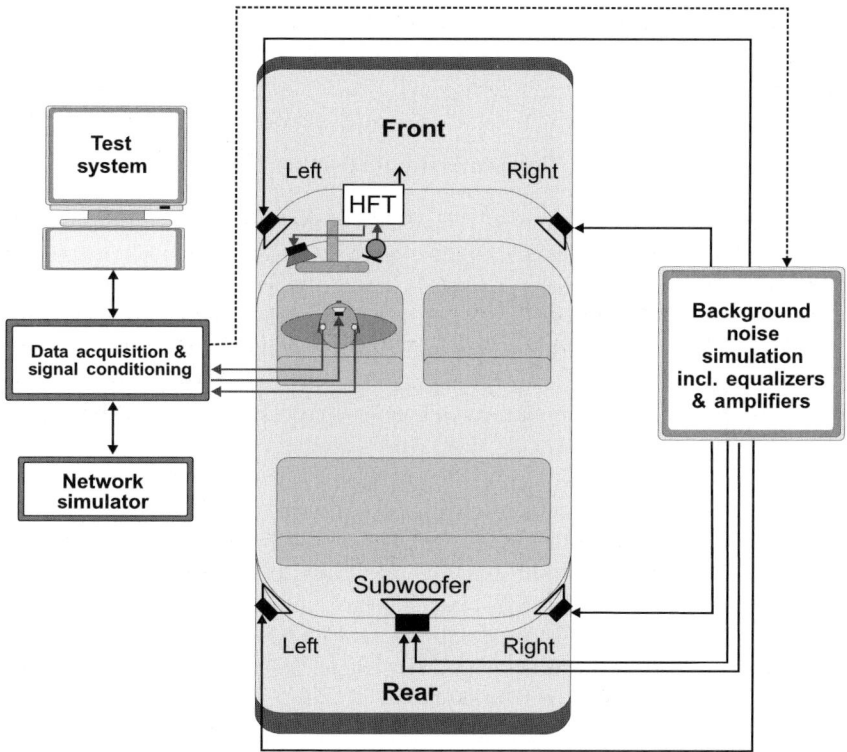

Fig. 10.3. Test arrangement with background noise simulation.

(see e.g. [25, 27]). In cars it is installed at the driver's seat for the measurement. The position has to be in line with the average user's position. Clearly there may be different locations for different users which may be taken into account in addition to the average users position. The position of the HATS (mouth/ears) within the placing arrangement is chosen individually for each type of car. The position used has to be described in detail by using suitable measures (marks in the car, relative position to A-, B-pillar, height from the floor etc.). The exact reproduction of the artificial head position must be possible at any later time. If no requirements for positioning are given, the distance from the microphone to the MRP [20, 21] is defined by the test lab.

The artificial head used should conform to ITU-T Recommendation P.58. Before conducting tests the artificial mouth is equalized at the MRP according to ITU-T Recommendation P.340 [27], the sound pressure level is calibrated at the HATS-HFRP (HATS-hands-free reference point) so that the average level at HATS-HFRP is -28.7 dB$_{Pa}$. The detailed description for equalization at the MRP and level correction at the HATS-HFRP can be found in ITU-T Recommendation P.581 [25]. For assessing the hands-free terminals in receiving direction the ear signal of the right ear of the artificial head is used (for cars where the steering wheel is on the right hand side, the left ear is used). The artificial head is free-field equalized as described in ITU-T Recommendation P.581.

10.4.5 Influence of the Transmission System

Measurements may be influenced by signal processing [1–4] (different speech codecs, DTX[2], comfort noise insertion. etc.) depending on the transmission system and the system simulator used in the test setup. In general, a network simulator (system simulator) is therefore used in the test in order to provide a mostly controlled network environment. All settings of the system simulator have to ensure that the audio signal is not disturbed by any processing and the transmission of the signal (in cases of mobile networks especially the radio signal) is error-free. DTX, VAD and other network signal processing is switched-off. In case of different speech coders available in a network the one providing the best audio performance is typically used. E.g. for measurements with AMR-codec [3] the highest bitrate of 12.2 kb/s is used. Nevertheless, there may be tests where lower bitrates providing less speech quality are used e.g. in order to evaluate the listening speech quality of the complete hands-free system in more detail. Except conditions which are targeted to investigate the influence of transmission errors such as packet loss or jitter no network impairments should influence the tests.

10.5 Test Signals and Analysis Methods

The choice of the test signal as well as the analysis method depends on the application. On the one hand, speech sequences are best suited as test signal for hands-free devices incorporating algorithms, which are optimized based on specific speech characteristics. But the dynamics of speech, the multitude of different languages with their specific characteristics, the directly related question of robustness and reproducibility of analyses make it difficult to come to a common agreement in standardization. On the other hand, artificial test signals providing speech-like properties have the advantage of not being limited to a specific language. These signals can be optimized to measure specific parameters and provide a high reproducibility of results, e.g. in different labs. However, it is also obvious, that typically a large number of test signals is needed – each designed for specific purposes. Furthermore, it is generally recommended to verify test results by speech recordings and listening examples. Test methods applicable for car hands-free evaluation can be separated in two main categories – the "traditional" analysis methods and the advanced test methods. The "traditional" analysis methods focus on the basic telephonometry parameters and include:

---

[2] The term *DTX* stands for *discontinuous transmission*.

- **Loudness Rating** calculations [26] which are the basis for setting the correct sensitivities in the hands-free terminals in order to ensure seamless interaction with the networks and the far end terminals.
  The Loudness Rating then is defined as:

$$LoudnessRating = -\frac{10}{m} \log_{10} \left\{ \sum_{i=1}^{N} 10^{-\frac{m}{10}(L_{\text{UME},i} - \overline{L_{\text{RME}}} + W_i)} \right\}. \quad (10.1)$$

  W_i are weighting factors as defined in [26], different for SLR, RLR, STMR, LSTR. L_{RME} represents the mouth-to-ear transmission loss of the reference speech path (IRS speech path [18]). m is a constant in the order of 0.2, different for the different loudness ratings. For a given telephone or transmission system, the values of L_{UME} can be derived from the measurement different sensitivities S_{MJ} (mouth-to-junction) for calculation of the SLR (sending loudness rating), from S_{JE} (junction-to-ear) for the calculation of the RLR (receiving loudness rating) or from S_{ME} (mouth-to-ear) for the overall loudness rating OLR.
  In a similar manner, the sidetone paths can be described: STMR is the Sidetone Masking Rating describing the perceived loudness of the user's own voice and LSTR (Listener Sidetone Rating) describes the perceived loudness of room noise coupled to the user's ear.
- Requirements for **frequency response characteristics** [27, 48] in sending and receiving in order to ensure a sufficient sound quality and intelligibility. In Sending a rising frequency response characteristics with a high pass characteristics at around 300 Hz is recommended to ensure sufficient intelligibility and the reduction of low frequency background noise. In receiving a most flat frequency response characteristics is advisable.
- **Echo loss requirements** [16] ensuring an echo free connection under different network conditions. Envisaging that delay is inserted
  - by the mobile network itself,
  - by the hands-free terminal where advanced signal processing leads to higher delay compared to standard handset terminals,
  - by connecting networks which increasingly insert VoIP transmission which adds additional delay in the transmission link,

  the echo loss requirements for car hands-free terminals are high. The terminal coupling loss (TCL) required is at least 40 dB, typically however higher (46 dB to 50 dB, see [36, 48]) in order to prevent the far end partner form the car hands-free echo. The basic information about transmission delay and the echo loss required can be found in [16].
- **Delay requirements** [48] referring to the processing delay introduced by the car hands-free terminal to ensure a minimum delay introduced by these terminals for the benefit of the overall conversational quality.

More details on these tests can be found in e.g. [36, 48].

10.5.1 Speech and Perceptual Speech Quality Measures

Hearing model based analysis methods like PESQ^{TM} [34,35] and TOSQA2001 [7,8] calculate estimated mean opinion scores. These objective scores represent the listening speech quality in a one-way transmission scenario with a high correlation to the results of a subjective listening test. The test signal used by such methods is speech. Due to the different characteristics of the different languages it is difficult to define an "average" speech signal to be used in conjunction with these methods. Therefore ITU-T has defined in Recommendation P.501 [22] a set of reference speech samples for different languages which can be used. The ITU-T recommended PESQ^{TM} has not been validated for acoustic terminal and handset testing, e.g. using HATS [34] and does therefore not play a practical role in testing hands-free implementations. TOSQA2001 is validated for terminal testing at acoustical interfaces [8] and is therefore also used for testing hand-free devices. The method estimates the listening speech quality (TMOS, TOSQA2001 mean opinion score) by using reference and degraded speech samples. Frequency content of the transmitted speech, loudness and noise, additive disturbances or non-linear coder distortions contribute to speech quality degradations and influence the TMOS score accordingly. These results are very useful in terminal testing, but provide only very limited information about the reason for unexpected quality degradations.

10.5.2 Speech-like Test Signals

Different test signals with different levels of complexity are available and have been evaluated for different types of applications. A comprehensive description of the most important signal can be found in ITU-T Recommendation P.501 [22]. ITU-T Recommendation P.502 [23] describes appropriate analysis methods for each signal. The most complex speech-like signal in telephonometry is the artificial voice as described in ITU-T Recommendation P.50 [19]. This signal provides a statistical representation of real speech. The signal duration amounts to 10 s, it is suited and often used to measure long-term or average parameters like frequency responses or loudness ratings [26].

An important signal for laboratory quality testing of hands-free telephones is the composite source signal (CSS) ([14,22], see Fig. 10.4). It is composed in the time domain and consists of different parts like voiced and unvoiced segments and a pause. Due to its short duration of the active signal part (approximately 250 ms) it is well suited to measure short term parameters, e.g. switching behaviour of AEC between single and double talk sequences. Parameters in the frequency domain such as frequency response, loudness ratings etc., as well as parameters in the time domain such as switch-on times can be determined.

Fig. 10.5 shows the combination of two uncorrelated CSS (composite source signals) to simulated a double talk sequence. The power density spectra are given in Fig. 10.6.

Evaluation of Hands-free Terminals 357

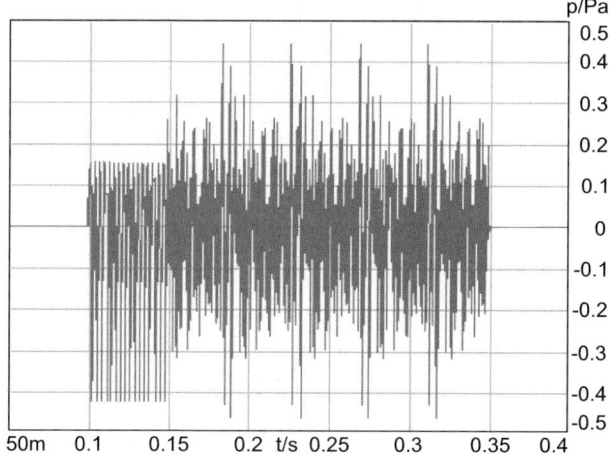

Fig. 10.4. Composite source signal. The signal consists of different parts like voiced and unvoiced segments and a pause.

The simulated double talk starts with a CSS burst applied in receiving direction (dark gray signal) followed by a near end double talk burst (light gray bursts). This sequence is then periodically repeated. The typical test signal levels are -4.7 dB$_{Pa}$ for the near end signal at the mouth reference point (MRP) and -16 dB$_{m0}$ in downlink direction. Measurements in sending direction of an HFT (hands-free terminal) typically analyze the transmitted

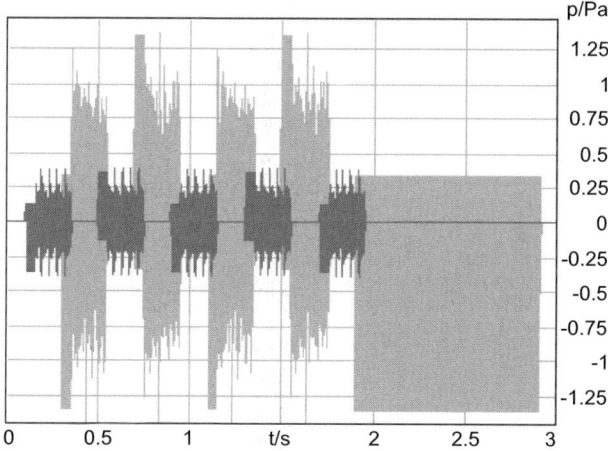

Fig. 10.5. Simulated double talk sequence. The simulated double talk starts with a CSS burst applied in receiving direction (dark gray signal) followed by a near end double talk burst (light gray bursts).

uplink signal referred to the near end test signal. The resulting sensitivity curve can be used to detect level modulations typically introduced by AEC post processing. Figs. 10.7 and 10.8 show two examples. The HFT implementation represented by the analysis curve in Fig. 10.7 introduces an attenuation of approximately 15 dB in the microphone path during the double talk sequence. The driver's voice is partly attenuated during a double talk sequence using real speech over this implementation. Vice versa the sending direction can be regarded as nearly transparent in the analysis curve in Fig. 10.8.

In the same way, the analysis can be carried out in receiving direction in order to verify if the implementations do not insert attenuation in this transmission path during double talk.

A third parameter determining the double talk capability of a hands-free implementation is the echo attenuation during double talk. Subjective test results are available comparing the echo attenuation during single and double periods [27, 40]. The challenge for measurement technique is to separate the near signal from the echo components in the send signal. A suitable test signal that provides this characteristic consists of two uncorrelated AM/FM modulated signals [22]. The time signal is shown in Fig. 10.9. The two signals show comb-filter spectra as given in Fig. 10.10, which are necessary to distinguish between the double talk signal (coming from the near end) and the echo signal (coming from the echo path as a reaction on the receive signal). The power density spectra of both signals calculated by Fourier transformation are given in Fig. 10.10. Echo components during double talk can be detected in the send signal by comparison of the uplink signal and the original downlink signal. The near signal components can easily be removed by appropriate filtering.

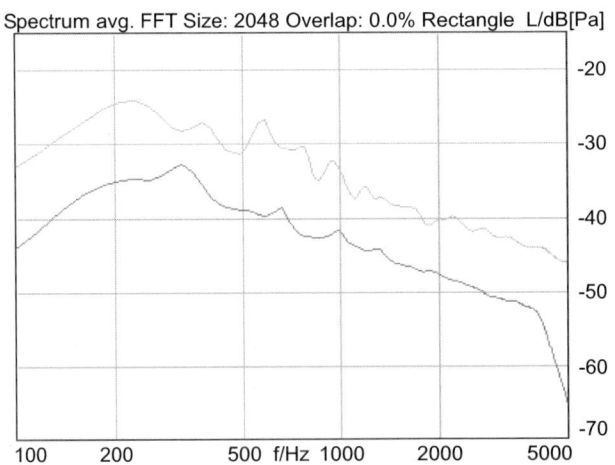

Fig. 10.6. Power density spectra of the signals shown in Fig. 10.5 (dark gray: far end, light gray: near end).

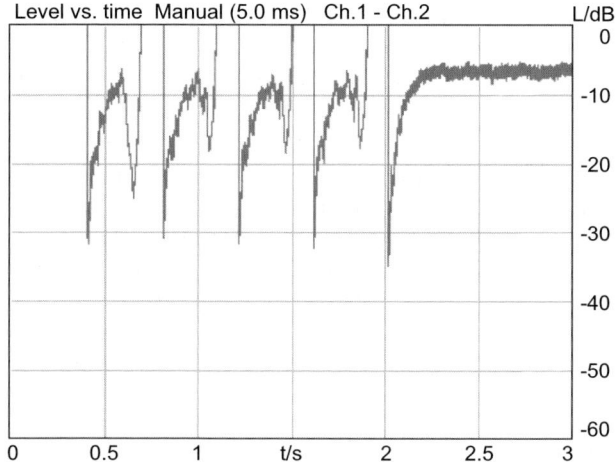

Fig. 10.7. Uplink sensitivity during double talk – Hands-free terminal 1. This terminal introduces an attenuation of approximately 15 dB in the microphone path during the double talk sequence.

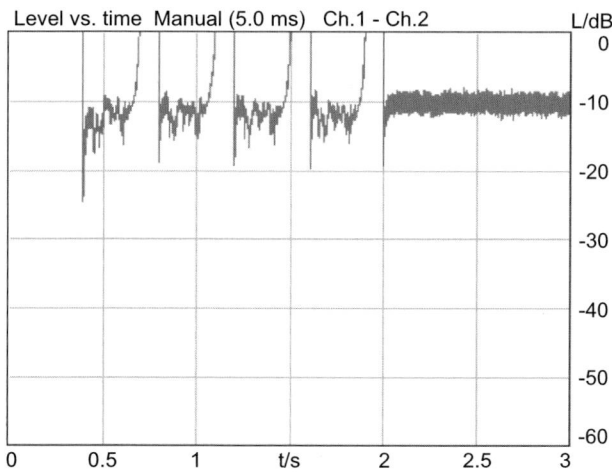

Fig. 10.8. Uplink sensitivity during double talk – Hands-free terminal 2. The sending direction can be regarded as nearly transparent in the analysis curve.

ITU-T Recommendation P.340 [27,40] defines different types of double talk performance for hands-free implementations. The characterization is based on the measured attenuation between the single and double talk situation inserted in sending and receiving direction (a_{HSDT}, a_{HRDT}). The third parameter is the echo attenuation during double talk (EL_{DT}).

The three parameters are measured independently. However, the worst result determines the characterization. A "type 1" implementation provides

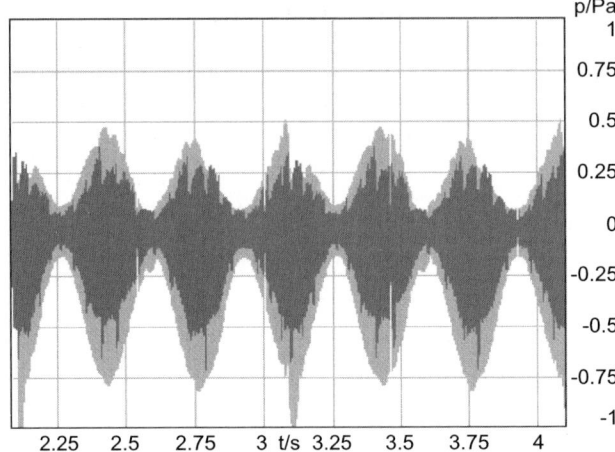

Fig. 10.9. AM/FM modulated test signals (dark gray: far end, light gray: near end) for determining the double talk capability of a hands-free implementation.

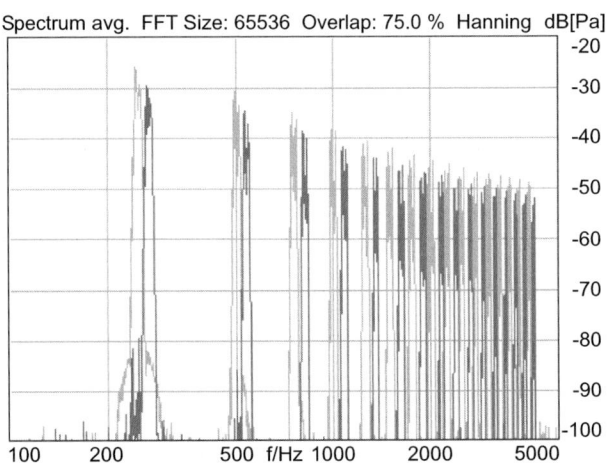

Fig. 10.10. Power density spectra of the signals presented in Fig. 10.9 (dark gray: far end, light gray: near end). Echo components during double talk can be detected in the send signal by comparison of the uplink signal (not depicted) and the original downlink signal (dark gray).

full duplex capability; "type 2a", "2b" and "2c" devices are partial duplex capable, a "type 3" characterization indicates no duplex capability.

10.5.3 Background Noise

Besides speech and artificial test signals the transmission quality of the background noise present e.g. in the driving vehicle needs to be evaluated in detail.

Table 10.1. Duplex capability.

Characterization	Type 1	Type 2a	Type 2b	Type 2c	Type 3
a_{HSDT}	≤ 3 dB	≤ 6 dB	≤ 9 dB	≤ 12 dB	> 12 dB
a_{HRDT}	≤ 3 dB	≤ 5 dB	≤ 8 dB	≤ 10 dB	> 10 dB
EL_{DT}	≥ 27 dB	≥ 23 dB	≥ 17 dB	≥ 11 dB	< 11 dB

The driving noise can not only be regarded as a disturbing signal for an HFT implementation, but carries important information for the far end subscriber. It is typically processed through noise reduction algorithms, might be modulated by echo suppression or partly substituted by comfort noise, if a downlink signal is applied. Related quality parameters range from D-value calculation, comparing the sensitivity of the microphone path on speech and on background noise, the signal to noise ratio, if near end signals are transmitted together with background noise or the modulation of transmitted background noise by echo cancellation or echo suppression.

A very promising method to analyze the performance of noise reduction algorithms is the *Relative Approach* [13, 47]. This method takes into account the sensitivity of the human ear on unexpected events both in the time and in the spectral domain. In contrary to all other methods the Relative Approach does not use any reference signal. The signal is band filtered (1/12 octave) and a forward estimation based on the signal history is calculated in order to predict the new back-ground noise signal value. Values between the frequency bands are interpolated.

The predicted signal pattern is compared to the actual signal characteristic and the deviation in time and frequency is displayed as an "estimation error". Thus instantaneous variations in time and dominant spectral structures are found based on the human ear sensitivity on these parameters. Typical disturbances produced by noise reduction algorithms like musical tones can be detected and verified, if these components lead to speech quality degradations. A typical example is shown in Fig. 10.11. It analyzes the adaptation phase of a noise reduction algorithm. Disturbing artefacts as detected by the Relative Approach are indicated by the arrows.

A more advanced test procedure is described in ETSI EG 202 396-3 [12]. The model described here is a perceptual model again based on the Relative Approach. The model is applicable for speech in background noise at the near end of a terminal and provides an estimation of the results that normally would be derived from a subjective test made using ITU-T recommendation P.835 [33]. Three MOS scores are predicted:

- S-MOS (speech MOS), describing the quality of the speech signal as perceived by the listener,

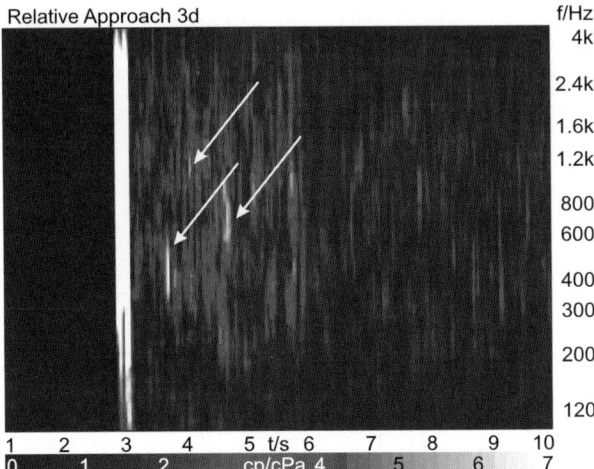

Fig. 10.11. Relative Approach analysis of an adaptation phase of noise reduction.

- N-MOS (noise MOS), describing the quality of the transmitted background noise, and
- G-MOS (global MOS), describing the perceived overall quality of the transmitted speech plus background noise signal.

Currently the test method is applicable for:

- Wideband handset and wideband hands-free devices (in sending direction),
- noisy environments (stationary or non-stationary noise),
- different noise reduction algorithms,
- AMR [3] and G.722 [17] wideband coders,
- VoIP networks introducing packet loss.

However the extension of this method to narrowband terminals and systems is already on ETSI's roadmap. Different input signals are required for the model and subsequently are used for the calculation of N-MOS, S-MOS and G-MOS. Beside the signals processed by the terminal or the near end device two additional signals are used as a priori knowledge for the calculation:

1. The "clean speech" signal, which is played back via the artificial mouth.
2. The "unprocessed signal", which is recorded close to the microphone position of the handset or the hands-free terminal.

Both signals are used in order to determine the degradation of speech and background noise due to the signal processing as the listeners did during the listening tests. The principle of the method is shown in Fig. 10.12. Further information and details about the algorithm can be found in [12].

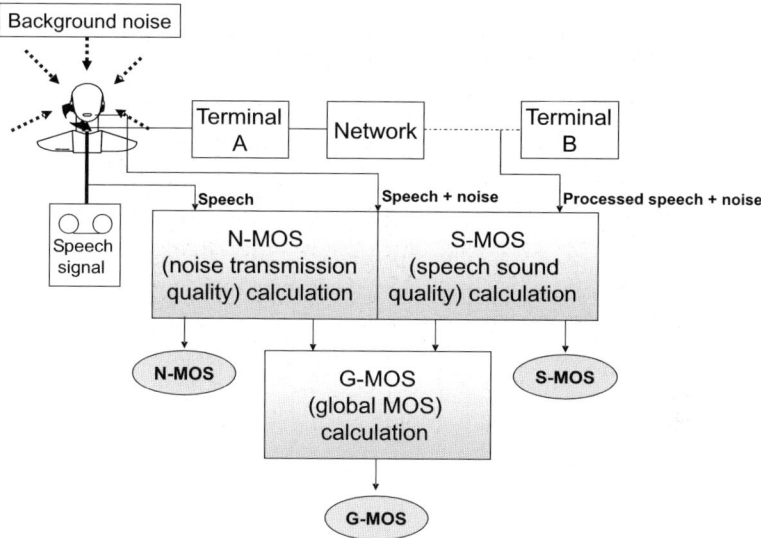

Fig. 10.12. Principle of the S-MOS, N-MOS and G-MOS prediction as described in [12].

10.5.4 Applications

A typical application example for these test signals and the interaction between the results is shown by a comparison analysis of two different HFT aftermarket implementations. Both devices are measured via Bluetooth connection to a commercially available 2G mobile phone. The GSM full rate speech coder is used.

The frequency response in Fig. 10.13 is relatively balanced without showing a strong high pass characteristic. The sending loudness rating [26] of 13.3 dB for this implementation absolutely meets the recommended range of 13 ± 4 dB according to [48]. The TMOS of 3.3 confirms the high uplink quality (recommended ≥ 3.0 TMOS [48]).

The signal-to-noise ratio estimated from the measurement result based on composite source signal bursts transmitted together with background noise (simulated 130 km/h background noise) is very high (approximately 27 dB, Fig. 10.14). The near end test signal is also affected by the uplink signal processing and attenuated. The D-value comparing the sensitivities of the uplink transmission path on speech and on background noise of +3.5 dB is also extremely high (recommended ≥ −10 dB). Both results are consistent. However, these parameters are extremely high especially when considering the balanced frequency response without a strong high pass characteristic. This indicates a very aggressive noise reduction algorithm. The undesired side effect is an unpleasant metallic speech sound and disturbing, artificial musical tones in the transmitted background noise.

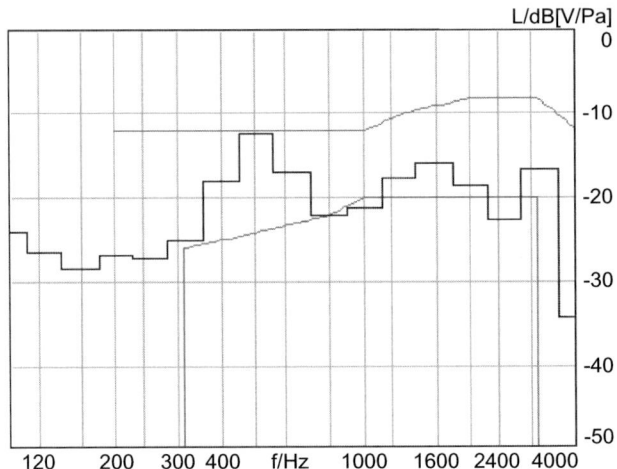

Fig. 10.13. Sending frequency response, hands-free terminal 1. The frequency response is relatively balanced without showing a strong high pass characteristics (compare with Fig. 10.15).

In comparison the analysis in Fig. 10.15 shows a frequency response providing a clear, distinct high order high pass around 300 Hz. The curve only slightly violates the tolerance scheme, which can practically be neglected. All other parameters like the SLR of 12.7 dB meet the requirements. The TMOS of 3.0 still indicates a sufficient listening speech quality – although the frequency response provides the strong high pass characteristic.

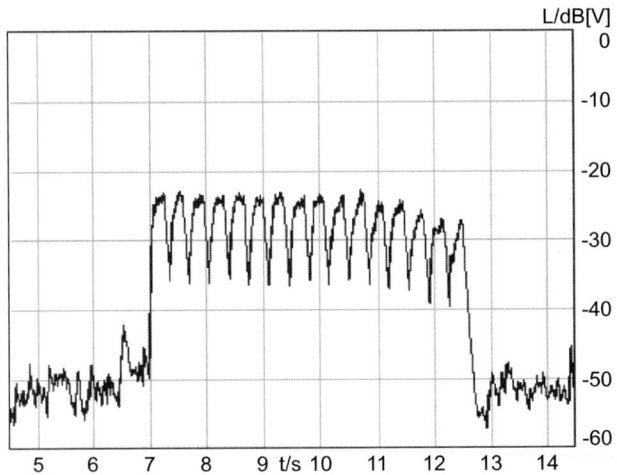

Fig. 10.14. Transmission of background noise and near end signal (level vs. time), hands-free terminal 1.

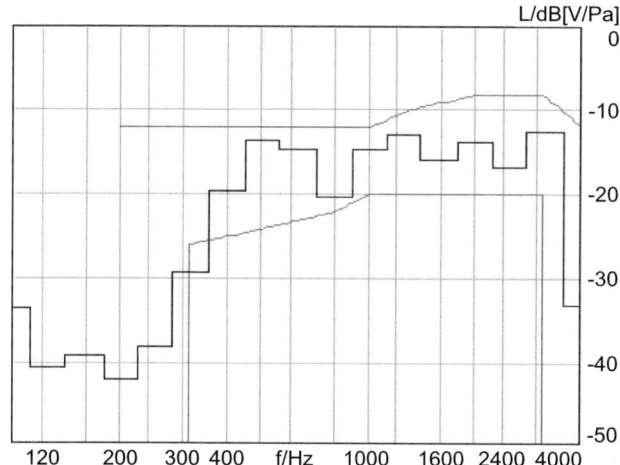

Fig. 10.15. Sending frequency response, hands-free terminal 2. This terminal has a frequency response providing a clear, distinct high order high pass around 300 Hz (compare with Fig. 10.13).

The signal-to-noise ratio as estimated from the analysis curve in Fig. 10.16 is 16 dB. The near end composite source signal bursts are only slightly distorted in this level analysis. This indicates that the noise cancellation algorithm does not significantly affect the near end test signal when transmitted together with background noise. The signal-to-noise ratio at the algorithm input seems to be high enough to clearly distinguish between both signals. The 16 dB signal-to-noise ratio estimated from this analysis and a reasonable D-value of -6.8 dB are consistent. The good quality for the uplink transmission can be confirmed by the listening example of the transmitted speech together with background noise.

The strong high pass as indicated above significantly contributes to a high signal-to-noise ratio in sending direction. Audible disturbances like musical tones are minimized, thus indicating that a high order microphone high pass significantly improves the performance in the presence of background noise. An acoustical tuning already at the microphone is a good compromise though even it might slightly degrade the listening speech quality under silent conditions.

10.6 Result Representation

The complexity of in-depth quality testing of hands-free implementations and the multitude of results acquired during laboratory tests require an appropriate result representation. An overall quality score that covers all conversational aspects is not yet available. Moreover, such a one-dimensional score

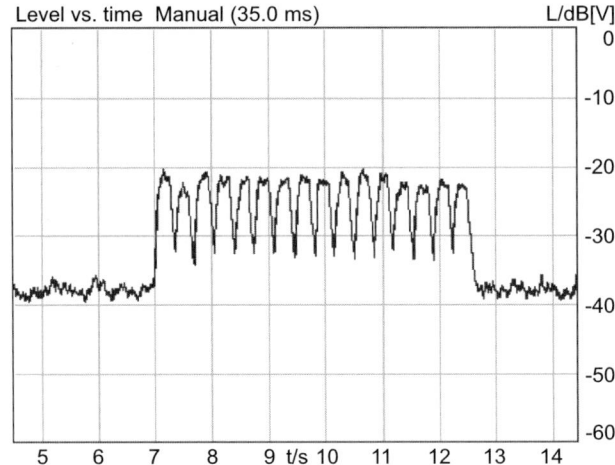

Fig. 10.16. Transmission of background noise and near end signal (level vs. time), hands-free terminal 2.

might even be misleading and therefore fail in practice, because completely different implementations might be represented by the same score. Such a score does not represent the acoustical "fingerprint" of an individual implementation.

The ITU-T Recommendation P.505 [24] provides a new representation methodology – best described as a "quality pie" – that bridges this gap. The circle segments and displayed parameters can be selected and adapted to the application, i.e. the device under test. An example of a hands-free implementation with parameter selection according to the VDA specification [48] is shown in Fig. 10.17.

The focus of this representation is to provide

- a "quick and easy to read" overview about the implementation for experts and non-experts including strengths and weaknesses,
- a comparison to limits, recommended values or average results from benchmarking tests, and
- detailed information for development to improve the performance.

10.6.1 Interpretation of HFT "Quality Pies"

The hands-free "quality pie" shown in Fig.10.17 does not represent an existing implementation. It is only used here for explanation purposes. In general the 12 segments – which can be regarded as a maximum suitable number being visualized in one diagram – can be subdivided into three groups covering different conversational aspects. The first 5 segments – clockwise arranged – represent one-way transmission parameters. The sending direction is

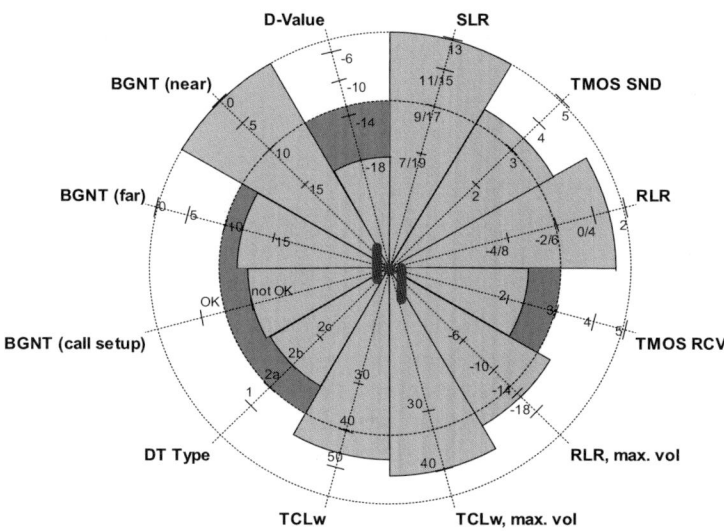

Fig. 10.17. Hands-free "quality pie" according to [24]. The following abbreviations were used: *SLR* means sending loudness rating, *TMOS SND* stands for TMOS value in sending direction, *RLR* abbreviates receiving loudness rating, *TMOS RCV* is short for the TMOS value in receiving direction, *RLR, max. vol* is the receiving loudness rating at maximum volume, *TCLw, max. vol* is the terminal coupling loss measured at maximum volume, *TCLw* abbreviates the terminal coupling loss measured during standard terminal operation, *DT type* means double-talk type, and the different *BGNT* slices show the background noise transmission in different situations.

covered by the sending loudness rating and the TMOS. The following two slices represent the receiving loudness rating (RLR) and the TMOS in receiving direction. The fifth segment represents the RLR value at maximum volume.

The following 3 segments indicate the echo attenuation expressed through the parameter weighted terminal coupling loss according to ITU-T Recommendation G.122 [16] measured at maximum volume ("TCL_W(max.vol.)"), at nominal volume ("TCL_W") and the double talk performance ("DT type"). The last 4 segments represent parameters concerning the quality of background noise transmission.

The following general assumptions are made for the quality pie representation: Each parameter is represented by a pie slice. The size of each slice directly correlates to quality. The gray color indicates a quality higher than the requirement for this specific parameter. Interaction aspects between single parameters are not considered. An inner circle (dark gray) indicates the minimum requirement for each parameter. For those parameters that should be within a range, like the sending loudness rating (SLR) of 13 ± 4 dB [48] the axis is double scaled. It raises from the origin of the diagram radial to the outside up to the recommended value (13 dB for the SLR in this example) and in addition radial to the inside. Other axes like the background noise transmission quality after call setup ("BGNT call setup") are scaled only between two states (ok, not ok).

10.6.2 Examples

The significance of this representation, e.g. in tracing different development phases can best be shown on a practical example. Fig. 10.18(a) represents an early quality status of a hands-free implementation during development.

The left pie chart points out the following:

- The SLR of 13 dB indicates a sufficient loudness, the TMOS score (parameter "TMOS SND") significantly exceeds the limit in sending direction.
- The D-value of -18 dB is too low, the inner dark gray circle represents the limit of -10 dB and gets visible. The sensitivity on background noise needs to be reduced or the sensitivity on speech increased.
- The echo attenuation is too low at maximum volume, the TCL_W requirement is violated under this condition (parameter "TCL_W(max. vol.)").
- Significant impairments could also be observed in background noise transmission during the application of far end signals (parameter "BGNT(far end)"). The background noise is completely attenuated by echo suppression, comfort noise is not inserted. The resulting modulation in the transmitted background is very high, gaps occur. The maximum acceptable level modulation of 10 dB for this parameter is exceeded.

The quality pie in Fig. 10.18(b) indicates a significantly improved performance compared to the previous status represented in part (a). However, the next step that should be addressed by tuning echo cancellation, echo suppression, double talk detection and the associated control parameters is the double talk performance. "Type 1" implementations, i.e. full duplex capable hands-free implementations are available today. It should be noted that it is not always recommended to tune the algorithms to full duplex capability, especially not for the price of lower robustness. Partial duplex capable HFTs ("type 2a" or even "2b") may sometimes be a preferable solution.

In summary, it can be stated that the quality pie representation simplifies the performance discussion. This representation can serve as a basis for commercial decisions, but still provides enough detailed information to discuss possible next optimization steps for speech quality. Important features like interaction aspects between single parameters are explicitly not considered yet and require further investigations.

10.7 Related Aspects

10.7.1 The Lombard Effect

The Lombard effect – also designated as Lombard reflex emphasizing more its intuitive character – describes the result of speech transformation under the

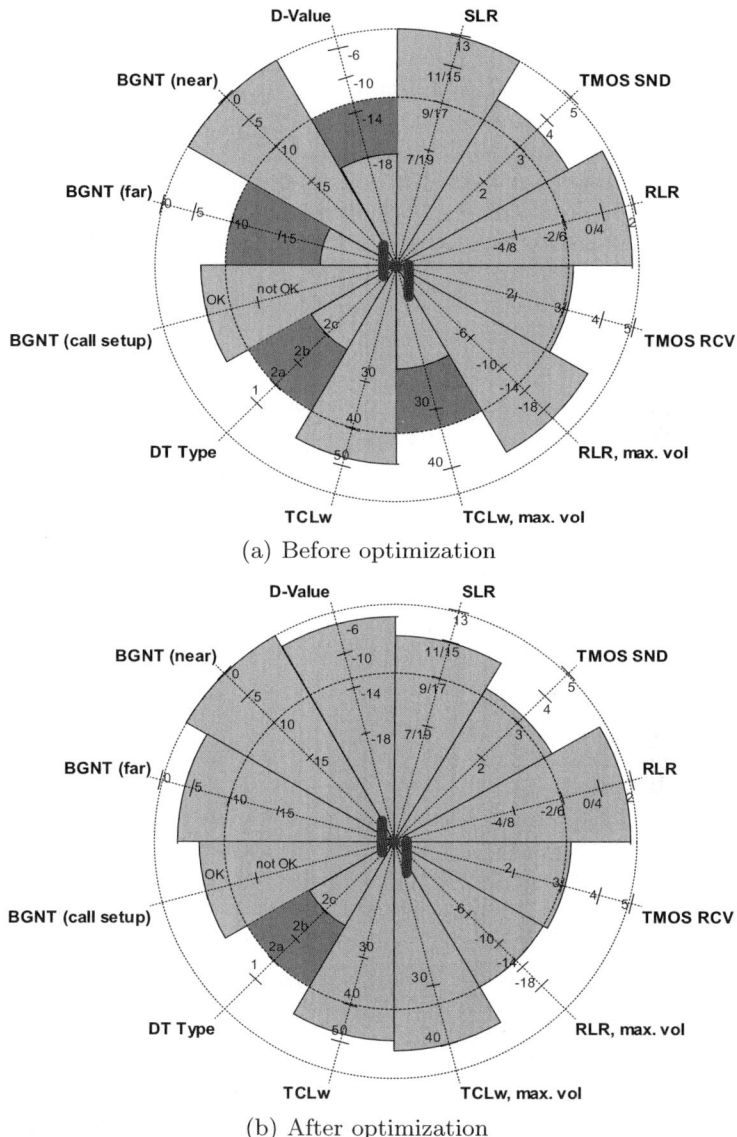

Fig. 10.18. Quality pie before (a) and after (b) first optimization step. For the meaning of the different abbreviations see the caption of Fig. 10.17.

influence of a reduced acoustical feedback, e.g. for hearing impaired people or under the influence of noise and stress. However, the Lombard effect is not only a physiological effect. In practice the main intention for the modification of speech production in a conversation is to be more intelligible to others. It can therefore be assumed that the "naturalness" of Lombard speech cannot be completely reproduced in laboratory testing by recording speech samples that are read from a list. Furthermore, different studies showed that multi-talker babble noise led to different Lombard speech characteristics, e.g. larger vowel duration as compared to stationary noise. In the same way, there is a dependence of the Lombard effect on the noise frequency distribution [38]. Lombard speech recorded under the influence of non-stationary noise provides a higher dynamic compared to Lombard speech produced under stationary noise conditions.

Databases available today do not always consider the mentioned aspects. They are typically recorded with test persons reading predefined sentences. These data are of course valid to be used for certain applications, however, the restrictions need to be known and considered.

It is obvious and reasonable to consider Lombard speech characteristics not only for testing speech recognition systems (e.g. [43]) but also for hands-free terminal testing instead of using neutral voice. An appropriate method is to play back these recordings via artificial head systems in a driving car or in a driving simulator [43]. Furthermore, it is important to analyze Lombard speech in order to verify, if important characteristics need to be considered in objective speech quality tests and analyses.

There are different simulation techniques in use providing a recording scenario for Lombard speech. Test persons typically wear equalized closed headphones during noise playback while their Lombard speech is recorded [10, 41]. These headphones lower the perception of the own voice, thus introducing already the Lombard effect. This can be minimized by introducing a feedback path between the microphone and the headphones itself, thus playing back simultaneously the recorded speech via the headphones [45]. Comparison tests with and without this feedback path indicated that the Lombard effect introduced by the headset itself can be neglected compared to the Lombard effect introduced by the background noise scenario e.g. simulating a driving car [45].

Recording scenarios for Lombard speech under the influence of driving noise are described e.g. in [10, 41]. The setup used during own tests is shown in Fig. 10.19. The recordings were carried out in a driving simulator consisting of a real car cabin equipped with an acoustical background noise simulation system.

It is important to reproduce not only the driving situation acoustically during this kind of speech recordings but also the concentration for a typical driving situation and the impression of having a real conversation over a hands-free system. The driving simulator is therefore operated interactively. The speed is indicated on a speedometer and the test persons are instructed to keep a constant speed. Furthermore, a typical hands-free microphone is installed visible near the interior mirror. The test persons were instructed that they should imagine the conversational situation of having a telephone conversation over a hands-free system.

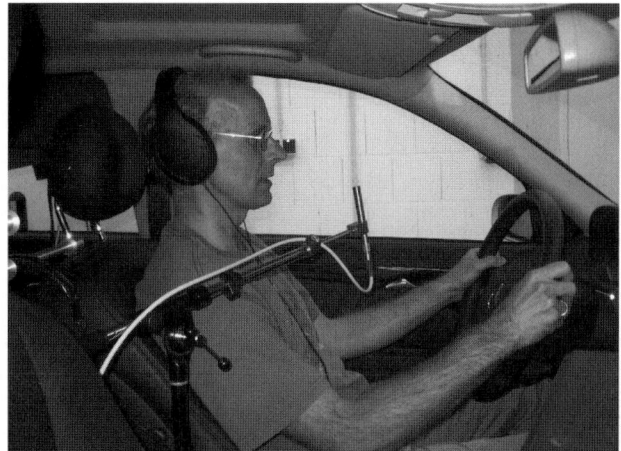

Fig. 10.19. Setup for Lombard speech recordings.

This suitability of this scenario was verified by recording Lombard speech of eight test persons, four male and four female speakers. Different speech material like free utterances, given test sentences to be read from a list and command words were recorded and analyzed. The speech level analyses first demonstrated that the influence of the headphones can practically be neglected. The average speech levels for the different speech materials increased by less than 1 dB if the test persons wore headphones.

Fig. 10.20 shows the average speech levels for free utterances (dark gray bars) and command words (light gray bars). The speech level under quiet conditions was determined to approximately -1 dB$_{Pa}$ at the mouth reference point (MRP) of the test persons for the free utterances. A standardized test signal level for objective terminal testing is -4.7 dB$_{Pa}$ at the MRP. However, it is reported that people tend to increase their speech level by approximately 3 dB when using hands-free devices [27]. The resulting level of approximately -1.7 dB$_{Pa}$ at the mouth reference point is rather accurately confirmed by the measured level of -1 dB$_{Pa}$ for the free utterances. These speech recordings confirm the tendency given in [27], the analyses of speech material recorded from eight test persons are not representative in a statistical sense.

The Lombard recordings were carried out for three different speeds and levels of 50 km/h (49 dB$_{SPL(A)}$), 130 km/h (69 dB$_{SPL(A)}$) and 200 km/h (79 dB$_{SPL(A)}$). Fig. 10.20 shows the average speech levels for the command words and the free utterances. An offset of approximately 2.5 dB can be measured for the two speech materials. The command words are more pronounced and therefore provide a higher level compared to the free speech.

The regression further points out that the speech level increases by approximately 0.4 dB/dB$_{(A)}$ for driving situations with a background noise level between approximately 55 dB$_{(A)}$ and 70 dB$_{(A)}$. Similar results are

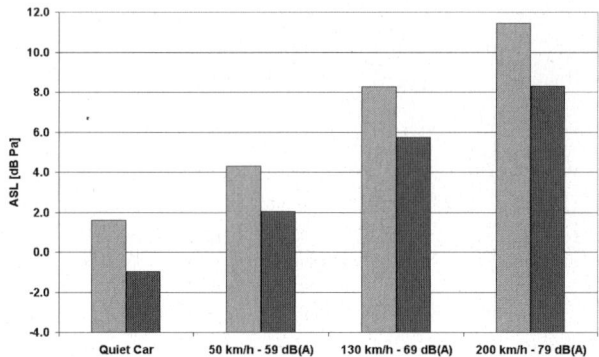

Fig. 10.20. Active speech levels (ASL) at different simulated conditions (light gray: command words, dark gray: free utterances).

reported in [10]. For higher speed the speech level increases by approximately 0.3 dB/dB$_{(A)}$ for the command words and 0.25 dB/dB$_{(A)}$ for the free utterances.

Important conclusions can be drawn from such investigations for objective laboratory tests because they again raise the question of adapting test signal levels during hands-free telephone tests. These results support the idea of increasing the test signal levels for all objective tests by approximately 3 dB at the artificial mouth of an artificial head measurement system simulating the driver's voice. Furthermore the Lombard effect depending on the different background noise scenarios simulated during laboratory tests should be considered and can be estimated from data as analyzed above.

10.7.2 Intelligibility Outside Vehicles

The intelligibility of telephone conversations outside the vehicle is a very important aspect but users are not always aware of this situation. The reason for this undesired effect is elementary: the downlink signal of a hands-free telephone conversation in a vehicle, typically played back via the built-in loudspeakers in the front door, exciting the door structure. The whole surface emits the audible sound outside the vehicle.

This implies, besides the privacy aspect, also a political aspect: a huge effort is taken by legislation in order to lower the external vehicle sound produced e.g. by motors, exhaust systems and tires [5]. The aspect of sound played back via the internal audio systems has – so far – not been addressed. The acoustical coupling between the loudspeakers and the chassis needs to be evaluated in detail in order to identify the transmission paths and individual contributions.

Combined electro-acoustic measures, intelligibility and perceptual analyses on the one hand and vibration analyses on the other hand are necessary in

Fig. 10.21. Intelligibility outside the vehicle.

order to document the status, evaluate the transmission paths and verify the effectiveness of modifications [42]. The acoustically relevant parameters can realistically be measured by using two artificial head measurement systems on the driver's seat and outside in a predefined distance and position, e.g. 1 and 2 m from the B-pillar.

The speech intelligibility index SII [6], can – in principle – be used and calculated for different noise scenarios. But the intelligibility of speech highly depends on the test corpus. The SII calculation is based on a weighted spectral distance between average speech and noise spectra. However, the sentence intelligibility is significantly higher than the SII due to its context information [46].

A more analytical analysis is given by the calculation of the attenuation provided by the car chassis. Fig. 10.22 shows the spectral attenuation between the inside HATS at the driver's position and outside in a distance of 1 m from the B-pillar. The curve indicates a strong low frequency coupling between the loudspeaker and the chassis. The attenuation of the high frequencies above approximately 1 kHz is around 20 dB to 25 dB higher.

Besides the speech-based analyses, a vibration analysis (laser scan of the driver's door, see Fig. 10.23) links the intelligibility to the technical source of the emitted signal. The oscillation amplitude is color coded. The complete door is excited by the acoustic signal. Further tests with different loudspeaker modifications (mechanically decoupling loudspeakers from door structure, use of damping material) in one test car showed that the main factor for the outside intelligibility is caused by airborne coupling between loudspeaker and door. Structure borne coupling played a minor role.

The efficiency of modifications is vehicle dependent. Acoustical coupling typically can be significantly reduced only by new loudspeaker positions and mountings or a complete encapsulation. Both would require an enormous effort in modifying vehicle design. The need and motivation for modifications is

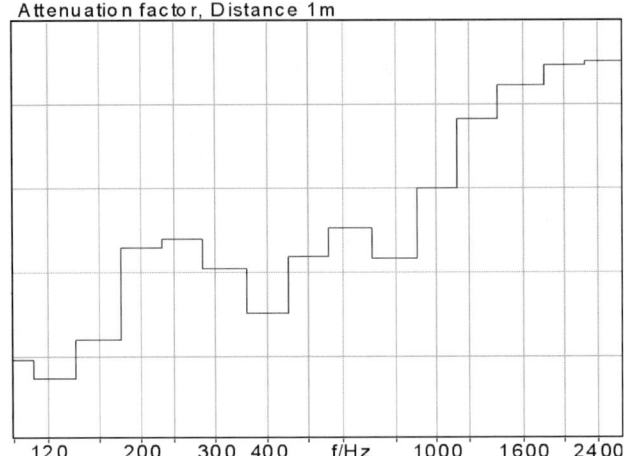

Fig. 10.22. Attenuation "Inside to outside" in 1 m distance, B-pillar.

probably driven by customer's expectations and complaints. A suggestion for reasonable limits for the outside intelligibility can be derived from practical approaches: the intelligibility of the driver's voice outside the vehicle or the intelligibility when using external loudspeakers for playback, e.g. positioned in the drivers and co-drivers footwell.

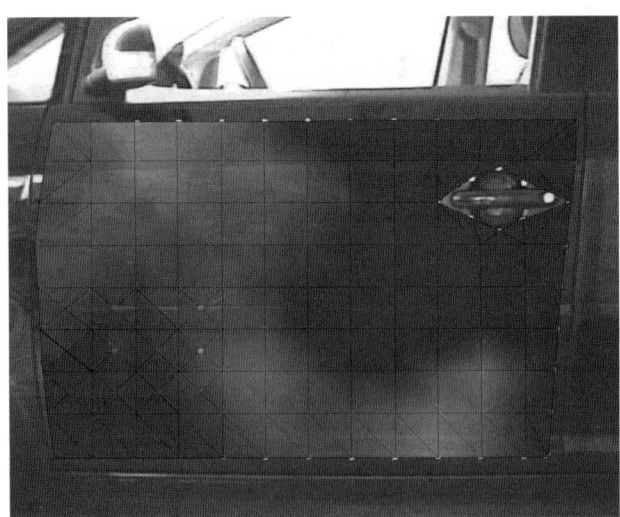

Fig. 10.23. Laser scan, vibration of door structure (excitation frequency 336 Hz (example)).

References

1. 3GPP:TS 46.010: Full rate speech encoding, *Third Generation Partnership Project (3GPP)*, 2002.
2. 3GPP:TS 46.051: GSM Enhanced full rate speech processing functions: General description, *Third Generation Partnership Project (3GPP)*, 2002.
3. 3GPP : TS 46.090: AMR speech codec: Transcoding functions, *Third Generation Partnership Project (3GPP)*, 2002.
4. 3GPP TS 26.077: Technical specification group services and system aspects; Minimum performance requirements for noise suppresser; Application to the adaptive multi-rate (AMR) speech encoder, *Third Generation Partnership Project (3GPP)*, 2003.
5. 70/157/EWG: Richtlinie des Rates zur Angleichung der Rechtsvorschriften der Mitgliedsstaaten über den zulässigen Geräuschpegel und die Auspuffvorrichtung von Kfz, 6. Feb. 1970 (in German).
6. ANSI S3.5-1997: Methods for calculation of the speech intelligibility index.
7. J. Berger: Instrumentelle Verfahren zur Sprachqualitätsschätzung – Modelle auditiver Tests, Ph.D. thesis,, Kiel, 1998 (in German).
8. J. Berger: Results of objective speech quality assessment including receiving terminals using the advanced TOSQA2001, ITU-T Contribution COM 12-20-E, Dec. 2000.
9. J. Blauert: *Spatial Hearing: The Psychophysics of Human Sound Localization*, Cambridge, MA, USA: MIT Press, 1997.
10. M. Buck, H.-J. Köpf, T. Haulick: Lombard-Sprache für Kfz-Anwendungen: eine Analyse verschiedener Aufnahmekonzepte, *Proc. DAGA '06*, Braunschweig, Germany, 2006 (in German).
11. ETSI EG 202 396-1: Speech quality performance in the presence of background noise; Part 1: Background noise simulation technique and background noise database, 2005.
12. ETSI EG 202 396-3: Speech quality performance in the presence of background noise; Part 3: Background noise transmission – objective test methods, 2007.
13. K. Genuit: Objective evaluation of acoustic quality based on a relative approach, *Proc. Internoise '96*, Liverpool, UK, 1996.
14. H. W. Gierlich: The auditory perceived quality of hands-free telephones: Auditory judgements, instrumental measurements and their relationship, *Speech Communication*, **20**, 241–254, October 1996.
15. H. W. Gierlich, F. Kettler, E. Diedrich: Proposal for the definition of different types of hands-free telephones based on double talk performance, *ITU-T SG 12 Meeting*, COM 12-103, Geneva, Switzerland, 1999.
16. ITU-T Recommendation G.122: Influence of national systems on stability and talker echo in international connections, *International Telecommunication Union*, Geneva, Switzerland, 1993.
17. ITU-T Recommendation G.722: 7 kHz audio-coding within 64 kbit/s, *International Telecommunication Union*, Geneva, Switzerland, 1988.
18. ITU-T Recommendation P.48: Specification for an intermediate refernce system, *International Telecommunication Union*, Geneva, Switzerland, 1993.
19. ITU T Recommendation P.50: Artificial voices, *International Telecommunication Union*, Geneva, Switzerland, 1999.
20. ITU T Recommendation P.51: Artificial mouth, *International Telecommunication Union*, Geneva, Switzerland, 1996.

21. ITU-T Recommendation P.58: Head and torso simulator for telephonometry, *International Telecommunication Union*, Geneva, Switzerland, 1996.
22. ITU T Recommendation P.501: Test signals for use in telephonometry, *International Telecommunication Union*, Geneva, Switzerland, 2000.
23. ITU T Recommendation P.502: Objective test methods for speech communication systems using complex test signals, *International Telecommunication Union*, Geneva, Switzerland, 2000.
24. ITU-T Recommendation P.505: One-view visualization of speech quality measurement results, *International Telecommunication Union*, Geneva, Switzerland, 2005.
25. ITU T Recommendation P.581: Use of head and torso simulator (HATS) for hands free terminal testing, *International Telecommunication Union*, Geneva, Switzerland, 2000.
26. ITU-T Recommendation P.79: Calculation of loudness ratings for telephone sets, *International Telecommunication Union*, Geneva, Switzerland, 2000.
27. ITU-T Recommendation P.340: Transmission characteristics and speech quality parameters of hands-free telephones, *International Telecommunication Union*, Geneva, Switzerland, 2000.
28. ITU-T Recommendation P.800.1: Mean opinion score (MOS terminology), *International Telecommunication Union*, Geneva, Switzerland, 2003.
29. ITU-T Recommendation P.800: Methods for subjective determination of speech quality, *International Telecommunication Union*, Geneva, Switzerland, 2003.
30. ITU-T Recommendation P.830: Subjective performance assessment of telephone-band and wideband digital codes, *International Telecommunication Union*, Geneva, Switzerland, 1996.
31. ITU-T Recommendation P.831: Subjective performance evaluation of network echo cancellers, *International Telecommunication Union*, Geneva, Switzerland, 1998.
32. ITU-T Recommendation P.832: Subjective performance evaluation of hands-free terminals, *International Telecommunication Union*, Geneva, Switzerland, 2000.
33. ITU-T Recommendation P.835: Subjective performance of noise suppression algorithms, *International Telecommunication Union*, Geneva, Switzerland, 2003.
34. ITU-T Recommendation P.862: Perceptual evaluation of speech quality (PESQ): An objective method for end-to-end speech quality assessment of narrow-band telephone networks and speech codecs, *International Telecommunication Union*, Geneva, Switzerland, 2001.
35. ITU-T Recommendation P.862.1: Mapping function for transforming P.862 raw result scores to MOS-LQO, *International Telecommunication Union*, Geneva, Switzerland, 2003.
36. ITU-T Focus Group FITcar: Draft specification for hands-free testing.
37. U. Jekosch: *Voice and Speech Quality Perception,* Berlin, Germany: Springer, 2005.
38. J.-C. Junqua: The influence of acoustics on speech production: A noise-induced stress phenomenon known as the Lombard reflex, *Speech Communication*, **20**, 13–22, 1996.
39. W. Kellermann: Acoustic echo cancellation for beamforming microphone arrays, in M. Brandstein, D. Ward (eds.), *Microphone Arrays*, Berlin, Germany: Springer: 2001.

40. F. Kettler, H. W. Gierlich, E. Diedrich: Echo and speech level variations during double talk influencing hands-free telephones transmission quality, *Proc. IWAENC '99*, Pocono Manor, PA, USA, 1999.
41. F. Kettler, M. Röber: Generierung von Sprachmaterial zum realitätsnahen Test von Freisprecheinrichtungen, *Proc. DAGA '03*, Aachen, Germany, 2003 (in German).
42. F. Kettler, F. Rohrer, C. Nettelbeck, H. W. Gierlich: Intelligibility of hands-free phone calls outside the vehicle, *Proc. DAGA '07*, Stuttgart, Germany, 2007.
43. M. Lieb: Evaluating speech recognition performance in the car, *Proc. CFA/DAGA '04*, Strasbourg, France, 2004.
44. S. Möller: *Assessment and Prediction of Speech Quality in Telecommunications*, Boston, MA, USA: Kluwer Academic Press, 2000.
45. C. Pörschmann: Eigenwahrnehmung der Stimme in virtuellen akustischen Umgebungen, *Proc. DAGA '98*, Zürich, Switzerland, 1998 (in German).
46. J. Sotschek: Methoden zur Messung der Sprachgüte I: Verfahren zur Bestimmung der Satz- und Wortverständlichkeit, *Der Fernmelde-ingenieur*, 1976 (in German).
47. R. Sottek, K. Genuit: Models of signal processing in human hearing, *International Journal of Electronics and Communications*, 157–165, 2005.
48. VDA-Specification for Car Hands-Free Terminals, Version 1.5, VDA, 2005.
49. N. Xiang, K. Genuit, H. W. Gierlich: Investigations on a new reproduction procedure for binaural recordings, *Proc. AES 95th Convention*, Preprint 3732 (B2-AM-9), New York, NY, USA, 1993.

Part IV

Multi-Channel Processing

11

Correlation-Based TDOA-Estimation for Multiple Sources in Reverberant Environments

Jan Scheuing and Bin Yang

University of Stuttgart, Germany

11.1 Introduction

Estimation of time difference of arrival (TDOA) using a microphone array is an important task in many speech related applications such as speaker localization or near-field beamforming for noise reduction. Basically, two different approaches for TDOA estimation have been used: search for the extrema in cross-correlation of microphone signals [7, 10] or blindly estimate the room impulse responses [3]. Both approaches have been approved in many single source scenarios. However, little research work has been done so far on simultaneous TDOA estimation of multiple sources in reverberant environments.

The main idea to extend single source room impulse response techniques to the multiple source case is to split the multi-input multi-output (MIMO) system into several independent single-input multi-output (SIMO) systems. Usually, this is achieved either under the ideal assumption that each speaker is active exclusively during some time intervals [8, 9, 13] or by blind source separation [1, 5].

The cross-correlation technique is both conceptually and computationally simple. Under ideal conditions of a single white source signal in additive white noise without multipath propagation, the TDOA estimate derived from the cross-correlation reaches the Cramer-Rao lower bound and is efficient [7]. In order to combat spurious peaks in the cross-correlation caused by a colored source signal spectrum or even nearly periodic signals like voiced speech segments, different generalized cross-correlations have been proposed [10].

However, two major problems of the cross-correlation based TDOA estimation remain unsolved. In a reverberant environment, the (generalized) cross-correlation will show a large number of peaks due to direct path and echo path propagations. While only the TDOAs originating from the direct paths contain useful information for source localization and beamforming, the cross-correlation peaks caused by echo paths make the TDOA estimation ambiguous. How can we distinguish between the direct path and echo path extrema in the cross-correlation?

A second TDOA ambiguity arises from multiple, simultaneously active sources. In this case, each cross-correlation between a pair of microphones contains both direct path and echo path extrema from all sources. Even if we could identify the direct path TDOAs, how can we assign them correctly to the different sources? Both source localization and beamforming require namely TDOA estimates of different sensor pairs for the same source. Combining TDOAs of different sources will cause a phantom source in localization and steer a beam to a wrong direction.

In this chapter, we present a novel approach to reduce and resolve both ambiguities [15–17]. Starting with the traditional cross-correlation, we find ways to identify the desired direct path TDOAs and to assign them correctly to the different sources. Actually, our approach is mainly based on two fairly simple observations of information redundancy which have not been exploited yet in the literature:

- First, the extremum positions of a cross-correlation between two microphone signals appear in well defined distances which can be predicted from the extremum positions of the corresponding autocorrelations of the microphone signals. Under ideal conditions, combining the cross-correlation with the autocorrelations will uniquely identify the desired direct path TDOA and reject all ambiguous cross-correlation extrema caused by echo paths.
- The second information redundancy is the zero cyclic sum of TDOAs over any number of microphones. Given TDOA estimates for various microphone pairs, the set of direct path TDOAs belonging to the same source clearly satisfies this zero cyclic sum condition. In contrast, TDOA estimates originating from direct and echo paths and from different sources usually violate this condition. This provides an additional mean to find the matching TDOAs.

In Sec. 11.2, we formulate the signal model and the TDOA estimation problem. Then we analyze different TDOA ambiguities and their origins. We show the information redundancy and present our basic ideas of TDOA disambiguation. In Sec. 11.3, we present an algorithm exploiting the information redundancy contained in the autocorrelation of the microphone signals. By using a so called raster matching approach, we show how to identify and reject the echo path cross-correlation extrema. Sec. 11.4 formulates the combination of TDOA estimates abstractly in a consistent TDOA graph. By using the zero cyclic sum condition, we search for matching TDOAs by a synthesis of consistent TDOA graphs. We present a very efficient synthesis algorithm based on consistent triples. Sec. 11.5 describes a real experiment of locating multiple sources in reverberant environments. It demonstrates the effectiveness, the real-time capability, and the high localization accuracy of our algorithms and system.

11.2 Analysis of TDOA Ambiguities

11.2.1 Signal Model

We assume N acoustic sources in a room recorded by M microphones. Neglecting noise and assuming omnidirectional characteristics of both the sources and the microphones, the discrete-time signal of the k-th microphone can be described by

$$y_k(n) = \sum_{a=1}^{N} (h_{a,k} * s_a)(n), \quad k \in \{1, \ldots, M\}. \tag{11.1}$$

Here $s_a(n)$ represents the signal of source $a \in \{1, \ldots, N\}$, "$*$" denotes the convolution sum, and $h_{a,k}(n)$ is the room impulse response between source a and microphone k. The latter consists of $\Lambda_{a,k}$ propagation paths $\mu \in \{0, \ldots, \Lambda_{a,k}-1\}$ characterized by the amplitudes $h_{a,k,\mu}$ and integer delays $\tau_{a,k,\mu}$. The room impulse response is

$$h_{a,k}(n) = \sum_{\mu=0}^{\Lambda_{a,k}-1} h_{a,k,\mu}\, \delta(n - \tau_{a,k,\mu}), \tag{11.2}$$

where $\delta(n)$ is the unit sample sequence. The microphone signal can then be written as

$$y_k(n) = \sum_{a=1}^{N} \sum_{\mu=0}^{\Lambda_{a,k}-1} h_{a,k,\mu}\, s_a(n - \tau_{a,k,\mu}). \tag{11.3}$$

All delays $\tau_{a,k,\mu}$ are sorted in ascending order, i.e. $\tau_{a,k,\mu} > \tau_{a,k,\nu}$ if $\mu > \nu$. The TDOA is defined as

$$n_{a,kl,\mu\nu} = \tau_{a,k,\mu} - \tau_{a,l,\nu}. \tag{11.4}$$

We assume that the *line of sight* condition is satisfied for all sources and microphones. The corresponding direct paths are denoted by $\mu = 0$; otherwise a localization based on the TDOAs would hardly be possible.

The subject of *TDOA estimation* is to estimate a *source TDOA vector*

$$\boldsymbol{n}_a = \begin{bmatrix} n_{a,12,00},\, n_{a,13,00},\, \ldots,\, n_{a,M-1\,M,00} \end{bmatrix}^{\mathrm{T}} \tag{11.5}$$

of length $\binom{M}{2}$ for each active source from the microphone signals. There are four requirements for the TDOA estimation task:

- All TDOAs in $\boldsymbol{n}_a$ should originate from direct paths only as indicated by the double zero index "$_{00}$".
- All TDOAs in $\boldsymbol{n}_a$ should originate from the same source a.
- The vector $\boldsymbol{n}_a$ should be as complete as possible (few missing elements).

- The estimation process should be computationally as efficient as possible.

While the last two requirements represent soft wishes, the first two requirements are mandatory because otherwise we would obtain a wrong source position estimation. Unfortunately, a number of reasons make the TDOA estimation ambiguous and difficult. In the following, three different types of *TDOA ambiguity* are analyzed using simple scenarios. They are multipath propagation, multiple sources, and periodic signals [15]. For notational convenience, if we consider only one source or only one path, we drop the corresponding index a or μ in Eq. 11.3.

11.2.2 Multipath Ambiguity

The first type of ambiguity is caused by the multipath propagation of signals. A single source ($N=1$) propagating on Λ_k different paths to the microphone k causes the microphone signal

$$y_k(n) = \sum_{\mu=0}^{\Lambda_k - 1} h_{k,\mu}\, s(n - \tau_{k,\mu}). \tag{11.6}$$

If the source signal $s(n)$ is white, the cross-correlation

$$r_{kl}(n) = \mathrm{E}\{y_k(m+n)\, y_l(m)\} \tag{11.7}$$

between the two microphone signals $y_k(n)$ and $y_l(n)$ will show $\Lambda_k \Lambda_l$ local extrema at the TDOA positions

$$n_{kl,\mu\nu} = \tau_{k,\mu} - \tau_{l,\nu} \quad \text{with} \quad \mu \in \{0,\ldots,\Lambda_k - 1\}, \nu \in \{0,\ldots,\Lambda_l - 1\}. \tag{11.8}$$

The only relevant and desired TDOA for source localization is the *direct path TDOA* $n_{kl,00} = \tau_{k,0} - \tau_{l,0}$, the difference between the two direct path delays. All other $\Lambda_k \Lambda_l - 1$ TDOA values involve at least one echo path. They are referred to as *echo path TDOA* and result in wrong hyperbola of possible source location, see Fig. 11.1. They appear as spurious peaks in the cross-correlation $r_{kl}(n)$. The problem is how to determine which of the $\Lambda_k \Lambda_l$ cross-correlation extrema represents the desired direct path TDOA.

11.2.3 Multiple Source Ambiguity

The second type of ambiguity is due to the presence of multiple simultaneously active sources. Assuming a direct path propagation of N source signals, the k-th microphone signal is

$$y_k(n) = \sum_{a=1}^{N} h_{a,k}\, s_a(n - \tau_{a,k}). \tag{11.9}$$

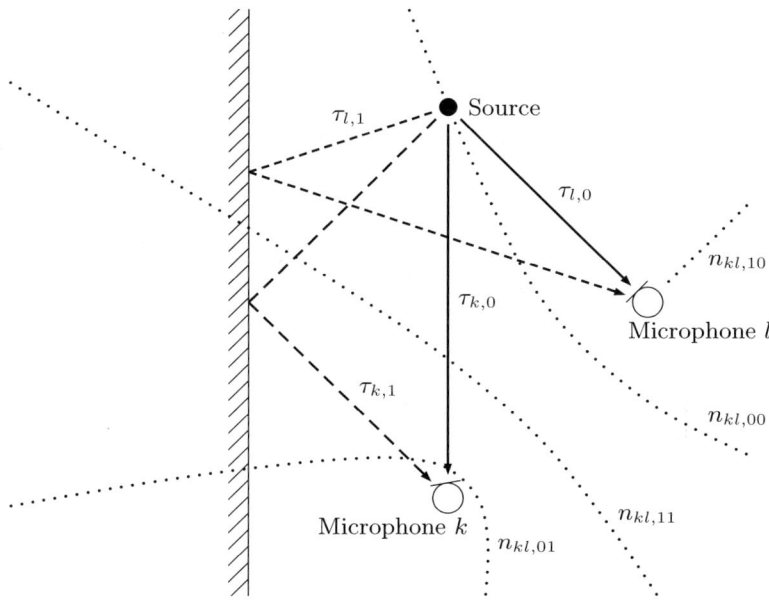

Fig. 11.1. Assuming a direct path (solid) and an echo path (dashed) from the source to each of the two microphones, four TDOA values are possible resulting in four hyperbola (dotted) of possible source location. Only the hyperbola corresponding to the direct path TDOA $n_{kl,00}$ passes the source location.

In this case, $\tau_{a,k}$ denotes the direct path delay from source a to microphone k. If the source signals $s_a(n)$ are uncorrelated and white, the cross-correlation $r_{kl}(n)$ will show N local extrema at the TDOA positions

$$n_{a,kl} = \tau_{a,k} - \tau_{a,l}. \tag{11.10}$$

The difficulty is to assign those N TDOA values correctly to the N sources such that all TDOAs of different microphone pairs for the same source are collected together. For unknown source positions, this assignment is ambiguous.

By using TDOAs of $L \leq \binom{M}{2}$ microphone pairs, there are $(N!)^{L-1}$ different possibilities of combining the N extremum positions of one cross-correlation with those of the $L-1$ remaining cross-correlations. Any erroneous combination of TDOAs like the combination of $n_{a,kl}$ and $n_{b,lm}$ in Fig. 11.2 causes a phantom source. Therefore, the output of a perfect multiple source TDOA estimation would be a set of N source TDOA vectors $\{\hat{\boldsymbol{n}}_{\sigma_1}, \ldots, \hat{\boldsymbol{n}}_{\sigma_N}\}$ corresponding to the N sources, with each $\hat{\boldsymbol{n}}_{\sigma_a}$ containing L matching TDOAs for one source. Obviously, there are $N!$ permutations in the enumeration of sources which can not be resolved.

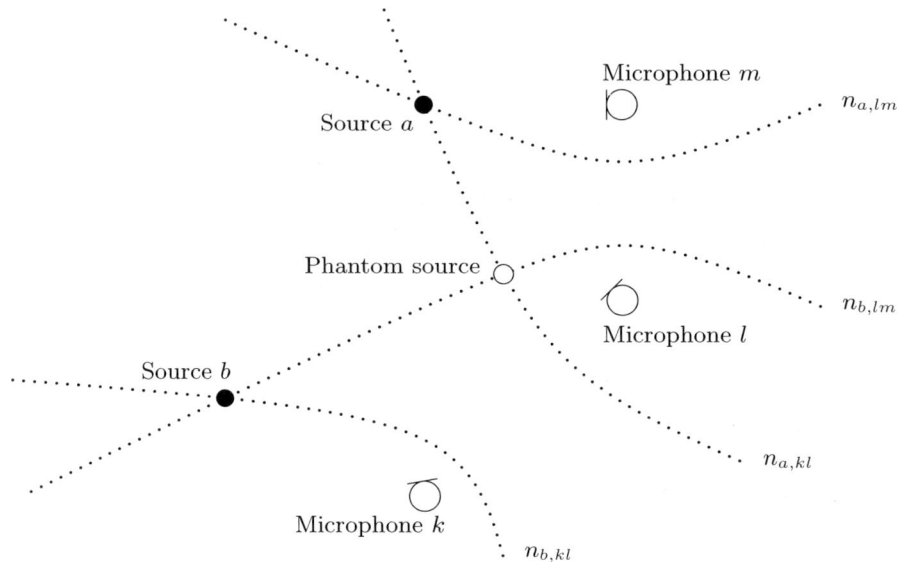

Fig. 11.2. A combination of TDOAs originating from different sources causes a phantom source.

11.2.4 Ambiguity due to Periodic Signals

Segments of natural speech signals can be categorized into three main types:
- silence for which neither localization nor beamforming makes sense,
- unvoiced speech which is more or less a colored noise signal,
- and voiced speech which shows a high periodicity [14].

Fig. 11.3 depicts the autocorrelation of a voiced speech signal. It looks like a modulated sinusoidal signal, indicating the presence of a pitch. Similarly, many natural sounds and signals of machine noise contain periodic parts as well. The periodic extrema of the autocorrelation of the source signals will also appear in the cross-correlation of the microphone signals, even in a single source scenario without multipath propagation. This makes the TDOA estimation ambiguous.

11.2.5 Principles of TDOA Disambiguation

In practice, all three types of ambiguity occur simultaneously, making the TDOA estimation even more difficult. Fig. 11.4 shows the cross-correlation of two microphone signals in a real experiment. Two speech sources are simultaneously active (see Sec. 11.5 for more details about the experiment). At a first look, it is impossible to determine the two direct path TDOAs from the cross-correlation. Due to directional characteristics of both sources and

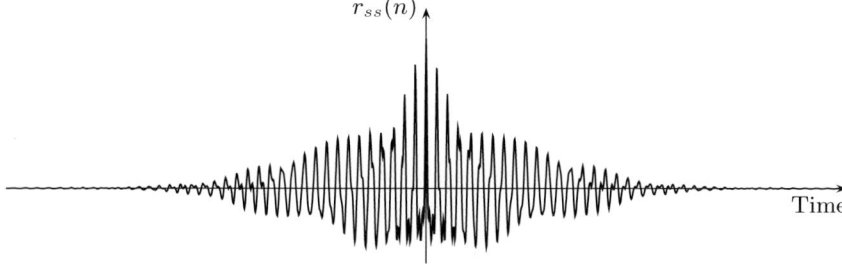

Fig. 11.3. Autocorrelation of a voiced speech signal recorded in an anechoic room.

microphones in reality, it is not always true that the direct path signal has the strongest amplitude. Even if we could find the direct path TDOAs, there seems to be no way to assign them correctly to the two sources. Below we will present some novel ideas to resolve these TDOA ambiguities. We call this process *TDOA disambiguation*.

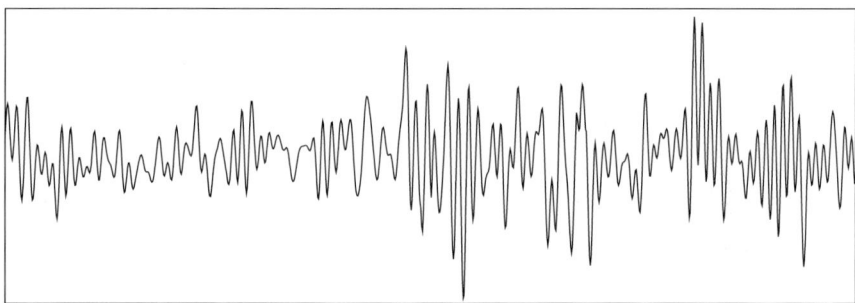

Fig. 11.4. Cross-correlation of two microphone signals in a real experiment.

As will be described in Sec. 11.3, we use a modified phase transform to prewhiten the microphone signals. In order to simplify the discussion in the sequel, we assume that the transformed microphone signals consist of approximately white source signals. This implies that each true TDOA appears as a local extremum in the cross-correlation.

Our approaches are based on two observations. The first one is the relationship between the extremum positions in the cross-correlation and autocorrelation of the microphone signals. For simplicity, we consider the single source and two-path (one direct path and one echo path) scenario in Fig. 11.1 again. Fig. 11.5 shows the four TDOA values corresponding to the four extremum positions in the cross-correlation $r_{kl}(n)$. It also shows the extremum positions of the two autocorrelations

$$r_{kk}(n) = \mathrm{E}\{y_k(m+n)\,y_k(m)\} \quad \text{and}$$
$$r_{ll}(n) = \mathrm{E}\{y_l(m+n)\,y_l(m)\}\,. \tag{11.11}$$

Obviously, the cross-correlation extrema appear in well defined distances like rasters. The difference between many pairs of cross-correlation TDOAs can be predicted by the extremum positions in the autocorrelations.

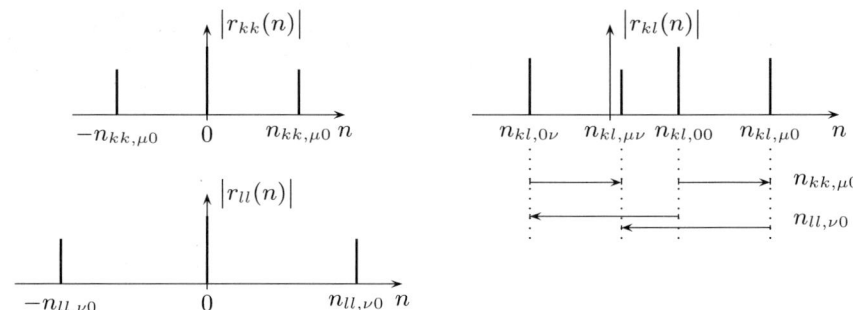

Fig. 11.5. Relationship between extremum positions in auto- and cross-correlation.

Let
$$y_k(n) = h_{k,0}\,s(n - \tau_{k,0}) + h_{k,\mu}\,s(n - \tau_{k,\mu}) \quad \text{and}$$
$$y_l(n) = h_{l,0}\,s(n - \tau_{l,0}) + h_{l,\nu}\,s(n - \tau_{l,\nu}) \tag{11.12}$$

be the microphone signals. If $s(n)$ is white, the autocorrelations $r_{kk}(n)$ and $r_{ll}(n)$ show, in addition to the zero-lag extrema $r_{kk}(0)$ and $r_{ll}(0)$, local extrema at the positions

$$n_{kk,\mu 0} = \tau_{k,\mu} - \tau_{k,0}\,,$$
$$n_{kk,0\mu} = -n_{kk,\mu 0} \tag{11.13}$$

and

$$n_{ll,\nu 0} = \tau_{l,\nu} - \tau_{l,0}\,,$$
$$n_{ll,0\nu} = -n_{ll,\nu 0}\,, \tag{11.14}$$

respectively. They coincide with the differences of cross-correlation extremum positions

$$n_{kk,\mu 0} = \tau_{k,\mu} - \tau_{k,0} = (\tau_{k,\mu} - \tau_{l,\eta}) - (\tau_{k,0} - \tau_{l,\eta}) = n_{kl,\mu\eta} - n_{kl,0\eta} \quad \text{and}$$
$$n_{ll,\nu 0} = \tau_{l,\nu} - \tau_{l,0} = (\tau_{k,\eta} - \tau_{l,0}) - (\tau_{k,\eta} - \tau_{l,\nu}) = n_{kl,\eta 0} - n_{kl,\eta\nu} \tag{11.15}$$

for any direct or echo path η. This condition is referred to as the *raster condition*. Since the direct path always has the shortest delay, $n_{kk,\mu 0}$ and $n_{ll,\nu 0}$

in Eq. 11.15 are positive. Thus, when the distance of two cross-correlation extrema in $r_{kl}(n)$ is $n_{kk,\mu 0}$, the cross-correlation extremum at the right hand side is caused by one echo path more than the left side extremum. For a distance of $n_{ll,\nu 0}$, the left side extremum involves more echo paths than the right side one. Based on this observation, we draw arrows below the cross-correlation function $r_{kl}(n)$ in Fig. 11.5 according to the following rules:

- The length of each arrow is given by the extremum position $n_{kk,\mu 0}$ and $n_{ll,\nu 0}$ in the autocorrelations.
- Each arrow representing an extremum from $r_{kk}(n)$ points from left to right and each arrow from $r_{ll}(n)$ points from right to left.

By using this convention, each arrowhead corresponds to a cross-correlation TDOA which involves at least one echo path. It is an echo path TDOA. The direct path TDOA $n_{kl,00}$ is that extremum in the cross-correlation $r_{kl}(n)$ which shows only arrow tails and no arrowheads. This *raster matching* approach combines the extremum positions of both auto- and cross-correlations and enables us to identify the desired direct path TDOA $n_{kl,00}$ even in a reverberant environment.

The second important observation is the redundancy of TDOAs. For each subset of microphones $\{k, l, \ldots, m, o\} \subseteq \{1, \ldots, M\}$ and their corresponding paths $\mu, \nu, \ldots, \eta, \kappa$, there exists a *zero cyclic sum condition*

$$n_{a,kl,\mu\nu} + \cdots + n_{a,mo,\eta\kappa} + n_{a,ok,\kappa\mu}$$
$$= (\tau_{a,k,\mu} - \tau_{a,l,\nu}) + \cdots + (\tau_{a,m,\eta} - \tau_{a,o,\kappa}) + (\tau_{a,o,\kappa} - \tau_{a,k,\mu}) = 0 \quad (11.16)$$

for TDOAs originating from the same source. Fig. 11.6 illustrates this observation for the direct path TDOAs $\mu = \nu = \ldots = 0$. If the cyclic sum does not disappear, either different paths or different sources are involved. By applying a *zero cyclic sum matching* of TDOA values, phantom sources like in Fig. 11.2 can be avoided.

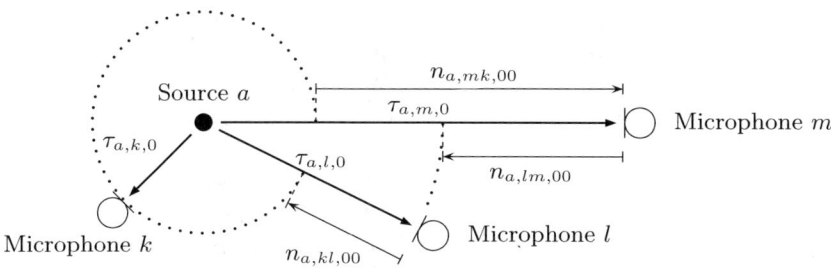

Fig. 11.6. Zero cyclic sum of direct path TDOAs from the same source.

In order to further reduce the number of cross-correlation extrema, we apply two additional disambiguation criteria:

- The direct path amplitudes $h_{a,k,0}$ in Eq. 11.3 are always positive. Hence we only search for the maxima instead of extrema in the cross-correlation $r_{kl}(n)$ [11].
- The triangular inequality posses a constraint on any direct path TDOAs. Each direct path TDOA between two microphones, multiplied by the sound speed, can never exceed the distance between the two microphones. Any TDOA estimates beyond this geometrical upper bound are discarded.

In Sec. 11.3, the estimation of direct path TDOAs based on the raster matching is described in detail. Sec. 11.4 discusses the synthesis of consistent TDOA graphs as an efficient implementation of the zero cyclic sum matching.

11.3 Estimation of Direct Path TDOAs

In the following, we consider the TDOA estimation for a single signal block containing all M microphone signals. For each signal block of length K, we compute the auto- and cross-correlations of the microphone signals. Then we determine the extremum positions of the autocorrelations and the maximum positions of the cross-correlations. Using this time information, a raster matching is performed on each cross-correlation in order to identify the direct path TDOAs.

11.3.1 Correlation and Extremum Positions

We use a modified version of the well known phase transform (PHAT) [10] to compute the *generalized cross-correlation* (GCC) between two microphone signals. In order to save computation time, we use the fast Fourier transform (FFT) combined with zero-padding to calculate the linear correlation.

Let the block length K be a power of two. We append K zeros to the k-th microphone signal in order to obtain a length $2K$ vector

$$\boldsymbol{y}_k = \left[y_k(n), y_k(n-1), \ldots, y_k(n-K+1), 0, \ldots, 0\right]^{\mathrm{T}}. \quad (11.17)$$

This finite-length sequence is transformed to the frequency domain by a radix-2 FFT

$$\boldsymbol{Y}_k = \left[Y_k(0), Y_k(1), \ldots, Y_k(2K-1)\right]^{\mathrm{T}} = \mathrm{FFT}\{\boldsymbol{y}_k\}, \quad (11.18)$$

where $Y_k(\Omega)$ represents the discrete Fourier transform of $\boldsymbol{y}_k$ in the Ω-th frequency bin.

In order to combat the TDOA ambiguity caused by (nearly) periodic signals, a prewhitening step is necessary. A simple but effective method is the *phase transform* (PHAT) which normalizes $Y_k(\Omega)$ by its magnitude [10]. After this operation, the phase information of all frequency bins contribute with equal weight to the cross-correlation. This approach has one disadvantage:

In frequency bins with a low signal spectrum, the noise has a pretty large influence to the generalized cross-correlation. For uniformly distributed phase noise, an additional maximum at $n = 0$ will occur in the cross-correlation. In order to reduce the noise effect in these frequency bins, we propose to modify the original PHAT by incorporating an empirical limiting parameter $\Gamma_{\text{PHAT}} > 0$ into the estimate of the generalized cross power spectral density

$$\widehat{S}_{kl}(\Omega) = \frac{Y_k(\Omega)\, Y_l^*(\Omega)}{\max\left\{\left|Y_k(\Omega)\, Y_l^*(\Omega)\right|, \Gamma_{\text{PHAT}}\right\}}. \quad (11.19)$$

The choice of Γ_{PHAT} depends on the noise level and the block length K. For a normalized signal dynamic range of $[-1, 1]$, we used the value $\Gamma_{\text{PHAT}} = K \cdot 10^{-3}$ in moderately reverberant environments.

After the prewhitening step in Eq. 11.19, we compute the inverse fast Fourier transform (IFFT) of $\widehat{S}_{kl}(\Omega)$

$$\begin{aligned}&\left[\tilde{r}_{kl}(0),\, \tilde{r}_{kl}(1),\, \ldots,\, \tilde{r}_{kl}(K-1),\, 0,\, \tilde{r}_{kl}(-K+1),\, \ldots,\, \tilde{r}_{kl}(-1)\right]^T \\ &= \text{IFFT}\left\{\left[\widehat{S}_{kl}(0),\, \widehat{S}_{kl}(1),\, \ldots,\, \widehat{S}_{kl}(2K-1)\right]^T\right\}. \end{aligned} \quad (11.20)$$

Since the rectangular window and zero-padding in Eq. 11.17 cause a triangular windowing of the sample cross-correlation function, we compensate this scaling by

$$\hat{r}_{kl}(n) = \frac{1}{K - |n|}\, \tilde{r}_{kl}(n) \quad \forall\, |n| \leq n_{\max}, \quad (11.21)$$

where $n_{\max}$ is the largest TDOA value to be considered. This scaling compensation has to be adjusted if we use other windows than the rectangular one in Eq. 11.17.

Now we determine a set $\mathbb{P}_{kl}$ of "relevant" maximum positions $n_{kl,\sigma}$ from the estimated cross-correlation $\hat{r}_{kl}(n)$

$$\mathbb{P}_{kl} = \left\{n_{kl,\sigma}\, \middle|\, n_{kl,\sigma} = \arg\max_n{}^\sigma\{\hat{r}_{kl}(n)\},\, \sigma \in \{1, \ldots, \Sigma\}\right\}. \quad (11.22)$$

The notation $\arg\max_n{}^\sigma\{\ldots\}$ determines the position of the σ-th highest maximum. The number of maxima Σ should be considerably larger than the number of sources N. The term "relevant" means that $\mathbb{P}_{kl}$ hopefully contains the direct path TDOA of all sources. In our experiments, we choose the 15 highest local maxima whose correlation values are above 20 % of the global maximum. Of course, these empirical parameters need to be adjusted in other environments.

For simplicity, we consider integer delays and TDOA values throughout this chapter. By using interpolation of the discrete-time correlation functions [4,6], real-valued TDOAs with a higher estimation accuracy are possible. Our disambiguation approach is applicable to both integer and real-valued TDOAs.

Relevant extremum positions of the autocorrelations are extracted in a similar way as for the cross-correlations. The main difference is that PHAT is not applicable to autocorrelations since it would return $\hat{S}_{kk}(\Omega) = 1$. Instead, we assume that the main part of the signal spectrum of each source is received by all microphones in the same block. This motivates an average of the power spectral densities of all microphone signals for each frequency bin. The resulting generalized auto power spectral density of the k-th microphone signal is

$$\hat{S}_{kk}(\Omega) = \frac{Y_k(\Omega)\, Y_k^*(\Omega)}{\max\left\{\dfrac{1}{M}\sum_{k=1}^{M}|Y_k(\Omega)\, Y_k^*(\Omega)|,\, \Gamma_{\text{PHAT}}\right\}}. \tag{11.23}$$

Again, we use the limiting parameter $\Gamma_{\text{PHAT}} > 0$ to avoid a close-to-zero denominator. In analogy to Eqs. 11.20 to 11.22, the extremum positions $n_{kk,\sigma}$ of the autocorrelation $\hat{r}_{kk}(n)$ except for the zero lag are determined and collected in the set $\mathbb{P}_{kk}$. Due to the symmetry of the autocorrelation function, we only need to store the positive extremum positions.

11.3.2 Raster Matching

Besides the desired direct path TDOAs, it is very likely that the set $\mathbb{P}_{kl}$ determined in Eq. 11.22 also contains echo path TDOAs. These spurious maxima in the cross-correlation have to be identified and rejected using the autocorrelation extrema in $\mathbb{P}_{kk}$ and $\mathbb{P}_{ll}$. In a first attempt, the raster condition in Eq. 11.15 motivates a search for all pairs of cross-correlation TDOAs whose differences match an autocorrelation extremum position

$$n_\mu - n_\nu = n_\eta \quad \text{with} \quad n_\eta \in (\mathbb{P}_{kk} \cup \mathbb{P}_{ll}),\ n_\mu, n_\nu \in \mathbb{P}_{kl},\ n_\mu > n_\nu. \tag{11.24}$$

If such a pair has been found, that cross-correlation maximum associated to the arrowhead can be rejected immediately as Fig. 11.5 illustrates. To be more specific, if $n_\eta \in \mathbb{P}_{kk}$, n_μ can not be a direct path TDOA, and if $n_\eta \in \mathbb{P}_{ll}$, we remove n_ν from $\mathbb{P}_{kl}$. After repeating this procedure for all autocorrelation extrema n_η from $\mathbb{P}_{kk} \cup \mathbb{P}_{ll}$, we hope that $\mathbb{P}_{kl}$ will finally contain all N direct path TDOAs and no echo path TDOAs. In other words, there is no *false detection* (echo path TDOA accepted) and no *miss detection* (direct path TDOA rejected).

Unfortunately, this is not true in the reality. A number of practical problems cause both false and miss detection. This makes it necessary to develop a more sophisticated strategy to distinguish between the direct path and echo path TDOAs. In general, a miss detection is more critical than a false detection. While a rejected direct path TDOA is lost for ever, echo path TDOAs which have not been rejected yet can still be identified during the subsequent steps like zero cyclic sum matching, see next section. Below we describe the main practical problems and discuss some implementation issues.

First of all, the raster condition 11.15 is necessary but not sufficient. It is possible that two cross-correlation maxima from different paths or different sources appear in a distance which is identical to one of the autocorrelation extremum positions. If we erroneously reject a direct path TDOA too early based on the matching of only one pair of cross-correlation maxima, this direct path TDOA is lost in all future steps resulting in a miss detection.

In order to prevent this from happening, we propose to first count all direct path and echo path hits according to the number of arrowheads and arrow tails, respectively, see Fig. 11.7. Obviously, those cross-correlation maxima with a high number of direct path hits and a low number of echo path hits are promising candidates for the direct path TDOAs.

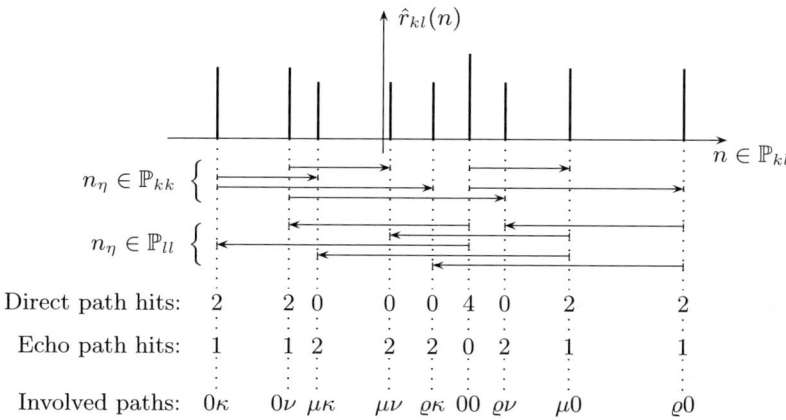

Fig. 11.7. Number of direct path and echo path hits in a scenario with one direct and two echo paths between one source and each of the two microphones. In this case, the desired direct path TDOA is that cross-correlation maximum position with zero echo path hits.

Second, due to sampling, noise effects, finite block length, and imperfect prewhitening of the source signals, the estimation of the maximum/extremum positions in the cross- and autocorrelation is not perfect. In particular, the raster condition 11.15 is not satisfied exactly for quantized time delays. Hence it is not a good idea to make a hard decision based on an exact raster match.

We introduce a soft decision instead of a hard decision. For each cross-correlation maximum $n_\mu \in \mathbb{P}_{kl}$, we define a quality measure $q(n_\mu)$. Its initial value is the positive cross-correlation amplitude $\hat{r}_{kl}(n_\mu)$. This quality value will be increased or decreased during the subsequent steps. Its final value determines whether n_μ represents a direct path or echo path TDOA. In particular, in order to cope with quantized delays and other estimation errors, we introduce a positive *tolerance function of raster match* (TFRM) $\Gamma_{\text{TFRM}}(n)$, see Fig. 11.8. Its width, the *tolerance width of raster match* (TWRM) Γ_{TWRM}, is typically in the order of a few samples. Its shape determines how smoothly the quality measure of a TDOA candidate decreases from a perfect raster match to a less accurate match.

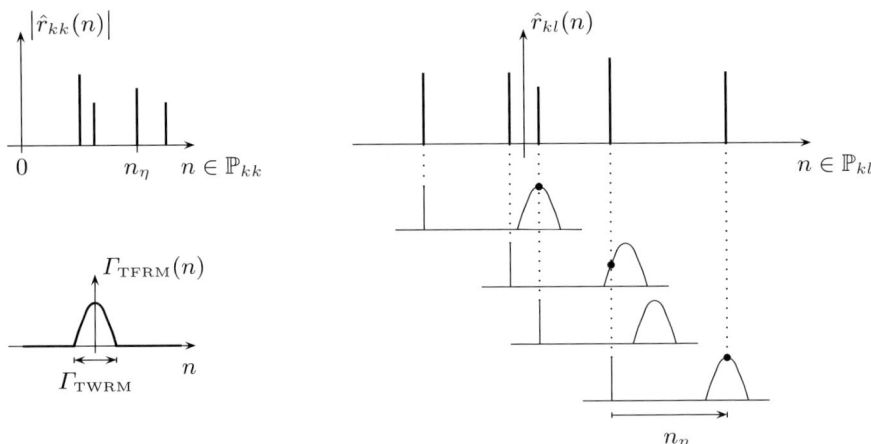

Fig. 11.8. Soft raster matching by using a tolerance function $\Gamma_{\text{TFRM}}(n)$. We look for pairs of cross-correlation maxima from $\mathbb{P}_{kl}$ whose distance matches an autocorrelation extremum n_η in $\mathbb{P}_{kk}$. The amplitude of the black dot • represents a quality measure of the match. If there is no black dot, there is no match at all.

Each time when we find a (soft) raster match for n_μ

$$n_\eta \approx |n_\mu - n_\nu|, \quad n_\eta \in (\mathbb{P}_{kk} \cup \mathbb{P}_{ll}),\, n_\mu, n_\nu \in \mathbb{P}_{kl} \qquad (11.25)$$

where n_μ is assigned to the arrow tail (likely a direct path TDOA), we increase the quality value $q(n_\mu)$ by the magnitude of the autocorrelation amplitude $|\hat{r}_{kk}(n_\eta)|$ or $|\hat{r}_{ll}(n_\eta)|$ which is involved in the raster match. If n_μ is assigned to the arrowhead (likely an echo path TDOA), $q(n_\mu)$ is decreased by the same value. In addition, we take a non-perfect match $n \neq 0$ into account and scale the correction value $|\hat{r}_{kk}(n_\eta)|$ or $|\hat{r}_{ll}(n_\eta)|$ by the tolerance function $\Gamma_{\text{TFRM}}(n)$. In total, we calculate the quality value $q(n_\mu)$ of the cross-correlation maximum n_μ by

$$q(n_\mu) = \hat{r}_{kl}(n_\mu) + \sum_{n_\eta \in \mathbb{P}_{kk}^+} |\hat{r}_{kk}(n_\eta)|\, \Gamma_{\text{TFRM}}(n_\eta + n_\mu - n_\nu)$$

$$- \sum_{n_\eta \in \mathbb{P}_{kk}^-} |\hat{r}_{kk}(n_\eta)|\, \Gamma_{\text{TFRM}}(n_\eta - n_\mu + n_\nu)$$

$$+ \sum_{n_\eta \in \mathbb{P}_{ll}^+} |\hat{r}_{ll}(n_\eta)|\, \Gamma_{\text{TFRM}}(n_\eta - n_\mu + n_\nu)$$

$$- \sum_{n_\eta \in \mathbb{P}_{ll}^-} |\hat{r}_{ll}(n_\eta)|\, \Gamma_{\text{TFRM}}(n_\eta + n_\mu - n_\nu) \qquad (11.26)$$

with

$$\mathbb{P}_{kk}^+ = \left\{ n_\eta \in \mathbb{P}_{kk} \,\middle|\, |n_\eta + n_\mu - n_\nu| < 0.5\,\Gamma_{\text{TWRM}} \text{ and } n_\mu < n_\nu \right\},$$

$$\mathbb{P}_{kk}^- = \left\{ n_\eta \in \mathbb{P}_{kk} \,\middle|\, |n_\eta - n_\mu + n_\nu| < 0.5\,\Gamma_{\text{TWRM}} \text{ and } n_\mu > n_\nu \right\},$$

$$\mathbb{P}_{ll}^+ = \left\{ n_\eta \in \mathbb{P}_{ll} \,\middle|\, |n_\eta - n_\mu + n_\nu| < 0.5\,\Gamma_{\text{TWRM}} \text{ and } n_\mu > n_\nu \right\},$$

$$\mathbb{P}_{ll}^- = \left\{ n_\eta \in \mathbb{P}_{ll} \,\middle|\, |n_\eta + n_\mu - n_\nu| < 0.5\,\Gamma_{\text{TWRM}} \text{ and } n_\mu < n_\nu \right\},$$

$$n_\nu \in \mathbb{P}_{kl}.$$

For each n_μ, the sets $\mathbb{P}_{kk}^+$ and $\mathbb{P}_{kk}^-$ contain those autocorrelation extremum positions from $\mathbb{P}_{kk}$ which match to a pair of cross-correlation TDOAs n_μ and n_ν. While $\mathbb{P}_{kk}^+$ denotes matches in which n_μ is assigned to an arrow tail, $\mathbb{P}_{kk}^-$ denotes matches in which n_μ corresponds to an arrowhead. $\mathbb{P}_{ll}^+$ and $\mathbb{P}_{ll}^-$ are defined in a similar way.

The final decision about the cross-correlation TDOA n_μ is based on the final value of $q(n_\mu)$:

$$n_\mu \text{ is } \begin{cases} \text{a direct path TDOA,} & \text{if } q(n_\mu) > t_{kl} \\ \text{an echo path TDOA,} & \text{otherwise.} \end{cases} \qquad (11.27)$$

All accepted TDOAs are collected in a reduced set $\mathbb{P}'_{kl}$. We used the threshold

$$t_{kl} = \min_{n_\nu \in \mathbb{P}_{kl}} \{r_{kl}(n_\nu)\} \qquad (11.28)$$

in Eq. 11.27. This choice is intentionally conservative in order to ensure that no direct path TDOAs are rejected at this early step. Echo path TDOAs which remain in $\mathbb{P}'_{kl}$ can be identified at the next step, the zero cyclic sum matching in the next section.

Note that, though the raster matching greatly reduces the TDOA ambiguity in cross-correlations, it is not able to resolve all ambiguities. A few difficult situations are given below:

- The raster condition in Eq. 11.15 is based on the assumption that each autocorrelation extremum used in the raster matching involves the direct path. If an autocorrelation extremum used in the raster matching arises from two echo paths, the relationship

$$n_{kk,\mu\nu} = n_{kl,\mu\eta} - n_{kl,\nu\eta}, \quad \mu \neq 0, \nu \neq 0. \tag{11.29}$$

holds. In this case, both quality values $q(n_{kl,\mu\eta})$ and $q(n_{kl,\nu\eta})$ have to be decreased since both $n_{kl,\mu\eta}$ and $n_{kl,\nu\eta}$ are echo path TDOAs. In our algorithm, however, one of the quality values will be decreased (arrowhead) and the other one will be increased (arrow tail) since we assume that one of the two (here path ν) is a direct path.
- If an echo path is involved in $\mathbb{P}_{kl}$ but neither in $\mathbb{P}_{kk}$ or in $\mathbb{P}_{ll}$, the raster matching would not work.
- It can happen that different sources may produce identical direct or echo path cross-correlation maxima. They sum to a larger maximum which can be associated to both arrowhead and arrow tail. Fig. 11.9 shows such a situation where a direct path maximum of source a overlaps an echo path maximum of source b (bold line). In this case, a simple binary decision about that maximum is impossible.

Nevertheless, these cases rarely occur in practice as we observed in our experiments.

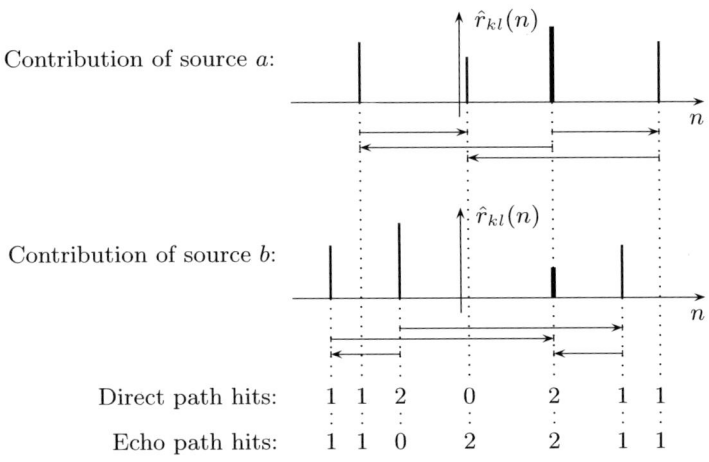

Fig. 11.9. Number of direct path and echo path hits for two sources with overlapping cross-correlation maxima.

11.4 Consistent TDOA Graphs

Starting from the sets $\mathbb{P}'_{kl}$ of direct path TDOA candidates for all microphone pairs determined in the previous section, we now apply the zero cyclic sum matching (Eq. 11.16) to estimate the source TDOA vectors $\hat{\boldsymbol{\tau}}_\sigma$. Clearly, some combinations of TDOA estimates satisfy the zero cyclic sum condition and the others not. The latter are caused by different propagation paths or different sources. We study this combination problem in the framework of consistent graphs. After a brief introduction into TDOA graphs, we discuss different strategies for consistency check, describe properties of TDOA graphs, and present a novel approach for the synthesis of consistent graphs as an efficient implementation of the zero cyclic sum matching [16].

11.4.1 TDOA Graph

As shown in Fig. 11.10, the content of a source TDOA vector $\boldsymbol{n}_a$ can be visualized by a weighted directed graph. It is called a *TDOA graph*. Each node represents a microphone and each directed edge between two nodes has a weight corresponding to the TDOA between these two microphones. The edge direction is given by the order of microphones in the cross-correlation. A change of the edge direction has the effect of a sign change of its weight. Again we use integer TDOA values for simple illustration. The concept of TDOA graphs applies to real-valued weights as well.

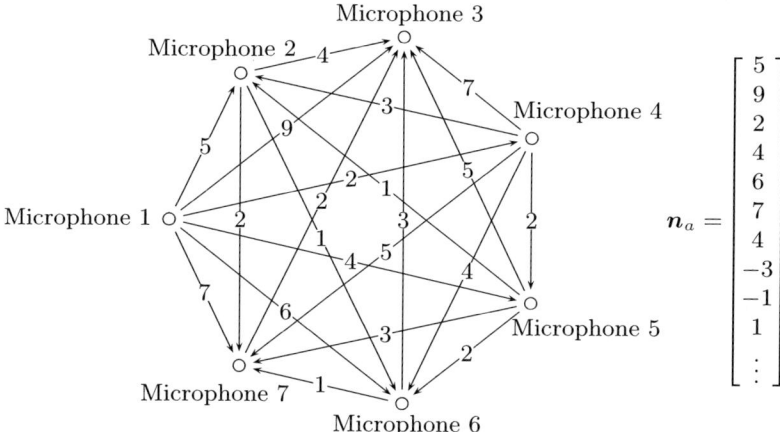

Fig. 11.10. A fully connected consistent TDOA graph with 7 nodes and the related source TDOA vector $\boldsymbol{n}_a$.

TDOA graphs have the topology of Hamilton graphs, see Fig. 11.11. In a *Hamilton graph*, there always exists a closed path through the graph containing each node once. This closed path is known as a *Hamilton cycle*. A fully connected graph corresponds to a source TDOA vector of length $\binom{M}{2}$. For partially connected graphs, some vector elements are missing. The aim of our synthesis algorithm is to compose Hamilton graphs that include the highest number of nodes and are maximally connected.

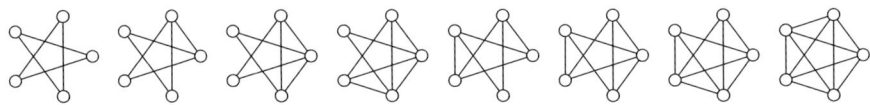

Fig. 11.11. All possible topologies of Hamilton graphs with 5 nodes [12].

A TDOA graph consisting of exact TDOA values (not estimates) is always *consistent* in the sense that the sum of all edge weights along any closed path is zero according to the zero cyclic sum condition (11.16). This is very similar to Kirchhoff's second law valid for electrical circuits (voltage graphs) except that we now replace voltages by TDOA values.

11.4.2 Strategies of Consistency Check

Below we analyze the complexity of different strategies to check the consistency of a given TDOA graph. We assume a fully connected graph with M nodes and $\binom{M}{2}$ edges. For simplicity, we count each addition or comparison as one operation.

In analogy to electrical voltage and potential, we can define a *time potential* at each node as the time of arrival with respect to a reference node. Let the time potential of the reference node be zero. The time potentials of the remaining $M-1$ nodes are then equal to their weights relative to the reference node. In order to check the consistency of the graph, we only need to compare the $\binom{M-1}{2}$ edge weights not involving the reference node with the corresponding potential differences. This leads to

$$C_{\text{pair}} = 2 \cdot \binom{M-1}{2} = (M-1)(M-2) \tag{11.30}$$

operations. It can be shown that this is also the minimum complexity of any consistency check.

Alternatively, we can analyze all $\binom{M}{k}$ subgraphs of k nodes with $k \geq 3$. Since there are $\frac{(k-1)!}{2}$ different closed paths combining the k nodes and each path causes $(k-1)$ operations for consistency check, this approach requires

$$C_{k\text{-tup}} = \binom{M}{k} \cdot \frac{(k-1)!}{2} \cdot (k-1) = \frac{k-1}{2k} M(M-1) \cdots (M-k+1) \tag{11.31}$$

operations. Interestingly, only a subset of the $\binom{M}{k}$ subgraphs are *independent* in the sense that they have to be checked for consistency separately. The consistency of the remaining subgraphs follows immediately from the consistency of these independent subgraphs.

A simple special case is $k = 3$. Here we only need to check the consistency of $\binom{M-1}{2}$ *independent triples*, see Fig. 11.12. This reduces the number of operations to the minimum

$$C_{\text{triple}} = \binom{M-1}{2} \cdot \frac{(k-1)!}{2} \cdot (k-1)\bigg|_{k=3} = (M-1)(M-2) = C_{\text{pair}}. \tag{11.32}$$

The advantage of this independent triple approach in comparison to time potential is the flexibility that there is no need to define a reference node. Instead of the six independent triples in Fig. 11.12 sharing a common reference node, we can also check the six independent triples in Fig. 11.13 without a reference node. The remaining $\binom{M}{3} - \binom{M-1}{2} = \binom{M-1}{3}$ dependent triples need not to be considered.

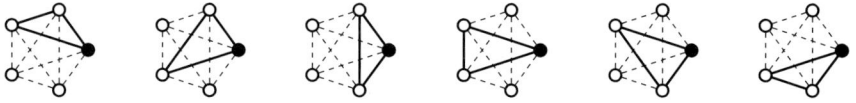

Fig. 11.12. Splitting a fully connected graph of 5 nodes into 6 independent triples by using a common reference node •.

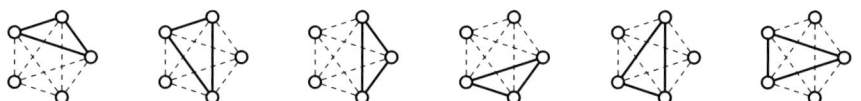

Fig. 11.13. Splitting a fully connected graph of 5 nodes into 6 independent triples without a reference node.

11.4.3 Properties of TDOA Graphs

In multiple source TDOA estimation, the aim is not the analysis but the synthesis of consistent TDOA graphs starting from the sets $\mathbb{P}'_{kl}$ of TDOA estimates. In the ideal case, the cardinal number $|\mathbb{P}'_{kl}|$ is equal to the number of sources N and each TDOA estimate from $\mathbb{P}'_{kl}$ is equal to one of the true direct path TDOAs. In reality, TDOA estimates might not exactly match their true values. Some true TDOAs may be lost (miss detection) due to other strong sources close to the microphones and some wrong TDOA estimates may have been produced by echo paths or other measurement errors (false detection). For different microphone pairs (k, l), $\mathbb{P}'_{kl}$ may have different cardinal numbers.

As it is not ensured that all N true TDOAs are contained in $\mathbb{P}'_{kl}$, the TDOA graph can be incomplete. In addition to the $|\mathbb{P}'_{kl}|$ possible edge weights, the edge between node k and l can be missing as well. By taking these $|\mathbb{P}'_{kl}| + 1$ possibilities for the microphone pair (k, l) into account, a brute force synthesis would need to check all

$$C_{\text{bf}} = \prod_{k=1}^{M-1} \prod_{l=k+1}^{M} \left(|\mathbb{P}'_{kl}| + 1\right) \tag{11.33}$$

possible graphs for consistency. For $M = 8$ microphones and $|\mathbb{P}'_{kl}| = 7$ TDOA estimates for each microphone pair (k, l), the total number of possible graphs is $C_{\text{bf}} = 8^{28} \approx 2 \cdot 10^{25}$. This is unacceptable for real-time applications.

One possibility to reduce the complexity is the use of time potentials as introduced in the previous subsection. This is, however, not practical as illustrated in Fig. 11.14 where two sources a and b result in two incomplete TDOA graphs. The missing edges are assumed to be not detectable. By considering node 1 as the reference node • for defining time potential, the graph corresponding to source a can be synthesized completely from TDOA estimates. In contrast, only the bold subgraph containing six edges for source b at the right hand side can be synthesized because the other nodes are not connected to the reference node at all. The situation will even become worse if we use other reference nodes than node 1. For this reason, we do not follow the time potential approach. Instead, we propose a synthesis algorithm based on consistent triples.

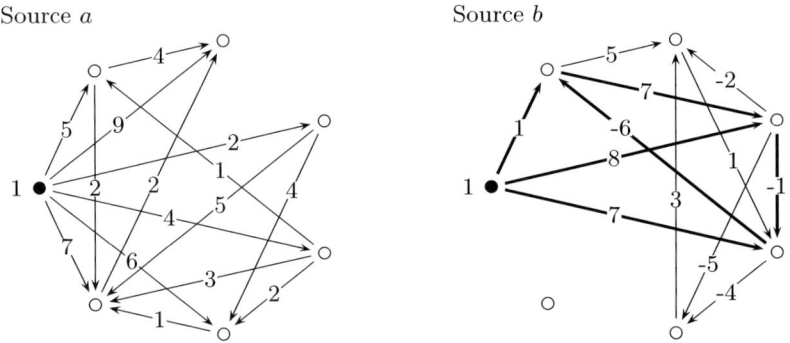

Fig. 11.14. Incomplete TDOA graphs make a choice of a common reference node for all sources difficult.

Note that besides the desired direct path TDOA graphs, other combinations of TDOAs can also form a consistent graph. There are two reasons for this phenomenon of *misleading consistency*. First, the zero cyclic sum condition 11.16 is necessary but not sufficient for TDOAs originating from a common source. Scenarios are possible, where TDOAs of different sources a, b, and c satisfy

$$n_{a,kl,00} + n_{b,lm,00} + n_{c,mk,00} = 0. \qquad (11.34)$$

Of course, the probability of this occurrence is quite small for randomly distributed sources. In practice, sound reflections frequently lead to misleading consistency because the zero cyclic sum condition is also fulfilled even if an echo path $\mu \neq 0$ is involved

$$n_{a,kl,0\mu} + n_{a,lm,\mu 0} + n_{a,mk,00} = 0 \quad (\mu \neq 0). \qquad (11.35)$$

Typically, microphone l is close to a wall. When we model sound propagation and reflection by acoustic rays like the image source method [2], a reflecting wall has the same effect on a microphone signal as a corresponding mirrored microphone, see Fig. 11.15. Clearly, both the direct path graph b) and the graph c) in Fig. 11.15 containing two echo path TDOAs are consistent. They can not be distinguished by the zero cyclic sum condition 11.16. We will see in the next section that the residual error in the source position estimation can help to resolve this ambiguity.

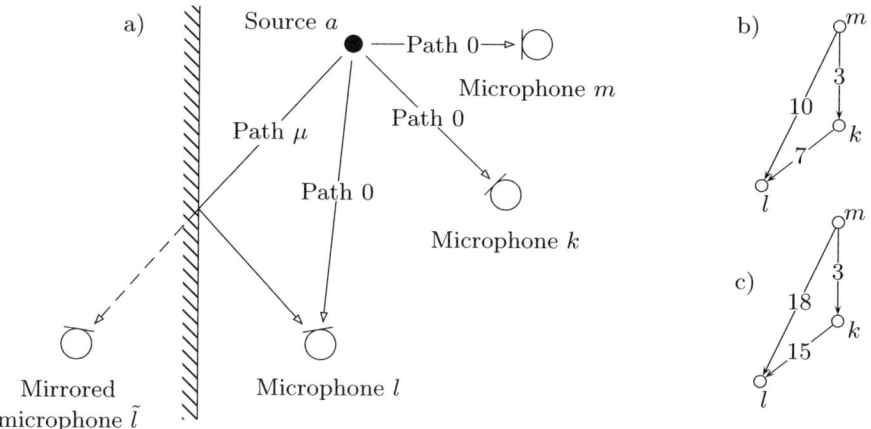

Fig. 11.15. A typical scenario of sound reflection where TDOA ambiguity occurs. Both paths 0 and μ to microphone l produce TDOAs which form consistent graphs.

Second, since TDOA estimates are derived from sampled and noisy microphone signals, the zero cyclic sum condition is only approximately fulfilled. We will accept a deviation of the zero cyclic sum

$$\left| n_{kl,\mu_1} + n_{lm,\mu_2} + \ldots + n_{pq,\mu_{\kappa-1}} + n_{qk,\mu_\kappa} \right| < \frac{\Gamma_{\text{TWTM}}}{2} \qquad (11.36)$$

where the *tolerance width of triple match* (TWTM) Γ_{TWTM} is in the order of a few samples. The choice of Γ_{TWTM} depends on both the magnitude of the TDOA estimation errors and the path length κ. In order to keep Γ_{TWTM} as small as possible, short paths are preferred for consistency check. In analogy to the tolerance function of raster match $\Gamma_{\text{TFRM}}(n)$ in Sec. 11.3, we also introduce a positive, smoothly decreasing *tolerance function of triple match* (TFTM) $\Gamma_{\text{TFTM}}(n)$ to measure how good the zero cyclic sum condition is satisfied for a TDOA triple.

11.4.4 Efficient Synthesis Algorithm

Our strategy to synthesize approximately consistent TDOA graphs is based on triples. For each microphone triple (k, l, m), let $\mathbb{T}_{klm}$ denote the set of approximately consistent TDOA triples

$$\left(n_{kl,\mu}, n_{lm,\nu}, n_{mk,\eta}\right) \quad \text{with} \quad \left|n_{kl,\mu} + n_{lm,\nu} + n_{mk,\eta}\right| < \frac{\Gamma_{\text{TWTM}}}{2}.$$

In total, the number of TDOA triples that need to be checked is

$$\sum_{k=1}^{M-2} \sum_{l=k+1}^{M-1} \sum_{m=l+1}^{M} \left|\mathbb{P}'_{kl}\right| \cdot \left|\mathbb{P}'_{lm}\right| \cdot \left|\mathbb{P}'_{mk}\right|. \tag{11.37}$$

This number can be reduced if the TDOA sets are stored as sorted lists. Those TDOA estimates from $\mathbb{P}'_{kl}, \mathbb{P}'_{lm}$, and $\mathbb{P}'_{mk}$ which do not contribute to consistent triples in $\mathbb{T}_{klm}$, will not be further considered. Since typically

$$\left|\mathbb{T}_{klm}\right| \ll \left|\mathbb{P}'_{kl}\right| \cdot \left|\mathbb{P}'_{lm}\right| \cdot \left|\mathbb{P}'_{mk}\right|, \tag{11.38}$$

the complexity of our synthesis algorithm is significantly reduced.

Starting with an initial TDOA triple $(n_{kl,\mu_1}, n_{lm,\nu_1}, n_{mk,\eta_1})$ from $\mathbb{T}_{klm}$, we now consider an additional microphone $o \in \{1, \ldots, M\} \setminus \{k, l, m\}$ and try to extend the TDOA triple to a consistent TDOA quadruple involving the four microphones k, l, m, o. We search for at least two other TDOA triples in the new sets $\mathbb{T}_{lmo}, \mathbb{T}_{mok}$, and $\mathbb{T}_{okl}$ with pairwise common edge weights. If, for example, the triples $(n_{mo,\sigma_2}, n_{ok,\varrho_2}, n_{km,\eta_2}) \in \mathbb{T}_{mok}$ and $(n_{ok,\varrho_3}, n_{kl,\mu_3}, n_{lo,\kappa_3}) \in \mathbb{T}_{okl}$ have common edge weights

$$n_{kl,\mu_1} = n_{kl,\mu_3}, \quad n_{mk,\eta_1} = -n_{km,\eta_2}, \quad n_{ok,\varrho_2} = n_{ok,\varrho_3},$$

we build a fully connected consistent TDOA quadruple by combining these three triples, see Fig. 11.16.

We repeat the synthesis of quadruple for any of the $M-3$ nodes $\{1, \ldots, M\} \setminus \{k, l, m\}$ which are not contained in the initial triple. All fourth nodes $o, p, \ldots$ for which a complete consistent quadruple can be successfully constructed are collected in a new set $\mathbb{K}$ with $|\mathbb{K}| \leq M-3$. Clearly, these consistent quadruples

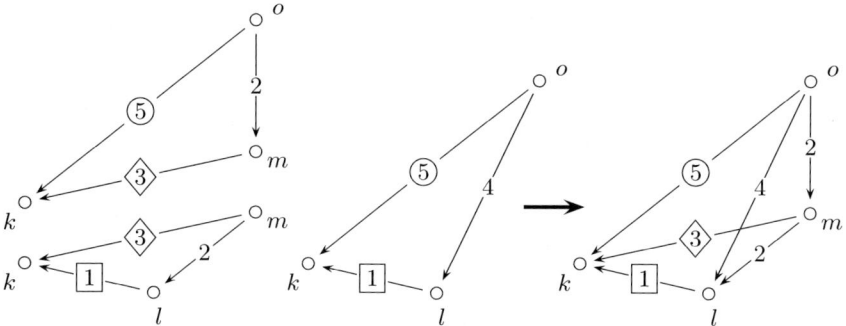

Fig. 11.16. Three consistent triples with pairwise common edge weights are combined to a consistent quadruple.

with the same initial triple (k, l, m) form a consistent but not fully connected *star graph* as shown on the left hand side of Fig. 11.17 for $|\mathbb{K}| = 2$. The missing $\binom{|\mathbb{K}|}{2}$ edges among the $|\mathbb{K}|$ new nodes can be completed by triples which have two edges in common with the star graph. On the right hand side of Fig. 11.17, one such completing triple from $\mathbb{T}_{mop}$ is shown. Obviously, any other matching triple from $\mathbb{T}_{lop}$ or $\mathbb{T}_{kop}$ can also be used for this purpose.

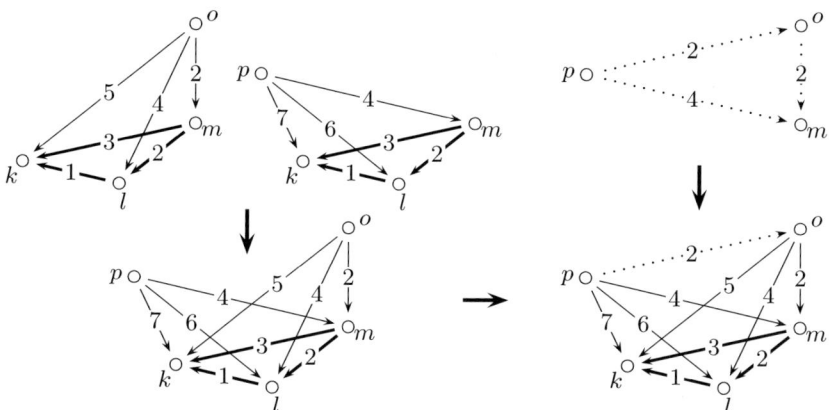

Fig. 11.17. Combining quadruples with a common initial triple (bold line) will result in a star graph. The star graph is completed by matching triples (dotted line).

If all sets of TDOA triples $\mathbb{T}$ contain only the direct path triples of all sources, any choice of the initial triple will result in a complete TDOA graph for a particular source. But if the sets $\mathbb{T}$ also contain echo path triples like in Fig. 11.15 c), different star graphs for the same initial triple are possible. Fig. 11.18 illustrates this phenomenon for $M = 6$ nodes. Starting with the boldface initial triple, four quadruples have been synthesized for the remaining 3 fourth nodes o, p, q. While the fourth nodes p and q each produce only one quadruple, the first two quadruples caused by the fourth node o may originate from the true microphone and its mirror. At this position, we are not able to decide which one is the correct one. Hence we accept all four quadruples and combine them into two star graphs as shown in the bottom row of Fig. 11.18. The final step is to complete these star graphs by looking for triples which connect the nodes o, p, q.

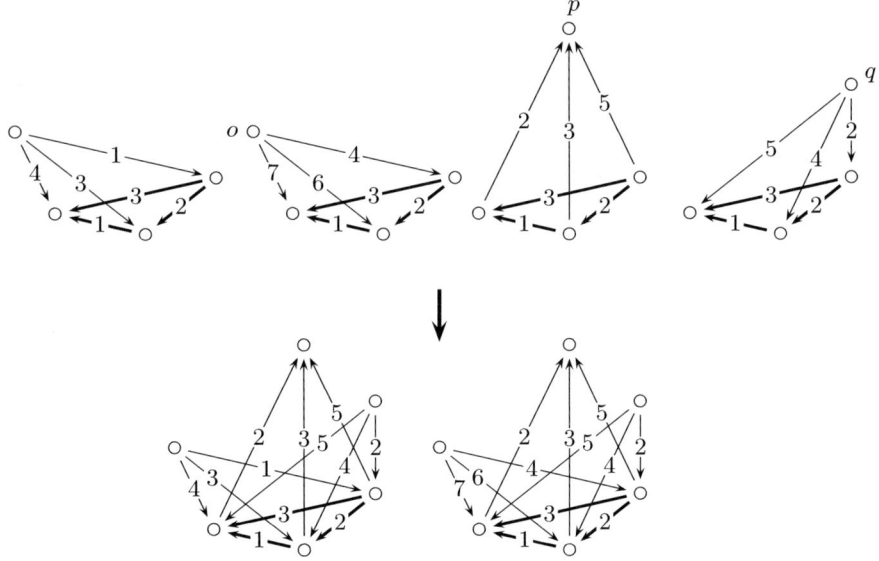

Fig. 11.18. When combining quadruples to a star graph, echo path triples result in different ambiguous star graphs.

One important feature of our synthesis algorithm is that no further consistency check is necessary during the synthesis process because each closed path and each subgraph of the resulting TDOA graph are (approximately) consistent by construction.

11.4.5 Initialization and Termination

The iterative synthesis process, i.e. extending an initial TDOA triple to a larger TDOA graph, is terminated if either the TDOA graph is already complete or no TDOA triples can be found to complete the star graph. All triples used in a synthesized TDOA graph σ are summarized in a set $\mathbb{G}_\sigma$ with $|\mathbb{G}_\sigma| \leq \binom{M}{3}$.

For multiple sources, the search for a new graph has to be initialized. In order to avoid the synthesis of identical graphs, those triples which have already been used in existing TDOA graphs are discarded from further consideration as initial triple. But we still use them to extend the next initial triple to a larger graph for a quite simple reason. The first graph we have synthesized might contain echo path TDOAs and thus will not be useful for localization. We should have a chance to reuse the triples contained in the first graph for a second attempt; otherwise these triples are lost for ever. The complete synthesis algorithm is terminated, if each triple has been used in a TDOA graph or if the remaining triples can not be combined even to quadruples. These isolated triples are rejected since a three-dimensional source localization requires at least four microphones.

The choice and the order of the initial triples have a great influence on the number and quality of the synthesized TDOA graphs and the speed of convergence. A high quality direct path initial triple would integrate much more other triples into a highly connected graph than a poor initial triple. The number of remaining triples is reduced significantly and the convergence is fast. We observed in our experiments that this is particularly the case if the initial triple has a small deviation from the zero cyclic sum and high quality values of TDOA estimates. For this reason, we define a quality value

$$q_\mathrm{T} = \Gamma_\mathrm{TFTM}\bigl(n_{kl,\mu}+n_{lm,\nu}+n_{mk,\eta}\bigr) \cdot \Bigl[q(n_{kl,\mu}) + q(n_{lm,\nu}) + q(n_{mk,\eta})\Bigr] \tag{11.39}$$

for each TDOA triple $(n_{kl,\mu}, n_{lm,\nu}, n_{mk,\eta})$. The initial triple is then that unused one with the highest triple quality q_T.

11.4.6 Estimating the Number of Active Sources

In practice, the above described synthesis algorithm returns a larger number of TDOA graphs than the number of active sources, mainly due to residual echo path triples. Typically, TDOA graphs corresponding to true source positions are highly connected while erroneous graphs caused by echo path triples have a small number of nodes and edges. This motivates the introduction of the *connectivity* w for each synthesized graph. It measures the degree of connection of the graph and how good the zero cyclic sum condition is satisfied for each valid triple

$$w = \sum_{(n_{kl,\mu},n_{lm,\nu},n_{mk,\eta})\in\mathbb{G}_\sigma} \Gamma_\mathrm{TFTM}\bigl(n_{kl,\mu} + n_{lm,\nu} + n_{mk,\eta}\bigr). \tag{11.40}$$

The maximum value of w is $\binom{M}{3} \cdot \max_n \{\Gamma_\mathrm{TFTM}(n)\}$. The larger w is, the higher the connectivity of the graph is. Our experiments show a significant gap in w between correct and erroneous TDOA graphs in most cases. Hence the number of high-connectivity graphs can be used to estimate the number of active sources if it is unknown.

Nevertheless, if a microphone is located close to a highly reflecting wall, an echo path graph caused by the mirrored microphone might show approximately the same graph connectivity as its corresponding direct path TDOA graph. In this case, a disambiguation is difficult on the graph level. As the position of the mirrored microphone does not match the true microphone position which is used in source position estimation, a mirrored microphone graph usually leads to a larger residual position error. This is an additional mean to identify echo path TDOA graphs on the level of source position estimation.

11.5 Experimental Results

We evaluated the proposed algorithms for TDOA estimation in a real-time demonstration system for multiple speaker localization. After a short introduction into the system, we describe in details the behavior and performance of our algorithms based on real experiments.

11.5.1 Localization System

Speech sources are recorded in a small rectangular acoustic lab of the size $4\,\text{m} \times 2\,\text{m} \times 2\,\text{m}$. The floor and the wooden walls are covered by a thin carpet. The ceiling is an acrylic glass. Fig. 11.19 shows a measured room impulse response. It shows a large number of echos. The reverberation time is $T_{60} \approx 300\,\text{ms}$. Due to fans and illumination, there is a weak background noise. The signal-to-noise ratio (SNR) is roughly 50 dB. For reproducible experiments, the speech signals are played back from loudspeakers instead of human speakers.

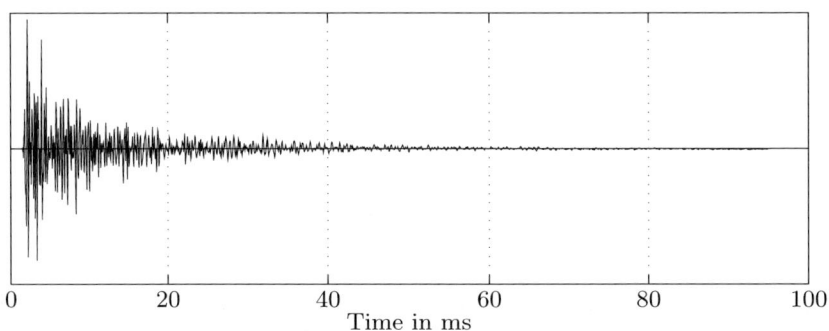

Fig. 11.19. Room impulse response measured in the acoustic lab.

Eight capacitive microphones and two loudspeakers are placed arbitrarily in the room. The positions of the microphones are

$$\begin{bmatrix}3.71\\0.49\\1.59\end{bmatrix}, \begin{bmatrix}3.69\\1.21\\1.57\end{bmatrix}, \begin{bmatrix}2.61\\0.01\\0.17\end{bmatrix}, \begin{bmatrix}1.86\\0.75\\1.68\end{bmatrix},$$
$$\begin{bmatrix}1.85\\1.63\\1.25\end{bmatrix}, \begin{bmatrix}0.82\\0.29\\1.78\end{bmatrix}, \begin{bmatrix}0.39\\0.40\\1.03\end{bmatrix}, \begin{bmatrix}0.30\\1.76\\0.97\end{bmatrix} \quad (11.41)$$

in meter. The positions of source a and b are

$$\boldsymbol{p}_a = \begin{bmatrix}1.67\\1.66\\0.71\end{bmatrix}, \quad \boldsymbol{p}_b = \begin{bmatrix}2.72\\0.65\\1.25\end{bmatrix}. \quad (11.42)$$

They are, together with their directions, illustrated in Fig. 11.20. The microphone signals are synchronously sampled at 96 kHz and processed by a Linux PC (kernel 2.6, dual core CPU at 2.8 GHz). Taking this high sampling rate and the sound speed of 343 m/s at the room temperature 20°C into account, one sampling interval corresponds to a range quantization of 3.6 mm.

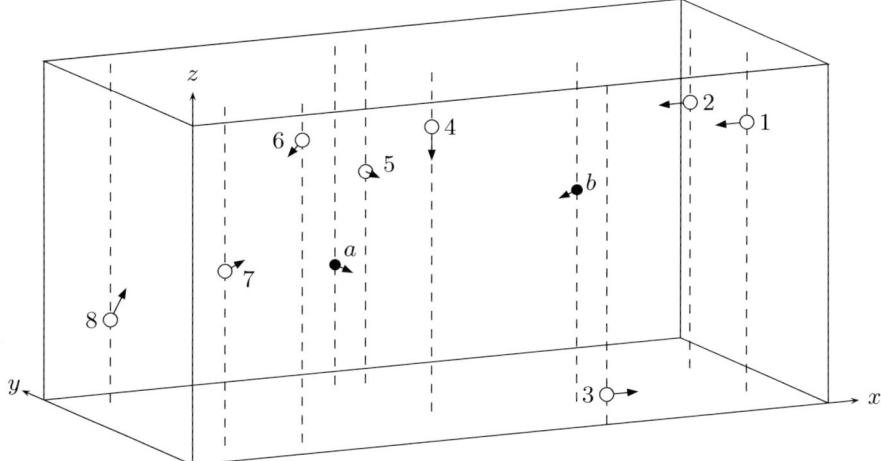

Fig. 11.20. Position and direction of microphones (○) and sources (●) in the lab.

The real-time ability of the system is mainly determined by its latency and throughput. Due to block processing, the latency time is determined by the block length. In order to reduce the overall latency time, we implemented a short data pipeline containing eight data buffers. The computation time for the cross- and autocorrelations of the microphone signals is fixed for one signal block. The time required for the raster matching and synthesis of consistent graphs, however, depends highly on the number of accepted cross-correlation maxima. It varies from block to block. Occasionally, we skip the processing of individual blocks if the processor load is too high. This is acceptable as long as the rate of evaluated blocks is still high enough. In particular, the blocks without speech activities need not be processed. We used a voice activity detector (VAD) for this purpose.

The experiments are based on a block length of $K = 4096$ samples. This corresponds to a time interval of approximately 43 ms. Due to the high sampling rate, we did not perform interpolation of the discrete-time correlations. The limiting parameter Γ_{PHAT} for the phase transform in Eqs. 11.19 and 11.23 is chosen to be $K \cdot 10^{-3} \approx 4$. The tolerance functions used are depicted in Fig. 11.21. Since we do not perform TDOA interpolation, these tolerance functions need only be defined for integer arguments n.

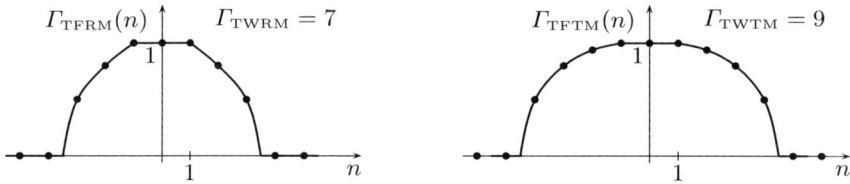

Fig. 11.21. Tolerance functions used for echo path detection (left side) and approximate triple consistency (right side).

11.5.2 TDOA Estimation of a Single Signal Block

In the following, one signal block of length 43 ms is analyzed. The eight microphone signals are shown in Fig. 11.22. During this time, the source signals contain a fricative [f] of one speaker and a diphthong [au] of the other speaker.

We calculated $M = 8$ autocorrelation and $\binom{M}{2} = 28$ cross-correlation signals. The total number of cross-correlation maxima we selected for further study is $28 \cdot 15 = 420$. After applying the raster matching as described in section 11.3, the number of direct path TDOA candidates is reduced to 148. The exact number of desired TDOAs for two sources is 56. The reason for our intentional overestimation is the conservative detection in Eq. 11.27 in order to avoid miss detections.

As an example, we consider below the cross-correlation between microphone 1 and 4. The microphones have a distance of 1.87 m, corresponding to an upper bound of $n_{\max} = 523$ samples for the TDOA. From the source and microphone positions, we calculated the true direct path TDOAs. They are 25.8 and 326.7 samples. The cross-correlation $\hat{r}_{14}(n)$ is shown in Fig. 11.23. Its 15 highest maxima are at the positions

$$\mathbb{P}_{14} = \{-81, -31, -4, 21, 48, 109, 162, 188, 267, 327, 337, 347, 358, 438, 448\}. \tag{11.43}$$

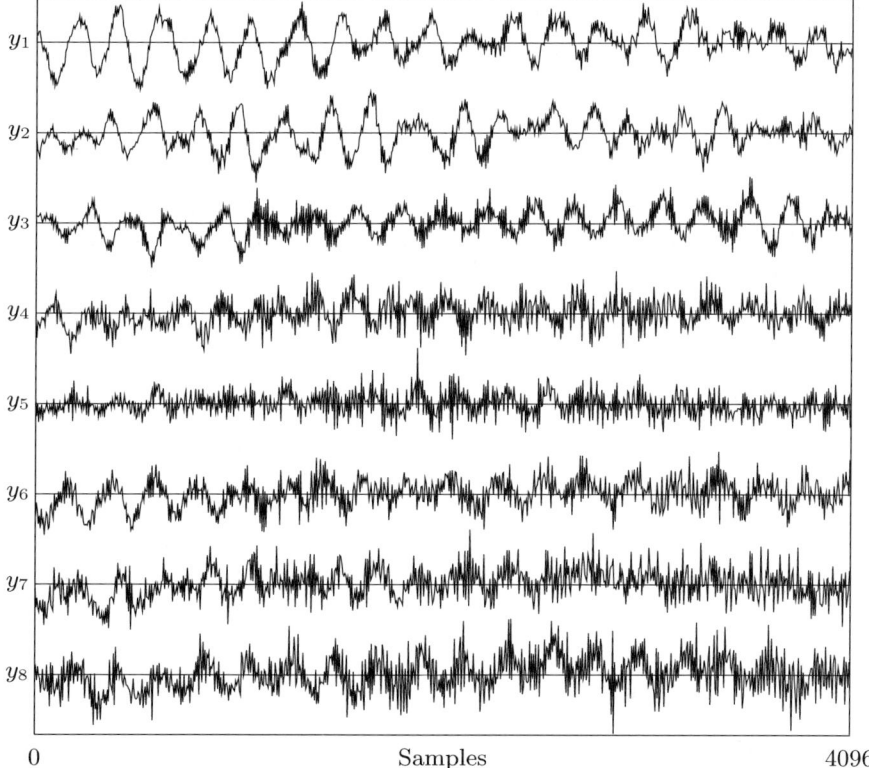

Fig. 11.22. One block of eight microphone signals

The autocorrelation extrema are located at

$$\mathbb{P}_{11} = \{10, 21, 28, 142\} \quad \text{and} \quad \mathbb{P}_{44} = \{35, 52, 79, 224\}. \tag{11.44}$$

Taking the tolerance width $\Gamma_{\text{TWRM}} = 7$ for the raster matching as shown in Fig. 11.21 into account, we found a number of cross-correlation TDOA pairs which match certain autocorrelation extremum positions from $\mathbb{P}_{11} \cup \mathbb{P}_{44}$. They are depicted in Fig. 11.24. If we only choose the cross-correlation maxima without arrowheads, there would be just one valid TDOA at the position 438 which does not correspond to any of the source positions. By applying the soft raster match as described in Sec. 11.3, the set $\mathbb{P}_{14}$ is reduced to

$$\mathbb{P}'_{14} = \{-31, 21, 48, 267, 327, 337, 438\}. \tag{11.45}$$

Obviously, eight echo path TDOAs have been successively rejected while the true direct path TDOAs 21 and 327 are still contained in $\mathbb{P}'_{14}$. We also analyzed the other signal blocks. In general, those cross-correlation maxima with more arrowheads than arrow tails were rejected.

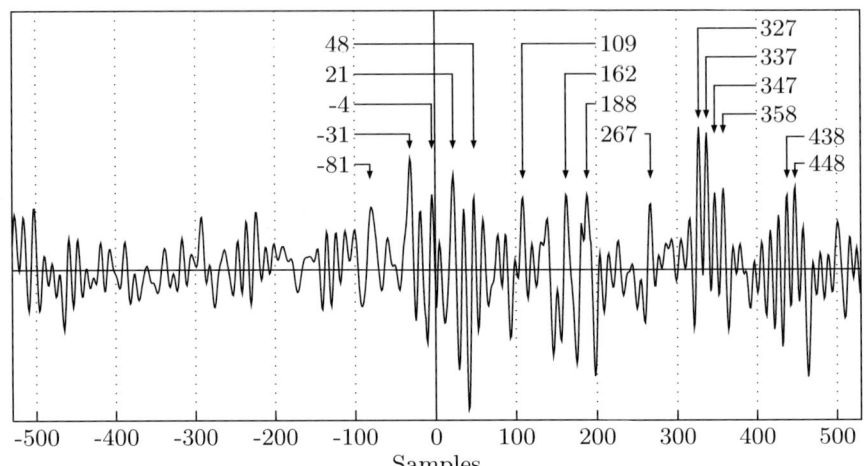

Fig. 11.23. Cross-correlation $\hat{r}_{14}(n)$ between microphone 1 and 4 and its 15 maxima.

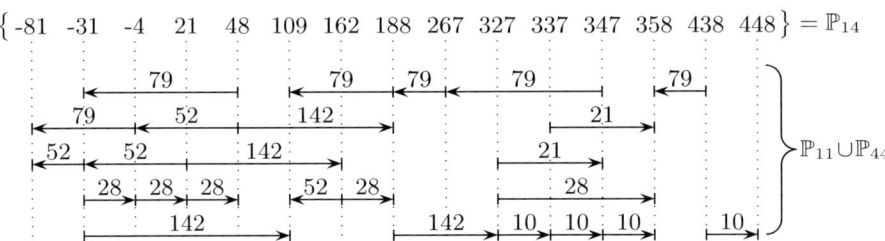

Fig. 11.24. Detected pairs of cross-correlation TDOAs whose distances match autocorrelation extremum positions.

Now we combine the 7 selected TDOA candidates from $\mathbb{P}'_{14}$ with those of the other microphone pairs by using consistent graphs. Our algorithm as proposed in Sec. 11.4 returned 17 approximately consistent TDOA graphs. Four of them are shown in Fig. 11.25. We only study graph I to III further. The other graphs contain only four connected nodes like graph IV in Fig. 11.25 and will not be further considered due to their low connectivity.

Graph I connects seven of the eight microphones. A total number of 21 TDOA estimates fit together to one big approximately consistent graph corresponding to one source position. The TDOA between microphone 1 and 4 (bold edge in Fig. 11.25) is the value 21 from $\mathbb{P}'_{14}$.

Graph II and III are quite similar. They connect 6 and 5 microphones, respectively. Interestingly, both graphs share the same TDOA values between microphone 1, 4, 6, and 7. They seem to originate from the same source, the second source. Only sensor 3 shows different TDOA values to other microphones in both graphs. The explanation of this phenomenon is that sensor 3

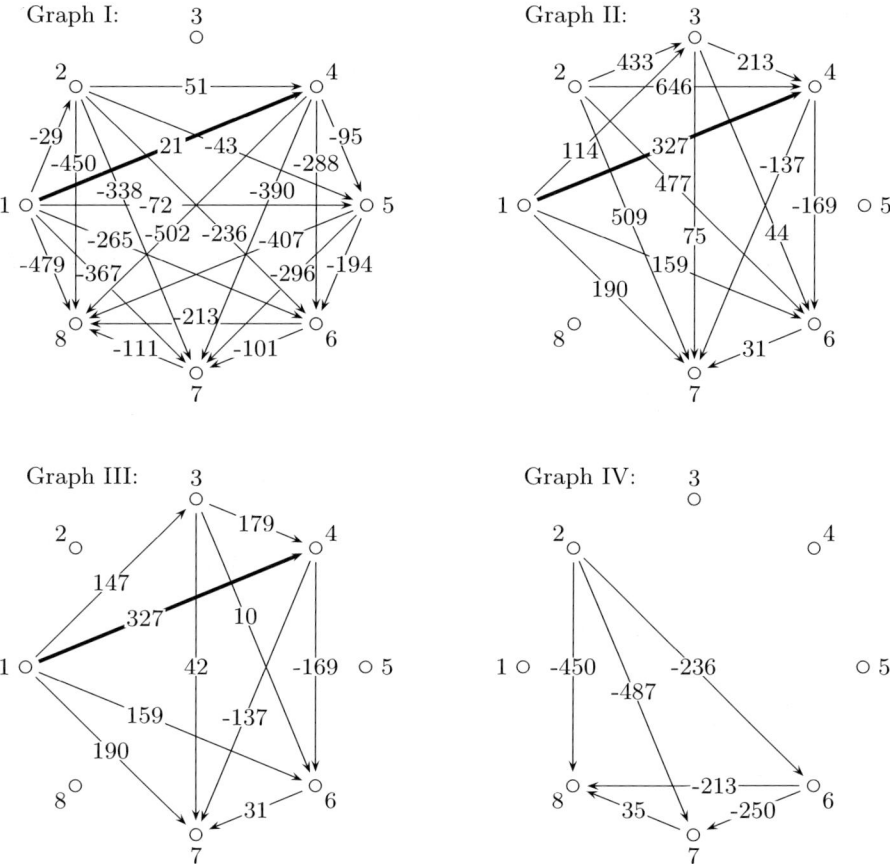

Fig. 11.25. Synthesized approximately consistent TDOA graphs based on the measured data.

represents in one graph the true microphone and in the other graph the mirrored microphone with respect to a wall. This is likely because microphone 3 is quite close to a wall. Which graph the correct one is can not be answered here. In the next section, we will resolve this ambiguity based on the residual error of the source position estimation.

A similar explanation applies to graph IV. It is a variation of graph I caused by a mirrored microphone. The triple (2, 6, 8) is identical in both graphs. The sensor 7 in graph IV represents a mirror of the true microphone 7 in graph I with respect to a wall.

11.5.3 Source Position Estimation

Taking the sound speed of 343 m/s into account, all TDOA estimates in graph I to III are converted to distances. We used the simple spherical interpolation (SI) method [18] to estimate the position of the two sources. Microphone 7 served as the reference sensor. Correspondingly, only 6, 5, and 4 TDOA estimates with respect to microphone 7 from graph I to III are used in source localization. The estimated source positions for the three TDOA graphs are

$$\hat{p}_\mathrm{I} = \begin{bmatrix} 2.725 \\ 0.667 \\ 1.278 \end{bmatrix}, \quad \hat{p}_\mathrm{II} = \begin{bmatrix} 1.712 \\ -0.688 \\ 1.291 \end{bmatrix}, \quad \hat{p}_\mathrm{III} = \begin{bmatrix} 1.671 \\ 1.693 \\ 0.713 \end{bmatrix}. \qquad (11.46)$$

The corresponding vector norm of the residual TDOA errors is

$$\|e_\mathrm{I}\| = 0.016, \quad \|e_\mathrm{II}\| = 1.027, \quad \|e_\mathrm{I}\| < 0.001. \qquad (11.47)$$

Clearly, $\hat{p}_\mathrm{I}$ and $\hat{p}_\mathrm{III}$ are quite good estimates for p_b and p_a in Eq. 11.42, although further improvement could be achieved by using the complete TDOA graph. The large residual error $\|e_\mathrm{II}\|$ shows that, though graph II is approximately consistent, the position estimation is not successful because the sensor 3 in graph II is actually a mirror of the true microphone 3 with respect to a wall. All TDOAs relative to sensor 3 in graph II are "direct path" TDOAs for the mirror microphone but echo path TDOAs for the true microphone. In contrast, all TDOAs in graph III originate from direct path propagations, thus leading to a very small residual error. We see that the residual error in source position estimation provides an additional mean to resolve the ambiguity of TDOA graphs.

11.5.4 Evaluation of Continuous Measurements

In this subsection, we describe the performance of our real-time multi-source localization system during a continuous operation in the same setup as before.

The average CPU load is about 40 %. For each signal block of length 43 ms, the average computation time of our complete localization algorithm consisting of

- preprocessing like VAD
- cross- and autocorrelations
- raster matching
- synthesis of consistent graphs
- source position estimation

is roughly 17 ms. The main computational effort is the calculation of cross- and autocorrelations. By using a pipeline structure containing eight signal blocks, the average block rejection rate of our system is in the order of a few blocks per minute. As expected, this low complexity is mainly due to the significant reduction of ambiguous TDOAs by raster matching and the synthesis of consistent graphs.

Fig. 11.26 illustrates the efficiency of the synthesis of consistent graphs. Each point represents one signal block. The abscissa denotes the total number of possible graphs C_{bf} for this block according to Eq. 11.33 if we perform a brute force search. The ordinate shows the number of consistent triples for this block which are then used in the synthesis of larger consistent graphs. For the particular signal block we studied in the previous subsections, the total number of 148 TDOA candidates would lead to $C_{\mathrm{bf}} \approx 2 \cdot 10^{21}$ different brute force graphs. Our approach finds only 153 consistent TDOA triples.

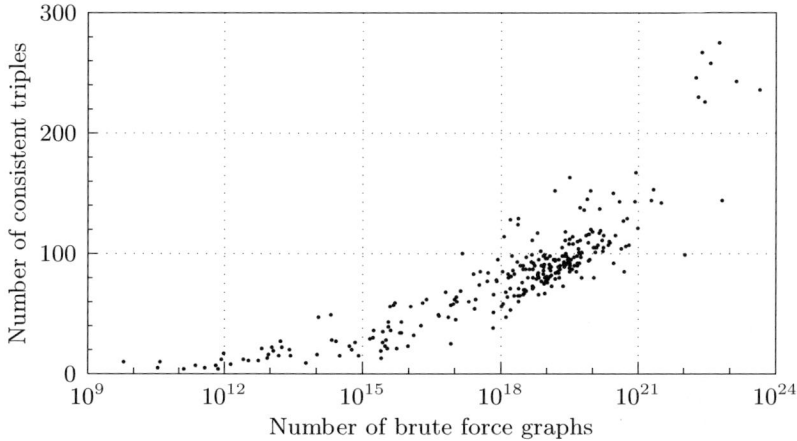

Fig. 11.26. Number of brute force graphs versus number of consistent triples.

In order to compare different synthesized TDOA graphs, we also computed a quality value q_{G} for each synthesized graph by summing up the quality values $q(n_\mu)$ of all involved TDOA estimates n_μ. As $q(n_\mu)$ and thus also q_{G} depend on the signal energy, we normalized each q_{G} to an exponentially weighted average (over signal blocks) of the maximum q_{G} of all graphs. In order to ensure for this experiment that exactly two speech sources are active in each block, we modified the transmitted speech signals slightly by cutting off all silence intervals from both source signals.

Fig. 11.27 shows the resulting graph qualities for a total number of 140 blocks. Each graph is assigned to either source a or b if the distance between the true and estimated source position is less than 10 cm. In 24 % and 30 % of the blocks, source a and b have no assigned graphs resulting in the value $q_{\mathrm{G}} = 0$. The sum of quality values of all unassigned graphs is shown in Fig. 11.27 c). A detailed analysis reveals the following three types of unassigned graphs:

- Mirror microphone graphs like graph II and IV in Fig. 11.25 are approximately consistent, but lead to wrong source position estimates.

- Some graphs result in source position estimates immediately in front of or behind the loudspeakers, especially for voiced speech blocks. One explanation of this phenomenon could be the occurrence of standing waves in the neighborhood of loudspeakers which cause virtual sources.
- Other graphs lead to random erroneous source position estimates which do not appear in adjacent blocks. They can be easily detected and removed by source tracking.

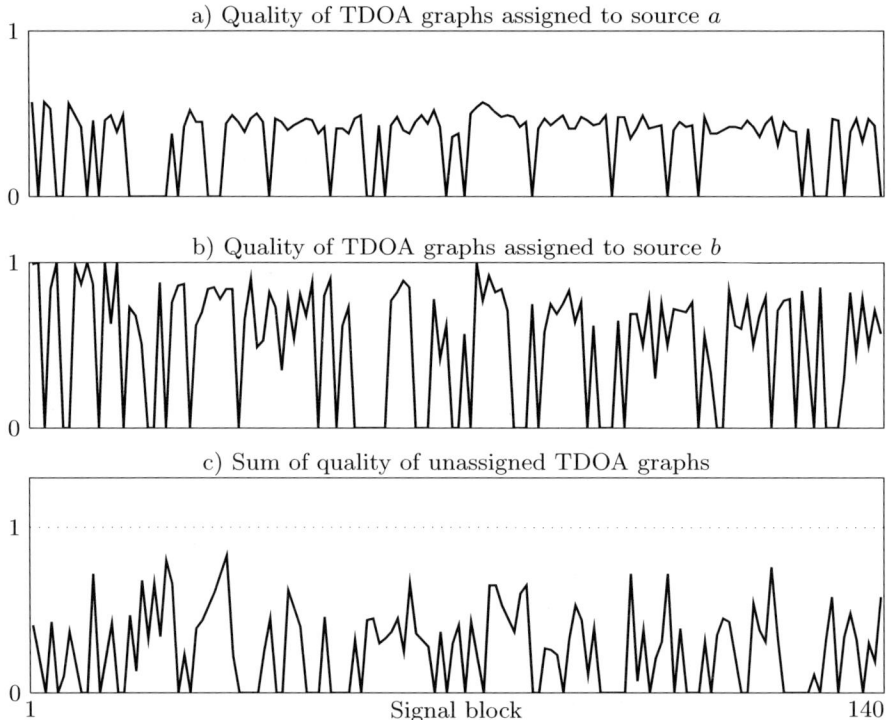

Fig. 11.27. Quality values of assigned and unassigned TDOA graphs.

11.6 Summary

In this chapter, we have studied the cross-correlation based TDOA estimation for locating multiple acoustic sources in reverberant environments. Due to echo path propagations, multiple sources, and periodic signals, a typical (generalized) cross-correlation between two microphone signals shows a large number of maxima. The TDOA estimation is thus ambiguous. How to identify the direct path TDOAs and how to assign them correctly to different sources remain difficult problems for a quite long time.

We have presented a novel approach to resolve these ambiguities. It is based on two fairly simple observations:

- Cross-correlation maxima resulting from a direct and an echo path of the same source signal always occur in a distance which also appears as an extremum position in the autocorrelation of the microphone signals.
- TDOAs of all microphone pairs are redundant because their cyclic sum is always zero.

The first observation motivates the use of additional extremum positions of autocorrelations to find raster matching in cross-correlations. By doing so, many of the echo path TDOAs can be identified and rejected.

In order to exploit the zero cyclic sum redundancy, we formulated the TDOA disambiguation problem in the framework of consistent graphs. The problem is then simplified to a synthesis of consistent graphs starting from many different TDOA candidates for each graph edge. We have developed a very efficient synthesis algorithm based on consistent triples. They are combined and extended to larger TDOA graphs according to certain simple rules. Each resulting TDOA graph is consistent by design and contains all TDOA estimates belonging to the same source.

We also addressed many practical issues like delay quantization, estimation error, mirrored microphone, and incomplete graphs. We introduced different levels of quality for TDOA estimate, TDOA triple, and TDOA graph. These quality values take approximate raster matching and approximate consistent graphs into account. In particular, the quality of a TDOA graph is useful for estimating the number of active sources if it is unknown in advance.

Finally, the efficiency and the real-time capability of our algorithms are demonstrated in a real experiment using simultaneously active speech sources. A single signal block is analyzed in details and continuous measurements are reported.

Note that our algorithms for TDOA disambiguation are generic. They can also be applied to TDOA estimates derived from other methods like impulse response techniques. A further improvement of the described TDOA estimation and localization system is an additional tracking on the TDOA or source position level. An interesting idea in this context is the tracking of detected echo sources that would allow the collection of information about sound reflecting walls. Also the synthesis of approximately consistent graphs can be applied to other signal processing areas.

References

1. R. Aichner, H. Buchner, S. Wehr, W. Kellermann: Robustness of acoustic multiple-source localization in adverse environments, *Proc. 7. ITG-Fachtagung Sprach-Kommunikation*, Kiel, Germany, 2006.

2. J. Allen, D. Berkley: Image method for efficiently simulating small-room acoustics, *Jasa*, **65**(4), 943–950, 1979.
3. J. Benesty: Adaptive eigenvalue decomposition algorithms for passive acoustic source localization, *JASA*, **107**, 384–391, 2000.
4. M. Brandstein: Time-delay estimation of reverberated speech exploiting harmonic structure, *JASA*, **105**(5), 2914–2919, 1999.
5. H. Buchner, R. Aichner, W. Kellermann: Blind source separation for convolutive mixtures: A unified treatment, in Y. Huang, J. Benesty (eds.), *Audio Signal Processing For Next-Generation Multimedia Communication Systems*, Kluwer Academic Publishers, Boston, MA, USA, 2004.
6. M. Drews: Time delay estimation for microphone array speech enhancement systems, *Proc. EUROSPEECH '95*, 2013–2016, Madrid, Spain, 1995.
7. W. R. Hahn, S. A. Tretter: Optimum processing for delay-vector estimation in passive signal arrays, *IEEE Trans. Information Theory*, **19**, 608–614, 1973.
8. Y. Huang, J. Benesty, J. Chen: A blind channel identification-based two-stage approach to separation and dereverberation of speech signals in a reverberant environment, *IEEE Trans. Speech Audio Process.*, **T-SA-13**(5), 882–895, 2005.
9. Y. Huang, J. Benesty and J. Chen: Speech acquisition and enhancement in a reverberant, cocktail-party-like environment, *Proc. ICASSP '06*, **5**, 25–29, Toulouse, France, 2006.
10. C. Knapp, C. Carter: The generalized correlation method for estimation of time delay, *IEEE Trans. Acoust., Speech, Sig. Process.* **T-ASSP-24**(4), 320–327, 1976.
11. Y. Lin, D. Lee, L. Saul: Nonnegative deconvolution for time of arrival estimation, *Proc. ICASSP '04*, **2**, 377–380, Montreal, Canada, 2004.
12. E. Weisstein: MathWorld *http://mathworld.wolfram.com*.
13. R. Nickel, A. Iyer: A novel approach to automated source separation in multi-speaker environments, *Proc. ICASSP '06*, **5**, 629–632, Toulouse, France, 2006.
14. S. Saito, K. Nakata: *Fundamentals of Speech Signal Processing*, Academic Press, New York, NY, USA, 1985.
15. J. Scheuing, B. Yang: Disambiguation of TDOA estimates in multi-path multi-source environments (DATEMM), *Proc. ICASSP '06*, **4**, 837–840, Toulouse, France, 2006.
16. J. Scheuing, B. Yang: Efficient synthesis of approximately consistent graphs for acoustic multi-source localization, *Proc. ICASSP '07*, **4**, 501–504, Honolulu, Hawaii, USA, 2007.
17. J. Scheuing, B. Yang: Klassifikation von Korrelationsextrema zur Laufzeitdifferenzschätzung, *Proc. 33. Deutsche Jahrestagung für Akustik - DAGA '07*, Stuttgart, Germany, 2007.
18. J. O. Smith, J. S. Abel: Closed-form least-squares source location estimation from range-difference measurements, *IEEE Trans. Acoust., Speech, Sig. Process.*, **T-ASSP-35**(12), 1661–1669, 1987.

12

Microphone Calibration for Multi-Channel Signal Processing

Markus Buck[1], Tim Haulick[1], and Hans-Jörg Pfleiderer[2]

[1] Harman/Becker, Ulm, Germany
[2] University Ulm, Ulm, Germany

12.1 Introduction

The application of microphone arrays and adaptive beamforming techniques promises significant improvements compared to systems operating with a single microphone as the spatial properties of the sound field are exploited. However, for real-world applications beamformers imply the risk of severe signal degradation due to mismatched microphones. Microphone mismatch naturally arises from production tolerances as well as from aging effects in the long run. The influence of microphone mismatch on beamforming techniques is barely addressed in literature. Mostly microphone mismatch is simply considered equivalent to deviations in the steering direction of a beamformer. In fact both of these imperfections may cause the well-known signal cancellation effect which usually comes along with a growth of the magnitudes of the filter coefficients. However, there are major differences as it will be pointed out in this chapter.

There are several well-known methods for improving the robustness of adaptive beamformers. These methods can be categorized into two classes: one class restricts the performance of the beamformer in such a way that the beamformer operates reasonably despite the present errors (combatting the symptoms). The other class aims to reduce the deviations themselves (combatting the causes), i. e. by performing an equalization or a calibration.

A widely used member of the first class is the *norm constraint* [11] which prevents an excessive increase of the filter coefficients by limiting their norm to an upper bound. Another method to prevent the filter coefficient magnitudes from growing too large is the application of a *leakage factor* [38]. In each time step the filter coefficients are scaled down by a factor that is slightly smaller than one. Especially for speech applications an *adaptation control* turns out to be an effective method to prevent signal cancellation. The beamformer filters are adapted during speech pauses only. However, during speech activity the filters are no longer able to track the optimal solution.

As mentioned above a different approach to avoid signal cancellation is to reduce its causes, i. e. to reduce microphone mismatch as well as mismatch of the room acoustics to the model assumed. A simple but costly method is to pre-select the microphones for an optimal match. It is also possible to pre-determine calibration filters by a special measurement and apply them afterwards. However, both solutions require additional and costly measurements. Furthermore, the microphones' characteristics usually change over time due to aging effects or environmental influences. For that reason, an adaptive matching that tracks these changes is desirable. Recently, the topic of microphone calibration has found increasing attention in literature. There were several publications on adaptive self-calibration [25, 31] as well as on approaches with fixed calibration filters [28, 30, 40].

This chapter gives an overview on calibration techniques with fixed as well as with adaptive filters. After a short introduction to beamforming techniques we will point out the problem of mismatched microphones by theoretical investigations under the assumption of a diffuse noise field. In the systematic investigations of adaptive beamformers an unknown characteristic was discovered which could be denoted as "pattern cancellation effect". In a third section different calibration techniques are introduced. These techniques are evaluated under more practical conditions by conducting Monte Carlo simulations on the basis of measured microphone characteristics. Here, the limits of the different calibration techniques are pointed out. In the following section, we will focus on systems which calibrate the microphones automatically. A class of efficient techniques for self-calibration is presented and compared to existing methods. Such self-calibration methods perform a calibration adaptively in the background during normal operation of the system and therefore save the need for an additional costly calibration procedure. The performance of the calibration techniques is examined using practical real-world scenarios.

12.2 Beamforming with Ideal Microphones

12.2.1 Principle of Beamforming

The term *beamformer* is used for systems which sample a propagating wave field with a certain number of sensors for achieving a spatial directionality by processing the sensor signals to a single output signal. This directionality can be used for noise reduction if desired signal components impinge from different directions than the noise signal components onto the sensor group. Particularly, for applications where signal and noise occur at the same time and for the same frequency beamformer approaches yield effective solutions for noise suppression.

In Fig. 12.1 a beamformer with M microphones is depicted. The sensor signals are considered preliminarily as continuous time signals. Their spectra $X_m^{\mathrm{M}}(\omega)$ are filtered by filters $W_m(\omega)$ and summed up to a single output signal with the spectrum $X^{\mathrm{BF}}(\omega)$. With vector notation

$$\boldsymbol{X}(\omega) = \left[X_1^{\mathrm{M}}(\omega), \ldots, X_M^{\mathrm{M}}(\omega)\right]^{\mathrm{T}},\qquad(12.1)$$

$$\boldsymbol{W}(\omega) = \left[W_1(\omega), \ldots, W_M(\omega)\right]^{\mathrm{T}}\qquad(12.2)$$

the processing can be written as an inner product of two vectors:[3]

$$X^{\mathrm{BF}}(\omega) = \boldsymbol{W}^{\mathrm{H}}(\omega)\,\boldsymbol{X}(\omega)\,.\qquad(12.3)$$

Using the power spectral density matrix of the input signals $\boldsymbol{\Phi}^{\mathrm{M}}(\omega)$ the power spectral density of the output signal can be expressed as a quadratic form

$$\Phi^{\mathrm{BF}}(\omega) = \boldsymbol{W}^{\mathrm{H}}(\omega)\,\boldsymbol{\Phi}^{\mathrm{M}}(\omega)\,\boldsymbol{W}(\omega)\qquad(12.4)$$

where $\boldsymbol{\Phi}^{\mathrm{M}}(\omega)\big|_{m,n} = \Phi^{\mathrm{M}}_{m,n}(\omega)$ denotes the cross power spectral density of the output signals of microphones m and n.

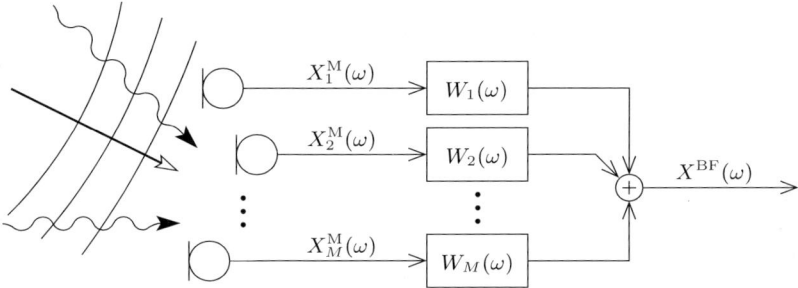

Fig. 12.1. Beamformer in direct structure. The sensor signals are filtered and summed up to a single output signal. The desired signal is indicated by the straight arrow.

The directionality of the beamformer is substantially determined by the positions of the microphones and the adjustment of the filters $W_m(\omega)$. For the design of the filters a *desired direction* $\mathbf{u}_0$ is specified which determines the orientation of the beamformer. The corresponding relative time delays $\tau_{m,n}(\mathbf{u}_0)$ related to a plane wave front coming from the desired direction have to be compensated by the filters. These time delays depend on the positions of the single microphones. The frequency domain representation of the time delays are described by the *steering vector* $\boldsymbol{E}_0(\omega) = \boldsymbol{E}(\omega, \mathbf{u}_0)$, where $\boldsymbol{E}(\omega, \mathbf{u})$ corresponds to the relative time delays for a plane wave from an arbitrary spatial direction $\mathbf{u}$:[4]

---

[3] The filter operation is formulated with conjugate complex values – as common in the literature. This leads to simplified formulas for the filter design.

[4] Here microphone 1 is used as reference.

$$\boldsymbol{E}(\omega, \mathbf{u}) = \Big[1, \exp\big\{ -j\omega\tau_{2,1}(\mathbf{u}) \big\}, \ldots, \exp\big\{ -j\omega\tau_{M,1}(\mathbf{u}) \big\} \Big]^{\mathrm{T}}. \qquad (12.5)$$

For the simplest case – the so-called *delay-and-sum beamformer* – the filters $\boldsymbol{W}(\omega) = \boldsymbol{E}_0(\omega)$ solely compensate for the time delays corresponding to the desired direction. Due to this phase alignment by the beamformer filters signal components arriving from the desired direction interfere constructively at the summation point at the output of the beamformer. In contrast, signal components arriving from off the desired direction interfere destructively. However, the suppression of interferers strongly depends on the direction of arrival and the wavelength of the interfering signal.

For the *filter-and-sum beamformer* the pure time delays are replaced by linear filters with arbitrary transfer functions. This generalization leads to some degrees of freedom over the delay-and-sum beamformer which can be used for a more effective noise suppression. In a consequence, signals coming from the desired direction are no more summed up in phase. The resulting partly destructive interference for the desired signal has to be compensated by higher filter amplifications in order to keep an undistorted transmission for the desired direction. Due to its enhanced directional properties this approach is also called *superdirective* beamforming.

Additional to the beamformer structure depicted in Fig. 12.1 the time delays can be realized as separate *time delay compensation* filters. These filters are connected in series to the residual filters. Such an implementation offers the opportunity to design the noise suppression on the basis of time aligned input signals which has the advantage of being simpler and more efficient.

A very important parameter of a beamformer is the spatial arrangement of the microphones – the array geometry. Commonly the microphones are positioned along an array axis $\mathbf{u}_A$ as linear array. Depending on the steering direction $\mathbf{u}_0$ broadside and endfire arrays are distinguished having the steering direction perpendicular or parallel to the array axis, respectively.

In case of *fixed* beamformers the filters $W_m(\omega)$ are designed a priori under the assumption of a known noise field or by specifying a desired beampattern [34]. On the other hand the filter settings for *adaptive* beamformers are changed permanently during operation in order to adapt to the present noise situation.

In this section only continuous time signals are considered in order to accommodate to the physical conditions. The analysis and the design of beamformer filters are done in the frequency domain by transfer functions. In order to implement the beamformers the resulting filter transfer functions have to be realized by an appropriate structure. For this, usually FIR filters in the time or frequency domain are utilized. Also sub-band implementations are very common, where an analysis filter bank decomposes the signal into independent sub-band signals. A synthesis filter bank recombines the signals. Depending on the chosen filter structure the specified transfer functions may be replicated only approximately.

12.2.2 Evaluation of Beamformers

In order to design optimum beamformers suitable performance criteria are necessary. As in the literature hardly any attention is payed to the influence of microphone characteristics on beamformers a comprehensive theoretical basis for the evaluation of beamformers only exists for the case of omnidirectional microphones, e. g. [8,12]. For this reason, in this section a generalization of the known criteria for beamformer evaluation is proposed by explicitly considering the influence of the single microphones.

For the evaluation as well as for the design of beamformers the acoustic situation is commonly described by a simplified representation. The signal spectra are separated into a desired component and a noise component. Often a simple signal model is considered where the desired component consists of one plane wave arriving from a single direction. The remaining signal portion is considered as noise assuming that the desired signal and the noise are statistically independent. In many practical applications this simplified signal model turns out to be strongly idealized. For example in reverberant environments the direct path signal is superposed by various reflections which arrive with some delay from arbitrary directions at the microphones. Depending on the objective the reverberant signal path may be classified as an interfering signal (i. e. for dereverberation) or as desired signal portion (i. e. for noise suppression). In practice the performance which is predicted on the basis of the simple signal model can mostly not be achieved completely.

12.2.2.1 Transfer Function

By its transfer function a beamformer is analyzed independently from a present sound field purely by its array geometry and filter settings. The transfer function $H^{BF}(\omega, \mathbf{u})$ describes the relationship between an impinging signal of a plane wave front and the output signal of the beamformer in dependence of the direction of arrival $\mathbf{u}$ and the angular frequency ω. The microphones are described by transfer functions $H_m^M(\omega, \mathbf{u})$ of dimension one which contain the directional characteristics in a normalized description:

$$H^{BF}(\omega, \mathbf{u}) = \sum_{m=1}^{M} W_m^*(\omega) \, H_m^M(\omega, \mathbf{u}) \, \exp\{-j\omega \tau_{m,1}(\mathbf{u})\} \qquad (12.6)$$

$$= \boldsymbol{W}^H(\omega) \, \boldsymbol{H}(\omega, \mathbf{u}) \, \boldsymbol{E}(\omega, \mathbf{u}) \, . \qquad (12.7)$$

The three expressions in Eq. 12.7 represent the characteristics of the sound field ($\boldsymbol{E}(\omega, \mathbf{u})$), of the microphone ($\boldsymbol{H}(\omega, \mathbf{u})$) and of the beamformer ($\boldsymbol{W}^H(\omega)$). According to Eq. 12.5 vector $\boldsymbol{E}(\omega, \mathbf{u})$ contains the relative time delays $\tau_{m,1}(\mathbf{u})$ and with it implicitly the *positioning* of the microphones. The *spatial orientation* of the microphones is taken into account by the diagonal matrix $\boldsymbol{H}(\omega, \mathbf{u})$:

$$\boldsymbol{H}(\omega, \mathbf{u}) = \text{diag}\{\, H_1^M(\omega, \mathbf{u}), \ldots, H_M^M(\omega, \mathbf{u}) \,\} \, . \qquad (12.8)$$

By introducing the matrix $\boldsymbol{H}(\omega, \mathbf{u})$ the well-known beamformer transfer function [8, 12] is generalized. Instead of the normally assumed identical and omnidirectional microphones now microphones with arbitrary directional characteristics can be considered.

12.2.2.2 Beampattern

The beampattern of a directional device is defined by the squared magnitude of its transfer function:

$$\Psi^{\mathrm{BF}}(\omega, \mathbf{u}) = \left| H^{\mathrm{BF}}(\omega, \mathbf{u}) \right|^2 . \tag{12.9}$$

Due to Eq. 12.6 the beampattern can also be expressed by

$$\Psi^{\mathrm{BF}}(\omega, \mathbf{u}) = \boldsymbol{W}^{\mathrm{H}}(\omega) \, \boldsymbol{\Phi}^{\mathrm{M,p}}(\omega) \, \boldsymbol{W}(\omega) \tag{12.10}$$

where the normalized power density matrix for a plane wavefront[5] $\boldsymbol{\Phi}^{\mathrm{M,p}}(\omega)$ is composed by the elements

$$\Phi_{m,n}^{\mathrm{M,p}}(\omega, \mathbf{u}) = H_m^{\mathrm{M}}(\omega, \mathbf{u}) \left(H_n^{\mathrm{M}}(\omega, \mathbf{u}) \right)^* \exp\left\{ j \frac{\omega}{c} (\mathbf{r}_m - \mathbf{r}_n)^{\mathrm{T}} \mathbf{u} \right\} . \tag{12.11}$$

Here, $\mathbf{r}_m$ and $\mathbf{r}_n$ denote the positions of the microphones m and n, respectively.

12.2.2.3 Directivity

On the basis of the beampattern the directivity $D^{\mathrm{BF}}(\omega)$ forms an integral measure which doesn't depend on the direction of arrival. For this purpose the beampattern for the desired direction $\mathbf{u}_0$ is normalized to the spatial integration of the beampattern over all spherical directions:[6]

$$D^{\mathrm{BF}}(\omega) = \frac{\Psi^{\mathrm{BF}}(\omega, \mathbf{u}_0)}{\frac{1}{4\pi} \int_0^{2\pi} \int_0^{\pi} \Psi^{\mathrm{BF}}(\omega, \mathbf{u}) \sin\theta \, d\theta \, d\varphi} . \tag{12.12}$$

The steering direction of the beamformer $\mathbf{u}_0$ and the orientation of the microphones $\mathbf{u}_m^{\mathrm{M}}$ do not need to be identical. However, in general a match of these directions is advantageous with respect to an optimal performance.

For the case of an isotropic sound field as noise component and a plane wave from direction $\mathbf{u}_0$ as desired signal component the directivity corresponds to the array gain [11]:

---

[5] Quantities corresponding to a plane wave sound field are labeled with an index p. Respectively, the index i is used for the diffuse (isotropic) sound field. Only sound signals with normalized power density spectra are considered here: $\Phi_{m,m}^{\mathrm{S,p}}(\omega, \mathbf{u}) = \Phi_{m,m}^{\mathrm{S,i}}(\omega) = 1$.

[6] For the reason of simplicity the dependance on $\mathbf{u}_0$ has been dropped for $D^{\mathrm{BF}}(\omega)$ here and in the following.

$$D^{\mathrm{BF}}(\omega) = \frac{\Phi^{\mathrm{BF,P}}(\omega, \mathbf{u}_0)}{\Phi^{\mathrm{BF,i}}(\omega)} . \tag{12.13}$$

To consider the influence of the microphones the correlation characteristics of the converted signals according to Eq. 12.4 have to be used.

Accordingly, the cross power spectral density for a diffuse sound field including the transducers' characteristics reads

$$\Phi^{\mathrm{M,i}}_{m,n}(\omega) = \frac{1}{4\pi} \int_0^{2\pi}\!\!\int_0^{\pi} H^{\mathrm{M}}_m(\omega, \mathbf{u}) \left(H^{\mathrm{M}}_n(\omega, \mathbf{u})\right)^* \exp\{j\tfrac{\omega}{c}(\mathbf{r}_m - \mathbf{r}_n)^{\mathrm{T}}\mathbf{u}\} \sin\theta\, d\theta\, d\varphi . \tag{12.14}$$

By using the two normalized power spectral density matrices $\boldsymbol{\Phi}^{\mathrm{M,P}}(\omega, \mathbf{u}_0)$ and $\boldsymbol{\Phi}^{\mathrm{M,i}}(\omega)$ the power spectral densities $\Phi^{\mathrm{BF,P}}(\omega, \mathbf{u}_0)$ and $\Phi^{\mathrm{BF,i}}(\omega)$ at the beamformer output can be determined using Eq. 12.4. In the case of a plane wavefront arriving from the desired direction $\Phi^{\mathrm{BF,P}}(\omega, \mathbf{u}_0)$ corresponds to $\Psi^{\mathrm{BF}}(\omega, \mathbf{u}_0)$. Thus, it is possible to determine the directivity of a beamformer according to Eq. 12.13 on the basis of the cross power spectrum density of the microphone signals and the beamformer filters:

$$D^{\mathrm{BF}}(\omega) = \frac{\Psi^{\mathrm{BF}}(\omega, \mathbf{u}_0)}{\boldsymbol{W}^{\mathrm{H}}(\omega)\, \boldsymbol{\Phi}^{\mathrm{M,i}}(\omega)\, \boldsymbol{W}(\omega)} . \tag{12.15}$$

This generalized expression will be used to investigate the influence of non-ideal microphone characteristics on beamformers. In that analysis the specific transfer function of each microphone depends on frequency as well as the angle of incidence.

12.2.2.4 Susceptibility

The evaluation criteria introduced so far all describe basically the *mechanism* of a beamformer. A measure for a beamformer's *susceptibility* to disturbances was defined in [18]:[7]

$$K(\omega) = \frac{\boldsymbol{W}^{\mathrm{H}}(\omega)\, \boldsymbol{W}(\omega)}{\Psi^{\mathrm{BF}}(\omega, \mathbf{u}_0)} . \tag{12.16}$$

It corresponds to the L_2 norm of the filter vector referring to the beampattern for the desired direction $\Psi^{\mathrm{BF}}(\omega, \mathbf{u}_0) = |\boldsymbol{W}^{\mathrm{H}}(\omega)\, \boldsymbol{H}(\omega, \mathbf{u}_0)\, \boldsymbol{E}_0(\omega)|^2$. Thus, a high value $K(\omega)$ also indicates a high internal signal amplification by the filters. The inverse of the susceptibility is also known as *white noise gain* [11].

---

[7] Deviating from the original definition in [18], here, the microphone influence according to the generalized evaluation after Eq. 12.7 is incorporated in the denominator.

12.2.3 Statistically Optimum Beamformers

This section describes the design of statistically optimum beamformers for known sound field characteristics on the basis of the generalized expression according to Sec. 12.2.2.

12.2.3.1 MVDR Criterion

A widely used design approach is the *minimum variance distortionless response* (MVDR) design. The beamformer filters $\boldsymbol{W}(\omega)$ are adjusted in such a way that the output signal of the beamformer has minimum power. In order to prevent the trivial solution $\boldsymbol{W}(\omega) = \boldsymbol{0}_{M \times 1}$ it is required that at the same time the transfer function for the desired direction equals $H^{\mathrm{BF}}(\omega, \mathbf{u}_0) = 1$. With the power spectral density of the output signal $\Phi^{\mathrm{BF}}(\omega)$ according to Eq. 12.4 and the beamformer transfer function $H^{\mathrm{BF}}(\omega, \mathbf{u})$ according to Eq. 12.7 the MVDR optimization criterion can be formulated as

$$\min_{\boldsymbol{W}(\omega)} \left\{ \boldsymbol{W}^{\mathrm{H}}(\omega) \, \boldsymbol{\Phi}^{\mathrm{M}}(\omega) \, \boldsymbol{W}(\omega) \right\} \tag{12.17}$$

subject to

$$\boldsymbol{W}^{\mathrm{H}}(\omega) \, \boldsymbol{H}(\omega, \mathbf{u}_0) \, \boldsymbol{E}_0(\omega) = 1 \ . \tag{12.18}$$

This is an optimization problem with linear constraints. The solution of this problem can be derived analytically by the use of Lagrange multipliers and reads:

$$\boldsymbol{W}_{\mathrm{MVDR}}(\omega) = \frac{\left(\boldsymbol{\Phi}^{\mathrm{M}}(\omega)\right)^{-1} \boldsymbol{H}(\omega, \mathbf{u}_0) \, \boldsymbol{E}_0(\omega)}{\boldsymbol{E}_0^{\mathrm{H}}(\omega) \, \boldsymbol{H}^{\mathrm{H}}(\omega, \mathbf{u}_0) \left(\boldsymbol{\Phi}^{\mathrm{M}}(\omega)\right)^{-1} \boldsymbol{H}(\omega, \mathbf{u}_0) \, \boldsymbol{E}_0(\omega)} \ . \tag{12.19}$$

The expression in the numerator can be interpreted as a two step process [9]: the inverse matrix affects a decorrelation of the microphone signals whereas the remaining part acts as a matched filter beamformer. Since after this filtering the desired signal components don't necessarily sum up in phase MVDR beamformers are superdirectional.

For the MVDR design the characteristics of the sound field as well as of the microphones have to be specified within the normalized power density matrix $\boldsymbol{\Phi}^{\mathrm{M}}(\omega)$. As sound field, commonly, a diffuse sound field is supposed. With this choice inherently the directivity of the beamformer is maximized.

The filters resulting from the MVDR design typically show strong amplifications for low frequencies. As a result the susceptibility $K(\omega)$ increases and the beamformer gets sensitive to disturbances. For this reason it is advantageous to limit the susceptibility to a maximum value $K_{\max}$. For this purpose the optimization criteria according to Eqs. 12.17 and 12.18 are extended by an additional quadratic constraint. The modified MVDR criterion reads

$$\min_{\boldsymbol{W}(\omega)} \left\{ \boldsymbol{W}^{\mathrm{H}}(\omega) \, \boldsymbol{\Phi}^{\mathrm{M}}(\omega) \, \boldsymbol{W}(\omega) \right\} \qquad (12.20)$$

subject to

$$\boldsymbol{W}^{\mathrm{H}}(\omega) \, \boldsymbol{H}(\omega, \mathbf{u}_0) \, \boldsymbol{E}_0(\omega) = 1 \quad \text{and} \quad \boldsymbol{W}^{\mathrm{H}}(\omega) \, \boldsymbol{W}(\omega) \le K_{\max} \, . \qquad (12.21)$$

In [12] it is shown that a regularization of the normalized power spectral density matrix by the term $\mu(\omega) \, \boldsymbol{I}_{M \times M}$ solves this optimization problem, where $\boldsymbol{I}_{M \times M}$ denotes the identity matrix:

$$\boldsymbol{W}^{(\mu)}_{\mathrm{MVDR}}(\omega) = \frac{\left(\boldsymbol{\Phi}^{\mathrm{M}}(\omega) + \mu(\omega) \, \boldsymbol{I}_{M \times M} \right)^{-1} \boldsymbol{H}(\omega, \mathbf{u}_0) \, \boldsymbol{E}_0(\omega)}{\boldsymbol{E}_0^{\mathrm{H}}(\omega) \, \boldsymbol{H}^{\mathrm{H}}(\omega, \mathbf{u}_0) \left(\boldsymbol{\Phi}^{\mathrm{M}}(\omega) + \mu(\omega) \, \boldsymbol{I}_{M \times M} \right)^{-1} \boldsymbol{H}(\omega, \mathbf{u}_0) \, \boldsymbol{E}_0(\omega)} \, . \qquad (12.22)$$

By the parameter $\mu(\omega) \in \mathbb{R}_0^+$ it is possible to influence the susceptibility $K^{(\mu)}(\omega)$ where $K^{(\mu)}(\omega)$ shows a monotonously decreasing run over $\mu(\omega)$ [18]. The optimal value for $\mu(\omega)$ cannot be calculated analytically. But it is possible to approximate the value $\mu(\omega)$ in an iterative manner for example using the bisection method [12].

12.2.3.2 Adaptive Beamformers

In case of the optimal beamformers investigated so far in this chapter it is assumed that the statistical properties of the sound field are known a priori. Since in general the acoustic situation may change over time in most cases the sound field is not known in advance. For this reason the statistical properties may be estimated on the basis of the present microphone signals. The beamformer filters are adjusted adaptively during operation in order to track the optimum filter settings. For the optimization the MVDR criterion according to Eqs. 12.20 and 12.21 may be used. An explicit consideration of the microphone characteristics is not necessary for adaptive beamformers as the estimate of the statistical properties bases on microphone signals rather than on sound signals.

An adaptive MVDR beamformer can be implemented directly as a filter-and-sum beamformer [15] as depicted in Fig. 12.2 a. Alternatively the MVDR beamformer can be realized in the structure of a *generalized sidelobe canceller* (GSC) [19] as shown in Fig. 12.2 b. In this approach the signal processing is divided into two paths after the time delay compensation. The upper path consisting of a fixed beamformer is non-adaptive. In the lower path a blocking matrix realizes the directional constraint by cancelling signals from the desired direction and thus in the ideal case generates pure noise reference signals. These noise reference signals are led to unconstrained adaptive filters $\boldsymbol{W}_{\mathrm{a}}(\omega)$ which aim to cancel the remaining noise in the output signal of the fixed beamformer $X^{\mathrm{FBF}}(\omega)$.

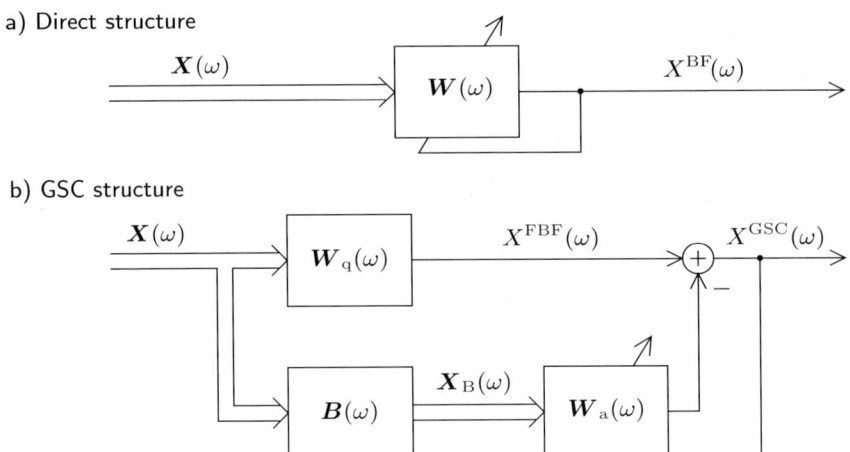

Fig. 12.2. Diagram of the a) direct structure and the b) GSC structure of a beamformer.

Obviously the optimization is based on the simple signal model, after which the desired signal impinges as a plane wavefront from one discrete direction $\mathbf{u}_0$ onto the microphone arrangement. For speech applications in closed rooms this model is violated substantially: on the one hand there are deviations to the *acoustic* model which may occur because of a mismatched steering direction $\mathbf{u}_0$, sound reflections or nearfield effects. On the other hand the assumption of an ideal *transducer* is not met because of microphone tolerances.

12.2.3.3 Signal Cancellation Effect

The effect of deviations from the signal model can be illustrated for the GSC shown in Fig. 12.2 b: because of the above listed non-idealities, portions of the desired signal may pour through the blocking matrix into the noise reference signals $\mathbf{X}_B(\omega)$. From these portions the adaptive filters are able to partly reconstruct the desired signal component in the output signal of the fixed beamformer $X^{\mathrm{FBF}}(\omega)$. Thus, with the subtraction in Fig. 12.2 b the desired signal is attenuated even though the directional constraint is formally fulfilled. This phenomenon is known as *signal cancellation* effect [37]. For the case of a mismatched steering direction this effect has been intensively studied [35,39].

Many methods have been proposed in order to prevent this effect. An overview over these methods can be found for example in [24] or also in Sec. 12.1. Here, only a few approaches which are particularly suitable for speech applications are outlined:

- By additional linear constraints the degrees of freedom for the beampattern can be reduced. For example so-called *derivative constraints* can be used

in order to broaden the main lobe of the beamformer [2, 14, 32]. In this way the beamformer gets more robust for small deviations of the steering direction.
- By specifying quadratic constraints the susceptibility of the adaptive beamformer can be limited. Thus, the robustness is enhanced. An example for a quadratic constraint is given with Eq. 12.21 where the norm of the filter vectors is bounded [11].
- Particularly for speech applications it is close at hand to control the adaptation step size depending on the acoustic situation. The signal cancellation effect can be avoided if the adaption is stopped during speech activity and performed only in speech pauses. One method for such a step size control will be described in Sec. 12.5.4.
- The approaches listed so far each have the drawback of degrading the performance of the beamformer. Instead, it can be attempted to compensate for the room influences adaptively [1, 16, 23, 33]. In this way the causes of signal cancellation are controlled without hampering the beamformer. In [22] this approach is denoted as optimization under *spatio-temporal constraints*.

12.3 Microphone Mismatch and its Effect on Beamforming

While in the last section mainly microphones with ideal characteristics have been considered, in this section the influence of non-ideal microphones is examined. Fixed and adaptive beamformers are handled separately in this investigation. The analyses are done in two steps: first, the filters $\boldsymbol{W}^\mathrm{H}(\omega)$ of the beamformer are determined, then, the resulting beamformer is evaluated taking into account the microphone characteristics. For this, instead of signal $\boldsymbol{X}^\mathrm{M}(\omega)$ obtained by ideal microphones now the signal $\tilde{\boldsymbol{X}}^\mathrm{M}(\omega)$ originating from imperfect microphones is used. Accordingly, the cross power spectral density of two microphone signals is $\tilde{\Phi}_{m,n}^\mathrm{M}(\omega)$.

There is a fundamental difference between designing the filters of a fixed beamformer and determining the adapted filters of an adaptive beamformer: the filters of a fixed beamformer are optimized a priori with the assumption of fictive boundary conditions and are applied independent of the specific microphone characteristic of the concrete array. On the contrary, the filters of the adaptive beamformer adjust to the real boundary conditions on hand and therefore depend among other things on the present microphone characteristics.

In this chapter only linear deviations like mismatch of the frequency response or the directional characteristics are considered. Non-linearities like saturation effects are not treated here.

12.3.1 Model for Non-Ideal Microphone Characteristics

In practical applications single microphones deviate from each other in their transfer characteristics because of production tolerances or by environmental influences. These deviations are regarded here as linear additive tolerances relative to an ideal nominal value $H_m^M(\omega, \mathbf{u})$:

$$\tilde{H}_m^M(\omega, \mathbf{u}) = H_m^M(\omega, \mathbf{u}) + \Delta H_m^M(\omega, \mathbf{u}) \ . \tag{12.23}$$

In the case of mismatched microphones an evaluation can be accomplished by the correlation properties. With $\tilde{H}_m^M(\omega, \mathbf{u})$ Eq. 12.11 for a propagating plane wave the spectral density now reads:

$$\tilde{\Phi}_{m,n}^{M,p}(\omega, \mathbf{u}) = \tilde{H}_m^M(\omega, \mathbf{u}) \left(\tilde{H}_n^M(\omega, \mathbf{u})\right)^* \exp\{j \tfrac{\omega}{c} (\mathbf{r}_m - \mathbf{r}_n)^T \mathbf{u}\} \ . \tag{12.24}$$

For a diffuse sound field Eq. 12.24 is averaged over all directions of a sphere $\mathbf{u} = \mathbf{u}(\theta, \varphi)$ according to Eq. 12.14:

$$\tilde{\Phi}_{m,n}^{M,i}(\omega) = \frac{1}{4\pi} \int_0^{2\pi}\int_0^{\pi} \tilde{H}_m^M(\omega, \mathbf{u}) \left(\tilde{H}_n^M(\omega, \mathbf{u})\right)^* \exp\{j \tfrac{\omega}{c} (\mathbf{r}_m - \mathbf{r}_n)^T \mathbf{u}\} \sin\theta \, d\theta \, d\varphi \ . \tag{12.25}$$

The cross power density spectra are determined pairwise for all microphone combinations of the array and gathered in the spectral density matrices $\tilde{\boldsymbol{\Phi}}^{M,p}(\omega, \mathbf{u})$ and $\tilde{\boldsymbol{\Phi}}^{M,i}(\omega)$, respectively. These matrices contain information on the sound field as well as on the microphone characteristics, implicitly including the microphone mismatch.

With a given set of beamformer filters $W_m(\omega)$ an analysis can be performed for actual non-idealities. The beampattern described by Eq. 12.10 reads for mismatched microphones:

$$\tilde{\Psi}^{BF}(\omega, \mathbf{u}) = \left| \sum_{m=1}^{M} W_m^*(\omega) \tilde{H}_m^M(\omega, \mathbf{u}) \exp\{j \tfrac{\omega}{c} \mathbf{r}_m^T \mathbf{u}\} \right|^2 \tag{12.26}$$

$$= \sum_{\substack{m=1 \\ }}^{M} \sum_{\substack{n=1 \\ n \neq m}}^{M} W_m^*(\omega) W_n(\omega) \left(H_m^M(\omega, \mathbf{u}) + \Delta H_m^M(\omega, \mathbf{u})\right)$$

$$\cdot \left(H_n^M(\omega, \mathbf{u}) + \Delta H_n^M(\omega, \mathbf{u})\right)^* \exp\{j \tfrac{\omega}{c} (\mathbf{r}_m - \mathbf{r}_n)^T \mathbf{u}\} \tag{12.27}$$

$$+ \sum_{m=1}^{M} |W_m(\omega)|^2 \left|H_m^M(\omega, \mathbf{u}) + \Delta H_m^M(\omega, \mathbf{u})\right|^2$$

and the directivity of Eq. 12.15 is now written as:

$$\tilde{D}^{BF}(\omega) = \frac{\tilde{\Psi}^{BF}(\omega, \mathbf{u}_0)}{\mathbf{W}^H(\omega) \tilde{\boldsymbol{\Phi}}^{M,i}(\omega) \mathbf{W}(\omega)} \ . \tag{12.28}$$

With a nominal beampattern that is normalized for the steering direction $\Psi^{\mathrm{BF}}(\omega, \mathbf{u}_0) = 1$ for the susceptibility of Eq. 12.16 it follows:

$$K(\omega) = \sum_{m=1}^{M} |W_m(\omega)|^2 = \|\mathbf{W}(\omega)\|^2 . \tag{12.29}$$

12.3.2 Effect of Microphone Mismatch on Fixed Beamformers

Independently of the input signals non-adaptive beamformers show a fixed directionality that is defined by the design of the beamformer filters. However, in practical applications with a real array the directional properties not only depend on the filters but also on the actual properties of the single microphones. That means that the nominal beampattern is not met exactly.

In this section, the imperfect beampattern according to Eq. 12.27 is statistically analyzed for given filter settings.[8] For this purpose it is assumed that the microphone array contains only microphones of the same nominal type aligned into the same direction: $H_m^{\mathrm{M}}(\omega, \mathbf{u}) = H^{\mathrm{M}}(\omega, \mathbf{u})$. The microphone deviations are considered as complex random variables that have zero-mean $\mathrm{E}\{\Delta H_m^{\mathrm{M}}(\omega, \mathbf{u})\} = 0$ and that are orthogonal $\mathrm{E}\{\Delta H_m^{\mathrm{M}}(\omega, \mathbf{u}) \Delta H_n^{\mathrm{M}}(\omega, \mathbf{u})\} = 0$ for $n \neq m$. Using these assumptions the expectation value of the beampattern can be determined:

$$\begin{aligned} \mathrm{E}\left\{\tilde{\Psi}^{\mathrm{BF}}(\omega, \mathbf{u})\right\} = & \sum_{m=1}^{M} \sum_{\substack{n=1 \\ n \neq m}}^{M} W_m^*(\omega) W_n(\omega) H^{\mathrm{M}}(\omega, \mathbf{u}) \left(H^{\mathrm{M}}(\omega, \mathbf{u})\right)^* \\ & \cdot \exp\{j \tfrac{\omega}{c} (\mathbf{r}_m - \mathbf{r}_n)^{\mathrm{T}} \mathbf{u}\} \\ & + \sum_{m=1}^{M} |W_m(\omega)|^2 |H^{\mathrm{M}}(\omega, \mathbf{u})|^2 \\ & + \sum_{m=1}^{M} |W_m(\omega)|^2 E\left\{|\Delta H_m^{\mathrm{M}}(\omega, \mathbf{u})|^2\right\} . \end{aligned} \tag{12.30}$$

For the succeeding investigation it is assumed that the deviations of the single microphones have identical statistical distributions. With $\varepsilon^2(\omega, \mathbf{u}) = E\{|\Delta H^{\mathrm{M}}(\omega, \mathbf{u})|^2\}$ the last equation can be reformulated:

$$E\left\{\tilde{\Psi}^{\mathrm{BF}}(\omega, \mathbf{u})\right\} = \Psi^{\mathrm{BF}}(\omega, \mathbf{u}) + \varepsilon^2(\omega, \mathbf{u}) \sum_{m=1}^{M} |W_m(\omega)|^2 \tag{12.31}$$

or

$$E\left\{\tilde{\Psi}^{\mathrm{BF}}(\omega, \mathbf{u})\right\} = \Psi^{\mathrm{BF}}(\omega, \mathbf{u}) + \varepsilon^2(\omega, \mathbf{u}) K(\omega) . \tag{12.32}$$

---

[8] This approach corresponds to the analyses in [18] and [12]. In contrast to those analyses, in the present approach a generalized model for the deviations of the sensor transfer functions is utilized.

Thus, in the statistical mean the microphone deviations cause the real beampattern to be expanded for all directions relative to the nominal beampattern. This amplification is proportional to the susceptibility of the beamformer. This means in particular that the nulls of the beampattern are shaped less clearly. For the case of a first-order differential array it has been shown in [3] that phase deviations of the microphone transfer functions move the positions of the nulls within the pattern whereas amplitude deviations reduce the depth of the "nulls".

12.3.3 Effect of Microphone Mismatch on Adaptive Beamformers

While non-adaptive beamformers are designed by predefined *sound field* properties without taking into account the microphone deviations of a present microphone array realization the optimization of adaptive beamformers is based directly on *microphone* signals. Thus, the adjustment of the filters depend on the microphone deviations on hand. According to Sec. 12.2.3.1 the power of the beamformer output signal is to be minimized:

$$\min_{\boldsymbol{W}(\omega)} \left\{ \boldsymbol{W}^{\mathrm{H}}(\omega)\, \tilde{\boldsymbol{\Phi}}^{\mathrm{M}}(\omega)\, \boldsymbol{W}(\omega) \right\} . \tag{12.33}$$

The power density matrix of the microphone signals $\tilde{\boldsymbol{\Phi}}^{\mathrm{M}}(\omega)$ implicitly includes the influence of the microphone deviations. The directional constraint according to Eq. 12.21 shall preserve an undistorted response for the desired direction $\mathbf{u}_0$:

$$\boldsymbol{W}^{\mathrm{H}}(\omega)\, \boldsymbol{H}(\omega, \mathbf{u}_0)\, \boldsymbol{E}_0(\omega) = 1 \quad \text{and} \quad \boldsymbol{W}^{\mathrm{H}}(\omega)\, \boldsymbol{W}(\omega) \leq K_{\max} . \tag{12.34}$$

This constraint does not depend on the input data since it is stiffly implemented into the algorithm. That implies that microphone deviations are taken into account for the power minimization in Eq. 12.33 but not for the constraint in Eq. 12.34. Besides the matrix $\tilde{\boldsymbol{\Phi}}^{\mathrm{M}}(\omega)$ the optimal solution for the filters reads equal to Eq. 12.22:

$$\tilde{\boldsymbol{W}}_{\mathrm{MVDR}}^{(\mu)}(\omega) = \frac{\left(\tilde{\boldsymbol{\Phi}}^{\mathrm{M}}(\omega) + \mu(\omega)\, \boldsymbol{I}_{M \times M} \right)^{-1} \boldsymbol{H}(\omega, \mathbf{u}_0)\, \boldsymbol{E}_0(\omega)}{\boldsymbol{E}_0^{\mathrm{H}}(\omega)\, \boldsymbol{H}^{\mathrm{H}}(\omega, \mathbf{u}_0) \left(\tilde{\boldsymbol{\Phi}}^{\mathrm{M}}(\omega) + \mu(\omega)\, \boldsymbol{I}_{M \times M} \right)^{-1} \boldsymbol{H}(\omega, \mathbf{u}_0)\, \boldsymbol{E}_0(\omega)} . \tag{12.35}$$

If the microphones' transfer characteristics deviate for the steering direction then despite keeping the directional constraint of Eq. 12.34 in general the constraint is not met for the actual microphone transfer functions $\tilde{\boldsymbol{H}}(\omega, \mathbf{u}_0)$:

$$\boldsymbol{W}^{\mathrm{H}}(\omega)\, \tilde{\boldsymbol{H}}(\omega, \mathbf{u}_0)\, \boldsymbol{E}_0(\omega) \neq 1 . \tag{12.36}$$

Thus, there is some room to move for the adjustment of the beamformer filters, that as a matter of principle is used for further reduction of the output power according to Eq. 12.33. Because of the microphone deviations the constraint doesn't take effect completely.

If the beamformer is adapted during speech activity the power density-matrix $\tilde{\boldsymbol{\Phi}}^{\mathrm{M}}(\omega)$ contains strongly correlated components. This may lead to a compensation of the speech signal according to the signal cancellation effect investigated in Sec. 12.2.3.3. However, in the further analysis we will restrict solely to the case where adaptation is performed in speech pauses only. Depending on the correlation properties of the noise sound field also a compensation effect occurs in this case. This effect may affect the beamformer transfer function for any direction of incidence. As the main lobe of the beamformer picks up a considerable portion of noise power, the steering direction and thus the response for the desired signal is affected in particular.

This effect has been studied in more detail in [4] by using a simple example for concrete mismatch: a broadside beamformer consisting of two omnidirectional microphones with spacing d is considered. The noise sound field is supposed to be diffuse. The transfer functions of the two microphones are specified by the real positive values η_1 and η_2 and shall not depend on frequency nor on the direction of incidence:

$$\tilde{H}_1^{\mathrm{M}}(\omega, \mathbf{u}) = \eta_1 \quad \text{and} \quad \tilde{H}_2^{\mathrm{M}}(\omega, \mathbf{u}) = \eta_2 \; . \tag{12.37}$$

For the diffuse sound field it is:

$$\tilde{\Phi}_{m,n}^{\mathrm{M,i}}(\omega) = \eta_m \, \eta_n \, \mathrm{si}\!\left(\omega \frac{(n-m)\,d}{c}\right), \tag{12.38}$$

with $\mathrm{si}(x) = \sin(x)/x$. With the acoustic delay vector for broadside steering $\boldsymbol{E}_0(\omega) = [1, 1]^\mathrm{T}$ the filters of an MVDR beamformer result to

$$\tilde{\boldsymbol{W}}_{\mathrm{MVDR}}(\omega) = \frac{\eta_1 \eta_2}{\eta_1^2 + \eta_2^2 - 2\,\eta_1\eta_2\,\mathrm{si}(\omega\frac{d}{c})} \begin{bmatrix} \frac{\eta_2}{\eta_1} - \mathrm{si}\!\left(\omega\frac{d}{c}\right) \\ \frac{\eta_1}{\eta_2} - \mathrm{si}\!\left(\omega\frac{d}{c}\right) \end{bmatrix} . \tag{12.39}$$

For the case of identical microphones ($\eta_1 = \eta_2$) a delay-and-sum beamformer with $\tilde{W}_{\mathrm{MVDR},1}(\omega) = \tilde{W}_{\mathrm{MVDR},2}(\omega) = \frac{1}{2}$ results. For this constellation no superdirective effect is achievable [7]. If the microphone characteristics differ ($\eta_1 \neq \eta_2$) filter curves result which deviate from the delay-and-sum beamformer and depend on frequency.

For the desired direction the beamformer's frequency response is depicted for the given example in Fig. 12.3. Obviously, for mismatched microphones a cancellation effect occurs which increases to lower frequencies. An attenuation of 6 dB is reached for [4]

$$f_{6\,\mathrm{dB}} \approx \frac{\sqrt{3}\,c}{2\pi\,d} \frac{|\eta_1 - \eta_2|}{\sqrt{\eta_1\,\eta_2}} . \tag{12.40}$$

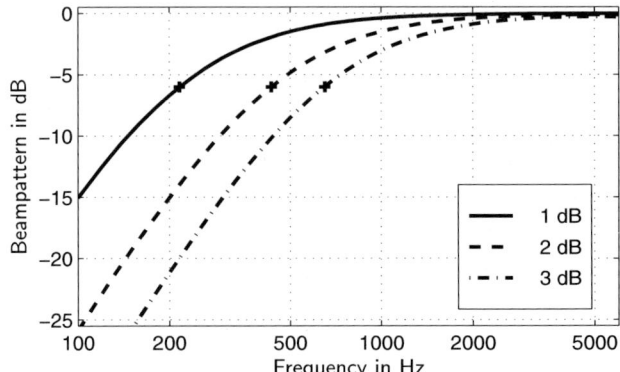

Fig. 12.3. Beampattern for the desired direction $\Psi^{\mathrm{BF}}(\omega, \mathbf{u}_0)$ for various relative deviations of the microphone sensitivities. The frequencies $f_{6\,\mathrm{dB}}$ are marked.

For a relative deviation $\frac{|\eta_1 - \eta_2|}{\sqrt{\eta_1 \eta_2}}$ of 1 dB and a microphone spacing of $d = 5$ cm a frequency $f_{6\,\mathrm{dB}} \approx 216$ Hz results.

In this simple example the mismatch of the microphones takes effect as an high-pass filter for the desired direction despite the flat response which has been specified by the directional constraint. The transfer characteristics of real microphones are much more complex: in general the transfer functions of the microphones depend on frequency as well as on the direction of incidence. The deviations concern magnitudes as well as phases.

12.3.4 Comparison of Fixed and Adaptive Beamformers

Fixed beamformers are designed on the basis of pre-defined characteristics of the *sound field* disregarding the actual microphone deviations. The optimal solution for an adaptive beamformer however, is based directly on the *microphone signals* and therefore depends on the present microphone deviations. Solely the directional constraint is specified a priori for ideal microphones without considering concrete deviations. The microphone deviations have contrary effects on fixed and adaptive beamformers: while fixed beamformers show an expansion of the beampattern in the statistical mean, adaptive beamformers tend to attenuate the output signal.

12.4 Calibration Techniques and their Limits for Real-World Applications

12.4.1 Calibration of Single Microphones

As discussed in the preceding section multi-channel noise reduction systems lose performance for non-ideal microphones. For array processing the magnitude as well as the phase of the frequency response of each single microphone are important.

In order to determine in which range microphone mismatch may occur in practical applications a series of microphones was measured and evaluated. All microphones had the same nominal characteristics and were of the type AKG Q400. This type of microphone is a differential microphone which has been designed particularly for speech signal acquisition in automobiles.

12.4.1.1 Model for Microphone Deviations

First, it has to be specified which quantities are investigated and how deviations are modeled.

Nominal Transfer Characteristics

The transfer functions of the microphones are considered to depend on the angular frequency ω and the direction of incidence $\mathbf{u}$. The nominal transfer function $G_{\text{ref}}^{\text{M}}(\omega, \mathbf{u})$ describes the conversion of the acoustic sound pressure signal into an electrical output signal for an ideal microphone. The transfer function can be separated into three factors: sensitivity $T_{\text{ref}}^{\text{M}}$, frequency response $E_{\text{ref}}^{\text{M}}(\omega)$ and directional response $H_{\text{ref}}^{\text{M}}(\omega, \mathbf{u})$:

$$G_{\text{ref}}^{\text{M}}(\omega, \mathbf{u}) = T_{\text{ref}}^{\text{M}} \, E_{\text{ref}}^{\text{M}}(\omega) \, H_{\text{ref}}^{\text{M}}(\omega, \mathbf{u}) \ . \tag{12.41}$$

The sensitivity is usually specified for the frequency $f = 1\,\text{kHz}$. For this reason, the nominal frequency response is normalized to this frequency:

$$E_{\text{ref}}^{\text{M}}(\omega = 2\pi \cdot 1\,\text{kHz}) = 1. \tag{12.42}$$

With this definition the nominal directional response $H_{\text{ref}}^{\text{M}}(\omega, \mathbf{u})$ is not distorted for the main direction $\mathbf{u}^{\text{M}}$ since it describes the relative directional properties relative to the main direction:

$$H_{\text{ref}}^{\text{M}}(\omega, \mathbf{u}^{\text{M}}) = 1 \ . \tag{12.43}$$

For analyzing the influence of mismatch the relative deviations of the microphones are important. Therefore, the mean value of the measured quantities are taken as nominal value for the microphones. The corresponding results are depicted in Figs. 12.4 and 12.5. The nominal sensitivity has been determined to $T_{\text{ref}}^{\text{M}} = 611\,\frac{\text{mV}}{\text{Pa}}$.

Microphones with Deviations

The transfer functions $G_n^{\text{M}}(\omega, \mathbf{u})$ of real-world microphones show deviations among one another. In this investigation these deviations are assigned completely to the directional responses $H_n^{\text{M}}(\omega, \mathbf{u})$, whereas the frequency response $E_n^{\text{M}}(\omega) = E_{\text{ref}}^{\text{M}}(\omega)$ and the sensitivity $T_n^{\text{M}} = T_{\text{ref}}^{\text{M}}$ are considered as ideal for all

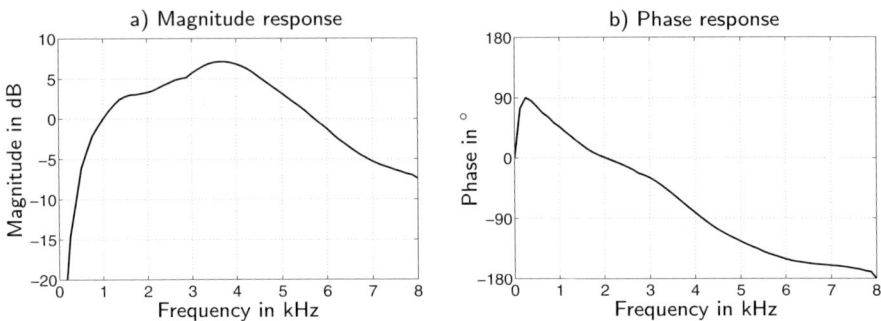

Fig. 12.4. Nominal frequency response $E_{\text{ref}}^{\text{M}}(\omega)$ for the microphone type analyzed. According to the definition there is $E_{\text{ref}}^{\text{M}}(\omega = 2\pi \cdot 1\,\text{kHz}) = 1$.

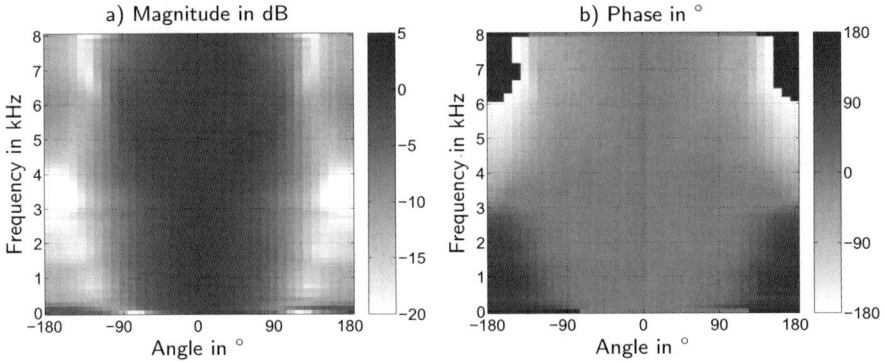

Fig. 12.5. Nominal directional response $H_{\text{ref}}^{\text{M}}(\omega, \mathbf{u})$ separated in magnitude and phase for the microphone type analyzed. According to the definition it is $H_{\text{ref}}^{\text{M}}(\omega, \mathbf{u}^{\text{M}}) = 1$.

microphones. Thus, the analysis of the microphone mismatch can be accomplished purely on the basis of the directional responses $H_n^{\text{M}}(\omega, \mathbf{u})$. Because of the equalized reference value according to Eq. 12.43 the results can be evaluated directly without a further normalization. In contrast to the preceding section here multiplicative deviations are considered instead of additive ones:

$$H_n^{\text{M}}(\omega, \mathbf{u}) = \eta_n^{\text{M}}(\omega, \mathbf{u}) \cdot H_{\text{ref}}^{\text{M}}(\omega, \mathbf{u}) \qquad (12.44)$$

The multiplicative deviations $\eta_n^{\text{M}}(\omega, \mathbf{u})$ are regarded as complex random values with

$$\mathcal{E}\{\eta_n^{\text{M}}(\omega, \mathbf{u})\} = 1 \ . \qquad (12.45)$$

From the multiplicative deviations the magnitude and phase mismatch can directly be determined.

12.4.1.2 Measurement of Microphone Deviations

For $N_{\text{Mic}} = 47$ microphones the transfer functions were measured in a plane using a steerable turntable. For each microphone transfer functions were determined for equidistant angles with distance $\Delta\theta = 10°$ at a sampling rate of $f_A = 16\,\text{kHz}$ with 65 frequency points.[9]

Equalization of Microphone Responses

In order to determine deviations, first, the nominal values have to be specified. As mentioned in Sec. 12.4.1.1 the average of the measured transfer functions $\overline{G^{\text{M}}}(\omega, \mathbf{u})$ can be taken as nominal value $G_{\text{ref}}^{\text{M}}(\omega, \mathbf{u})$.[10] In order to accomplish an equalization for the main direction $\mathbf{u}^{\text{M}}$ the measured microphone transfer functions are normalized:

$$H_n^{\text{M}}(\omega, \mathbf{u}) = \frac{G_n^{\text{M}}(\omega, \mathbf{u})}{\overline{G^{\text{M}}}(\omega, \mathbf{u}^{\text{M}})} . \qquad (12.46)$$

Then, the nominal directional response $H_{\text{ref}}^{\text{M}}(\omega, \mathbf{u})$ corresponds to the arithmetic mean over all measured microphones:

$$\overline{H^{\text{M}}}(\omega, \mathbf{u}) = \frac{1}{N_{\text{Mic}}} \sum_{n=1}^{N_{\text{Mic}}} H_n^{\text{M}}(\omega, \mathbf{u}) . \qquad (12.47)$$

Determination of Deviations

The deviations are calculated according to Eq. 12.44:

$$\eta_n^{\text{M}}(\omega, \mathbf{u}) = \frac{H_n^{\text{M}}(\omega, \mathbf{u})}{\overline{H^{\text{M}}}(\omega, \mathbf{u})} . \qquad (12.48)$$

By these definitions the following conditions are ensured:[11]

$$\overline{H^{\text{M}}}(\omega, \mathbf{u}^{\text{M}}) = 1 , \qquad (12.49)$$

$$\overline{\eta^{\text{M}}}(\omega, \mathbf{u}) = 1 . \qquad (12.50)$$

For the analysis of the multiplicative deviations $\eta_n^{\text{M}}(\omega, \theta)$ the magnitudes

---

[9] For the sake of a better readability in the further descriptions continuous quantities ω and $\mathbf{u}$ are used instead of the discrete quantities ω_ν and θ_μ of the measurement points.

[10] The arithmetic mean of a quantity is marked by an overline.

[11] Within Eq. 12.48 the denominator can take on very small values for example near a null of the directional response. By averaging over many measurements the exact value zero doesn't occur any more – however, very small values of the denominator lead to high values for the deviations $\eta_n^{\text{M}}(\omega, \mathbf{u})$. In these areas the phases for the respective angles are not very reliable.

$$20 \log_{10} \left| \eta_n^{\mathrm{M}}(\omega, \theta) \right| \mathrm{dB} \tag{12.51}$$

and the phases

$$\arg\{\eta_n^{\mathrm{M}}(\omega, \theta)\} \tag{12.52}$$

are evaluated separately. The results are presented in Sec. 12.4.1.4.

12.4.1.3 Calibration of Microphones

The objective of calibration is to measure existing deviations and to compensate for them. In order to investigate the effect of a microphone calibration residual deviations after the calibration are determined, evaluated, and compared to the deviation of the uncalibrated microphones.

For the calibration each microphone is measured and a calibration filter $A_n(\omega)$ is determined on the basis of this measurement. Since calibration is accomplished by filtering the microphone output signal a calibration can only correct for a frequency dependent deviation but not for deviations which depend on the angle of incidence. As a matter of principle a complete match of microphone transfer functions – i. e. for all frequencies and for all angles of incidence – is not possible. Therefore, a calibration is mostly performed for one specific direction of incidence, which typically is the direction of the desired signal.

In this section, two calibration methods are compared:

- *Calibration of the sensitivity*
  The deviation of the directional pattern at one discrete frequency f_s in the main direction $\mathbf{u}^{\mathrm{M}}$ is compensated by a real-valued scalar factor A_n^s:

$$\tilde{H}_n^{\mathrm{M,s}}(\omega, \mathbf{u}) = A_n^\mathrm{s}\, H_n^{\mathrm{M}}(\omega, \mathbf{u})\,, \quad \text{with } A_n^\mathrm{s} = \frac{1}{\left|H_n^{\mathrm{M}}(2\pi f_\mathrm{s}, \mathbf{u}^{\mathrm{M}})\right|} . \tag{12.53}$$

This is equivalent to a calibration of the microphone's sensitivity. After this kind of calibration the condition according to Eq. 12.49 doesn't hold any more. A re-normalization becomes necessary which may shift the nominal values marginally:

$$H_n^{\mathrm{M,s}}(\omega, \mathbf{u}) = \frac{\tilde{H}_n^{\mathrm{M,s}}(\omega, \mathbf{u})}{\overline{\tilde{H}^{\mathrm{M,s}}}(\omega, \mathbf{u}^{\mathrm{M}})} . \tag{12.54}$$

- *Calibration of the frequency response*
  The directional response in main direction $\mathbf{u}^{\mathrm{M}}$ is equalized for the whole frequency range by a complex-valued filter $A_n^\mathrm{r}(\omega)$:

$$H_n^{\mathrm{M,r}}(\omega, \mathbf{u}) = A_n^\mathrm{r}(\omega)\, H_n^{\mathrm{M}}(\omega, \mathbf{u})\,, \quad \text{with } A_n^\mathrm{r}(\omega) = \frac{1}{H_n^{\mathrm{M}}(\omega, \mathbf{u}^{\mathrm{M}})} . \tag{12.55}$$

Thus, the frequency response in main direction is equalized for angle and phase deviations for each microphone. After having performed the calibration the residual deviations are determined according to Eq. 12.48 with x ∈ {s,r}:

$$\eta_n^{M,x}(\omega, \mathbf{u}) = \frac{H_n^{M,x}(\omega, \mathbf{u})}{\overline{H^{M,x}(\omega, \mathbf{u})}} \ . \tag{12.56}$$

12.4.1.4 Results of the Measurements

By measuring the transfer functions $G_n^M(\omega, \theta)$ for the $N_{Mic} = 47$ microphones the random variables were determined according to Eqs. 12.51 and 12.52. The mean value and the standard deviation of these random variables were calculated, respectively.

The measured data sets were used to investigate the effect of the methods for microphone calibration presented in Sec. 12.4.1.3. Since in this section we are focussing on deviations, only standard deviations are depicted. The nominal microphone transfer functions are shown in Sec. 12.4.1.1 on page 434.

Directional Pattern for Uncalibrated Microphones

For the uncalibrated microphones the resulting standard deviations of the logarithmic magnitudes and of the phases are shown in Fig. 12.6. The statistical analysis of magnitudes and phases according to Eqs. 12.51 and 12.52 is problematic in the vicinity of nulls of the transfer function. There, the phase is strongly varying and the logarithmic magnitude is showing very small values with high dynamical range. Thus, the standard deviation is taking relative high values in these areas. For the main direction uncalibrated microphones show a standard deviation of about 0.5 dB to 1 dB for the logarithmic magnitudes and a standard deviation of about 4° to 5° for the phases.

Directional Pattern for Sensitivity-Calibrated Microphones

For the measured data base of transfer functions a calibration of the sensitivities was performed according to Sec. 12.4.1.3. For this purpose, each microphone transfer function was matched for the frequency $f_s = 1\,\text{kHz}$ and the main direction $\theta = 0°$ by a real-valued scalar factor. After this calibration the statistical analysis was accomplished. The results are depicted in Fig. 12.7. Using the multiplicative model for the deviations the effect of the sensitivity calibration gets obvious: in Fig. 12.7 b the standard deviation of the logarithmic magnitude for $f = 1\,\text{kHz}$ in main direction is 0 dB, whereas the standard deviation of the phase at this point in the pattern doesn't change significantly compared to the uncalibrated case. This becomes evident by comparing Fig. 12.7 d for the sensitivity calibration to Fig. 12.6 d for the case without any calibration.

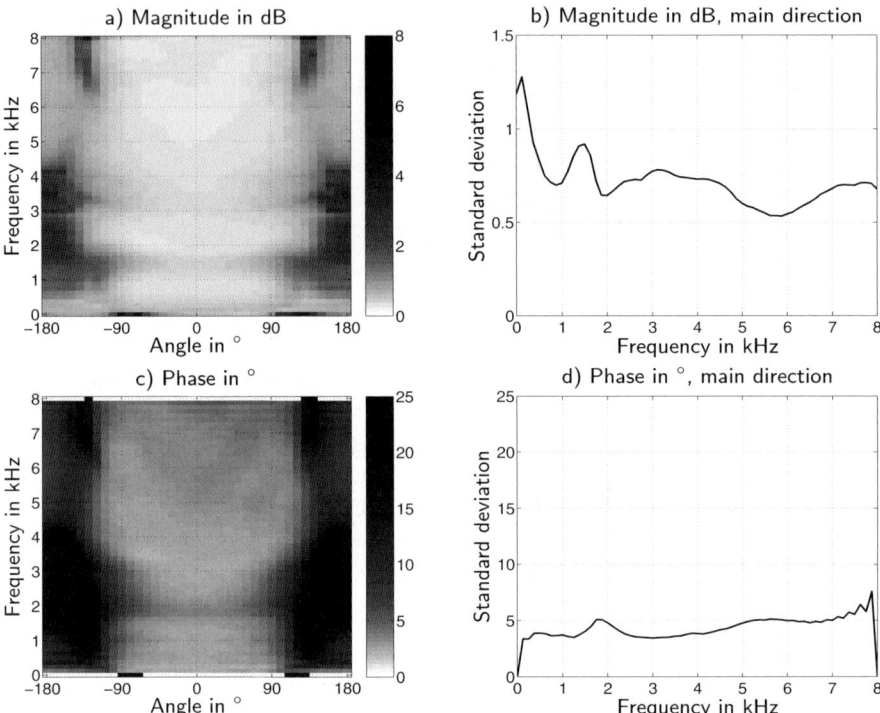

Fig. 12.6. Results for the multiplicative deviations $\eta_n^{\mathrm{M}}(\omega,\mathbf{u})$ for uncalibrated microphones. a), b) Standard deviation of the logarithmic magnitude $20\log_{10}|\eta_n^{\mathrm{M}}(\omega,\mathbf{u})|$ dB. c), d) Standard deviation of the phase $\arg\{\eta_n^{\mathrm{M}}(\omega,\mathbf{u})\}$.

Directional Pattern for Response-Calibrated Microphones

According to Sec. 12.4.1.3 also a frequency response calibration has been investigated by using the measured data base. For the main direction $\theta=0°$ the magnitudes as well as the phases of the microphone transfer functions were equalized by calibration filters. The corresponding results of the statistical analysis are shown in Fig. 12.8.

After a response calibration (in the ideal case) no deviation should remain for the main direction $\theta=0°$. Therefore, the standard deviations of the investigated random variables are exactly zero at these points. For this reason the corresponding figures for the main direction are not depicted.

Analyzing Figs. 12.8 a and 12.8 b it is obvious that despite the response calibration considerable deviations remain especially in the backward half-space of the directional pattern of the microphones.

Discussion of the Different Calibration Methods

In Fig. 12.9 the statistical results of uncalibrated and calibrated microphones are directly compared at different frequencies. Instead of multiplicative

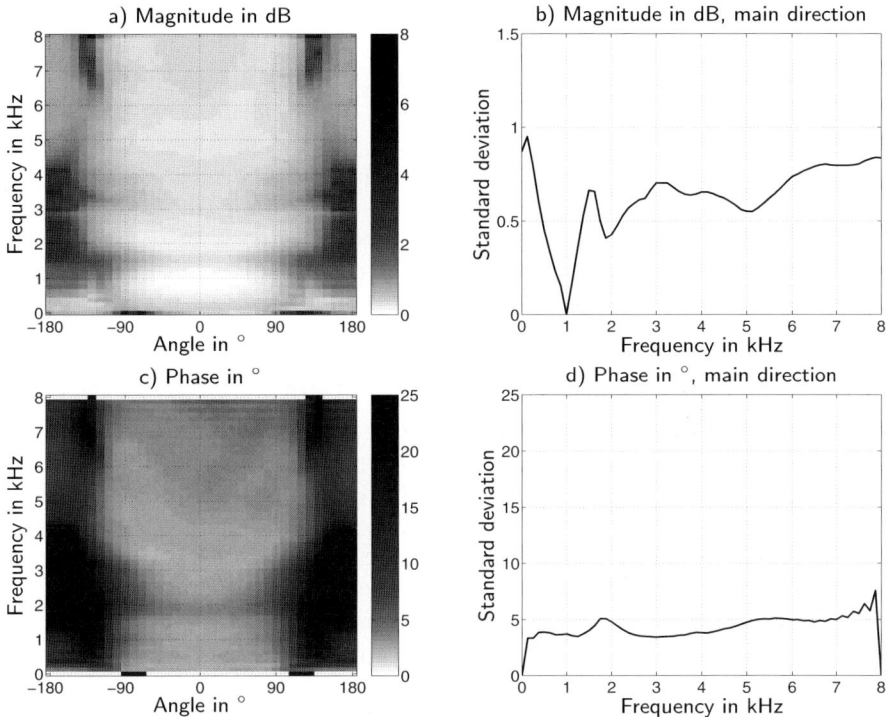

Fig. 12.7. Results for the multiplicative deviations $\eta_n^{M,s}(\omega, \mathbf{u})$ for sensitivity-calibrated microphones. a), b) Standard deviation of the logarithmic magnitude $20\log_{10}|\eta_n^{M,s}(\omega, \mathbf{u})|$ dB. c), d) Standard deviation of the phase $\arg\{\eta_n^{M,s}(\omega, \mathbf{u})\}$.

deviations again complex valued additive deviations $\Delta H^M(\omega, \mathbf{u})$ are considered here in order to get a more compact representation.

It is evident that calibration cannot improve the deviations at all angles. Regarding the directional pattern a calibration is most effective in the local surrounding of the direction for which the calibration has been performed. Particularly for higher frequencies neither the sensitivity calibration nor the response calibration are capable to reduce the microphone deviations significantly in the backward half-space.

For all depictions in Fig. 12.9 strong deviations occur around an angle of $\theta = 180°$, despite a zero is expected in this range. Reasons for that phenomenon may be the fact that the position of the spatial nulls are varying from microphone to microphone and that the depths of these "nulls" are different.

Overall, this investigation gives an indication what kind of microphone specific deviations may occur in practical applications.

12.4.2 Analysis of Fixed Beamformers

The data base of measured single microphones offers the possibility of investigating the effect of microphone mismatch on beamforming by Monte Carlo simulations of concrete array realizations. In these simulations the deviations are no more considered as *statistical* quantities but as *deterministic* ones.

For different beamformer parametrizations simulation runs were conducted with $I = 500$ concrete array "realizations" each. For each array "realization" the transfer characteristics of the single microphones have been specified by randomly selecting M different microphones and taking the corresponding measurement data out of the data base. Thus, fictive microphone arrays are combined on the basis of measured directional patterns.

For each combined array $i \in \{1,\ldots,I\}$ beamformers were calculated for which the beampattern $\tilde{\Psi}_i^{\mathrm{BF}}(\omega,\mathbf{u})$ and the directivity $\tilde{D}_i^{\mathrm{BF}}(\omega)$ were evaluated.[12]

As array configuration a linear broadside geometry is considered, which consists of $M = 4$ microphones arranged equidistantly on the array axis $\mathbf{u}^{\mathrm{A}}$ with a spacing of $d = 5\,\mathrm{cm}$. The orientations of the single microphones are identical to the desired direction of the beamformer: $\mathbf{u}_m^{\mathrm{M}} = \mathbf{u}_0$.

The calibration methods described in the preceding section are applied, here: the simulations are carried out for uncalibrated, for sensitivity-calibrated and for response-calibrated microphone data.

For the design of fixed beamformers according to Sec. 12.2.3 ideal microphone characteristics are supposed. As constraint the susceptibility of the beamformer was limited to K_{max}. The actual value of K_{max} is a trade-off

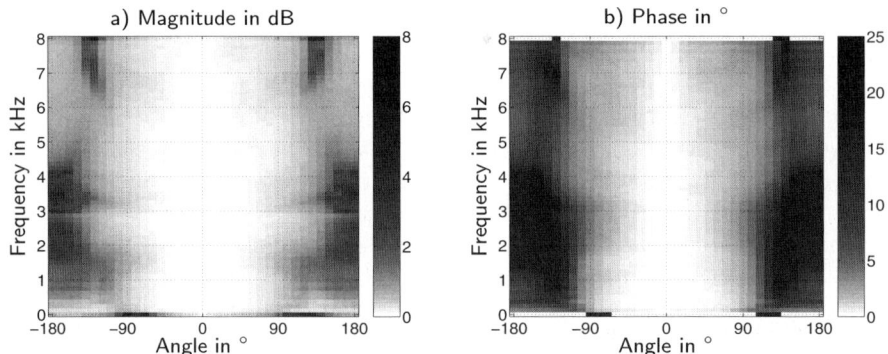

Fig. 12.8. Results for the multiplicative deviations $\eta_n^{\mathrm{M,r}}(\omega,\mathbf{u})$ for response-calibrated microphones. a) Standard deviation of the logarithmic magnitude $20\log_{10}|\eta_n^{\mathrm{M,r}}(\omega,\mathbf{u})|\,\mathrm{dB}$. b) Standard deviation of the phase $\arg\{\eta_n^{\mathrm{M,r}}(\omega,\mathbf{u})\}$. The standard deviations in main direction $\mathbf{u}^{\mathrm{M}}$ are both zero in accordance with the definition.

---

[12] Quantities which depend on concrete microphone deviations are marked with a tilde ($\sim$).

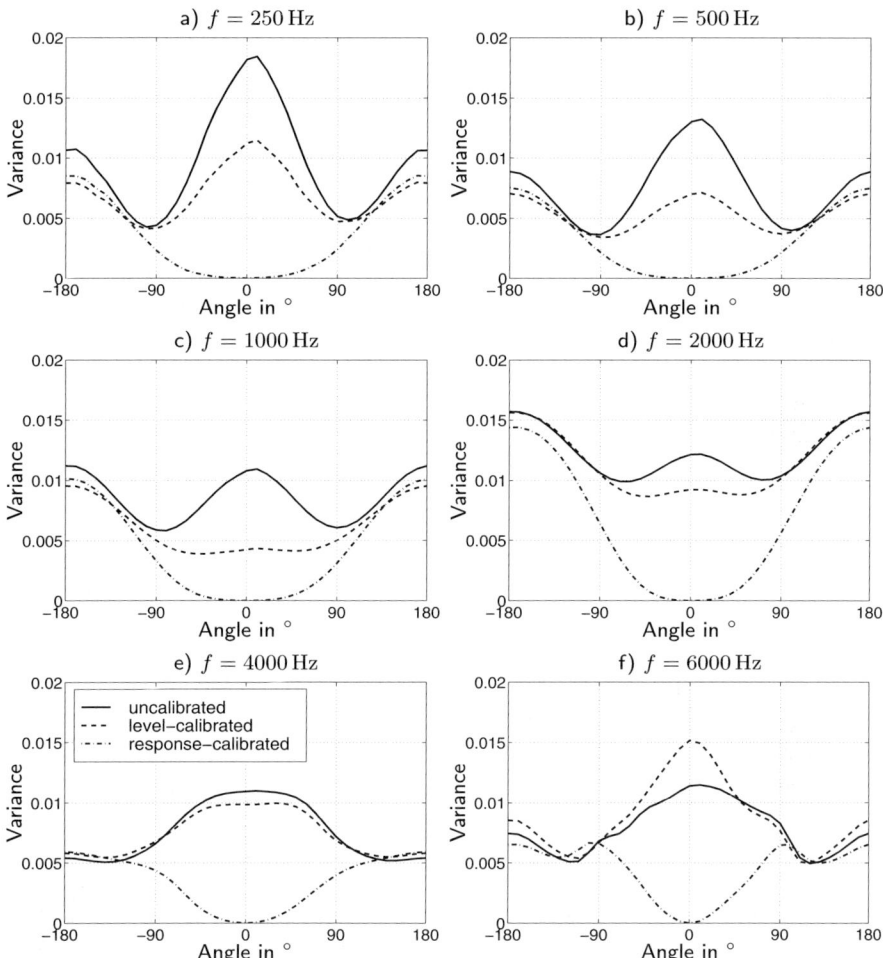

Fig. 12.9. Variance of the additive microphone deviations for the different calibration methods.

between the robustness against microphone mismatch on the one hand and the potential performance for noise reduction on the other hand. Thus, $K_{\max}$ should be chosen corresponding to the variance of the expected microphone deviations. Within the simulations different values $K_{\max} \in \{\frac{1}{M}, 1, 10, 100\}$ were analyzed. For the beamformer design ideal differential microphones of the type supercardioid [13] were assumed. The filters were determined independently of each calibration.

The beamformers were designed as MVDR beamformers according to Sec. 12.2.3.1 under the assumption of a diffuse sound field. Thus, the sound fields which are utilized for the design as well as for the evaluation agree.

From the evaluated directivity curves the effect of microphone deviations on the noise suppression performance can be deduced.

As result of the Monte Carlo simulations Fig. 12.10 shows the mean values of the directivities and beampatterns for the desired direction for uncalibrated and response-calibrated microphones. Compared to the delay-and-sum beam-

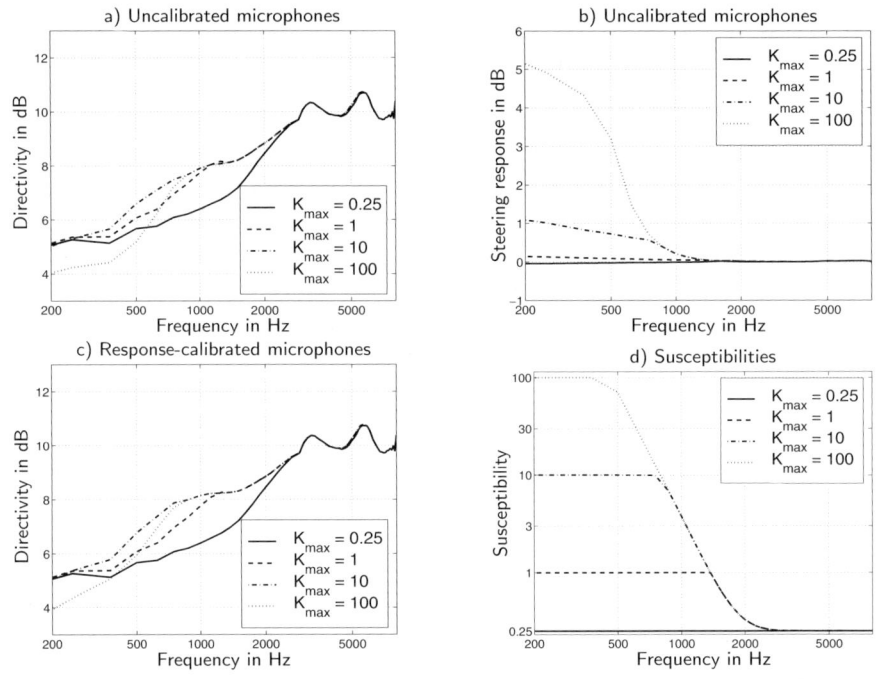

Fig. 12.10. Average directivity $\overline{\tilde{D}^{\mathrm{BF}}}(\omega)$ and average steering response $\overline{\tilde{\Psi}^{\mathrm{BF}}}(\omega, \mathbf{u}_0)$ for fixed beamformers for different limits for the susceptibilities. a, b) uncalibrated mirophones, c) response-calibrated microphones, d) susceptibility $K(\omega)$ of the investigated beamformers.

former ($K_{\max} = 0.25$) superdirective beamformers ($K_{\max} > 0.25$) show higher directivities which are increased by up to 2 dB. According to Figs. 12.10 a and c this value is achieved for frequencies of about 1000 Hz. As expected, in the frequency range above 3000 Hz no significant differences result for the different beamformers because the allowed maximum value of the susceptibility $K_{\max}$ is not exploited there (Fig. 12.10 d). In the frequency range below 500 Hz a high susceptibility has a negative effect:[13] due to the high filter amplifications the effect of microphone deviation is magnified. In Sec. 12.3.2 it has already been shown that the directionality of a fixed beamformer is reduced for high

---

[13] A similar behaviour has been observed in [12].

susceptibilities in the statistical mean. The average directivity of the single microphones is about 5 dB (see Fig. 12.14 on page 449).

According to the theoretical analysis in Sec. 12.3.2 the steering response $\tilde{\Psi}^{BF}(\omega, \mathbf{u}_0)$ equals the co-variance of the microphone transfer functions multiplied by the susceptibility of the beamformer. The results obtained by the Monte Carlo simulations in Fig. 12.10 b confirm this effect.

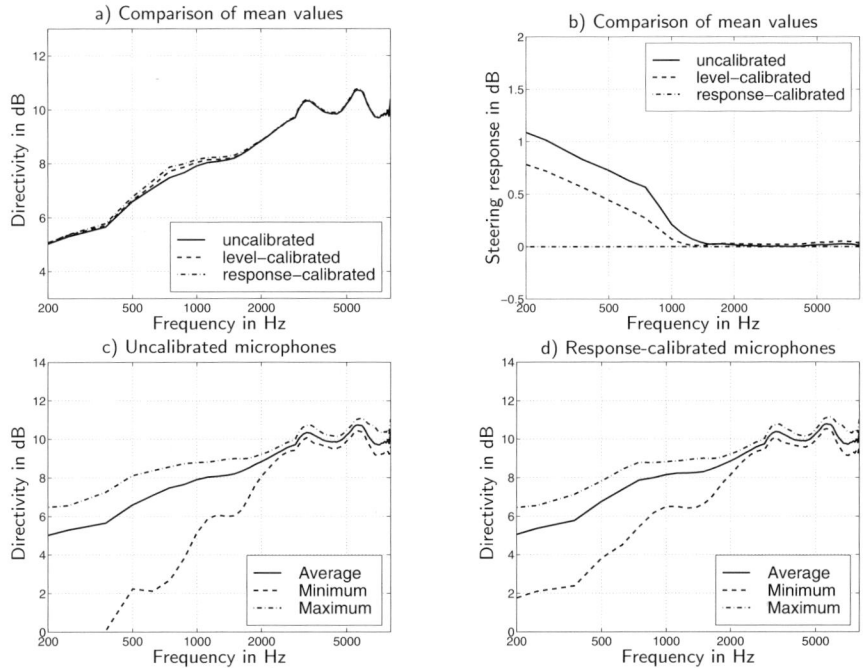

Fig. 12.11. Results for fixed beamformers with $K_{\max} = 10$. Comparison of a) average directivity and b) average steering response for different calibration methods. c, d) mean values and extreme values of the directivities for different calibration methods.

In Fig. 12.11 the results for the different calibration methods are compared for $K_{\max} = 10$. According to Fig. 12.11 a the directivity cannot be increased significantly by calibration of the microphones. This refers to the problem of matching single microphones as pointed out in Sec. 12.4.1.4. There, it has been noticed that microphone deviations cannot completely be compensated for (see Fig. 12.9).

For the case of the response-calibrated microphones – as expected – no distortion of the steering direction occurs (Fig. 12.11 b) since for the desired direction $\mathbf{u}_0$ the microphone deviations were equalized. By means of a sensitivity-calibration only a slight improvement compared to the uncalibrated case can be achieved. This is due to the fact that solely the magnitude

at a specific frequency ($f_\mathrm{s} = 1\,\mathrm{kHz}$) was matched and the phases were not considered at all.
As illustrated in Figs. 12.11 c and d the spread of the directivities is significantly reduced by the response calibration.

12.4.3 Analysis of Adaptive Beamformers

Adaptive beamformers permanently adjust their filter coefficients in order to adapt to varying acoustic situations. For stationary sound field properties the beamformer filters converge against the optimal Wiener filter solution [11].

This optimal solution depends on the sound field characteristics as well as on the microphone properties which are comprised in the power density matrix $\boldsymbol{\Phi}^\mathrm{M}(\omega)$ according to Sec. 12.2.3. With the modification for mismatched microphones introduced in Sec. 12.3.3 the power density matrix $\tilde{\boldsymbol{\Phi}}^\mathrm{M}(\omega)$ results. Thus, the optimal solution for the beamformer filters does not depend only on the properties of the sound field but also on the actual deviations of the microphone transfer functions. In the Monte Carlo simulations for each array "realization" a specific power density matrix $\tilde{\boldsymbol{\Phi}}_i^\mathrm{M}(\omega)$ has to be determined by using specific transfer functions of single microphones. Using this specific matrix the corresponding optimal beamformer filters $\tilde{\boldsymbol{W}}_{\mathrm{MVDR},i}^{(\mu)}(\omega)$ can be calculated which are in the same way specific for each array "realization".

For speech applications the adaptation of the beamformer filters is performed only during speech pauses in order to prevent the signal cancellation effect. Accordingly, the analysis of the simulated beamformers is accomplished by considering the noise sound field only, i.e. in the absence of the desired signal. For the investigation of the adaptive beamformers – analogous to the analysis of the fixed beamformers in the preceding section – a diffuse sound field is assumed. A method to determine the power density matrix $\tilde{\boldsymbol{\Phi}}_i^{\mathrm{M,i}}(\omega)$ in a diffuse noise field for given microphone transfer functions has been sketched in the preceding section. This method is documented in more detail in the appendix 12.A.

With the resulting normalized spectral density matrix $\tilde{\boldsymbol{\Phi}}_i^{\mathrm{M,i}}(\omega)$ the optimal solution for the beamformer filters can be calculated in accordance to Eq. 12.35 [8, 11]:

$$\tilde{\boldsymbol{W}}_{\mathrm{MVDR},i}^{(\mu)}(\omega) = \frac{\left(\tilde{\boldsymbol{\Phi}}_i^{\mathrm{M,i}}(\omega) + \mu(\omega, K_{\max})\,\boldsymbol{I}_{M\times M}\right)^{-1} \boldsymbol{E}_0(\omega)}{\boldsymbol{E}_0^\mathrm{H}(\omega)\left(\tilde{\boldsymbol{\Phi}}_i^{\mathrm{M,i}}(\omega) + \mu(\omega, K_{\max})\,\boldsymbol{I}_{M\times M}\right)^{-1} \boldsymbol{E}_0(\omega)} . \quad (12.57)$$

Using the parameter $\mu(\omega, K_{\max})$ and the identity matrix $\boldsymbol{I}_{M\times M}$ the normalized power density matrix $\tilde{\boldsymbol{\Phi}}_i^{\mathrm{M,i}}(\omega)$ can be regularized for any frequency ω, in such a way that the maximum value for the susceptibility $K_{\max}$ is satisfied:

$$\left\|\tilde{\boldsymbol{W}}_{\mathrm{MVDR},i}^{(\mu)}(\omega)\right\|^2 \leq K_{\max} . \quad (12.58)$$

In contrast to the design of fixed beamformers no assumptions about the microphone type have to be made for adaptive beamformers. As the power density matrix $\tilde{\boldsymbol{\Phi}}_i^{\mathrm{M},\mathrm{i}}(\omega)$ refers to ("real") microphone signals the directional properties of the single microphones – including the specific deviations – are already considered.

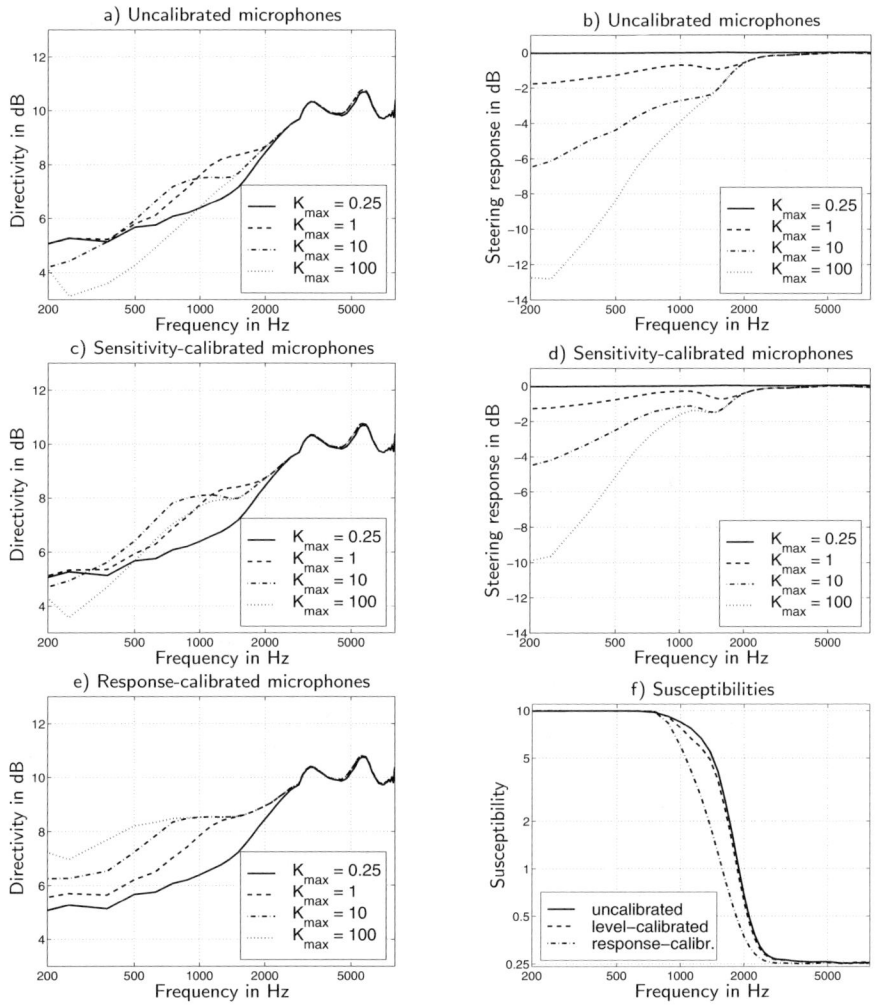

Fig. 12.12. Average directivity $\overline{\tilde{D}^{\mathrm{BF}}}(\omega)$ and average steering response $\overline{\tilde{\Psi}^{\mathrm{BF}}}(\omega, \mathbf{u}_0)$ for adaptive beamformers for different values $K_{\max}$. a, b) uncalibrated microphones, c, d) sensitivity-calibrated microphones, e) response-calibrated microphones, f) average susceptibility $K(\omega)$ for the different calibration methods.

As shown in Figs. 12.12 a and c high values of the susceptibility can have a negative effect on adaptive beamformers as well. However, this holds only for the cases where the microphone transfer functions $\tilde{H}_m^{\mathrm{M}}(\omega, \mathbf{u})$ show deviations for the steering direction $\mathbf{u}_0$: for response-calibrated microphones the directivity shows a monotonously increasing run over the susceptibility. In these cases a higher susceptibility never means a loss in directivity (see Fig. 12.12 e). Particularly for the lower frequencies the advantage of the calibration is evident.

In Figs. 12.12 b and d a strong attenuation for the steering direction in the low frequency range is conspicuous. At the first glance this does not seem plausible since the optimal beamformer filters $\tilde{\boldsymbol{W}}_{\mathrm{MVDR},i}^{(\mu)}(\omega)$ have been designed by implicitly considering the specific microphone deviations. As discussed in Sec. 12.3.3 the reason is that the concrete microphone deviations are considered of course for the power minimization at the beamformer output but not in the formulation of the directional constraint. However, the directional constraint was introduced with the aim to ensure an undistorted steering response.

The directional constraint for the beamformer filters $\boldsymbol{W}(\omega)$ is formulated for ideal microphones independently of the actual sound field or microphone properties:

$$H^{\mathrm{BF}}(\omega, \mathbf{u}_0) = \boldsymbol{W}^{\mathrm{H}}(\omega)\, \boldsymbol{H}(\omega, \mathbf{u}_0)\, \boldsymbol{E}_0(\omega) \stackrel{!}{=} 1 \,. \tag{12.59}$$

If deviations from the ideal microphone transfer functions occur it is not ensured that the a priori specified constraint according to Eq. 12.59 is met:

$$\tilde{H}^{\mathrm{BF}}(\omega, \mathbf{u}_0) = \boldsymbol{W}^{\mathrm{H}}(\omega)\, \tilde{\boldsymbol{H}}(\omega, \mathbf{u}_0)\, \boldsymbol{E}_0(\omega) \stackrel{?}{=} 1 \,. \tag{12.60}$$

The adaptive filters $\boldsymbol{W}(\omega)$ are adjusted in such a way that the output signal power of the beamformer is minimized while keeping the constraint of Eq. 12.59. If the microphone transfer functions deviate for the steering direction a margin results which as a matter of principle is used by the adaptive filters to further reduce the output power. Since for a diffuse sound field the noise is spatial equally distributed any spatial direction may be subject of this attenuation effect. This effect particularly affects the steering direction as the main lobe of the beamformer is aiming in this direction and is receiving a considerable portion of the total signal energy. With that, the reduction in directivity shown in Figs. 12.12 a and c can be explained.

For the response-calibration $\tilde{H}_m^{\mathrm{M}}(\omega, \mathbf{u}_0) = H_m^{\mathrm{M}}(\omega, \mathbf{u}_0)$ is valid. Thus, Eqs. 12.59 and 12.60 are equivalent. No distortion for the steering direction can occur and high filter amplifications – i.e. high susceptibilities – cause an increase of the directivity (see Fig. 12.12 e). In the statistical mean there are higher filter amplifications for uncalibrated microphones compared to calibrated microphones (see Fig. 12.12 f). This is again caused by the attenuation effect described above which requires higher filter amplifications for the signal attenuation. In Sec. 12.3.3 this phenomenon has been discussed theoretically by means of a simple example.

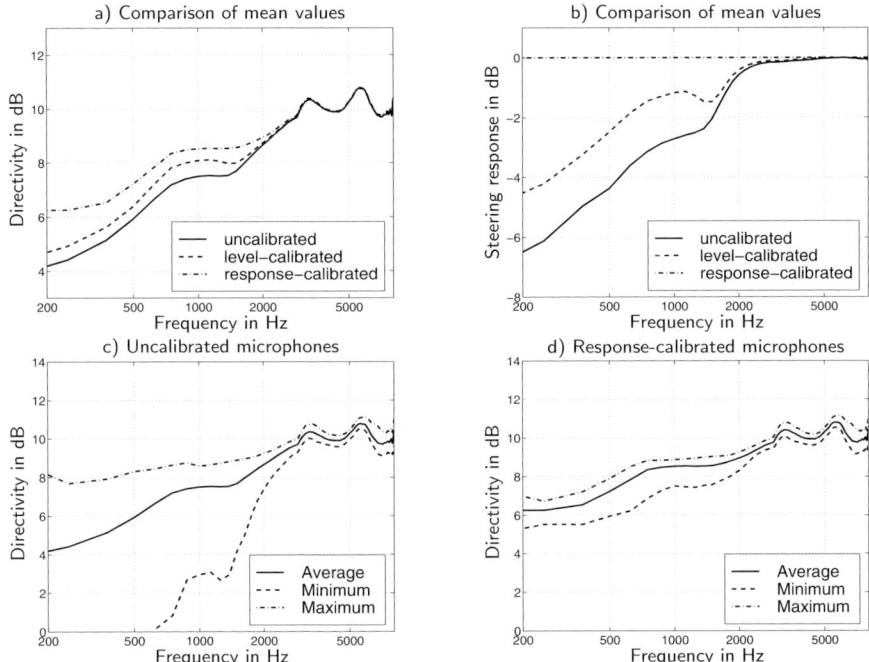

Fig. 12.13. Results for $K_{\max} = 10$ for adaptive beamformers. Comparison of a) average directivity and b) average steering response for different calibration methods. c, d) mean values and extreme values of the directivities for different calibration methods.

It has to be emphasized that the previously discussed attenuation for the steering direction is different to the well-known "signal cancellation" effect. Signal cancellation can only occur when signal components of the *desired* signal impinge onto the microphone array from directions other than the steering direction. This may occur due to an inaccurate time delay compensation or due to reflections within a room [23, 37]. However, in the present case any effect of the desired signal can be excluded as the beamformer filters are adapted solely on the basis of the *noise* signal. Consequently, the attenuation of the desired signal is caused only by the microphone deviations, as already described above. In the literature very few contributions exist which analyze the effect of microphone deviations on beamformers in the absence of the desired signal. In [10] and [29] the influence of an incorrect steering direction is analyzed in the absence of the desired signal. However, the respective approaches suppose ideal microphones and are therefore not suitable to consider uncalibrated microphones.

Fig. 12.13 illustrates the effect of the different calibration methods on adaptive beamformers with a maximum susceptibility of $K_{\max} = 10$. In contrast to fixed beamformers the directivity can be increased by up to 2 dB by means of microphone calibration (Fig. 12.13 a). Considering the steering response in Fig. 12.13 b the necessity of a calibration becomes evident. The sensitivity calibration shows good performance around $f=1\,\mathrm{kHz}$. But the response calibration performs clearly best for all frequencies $f<2\,\mathrm{kHz}$. It is also evident that the range of variation of the directivities $\tilde{D}_i^{\mathrm{BF}}(\omega)$ are significantly smaller in Fig. 12.13 d than in Fig. 12.13 c.

12.4.4 Comparison of Fixed and Adaptive Beamformers

According to Figs. 12.14 a and c fixed beamformers show an advantage over adaptive beamformers in the case of uncalibrated microphones and a diffuse noise field. In the low frequency range the average directivity of the adaptive beamformer is even lower than the average directivity of the single microphones. Due to microphone deviations the signal attenuation effect described in the preceding section as well as in Sec. 12.3.3 occurs. For the average steering responses contrary curves result for fixed and adaptive beamformers. In case of fixed beamformers the filters have been designed a priori without considering microphone deviations. In the statistical mean a moderate amplification of the steering direction $\mathbf{u}_0$ results for mismatched microphones. By contrast to this the filters of the adaptive beamformer were adapted specifically for the existing microphone deviations. This specific design induces an attenuation of the signals from the desired direction which may be more or less severe depending on the actual deviations.

A response calibration shows much more effect in the case of adaptive beamformers than in the case of fixed beamformers. Especially in the low frequency range the directivity of adaptive beamformers exceeds the ones of the fixed beamformers (Fig. 12.14 b).

When comparing the susceptibilities according to Fig. 12.14 d, the filters of fixed beamformers show in the average smaller amplifications than the (specific) filters of the adaptive beamformers. Particularly the specific optimization towards the existing microphone transfer functions requires higher filter amplifications in the case of the adaptive beamformer. If the microphones show deviations for the steering direction $\mathbf{u}_0$ additional freedom for the filter adjustment occurs within the specific design which is misused for an attenuation of the desired signal and a higher susceptibility.

In Fig. 12.15 the average beampatterns for $K_{\max}=10$ are depicted for the frequency $f=1\,\mathrm{kHz}$. Comparing Figs. 12.15 a and c the described attenuation effect for adaptive beamformers with uncalibrated microphones is evident: for almost all spatial directions adaptive beamformers show smaller values than fixed beamformers do. This effect increases towards lower frequencies. In contrast to the well-known "signal cancellation" effect this phenomenon could

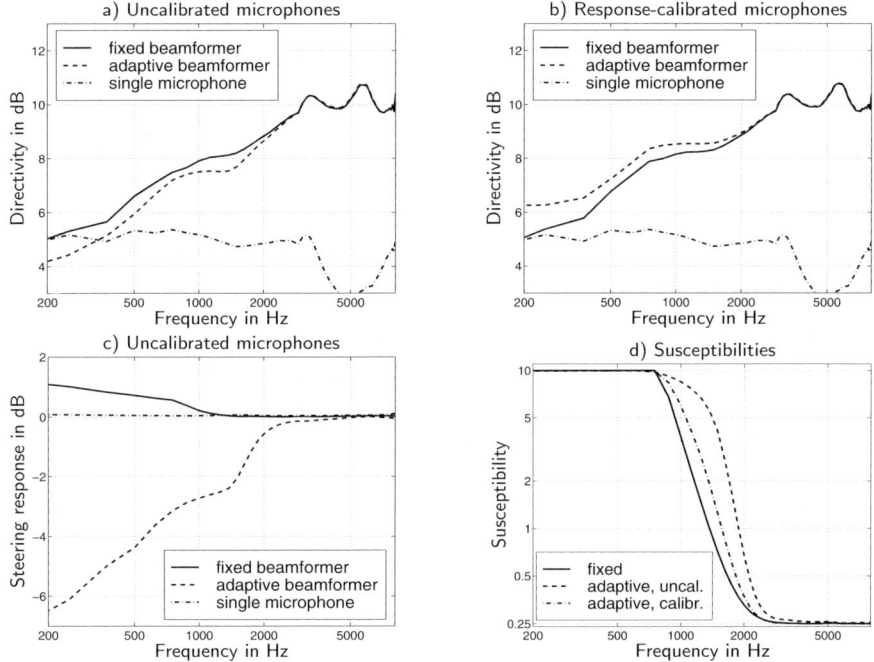

Fig. 12.14. Comparison of fixed and adaptive broadside beamformers for $K_{\max} = 10$. a) average directivity for uncalibrated microphones, b) average directivity for response-calibrated microphones, c) average steering response for uncalibrated microphones, d) (average) susceptibility.

more precisely be named as "pattern cancellation" since no desired signal was considered for the filter adjustment and the whole beampattern is affected.[14]

For calibrated microphones (Figs. 12.15 b and d) the "nulls" in the beampattern are more distinctive. Furthermore the constraint of an undistorted response for the desired direction $\mathbf{u}_0$ is fulfilled on principle.

12.5 Self-Calibration Techniques

As discussed in the preceding section the signal model that has been used for the beamformer design in Sec. 12.2.3 has turned out to be too simple for most practical applications. On the one hand a more complex acoustic situation is present: instead of a pure plane wave from the steering direction the desired signal may impinge also from other directions due to sound reflections. Also a mismatch between the steered direction and the real direction of incidence

---

[14] The "signal cancellation" effect occurs due to signal components which are correlated to the desired signal and impinge from directions other than the desired direction.

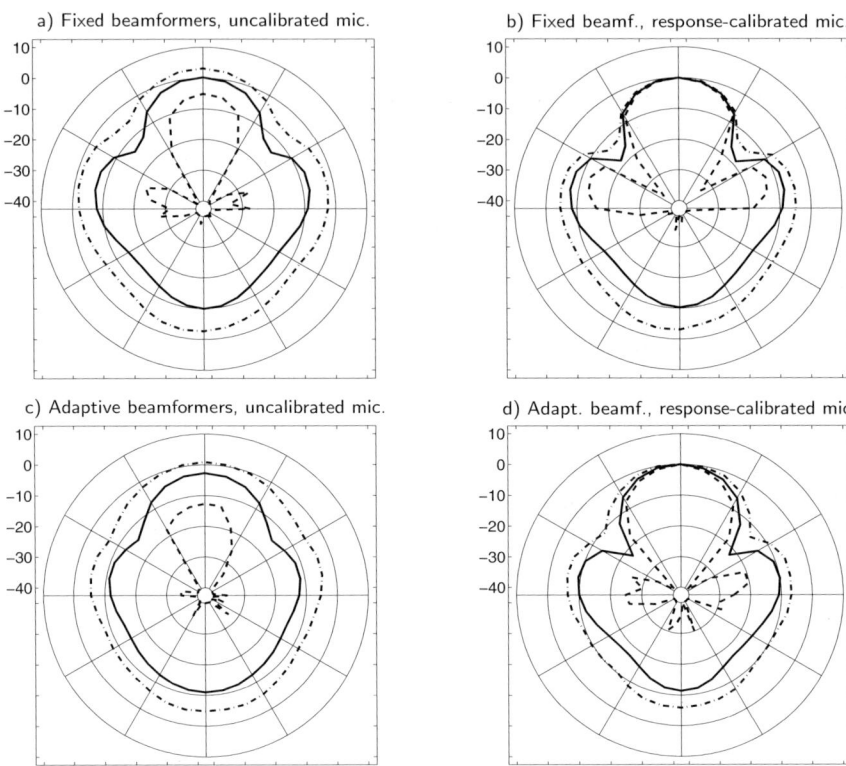

Fig. 12.15. Beampatterns for fixed and adaptive broadside beamformers for the frequency $f = 1\,\text{kHz}$ and the maximum susceptibility $K_{\max} = 10$: average values (——), maximum values (— ·) and minimum values (— —) each in dB. a) fixed beamformers with uncalibrated microphones, b) fixed beamformers with response-calibrated microphones, c) adaptive beamformers with uncalibrated microphones, d) adaptive beamformers with response-calibrated microphones.

of the desired signal may occur. On the other hand in this signal model ideal microphones are supposed. However, in practical applications tolerances resulting from the production process or caused by environmental stress occur by nature. Furthermore, microphone deviations may change over time due to aging effects. They therefore cannot be considered as constant over the whole life cycle.

Both, real room acoustics and microphone mismatch mean deviations form the signal model and therefore may impair the performance of a beamformer. This chapter is mainly focussed on deviations originating from microphones. The effects of microphone mismatch has been studied theoretically in Sec. 12.3 and by simulations with real measured data in Sec. 12.4. Particularly adaptive beamformers show a high susceptibility to microphone mismatch ("pattern cancellation" effect in Sec. 12.4.4). In this section a class of methods for performing a kind of response-calibration is presented. These methods work adaptively during operation of the beamformer in the background. Whereas in the preceding sections mainly continuous time signals have been considered, now, sampled time signals are assumed. The signal processing is performed in the sub-band domain with complex valued sub-band signals where the time index of the sub-sampled signals is denoted with n. It is supposed that the time domain input signals have already been decomposed by an analysis filter-bank [20]. With an appropriate synthesis filter-bank the sub-band signals at the output are combined to a time domain output signal.

12.5.1 Basic Unit

The methods proposed in this section are based on a simple basic unit. For the multi-channel case one or more instances of this basic unit are necessary to build up a self-calibrating stage. The basic unit relates two input signals to two output signals as depicted in Fig. 12.16. For each sub-band μ the reference signal $x_{\text{IC},\mu}(n)$ is filtered by an adaptive FIR filter having N coefficients $v_{\text{IC},\mu}(l,n)$ with $0 \leq l < N$ and n denoting the subsampled time index. As a result the calibrated output signal $x^{\text{c}}_{\text{IC},\mu}(n)$ is obtained:[15]

$$x^{\text{c}}_{\text{IC},\mu}(n) = \sum_{l=0}^{N-1} v^{*}_{\text{IC},\mu}(l,n)\, x_{\text{IC},\mu}(n-l)\,. \tag{12.61}$$

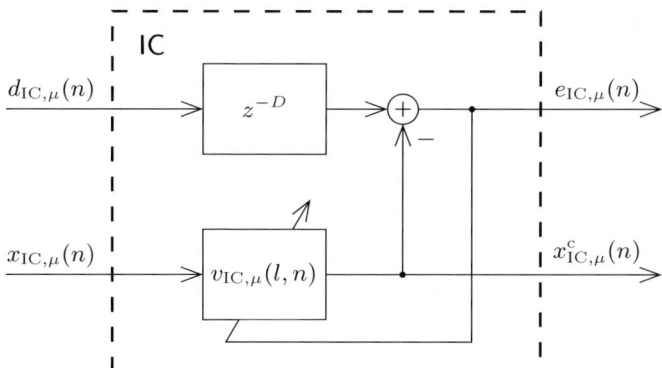

Fig. 12.16. Basic unit for self-calibration. The reference signal $x_{\text{IC},\mu}(n)$ is matched to the desired signal $d_{\text{IC},\mu}(n)$ by the adaptive filter $v_{\text{IC},\mu}(l,n)$. Output signals of the basic unit are the calibrated signal $x^{\text{c}}_{\text{IC},\mu}(n)$ and the error signal $e_{\text{IC},\mu}(n)$. The abbreviation *IC* stands for *interference canceller*.

The other input signal $d_{\text{IC},\mu}(n)$ is used as target signal. For adjusting the adaptive filter $v_{\text{IC},\mu}(l,n)$ the error signal $e_{\text{IC},\mu}(n)$ is calculated as the difference

---

[15] The raised index c indicates the calibrated signal.

of the target signal and the calibrated signal

$$e_{\text{IC},\mu}(n) = d_{\text{IC},\mu}(n-D) - x^c_{\text{IC},\mu}(n) \ . \tag{12.62}$$

In order to model non-causal portions a delay of D samples may be introduced for the target signal. The filter coefficients $v_{\text{IC}}(l,n)$ are adjusted such that the mean-squared error is minimized. This can be accomplished adaptively by the normalized least-mean-square algorithm (NLMS) [21]:

$$v_{\text{IC},\mu}(l,n+1) = v_{\text{IC},\mu}(l,n) + \frac{\beta(n)}{\sum_{p=0}^{N-1} |x_{\text{IC},\mu}(n-p)|^2} \, e^*_{\text{IC},\mu}(n) \, x_{\text{IC},\mu}(n-l) \ . \tag{12.63}$$

With the stepsize parameter $\beta(n)$ the adaptation can be controlled.

The structure in Fig. 12.16 is also known as "Interference Canceller" (IC) [36]. Originally, its task was to cancel out correlated signals from the input signal $d_{\text{IC},\mu}(n)$ and it was the error signal $e_{\text{IC},\mu}(n)$ which was of interest. However, in this contribution this basic unit is used to perform a matching of the two input signals. Therefore the calibrated output signal $x^c_{\text{IC},\mu}(n)$ is important, too. The filter coefficients are adapted only in time frames where the desired signal has sufficient signal power over the background noise.

12.5.2 Configurations for Array Processing

For the multi-channel case with more than two microphones several instances of the basic unit are necessary to build up a self-calibration stage. There are different possibilities to choose the input signals of these units [6]. Four different configurations can be distinguished which are depicted in Fig. 12.17. These structures are marked by letters A, B, C and D, respectively. It is supposed that the input signals are time aligned with respect to the desired direction. In array configurations other than broadside this can be accomplished by using fractional delay filters [27] that compensate for relative time delays.

12.5.2.1 Configuration A

The output signal of one arbitrarily chosen microphone from the array is used as target signal. The signals of the $M-1$ other microphones are utilized as reference signals for $M-1$ basic units, respectively. These signals are filtered and thus calibrated with respect to the target signal. The calibrated output signals can be used as input signals of a succeeding beamformer. The signal of the target microphone must be delayed by D samples to enable the adaptive filters to model positive as well as negative relative time delays.

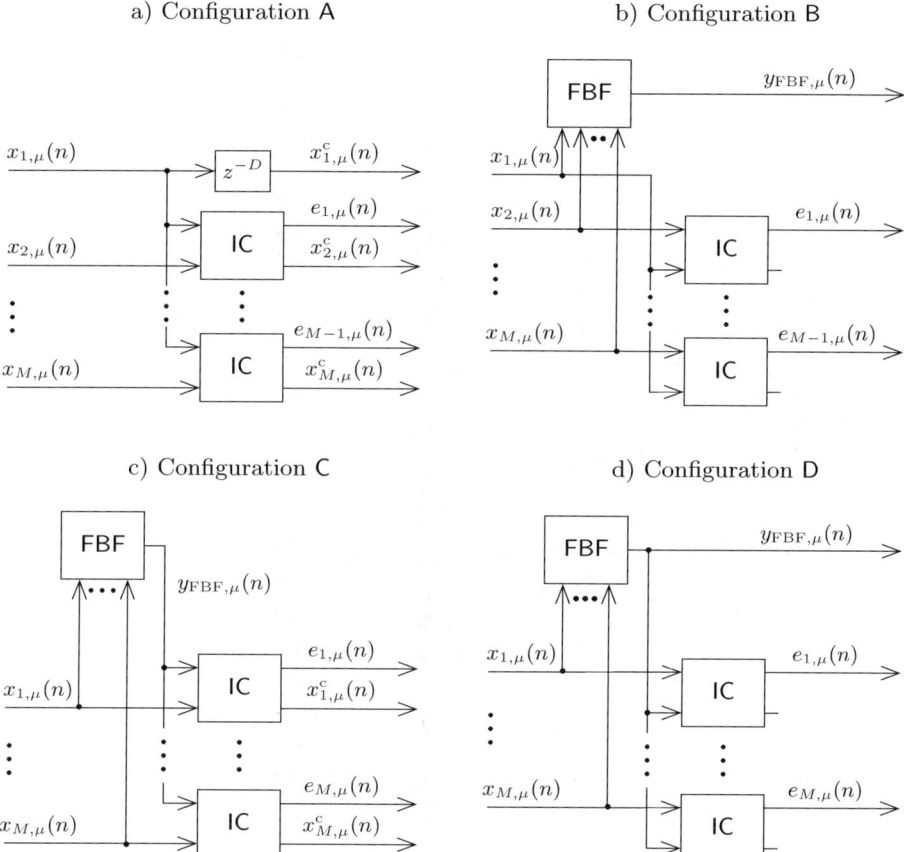

Fig. 12.17. Possible configurations of self-calibration for multi-channel signal processing. For microphone arrangements other than linear broadside an additional time delay compensation has to be applied in front of the self-calibration stage. The abbreviation *FBF* stands for *fixed beamformer*.

12.5.2.2 Configuration B

A new configuration results if the input signals of each basic unit in configuration A are exchanged pairwise. Again $M-1$ basic units are utilized but now $M-1$ microphones supply $M-1$ different target signals. The signal of the remaining microphone is taken as reference signal for each basic unit. In Fig. 12.17 b the output of microphone 1 is chosen as reference signal. The calibrated output signals of the units are filtered versions of the reference signal which originates from the same microphone for each basic unit. For this reason they are not suitable as input signals for a succeeding beamformer because they don't contain useful spatial information. The ensemble of error signals, however, do contain spatial information because they each depend on the corresponding target input signal. The error signals can be utilized as noise reference signals within a GSC beamformer. In this case configuration B replaces the blocking matrix and can be regarded as part of the beamformer itself. In Fig. 12.17 b the fixed beamformer (FBF) is already indicated.

12.5.2.3 Configuration C

The configurations A and B, both, make use of single microphone signals only. But employing a single microphone's signal as reference signal bears the risk of choosing a faulty microphone. Instead, the output signal of a fixed beamformer can be chosen as an enhanced target signal for the basic units:

$$y_{\text{FBF},\mu}(n) = \sum_{m=1}^{M} a_{m,\mu}\, x_{m,\mu}(n)\,. \tag{12.64}$$

With coefficients chosen to $a_{m,\mu} = \frac{1}{M}$ a delay-and-sum beamformer results. The fixed beamformer averages the microphone characteristics which makes this configuration more robust. The reference signals of the basic units are chosen similar to configuration A. However, M basic units have to be applied now. Configuration C is depicted in Fig. 12.17 c.

12.5.2.4 Configuration D

A further configuration can be derived by exchanging the input signals of the basic units of configuration C. In this case, the calibrated output signals of the basic units contain no spatial information that could be exploited by a beamformer analogous to configuration B. Therefore, the error signals are utilized as noise reference signals within a GSC beamformer. Correspondingly to configuration B the self-calibration stage replaces again the blocking matrix and can be interpreted as part of the GSC. One important advantage of configuration D is the fact that the reference signal shows a better SNR compared to the other configuration as it is preprocessed by a fixed beamformer. However, in the GSC beamformer M adaptive noise cancelling filters are necessary as configuration D generates M output signals $e_{m,\mu}(n)$. In contrast the other configurations do need only $M-1$ noise cancelling filters in the beamformer and are therefore computationally more efficient.

12.5.2.5 Relationship to Known Solutions

Some of the configurations introduced in this section correspond to known approaches that have been sketched in Sec. 12.1: Configuration A agrees with the system proposed by Van Compernolle [33], whereas configuration B is very similar to the system of Gannot et al. [17]. Furthermore configuration D corresponds to the adaptive blocking matrix in the approach of

Hoshuyama et al. [23] except for the constraints on the adaptive filters which have been skipped here. All these methods have been developed in order to compensate for a mismatch between the acoustic model and the real acoustic situation (e. g. an inaccurate steering direction). Thus, several well-known methods are closely related to the methods for microphone calibration that have been presented in this section. In fact, a mismatched steering direction causes a similar effect as phase deviations of the microphones [3]. As deviations of the steering direction and deviations of the microphone transfer functions can hardly be distinguished from each other an adaptive self-calibration jointly compensates for both kind of deviation.

12.5.3 Recursive Configuration

Configuration C can be extended by a recursive implementation. Instead of the original microphone signals $x_{m,\mu}(n)$ the calibrated microphone signals $x^c_{m,\mu}(n)$ are applied to a fixed beamformer:

$$x^c_{m,\mu}(n) = \sum_{l=0}^{N-1} v^*_{m,\mu}(l,n)\, x_{m,\mu}(n-l) \,, \tag{12.65}$$

$$y^c_{\text{FBF},\mu}(n) = \sum_{m=1}^{M} a_{m,\mu}\, x^c_{m,\mu}(n) \,. \tag{12.66}$$

The output signal of this fixed beamformer $y^c_{\text{FBF},\mu}(n)$ serves as an enhanced target signal for the adaptive filters of the basic units

$$e_{m,\mu}(n) = y^c_{\text{FBF},\mu}(n) - x^c_{m,\mu}(n) \,. \tag{12.67}$$

Compared to configuration C the quality of the target signal is further enhanced because the calibrated signals are used for generating the target signal instead of the uncalibrated signals. This modified structure is referred to as configuration Cc. The corresponding processing structure is depicted in Fig. 12.18.

By taking the calibrated signals $x^c_{m,\mu}(n)$ for generating the target signal, a closed loop has been introduced. Because of this recursive structure a constraint has to be applied in order to prevent the filters from converging towards zero. The arithmetic mean value of the calibration filters over the microphones has to correspond to a pure delay of D samples[16]:

$$\frac{1}{M} \sum_{m=1}^{M} v_{m,\mu}(l,n) = f_l \tag{12.68}$$

with

---

[16] With the delay D ($0 \leq D < N$) the adaptive calibration filters can model non-causal parts.

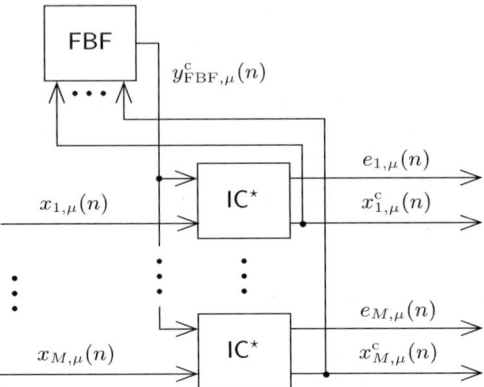

Fig. 12.18. Configuration Cc as extension of configuration C. The target signal $y_{\mathrm{FBF},\mu}^{\mathrm{c}}(n)$ is generated from the calibrated microphone signals. Because of the closed loop the basic unit IC* has to be stabilized by additional measures.

$$f_l = \begin{cases} 1, \text{ for } l = D, \\ 0, \text{ for } l \neq D. \end{cases} \quad (12.69)$$

For the NLMS algorithm a method which accomplishes this constrained adaptation of the filter coefficients was derived in [4]. As cost function $J_\mu(n)$ the sum of the mean-squared error signals is to be minimized:

$$J_\mu(n) = E\left\{ \sum_{m=1}^{M} |e_{m,\mu}(n)|^2 \right\}. \quad (12.70)$$

The expression $E\{.\}$ denotes the expectation value. The proposed algorithm consists of two steps. In the first step intermediate filter coefficients $\tilde{v}_{m,\mu}(l,n+1)$ are calculated by unconstrained adaptation:

$$\tilde{v}_{m,\mu}(l,n+1) = v_{m,\mu}(l,n) + $$
$$\gamma_{\mathrm{SC},\mu}(n) \left[e_{m,\mu}(n) - a_{m,\mu} \sum_{i=1}^{M} e_{i,\mu}(n) \right]^* x_{m,\mu}(n-l). $$
$$(12.71)$$

Afterwards, in the second step the constraint from Eq. 12.68 is applied to the intermediate filter coefficients:

$$v_{m,\mu}(l,n+1) = f_l + \tilde{v}_{m,\mu}(l,n+1) - \frac{1}{M} \sum_{i=1}^{M} \tilde{v}_{i,\mu}(l,n+1). \quad (12.72)$$

For the adaptation step in Eq. 12.71 the step-size is normalized to the magnitude square of the input data of the adaptive filters

$$\gamma_{\text{SC},\mu}(n) = \frac{\beta_{\text{SC},\mu}(n)}{\sum_{m=1}^{M}\sum_{l=0}^{N-1}\left|x_{m,\mu}(n-l)\right|^{2}}. \qquad (12.73)$$

With $\beta_{\text{SC},\mu}(n)$ the adaptation speed can be controlled. As the calibration is accomplished exclusively for the steering direction, the adaption is only permitted in phases when the desired signal is dominant. During speech pauses the step-size is set to $\beta_{\text{SC},\mu}(n) = 0$. In order to avoid instability $\beta_{\text{SC},\mu}(n)$ must be chosen in the range of $0 \leq \beta_{\text{SC},\mu}(n) \leq 2$.

12.5.4 Adaptation Control

The self-calibration systems presented in the preceding sections are especially suited for use combined with an adaptive beamformer. The adaptive filters of a self-calibration system are only updated when the desired signal is active, whereas the adaptive filters of the beamformer should only be adapted during speech pauses. Therefore an adaptation control is required which analyzes the microphone signals and determines step-sizes for the self-calibration stage and the beamformer, respectively.

The adaptation step-sizes for both the adaptive beamformer as well as the proposed self-calibration system are controlled by a criterion which detects signal activity from the desired direction. This is accomplished by comparing the output signal of a fixed beamformer

$$y_{\text{FBF},\mu}(n) = \sum_{m=1}^{M} a_{m,\mu} x_{m,\mu}(n) \qquad (12.74)$$

which contains a strong portion of the desired signal and the output signals of the blocking matrix which can be calculated for instance by

$$b_{m,\mu}(n) = x_{m,\mu}(n) - x_{m+1,\mu}(n), \quad m = 1, \ldots, M-1 \qquad (12.75)$$

containing only small portions of the desired signal. The power values of these signals are smoothed over time by a first-order IIR filter, where α is a constant smoothing parameter:

$$P_{\text{sum},\mu}(n) = \alpha\, P_{\text{sum},\mu}(n-1) + (1-\alpha)\left|y_{\text{FBF},\mu}(n)\right|^{2}, \qquad (12.76)$$

$$P_{\text{dif},\mu}(n) = \alpha\, P_{\text{dif},\mu}(n-1) + \frac{1-\alpha}{M-1}\sum_{m=1}^{M-1}\left|b_{m,\mu}(n)\right|^{2}. \qquad (12.77)$$

During speech activity $P_{\text{sum},\mu}(n)$ will exceed $P_{\text{dif},\mu}(n)$ because the summation emphasizes signal components arriving from the steering direction whereas the blocking operation suppresses them. Therefore, a ratio of both powers can be used to define a spatial criterion for detecting signals from the target direction:

$$q_{\text{sd},\mu}(n) = \frac{P_{\text{sum},\mu}(n)}{P_{\text{dif},\mu}(n)\, w_{\text{eq},\mu}(n)}\ . \tag{12.78}$$

The adaptive equalization factors $w_{\text{eq},\mu}(n)$ serve to normalize the ratios in each sub-band. These factors are adjusted in speech pauses in such a way that the ratio satisfies $q_{\text{sd},\mu}(n) \approx 1$ during stationary background noise. The adjustment can be accomplished with a multiplicative correction

$$w_{\text{eq},\mu}(n+1) = w_{\text{eq},\mu}(n) \cdot \eta_{\text{eq},\mu}(n) \tag{12.79}$$

where the correction $\eta_{\text{eq},\mu}(n)$ is controlled by a voice activity detection:

$$\eta_{\text{eq},\mu}(n) = \begin{cases} \eta_0, & \text{for speech pauses and } q_{\text{sd},\mu}(n) > 1\ , \\ \eta_0^{-1}, & \text{for speech pauses and } q_{\text{sd},\mu}(n) <= 1\ , \\ 1, & \text{for speech activity}\ . \end{cases} \tag{12.80}$$

The constant η_0 has to be chosen slightly larger than one, for example $\eta_0 = 1.001$ in order to ensure a smooth transition of the equalization factor.

Thus, values of $q_{\text{sd},\mu}(n)$ larger than one indicate signals coming from the steering direction. To further improve robustness of this criterion the ratio $q_{\text{sd},\mu}(n)$ can be averaged or smoothed over several sub-bands.

The beamformer filters are adjusted only when $q_{\text{sd},\mu}(n)$ falls below a preset threshold $Q_{\text{BF},\mu}$. The calibration filters are adapted only if this ratio exceeds a threshold $Q_{\text{SC},\mu}$ (with $Q_{\text{BF},\mu} < Q_{\text{SC},\mu}$) and the signal power is sufficiently high above the power of the background noise.

12.5.5 Experimental Results

The different realizations for self-calibration presented in this section were evaluated in a real-life car environment with human speakers and driving noise. As microphone array the system equipped as standard in Mercedes sedans was used which consists of $M=4$ differential microphones integrated at the bottom of the rear view mirror. The microphone arrangement is approximately linear with a microphone spacing of about 5 cm. The microphone signals were sampled at a rate of $f_s = 11025\,\text{Hz}$ and were decomposed each into $P = 256$ complex sub-bands by using a poly-phase filter bank with a sub-sampling rate of $R = 64$. As prototype low-pass filter a Hann window of length $N_{\text{win}} = 256$ was used.

The performance for speech recognition was examined by using a Lombard database. This database includes various utterances from a total of 53 speakers. To induce the Lombard effect [26] different driving noises were played back over headphones during recording the speech data with a headset. For the evaluation these speech recordings were convolved with impulse responses, which had been measured in a Mercedes S class sedan between the mouth reference point (MRP) and the single microphones of the array. Driving noise which had been measured in the same car with the same microphones was added to the convolved signals in order to generate realistic microphone signals [5]. Afterwards the signals were processed off-line by the multi-channel algorithms as discussed in the previous sections. For the evaluation a GSC beamformer without any calibration serves as reference which is compared to GSC beamformers with different configurations for self-calibration as preprocessing stage.

For the recognition tests a speaker-independent continuous-word recognition engine was utilized which had been trained for single channel operation in cars. The speech data consisted of digit loops of variable lengths. The test set consisted of 1373 utterances with in total 12791 digits spoken. In Tab. 12.1 the results for a driving speed of 130 km/h are listed. Configurations B, Cc and D were tested each in combination with a GSC beamformer. For comparison also the results for a single microphone and for a fixed beamformer are given. By self-calibration the word error rate reduces considerably. In practical real-life applications the self-calibration algorithms have proven to be very effective and robust to noise. Typically the computational effort for the self-calibration stage is significantly less compared to the pure GSC since the length of the calibration filters should be chosen considerably shorter than the length of the noise cancelling filters of the GSC.

Table 12.1. Results of speech recognition tests with digit loops at a driving speed of 130 km/h.

	Mik. 1	FBF	GSC	B+GSC	Cc+GSC	D+GSC
Word error rate	5.44%	3.17%	2.81%	2.30%	2.18%	2.21%
Relative reduction of the word error rate	-93.6%	-12.8%	0%	18.1%	22.4%	21.4%

12.6 Summary

In this chapter the effect of microphone mismatch on multi-channel noise reduction systems was studied by investigating MVDR beamformers. For this, the well-known signal model was extended by considering the influence of each single microphone in the equations for beamformer evaluation. This generalization leads to a modified power density matrix which includes the microphone characteristics implicitly. On the basis of the new model the effect of microphone deviations was studied theoretically and "practically" by simulating beamformers with real measured data.

Several well-known algorithmic approaches to improve the robustness of beamformers against microphone mismatch were presented. These methods mainly limit the performance of the beamformer in such a way that the beamformer operates reasonably despite the existing microphone deviations. In contrast, calibration aims to compensate for existing microphone mismatch without impairing the beamformer's performance. However, it turned out that a calibration is not able to compensate perfectly for microphone mismatch. In particular, when the deviations depend on the angle of incidence considerable residual deviations may remain after a calibration.

In this chapter several approaches for fixed calibration as well as for adaptive calibration have been presented. The adaptive calibration adjusts its filters during operation of the system by using the speech signal as target signal for the adaptation. The adaptive calibration is especially valuable for mass-produced products, where an additional off-line calibration would mean an increase in production costs.

12.A Experimental Determination of the Directivity Index

This appendix describes how the directivity of a beamformer can be calculated on the basis of free-field measurements. In general all spatial directions have to be considered in order to determine the directivity. Within the mathematical description according to Eq. 12.12 this is done by integrating over a spherical surface.

First, in the following subsection it is shown how this integral over a sphere can be approximated numerically with only a limited number of available measurement points. In the succeeding subsections this result is used to determine the directivity for the present problem.

12.A.1 Numerical Integration over a Spherical Surface

For a scalar function $F(\theta, \varphi): S \mapsto \mathbb{C}$, which is defined on a surface S of a sphere with radius 1 the integral

$$\frac{1}{4\pi} \int_0^{2\pi} \int_0^{\pi} F(\theta, \varphi) \sin\theta \, d\theta \, d\varphi \qquad (12.81)$$

has to be solved numerically. θ and φ denote the angle of elevation and azimuth of a spherical coordinate system, respectively.

The functional values $F(\theta_p, \varphi_q)$ are available only for discrete points (θ_p, φ_q) on an angular grid. It is assumed that the grid points are equidistantly spaced over the azimuth angle θ as well as over the elevation angle φ:

$$\theta_p = \Delta\theta \cdot p, \quad \text{with} \quad \Delta\theta = \frac{\pi}{P} \quad \text{and} \quad p \in \{0, 1, \ldots, P\}, \tag{12.82}$$

$$\varphi_q = \Delta\varphi \cdot q, \quad \text{with} \quad \Delta\varphi = \frac{2\pi}{Q} \quad \text{and} \quad q \in \{0, 1, \ldots, Q-1\}. \tag{12.83}$$

Now, the spherical surface S is divided into small segments $S_{p,q}$, which are each assigned to one grid point (θ_p, φ_q). For the numerical integration the functional values of the points (θ_p, φ_q) are each passed on the complete area of the corresponding segment $S_{p,q}$. The segments are specified by angular sectors of equal width, which are chosen in such a way that the grid points lay in the middle of the sectors. The borders of the surface segments are specified by

$$\theta = \theta_p + \frac{\Delta\theta}{2} \quad \text{with} \quad p \in \{0, 1, \ldots, P-1\} \tag{12.84}$$

and

$$\varphi = \varphi_q + \frac{\Delta\varphi}{2} \quad \text{with} \quad q \in \{0, 1, \ldots, Q-1\}. \tag{12.85}$$

Because the range of integration for the azimuth angle θ reaches from 0 to π the boundary sectors have half width compared to the other sectors.

The double integral over θ and φ of Eq. 12.81 can be divided into a double sum of sub-integrals:

$$\frac{1}{4\pi} \int_0^{2\pi} \int_0^{\pi} F(\theta, \varphi) \sin\theta \, d\theta \, d\varphi \approx \frac{1}{2\pi} \sum_{q=0}^{Q-1} \int_{\varphi_q - \frac{\Delta\varphi}{2}}^{\varphi_q + \frac{\Delta\varphi}{2}} \frac{1}{2} \left[\int_0^{\frac{\Delta\theta}{2}} F(\theta_0, \varphi_q) \sin\theta \, d\theta \right.$$

$$\left. + \sum_{p=1}^{P-1} \int_{\theta_p - \frac{\Delta\theta}{2}}^{\theta_p + \frac{\Delta\theta}{2}} F(\theta_p, \varphi_q) \sin\theta \, d\theta + \int_{\pi - \frac{\Delta\theta}{2}}^{\pi} F(\theta_P, \varphi_q) \sin\theta \, d\theta \right] d\varphi . \tag{12.86}$$

As the functional value F is considered to be constant within each subintegral it can be put out of the integrals. Therefore, the value of each subintegral is equivalent to the corresponding area of integration $S_{p,q}$ multiplied by the corresponding functional value $F(\theta_p, \varphi_q)$:

$$\frac{\Delta\varphi}{2\pi} \sum_{q=0}^{Q-1} \frac{1}{2} \left[F(\theta_0, \varphi_q) \cdot \left(1 - \cos\tfrac{\Delta\theta}{2}\right) \right.$$

$$+ \sum_{p=1}^{P-1} F(\theta_p, \varphi_q) \cdot \left(\cos(\theta_p - \tfrac{\Delta\theta}{2}) - \cos(\theta_p + \tfrac{\Delta\theta}{2})\right) \tag{12.87}$$

$$\left. + F(\theta_P, \varphi_q) \cdot \left(\cos(\pi - \tfrac{\Delta\theta}{2}) + 1\right) \right] .$$

By applying the addition theorems and resorting the terms there is

$$\frac{\Delta\varphi}{2\pi}\sum_{q=0}^{Q-1}\left[\sin\frac{\Delta\theta}{2}\sum_{p=1}^{P-1}\sin\theta_p\,F(\theta_p,\varphi_q)+\frac{1}{2}\left(1-\cos\frac{\Delta\theta}{2}\right)\left(F(\theta_0,\varphi_q)+F(\theta_P,\varphi_q)\right)\right].$$
(12.88)

This expression can be formulated in matrix notation. With the weighting factors for the elevation angle

$$\boldsymbol{g}_\theta = \begin{bmatrix} g_{\theta,0}, & \ldots, & g_{\theta,P} \end{bmatrix}^{\mathrm{T}},$$
(12.89)

with

$$g_{\theta,p} = \begin{cases} \frac{1}{2}\left(1-\cos(\frac{\pi}{2P})\right), & \text{for } p=0 \text{ and } p=P, \\ \sin(\frac{\pi}{2P})\sin(\frac{\pi p}{P}), & \text{for } p \in \{1,\ldots,P-1\} \end{cases}$$
(12.90)

and the weighting factors for the azimuth angle

$$\boldsymbol{g}_\varphi = \begin{bmatrix} g_{\varphi,0}, & \ldots, & g_{\varphi,Q-1} \end{bmatrix}^{\mathrm{T}},$$
(12.91)

with

$$g_{\varphi,q} = \frac{1}{Q}, \text{ for } q \in \{0, 1, \ldots, Q-1\}$$
(12.92)

and a matrix of size $(P+1) \times Q$ containing the functional values

$$\boldsymbol{F} = \begin{bmatrix} F(\theta_0,\varphi_0) & \cdots & F(\theta_0,\varphi_{Q-1}) \\ \vdots & \ddots & \vdots \\ F(\theta_P,\varphi_0) & \cdots & F(\theta_P,\varphi_{Q-1}) \end{bmatrix}$$
(12.93)

the numerical integration of Eq. 12.81 can be written as a product:

$$\frac{1}{4\pi}\int_0^{2\pi}\int_0^{\pi} F(\theta,\varphi)\sin\theta\,d\theta\,d\varphi \approx \boldsymbol{g}_\theta^{\mathrm{T}}\,\boldsymbol{F}\,\boldsymbol{g}_\varphi.$$
(12.94)

In this way the integration of a scalar function over a spherical surface can be approximated numerically by means of a matrix multiplication.

Eq. 12.94 is utilized in Secs. 12.4.2 and 12.4.3 in order to calculate cross power density spectra of microphone signals in a diffuse sound field. There, the parameters are set to $P = 36$ and $Q = 36$.

12.A.2 Definition of the Coordinate System

The directivity of a beamformer is to be determined on the basis of the measurement values which are available for the transfer functions of the single microphones. In order to apply the numerical integration according to Eq. 12.94, first, a coordinate system has to be defined. An appropriate choice of the coordinate system can simplify the calculations considerably.

For array geometries where the microphones are all oriented into the same direction $\mathbf{u}_m^{\mathrm{M}} = \mathbf{u}^{\mathrm{M}}$ it is advantageous to set the z axis of the coordinate system equal to the main direction of the microphone: $\mathbf{e}_z = \mathbf{u}^M$. The azimuth angle $\varphi = 0$ of the spherical coordinate system is specified by the array axis $\mathbf{u}^A$.[17] For the microphone arrangements considered in this chapter the grid points of the numerical integration correspond to the discrete angles for which the microphone transfer functions have been measured as described in Sec. 12.4.1.2.

The angle enclosed by the microphone axis $\mathbf{u}^M$ and the array axis $\mathbf{u}^A$ is denoted θ_A according to Fig. 12.19. Thus, the array axis can be expressed in cartesian coordinates:

$$\mathbf{u}^A = \sin\theta_A\,\mathbf{e}_x + \cos\theta_A\,\mathbf{e}_z \ . \tag{12.95}$$

For an arbitrary angle of incidence $\mathbf{u}(\theta, \varphi)$ it is

$$\mathbf{u}(\theta,\varphi) = \sin\theta\,\cos\varphi\,\mathbf{e}_x + \sin\theta\,\sin\varphi\,\mathbf{e}_y + \cos\theta\,\mathbf{e}_z \ . \tag{12.96}$$

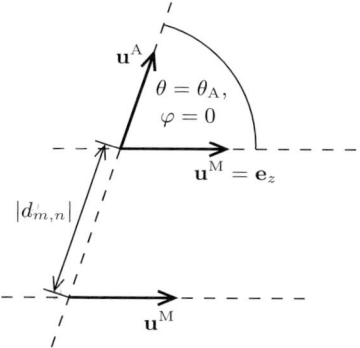

Fig. 12.19. Position of the directional vectors for microphone and array.

In order to calculate the directivity the relative time delays between the single microphone positions have to be determined. Considering two positions $\mathbf{r}_m$ and $\mathbf{r}_n = \mathbf{r}_m + d_{m,n}\,\mathbf{u}^A$ with spacing $|d_{m,n}|$ on the array axis the relative time delay between these two positions for a plane wave coming from direction $\mathbf{u}(\theta, \varphi)$ can be expressed by

$$\tau_{m,n}(\theta,\varphi) = \frac{d_{m,n}}{c}\cos\bigl(\psi(\theta,\varphi)\bigr) \ . \tag{12.97}$$

$\psi(\theta,\varphi)$ denotes the angle which is enclosed by the two directional vectors $\mathbf{u}(\theta,\varphi)$ and $\mathbf{u}^A$. The cosine can be expressed by the scalar product

$$\cos\bigl(\psi(\theta,\varphi)\bigr) = \mathbf{u}^T(\theta,\varphi) \cdot \mathbf{u}^A \ . \tag{12.98}$$

---

[17] For endfire arrays ($\mathbf{u}^A = \mathbf{u}^M = \mathbf{e}_z$) the expression $\varphi = 0$ is ambiguous. In this case $\varphi = 0$ can be set arbitrarily.

Thus, the relative time delay can be determined for the geometric arrangements by using Eq. 12.95 and 12.96:

$$\tau_{m,n}(\theta, \varphi) = \frac{d_{m,n}}{c} \left(\sin\theta \cos\varphi \cdot \sin\theta_A + \cos\theta \cdot \cos\theta_A \right). \tag{12.99}$$

With the coordinate system specified in this way the grid for the elevation angle θ_p can be chosen according to the angular grid which has been used for the microphone measurements. However, the microphone measurements described in Sec. 12.4.1.4 have only be done in a plane and not for a sphere. Thus, for the numerical integration measurement values for the remaining spherical grid points are necessary. These values are generated by taking the values which have been measured in the plane and assuming them identically for all azimuth angles. Thus, rotational symmetric microphone transfer functions result.

12.A.3 Determination of the Directivity using the Normalized Cross Power Spectral Densities of Microphone Signals

The directivity can be determined on the basis of the normalized cross power spectral density of the microphone signals in a diffuse noise field $\tilde{\boldsymbol{\Phi}}^{M,i}(\omega)$ according to Sec. 12.2.2.3. According to Eq. 12.15 the directivity results to:

$$\tilde{D}^{BF}(\omega) = \frac{\tilde{\Psi}^{BF}(\omega, \mathbf{u}_0)}{\boldsymbol{W}^H(\omega) \tilde{\boldsymbol{\Phi}}^{M,i}(\omega) \boldsymbol{W}(\omega)}. \tag{12.100}$$

The cross power density of two microphone signals in a diffuse sound field can be determined according to Eq. 12.14 by integration over a sphere. Considering deviations of the microphone transfer functions the formula reads

$$\tilde{\Phi}_{m,n}^{M,i}(\omega) = \frac{1}{4\pi} \int_0^{2\pi} \int_0^{\pi} \tilde{H}_m^M(\omega, \mathbf{u}(\theta, \varphi)) \left(\tilde{H}_n^M(\omega, \mathbf{u}(\theta, \varphi)) \right)^* \tag{12.101}$$

$$\cdot \exp\{-j\omega\tau_{m,n}(\theta, \varphi)\} \sin\theta \, d\theta \, d\varphi.$$

This integral can be solved numerically by matrix multiplication using Eq. 12.94. For this purpose, a coordinate system according to Sec. 12.A.2 is specified and the relative time delays $\tau_{m,n}(\theta_p, \varphi_q)$ are calculated. Then, the elements of the matrix $\boldsymbol{F}$ can be determined on the basis of the present measurement values for the microphone transfer functions:

$$F(\theta_p, \varphi_q) = \tilde{H}_m^M(\omega, \mathbf{u}(\theta_p, \varphi_q)) \left(\tilde{H}_n^M(\omega, \mathbf{u}(\theta_p, \varphi_q)) \right)^* \exp\{-j\omega\tau_{m,n}(\theta_p, \varphi_q)\}. \tag{12.102}$$

The numerical integration has to be accomplished for each pair of microphones (m, n) of the array. The resulting cross power spectral densities $\tilde{\Phi}_{m,n}^{M,i}(\omega)$ are gathered within the matrix $\tilde{\boldsymbol{\Phi}}^{M,i}(\omega)$.

The steering response $\tilde{\Psi}^{\mathrm{BF}}(\omega, \mathbf{u}_0)$ is calculated as specified in Sec. 12.2.2.2:

$$\tilde{\Psi}^{\mathrm{BF}}(\omega, \mathbf{u}_0) = \left| \sum_{m=1}^{M} W_m^*(\omega)\, \tilde{H}_m^{\mathrm{M}}\big(\omega, \mathbf{u}(\theta_0, \varphi_0)\big) \exp\big\{ -j\omega \tau_{m,n}(\theta_0, \varphi_0) \big\} \right|^2 .$$

(12.103)

In this way, the directivity according to Eq. 12.100 can be determined approximately. This method has been used in Secs. 12.3.2 and 12.3.3 within the Monte Carlo simulations to evaluate the effect of microphone mismatch on beamforming.

References

1. S. Affes, Y. Grenier: A signal subspace tracking algorithm for microphone array processing of speech, *IEEE Trans. Speech and Audio Process.*, **5**(5), 425–437, 1997.
2. K. M. Buckley, L. J. Griffiths: An adaptive generalized sidelobe canceller with derivative constraints, *IEEE Trans. Antennas and Propagation*, **AP-34**(3), 311–319, 1986.
3. M. Buck: Aspects of first-order differential microphone arrays in the presence of sensor imperfections, *Europ. Trans. on Telecommunication*, **13**(2), 115–122, 2002.
4. M. Buck: *Mehrkanalige Systeme zur Geräuschunterdrückung für Sprachanwendungen unter Berücksichtigung von Mikrofoneigenschaften*, Aachen, Germany: Shaker Verlag, 2004 (in German).
5. M. Buck, H.-J. Köpf, T. Haulick: Lombard-Sprache für Kfz-Anwendungen: eine Analyse verschiedener Aufnahmekonzepte. *Proc. DAGA '06*, 217–218, Braunschweig, Germany, 2006 (in German).
6. M. Buck, T. Haulick, H.-J. Pfleiderer: Self-calibrating microphone arrays for speech signal acquisition: A systematic approach, *Signal Processing*, **86**(6), 1230–1238, 2006.
7. J. Bitzer, K. U. Simmer, K.-D. Kammeyer: An alternative implementation of the superdirective beamformer, *Proc. WASPAA '99*, 7–10, New Paltz, NY, USA, 1999.
8. J. Bitzer, K. U. Simmer: Superdirective microphone arrays, in M. Brandstein, D. Ward (eds.), *Microphone Arrays – Signal Processing Techniques and Applications*, 19–38, Berlin, Germany: Springer, 2001.
9. P. L. Chu: Superdirective microphone array for a set-top videoconferencing system, *Proc. ICASSP '97*, 235–238, Munich, Germany, 1997.
10. H. Cox: Resolving power and sensitivity to mismatch of optimum array processors, *JASA*, **54**, 771–785, 1973.
11. H. Cox, R. Zeskind, M. Oven: Robust adaptive beamforming, *IEEE Trans. Acoust. Speech Signal Process.*, **ASSP-35**(10), 1365–1376, 1987.
12. M. Dörbecker: *Mehrkanalige Signalverarbeitung zur Verbesserung akustisch gestörter Sprachsignale am Beispiel elektonischer Hörhilfen*, Aachen, Germany: Verlag der Augustinus Buchhandlung, 10, 1998 (in German).

13. G. W. Elko: Superdirectional microphone arrays, in S. L. Gay, J. Benesty (eds.), *Acoustic Signal Processing for Telecommunication*, Boston, MA, USA: Kluwer, 2000, 181–237.
14. M. H. Er, A. Cantoni: Derivative constraints for broad-band element space antenna array processors, *IEEE Trans. Acoust. Speech Signal Process.*, **ASSP-31**(6), 1378–1393, 1983.
15. O. L. Frost, III: An algorithm for linearily constrained adaptive array processing, *Proc. IEEE*, **60**(8), 926–935, 1972.
16. S. Gannot, D. Burshstein, E. Weinstein: Beamforming methods for multichannel speech enhancement, *Proc. IWAENC '99*, 96–99, Pocono Manor, PA, USA, 1999.
17. S. Gannot, D. Burshstein, E. Weinstein: Signal enhancement using beamforming and nonstationarity with applications to speech, *IEEE Trans. on Signal Process.*, **SP-49**(8), 1614–1626, 2001.
18. E. N. Gilbert, S. P. Morgan: Optimum design of directive antenna arrays subject to random variation, *The Bell System Technical Journal*, **34**(5), 637–663, 1955.
19. L. J. Griffiths, C. W. Jim: An alternative approach to linearly constrained adaptive beamforming, *IEEE Tans. Antennas and Propagation*, **AP-30**(1), 24–34, 1982.
20. E. Hänsler, G. Schmidt: *Acoustic Echo and Noise Control: A Practical approach*, Hoboken, NJ, USA: Wiley, 2004.
21. S. Haykin: *Adaptive Filter Theory*, 4th ed., Englewood Cliffs, NJ, USA: Prentice Hall, 2002.
22. W. Herbordt, W. Kellermann: Adaptive beamforming for audio signal acquisition, in J. Benesty, Y. Huang (eds.), *Adaptive signal processing: applications to real-world problems*, Berlin, Germany: Springer, 2003, 155–194.
23. O. Hoshuyama, A. Sugiyama, A. Hirano: A robust adaptive beamformer for microphone arrays with a blocking matrix using constrained adaptive filters, *IEEE Trans. Signal Process.*, **47**(10), 2677–2684, 1999.
24. O. Hoshuyama, A. Sugiyama: Robust adaptive beamforming, in M. Brandstein, D. Ward (eds.), *Microphone Arrays – Signal Processing Techniques and Applications*, Berlin, Germany: Springer, 2001, 87–109.
25. T. P. Hua, A. Sugiyama, G. Faucon: A new self-calibration technique for adaptive microphone arrays, *Proc. IWAENC '05*, 237–240, Eindhoven, Netherlands, 2005.
26. J.-C. Juncqua: The influence of acoustics on speech production: a noise-induced stress phenomenon known as the Lombard reflex. *Speech Communication*, **20**, 13–22, 1996.
27. T. I. Laakso, V. Välimäki, M. Karjalainen, U. K. Laine: Splitting the unit delay – tools for fractional delay filter design, *IEEE Signal Process. Mag.*, **13**(1), 30–60, 1996.
28. Z. Liu, M. L. Seltzer, A. Acero, I. Tashev, Z. Zhang, M. Sinclair: A compact multi-sensor headset for hands-free communication, *Proc. WASPAA '05*, 138–141, New Paltz, NY, USA, 2003.
29. D. G. Manolakis, V. K. Ingle, S. M. Kogon: *Statistical and adaptive signal processing: spectral estimation, signal modeling, adaptive filtering and array processing*, Boston, MA, USA: McGraw-Hill, 2000.

30. S. Nordholm, I. Claesson, M. Dahl: Adaptive microphone array employing calibration signals: an analytical evaluation, *IEEE Trans. on Speech and Audio Processing*, **7**(3), 241–252, 1999.
31. P. Oak, W. Kellermann: A calibration algorithm for robust generalized sidelobe cancelling beamformers, *Proc. IWAENC '05*, 97–100, Eindhoven, Netherlands, 2005.
32. A. K. Steele: Comparison of directional and derivative constraints for beamformers subject to multiple linear constraints, *IEE Proceedings part H*, **130**(2), 41–45, 1983.
33. D. Van Compernolle: Switching adaptive filters for enhancing noisy and reverberant speech from microphone array processing, *Proc. ICASSP '90*, 833–836, Albuquerque, MN, USA, 1990.
34. B. D. Van Veen, K. M. Buckley: Beamforming: A versatile approach to spatial filtering, *IEEE ASSP Mag.*, **5**(2), 4–24, 1988.
35. E. Wallach: On superresolution effects in maximum likelihood adaptive antenna arrays, *IEEE Trans. on Antennas and Propagation*, **32**(3), 259–263, 1984.
36. B. Widrow, J. R. Glover, F. M. McCool, J. Kaunitz, C. S. Williams, R. H. Hearn, J. R. Zeidler, E. Dong, R. C. Goodlin: Adaptive noise cancellation: principles and applications, *Proc. IEEE*, **63**(12), 1692–1718, 1975.
37. B. Widrow, K. M. Duvall, R. P. Gooch, W. C. Newman: Signal cancellation phenomena in adaptive antennas: causes and cures, *IEEE Trans. on Antennas and Propagation*, **30**(3), 469–478, 1982.
38. B. Widrow, S. Stearns: *Adaptive Signal Processing*, Englewood Cliffs, NJ, USA: Prentice Hall, 1985.
39. C. L. Zahm: Effects of errors in the direction of incidence on the performance of an adaptive array, *Proc. IEEE*, **60**(8), 1008–1009, 1972.
40. X. Zhang, J. H. L. Hansen: CSA-BF: Novel constrained switched adaptive beamforming for speech enhancement and recognition in real car environments, *Proc. ICASSP '03*, **2**, 125–128, Hong Kong, 2003.

13

Convolutive Blind Source Separation for Noisy Mixtures

Robert Aichner[1], Herbert Buchner[2], and Walter Kellermann[3]

[1] Microsoft Corporation, Redmond, WA, USA[†]
[2] Deutsche Telekom Laboratories, Technical University Berlin, Germany[†]
[3] University of Erlangen-Nuremberg, Germany

Convolutive blind source separation (BSS) is a promising technique for separating acoustic mixtures acquired by multiple microphones in reverberant environments. In contrast to conventional beamforming methods no a-priori knowledge about the source positions or sensor arrangement is necessary resulting in a greater versatility of the algorithms. In this contribution we will first review a general BSS framework called TRINICON which allows a unified treatment of broadband and narrowband BSS algorithms. Efficient algorithms will be presented and their high performance will be confirmed by experimental results in reverberant rooms. Subsequently, the BSS model will be extended by incorporating background noise. Commonly encountered realistic noise types are examined and, based on the resulting model, pre-processing methods for noise-robust BSS adaptation are investigated. Additionally, an efficient post-processing technique following the BSS stage, will be presented, which aims at simultaneous suppression of background noise and residual cross-talk. Combining these pre- or post-processing approaches with the algorithms obtained by the TRINICON framework yield versatile BSS systems which can be applied in adverse environments as will be demonstrated by experimental results.

13.1 Introduction

Acoustic blind source separation can be applied to scenarios where there are a number of point sources whose signals are picked up by several microphones. As each microphone is located at a different position, each sensor acquires a slightly different mixture of the original source signals. The goal of blind source separation is to recover the separated source signals from this set of

---

[†] The research underlying this work was performed while the authors were with Multimedia Communications and Signal Processing, University of Erlangen-Nuremberg.

sensor signals. The term "blind" stresses the fact that the source signals and the mixing system are assumed to be unknown and no information about the source positions and sensor arrangement is necessary. The fundamental assumption for BSS methods is that the original source signals are mutually statistically independent. In reality this assumption holds for a variety of signals, such as multiple speakers. Therefore, the problem of BSS refers to finding a demixing system whose outputs are statistically independent.

In reverberant environments delayed and attenuated versions of the source signals $s_q(n)$ are picked up by the microphones. Assuming point sources, this can be modeled by a mixing system consisting of finite impulse response (FIR) filters of length M given as

$$x_p(n) = \sum_{q=1}^{Q} \sum_{\kappa=0}^{M-1} h_{qp,\kappa} s_q(n-\kappa) + n_p(n), \qquad (13.1)$$

where $h_{qp,\kappa}$, $\kappa = 0, \ldots, M-1$ denote the coefficients of the FIR filter model from the q-th source to the p-th sensor. In addition to the source signals, a noise signal $n_p(n)$ may be picked up by each sensor which contains both, background noise and sensor noise. In blind source separation, we are interested in finding a corresponding demixing system whose output signals $y_q(n)$ are described by

$$y_q(n) = \sum_{p=1}^{P} \sum_{\kappa=0}^{L-1} w_{pq,\kappa} x_p(n-\kappa). \qquad (13.2)$$

The parameter L denotes the FIR filter length of the demixing filters $w_{pq,\kappa}$. The convolutive mixing model together with the demixing system is depicted as a block diagram in Fig. 13.1. From this it is obvious that BSS can be classified as a blind multiple-input multiple-output (MIMO) technique. Throughout this chapter, we regard the standard BSS model where the number Q of *potentially simultaneously active source signals* $s_q(n)$ is equal to the number of

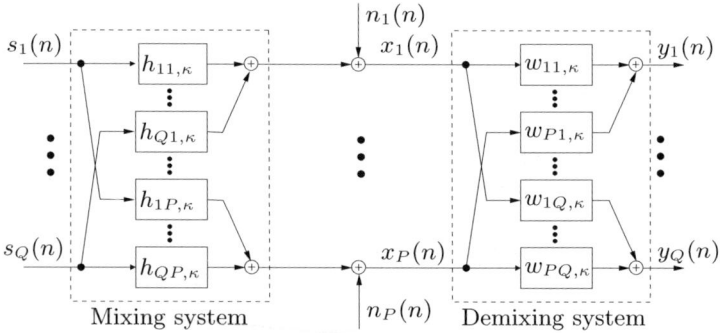

Fig. 13.1. Convolutive MIMO model for BSS.

sensor signals $x_p(n)$, i.e., $Q = P$. It should be noted that in contrast to other BSS algorithms we do not assume prior knowledge about the exact number of active sources. Thus, even if the algorithms will be derived for the case $Q = P$, the number of simultaneously active sources may change throughout the application of the BSS algorithm and only the condition $Q \leq P$ has to be fulfilled. For $Q > P$ the demixing system cannot be computed directly so that usually the sparseness of the sources in transform domains, such as the discrete Fourier transform (DFT) domain, is exploited and subsequently time-frequency masking are applied to separate the sources. This research field is termed computational auditory scene analysis (CASA) and a recent overview on the state-of-the-art can be found, e.g., in [29, 84] or in Chap. 14 of this book. Alternative statistical approaches for the case of $P < Q$ are still in an early stage [87].

As pointed out above, the source signals $s_q(n)$ are assumed to be mutually statistically independent. For the case $P = Q$ considered in this chapter it was shown in [85] that merely utilizing second-order statistics (SOS) by decorrelating the output signals $y_q(n)$ does not lead to a separation of the sources. This implies that we have to force the output signals to become statistically decoupled up to joint moments of a certain order by using additional conditions. This can be realized by exploiting one of the following source signal properties:

(a) **Nongaussianity.** The probability density function (PDF) of an acoustic source signal $s_q(n)$ is in general not Gaussian. Thus, the nongaussianity can be exploited by using higher-order statistics (HOS) yielding a statistical decoupling of higher-order joint moments of the BSS output signals. BSS algorithms utilizing HOS are also termed independent component analysis (ICA) algorithms (e.g., [48, 77]).

(b) **Nonwhiteness.** Audio signals exhibit temporal dependencies which can be exploited by the BSS criterion. This means that the samples of *each* source signal are not independent along the time axis. However, the signal samples from different sources are *mutually* independent. Based on the assumption of mutual statistical independence for non-white sources, several algorithms can be found in the literature. There, mainly the nonwhiteness is exploited by simultaneous diagonalization of output correlation matrices over multiple time-lags, (e.g., [54, 67, 78, 81]).

(c) **Nonstationarity.** Audio signals are in general assumed to be nonstationary. Therefore, in most acoustic BSS applications nonstationarity of the source signals is exploited by simultaneous diagonalization of short-time output correlation matrices at different time instants (e.g., [50, 64, 73, 85]). The signals within each block, as necessary for estimating the correlation matrices, are usually assumed to be wide-sense stationary.

A simultaneous exploitation of two or even all three signal properties leads to improved results as was shown within the TRINICON framework [14, 16] which will be reviewed in the next section.

It should be pointed out that, as long as the concept of BSS is solely based on the assumption of mutual independence of the source signals, some ambiguities are unavoidable:

- Permutation ambiguity: The ordering of the separated sources cannot be determined.
- Filtering ambiguity: The estimated separated source signals can only be determined up to an arbitrary filtering operation.

The permutation ambiguity cannot be resolved without additional a-priori information. However, if, e.g., the sensor positions are known, then the position of each separated source can be determined from the demixing system [18]. For some applications this may be sufficient for solving the permutation problem.

The filtering ambiguity is caused by the fact that in general, BSS approaches do not aim at blind dereverberation which would lead to a deconvolution of the mixing system, i.e., at a recovery of the original source signals up to an arbitrary scaling factor and a constant delay. Blind dereverberation is a more challenging task as it requires to distinguish between the temporal correlations introduced by the vocal tract of the human speaker and the correlations originating from the reverberation of the room. This was addressed in the extension of the TRINICON framework to blind dereverberation in [15]. However, even if we do not strive for solving the dereverberation problem in BSS it is still desirable to avoid the arbitrariness of the filtering operation in blind source separation. Fortunately, it can be shown [8, 20] that the filtering ambiguity reduces to a scaling ambiguity, if the demixing filter length L is chosen less or equal to the optimum BSS demixing filter length $L_{\mathrm{opt}} = \frac{(Q-1)(M-1)+1}{P-Q+1}$. Another popular approach to avoid the arbitrary filtering is to apply a constraint which minimizes the distortion introduced by the demixing system of the BSS algorithm. Thereby, the q-th separated source $y_q(n)$ is constrained to be equal to the component of the desired source $s_q(n)$ picked up, e.g., at the q-th microphone. This is done by back-projecting the estimated sources to the sensors or by introducing a constrained optimization scheme [49, 65]. In the following we disregard these ambiguities and first concentrate on the fundamental BSS problem for convolutive acoustic mixtures and then extend our treatment to noisy mixtures.

The rest of the chapter is structured as follows: In the next section the TRINICON framework which is based on a generic time-domain optimization criterion accounting for all three signal properties is reviewed. The minimization of the criterion leads to a natural gradient algorithm which exhibits a so-called Sylvester constraint. Subsequently, several approximations are discussed yielding various efficient BSS algorithms and experimental results in reverberant environments are given. In Sec. 13.3 the framework is extended to noisy environments. First, a model for background noise is discussed. Based on this model several pre-processing methods and a post-processing approach are presented which complement the BSS algorithms derived from the TRINICON framework. Especially the most promising post-processing scheme is discussed in detail and experimental results demonstrate the increased versatility of the complemented BSS algorithms.

13.2 Blind Source Separation for Acoustic Mixtures Based on the TRINICON Framework

In this section, we introduce, based on a compact matrix notation, a generic convolutive BSS framework which allows the simultaneous exploitation of the three signal properties. Several efficient algorithms are presented which can be derived from the optimization criterion of the framework and which allow real-time separation of multiple sources in reverberant environments. Moreover, links to well-known algorithms in the literature are illustrated.

13.2.1 Matrix Formulation

From the convolutive MIMO model illustrated in Fig. 13.1 it can be seen that the output signals $y_q(n)$ are obtained by convolving the input signals $x_p(n)$ with the demixing filter coefficients $w_{pq,\kappa}$, $\kappa = 0, \ldots, L-1$. For an algorithm which utilizes the nonwhiteness property of the source signals accounting for $D-1$ time-lags, a memory containing the current and the previous $D-1$ output signal values $y_q(n), \ldots, y_q(n-D+1)$ has to be introduced. The linear convolution yielding the D output signal values can be formulated using matrix-vector notation as

$$\boldsymbol{y}_q(n) = \sum_{p=1}^{P} \boldsymbol{W}_{pq}^{\mathrm{T}} \boldsymbol{x}_p(n), \tag{13.3}$$

with the column vectors $\boldsymbol{x}_p$ and $\boldsymbol{y}_q$ given as[1]

$$\boldsymbol{x}_p(n) = \begin{bmatrix} x_p(n), \ldots, x_p(n-2L+1) \end{bmatrix}^{\mathrm{T}}, \tag{13.4}$$

$$\boldsymbol{y}_q(n) = \begin{bmatrix} y_q(n), \ldots, y_q(n-D+1) \end{bmatrix}^{\mathrm{T}}. \tag{13.5}$$

To express the linear convolution as a matrix-vector product, the $2L \times D$ matrix $\boldsymbol{W}_{pq}$ exhibits a Sylvester structure that contains all L coefficients of the respective demixing filter in each column:

---

[1] With respect to efficient DFT-domain implementations the vector $\boldsymbol{x}_p$ contains $2L$ sensor signal samples instead of the $L+D-1$ samples required for the linear convolution ($1 \leq D \leq L$).

$$\boldsymbol{W}_{pq} = \begin{bmatrix} w_{pq,0} & 0 & \cdots & 0 \\ w_{pq,1} & w_{pq,0} & \ddots & \vdots \\ \vdots & w_{pq,1} & \ddots & 0 \\ w_{pq,L-1} & \vdots & \ddots & w_{pq,0} \\ 0 & w_{pq,L-1} & \ddots & w_{pq,1} \\ \vdots & & \ddots & \vdots \\ 0 & \cdots & 0 & w_{pq,L-1} \\ 0 & \cdots & 0 & 0 \\ \vdots & \cdots & \vdots & \vdots \\ 0 & \cdots & 0 & 0 \end{bmatrix}. \qquad (13.6)$$

It can be seen that for the general case, $1 \leq D \leq L$, the last $L - D + 1$ rows of $\boldsymbol{W}_{pq}$ are padded with zeros to ensure compatibility with the length of $\boldsymbol{x}_p(n)$ which was chosen to $2L$. Note that for $D = 1$, Eq. 13.3 simplifies to the well-known vector formulation of a convolution, as it is used extensively in the literature on supervised adaptive filtering, e.g., [41]. Finally, to allow a convenient notation we combine all channels and thus, we can write Eq. 13.3 compactly as

$$\boldsymbol{y}(n) = \boldsymbol{W}^{\mathrm{T}} \boldsymbol{x}(n), \qquad (13.7)$$

with

$$\boldsymbol{x}(n) = \begin{bmatrix} \boldsymbol{x}_1^{\mathrm{T}}(n), \ldots, \boldsymbol{x}_P^{\mathrm{T}}(n) \end{bmatrix}^{\mathrm{T}}, \qquad (13.8)$$

$$\boldsymbol{y}(n) = \begin{bmatrix} \boldsymbol{y}_1^{\mathrm{T}}(n), \ldots, \boldsymbol{y}_P^{\mathrm{T}}(n) \end{bmatrix}^{\mathrm{T}}, \qquad (13.9)$$

$$\boldsymbol{W} = \begin{bmatrix} \boldsymbol{W}_{11} & \cdots & \boldsymbol{W}_{1P} \\ \vdots & \ddots & \vdots \\ \boldsymbol{W}_{P1} & \cdots & \boldsymbol{W}_{PP} \end{bmatrix}, \qquad (13.10)$$

with $\boldsymbol{W}$ exhibiting a blockwise Sylvester structure.

13.2.2 Optimization Criterion and Coefficient Update

As pointed out before, we aim at an optimization criterion simultaneously exploiting the three signal properties nonstationarity, nonwhiteness, and nongaussianity. Therefore, based on a generalization of Shannon's mutual information [27], the following optimization criterion was defined in [14] and was termed "**TRI**ple-**N**-**I**ndependent component analysis for **CON**volutive mixtures" (**TRINICON**) as it simultaneously accounts for the three fundamental properties **N**onwhiteness, **N**onstationarity, and **N**ongaussianity:

$$\mathcal{J}(m, \boldsymbol{W}) = \sum_{i=0}^{\infty} \beta(i,m) \frac{1}{N} \sum_{j=0}^{N-1} \left\{ \log \frac{\hat{p}_{y,PD}(\boldsymbol{y}(iL+j))}{\prod_{q=1}^{P} \hat{p}_{y_q,D}(\boldsymbol{y}_q(iL+j))} \right\}. \quad (13.11)$$

The variable $\hat{p}_{y_q,D}(\cdot)$ is the estimated or assumed multivariate probability density function (PDF) for channel q of dimension D and $\hat{p}_{y,PD}(\cdot)$ is the joint PDF of dimension PD over all channels. The usage of PDFs allows to exploit the *nongaussianity* of the signals. Furthermore, the multivariate structure of the PDFs, which is given by the memory length D, i.e., the number of time-lags, models the *nonwhiteness* of the P signals with D chosen to $1 \leq D \leq L$. The expectation operator of the mutual information [27] is replaced in Eq. 13.11 by a short-time estimate of the multivariate PDFs using N time instants. To allow for a proper estimation of the multivariate PDFs the averaging has to be done in general for $N > PD$ time instants. The block indices i, m refer to the blocks which are underlying to the statistical estimation of the multivariate PDFs. For each output signal block $\boldsymbol{y}_q(iL+j)$ containing D samples a sensor signal block of length $2L$ is required according to Eq. 13.4. The *nonstationarity* is taken into account by a weighting function $\beta(i,m)$ with the block indices i, m and with finite support. The weighting function is normalized according to

$$\sum_{i=0}^{\infty} \beta(i,m) = 1, \quad (13.12)$$

and allows offline, online, and block-online implementations of the algorithms [16]. As an example,

$$\beta(i,m) = \begin{cases} (1-\lambda)\lambda^{m-i}, & \text{for } 0 \leq i \leq m, \\ 0, & \text{else,} \end{cases} \quad (13.13)$$

leads to an efficient online version allowing for tracking in time-variant environments. The forgetting factor λ is usually chosen close to, but less than 1. A robust block-online adaptation was discussed in detail in [6].

The approach followed here is carried out with overlapping data blocks as the sensor signal blocks of length $2L$ are shifted only by L samples due to the time index iL in Eq. 13.11. Analogously to supervised block-based adaptive filtering [41], this increases the convergence rate and reduces the signal delay. If further overlapping is desired, then the time index iL in Eq. 13.11 is simply replaced by iL/α. The overlap factor α with $1 \leq \alpha \leq L$ should be chosen suitably to obtain integer values for the time index.

The derivation of the gradient with respect to the demixing filter weights $w_{pq,\kappa}$ for $p, q = 1, \ldots, P$ and $\kappa = 0, \ldots, L-1$ can be expressed compactly in matrix notation by defining the matrix $\check{\boldsymbol{W}}$ given as

$$\check{\boldsymbol{W}} = \begin{bmatrix} \boldsymbol{w}_{11} & \cdots & \boldsymbol{w}_{1P} \\ \vdots & \ddots & \vdots \\ \boldsymbol{w}_{P1} & \cdots & \boldsymbol{w}_{PP} \end{bmatrix},$$

which is composed of the column vectors $\boldsymbol{w}_{pq}$ containing the demixing filter coefficients

$$\boldsymbol{w}_{pq} = \begin{bmatrix} w_{pq,0}, \ldots, w_{pq,L-1} \end{bmatrix}^{\mathrm{T}}. \tag{13.14}$$

Then the gradient with respect to the $P^2 L$ demixing filter coefficients can be expressed compactly as

$$\nabla_{\check{\boldsymbol{W}}} \mathcal{J}(m, \boldsymbol{W}) = \frac{\partial \mathcal{J}(m, \boldsymbol{W})}{\partial \check{\boldsymbol{W}}}. \tag{13.15}$$

With an iterative optimization procedure, the current demixing matrix is obtained by the recursive update equation

$$\check{\boldsymbol{W}}(m) = \check{\boldsymbol{W}}(m-1) - \mu \Delta \check{\boldsymbol{W}}(m), \tag{13.16}$$

where μ is a stepsize parameter, and $\Delta \check{\boldsymbol{W}}(m)$ is the update which is set equal to $\nabla_{\check{\boldsymbol{W}}} \mathcal{J}(m, \boldsymbol{W})$ for gradient descent adaptation.

In order to calculate the gradient (Eq. 13.15), the TRINICON optimization criterion $\mathcal{J}(m, \boldsymbol{W})$ given in Eq. 13.11 has to be expressed in terms of the demixing filter coefficients $w_{pq,\kappa}$. This can be done by inserting the definition of the linear convolution $\boldsymbol{y} = \boldsymbol{W}^{\mathrm{T}} \boldsymbol{x}$ given in Eq. 13.7 into $\mathcal{J}(m, \boldsymbol{W})$ and subsequently transforming the output signal PDF $\hat{p}_{y,PD}(\boldsymbol{y}(iL+j))$ into the PD-dimensional input signal PDF $\hat{p}_{x,PD}(\cdot)$ using the Sylvester matrix $\boldsymbol{W}$, which is considered as a mapping matrix for this linear transformation [71]. This leads to an expression of the optimization criterion 13.11 with respect to the Sylvester matrix $\boldsymbol{W}$. To be able to take the derivative with respect to $\check{\boldsymbol{W}}$ instead of the Sylvester matrix $\boldsymbol{W}$, the chain rule for the derivative of a scalar function with respect to a matrix [40] was applied to the gradient (Eq. 13.15) in [20]. There, it was shown that the chain rule leads to a $\mathcal{S}$ylvester $\mathcal{C}$onstraint operator ($\mathcal{SC}$) which relates the gradient with respect to $\check{\boldsymbol{W}}$ and with respect to $\boldsymbol{W}$ as

$$\nabla_{\check{\boldsymbol{W}}} \mathcal{J}(m, \boldsymbol{W}) = \mathcal{SC}\{\nabla_{\boldsymbol{W}} \mathcal{J}(m, \boldsymbol{W})\}. \tag{13.17}$$

The Sylvester constraint operator $\mathcal{SC}$ is illustrated for the pq-th submatrix of $\nabla_{\boldsymbol{W}} \mathcal{J}(m, \boldsymbol{W})$ in the left plot of Fig. 13.2 where it can be seen that it corresponds (up to a scaling by the constant factor D) to an arithmetic average over the elements on each diagonal of the $2L \times D$ submatrices of the gradient $\nabla_{\boldsymbol{W}} \mathcal{J}(m, \boldsymbol{W})$. Thus, the $2PL \times PD$ gradient $\nabla_{\boldsymbol{W}} \mathcal{J}(m, \boldsymbol{W})$ will be reduced to the $PL \times P$ gradient $\nabla_{\check{\boldsymbol{W}}} \mathcal{J}(m, \boldsymbol{W})$.

To reduce computational complexity, two efficient approximated versions of the Sylvester constraint $\mathcal{SC}$ (see Fig. 13.2) were discussed in [6] leading to two different classes of algorithms:

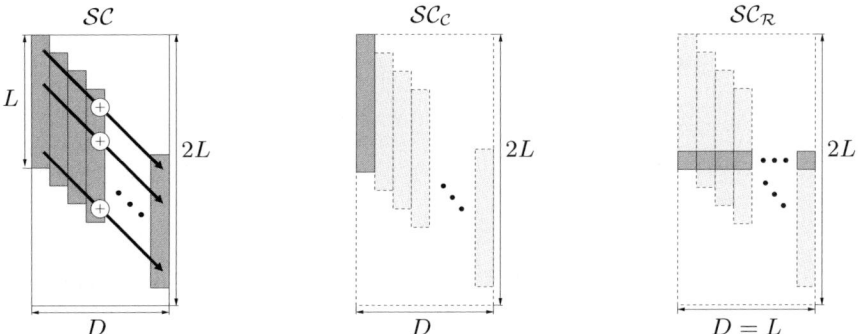

Fig. 13.2. The Sylvester constraint ($\mathcal{SC}$) and two popular approximations denoted as the column Sylvester constraint $\mathcal{SC}_\mathcal{C}$ and row Sylvester constraint $\mathcal{SC}_\mathcal{R}$ all illustrated for the gradient $\nabla_{\boldsymbol{W}_{pq}} \mathcal{J}(m, \boldsymbol{W})$ with respect to the pq-th submatrix $\boldsymbol{W}_{pq}$.

(1) Computing only the *first column* of each channel of the update matrix to obtain the new coefficient matrix $\boldsymbol{\check{W}}$. This method is denoted as $\mathcal{SC}_\mathcal{C}$.
(2) Computing only the *L-th row* of each channel of the update matrix to obtain the new coefficient matrix $\boldsymbol{\check{W}}$. This method is denoted as $\mathcal{SC}_\mathcal{R}$.

It can be shown that in both cases the update process is considerably simplified [6]. However, in general, both choices require some tradeoff in the algorithm performance. While simulations showed [4] that $\mathcal{SC}_\mathcal{C}$ may provide a potentially more robust convergence behaviour, it will not work for arbitrary source positions (e.g., in the case of two sources, they are required to be located in different half-planes with respect to the orientation of the microphone array), or for $P > 2$, which is in contrast to the more versatile $\mathcal{SC}_\mathcal{R}$ [4,6]. Note that the choice of $\mathcal{SC}$ also determines the appropriate coefficient initialization [4,6].

It is known that stochastic gradient descent, i.e., $\Delta \boldsymbol{\check{W}}(m) = \nabla_{\boldsymbol{\check{W}}} \mathcal{J}(m, \boldsymbol{W})$ suffers from slow convergence in many practical problems. In the BSS application the gradient and thus, the separation performance depends on the MIMO mixing system. Fortunately, a modification of the ordinary gradient, termed the *natural gradient* by Amari [9] and the *relative gradient* by Cardoso [21] (which is equivalent to the natural gradient in the BSS application) has been developed that largely removes all effects of an ill-conditioned mixing matrix, assuming an appropriate initialization of $\boldsymbol{W}$ and thus leads to better performance compared to the stochastic gradient descent. The idea of the relative gradient is based on the equivariance property. Generally speaking, an estimator behaves equivariantly if it produces estimates that, under data transformation, are transformed in the same way as the data [21]. In the context of BSS the key property of equivariant estimators is that they exhibit uniform performance, e.g., in terms of bias and variance, independently of the mixing system. In [17] the natural/relative gradient has been extended to the case of Sylvester matrices $\boldsymbol{W}$ which together with the Sylvester constraint

yields
$$\nabla_{\boldsymbol{W}}^{\mathrm{NG}} \mathcal{J}(m, \boldsymbol{W}) = \mathcal{SC}\left\{\boldsymbol{W}\boldsymbol{W}^{\mathrm{T}} \nabla_{\boldsymbol{W}} \mathcal{J}(m, \boldsymbol{W})\right\}. \tag{13.18}$$

This leads to the following expression for the *HOS natural gradient*

$$\nabla_{\boldsymbol{W}}^{\mathrm{NG}} \mathcal{J}(m, \boldsymbol{W})$$
$$= \mathcal{SC}\left\{\sum_{i=0}^{\infty} \beta(i,m)\boldsymbol{W}(i)\frac{1}{N}\sum_{j=0}^{N-1}\left\{\boldsymbol{y}(iL+j)\boldsymbol{\Phi}^{\mathrm{T}}\left(\boldsymbol{y}(iL+j)\right) - \boldsymbol{I}\right\}\right\}, \tag{13.19}$$

with the general weighting function $\beta(i,m)$ and the *multivariate score function* $\boldsymbol{\Phi}(\boldsymbol{y}(.))$ consisting of the stacked channel-wise multivariate score functions $\boldsymbol{\Phi}_q(\boldsymbol{y}_q(.))$, $q = 1, \ldots, P$ defined as

$$\boldsymbol{\Phi}\bigl(\boldsymbol{y}(iL+j)\bigr) = \left[\left(-\frac{\frac{\partial \hat{p}_{\boldsymbol{y}_1,D}(\boldsymbol{y}_1(iL+j))}{\partial \boldsymbol{y}_1(iL+j)}}{\hat{p}_{\boldsymbol{y}_1,D}(\boldsymbol{y}_1(iL+j))}\right)^{\mathrm{T}}, \ldots, \left(-\frac{\frac{\partial \hat{p}_{\boldsymbol{y}_P,D}(\boldsymbol{y}_P(iL+j))}{\partial \boldsymbol{y}_P(iL+j)}}{\hat{p}_{\boldsymbol{y}_P,D}(\boldsymbol{y}_P(iL+j))}\right)^{\mathrm{T}}\right]^{\mathrm{T}}$$
$$:= \left[\boldsymbol{\Phi}_1^{\mathrm{T}}\bigl(\boldsymbol{y}_1(iL+j)\bigr), \ldots, \boldsymbol{\Phi}_P^{\mathrm{T}}\bigl(\boldsymbol{y}_P(iL+j)\bigr)\right]^{\mathrm{T}}. \tag{13.20}$$

The update in Eq. 13.19 represents a so-called holonomic algorithm as it imposes the constraint $\boldsymbol{y}(iL+j)\boldsymbol{\Phi}^{\mathrm{T}}(\boldsymbol{y}(iL+j)) = \boldsymbol{I}$ on the magnitudes of the recovered signals. However, when the source signals are nonstationary, these constraints may force a rapid change in the magnitude of the demixing matrix which in turn leads to numerical instabilities in some cases (see, e.g., [25]). By replacing $\boldsymbol{I}$ in Eq. 13.19 with the term bdiag$\{\boldsymbol{y}(iL+j)\boldsymbol{\Phi}^{\mathrm{T}}(\boldsymbol{y}(iL+j))\}$ the constraint on the magnitude of the recovered signals can be avoided. This is termed the *nonholonomic natural gradient* algorithm which is given as

$$\nabla_{\boldsymbol{W}}^{\mathrm{NG}} \mathcal{J}(m, \boldsymbol{W}) = \mathcal{SC}\Bigg\{\sum_{i=0}^{\infty} \beta(i,m)\boldsymbol{W}(i)\frac{1}{N}\sum_{j=0}^{N-1}\Big\{\boldsymbol{y}(iL+j)\boldsymbol{\Phi}^{\mathrm{T}}\bigl(\boldsymbol{y}(iL+j)\bigr)$$
$$- \mathrm{bdiag}\Big\{\boldsymbol{y}(iL+j)\boldsymbol{\Phi}^{\mathrm{T}}\bigl(\boldsymbol{y}(iL+j)\bigr)\Big\}\Big\}\Bigg\}. \tag{13.21}$$

Here, the bdiag operator sets all cross-channel terms to zero. Due to the improved convergence behaviour and the nonstationary nature of acoustic signals the remainder of this chapter will focus on the nonholonomic algorithm (Eq. 13.21) based on the natural gradient.

13.2.3 Approximations Leading to Special Cases

The natural gradient update (Eq. 13.21) rule provides a very general basis for BSS of convolutive mixtures. However, to apply it to real-world scenarios, the multivariate score function (Eq. 13.20) has to be estimated, i.e., we have to estimate P multivariate PDFs $\hat{p}_{y_q,D}(\boldsymbol{y}_q(iL+j))$, $q = 1, \ldots, P$ of dimension D. In general, this is a very challenging task, as it effectively requires estimation of all possible higher-order cumulants for a set of D output samples, where D may be on the order of several hundred or thousand in real acoustic environments.

As first shown in [14] we will present in Sec. 13.2.3.1 an efficient solution for the problem of estimating the multivariate score function by assuming so-called *spherically invariant random processes* (SIRPs). Moreover, efficient realizations based on second-order statistics will be derived in Sec. 13.2.3.2 by utilization of the multivariate Gaussian PDF.

13.2.3.1 Higher-Order Statistics Realization Based on Multivariate PDFs

Early experimental measurements [28] indicated that the PDF of speech signals in the time domain can be approximated by exponential distributions such as the Gamma or Laplacian PDF. Later on, a special class of multivariate PDFs based on the assumption of SIRPs was used in [13] to model bandlimited telephone speech. The SIRP model is representative for a wide class of stochastic processes [35,74,89] and is very attractive since multivariate PDFs can be derived analytically from the corresponding univariate probability density function together with the correlation matrices covering multiple time-lags. The correlation matrices can be estimated from the data while for the univariate PDF appropriate models can be assumed or the univariate PDF can be estimated based on parameterized representations, such as the Gram-Charlier or Edgeworth expansions [48].

The general model of a zero-mean non-white SIRP of D-th order for channel q is given by [13]

$$\hat{p}_{y_q,D}\big(\boldsymbol{y}_q(iL+j)\big)
= \frac{1}{\sqrt{\pi^D \det(\boldsymbol{R}_{\boldsymbol{y}_q\boldsymbol{y}_q}(i))}} f_{y_q,D}\left(\boldsymbol{y}_q^{\mathrm{T}}(iL+j)\boldsymbol{R}_{\boldsymbol{y}_q\boldsymbol{y}_q}^{-1}(i)\boldsymbol{y}_q(iL+j)\right) \quad (13.22)$$

with the $D \times D$ correlation matrix given as

$$\boldsymbol{R}_{\boldsymbol{y}_p\boldsymbol{y}_q}(i) = \frac{1}{N}\sum_{j=0}^{N-1} \boldsymbol{y}_p(iL+j)\,\boldsymbol{y}_q^{\mathrm{T}}(iL+j), \quad (13.23)$$

and the function $f_{y_q,D}(\cdot)$ depending on the chosen univariate PDF. As the best known example, the multivariate Gaussian can be viewed as a special case of the class of SIRPs. The multivariate PDFs are completely characterized by the scalar function $f_{y_q,D}(\cdot)$ and $\boldsymbol{R}_{\boldsymbol{y}_q\boldsymbol{y}_q}$. Due to the quadratic form $\boldsymbol{y}_q^{\mathrm{T}}\boldsymbol{R}_{\boldsymbol{y}_q\boldsymbol{y}_q}^{-1}\boldsymbol{y}_q$, the PDF is spherically invariant which means for the bivariate case ($D = 2$)

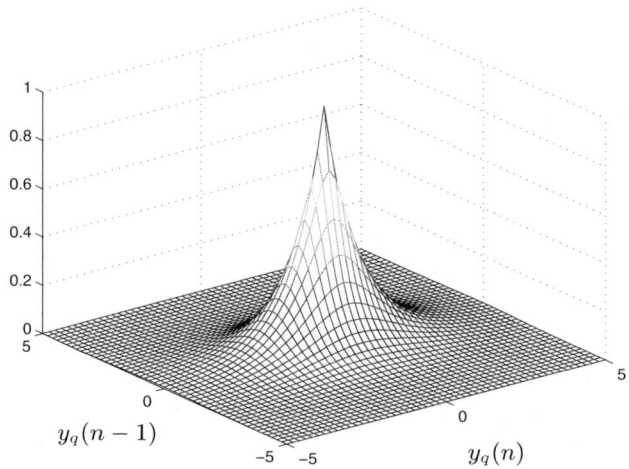

Fig. 13.3. Illustration of a bivariate SIRP PDF (i.e., $D = 2$).

that independent of the choice of $f_{y_q,D}(\cdot)$ the bivariate PDFs based on the SIRP model exhibit ellipsoidal or circular contour lines (see Fig. 13.3). The function $f_{y_q,D}(\cdot)$ is determined by the choice of the univariate PDF and can be calculated by using the so-called Meijer's G-functions as detailed in [13].

By introducing SIRPs into the BSS optimization criterion we obtain a considerably simplified expression for the multivariate score function (Eq. 13.20) as first presented in [14]. After applying the chain rule to Eq. 13.22, the multivariate score function for the q-th channel can be expressed as

$$\boldsymbol{\Phi}_q\big(\boldsymbol{y}_q(iL+j)\big) = -\frac{\frac{\partial \hat{p}_{y_q,D}(\boldsymbol{y}_q(iL+j))}{\partial \boldsymbol{y}_q(iL+j)}}{\hat{p}_{y_q,D}(\boldsymbol{y}_q(iL+j))}$$

$$= 2 \underbrace{\left[-\frac{\frac{\partial f_{y_q,D}(u_q(iL+j))}{\partial u_q(iL+j)}}{f_{y_q,D}(u_q(iL+j))}\right]}_{:=\phi_{y_q,D}(u_q(iL+j))} \boldsymbol{R}^{-1}_{\boldsymbol{y}_q\boldsymbol{y}_q}(i)\,\boldsymbol{y}_q(iL+j). \quad (13.24)$$

For convenience, we call the scalar function $\phi_{y_q,D}(u_q(iL+j))$ the *SIRP score* of channel q and the scalar argument given as the quadratic form is defined as

$$u_q(iL+j) = \boldsymbol{y}_q^T(iL+j)\,\boldsymbol{R}^{-1}_{\boldsymbol{y}_q\boldsymbol{y}_q}(i)\,\boldsymbol{y}_q(iL+j). \quad (13.25)$$

From Eq. 13.24 it can be seen that the estimation of multivariate PDFs reduces to an estimation of the correlation matrix together with a computation of the

SIRP score which can be determined by choosing suitable models for the multivariate SIRP PDF.

In [13, 34] it was shown that the *spherically symmetric multivariate Laplacian PDF* which exhibits Laplacian marginals is a good model for long-term properties of speech signals in the time-domain. A derivation of the multivariate Laplacian based on SIRPs can be found in, e.g., [13, 31, 53] and leads to Eq. 13.22 with the function $f_{y_q,D}(u_q(iL+j))$ given as

$$f_{y_q,D}(u_q(iL+j)) = \left(\frac{1}{\sqrt{2u_q(iL+j)}}\right)^{D/2-1} K_{D/2-1}\left(\sqrt{2u_q(iL+j)}\right). \tag{13.26}$$

where $K_\nu(\cdot)$ denotes the ν-th order modified Bessel function of the second kind. The SIRP score for the multivariate Laplacian SIRP PDF can be straightforwardly derived by using the relation for the derivative of a ν-th order modified Bessel function of the second kind given as [1]

$$\frac{\partial K_\nu\left(\sqrt{2u_q}\right)}{\partial \sqrt{2u_q}} = \frac{\nu}{\sqrt{2u_q}} K_\nu\left(\sqrt{2u_q}\right) - K_{\nu+1}\left(\sqrt{2u_q}\right), \tag{13.27}$$

and is obtained as

$$\phi_{y_q,D}(u_q(iL+j)) = \frac{1}{\sqrt{2u_q(iL+j)}} \frac{K_{D/2}\left(\sqrt{2u_q(iL+j)}\right)}{K_{D/2-1}\left(\sqrt{2u_q(iL+j)}\right)}. \tag{13.28}$$

It should be noted that the formulation of Eq. 13.28 in [14] is slightly different but equivalent. In practical implementations the ν-th order modified Bessel function of the second kind $K_\nu(\sqrt{2u_q})$ may be approximated by [1]

$$K_\nu\left(\sqrt{2u_q}\right) = \sqrt{\frac{\pi}{2\sqrt{2u_q}}} e^{-\sqrt{2u_q}} \left(1 + \frac{4\nu^2-1}{8\sqrt{2u_q}} + \frac{(4\nu^2-1)(4\nu^2-9)}{2!(8\sqrt{2u_q})^2} + \ldots\right). \tag{13.29}$$

Having derived the multivariate score function (Eq. 13.24) for the SIRP model, we can now insert it into the generic HOS natural gradient update equation with its nonholonomic extension (Eq. 13.21) and will find several attractive properties that lead to significant reductions in computational complexity relative to the general case. Considering the fact that the autocorrelation matrices are symmetric so that $(\boldsymbol{R}_{\boldsymbol{y}_q\boldsymbol{y}_q}^{-1})^{\mathrm{T}} = \boldsymbol{R}_{\boldsymbol{y}_q\boldsymbol{y}_q}^{-1}$ leads to the following expression for the *nonholonomic HOS-SIRP natural gradient*:

$$\nabla_{\boldsymbol{W}}^{\mathrm{NG}} \mathcal{J}(m,\boldsymbol{W})$$
$$= \mathcal{SC}\left\{2\sum_{i=0}^{\infty} \beta(i,m)\boldsymbol{W}(i)\left[\boldsymbol{R}_{\boldsymbol{y}\phi(\boldsymbol{y})}(i) - \mathrm{bdiag}\left\{\boldsymbol{R}_{\boldsymbol{y}\phi(\boldsymbol{y})}(i)\right\}\right]\mathrm{bdiag}^{-1}\left\{\boldsymbol{R}_{\boldsymbol{y}\boldsymbol{y}}(i)\right\}\right\} \tag{13.30}$$

with the second-order correlation matrix $\boldsymbol{R_{yy}}$ consisting of the channel-wise submatrices $\boldsymbol{R_{y_p y_q}}$ defined in Eq. 13.23 and $\boldsymbol{R_{y\phi(y)}}$ consisting of the channel-wise submatrices $\boldsymbol{R_{y_p \phi(y_q)}}$ given as

$$\boldsymbol{R}_{\boldsymbol{y}_p\phi(\boldsymbol{y}_q)}(i) = \frac{1}{N}\sum_{j=0}^{N-1} \boldsymbol{y}_p(iL+j)\phi_{\boldsymbol{y}_q,D}\bigl(\boldsymbol{u}_q(iL+j)\bigr)\boldsymbol{y}_q^{\mathrm{T}}(iL+j). \quad (13.31)$$

The SIRP score $\phi_{\boldsymbol{y}_q,D}(\cdot)$ of channel q which is a scalar value function causes a weighting of the correlation matrix in Eq. 13.31. In Eq. 13.30 only channel-wise submatrices have to be inverted so that it is sufficient to choose $N > D$ instead of $N > PD$ for the estimation of $\boldsymbol{R_{yy}}(i)$ and $\boldsymbol{R_{y\phi(y)}}$. Moreover, from the update equation 13.30, it can be seen that the SIRP model leads to an inherent normalization by the auto-correlation submatrices. This becomes especially obvious if the update (Eq. 13.30) is written explicitly for a 2-by-2 MIMO system leading to

$$\nabla_{\boldsymbol{W}}^{\mathrm{NG}} \mathcal{J}(m, \boldsymbol{W})$$
$$= \mathcal{SC}\left\{ 2\sum_{i=0}^{\infty} \beta(i,m)\boldsymbol{W}(i) \begin{bmatrix} 0 & \boldsymbol{R}_{\boldsymbol{y}_1\phi(\boldsymbol{y}_2)}(i)\boldsymbol{R}_{\boldsymbol{y}_2\boldsymbol{y}_2}^{-1}(i) \\ \boldsymbol{R}_{\boldsymbol{y}_2\phi(\boldsymbol{y}_1)}(i)\boldsymbol{R}_{\boldsymbol{y}_1\boldsymbol{y}_1}^{-1}(i) & 0 \end{bmatrix} \right\}.$$
$$(13.32)$$

The normalization is important as it provides good convergence even for correlated signals such as speech and also for a large number of filter taps. The normalization is similar as in the recursive least-squares (RLS) algorithm in supervised adaptive filtering where also the inverse of the auto-correlation matrix is computed [41]. To obtain efficient implementations, the normalization by the computationally demanding inverse of the $D \times D$ matrix can be approximated in several ways as shown in Sec. 13.2.4 and outlined in Sec. 13.2.5.

13.2.3.2 Second-Order Statistics Realization Based on the Multivariate Gaussian PDF

Using the model of the multivariate Gaussian PDF leads to a second-order realization of the BSS algorithm utilizing the nonstationarity and the non-whiteness of the source signals. The multivariate Gaussian PDF

$$\hat{p}_{\boldsymbol{y}_q,D}\bigl(\boldsymbol{y}_q(iL+j)\bigr) = \frac{1}{\sqrt{(2\pi)^D \det\bigl(\boldsymbol{R}_{\boldsymbol{y}_q\boldsymbol{y}_q}(i)\bigr)}}\, e^{-\frac{1}{2}\boldsymbol{y}_q^{\mathrm{T}}(iL+j)\boldsymbol{R}_{\boldsymbol{y}_q\boldsymbol{y}_q}^{-1}(i)\boldsymbol{y}_q(iL+j)}$$
$$(13.33)$$

is inserted in the expression for the multivariate score function (Eq. 13.20) whose elements reduce to

$$\boldsymbol{\Phi}_q\bigl(\boldsymbol{y}_q(iL+j)\bigr) = \boldsymbol{R}_{\boldsymbol{y}_q\boldsymbol{y}_q}^{-1}(i)\,\boldsymbol{y}_q(iL+j). \quad (13.34)$$

Inserting Eq. 13.34 into the natural gradient update (Eq. 13.19) yields the *SOS natural gradient*:

$$\nabla_{\boldsymbol{W}}^{\mathrm{NG}} \mathcal{J}(m, \boldsymbol{W})$$
$$= \mathcal{SC} \left\{ \sum_{i=0}^{\infty} \beta(i, m) \boldsymbol{W}(i) \Big[\boldsymbol{R}_{\boldsymbol{yy}}(i) - \mathrm{bdiag}\,\{\boldsymbol{R}_{\boldsymbol{yy}}(i)\} \Big] \mathrm{bdiag}^{-1}\{\boldsymbol{R}_{\boldsymbol{yy}}(i)\} \right\}. \tag{13.35}$$

Comparing Eq. 13.35 to the HOS-SIRP update (Eq. 13.30) shows that due to the fact that only SOS are utilized, we obtain the same update with the nonlinearity (Eq. 13.28) omitted, i.e., $\phi_{y_q, D}(u_q(iL + j)) = 1$, $q = 1, \ldots, P$. Therefore, the SOS natural gradient update also exhibits the inherent normalization by the auto-correlation matrices which leads to very robust convergence behaviour in real-world environments. Moreover, due to the inversion of channel-wise $D \times D$ submatrices, $N > D$ instead of $N > PD$ is again sufficient for the estimation of the correlation matrices.

In Fig. 13.4 the structure of the cost function in the case of SOS and idealized/simplified mechanism of the adaptation update (Eq. 13.35) is illustrated. By assuming the multivariate Gaussian PDF (Eq. 13.33) and then minimizing $\mathcal{J}(m, \boldsymbol{W})$, all cross-correlations for D time-lags are reduced and thus the algorithm exploits nonwhiteness. Nonstationarity is utilized by minimizing the correlation matrices simultaneously for several blocks i. Ideally, the cross-correlations will be equal to zero upon convergence which causes the update term to be zero because then $\boldsymbol{R}_{\boldsymbol{yy}}(i) - \mathrm{bdiag}\,\{\boldsymbol{R}_{\boldsymbol{yy}}(i)\} = \boldsymbol{0}$.

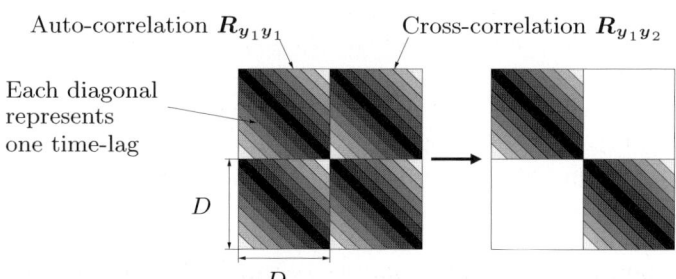

Fig. 13.4. Illustration of the diagonalization of the correlation matrices $\boldsymbol{R}_{\boldsymbol{yy}}(i)$ performed by the natural gradient update (Eq. 13.35) for the 2×2 case.

An alternative derivation of a SOS BSS algorithm leading to the same natural gradient update as given in Eq. 13.35 was presented in [17]. There, the derivation was based on a generalized version of the cost function used in [64], which also simultaneously exploits nonwhiteness and nonstationarity of the sources.

13.2.4 Estimation of the Correlation Matrices and an Efficient Normalization Strategy

In this section some implementation aspects are addressed which allow to reduce the computational complexity. The first aspect is the block-based estimation of the short-time output correlation matrices $\boldsymbol{R}_{\boldsymbol{y}_p \boldsymbol{y}_q}(i)$ for nonstationary signals for which two basic methods exist: The so-called *covariance method* and the *correlation method* as they are known from linear prediction problems [58]. It should be emphasized that the terms covariance method and correlation method are not based upon the standard usage of the covariance function as the correlation function with the means removed. In the definition of the correlation matrices in Eq. 13.23 the more accurate covariance method was introduced. To obtain more efficient implementations, the computationally less complex correlation method can be used which is obtained by assuming stationarity within each block i. This leads to a Toeplitz structure of the matrices $\boldsymbol{R}_{\boldsymbol{y}_p \boldsymbol{y}_q}(i)$ and thus simplifies the computation of the matrix products [6]. Furthermore, it is important to note that regardless of the estimation of the correlation matrices, the matrix product of Sylvester matrices $\boldsymbol{W}_{pq}$ and the remaining matrices in the update Eqs. 13.30 and 13.35 can be described by linear convolutions due to the Sylvester structure involved.

As a second aspect, we discuss the inherent normalization by the autocorrelation matrices in Eqs. 13.30 and 13.35 which is introduced by the usage of multivariate PDFs as pointed out in the previous section. The normalization is desirable as it guarantees fast convergence of the adaptive filters even for large filter lengths and correlated input signals. On the other hand this poses the problem of large computational complexity due to the required matrix inversion of P matrices of size $D \times D$. The complexity of a straightforward implementation is $\mathcal{O}(D^3)$ for using the covariance method and $\mathcal{O}(D^2)$ for the correlation method due to the Toeplitz structure involved. However, as D may be even larger than 1000 for realistic environments this is still prohibitive for a real-time implementation on regular PC platforms. Therefore, approximations are desirable which reduce the complexity with minimum degradation of the separation performance.

One possible solution is to approximate the auto-correlation matrices $\boldsymbol{R}_{\boldsymbol{y}_q \boldsymbol{y}_q}(i)$ by a diagonal matrix, i.e., by the output signal powers

$$\boldsymbol{R}_{\boldsymbol{y}_q \boldsymbol{y}_q}(i) \approx \frac{1}{N} \sum_{j=0}^{N-1} \operatorname{diag}\left\{\boldsymbol{y}_q(iL+j)\boldsymbol{y}_q^{\mathrm{T}}(iL+j)\right\}. \tag{13.36}$$

for $q = 1, \ldots, P$, where the operator $\operatorname{diag}\{\boldsymbol{A}\}$ sets all off-diagonal elements of matrix $\boldsymbol{A}$ to zero. This approximation is comparable to the one in the well-known normalized least mean squares (NLMS) algorithm in supervised adaptive filtering approximating the RLS algorithm [41]. It should be noted that the SOS natural gradient algorithm based on Eq. 13.35 together with the approximation 13.36 was also heuristically introduced for the case $D = L$

in [2,70] as an extension of [50] incorporating several time-lags. It should be pointed out that also a more sophisticated approximation of the normalization is possible. One approach which exploits the efficiency of computations in the DFT domain is outlined in the next section.

For blocks with speech pauses and low background noise the normalization by the auto-correlation matrix $\boldsymbol{R}_{\boldsymbol{y}_q \boldsymbol{y}_q}$ leads to the inversion of an ill-conditioned matrix or in the case of the approximation (Eq. 13.36) to a division by very small output powers or even by zero becomes likely and thus, the estimation of the filter coefficients becomes very sensitive. For a robust adaptation $\boldsymbol{R}_{\boldsymbol{y}_q \boldsymbol{y}_q}$ is replaced by a regularized version $\boldsymbol{R}_{\boldsymbol{y}_q \boldsymbol{y}_q} + \delta_{y_q} \boldsymbol{I}$. The basic feature of the regularization is a compromise between fidelity to data and fidelity to prior information about the solution [23]. As the latter increases robustness but leads to biased solutions, similarly to supervised adaptive filtering [19], a dynamical regularization

$$\delta_{y_q} = \delta_{\max} e^{-\sigma_{y_q}^2 / \sigma_0^2} \qquad (13.37)$$

can be used with two parameters $\delta_{\max}$ and σ_0^2. This exponential method provides a smooth transition between regularization for low output power $\sigma_{y_q}^2$ and data fidelity whenever the output power is large enough. Other popular strategies are the fixed regularization which simply adds a constant value to the output power $\delta_{y_q} = $ const and the approach of choosing the maximum out of the output signal power $\sigma_{y_q}^2$ and a fixed threshold δ_{th}.

13.2.5 On Broadband and Narrowband BSS Algorithms in the DFT Domain

In the previous sections it was shown how different time-domain algorithms can be derived from the TRINICON framework. On the other hand, for convolutive mixtures the classical approach of frequency-domain BSS appears to be an attractive alternative because all techniques originally developed for instantaneous BSS can typically be applied independently in each frequency bin, e.g., [48]. Unfortunately, this traditional narrowband approach exhibits several limitations as identified in, e.g., [10,55,75]. In particular, the permutation problem pointed out in Sec. 13.1, which is inherent in BSS may then also appear independently in each frequency bin so that extra repair measures have to be taken to address this *internal* permutation. Moreover, problems caused by circular convolution effects due to the narrowband approximation are reported in, e.g., [75].

To exploit the computational efficiency it is desirable to derive approaches in the DFT domain, but on the other hand the above-mentioned problems of the narrowband approach should be avoided. This can be achieved by transforming the equations of the TRINICON framework into the DFT domain in a rigorous way (i.e., without any approximations) as was shown in [16, 17]. As in the case of time-domain algorithms, the resulting generic DFT-domain broadband BSS may serve both as a unifying framework for existing algorithms, and also as a starting point for developing new improved algorithms by a considerate choice of *selective* approximations as shown in, e.g., [7, 16]. Fig. 13.5 gives an overview on the most important classes of DFT-domain BSS algorithms known so far (various more special cases may be developed in the future). A very important observation from this framework using multivariate PDFs is that the internal permutation problem is avoided. This is achieved by the following two elements:

1. Constraint matrices (consisting of an inverse DFT followed by a zeroing of several elements in the time domain and a subsequent DFT) appear in the generic DFT-domain formulation (see, e.g., [16, 17]) and describe the inter-frequency correlation between DFT components.
2. The coupling between the DFT bins is additionally ensured by the multivariate score function which is derived from the multivariate PDF [16]. As an example, for SIRPs the argument of the multivariate score function (which is in general a nonlinear function) is $\boldsymbol{y}_q^T(iL+j)\boldsymbol{R}_{\boldsymbol{y}_q\boldsymbol{y}_q}^{-1}(i)\boldsymbol{y}_q(iL+j)$ according to Eq. 13.22. Even for the simple case $\boldsymbol{R}_{\boldsymbol{y}_q\boldsymbol{y}_q}^{-1}(i) = \boldsymbol{I}$, where we have $\boldsymbol{y}_q^T(iL+j)\boldsymbol{y}_q(iL+j) = \|\boldsymbol{y}_q(iL+j)\|^2$, i.e., the quadratic norm, and – due to the Parseval theorem – the same in the DFT domain, i.e., the quadratic norm over all DFT components ensures a coupling between all DFT bins. From this we immediately see that for the adaptation of an individual DFT bin all DFT bins are taken into account simultaneously so that the internal permutation problem is at least mitigated if not completely avoided.

This illustrates that the dependencies among all DFT components (including higher-order dependencies) are inherently taken into account in the TRINICON framework. The traditional narrowband approach (with the internal permutation problem) would result as a special case if we assume all DFT components to be statistically independent from each other which is of course not the case for real-world broadband signals such as speech and audio signals. Actually, in the traditional narrowband approach, the additionally required repair mechanisms for permutation alignment try to exploit such inter-frequency dependencies [51].

In Fig. 13.5 it can be seen that the TRINICON framework allows to introduce several *selective* approximations which can be used to cover several well-known algorithms and also to derive new algorithms. Among these algorithms, the system described in [7] has turned out to be very efficient. There, only the normalization by the auto-correlation matrix in the SOS BSS update (Eq. 13.35) has been approximated by a narrowband inverse which allows to perform for each channel a scalar inversion for each DFT bin instead of a $D \times D$ matrix inverse. Therefore, this algorithm is included in the experimental evaluation in the next section. More details and a pseudo-code of the algorithm can be found in [7].

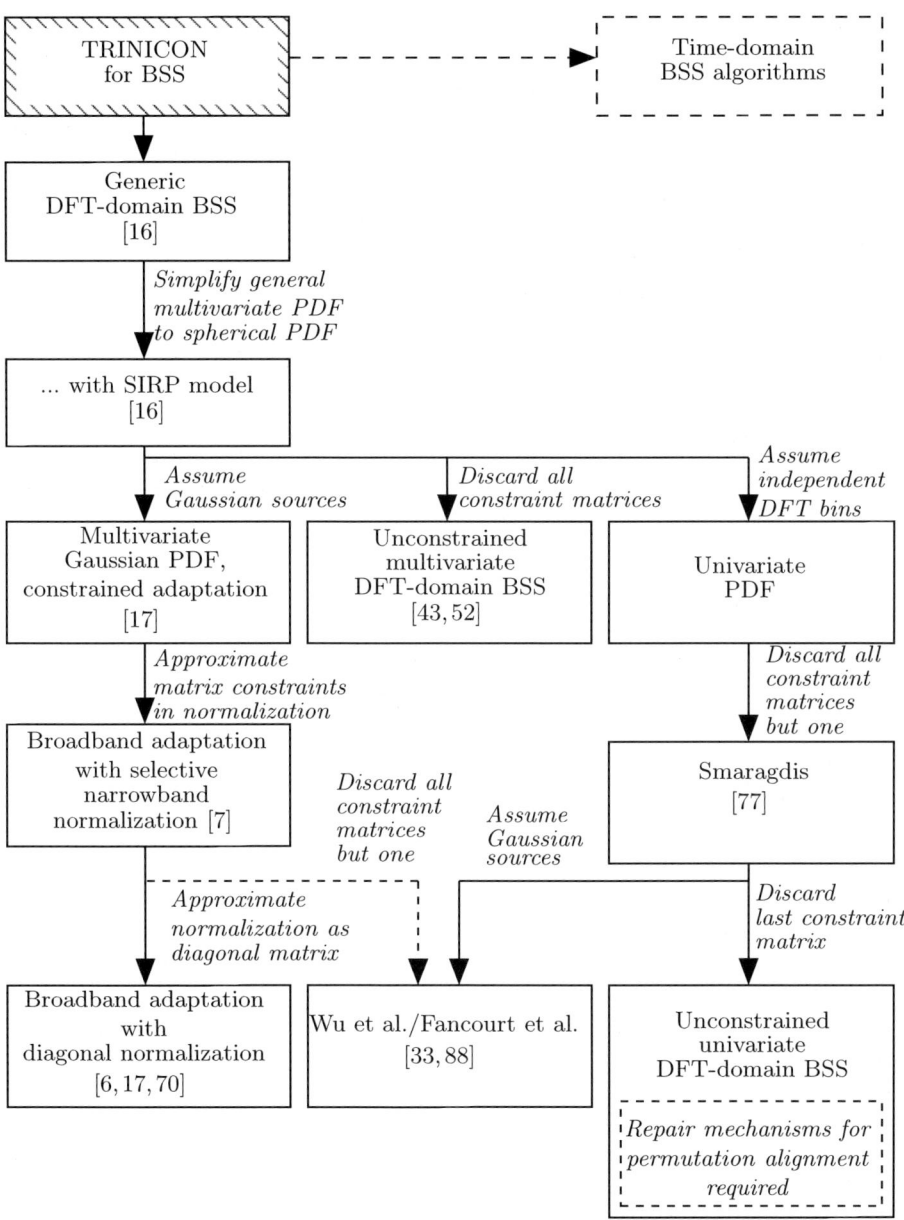

Fig. 13.5. Overview of BSS algorithms in the DFT domain.

13.2.6 Experimental Results for Reverberant Environments

The separation performance of various BSS algorithms derived from the TRINICON framework is shown for a living room scenario with a reverberation time $T_{60} = 200$ ms. Two sources have been placed at a distance of 1 m at $-20°, 40°$ from a microphone pair with omnidirectional sensors and the signals have been sampled at $f_s = 16$ kHz. To cope well with the reverberation, the demixing filter length has been chosen to $L = 1024$ taps. The nonwhiteness is exploited by the memory of $D = L = 1024$ introduced in the multivariate PDFs and in the correlation matrices. For accurate estimation of these quantities a block length of $N = 2048$ was chosen. For all examined algorithms the Sylvester constraint $\mathcal{SC_R}$ together with the initialization $w_{pp,15} = 1$, $p = 1, 2$ (all other taps are set to zero) is used due to the increased versatility [4,6] and the correlation method is used for the estimation of the correlation matrices to reduce computational complexity. For the iterative adaptation procedure the block-online update with $\ell_{\max} = 5$ offline iterations together with an adaptive stepsize is used (for details, see [6]). This allows online processing of the sensor signals and fast convergence when iterating $\ell_{\max}$ times on the same data block.

The evaluated algorithms include the computationally complex second-order statistics algorithm (Eq. 13.35) as well as efficient algorithms obtained by applying several approximations to the generic algorithm. A list of all algorithms is given in Tab. 13.1. The performance of the algorithms is measured in terms of the segmental signal-to-interference ratio (SIR) improvement ΔSIR_{seg}. The segmental SIR measures the ratio of the power of the desired signal versus the power of the interfering signals and then averages this quantity over all P channels.

Table 13.1. List of algorithms evaluated in the reverberant living room scenario.

Identifier	Algorithmic description
(A)	Broadband SOS algorithm (Eq. 13.35) based on the multivariate Gaussian PDF.
(B)	Broadband SOS algorithm based on multivariate Gaussian PDF (Eq. 13.35) with normalization approximated as a narrowband inverse [7].
(C)	Broadband SOS algorithm based on multivariate Gaussian PDF (Eq. 13.35) with normalization approximated as a scaling by the output signal variance (Eq. 13.36) [6, 17].
(D)	Narrowband SOS algorithm based on multivariate Gaussian PDF where the coupling between the DFT bins is ensured by one remaining constraint matrix [17].

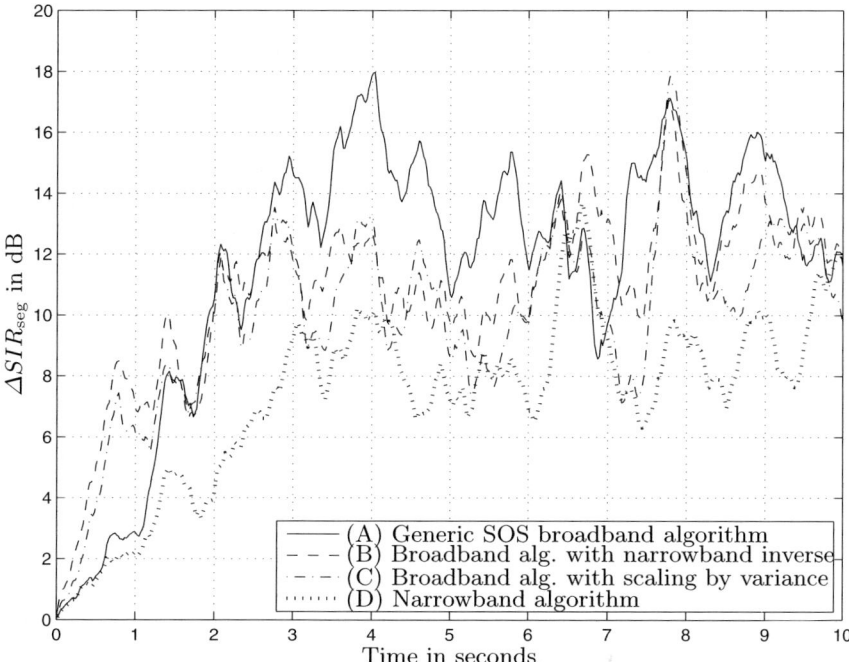

Fig. 13.6. Segmental SIR improvement $\Delta\overline{SIR}_{\text{seg}}$ for the second-order statistics algorithms (A)-(D) evaluated in the living room scenario ($T_{60} = 200\,\text{ms}$) for two source position setups.

The experimental results in Fig. 13.6 show that the SOS algorithm (A) provides the best performance. However, its high computational complexity prevents a real-time implementation on current state-of-the-art hardware platforms for such large demixing filter lengths. Therefore approximations are needed which minimally affect the separation performance but result in computationally efficient algorithms. As pointed out above, the main complexity in the second-order statistics algorithms is caused by the inverse of the autocorrelation matrix for each output channel. This inverse is approximated in the broadband algorithm (B) by a narrowband inverse which leads to a scalar inversion in each DFT bin [7]. The algorithm (B) can be implemented in real-time on regular PC hardware and it can be seen in Fig. 13.6 that the separation performance is only slightly reduced. In the broadband algorithm (C) the normalization is further simplified by using the variance of each output signal [6,17] as shown in Sec. 13.2.4. This means that the normalization is not frequency-dependent anymore. In the narrowband algorithm (D) all constraint matrices except one are approximated [17]. This means that the narrowband normalization is done analogously to algorithm (B), however, due to discarding all constraint matrices except for one, the complete decoupling of the

DFT bins is only prevented by the last remaining constraint matrix. Thus, the algorithm (D) already suffers from the permutation and scaling problem occuring in each DFT bin. This explains the inferior separation performance compared to the broadband algorithms (A)-(C). More extensive simulations for algorithms derived from the TRINICON framework including different source positions and reverberation times can be found in [8].

13.3 Extensions for Blind Source Separation in Noisy Environments

In the previous section only noiseless reverberant environments were considered with the maximum number of simultaneously active point sources Q assumed to be equal to the number of sensors P. However, in realistic scenarios, in addition to the point sources to be separated, some background noise will usually be present. Thus, in general BSS faces two different challenges in noisy environments:

1. The *adaptation* of the demixing BSS filters should be *robust* to the *noise signals* $n_1(n), \ldots, n_P(n)$ to ensure high separation performance of the desired point sources $s_1(n), \ldots, s_P(n)$. This means that the signal-to-interference ratio (SIR) should not deteriorate compared to the noiseless case.
2. The *noise* contribution contained in the separated BSS output signals *should be suppressed*, i.e., the signal-to-noise ratio (SNR) should be maximized.

Both requirements must be met if BSS should be attractive for noisy environments.

According to the literature (see e.g., [25, 48] and references therein), it has been tried to address the first point by developing noise-robust BSS algorithms. However, so far this has been considered only for the instantaneous BSS case. Additionally, several assumptions such as spatial or temporal uncorrelatedness are usually imposed on the noise signals allowing to generate optimization criteria which are not affected by the noise signals. However, in the case of convolutive BSS for acoustic signals these assumptions for the noise signals are too restrictive. In the next section when discussing a model for background noise, we will see that in realistic scenarios the background noise at the sensors is *temporally correlated* and for low frequencies and/or small microphone spacings it will also be *spatially correlated*. This does not allow the application of the noise-robust instantaneous BSS algorithms presented in the literature to the noisy convolutive BSS problem.

A more promising approach for increasing the robustness of the BSS adaptation are pre-processing methods. In Sec. 13.3.2 we will describe single-channel and multi-channel methods in order to remove the bias of the second-order correlation matrices caused by the noise. This will lead to a better performance of previously discussed BSS algorithms.

Another approach to the noisy convolutive BSS problem is the application of post-processing methods to the outputs of the BSS system. Without preprocessing the separation performance of the BSS algorithms will decrease in noisy environments. Therefore, the post-processing technique has to aim at the suppression of both, background noise and residual crosstalk from interfering point sources which could not be cancelled by the BSS demixing filters. This will be discussed in detail in Sec. 13.3.3.

13.3.1 Model for Background Noise in Realistic Environments

A model often used to describe background noise is the 3-dimensional isotropic sound field which is also termed diffuse sound field [56]. It can be modeled by an infinite number of statistically independent point sources which are uniformly distributed on a sphere. The phases of the emitted background noise signals are uniformly distributed between 0 and 2π. If the radius of the sphere is $r \to \infty$, then the propagating waves from each point source picked up be the microphones x_p can be assumed to be plane waves.

The diffuse sound field allows to describe, e.g., speech babble noise in a cafeteria, which is generated by a large number of background speakers or exterior noise recorded in the passenger compartment of a car which is a superposition of many different sources such as, e.g., motor, wind, or street noise. Moreover, the diffuse sound field is also often used to model reverberation [56]. This requires that the direct sound and the reflections are assumed to be mutually incoherent, i.e., the phase relations between the sound waves are neglected and thus, a superposition of the sound waves only results in a summation of the sound intensities. As the convolutive BSS demixing system accounts for the phase relations by the FIR filters of length L only the reflections exceeding the time-delay covered by L filter taps can be considered as being of diffuse nature. This case applies to highly reverberant environments such as, e.g., lecture rooms, or train stations.

In the convolutive BSS model depicted in Fig. 13.1 we assumed that the number of simultaneously active point sources Q is less or equal to the number of sensors P. Due to the limited number of point sources Q in the BSS scenario, we thus cannot model the diffuse sound field by an infinite number of point sources. Therefore, they are included in the BSS model in Fig. 13.1 as noise components $n_p(n)$, $p = 1, \ldots, P$ which are additively mixed to each microphone signal $x_p(n)$.

An adequate quantity to classify the sound field at the sensors is the magnitude-squared coherence (MSC) function whose estimate for the m-th data block and ν-th DFT bin is given by

$$\left|\underline{\Gamma}^{(\nu)}_{x_1 x_2}(m)\right|^2 = \frac{\left|\underline{S}^{(\nu)}_{x_1 x_2}(m)\right|^2}{\underline{S}^{(\nu)}_{x_1 x_1}(m)\underline{S}^{(\nu)}_{x_2 x_2}(m)}. \tag{13.38}$$

The estimation of the power-spectral densities $\underline{S}^{(\nu)}_{x_p x_q}(m)$ for nonstationary signals such as speech is usually performed using recursive averaging with a forgetting factor γ given as

$$\underline{S}^{(\nu)}_{x_p x_q}(m) = \gamma \underline{S}^{(\nu)}_{x_p x_q}(m-1) + (1-\gamma)\underline{X}^{(\nu)}_p(m)\underline{X}^{(\nu)*}_q(m). \quad (13.39)$$

A long-term estimate of the MSC can be obtained by averaging the short-term MSC $|\underline{\Gamma}^{(\nu)}_{x_1 x_2}(m)|^2$ over all blocks.

In an ideally diffuse sound field the MSC between the microphone signals $x_1(n)$ and $x_2(n)$ is given by

$$|\underline{\Gamma}^{(\nu)}_{x_1 x_2}|^2 = \frac{\sin^2\left(2\pi\nu R^{-1} f_s d\, c^{-1}\right)}{(2\pi\nu R^{-1} f_s d\, c^{-1})^2}, \quad (13.40)$$

where d denotes the distance between the microphones and R is the DFT length. This result assumes omnidirectional sensor characteristics and was first presented in [26] (a detailed derivation can be found, e.g., in [8, 60]). Eq. 13.40 reflects that the noise components $n_p(n)$ which originated from a diffuse sound field are strongly correlated between the sensors at low frequencies but less correlated for higher frequencies. Additionally, each $n_p(n)$, $p = 1, \ldots, P$ may also contain sensor noise which is usually assumed independent across different sensors. For comparison, the MSC for a point source $s_q(n)$ is equal to one. In Fig. 13.7 the estimated MSC of car noise is shown. A two-element omnidirectional microphone array was positioned in the passenger compartment at the interior mirror and two different spacings of $d = 4\,\mathrm{cm}$ and $d = 16\,\mathrm{cm}$ have been examined. The car noise was measured while driving through a suburban area. The estimate of the MSC (Eq. 13.38) was obtained by using the recursive averaging procedure (Eq. 13.39) with $\gamma = 0.9$, DFT length $R = 512$, and using Hann windowing. The long-term estimate $|\Gamma^{(\nu)}_{x_1 x_2}|^2$

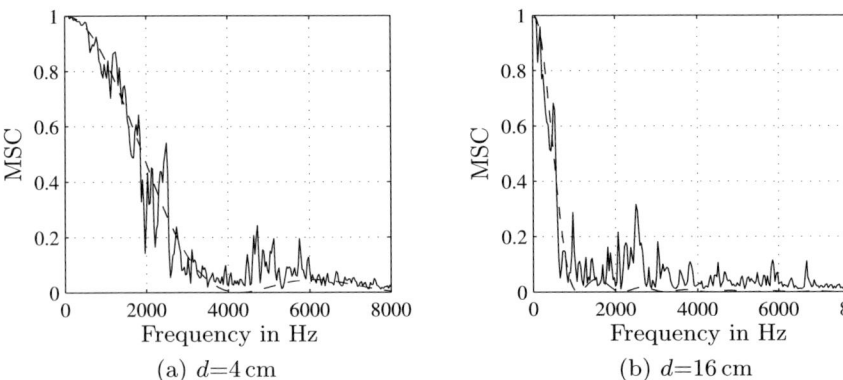

Fig. 13.7. MSC $|\Gamma^{(\nu)}_{x_1 x_2}|^2$ of car noise measured at two sensors $x_1(n)$ and $x_2(n)$ positioned at the interior mirror in a car compartment for different sensor spacings d.

was calculated by averaging over all blocks for a signal length of 20 sec. It can be seen that the MSC of the measured data (solid) corresponds very well to the $\sin^2(x)/x^2$ characteristic of the MSC of an ideal diffuse sound field (dashed) for both microphone spacings. Therefore, it can be concluded that the MSC of car noise can be approximated by the MSC of a diffuse sound field. Additionally, in [62] it was shown experimentally that also office noise originating from computer fans and hard disk drives can be assumed to exhibit the MSC of a diffuse noise field.

13.3.2 Pre-Processing for Noise-Robust Adaptation

From the literature only few pre-processing approaches for BSS in noisy environments are known. If the number of sensors P is equal to the number of sources Q, as considered in this chapter, then usually so-called *bias removal techniques* are used which aim at estimating and subtracting the contribution of the noise in the sensor signal itself or in the second-order correlation matrix and possibly also in the higher-order relation matrix of the sensor signals. These techniques will be discussed in the following. If more sensors than sources are available, i.e., $P > Q$, then also *subspace techniques* can be used as a pre-processing step to achieve a suppression of the background noise. As we restricted ourselves in this chapter to the case $P = Q$ the subspace approaches will not be treated here, but a summary and an outline of possible directions of future research can be found in [8].

The signal model in matrix-vector notation (Eq. 13.7) yields the BSS output signals $\boldsymbol{y}(n)$ containing D output signal samples for each of the $Q = P$ channels. If background noise $n_p(n)$ is superimposed at each sensor $p = 1, \ldots, P$, the signal model can be decomposed as

$$\boldsymbol{y}(n) = \boldsymbol{W}^\mathrm{T} \boldsymbol{x}(n)$$
$$= \boldsymbol{W}^\mathrm{T} \left(\boldsymbol{H}^\mathrm{T} \boldsymbol{s}(n) + \boldsymbol{n}(n) \right) \tag{13.41}$$

where the background noise and speech samples are contained in the column vectors

$$\boldsymbol{s}(n) = \left[\boldsymbol{s}_1^\mathrm{T}(n), \ldots, \boldsymbol{s}_P^\mathrm{T}(n) \right]^\mathrm{T}, \tag{13.42}$$

$$\boldsymbol{s}_p(n) = \left[s_p(n), \ldots, s_p(n - 2L - M + 2) \right]^\mathrm{T}, \tag{13.43}$$

$$\boldsymbol{n}(n) = \left[\boldsymbol{n}_1^\mathrm{T}(n), \ldots, \boldsymbol{n}_P^\mathrm{T}(n) \right]^\mathrm{T}, \tag{13.44}$$

$$\boldsymbol{n}_p(n) = \left[n_p(n), \ldots, n_p(n - 2L + 1) \right]^\mathrm{T}, \tag{13.45}$$

and the matrix $\boldsymbol{H}$ is composed of channel-wise Sylvester matrices $\boldsymbol{H}_{qp}$ of size $(M + 2L - 1) \times 2L$ containing the mixing FIR filters $h_{qp,\kappa}$, $\kappa = 0, \ldots, M - 1$. It can be seen from the noisy signal model (Eq. 13.41) that the second-order correlation matrix $\boldsymbol{R}_{\boldsymbol{yy}}(n)$ and also the higher-order relation matrix $\boldsymbol{R}_{\boldsymbol{y}\phi(\boldsymbol{y})}(n)$

will contain a bias due to the background noise. Due to the central limit theorem, the distribution of the diffuse background noise can be assumed to be closer to a Gaussian than the distribution of the speech signals. Therefore, the bias will be larger for the estimation of the cross-correlation matrix $\boldsymbol{R}_{yy}(n)$ and the background noise will affect the estimation of higher-order moments less. Therefore, we will focus on bias removal for second-order correlation matrices. The background noise $\boldsymbol{n}(n)$ and the point-source signals $\boldsymbol{s}(n)$ are assumed to be mutually uncorrelated so that the second-order correlation matrix $\boldsymbol{R}_{yy}(n)$ with its channel-wise submatrices defined in Eq. 13.23 can be decomposed as

$$\boldsymbol{R}_{yy}(n) = \boldsymbol{W}^{\mathrm{T}} \left(\boldsymbol{H}^{\mathrm{T}} \boldsymbol{R}_{ss}(n) \boldsymbol{H} + \boldsymbol{R}_{nn}(n) \right) \boldsymbol{W} \qquad (13.46)$$

with the source correlation matrix $\boldsymbol{R}_{ss}(n)$ and noise correlation matrix $\boldsymbol{R}_{nn}(n)$ defined as

$$\boldsymbol{R}_{ss}(n) = \frac{1}{N} \sum_{j=0}^{N-1} \boldsymbol{s}(n+j)\, \boldsymbol{s}^{\mathrm{T}}(n+j), \qquad (13.47)$$

$$\boldsymbol{R}_{nn}(n) = \frac{1}{N} \sum_{j=0}^{N-1} \boldsymbol{n}(n+j) \boldsymbol{n}^{\mathrm{T}}(n+j). \qquad (13.48)$$

To remove the bias introduced by the background noise it is possible to either aim at estimating and subsequently removing the noise component in Eq. 13.41, e.g., by using single-channel noise reduction techniques, or to estimate and remove the noise correlation matrix $\boldsymbol{R}_{nn}(n)$. The latter approach is already known from the literature on instantaneous BSS. There, usually spatially and temporally uncorrelated Gaussian noise is assumed, i.e., $\boldsymbol{R}_{nn}(n)$ is a diagonal matrix (see, e.g., [24,25,30,48]). Moreover, most approaches assume that $\boldsymbol{R}_{nn}(n)$ is known a-priori and stationary. However, in realistic scenarios usually temporally correlated background noise is present at the sensors. This noise can often be described by a diffuse sound field, leading to noise signals which are also spatially correlated for low frequencies (see Sec. 13.3.1). Additionally, background noise is in general nonstationary and its stochastic properties can at best be assumed slowly time-variant which thus requires a continuous estimation of the correlation matrix $\boldsymbol{R}_{nn}(n)$ based on short-time stationarity according to Eq. 13.48. The following bias removal techniques, aiming at the noise signal $\boldsymbol{n}(n)$ or the noise correlation matrix $\boldsymbol{R}_{nn}(n)$, will be examined under these conditions.

13.3.2.1 Single-Channel Noise Reduction

If the estimation and suppression of the noise components $\boldsymbol{n}_p(n)$ is desired for each sensor signal $\boldsymbol{x}_p(n)$ individually, then for each channel $p = 1, \ldots, P$ a single-channel noise reduction algorithm can be used. The estimation and suppression of background noise using one channel is already a long-standing research topic. In general, all algorithms consist of two main building blocks:

- the estimation of the noise contribution and
- the computation of a weighting rule to suppress the noise and enhance the desired signal.

An overview of various methods can be found, e.g., in [39].

In all well-known noise estimation methods in the literature usually the noise power spectral density (PSD) is estimated without recovering the phase of the clean signal but using the phase of the noisy signal instead. This is motivated by the fact that the human perception of speech is not much affected by a modification of the phase of the clean signal [83]. However, for BSS algorithms the relative phase of the signals acquired by different microphones is crucial as this information is implicitly used to suppress signals depending on their different directions of arrival. To evaluate the importance of amplitude and phase for pre-processing techniques applied to BSS algorithms, we will in the following generate pre-processed sensor signals by using the DFT-domain amplitude of the clean mixture signals and the phase of the noisy mixture signals. This corresponds to an optimum single-channel speech enhancement algorithm which perfectly estimates the amplitude of the clean mixture signal and thus, suppresses the background noise completely. These signals are then used as inputs for the second-order statistics BSS algorithm described in [7].

For this experiment we use two noisy scenarios. The first one is a car environment where a pair of omnidirectional microphones with a spacing of 20 cm was mounted to the interior mirror. The long-term SNR was adjusted to 0 dB which is a realistic value commonly encountered inside car compartments. Analogously to the BSS experiments in Sec. 13.2.6 a male and a female speech signal were convolved with the acoustic impulse response measured for the driver and co-driver positions. The second scenario corresponds to the cocktail party problem which is usually described by the task of listening to one desired point source in the presence of speech babble noise consisting of the utterances of many other speakers. The long-term statistics of speech babble are well described by a diffuse sound field, however, there may also be several other distinct noise point sources present. In our experiments we simulated such a cocktail party scenario inside a living room environment where speech babble noise was generated by a circular loudspeaker array with a diameter of 3 m. The two omnidirectional microphones with a spacing of 20 cm were placed in the center of the loudspeaker array from which 16 speech signals were reproduced to simulate the speech babble noise. Additionally, two distinct point sources at a distance of 1 m and at the angles of $0°$ and $-80°$ were used to simulate the desired and one interfering point source, respectively. The long-term input SNR at the microphones has been adjusted for the living room scenario to 10 dB. This is realistic, as due to the speech-like spectrum of the background noise the microphone signals exhibiting higher SNR values are perceptually already as annoying as those with significantly lower SNR values for lowpass car noise.

Due to the perfect estimation of the clean signal amplitude the background noise is almost inaudible. However, the results in Fig. 13.8 show that due to the noisy phase for the car environment no improvement in terms of separa-

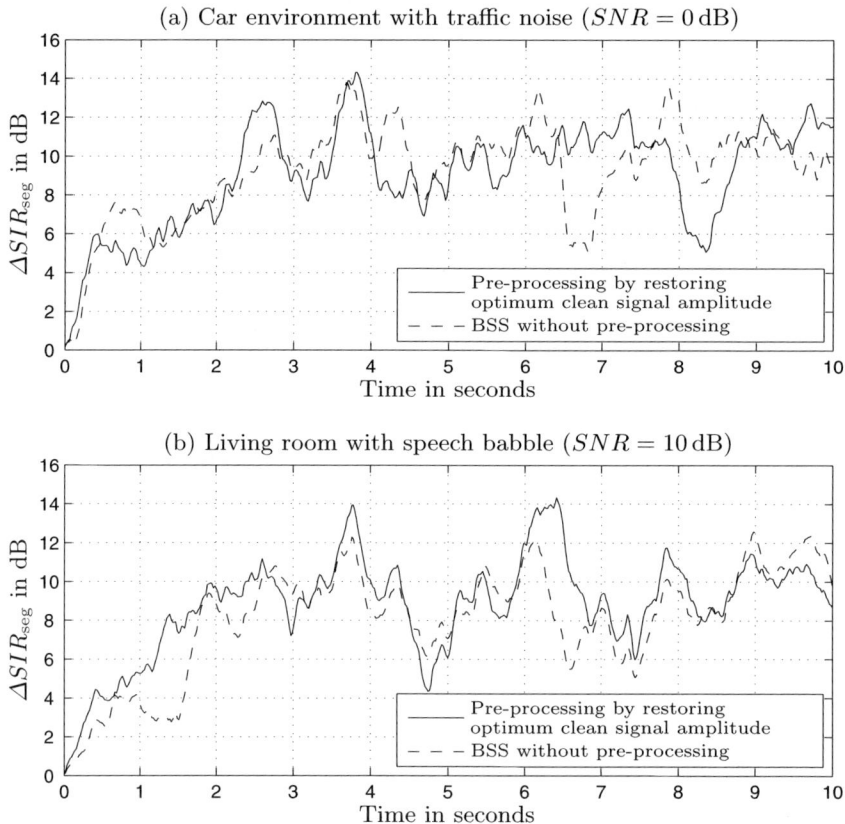

Fig. 13.8. Segmental SIR improvement ΔSIR_seg depicted over time for two noisy environments. Speech separation results are shown for the BSS outputs adapted with the noisy mixtures and for BSS with pre-processing by restoring the magnitude of the clean mixture signals but with the phase of the noisy mixtures.

tion performance can be obtained. Similarly, for the cocktail party scenario only a small improvement in terms of separation of the point sources can be achieved. Further experiments also indicated that when using a realistic state-of-the-art noise reduction algorithm as, e.g., proposed in [63], then also the improvements shown in Fig. 13.8(b) disappear. Therefore, it is concluded that pre-processing by single-channel noise reduction algorithms only suppresses the background noise, but does not improve the degraded separation performance of the subsequent BSS algorithm. To improve the separation, it is crucial that both, amplitude and phase of the clean mixture signals are estimated. This usually requires multi-channel methods as presented in the next section.

13.3.2.2 Multi-Channel Bias Removal

To also account for the phase contribution of the background noise we first review briefly some methods initially proposed for instantaneous BSS which aim at estimating and subsequently removing the noise correlation matrix $\boldsymbol{R_{nn}}(n)$. For convolutive BSS only a few approaches have been proposed so far: In [45] the special case of spatio-temporally white noise was addressed and has been extended to the diffuse noise case in [46]. There, stationarity of the noise was assumed and the preceding noise-only segments have been used for the estimation of the correlation matrix. Already earlier in [3, 6] a similar procedure was proposed where the minimum statistics approach [63] was used for the estimation of the noise characteristics. This method operates in the DFT domain and is based on the observation that the power of a noisy speech signal frequently decays to the power of the background noise. Hence by tracking the minima an estimate for the auto-power spectral density of the noise is obtained. However, due to the spatial correlation not only the auto- but also the cross-power spectral densities of the noisy signal $x_p(n)$ and the background noise $n_p(n)$ are required. They are estimated and averaged recursively for each DFT bin whenever we detect a minimum (i.e. speech pause) of the noisy speech signals. Thus, for slowly time-varying noise statistics this method gives an accurate estimate of the noise spectral density matrix used for the bias removal. In Fig. 13.9 the results of the approach in [3] are shown in terms of the segmental SIR improvement for the separation of the two point sources. It can be seen that the pre-processing slightly improves the separation performance for the noisy car environment described in the previous section.

In the cocktail party scenario this approach did not achieve good results as the noise statistics is more time-variant and due to only few speech pauses of the point sources the noise PSD cannot be estimated very well.

In contrast to the single-channel bias removal techniques, the multi-channel approaches do not achieve any background noise reduction as they merely aim at providing a better estimate of the correlation matrix of the point sources which is then used for the adaptation of the demixing filter weights. To additionally suppress the background noise, this approach would have to be complemented by a post-processing technique. Note also that due to fewer speech pauses it is more difficult to estimate the noise correlation matrix for multiple active speakers compared to a single speaker as typically encountered in single-channel speech enhancement applications. Therefore, the estimation of the noise contribution may be done more reliably after the BSS stage where already a partial suppression of the interfering point sources is achieved. This

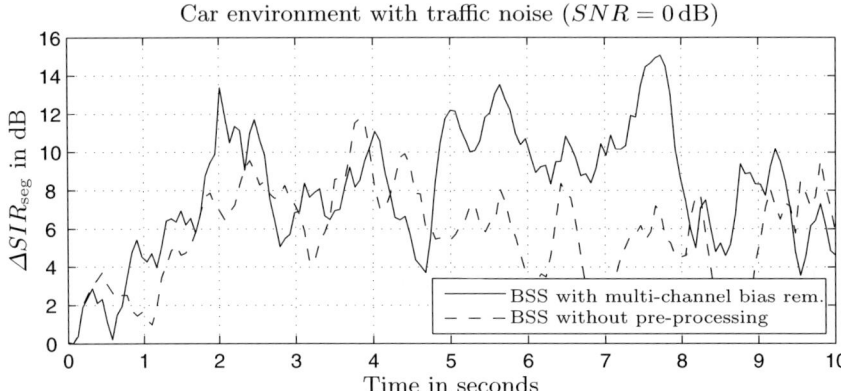

Fig. 13.9. Segmental SIR improvement ΔSIR_{seg} depicted over time for the noisy car environment. Speech separation results are shown for the BSS outputs adapted with the noisy mixtures and for BSS with pre-processing by multi-channel bias removal.

will be investigated in detail for the post-processing approach discussed in Sec. 13.3.3

13.3.3 Post-Processing for Suppression of Residual Crosstalk and Background Noise

In Sec. 13.3.2 several pre-processing approaches have been discussed. It could be seen that for the case $P = Q$ only multi-channel bias removal methods achieved some noise robustness of the BSS algorithm. For these methods a reliable voice activity detection is crucial but might be difficult to realize in environments with several speech point sources so that in such cases post-processing methods are a preferable alternative. Post-processing methods have the advantage that the BSS system already achieves a suppression of the interfering point sources so that in each BSS output channel only some remaining interference of the other point sources is present. As will be shown later, this simplifies the estimation of the quantities required by the post-processing method. A suitable post-processing scheme is given by a single-channel post-filter $g_{q,\kappa}$ applied to each BSS output channel as shown in Fig. 13.10. The motivation of using a single-channel postfilter for each BSS output channel is twofold:

Firstly, it is desired that the remaining background noise is reduced at the BSS output channels. In [20] it was shown that the optimum solution for BSS leads to blind MIMO identification and thus, the BSS demixing system can be interpreted for each output channel as a blind adaptive interference canceller aiming at the suppression of the interfering point sources. As the background noise is usually described by a diffuse sound field, the BSS system achieves only limited noise suppression. However, from adaptive beamforming

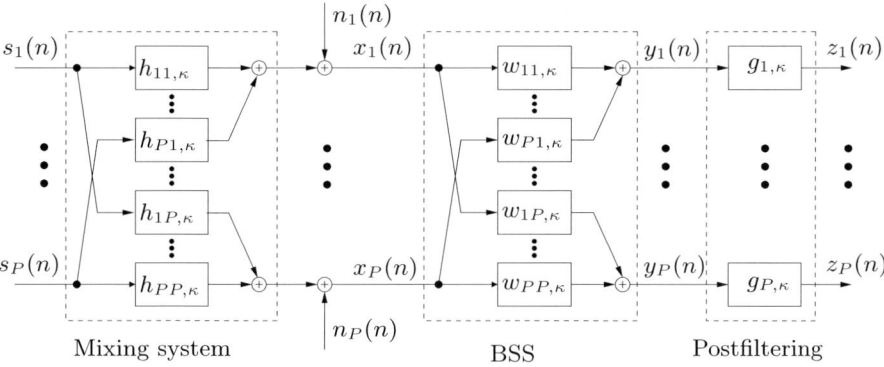

Fig. 13.10. Noisy BSS model combined with postfiltering.

(e.g., [76]) it is known that in such environments the concatenation of an adaptive interference canceller with a single-channel postfilter can improve the noise reduction.

Secondly, BSS algorithms are in noisy environments usually not able to converge to the optimum solution due to the bias introduced by the background noise. Moreover, moving point sources or an insufficient demixing filter length, which only partly covers the existing room reverberation, may lead to reduced signal separation performance and thus, to the presence of residual crosstalk from interfering point sources at the BSS output channels. In such situations, the single-channel postfilter should be designed such that it also provides additional separation performance. Analogously, similar considerations have led to a single-channel postfilter in acoustic echo cancellation which was first proposed in [11,59].

The reduced separation quality due to an insufficient demixing filter length in realistic environments was the motivation of several single-channel postfilter approaches that have been previously proposed in the BSS literature [22,68,69,72,80,82]. Nevertheless, a comprehensive treatment of the simultaneous suppression of residual crosstalk and background noise is still missing and will be presented in the following sections. We will first discuss in Sec. 13.3.3.1 the advantages of the implementation of the single-channel postfilter in the DFT domain and will introduce a spectral gain function requiring the power spectral density (PSD) estimates of the residual crosstalk and background noise. Then, the signal model for the residual crosstalk and the background noise will be discussed in Sec. 13.3.3.2 allowing to point out the relationships to previous post-processing approaches. The chosen signal model will lead to the derivation of a novel residual crosstalk PSD estimation and additionally the estimation of the background noise will be addressed. Subsequently, experimental results will be presented which illustrate the improvements that can be obtained by the application of single-channel postfilters both, in terms of SIR and SNR.

13.3.3.1 Spectral Weighting Function for a Single-Channel Postfilter

The BSS output signals $y_q(n)$, $q = 1, \ldots, P$ can be decomposed for the q-th channel as

$$y_q(n) = y_{s_r,q}(n) + y_{c,q}(n) + y_{n,q}(n), \tag{13.49}$$

where $y_{s_r,q}(n)$ is the component containing the desired source $s_r(n)$. As a possible permutation of the separated sources at the BSS outputs, i.e., $r \neq q$ does not affect the post-processing approach we will simplify the notation and denote in the following the desired signal component in the q-th channel as $y_{s,q}(n)$. The quantity $y_{c,q}(n)$ is the residual crosstalk component from the remaining point sources that could not be suppressed by the BSS algorithm and $y_{n,q}(n)$ denotes the contribution of the background noise.

From single-channel speech enhancement (e.g., [12]) or from the literature on single-channel postfiltering for beamforming (e.g., [76]) it is well-known that it is beneficial to utilize the DFT-domain representation of the signals and estimate the single-channel postfilter in the DFT domain. Thus, N_{post} samples are combined to an output signal block which is, after applying a windowing operation, transformed by the DFT of length $R_{\text{post}} \geq N_{\text{post}}$ yielding the DFT-domain representation of the output signals as

$$\underline{Y}_q^{(\nu)}(m) = \underline{Y}_{s,q}^{(\nu)}(m) + \underline{Y}_{c,q}^{(\nu)}(m) + \underline{Y}_{n,q}^{(\nu)}(m) \tag{13.50}$$

where $\nu = 0, \ldots, R_{\text{post}} - 1$ is the index of the DFT bin and m denotes the block time index. The advantage is that in the DFT domain speech signals are sparser, i.e., we can find regions in the time-frequency plane where the individual speech sources do not overlap (see e.g., [90]). This property is often exploited in underdetermined blind source separation where there are more simultaneously active sources than sensors (e.g., [29,84]). Here, this sparseness is used for the estimation of the quantities necessary for the implementation of the spectral gain function. A block diagram showing the main building blocks of a DFT-based postfilter is given in Fig. 13.11. There it can already be seen that analogously to single-channel speech enhancement or post-filtering applied to beamforming or acoustic echo cancellation, the DFT bins are treated in a narrowband manner as all computations are carried out independently in each DFT bin. Because of the narrowband treatment we have to ensure that circular convolution effects, appearing due to the signal modification by the spectral weighting, are not audible. Thus, the enhanced output signal z_q, which is the estimate $\hat{y}_{s,q}(n)$ of the clean desired source component, is computed by the means of an inverse DFT using a weighted overlap-add method including a tapered analysis and synthesis windows as suggested in [38]. This is in contrast to the BSS algorithms derived from the TRINICON framework where the linear convolution of the sensor signals with the estimated FIR demixing system is implemented without approximations equivalently in the DFT domain by the overlap-save method. In contrast to postfiltering, the

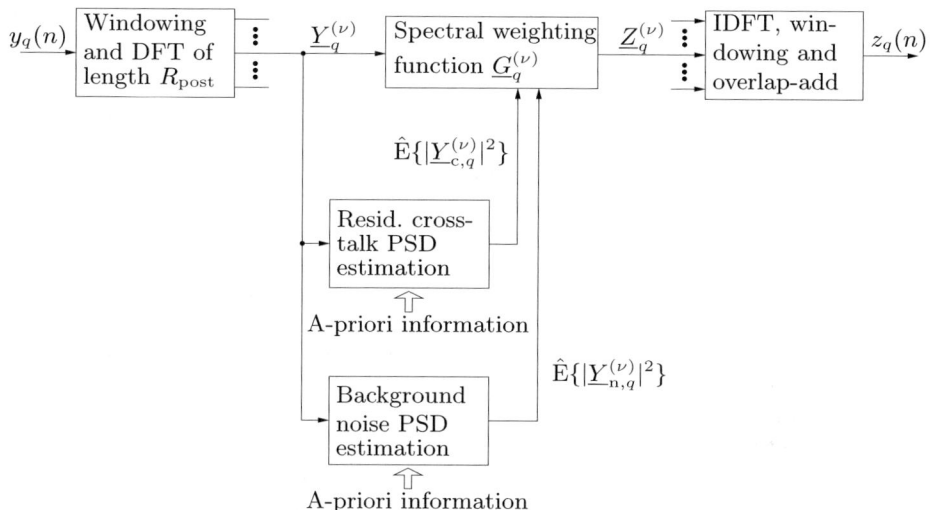

Fig. 13.11. DFT-based single-channel postfiltering depicted for the ν-th DFT bin in the q-th channel.

selective narrowband approximations which are applied in the TRINICON framework and have been outlined in Sec. 13.2.5 have only been made in the adaptation process of the demixing filters to obtain efficient BSS algorithms.

According to Fig. 13.11 a spectral gain function $\underline{G}_q^{(\nu)}(m)$ in the ν-th DFT bin aiming at simultaneous suppression of residual crosstalk and background noise has to be derived. The output signal of the post-processing scheme is the estimate of the clean desired source signal

$$\underline{Z}_q^{(\nu)} = \hat{\underline{Y}}_{s,q}^{(\nu)} \tag{13.51}$$

and is given as

$$\underline{Z}_q^{(\nu)}(m) = \underline{G}_q^{(\nu)}(m)\underline{Y}_q^{(\nu)}(m). \tag{13.52}$$

According to [5] we choose in this chapter to minimize the mean-squared error $\mathrm{E}\{(\underline{Z}_q^{(\nu)}(m) - \underline{Y}_{s,q}^{(\nu)}(m))^2\}$ with respect to $\underline{G}_q^{(\nu)}(m)$. This leads to the ν-th bin of the well-known Wiener filter for the q-th channel given as

$$\underline{G}_q^{(\nu)}(m) = \frac{\mathrm{E}\left\{\left|\underline{Y}_{s,q}^{(\nu)}(m)\right|^2\right\}}{\mathrm{E}\left\{\left|\underline{Y}_q^{(\nu)}(m)\right|^2\right\}}. \tag{13.53}$$

With the assumption that the desired signal component, the interfering signal components and the background noise in the q-th channel are all mutually uncorrelated, Eq. 13.53 can be expressed as

$$\underline{G}_q^{(\nu)}(m) = \frac{\mathrm{E}\left\{\left|\underline{Y}_{\mathrm{s},q}^{(\nu)}(m)\right|^2\right\}}{\mathrm{E}\left\{\left|\underline{Y}_{\mathrm{s},q}^{(\nu)}(m)\right|^2\right\} + \mathrm{E}\left\{\left|\underline{Y}_{\mathrm{c},q}^{(\nu)}(m)\right|^2\right\} + \mathrm{E}\left\{\left|\underline{Y}_{\mathrm{n},q}^{(\nu)}(m)\right|^2\right\}}. \quad (13.54)$$

From this equation it can be seen that for regions with desired signal *and* residual crosstalk or background noise components the output signal spectrum is reduced, whereas in regions without crosstalk or background noise the signal passed through. On the one hand this fulfills the requirement that an undisturbed desired source signal passes through the Wiener filter without any distortion. On the other hand, if crosstalk or noise is present, the magnitude spectrum of the noise or crosstalk attains a shape similar to that of the desired source signal, so that noise and crosstalk are therefore partially masked by the desired source signal. This effect was already exploited in postfiltering for acoustic echo cancellation aiming at the suppression of residual echo. There, this effect has been termed "echo shaping" [61]. Moreover, it can be observed in Eq. 13.54 that if the BSS system achieves the optimum solution, i.e., the residual crosstalk in the q-th channel $\underline{Y}_{\mathrm{c},q}^{(\nu)}(m) = 0$, then Eq. 13.54 reduces to the well-known Wiener filter for a signal with additive noise used in single-channel speech enhancement. To realize Eq. 13.54 in a practical system, the ensemble average $\mathrm{E}\{\cdot\}$ has to be estimated and thus, it is usually replaced by a time average $\hat{\mathrm{E}}\{\cdot\}$. Thereby, the Wiener filter is approximated by

$$\underline{G}_q^{(\nu)}(m) \approx \frac{\hat{\mathrm{E}}\left\{\left|\underline{Y}_q^{(\nu)}(m)\right|^2\right\} - \hat{\mathrm{E}}\left\{\left|\underline{Y}_{\mathrm{c},q}^{(\nu)}(m)\right|^2\right\} - \hat{\mathrm{E}}\left\{\left|\underline{Y}_{\mathrm{n},q}^{(\nu)}(m)\right|^2\right\}}{\hat{\mathrm{E}}\left\{\left|\underline{Y}_q^{(\nu)}(m)\right|^2\right\}}, \quad (13.55)$$

where $\hat{\mathrm{E}}\{|\underline{Y}_q^{(\nu)}(m)|^2\}$, $\hat{\mathrm{E}}\{|\underline{Y}_{\mathrm{c},q}^{(\nu)}(m)|^2\}$, and $\hat{\mathrm{E}}\{|\underline{Y}_{\mathrm{n},q}^{(\nu)}(m|^2\}$ are the PSD estimates of the BSS output signal, residual crosstalk, and background noise, respectively. Due to the reformulation in Eq. 13.55 the unobservable desired signal PSD $\mathrm{E}\{|\underline{Y}_{\mathrm{s},q}^{(\nu)}(m)|^2\}$ does not have to be estimated. However, the main difficulty is still to obtain reliable estimates of the unobservable residual crosstalk and background noise PSDs. A novel method for this estimation process leading to high noise reduction with little signal distortion will be shown in the next section.

Moreover, an estimate of the observable BSS output signal PSD is required. The PSD estimates can be used to implement spectral weighting algorithms other than the Wiener filter as described, e.g., in [39].

13.3.3.2 Estimation of Residual Crosstalk and Background Noise

In this section a model for the residual crosstalk and background noise is introduced. Subsequently, based on the residual crosstalk model an estimation procedure will be given which relies on an adaptation control. Different adaptation control strategies will be outlined. Moreover, the estimation of the background noise PSD will be discussed.

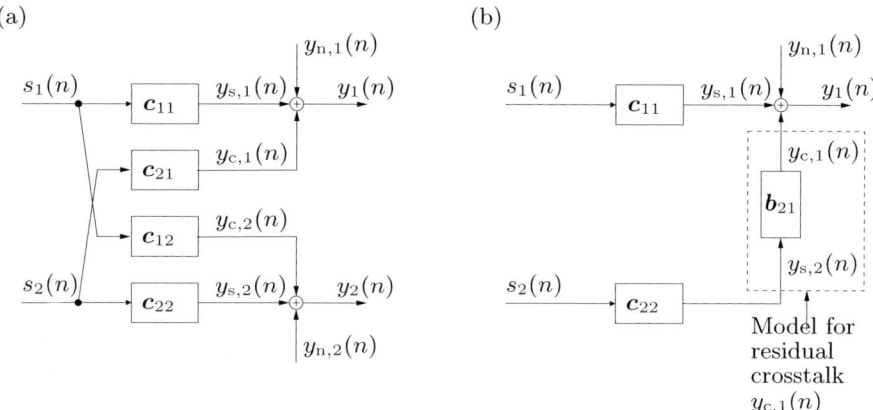

Fig. 13.12. (a) Representation of mixing and demixing system for the case $P = 2$ by using the overall system FIR filters c_{qr}. (b) Resulting model for the residual crosstalk $y_{c,1}(n)$.

Model of Residual Crosstalk and Background Noise

We restricted our scenario to the case that the number of microphones equals the maximum number of simultaneously active point sources. Therefore, the BSS algorithm is able to provide an estimate of one separated point source at each output $y_q(n)$. As pointed out above, due to movement of sources or long reverberation, the BSS algorithm might not converge fast enough to the optimum solution and thus, some residual crosstalk from point source interferers, denoted in the DFT domain by $\underline{Y}_{c,q}^{(\nu)}(m)$, remains in the BSS output. To obtain a good estimate of the residual crosstalk PSD $E\{|\underline{Y}_{c,q}^{(\nu)}(m)|^2\}$ as needed for the post-filter in the q-th channel, we first need to set up an appropriate model.

In Fig. 13.12(a) the concatenation of the mixing and demixing systems are expressed by the overall filters c_{qr} of length $M + L - 1$ which denote the path from the q-th source to the r-th output. For simplicity, we have depicted the case $Q = P = 2$ in Fig. 13.12. As can be seen in Fig. 13.12(a), the crosstalk component $y_{c,1}(n)$ of the first output channel is determined in the case $Q = P = 2$ by the source signal $s_2(n)$ and the filter c_{21}. However, as neither the original source signals nor the overall system matrix are observable, the crosstalk component $y_{c,1}(n)$ is expressed in Fig. 13.12(b) in terms of the desired source signal component $y_{s,2}(n)$ at the second output. This residual crosstalk model could be used if a good estimate of $y_{s,2}(n)$ is provided by the BSS system, i.e., if the source in the second channel is well-separated.

It should also be noted that even if $y_{s,2}(n)$ is available, then this model does not allow a perfect estimation of the residual crosstalk $y_{c,1}(n)$. This is due to the fact that for a perfect replica of $y_{c,1}(n)$ based on the input signal $y_{s,2}(n)$, the filter b_{21} has to model the combined system of c_{21} and the inverse of c_{22}. However, c_{22} is in general a non-minimum phase FIR filter

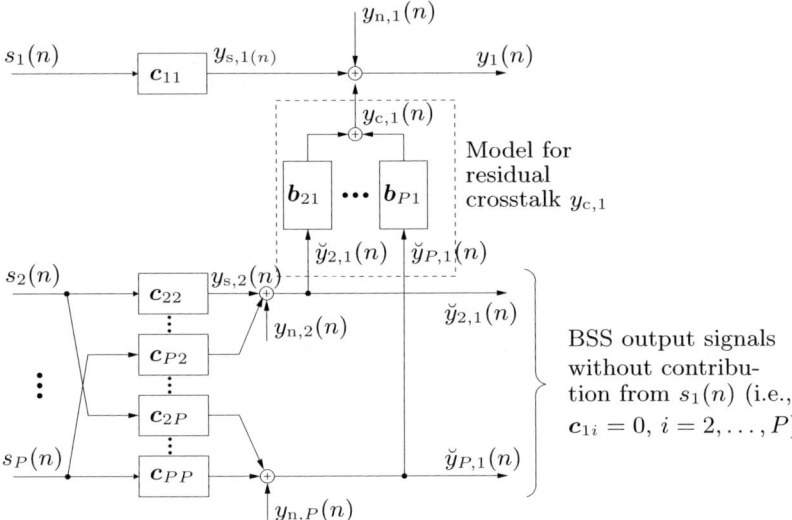

Fig. 13.13. Model of the residual crosstalk component $y_{c,q}$ contained in the q-th BSS output channel y_q illustrated for the first channel, i.e., $q = 1$. In contrast to Fig. 13.12(b) this model is solely based on observable quantities.

and thus, cannot be inverted in an exact manner by a single-input single-output system as was shown in [66]. Hence, analogously to single-channel blind dereverberation approaches, it is only possible to obtain an optimum filter $\mathbf{b}_{21}$ in the least-squares sense [66]. We will see in the following that due to the usage of additional a-priori information this model is nevertheless suitable for the estimation of the residual-cross talk PSD.

The model in Fig. 13.12(b) requires the desired source signal component $y_{s,2}(n)$ in the second BSS output. However, in practice it cannot be assumed that the BSS system always achieves perfect source separation. Especially in the initial convergence phase or with moving sources, there is some residual crosstalk remaining in all outputs. Therefore, we have to modify the residual crosstalk model so that only observable quantities are used. Hence, in Fig. 13.13 the desired signal component $y_{s,i}(n)$ for the i-th channel is replaced by the signal $\breve{y}_{i,q}(n)$ which denotes the BSS output signal of the i-th channel but without any interfering crosstalk components from the q-th point source (i.e., desired source $s_q(n)$). This means that the overall filters c_{qi} from the q-th source to the i-th output ($i = 1, \ldots, P$, $i \neq q$) are assumed to be zero. In practice, this condition is fulfilled by an adaptation control which determines time-frequency points where the desired source $s_q(n)$ is inactive. This a-priori information about desired source absence is important for a good estimation of the residual crosstalk PSD and thus, for achieving additional residual crosstalk cancellation. A detailed discussion of the adaptation control will be given in Sec. 13.3.3.2. Due to the bin-wise application of the single-channel postfilter we will in the following formulate the model in the DFT domain. Consequently, the model for the residual crosstalk in the q-th channel based on observable quantities is expressed for the ν-th DFT bin ($\nu = 1, \ldots, R_{\text{post}}$) as

$$\underline{Y}_{c,q}^{(\nu)}(m) = \sum_{i=1, i \neq q}^{P} \underline{\check{Y}}_{i,q}^{(\nu)}(m)\, \underline{B}_{i,q}^{(\nu)}(m)$$

$$= \underline{\check{Y}}_{q}^{(\nu)^{\mathrm{T}}}(m)\, \underline{B}_{q}^{(\nu)}(m), \tag{13.56}$$

where $\underline{\check{Y}}_{i,q}^{(\nu)}(m)$ and $\underline{B}_{i,q}^{(\nu)}(m)$ are the DFT-domain representations of $\check{Y}_{i,q}(m)$ and $b_{iq}(m)$, respectively. The variable $\underline{\check{Y}}_{q}^{(\nu)}(m)$ is the $P-1$ dimensional DFT-domain column vector containing $\underline{\check{Y}}_{i,q}^{(\nu)}(m)$ for $i = 1, \ldots, P$, $i \neq q$, and $\underline{B}_{q}^{(\nu)}(m)$ is the column vector containing the unknown filter weights $\underline{B}_{i,q}^{(\nu)}(m)$ for $i = 1, \ldots, P$, $i \neq q$.

It should be pointed out that the adaptation control only ensures that the desired source $s_q(n)$ is absent in the i-th BSS output channel $\underline{\check{Y}}_{i,q}^{(\nu)}(m)$. However, the background noise $\underline{Y}_{\mathrm{n},i}^{(\nu)}(m)$ is still present in the i-th BSS output channel as can also be seen in Fig. 13.13. If the background noise is spatially correlated between the q-th and i-th BSS output channel, then the coefficient $\underline{B}_{i,q}^{(\nu)}(m)$ would not only model the leakage from the separated source in the i-th channel, but $\underline{B}_{i,q}^{(\nu)}(m)$ would also be affected by the spatially correlated background noise. However, as an additional measure, the noise PSD $\mathrm{E}\{|\underline{Y}_{\mathrm{n},q}^{(\nu)}(m)|^2\}$ is estimated individually in each channel by one of the noise estimation methods known from single-channel speech enhancement. Therefore, if the background noise is already included in the residual crosstalk model, this would lead to an overestimation of the noise PSD. In Sec. 13.3.1 the character of the background noise such as car or babble noise was examined and the correlation of the noise sources between the sensors was evaluated using the magnitude squared coherence (MSC). It was concluded that the MSC of such background noise exhibits the same characteristics as a diffuse sound field leading to strong spatial correlation for low frequencies but to very small spatial correlation at higher frequencies. The model of the residual crosstalk is based on the BSS output signals and hence, it is of interest how the BSS system changes the MSC of the noise signals. In Fig. 13.14(a) and Fig. 13.14(b) the MSC of car noise and babble noise, which was estimated recursively according to Eq. 13.39 with the parameters $R = 512$, $\gamma = 0.9$ is plotted. For the car scenario a two-microphone array with a spacing of 4 cm was mounted at the interior mirror and the driver and co-driver were speaking simultaneously. Then the block-online BSS algorithm given in Eq. 13.35 together with the narrowband normalization [7] was applied. The same experiment was performed with two sources in a reverberant room where the babble noise was generated by a circular loudspeaker array using 16 individual speech signals. As pointed

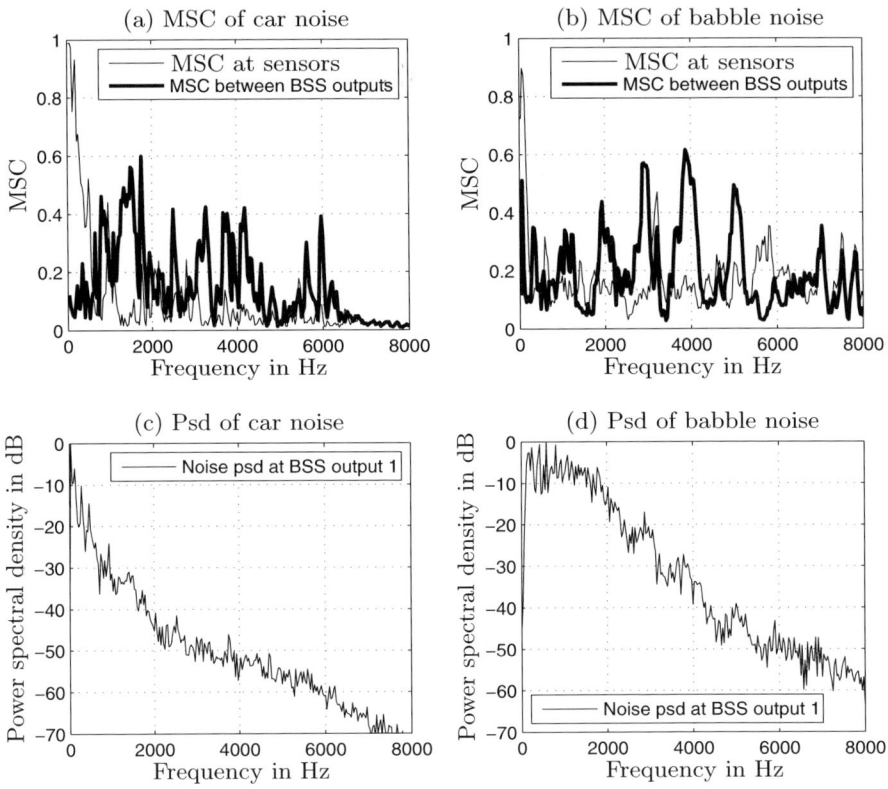

Fig. 13.14. Magnitude-squared coherence (MSC) of the car (a) and speech babble (b) background noise between the sensors and between the BSS outputs. The long-term noise PSDs of car noise and speech babble are shown in (c) and (d), respectively.

out in the beginning, the BSS algorithm tries to achieve source separation by aiming at mutual independence of the BSS output signals. From Fig. 13.14 it can be seen that in the presence of background noise this also leads to a *spatial decorrelation of the noise signals* at the BSS outputs. The car noise which is dominant at low frequencies (see Fig. 13.14(c)) has a MSC close to zero at these frequencies (see Fig. 13.14(a)). Only at higher frequencies, where the noise signal has much less energy, a larger MSC can be observed. The reduction of the MSC for the relevant frequencies can analogously also be observed for the babble noise. This observation shows that the background noise is spatially decorrelated at the BSS outputs and thus, confirms that the model for the residual crosstalk introduced in Eq. 13.56 is valid even in the case of background noise. This also justifies the independent estimations of the background noise in each channel and thus, we can apply noise estimation methods previously derived for single-channel speech enhancement algorithms. The residual cross-talk, however, is correlated across the output channels. These characteristics of residual cross-talk and background noise will be exploited in the next section to derive suitable estimation procedures.

After introducing the residual crosstalk model and validating it for the case of existing background noise, we briefly discuss the relationships to the models used in previous publications on post-processing for BSS. In [72] a Wiener-based approach for residual crosstalk cancellation is presented for the case $P = 2$. There, a greatly simplified model is used where all coefficients $\underline{B}_{i,q}^{(\nu)}(m)$ ($i, q = 1, 2, i \neq q$) are assumed to be equal to one. Similarly, in [80] one constant factor was chosen for all $\underline{B}_{i,q}^{(\nu)}(m)$. A model closer to Eq. 13.56, but based on magnitude spectra, was given in [22] which was then used for the implementation of a spectral subtraction rule. In contrast to the estimation method presented in the next section, the frequency-dependent coefficients of the model were learned by a modified least-mean-squares (LMS) algorithm. In [68] and [69] more sophisticated models were proposed allowing for time-delays or FIR filtering in each DFT bin. The model parameters were estimated by exploiting correlations between the channels or by using an NLMS algorithm. In all of these single-channel approaches the information of the multiple channels is only exploited to estimate the PSDs necessary for the spectral weighting rule. Alternatively, if also the phase of $\underline{Y}_{c,q}^{(\nu)}(m)$ is estimated, then it is also possible to directly subtract the estimate of the crosstalk component $\underline{Y}_{c,q}^{(\nu)}(m)$ from the q-th channel. This was proposed in [57] resulting in an adaptive noise canceller (ANC) structure [86]. The ANC was adapted by a leaky LMS algorithm [37] which includes a variable step size to allow also for strong desired signal activity without the necessity of an adaptation control.

The background noise component in the q-th channel $\underline{Y}_{n,q}^{(\nu)}(m)$ is usually assumed to be more stationary than the desired signal component $\underline{Y}_{s,q}^{(\nu)}(m)$. This assumption is necessary for the noise estimation methods known from single-channel speech enhancement which will be used to estimate the noise PSD $\hat{\mathrm{E}}\{|\underline{Y}_{n,q}^{(\nu)}(m)|^2\}$ in each channel and which are briefly discussed in the next section.

Estimation of Residual Crosstalk and Background Noise Power Spectral Densities

After introducing the residual crosstalk model (Eq. 13.56) we need to estimate the PSDs $\mathrm{E}\{|\underline{Y}_{c,q}^{(\nu)}(m)|^2\}$ of the residual crosstalk and $\mathrm{E}\{|\underline{Y}_{n,q}^{(\nu)}(m)|^2\}$ of the background noise for evaluating Eq. 13.55. To obtain an estimation procedure based on observable quantities we first calculate the cross-power spectral density vector $\underline{\boldsymbol{S}}_{\boldsymbol{\breve{Y}}_q Y_{c,q}}^{(\nu)}(m)$ between $\underline{\boldsymbol{\breve{Y}}}_q^{(\nu)}(m)$ and $\underline{Y}_{c,q}^{(\nu)}(m)$ given as

$$\begin{aligned}\underline{\boldsymbol{S}}^{(\nu)}_{\check{\boldsymbol{Y}}_q Y_{c,q}}(m) &= \hat{\mathrm{E}}\bigl\{\underline{\check{\boldsymbol{Y}}}_q^{(\nu)*}(m)\,\underline{Y}^{(\nu)}_{c,q}(m)\bigr\} \\ &= \hat{\mathrm{E}}\bigl\{\underline{\check{\boldsymbol{Y}}}_q^{(\nu)*}(m)\,\underline{\check{\boldsymbol{Y}}}_q^{(\nu)\mathrm{T}}(m)\bigr\}\,\underline{\boldsymbol{B}}^{(\nu)}_q(m) \\ &=: \underline{\boldsymbol{S}}^{(\nu)}_{\check{\boldsymbol{Y}}_q \check{\boldsymbol{Y}}_q}(m)\,\underline{\boldsymbol{B}}^{(\nu)}_q(m)\,, \end{aligned} \qquad (13.57)$$

where in the first step $\underline{\boldsymbol{B}}^{(\nu)}_q(m)$ was assumed to be slowly time-varying. Using Eq. 13.56, the power spectral density estimate $\hat{\mathrm{E}}\{|\underline{Y}^{(\nu)}_{c,q}(m)|^2\}$ can be expressed as

$$\begin{aligned}\hat{\mathrm{E}}\bigl\{|\underline{Y}^{(\nu)}_{c,q}(m)|^2\bigr\} &= \hat{\mathrm{E}}\bigl\{\underline{Y}^{(\nu)\mathrm{H}}_{c,q}(m)\,\underline{Y}^{(\nu)}_{c,q}(m)\bigr\} \\ &= \underline{\boldsymbol{B}}^{(\nu)\mathrm{H}}_q(m)\,\underline{\boldsymbol{S}}^{(\nu)}_{\check{\boldsymbol{Y}}_q \check{\boldsymbol{Y}}_q}(m)\,\underline{\boldsymbol{B}}^{(\nu)}_q(m)\,. \end{aligned} \qquad (13.58)$$

Solving Eq. 13.57 for $\underline{\boldsymbol{B}}^{(\nu)}_q(m)$ and inserting it into Eq. 13.58 leads to

$$\hat{\mathrm{E}}\bigl\{|\underline{Y}^{(\nu)}_{c,q}(m)|^2\bigr\} = \underline{\boldsymbol{S}}^{(\nu)\mathrm{H}}_{\check{\boldsymbol{Y}}_q Y_{c,q}}(m)\,\Bigl(\underline{\boldsymbol{S}}^{(\nu)}_{\check{\boldsymbol{Y}}_q \check{\boldsymbol{Y}}_q}(m)\Bigr)^{-1}\underline{\boldsymbol{S}}^{(\nu)}_{\check{\boldsymbol{Y}}_q Y_{c,q}}(m)\,. \qquad (13.59)$$

As $\underline{Y}^{(\nu)}_{c,q}(m)$, $\underline{Y}^{(\nu)}_{s,q}(m)$, and $\underline{Y}^{(\nu)}_{n,q}(m)$ in Fig. 13.13 are assumed to be mutually uncorrelated, $\underline{\boldsymbol{S}}^{(\nu)}_{\check{\boldsymbol{Y}}_q Y_{c,q}}(m)$ can also be estimated as the cross-power spectral density $\underline{\boldsymbol{S}}^{(\nu)}_{\check{\boldsymbol{Y}}_q Y_q}(m)$ between $\underline{\check{\boldsymbol{Y}}}^{(\nu)}_q(m)$ and q-th output of the BSS system $\underline{Y}^{(\nu)}_q(m)$ leading to the final estimation procedure:

$$\hat{\mathrm{E}}\bigl\{|\underline{Y}^{(\nu)}_{c,q}(m)|^2\bigr\} = \underline{\boldsymbol{S}}^{(\nu)\mathrm{H}}_{\check{\boldsymbol{Y}}_q Y_q}(m)\,\Bigl(\underline{\boldsymbol{S}}^{(\nu)}_{\check{\boldsymbol{Y}}_q \check{\boldsymbol{Y}}_q}(m)\Bigr)^{-1}\underline{\boldsymbol{S}}^{(\nu)}_{\check{\boldsymbol{Y}}_q Y_q}(m)\,. \qquad (13.60)$$

Thus, the power spectral density of the residual crosstalk for the q-th channel can be efficiently estimated in each DFT bin $\nu = 0,\ldots,R-1$ by computing the $1 \times P-1$ cross-power spectral density vector $\underline{\boldsymbol{S}}^{(\nu)}_{\check{\boldsymbol{Y}}_q Y_q}(m)$ between input and output of the model shown in Fig. 13.13 and calculating the $P-1 \times P-1$ cross-power spectral density matrix $\underline{\boldsymbol{S}}^{(\nu)}_{\check{\boldsymbol{Y}}_q \check{\boldsymbol{Y}}_q}(m)$ of the inputs. One possible implementation for estimating this expectation is given by an exponentially weighted average

$$\hat{\mathrm{E}}\{a(m)\} = (1-\gamma)\sum_i \gamma^{m-i} a(i)\,, \qquad (13.61)$$

where $a(m)$ is the quantity to be averaged. The advantage is that this can also be formulated recursively leading to

$$\underline{\boldsymbol{S}}^{(\nu)}_{\check{\boldsymbol{Y}}_q \check{\boldsymbol{Y}}_q}(m) = \gamma\,\underline{\boldsymbol{S}}^{(\nu)}_{\check{\boldsymbol{Y}}_q \check{\boldsymbol{Y}}_q}(m-1) + (1-\gamma)\,\underline{\check{\boldsymbol{Y}}}_q^{(\nu)*}(m)\,\underline{\check{\boldsymbol{Y}}}_q^{(\nu)\mathrm{T}}(m)\,, \qquad (13.62)$$

$$\underline{\boldsymbol{S}}^{(\nu)}_{\check{\boldsymbol{Y}}_q Y_q}(m) = \gamma\,\underline{\boldsymbol{S}}^{(\nu)}_{\check{\boldsymbol{Y}}_q Y_q}(m-1) + (1-\gamma)\,\underline{\check{\boldsymbol{Y}}}_q^{(\nu)*}(m)\,\underline{Y}^{(\nu)}_q(m)\,. \qquad (13.63)$$

In summary, the power spectral density of the residual crosstalk for the q-th channel can be efficiently estimated in each DFT bin $\nu = 0,\ldots,R-1$

using Eq. 13.60 together with the recursive calculation of the $P-1 \times P-1$ cross-power spectral density matrix (Eq. 13.62) and the $P-1 \times 1$ cross-power spectral density vector (Eq. 13.63). It should be noted that similar estimation techniques have been used to determine a post-filter for residual echo suppression in the context of acoustic echo cancellation (AEC) [32, 79]. However, the methods presented in [32, 79] are different in two ways: Firstly, in contrast to BSS where several interfering point sources may be active, the AEC post-filter was derived for a single channel, i.e., the residual echo originates from only one point source and thus all quantities in Eq. 13.60 reduce to scalar values. Secondly, in the AEC problem a reference signal for the echo is available. In BSS however, $\check{\underline{Y}}_q^{(\nu)}(m)$ is not immediately available as it can only be estimated if the desired source signal in the q-th channel is currently inactive. Strategies how to determine such time intervals are discussed in the next section.

The estimation of the PSD of the background noise $\hat{\mathrm{E}}\{|\underline{Y}_{\mathrm{n},q}^{(\nu)}(m)|^2\}$ is already a long-standing research topic in single-channel speech enhancement and an overview on the various methods can be found, e.g., in [39]. Usually it is assumed that the noise PSD is at least more stationary than the desired speech PSD. The noise estimation can be performed during speech pauses, which have to be detected properly by a voice activity detector. As voice activity detection algorithms are rather unreliable in low SNR conditions, several methods have been proposed which can track the noise PSD continuously. One of the most prominent methods is the minimum statistics approach which is based on the observation that the power of a noisy speech signal frequently decays to the power of the background noise. Hence, by tracking the minima the power spectral density of the noise is obtained. In [63] a minima tracking algorithm was proposed which includes an optimal smoothing of the noise PSD together with a bias correction and which will be applied in the experiments in Sec. 13.3.3.3. An overview of other methods providing continuous noise PSD estimates can be found, e.g., in [39].

Adaptation Control Based on SIR Estimation

In the previous sections it was shown that the estimation of the residual crosstalk power spectral density in the q-th channel is only possible at time instants when the desired point source of the q-th channel is inactive. As pointed out already above, speech signals can be assumed to be sufficiently sparse in the time-frequency domain so that even in reverberant environments regions can be found where one or more sources are inactive (see, e.g., [90]). This fact will be exploited by constructing a DFT-based adaptation control necessary for the estimation of the residual cross-talk PSD. In this section we will first briefly review an adaptation control approach which is already known from the literature on post-processing for BSS. Due to the similarity of the adaptation control necessary for estimating the residual crosstalk and the control necessary for adaptive beamformers applied to acoustic signals, also the existing approaches in the beamforming literature will be briefly summarized. A sophisticated bin-wise adaptation control proposed in [42] in the context of adaptive beamforming will then be applied in a slightly modified version to the post-processing scheme.

In general, all adaptation controls aim at estimating the SIR in the time domain or in a bin-wise fashion in the DFT domain. For the latter, the SIR estimate is given for the ν-th DFT bin as the ratio of the desired signal PSD and the PSD of the interfering signals. Thus, the SIR estimate at the q-th BSS output is given as

$$\widehat{SIR}_q^{(\nu)}(m) = \frac{\hat{\mathrm{E}}\{|\underline{Y}_{s,q}^{(\nu)}(m)|^2\}}{\hat{\mathrm{E}}\{|\underline{Y}_{c,q}^{(\nu)}(m)|^2\}}. \tag{13.64}$$

For the case of a BSS system with two output channels ($P = 2$) together with the assumption that for the number of simultaneously active point sources $Q \leq P$ holds, a simple SIR estimate is given by approximating the desired signal component $\underline{Y}_{s,q}^{(\nu)}(m)$ with the BSS output signal $\underline{Y}_q^{(\nu)}(m)$ of the q-th channel and approximating the interfering signal component by the BSS output signal of the other channel. This yields, e.g., for the approximated SIR estimate in the first BSS output channel

$$\widehat{SIR}_1^{(\nu)}(m) \approx \frac{\hat{\mathrm{E}}\{|\underline{Y}_1^{(\nu)}(m)|^2\}}{\hat{\mathrm{E}}\{|\underline{Y}_2^{(\nu)}(m)|^2\}}. \tag{13.65}$$

This approximation is justified if the BSS system already provides enough separation performance so that the BSS output signals can be seen as estimates of the point sources. In [68, 69] the time-average $\hat{\mathrm{E}}\{\cdot\}$ in Eq. 13.65 has been approximated by taking the instantaneous PSD values and the resulting approximated SIR was used successfully as a decision variable for controlling the estimation of the residual crosstalk. If $\widehat{SIR}_1^{(\nu)}(m) < 1$, then the crosstalk $\underline{Y}_{c,1}^{(\nu)}(m)$ was estimated and for $\widehat{SIR}_1^{(\nu)}(m) > 1$ the crosstalk of the second channel $\underline{Y}_{c,2}^{(\nu)}(m)$ was determined. In [5] this adaptation control was refined by the introduction of a safety margin Υ to improve reliability. By comparing Eq. 13.65 to a fixed threshold Υ it is ensured that a certain SIR value $\widehat{SIR}_1^{(\nu)}(m) < \Upsilon$ has to be attained to allow the conclusion that the desired signal is absent and thus, allow estimation of the residual crosstalk $\underline{Y}_{c,1}^{(\nu)}(m)$. The safety margin Υ has to be chosen between $0 < \Upsilon \leq 1$ and was set in [5] to $\Upsilon = 0.9$. For an extension of this mechanism to $P, Q > 2$ a suitable approximation for $\hat{\mathrm{E}}\{|\underline{Y}_{c,q}^{(\nu)}(m)|^2\}$ in the SIR estimate (Eq. 13.64) is important. In [5] it was suggested for $P, Q > 2$ to use the maximum PSD of the remaining channels $\hat{\mathrm{E}}\{|\underline{Y}_i^{(\nu)}(m)|^2\}$, $i \neq q$. For increasing P, Q this requires a very careful choice of Υ. In such scenarios, it is advantageous to replace the fixed threshold

$\varUpsilon$ by adaptive thresholding. As we will see in the following, such sophisticated adaptation controls were treated in the beamforming literature and will now be applied to the BSS post-processing scheme.

If adaptive beamformers, such as the generalized sidelobe canceller (GSC) (see, e.g., [42]), are applied to acoustic signals, then usually an adaptation control is required for the adaptive filters aiming at interference cancellation. Analogously to the residual crosstalk estimation procedure discussed in Sec. 13.3.3.2, the adaptation of the adaptive interference canceller has to be stalled in the case of a strong desired signal. This analogy allows to apply the approaches in the literature on adaptive beamforming to the post-processing of the BSS output signals. To control the adaptation of beamformers, a correlation-based method was proposed in [36] and recently in a modified form also in [47]. Another approach relies on the comparison of the outputs of a fixed beamformer with the main lobe steered towards the desired source and a complementary beamformer which steers a spatial null towards the desired source [44]. The ratio of the output signal powers, which constitutes an estimate of the SIR is then compared to a threshold to decide if the adaptation should be stopped. As both methods were suggested in the time-domain, this corresponds to a full-band adaptation control, so that in case of a strong desired signal the adaptation is stopped for all DFT bins. It has been pointed out before that speech signals are sparse in the DFT domain and thus, better performance of the adaptation algorithm can be expected when using a bin-wise adaptation control. This was the motivation in [42] for transferring the approach based on two fixed beamformer outputs to the DFT domain leading to a frequency-dependent SIR estimate. Instead of a fixed threshold $\varUpsilon$, additionally an adaptive threshold $\varUpsilon_q^{(\nu)}(m)$ for each channel and DFT bin has been proposed leading to a more robust decision. The application of this adaptation control to the estimation of the residual crosstalk, which is required for the post-processing algorithm, will be discussed in the following.

In [42] the estimate $\hat{\mathrm{E}}\{|\underline{Y}_{s,q}^{(\nu)}(m)|^2\}$ of the desired signal required for the SIR estimate (Eq. 13.64) is obtained by a delay-and-sum beamformer. This requires an array of several microphones which should have a spacing that is sufficiently large to allow the suppression of the interfering signals also at low frequencies. Moreover, the positions of the microphones are assumed to be known. This is in contrast to the BSS application where the sensors can be arbitrarily positioned and where there might be only a small number of sensors available (e.g., $P = 2$). Therefore, instead of a fixed beamformer output we will use the q-th BSS output signal PSD $\hat{\mathrm{E}}\{|\underline{Y}_q^{(\nu)}(m)|^2\}$ as an estimate of the desired signal PSD $\hat{\mathrm{E}}\{|\underline{Y}_{s,q}^{(\nu)}(m)|^2\}$.

The estimate of the interfering signal components required for the SIR estimate (Eq. 13.64) are obtained in [42] by a complementary beamformer which places a spatial null towards the desired source. The difference to the procedure in [44] is that this is done in a bin-wise manner. In our application we will use the PSD of a complementary BSS signal $\underline{\bar{Y}}_q^{(\nu)}$ which is obtained analogously to [42] as

$$\hat{\mathrm{E}}\{|\underline{\bar{Y}}_q^{(\nu)}(m)|^2\} = \hat{\mathrm{E}}\{|\underline{X}_q^{(\nu)}(m)|^2\} - \hat{\mathrm{E}}\{|\underline{Y}_q^{(\nu)}(m)|^2\}. \qquad (13.66)$$

Here it is assumed that the filtering due to the BSS demixing system is approximately linear phase and that the BSS output signal and the microphone signal have been properly time-aligned before subtracting their PSD estimates. It should be noted that due to the usage of a broadband BSS algorithm, the permutation at the BSS output signals is not frequency-dependent. Therefore, a possible permutation of the BSS output channels has no effect on the calculation of the complementary BSS signal.

Usually a recursive average is used for the time-average indicated by the operator $\hat{\mathrm{E}}\{\cdot\}$ which leads to the PSD estimates

$$\underline{S}_{x_q x_q}^{(\nu)}(m) = \gamma \underline{S}_{x_q x_q}^{(\nu)}(m-1) + (1-\gamma)|\underline{X}_q^{(\nu)}(m)|^2, \qquad (13.67)$$

$$\underline{S}_{y_q y_q}^{(\nu)}(m) = \gamma \underline{S}_{y_q y_q}^{(\nu)}(m-1) + (1-\gamma)|\underline{Y}_q^{(\nu)}(m)|^2, \qquad (13.68)$$

necessary for the estimation of the SIR in the q-th BSS output channel. The SIR estimate (Eq. 13.64) can thus be expressed as

$$\widehat{SIR}_q^{(\nu)}(m) \approx \frac{\underline{S}_{y_q y_q}^{(\nu)}(m)}{\underline{S}_{x_q x_q}^{(\nu)}(m) - \underline{S}_{y_q y_q}^{(\nu)}(m)}. \qquad (13.69)$$

The SIR estimate (Eq. 13.69) is then compared to an adaptive threshold $\underline{\Upsilon}_q^{(\nu)}(m)$. If $\widehat{SIR}_q^{(\nu)}(m) < \underline{\Upsilon}_q^{(\nu)}(m)$, then the absence of the desired signal in the q-th channel can be assumed. The adaptive threshold is given as the minimum of SIR estimate $\widehat{SIR}_q^{(\nu)}(m)$ which is determined for each DFT bin by taking into account the last D_Υ blocks [63]. In practice D_Υ must be large enough to bridge any peak of desired signal activity but short enough to track the nonstationary SIR variations in case of absence of the desired signal. Here, we choose an interval equivalent to a time period of 1.5 sec. Moreover, for small variations

$$\left| \frac{\widehat{SIR}_q^{(\nu)}(m) - \underline{\Upsilon}_q^{(\nu)}(m)}{\underline{\Upsilon}_q^{(\nu)}(m)} \right| \leq \Delta \Upsilon \qquad (13.70)$$

the threshold $\underline{\Upsilon}_q^{(\nu)}(m)$ is updated immediately. In Fig. 13.15 the SIR estimate $\widehat{SIR}_q^{(\nu)}$ and the adaptive threshold $\underline{\Upsilon}_q^{(\nu)}$ determined by minimum tracking are illustrated for the DFT bin corresponding to 1 kHz. The results are based on the output signals of the BSS system applied to the car environment. It can be seen that due to the parameter $\Delta \Upsilon = 0.3$ the threshold follows small changes of the SIR estimate immediately. Moreover, it should be pointed out that the SIR estimate in Fig. 13.15 exhibits high positive values due to the

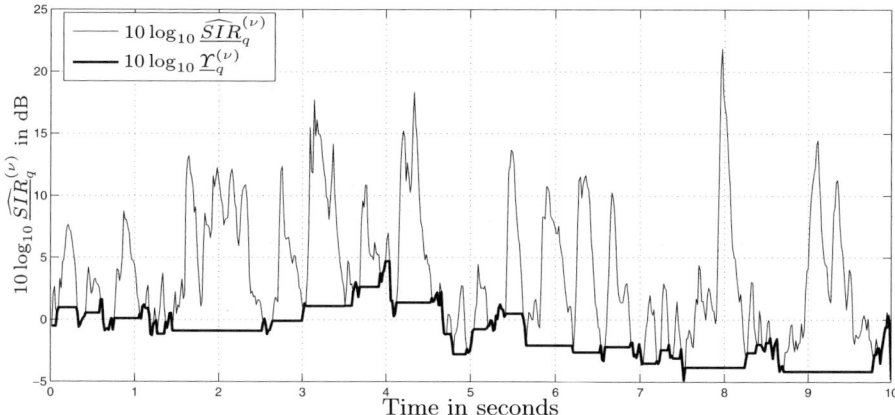

Fig. 13.15. Estimate $10\log_{10}\widehat{SIR}_q^{(\nu)}$ of the SIR and adaptive threshold $\underline{\Upsilon}_q^{(\nu)}$ determined by minimum tracking illustrated for the DFT bin corresponding to 1 kHz.

good convergence of the BSS algorithm. This is the reason why even in speech pauses of the desired signal, the SIR estimate does rarely exhibit negative SIR values.

In Fig. 13.16 the decision of the adaptation control is illustrated for the first output channel of the BSS system applied to the car environment ($P = Q = 2$). The desired component, residual crosstalk, and background noise component at the first BSS output are depicted in (a)-(c). The decision of the adaptation control is obtained by estimating the SIR according to Eq. 13.69 solely based on observable quantities. Especially due to the existence of background noise $y_{\mathrm{n},q}(n)$ this leads to a biased SIR. Nevertheless, the adaptation control is very robust due to the adaptive threshold $\underline{\Upsilon}_1^{(\nu)}$ based on minimum tracking and the parameter $\Delta\Upsilon = 0.3$ which allows for small variation of the threshold. This can be seen, when comparing the results of the adaptation control with the true SIR illustrated in (e) which is estimated based on unobservable quantities according to Eq. 13.64. In case of high SIR values $\widehat{SIR}_1^{(\nu)}(m)$, the desired signal in the first channel is present and the residual crosstalk PSD $\hat{\mathrm{E}}\{|\underline{Y}_{\mathrm{c},2}^{(\nu)}(m)|^2\}$ of the other channel is estimated. Vice versa, a low SIR in the first channel allows to adapt $\hat{\mathrm{E}}\{|\underline{Y}_{\mathrm{c},1}^{(\nu)}(m)|^2\}$.

In case that the adaptation control stalls the estimation of the residual crosstalk for the ν-th DFT bin in one of the P BSS output channels, the residual crosstalk estimate from the previous block has to be used. As speech is a nonstationary process and therefore, the statistics of the residual crosstalk are quickly time-varying, this would deteriorate the performance of the postfilter $\underline{G}_q^{(\nu)}(m)$. On the other hand, as pointed out above, the minimum statistics algorithm can provide continuous noise PSD estimates even in periods with desired signal activity. Therefore, for those time instants where the estimate of residual crosstalk cannot be updated, i.e., where the desired source signal

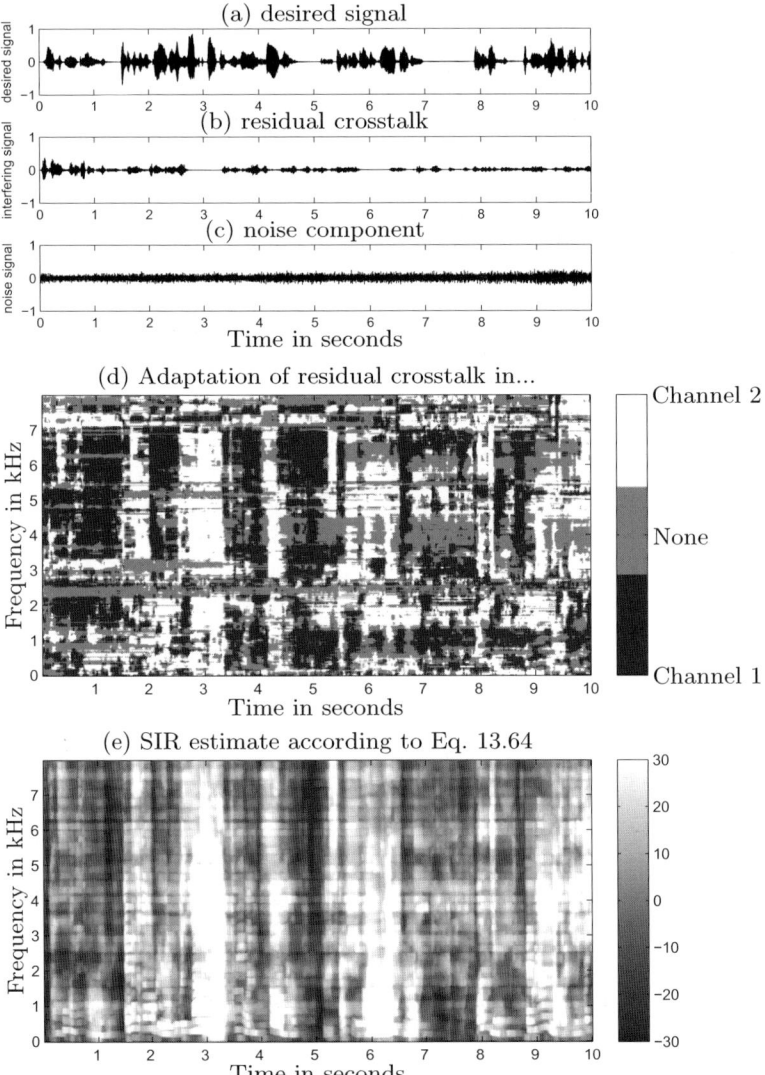

Fig. 13.16. BSS output signal components for the car environment with an input SNR at the sensors of 0 dB showing the desired signal (a), residual crosstalk (b) and background noise (c) in the first channel. Based on the SIR estimate (Eq. 13.69) and the adaptive threshold $\underline{\Upsilon}_1^{(\nu)}(m)$ the decision of the adaptation control is shown in (d). For comparison, the SIR (Eq. 13.64) computed for the true signal components in the first channel is illustrated in (e).

is dominant, a postfilter

$$\underline{G}_{\mathrm{n},q}^{(\nu)}(m) = \frac{\hat{\mathrm{E}}\{|\underline{Y}_q^{(\nu)}(m)|^2\} - \hat{\mathrm{E}}\{|\underline{Y}_{\mathrm{n},q}^{(\nu)}(m)|^2\}}{\hat{\mathrm{E}}\{|\underline{Y}_q^{(\nu)}(m)|^2\}} \qquad (13.71)$$

merely aiming at suppression of the background noise is applied.

In Tab. 13.2 the adaptation control and the resulting application of the postfilters is outlined for the q-th BSS output channel.

Table 13.2. Adaptation control and application of the postfilter for the q-th BSS output channel and ν-th DFT bin.

Number	Algorithmic part
1.	Estimate $\underline{S}_{x_q x_q}^{(\nu)}(m)$ and $\underline{S}_{y_q y_q}^{(\nu)}(m)$ according to Eqs. 13.67 and 13.68
2.	Estimate $\widehat{SIR}_q^{(\nu)}(m)$ according to Eq. 13.69
3.	Estimate $\hat{\mathrm{E}}\{|\underline{Y}_{\mathrm{n},q}^{(\nu)}(m)|^2\}$ by minimum statistics algorithm

4.	Tracking of minima of $\widehat{SIR}_q^{(\nu)}(m)$:
	If $|(\widehat{SIR}_q^{(\nu)}(m) - \underline{\varUpsilon}_q^{(\nu)}(m))/\underline{\varUpsilon}_q^{(\nu)}(m)| \leq \Delta \varUpsilon$
	   Replace all values of $\underline{\varUpsilon}_q^{(\nu)}(i)$ inside the buffer, i.e.,
	   $\underline{\varUpsilon}_q^{(\nu)}(i) = \widehat{SIR}_q^{(\nu)}(m)$, $i = m, \ldots, m - D_\varUpsilon + 1$
	If $\widehat{SIR}_q^{(\nu)}(m)$ is the minimum of $\underline{\varUpsilon}_q^{(\nu)}(m-i)$, $i = 0, \ldots, D_\varUpsilon - 1$
	   Set current value of buffer $\underline{\varUpsilon}_q^{(\nu)}(m) = \widehat{SIR}_q^{(\nu)}(m)$

5.	If minimum is detected, i.e., $\widehat{SIR}_q^{(\nu)}(m) \leq \underline{\varUpsilon}_q^{(\nu)}(m)$: Calculate residual crosstalk $\hat{\mathrm{E}}\{|\underline{Y}_{\mathrm{c},q}^{(\nu)}(m)|^2\}$ according to Eq. 13.60 Compute postfilter (Eq. 13.55) for residual crosstalk and noise
6.	If no minimum is detected, i.e., $\widehat{SIR}_q^{(\nu)}(m) > \underline{\varUpsilon}_q^{(\nu)}(m)$: Compute postfilter (Eq. 13.71) for noise only

13.3.3.3 Experimental Results for Reverberant and Noisy Environments

In the evaluation of the postfiltering algorithm summarized in Tab. 13.2 the same two noisy scenarios have been considered as in Sec. 13.3.2 and their description is briefly summarized. The first one is a car environment where a pair of omnidirectional microphones with a spacing of 20 cm was mounted at the interior mirror and recorded a male and female speaker at the driver and co-driver positions, respectively using a sampling rate of $f_\mathrm{s} = 16\,\mathrm{kHz}$.

The long-term SNR was adjusted to 0 dB which is a realistic value commonly encountered inside car compartments. The second scenario corresponds to the cocktail party problem which is usually described by the task of listening to one desired point source in the presence of speech babble noise consisting of the utterances of many other speakers. Speech babble is well described by a diffuse sound field, however, there may also be several other distinct noise point sources present. In our experiments we simulated such a cocktail party scenario inside a living room environment where speech babble noise was generated by a circular loudspeaker array with a diameter of 3 m. The two omnidirectional microphones with a spacing of 20 cm were placed in the center of the loudspeaker array from which 16 speech signals were reproduced to simulate the speech babble noise. Additionally, two distinct point sources at a distance of 1 m and at the angles of 0° and −80° were used to simulate the desired and one interfering point source, respectively. The long-term input SNR at the microphones has been adjusted for the living room scenario to 10 dB. This is realistic, as due to the speech-like spectrum of the background noise the microphone signals which exhibit higher SNR values are perceptually already as annoying as those with significantly lower SNR values for lowpass car noise.

The second-order statistics BSS algorithm with the narrowband normalization described in [7] is applied to the two noisy scenarios. To evaluate the performance two measures have been used: The segmental SIR which is defined as the ratio of the signal power of the desired signal to the signal power of the residual crosstalk stemming from point source interferers and the segmental SNR defined as the ratio of the signal power of the desired signal to the signal power of the possibly diffuse background noise. In both cases, the SIR and SNR improvement due to the application of the postfilter is measured and averaged over both channels. The segmental SIR improvement $\Delta SIR_{\text{seg}}(m)$ is plotted as a function of the block index m to illustrate the convergence effect of the BSS system. The channel-averaged segmental SNR improvement ΔSNR_{seg} is given as the average over all blocks. To assess the desired signal distortion, the unweighted log-spectral distance (SD) which describes the Euclidean distance between logarithmic short-time magnitude spectra has been measured between the desired signal at the input and the output of the postfilter and is given as

$$SD_{s_r,q} = \frac{1}{K_S} \sum_{m=1}^{K_S} \sqrt{\frac{1}{R} \sum_{\nu=0}^{R-1} \left(20 \lg \frac{\left| \underline{Z}_{s,q}^{(\nu)}(m) \right|}{\left| \underline{Y}_{s,q}^{(\nu)}(m) \right|} \right)^2}. \qquad (13.72)$$

The DFT length R for computing $SD_{s_r,q}$ is usually set to be small so that speech can be assumed stationary. In our experiments we used $R = 256$ and set K_S large enough to cover the whole signal length. To reduce artifacts such as, e.g., musical noise, the postfilter (Eq. 13.55) is usually calculated using an adaptive oversubtraction factor $\xi_q^{(\nu)}$ as proposed in [12]. Moreover, negative

Table 13.3. Segmental SNR and unweighted log-spectral distortion for both scenarios

Scenario	ΔSNR_seg at BSS outputs	ΔSNR_seg at postfilter outputs	SD at postfilter outputs
Car	3.0 dB	4.9 dB	1.0 dB
Cocktail party	0.2 dB	1.3 dB	1.6 dB

gains of the postfilters are set to zero. Hence in the experiments the postfilter

$$\underline{G}_q^{(\nu)}(m) = \frac{\max\left\{\left(\hat{\mathrm{E}}\left\{|\underline{Y}_q^{(\nu)}(m)|^2\right\} - \xi_q^{(\nu)}\left(\hat{\mathrm{E}}\left\{|\underline{Y}_{\mathrm{n},q}^{(\nu)}(m)|^2\right\} + \hat{\mathrm{E}}\left\{|\underline{Y}_{\mathrm{c},q}^{(\nu)}(m)|^2\right\}\right)\right), 0\right\}}{\hat{\mathrm{E}}\left\{|\underline{Y}_q^{(\nu)}(m)|^2\right\}}$$
(13.73)

was used. For the post-processing algorithm, $\gamma = 0.9$ and a DFT length of $R_\text{post} = 2048$ was chosen. The block length N_post was equal to the DFT length and an overlap factor $\alpha = 4$ was used. The parameters of the adaptation control are given as $\Delta \Upsilon = 0.3$ and $D_\Upsilon = 94$ corresponding to a period of 1.5 sec over which the minimum is tracked.

In Fig. 13.17 the results for the separation of the two speech point sources can be seen. For both scenarios the separation performance of the combined system of BSS and single-channel postfilter (solid) outperforms the BSS performance (dashed). In contrast to the BSS system which possesses an inherent adaptation control implied by the normalization term in the update equation, the postfilter relies on a-priori information provided by the adaptation control. Hence, it is possible to accurately estimate the residual crosstalk at the BSS outputs for further improvement of the speech separation performance. The reduced absolute level of the SIR improvement in the cocktail party scenario, i.e., in the reverberant living room (Fig. 13.17(b)) is due to longer reverberation and especially due to the background babble noise which exhibits a speech-like long-term spectrum.

Moreover, in both scenarios also the background noise could be partially suppressed. In Tab. 13.3 the segmental SNR averaged over all output channels of the BSS system and of the postfilter is shown. It can be observed that the postfilter achieves an additional SNR gain. As the car noise is more stationary compared to the speech babble noise, the minimum statistics algorithm can better estimate the noise PSD and thus a higher SNR improvement can be achieved by the postfilter.

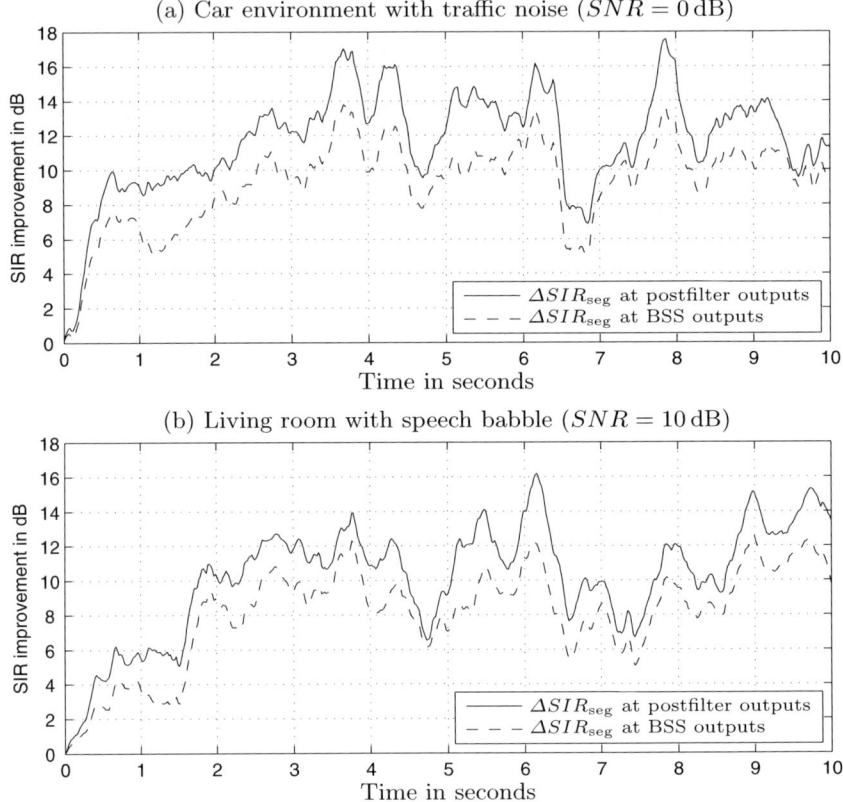

Fig. 13.17. Segmental SIR improvement $\Delta SIR_{\text{seg}}(m)$ depicted over time for two environments containing two speech point-source and additional background noise: (a) car compartment with background noise consisting of car and traffic noise ($SNR = 0\,\text{dB}$) and (b) living room scenario with speech babble background noise from 16 speakers ($SNR = 10\,\text{dB}$). Speech separation results are shown for BSS outputs and postfilter outputs.

To assess the speech quality, the SD (Eq. 13.72) between the desired signal at the input and output of the post-filter was calculated and averaged over both output channels. The small values in Tab. 13.3 indicate that the quality of the desired signal is preserved. This was also confirmed by informal listening tests where no musical noise was observed.

13.4 Conclusions

In this chapter we have presented a review of the TRINICON framework which allows to derive BSS algorithms simultaneously exploiting the signal properties nongaussianity, nonwhiteness, and nonstationarity. After the introduction of a generic natural gradient algorithm several special cases leading to efficient implementations have been discussed. It was also outlined how broadband BSS algorithms can be obtained from the TRINICON framework without introducing ambiguities appearing in narrowband algorithms. This has been supported by experimental results in a reverberant room. Subsequently, the application of BSS in noisy environments has been discussed. First, it has been shown that realistic background noise can often be described by the diffuse sound field. As such sound fields have to be modeled by an infinite or at least large number of point sources, the BSS approach only achieves limited noise reduction. Therefore, the extension of the TRINICON framework with pre- and post-processing approaches has been examined. It was shown that a single-channel postfilter applied to each BSS output signal can yield better results than bias-removal techniques used for pre-processing. The postfilter allowed to simultaneously address the cancellation of the residual crosstalk from point source interferers and the suppression of background noise. This was achieved by developing a model for the residual crosstalk and by using a-priori information provided by an adaptation control. The experiments in a car environment and a cocktail-party scenario with background babble noise showed good results for the complemented BSS algorithm. Thus, it can be concluded that by applying the presented post-processing approach, the versatility of the TRINICON BSS algorithms can be extended, resulting in a simultaneous separation of point sources and attenuation of background noise.

References

1. M. Abramowitz, I.A. Stegun (eds.): *Handbook of Mathematical Functions,* New York, NY, USA: Dover Publications, 1972.
2. R. Aichner, S. Araki, S. Makino, T. Nishikawa, H. Saruwatari: Time-domain blind source separation of non-stationary convolved signals by utilizing geometric beamforming, *Proc. NNSP '02*, 445–454, Martigny, Switzerland, September 2002.
3. R. Aichner, H. Buchner, W. Kellermann: Convolutive blind source separation for noisy mixtures, *Proc. CFA/DAGA '04*, 583–584, Strasbourg, France, March 2004.
4. R. Aichner, H. Buchner, W. Kellermann: On the causality problem in time-domain blind source separation and deconvolution algorithms, *Proc. ICASSP '05*, **5**, 181–184, Philadelphia, PA, USA, March 2005.
5. R. Aichner, M. Zourub, H. Buchner, W. Kellermann: Post-processing for convolutive blind source separation, *Proc. ICASSP '06*, **5**, 37–40, Toulouse, France, May 2006.
6. R. Aichner, H. Buchner, F. Yan, W. Kellermann: A real-time blind source separation scheme and its application to reverberant and noisy acoustic environments, *Signal Processing*, **86**(6), 1260–1277, June 2006.

7. R. Aichner, H. Buchner, W. Kellermann: Exploiting narrowband efficiency for broadband convolutive blind source separation, *EURASIP Journal on Applied Signal Processing*, 1–9, September 2006.
8. R. Aichner: *Acoustic Blind Source Separation in Reverberant and Noisy Environments*, PhD thesis, Universität Erlangen-Nürnberg, Erlangen, Germany, 2007.
9. S.-I. Amari: Natural gradient works efficiently in learning, *Neural Computation*, **10**, 251–276, 1998.
10. S. Araki, R. Mukai, S. Makino, T. Nishikawa, H. Saruwatari: The fundamental limitation of frequency-domain blind source separation for convolutive mixtures of speech, *IEEE Trans. Speech Audio Processing*, **11**(2), 109–116, March 2003.
11. B. Ayad, G. Faucon: Acoustic echo and noise cancelling for hands-free communication systems, *Proc. IWAENC '95*, 91–94, Røros, Norway, June 1995.
12. M. Berouti, R. Schwartz, J. Makhoul: Enhancement of speech corrupted by acoustic noise, *Proc. ICASSP '79*, 208–211, April 1979.
13. H. Brehm, W. Stammler: Description and generation of spherically invariant speech-model signals, *Signal Processing*, **12**, 119–141, 1987.
14. H. Buchner, R. Aichner, W. Kellermann: Blind source separation algorithms for convolutive mixtures exploiting nongaussianity, nonwhiteness, and nonstationarity, *Proc. IWAENC '03*, 275–278, Kyoto, Japan, September 2003.
15. H. Buchner, R. Aichner, W. Kellermann: TRINICON: A versatile framework for multichannel blind signal processing, *Proc. ICASSP' 04*, **3**, 889–892, Montreal, Canada, May 2004.
16. H. Buchner, R. Aichner, W. Kellermann: Blind source separation for convolutive mixtures: A unified treatment, in J. Benesty, Y. Huang (eds.), *Audio Signal Processing for Next-Generation Multimedia Communication Systems*, 255–293, Boston, MA, USA: Kluwer, 2004.
17. H. Buchner, R. Aichner, W. Kellermann: A generalization of blind source separation algorithms for convolutive mixtures based on second-order statistics, *IEEE Trans. Speech Audio Processing*, **13**(1), 120–134, January 2005.
18. H. Buchner, R. Aichner, J. Stenglein, H. Teutsch, W. Kellermann: Simultaneous localization of multiple sound sources using blind adaptive MIMO filtering, *Proc. ICASSP '05*, **3**, 97–100, Philadelphia, PA, USA, March 2005.
19. H. Buchner, J. Benesty, W. Kellermann: Generalized multichannel frequency-domain adaptive filtering: Efficient realization and application to hands-free speech communication, *Signal Processing*, **85**, 549–570, 2005.
20. H. Buchner, R. Aichner, W. Kellermann: TRINICON-based blind system identification with application to multiple-source localization and separation, in S. Makino, T.-W. Lee, S. Sawada (eds.), *Blind Speech Separation*, Berlin, Germany: Springer, 2007.
21. J.-F. Cardoso, B.H. Laheld: Equivariant adaptive source separation, *IEEE Trans. Signal Processing*, **44**(12), 3017–3030, December 1996.
22. C. Choi, G.-J. Jang, Y. Lee, S. R. Kim: Adaptive cross-channel interference cancellation on blind source separation outputs, *Proc. ICA '04*, 857–864, Granada, Spain, September 2004.
23. A. Cichocki, R. Unbehauen: *Neural Networks for Optimization and Signal Processing*, Chichester, USA: Wiley, 1994.
24. A. Cichocki, S. Douglas, S.-I. Amari: Robust techniques for independent component analysis (ICA) with noisy data, *Neurocomputing*, **22**, 113–129, 1998.

25. A. Cichocki, S.-I. Amari: *Adaptive Blind Signal and Image Processing*, Chichester, USA: Wiley, 2002.
26. R. K. Cook, R. V. Waterhouse, R. D. Berendt, S. Edelman, M.C. Thompson, Jr.: Measurement of correlation coefficients in reverberant sound fields, *JASA*, **27**(6), 1072–1077, November 1955.
27. T. M. Cover, J. A. Thomas: *Elements of Information Theory*, New York, NY, USA: Wiley, 1991.
28. W. B. Davenport: An experimental study of speech wave propability distribution, *JASA*, **24**(4), 390–399, 1952.
29. P. Divenyi (ed.): *Speech Separation by Humans and Machines*, Norwell, MA, USA: Kluwer, 2005.
30. S. C. Douglas, A. Cichocki, S.-I. Amari: A bias removal technique for blind source separation with noisy measurements, *Electronic Letters*, **34**(14), 1379–1380, July 1998.
31. T. Eltoft, T. Kim, T.-W. Lee: On the multivariate Laplace distribution, *IEEE Signal Processing Lett.*, **13**(5), 300–303, May 2006.
32. G. Enzner, R. Martin, P. Vary: Partitioned residual echo power estimation for frequency-domain acoustic echo cancellation and postfiltering, *Eur. Trans. Telecommun.*, **13**(2), 103–114, 2002.
33. C. L. Fancourt, L. Parra: The coherence function in blind source separation of convolutive mixtures of non-stationary signals, *Proc. NNSP '01*, 303–312, 2001.
34. S. Gazor, W. Zhang: Speech propability distribution, *IEEE Signal Processing Lett.*, **10**(7), 204–207, July 2003.
35. J. Goldman: Detection in the presence of spherically symmetric random vectors, *IEEE Trans. Inform. Theory*, **22**(1), 52–59, January 1976.
36. J. E. Greenberg, P. M. Zurek: Evaluation of an adaptive beamforming method for hearing aids, *JASA*, **91**(3), 1662–1676, March 1992.
37. J. E. Greenberg: Modified LMS algorithms for speech processing with an adaptive noise canceller, *IEEE Trans. Speech Audio Processing*, **6**(4), 338–351, 1998.
38. D. W. Griffin, J. S. Lim: Signal estimation from modified short-time fourier transform, *IEEE Trans. Acoust., Speech, Signal Processing*, **ASSP-32**(2), 236–243, April 1984.
39. E. Hänsler, G. Schmidt: *Acoustic Echo and Noise Control: A Practical Approach*, Hoboken, NJ, USA: Wiley, 2004.
40. D. A. Harville: *Matrix Algebra from a Statistician's Perspective*, Berlin, Germany: Springer, 1997.
41. S. Haykin: *Adaptive Filter Theory*, 4th ed., Englewood Cliffs, NJ, USA: Prentice-Hall, 2002.
42. W. Herbordt: *Sound Capture for Human/Machine Interfaces – Practical Aspects of Microphone Array Signal Processing*, volume 315 of *Lecture Notes in Control and Information Sciences*, Berlin, Germany: Springer, 2005.
43. A. Hiroe: Solution of permutation problem in frequency domain ICA, using multivariate probability density functions. *Proc. ICA '06*, 601–608, Charleston, SC, USA, March 2006.
44. O. Hoshuyama, A. Sugiyama: An adaptive microphone array with good sound quality using auxiliary fixed beamformers and its DSP implementation, *Proc. ICASSP '99*, 949–952, Phoenix, AZ, USA, March 1999.
45. R. Hu, Y. Zhao: Adaptive decorrelation filtering algorithm for speech source separation in uncorrelated noises, *Proc. ICASSP '05*, **1**, 1113–1115, Philadelphia, PA, USA, May 2005.

46. R. Hu, Y. Zhao: Fast noise compensation for speech separation in diffuse noise, *Proc. ICASSP '06*, **5**, 865–868, Toulouse, France, May 2006.
47. T. P. Hua, A. Sugiyama, R. Le Bouquin Jeannes, G. Faucon: Estimation of the signal-to-interference ratio based on normalized cross-correlation with symmetric leaky blocking matrices in adaptive microphone arrays, *Proc. IWAENC '06*, 1–4, Paris, France, September 2006.
48. A. Hyvaerinen, J. Karhunen, E. Oja: *Independent Component Analysis*, New York, NY, USA: Wiley, 2001.
49. S. Ikeda, N. Murata: A method of ICA in time-frequency-domain, *Proc. ICA '99*, 365–371, January 1999.
50. M. Kawamoto, K. Matsuoka, N. Ohnishi: A method of blind separation for convolved non-stationary signals, *Neurocomputing*, **22**, 157–171, 1998.
51. W. Kellermann, H. Buchner, R. Aichner: Separating convolutive mixtures with TRINICON, *Proc. ICASSP '06*, **5**, 961–964, Toulouse, France, May 2006.
52. T. Kim, T. Eltoft, T.-W. Lee: Independent vector analysis: An extension of ICA to multivariate components, *Proc.ICA '06*, 175–172, Charleston, SC, USA, March 2006.
53. S. Kotz, T. Kozubowski, K. Podgorski: *The Laplace Distribution and Generalizations*, Basel, Switzerland: Birkhäuser Verlag, 2001.
54. B. S. Krongold, D.L. Jones: Blind source separation of nonstationary convolutively mixed signals, *Proc. SSAP '00*, 53–57, Pocono Manor, PA, USA, August 2000.
55. S. Kurita, H. Saruwatari, S. Kajita, K. Takeda, F. Itakura: Evaluation of blind signal separation method using directivity pattern under reverberant conditions, *Proc. ICASSP '00*, **5**, 3140–3143, Istanbul, Turkey, June 2000.
56. H. Kuttruff: *Room Acoustics*, 4th ed., London, GB: Spon Press, 2000.
57. S. Y. Low, S. Nordholm, R. Tognieri: Convolutive blind signal separation with post-processing, *IEEE Trans. Speech Audio Processing*, **12**(5), 539–548, September 2004.
58. J. D. Markel, A. H. Gray: *Linear Prediction of Speech*, Berlin, Germany: Springer, 1976.
59. R. Martin, J. Altenhöner: Coupled adaptive filters for acoustic echo control and noise reduction, *Proc. ICASSP '95*, 3043–3046, Detroit, MI, USA, May 1995.
60. R. Martin: *Freisprecheinrichtungen mit mehrkanaliger Echokompensation und Störgeräuschreduktion*, PhD thesis, RWTH Aachen, Aachen, Germany, June 1995 (in German).
61. R. Martin: The echo shaping approach to acoustic echo control, *Speech Communication*, **20**, 181–190, 1996.
62. R. Martin: Small microphone arrays with postfilters for noise and acoustic echo reduction, in M. Brandstein, D. Ward (eds.), *Microphone Arrays: Signal Processing Techniques and Applications*, 255–279, Berlin, Germany: Springer, 2001.
63. R. Martin: Noise power spectral density estimation based on optimal smoothing and minimum statistics, *IEEE Trans. Speech Audio Processing*, **9**(5), 504–512, July 2001.
64. K. Matsuoka, M. Ohya, M. Kawamoto: Neural net for blind separation of nonstationary signals, *IEEE Trans. Neural Networks*, **8**(3), 411–419, 1995.
65. K. Matsuoka, S. Nakashima: Minimal distortion principle for blind source separation, *Proc. ICA '01*, 722–727, San Diego, CA, USA, December 2001.
66. M. Miyoshi, Y. Kaneda: Inverse filtering of room acoustics, *IEEE Trans. Acoust., Speech, Signal Processing*, **36**(2), 145–152, February 1988.

67. L. Molgedey, H. G. Schuster: Separation of a mixture of independent signals using time delayed correlations, *Physical Review Letters*, **72**, 3634–3636, 1994.
68. R. Mukai, S. Araki, H. Sawada, S. Makino: Removal of residual cross-talk components in blind source separation using time-delayed spectral subtraction, *Proc. ICASSP '02*, **2**, 1789–1792, Orlando, FL, USA, May 2002.
69. R. Mukai, S. Araki, H. Sawada, S. Makino: Removal of residual cross-talk components in blind source separation using LMS filters, *Proc. NNSP '02*, 435–444, Martigny, Switzerland, September 2002.
70. T. Nishikawa, H. Saruwatari, K. Shikano: Comparison of time-domain ICA, frequency-domain ICA and multistage ICA for blind source separation, *Proc. EUSIPCO 03*, **2**, 15–18, September 2002.
71. A. Papoulis: *Probability, Random Variables, and Stochastic Processes*, 4th ed., Boston, MA, USA: McGraw-Hill, 2002.
72. K. S. Park, J. S. Park, K. S. Son, H. T. Kim: Postprocessing with Wiener filtering technique for reducing residual crosstalk in blind source separation, *IEEE Signal Processing Lett.*, **13**(12), 749–751, December 2006.
73. L. Parra, C. Spence: Convolutive blind source separation of non-stationary sources, *IEEE Trans. Speech Audio Processing*, **8**(3), 320–327, May 2000.
74. L. Parra, C. Spence, P. Sajda: Higher-order statistical properties arising from the non-stationarity of natural signals, *Advances in Neural Information Processing Systems*, **13**, 786–792, Cambridge, MA, USA: MIT Press, 2000.
75. H. Sawada, R. Mukai, S. de la Kethulle de Ryhove, S. Araki, S. Makino: Spectral smoothing for frequency-domain blind source separation, *Proc. IWAENC '03*, 311–314, Kyoto, Japan, September 2003.
76. K. U. Simmer, J. Bitzer, C. Marro: Post-filtering techniques, in M. Brandstein, D. Ward (eds.), *Microphone Arrays: Signal Processing Techniques and Applications*, 39–60, Berlin, Germany: Springer, 2001.
77. P. Smaragdis: Blind separation of convolved mixtures in the frequency domain, *Neurocomputing*, **22**, 21–34, 1998.
78. L. Tong, R.-W. Liu, V.C. Soon, Y.-F. Huang: Indeterminacy and identifiability of blind identification, *IEEE Trans. on Circuits and Systems*, **38**(5), 499–509, May 1991.
79. V. Turbin, A. Gilloire, P. Scalart, C. Beaugeant: Using psychoacoustic criteria in acoustic echo cancellation algorithms, *Proc. IWAENC '97*, 53–56, London, UK, September 1997.
80. J.-M. Valin, J. Rouat, F. Michaud: Microphone array post-filter for separation of simultaneous non-stationary sources, *Proc. ICASSP '04*, **1**, 221–224, Montreal, Canada, May 2004.
81. S. Van Gerven, D. Van Compernolle: Signal separation by symmetric adaptive decorrelation: Stability, convergence and uniqueness, *IEEE Trans. Signal Processing*, **43**(7), 1602–1612, July 1995.
82. E. Visser, T.-W. Lee: Speech enhancement using blind source separation and two-channel energy based speaker detection, *Proc. ICASSP '03*, **1**, 836–839, HongKong, April 2003.
83. D. Wang and J. Lim: The unimportance of phase in speech enhancement, *IEEE Trans. Acoust., Speech, Signal Processing*, **ASSP-30**(4), 679–681, August 1982.
84. D. Wang, G. J. Brown (eds.): *Computational Auditory Scene Analysis: Principles, Algorithms, and Applications*, New York, NY, USA: Wiley, 2006.

85. E. Weinstein, M. Feder, A. Oppenheim: Multi-channel signal separation by decorrelation, *IEEE Trans. Speech Audio Processing*, **1**(4), 405–413, October 1993.
86. B. Widrow, J. Glover, J. MacCool, J. Kautnitz, C. Williams, R. Hearn, J. Zeidler, E. Dong, R. Goodlin: Adaptive noise cancelling: principles and applications, *Proc. IEEE*, **63**, 1692–1716, 1975.
87. S. Winter, W. Kellermann, H. Sawada, S. Makino: MAP-based underdetermined blind source separation of convolutive mixtures by hierarchical clustering and ℓ_1-norm minimization, *EURASIP Journal on Applied Signal Processing*, 1–12, 2007.
88. H.-C. Wu, J. C. Principe: Simultaneous diagonalization in the frequency domain (SDIF) for source separation, *Proc. ICA '99*, 245–250, Aussois, France, December 1999.
89. K. Yao: A representation theorem and its applications to spherically-invariant random processes, *IEEE Trans. Inform. Theory*, **19**(5), 600–608, September 1973.
90. O. Yilmaz, S. Rickard: Blind separation of speech mixtures via time-frequency masking, *IEEE Trans. Signal Processing*, **52**(7), 1830–1847, July 2004.

14
Binaural Speech Segregation

Nicoleta Roman[1] and DeLiang Wang[2]

[1] Ohio State University at Lima, Lima, USA
[2] Ohio State University, Columbus, USA

It is relatively easy for a human listener to attend to a particular speaker at a cocktail party in the presence of other speakers, music and environmental sounds. To perform this task, the human listener needs to separate the target speech from a mixture of multiple concurrent sources reflected by various surfaces. This process is referred to as *auditory scene analysis*. While humans excel at this task using only two ears, machine separation based on two-microphone recordings has proven to be extremely challenging. By incorporating the mechanisms underlying the perception of sound by human listeners, *computational auditory scene analysis* (CASA) offers a new approach to sound segregation. Binaural hearing – hearing with two ears – employs the difference in sound source locations to improve sound segregation. In this chapter, we describe the principles of binaural processing and review the state-of-the-art in binaural CASA, particularly for speech segregation.

14.1 Introduction

Human listeners are able to effectively process the multitude of acoustic events that surrounds them at all times. Each acoustic source generates a vibration of the medium (air) and our hearing is confronted by the superposition of all vibrations impinging on our eardrums. As Helmholtz noted in 1863, the final waveform is "complicated beyond conception" [26]. Nonetheless, at a cocktail party, we are able to attend to and understand a particular talker. This perceptual ability is known as the "cocktail-party effect" – a term introduced by Cherry in 1953 [15]. Cherry's original experiments have triggered research in widely different areas including speech perception in noise, selective attention, neural modeling, speech enhancement and source separation. Of special interest is a machine solution to the problem of sound separation in realistic environments, which is essential to many important applications including automatic speech and speaker recognition, hearing aid design and audio information retrieval. The field of automatic speech recognition (ASR), for example, has seen much progress in recent years. However, the performance of current recognition systems degrades rapidly in the presence of noise and reverberation and the degradation is much faster compared to human performance in similar conditions [30,33].

The sound separation problem has been investigated in the signal processing field for many years for both one-microphone recordings and multi-microphone ones (for recent reviews see [8,19]). One-microphone speech enhancement techniques include spectral subtraction [40], Kalman filtering [38], subspace analysis [20] and autoregressive modeling [5]. While requiring only one sensor benefits many applications, these algorithms make strong assumptions about interference and thus have difficulty in dealing with general acoustic mixtures. Microphone array algorithms include beamforming [8] and blind source separation (BSS) through independent component analysis (ICA) [31]. To separate multiple sound sources, beamforming takes advantage of their different directions of arrival while ICA relies on their statistical independence. The main drawback of these approaches is that they generally require the number of microphones equal or exceed the number of sound sources. In the case of two-microphone recordings typically only one wideband interfering source can be canceled out by steering a null towards its location. To address this problem, it has been proposed in [35] a subband adaptive beamformer which steers independent nulls in each time-frequency (T-F) unit to suppress the strongest interference. Another approach to this underdetermined problem – more sources than sensors – is sparse signal representation based ICA [64]. The performance of these approaches is still limited in realistic multi-source reverberant conditions.

Since the natural solution provided by human hearing is robust to noise and reverberation, one can expect that a solution to the sound separation problem can be devised using up to two microphones. While human listeners can separate speech monaurally, binaural hearing adds to this ability when sources are spatially separated [10]. A coherent theory on the human ability to segregate signals from noisy mixtures was presented by Bregman in 1990 [9]. He argues that humans perform an auditory scene analysis (ASA) of the acoustic input in order to form perceptual representations of individual sources called streams. ASA takes place in two stages: the first stage decomposes the input into a collection of sensory elements while the second stage selectively groups the elements into streams that correspond to individual sound sources. According to Bregman, stream segregation is guided by a variety of grouping cues including proximity in frequency and time, pitch, onset/offset, and spatial location. The ASA account has inspired a series of computational ASA (CASA) systems that have significantly advanced the state-of-the-art performance in monaural separation as well as binaural separation [12,28,60]. Mirroring the ASA processing described above, CASA systems generally employ two stages: segmentation (analysis) and grouping (synthesis).

In segmentation, the acoustic input is decomposed into sensory segments, or contiguous T-F regions, each of which mainly originates from a single source. In grouping, segments that are likely to come from the same source are put together. Most monaural segregation algorithms rely on pitch as the main grouping cue and therefore can operate only on voiced speech or other periodic sounds (e.g., [27]; see also [29] for an exception). On the other hand, binaural algorithms employ location cues which are independent of signal content and thus can be used to track both voiced and unvoiced speech. Compared with signal processing techniques, CASA systems make relatively few assumptions about the acoustic properties of the interference and the environment.

CASA systems typically employ T-F masking to segregate target signal from mixture signal [11, 59, 60]. Specifically, T-F units in the acoustic mixture are selectively weighted in order to enhance the desired signal. The weights can be binary or real [53]. T-F binary masking is motivated by the masking phenomenon in human audition, in which a weaker signal is masked by a stronger one in the same critical band [41]. Subsequently, the computational goal of a CASA system has been argued to be an ideal T-F binary mask, which selects the target if it is stronger than the interference in a local T-F unit [47, 49, 58]. Speech extracted from such masks has been shown to be highly intelligible in multi-source mixtures [14, 49], as well as to produce substantial improvements in robust speech recognition [17, 49]. Following the ASA account, the binary mask is estimated in a CASA system by grouping T-F units using various perceptual cues. This binary masking framework has recently become popular in the underdetermined BSS field as well, as it has been observed that different speech signals can be approximately orthogonal in a high-resolution T-F representation [3, 44, 62]. In [44] for example, ICA is combined with T-F binary masking to iteratively extract speech signals from underdetermined anechoic mixtures.

The binaural cues used by the auditory system for source localization are interaural time difference (ITD) and interaural intensity difference (IID) between the two ears [6]. While filtering produced by head, torso and external ear introduce only a weak frequency dependency for ITD [39], IID varies widely across frequencies ranging from 0.5-1 dB at low frequencies to as much as 30 dB at high frequencies. Consequently, while ITD is the main localization cue employed by the auditory system at lower frequencies (<1.5 kHz), both binaural cues are used at higher frequencies. A series of psychoacoustically inspired binaural processors have shown that these location cues can be used to substantially enhance target speech in binaural mixtures [7, 37] [61]. Recently, binaural CASA systems have employed supervised learning in the ITD-IID feature space to optimally extract the target signal [13, 49, 53]. The main observation is that, in a given T-F unit, there exists a systematic relationship between the *a priori* local SNR and the deviation of ITD/IID features [49]. Moreover, in the case of multiple concurrent sources, there exists a characteristic clustering in the ITD/IID space which can be used to estimate the ideal T-F binary mask. Systematic evaluations have shown that systems developed based on these observations perform very well under multi-source anechoic conditions [49, 53].

Reverberation presents an additional challenge to a binaural system as it introduces potentially an infinite number of additional sources due to reflections from hard surfaces. This smears considerably the ITD/IID statistics. As a result, the performance of the above location-based segregation systems degrades as reverberation level increases. Inspired by psychoacoustical studies, many systems use a model of precedence effect prior to binaural processing to emphasize the cues in the direct wavefront over the cues in the later reflections [13,43]. Alternatively, we have proposed to replace the anechoic modeling of ITD/IID with an adaptive filter to better characterize the target location in reverberant conditions [50]. The system in [50] performs target cancellation through adaptive filtering followed by an analysis of the output-to-input attenuation level to estimate the ideal binary mask. A systematic evaluation shows that the system results in large SNR gains and it outperforms standard two-microphone beamforming algorithms as well as a recent binaural processor.

The rest of the chapter is organized as follows. The next section describes T-F masks for CASA systems. Sec. 14.3 describes a binaural system for multi-source anechoic conditions. Sec. 14.4 describes a binaural system for multi-source reverberant conditions. Sec. 14.5 gives evaluation data for the systems described in Sec. 14.3 and Sec. 14.4. The last section concludes the chapter.

14.2 T–F Masks for CASA

The first stage of a CASA system is usually a T-F analysis of the input signal using either a physiologically motivated filterbank that mimics cochlear filtering [16] or a short-time Fourier transform (STFT). In this paper, we use an STFT representation to illustrate the concepts of T-F masking and binaural processing (see also the next two sections) but a similar description can be made using an auditory filterbank. Given a T-F decomposition of the acoustic mixture, the target source can be recovered by applying independent weights to individual T-F units. This type of T-F masking can be viewed as a nonstationary Wiener filter. The authors in [21] have shown that the minimum mean-square error estimate of the target signal amplitude in a T-F unit is related to the *a priori* local SNR. Hence, we define an ideal ratio mask using the *a priori* energy ratio as follows:

$$R(\Omega, t) = \frac{\left|S\left(e^{j\Omega}, t\right)\right|^2}{\left|S\left(e^{j\Omega}, t\right)\right|^2 + \left|N\left(e^{j\Omega}, t\right)\right|^2}, \qquad (14.1)$$

where $S(e^{j\Omega}, t)$ is the spectral value for the target signal and $N(e^{j\Omega}, t)$ is the spectral value for the interference at frequency Ω and frame index t.

As described previously, a number of researchers have shown the potential of binary T-F masking in speech segregation. The upper limit for a CASA system that uses binary masking is an ideal binary mask, which selects the T-F units where the target energy is stronger than the interference energy. Formally, this ideal binary mask is defined as follows:

$$M_{\text{IBM}}(\Omega, t) = \begin{cases} 1, & \text{if } \left|S\left(e^{j\Omega}, t\right)\right| > \left|N\left(e^{j\Omega}, t\right)\right|, \\ 0, & \text{otherwise.} \end{cases} \quad (14.2)$$

This is equivalent to applying a threshold of 0.5 on the energy ratio $R(\Omega, t)$. By selecting the T-F units where the target is stronger than the interference, this definition results in the optimal SNR gain among all possible binary masks because the SNR in each T-F unit is positive if the unit is retained and negative if the unit is discarded [27]. Although an ideal ratio mask will outperform an ideal binary mask [53], the estimation of an ideal ratio mask has turned out to be more sensitive to corruptions by noise and reverberation than estimating the ideal binary mask. This chapter will therefore focus on the estimation of an ideal binary mask.

An important application for CASA systems is to provide a robust front-end for ASR. Given a T-F mask (binary or ratio), the target signal can be reconstructed using the element-wise multiplication of the mask and the spectral energy of the mixture. While the signal obtained from a ratio mask can be used directly as input to a speech recognizer, conventional ASR systems are highly sensitive to the distortions introduced by binary masks. Cooke et al. [17] have proposed a missing-data approach to ASR which performs recognition using only the reliable (clean) components. Hence, a binary mask which labels the T-F units where target dominates interference is therefore an ideal front-end for this approach. A number of authors have shown that the ideal binary masks used as front-ends to a missing-data ASR provide impressive results even under very low SNR conditions [17, 49]. Alternatively, Raj et al. [46] have proposed a spectral reconstruction method for the T-F units dominated by noise to alleviate the mismatch introduced by binary masking. The reconstructed signal is then used as input to a conventional ASR system. While the missing-data ASR requires spectral features, a conventional ASR usually employs cepstral features which are known to be more effective than the spectral ones.

14.3 Anechoic Binaural Segregation

Under anechoic conditions, the signal emitted by an acoustic source arrives at the ear further away from the source at a later time and attenuated compared to the signal arriving at the ear closer to the emitting source. Inspired by psychoacoustical studies of sound localization, binaural sound separation systems have typically employed the binaural cues of ITD and IID for localization and further segregation of target source [6, 7, 49, 53]. Specifically, the filtering due to head, pinna and torso introduces at each frequency natural combinations of ITD and IID which are location dependent. When the target source dominates a particular frequency bin, the observed ITD and IID correspond to the target values. When an interfering source overlaps with the target one in the same frequency bin, the observed ITD and IID undergo systematic shifts as the energy ratio between the two sources changes [49]. This relationship is used to estimate the weights independently in each T-F unit in order to extract target signal from noisy mixture.

The presentation here is based on the binaural system proposed in [53]. The ITD and IID estimates are computed based on the spectral ratio at the left and right ears:

$$\left(\widehat{ITD}, \widehat{IID}\right)(\Omega, t) = \left[-\frac{1}{\Omega} A \left(\frac{X_L\left(e^{j\Omega}, t\right)}{X_R\left(e^{j\Omega}, t\right)}\right), \left|\frac{X_L\left(e^{j\Omega}, t\right)}{X_R\left(e^{j\Omega}, t\right)}\right|\right] \quad (14.3)$$

where $X_L(e^{j\Omega}, t)$ and $X_R(e^{j\Omega}, t)$ are the left and right ear spectral values of the mixture signal at frequency Ω and frame t and $A(re^{j\phi})=\phi$, $-\pi < \phi < \pi$. The function A computes the phase angle, in radians, of a complex number with magnitude r and phase angle ϕ. The phase is ambiguous corresponding to integer multiples of 2π. We therefore consider ITD in the range $2\pi/\Omega$ centered at zero delay.

A corpus of 10 speech signals from the TIMIT[3] database [23] is used for training. Five sentences correspond to the target location set and the rest belong to the interference location set. Binaural signals are obtained by convolving monaural signals with measured head-related impulse responses (HRIRs) corresponding to the direction of sound incidence. The responses to multiple sources are added at each ear. The HRIR measurements consist of left/right responses of a KEMAR[4] manikin from a distance of 1.4 m in the horizontal plane, resulting in 128 point impulse responses at a sampling rate of 44.1 kHz [22].

Fig. 14.1 shows empirical results from the above corpus for a two-source configuration: target source in the median plane and interference at 30°. The T-F resolution is 512 discrete Fourier transform (DFT) coefficients extracted every 20 ms with a 10 ms overlap. ITD/IID and energy ratio estimates are computed every frame using the formulas in Eqs. 14.1 and 14.3. The scatter plot in Fig. 14.1(a) shows samples of $\widehat{ITD}(\Omega, t)$ and $R(\Omega, t)$ for a frequency bin at 1 kHz. Similarly, Fig. 14.1(b) shows the results that describe the variation of $\widehat{IID}(\Omega, t)$ and $R(\Omega, t)$ for a frequency bin at 3.4 kHz. Note that the scatter plots in Fig. 14.1 exhibit a systematic shift of the estimated ITD and IID with respect to R. Moreover, a location-based clustering is observed in the joint ITD-IID space as shown in Fig. 14.1(c). Each peak in the histogram corresponds to a distinct active source.

---

[3] The term *TIMIT* results from *Texas Instruments (TI)* and *Massachusetts Institute of Technology (MIT)*.
[4] *KEMAR* abbreviates *Knowles electronic manikin for acoustic research*.

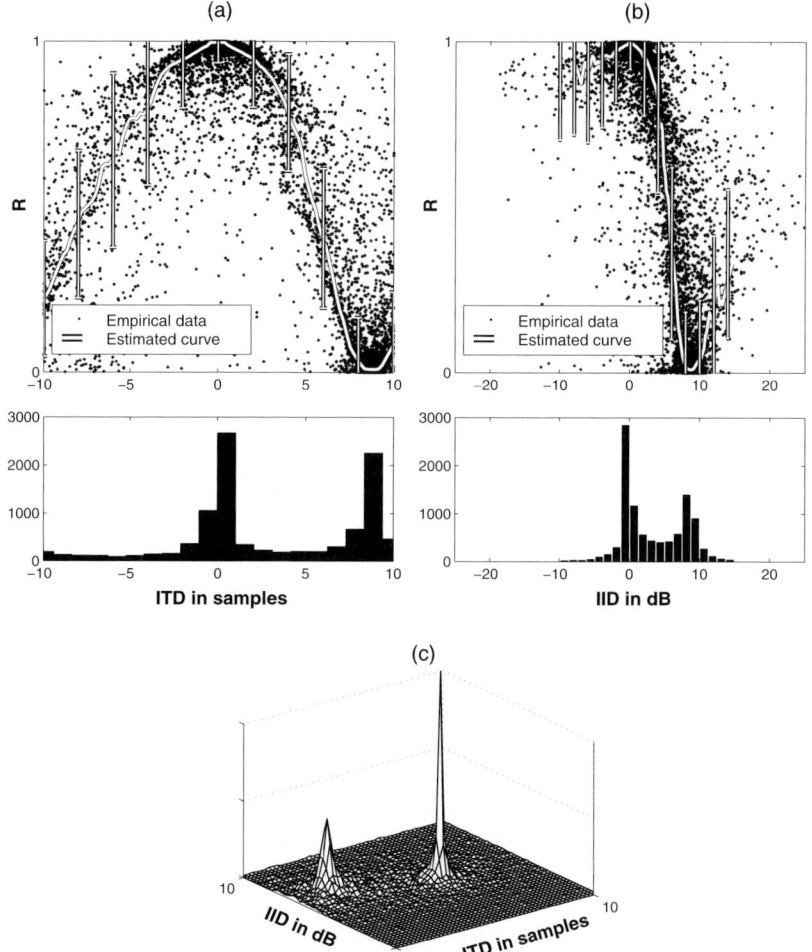

Fig. 14.1. Relationship between ITD/IID and the energy ratio R (from [53]). Statistics are obtained with target in the median plane and interference on the right side at $30°$. (a) The top panel shows the scatter plot for the distribution of R with respect to ITD for a frequency bin at 1 kHz. The solid white curve shows the mean curve fitted to the data. The vertical bars represent the standard deviation. The bottom panel shows the histogram of ITD samples. (b) Corresponding results for IID for a frequency bin at 3.4 kHz. (c) Histogram of ITD and IID samples for a frequency bin at 2 kHz.

To estimate the ideal binary mask $M_{\text{IBM}}(\Omega, t)$ we employ a non-parametric classification in the joint ITD-IID feature space. There are two hypotheses for the binary decision:

- H_1 – target is stronger or $R(\Omega, t) \geq 0.5$ and

- H_2 – interference is stronger or $R(\Omega, t) < 0.5$.

The estimated binary mask, $\widehat{M}_{\text{IBM}}(\Omega, t)$, is obtained using the maximum a posteriori (MAP) decision rule:

$$\widehat{M}_{\text{IBM}}(\Omega, t) = \begin{cases} 1, & \text{if } p(H_1)p(x|H_1) > p(H_2)p(x|H_2), \\ 0, & \text{otherwise,} \end{cases} \quad (14.4)$$

where x corresponds to the ITD and IID estimates. The prior probabilities, $p(H_i)$, are computed as the ratio of the number of samples in each class to the total number of samples. The conditional probabilities, $p(x|H_i)$ are estimated from the training data using the kernel density estimation method (see also [49]). Alternatively, the ITD/IID statistics can be used to derive ratio/soft masks [13,53]. In [53] for example, the empirical mean curves shown in Fig. 14.1 are used to estimate the energy ratio from the observed ITD/IID.

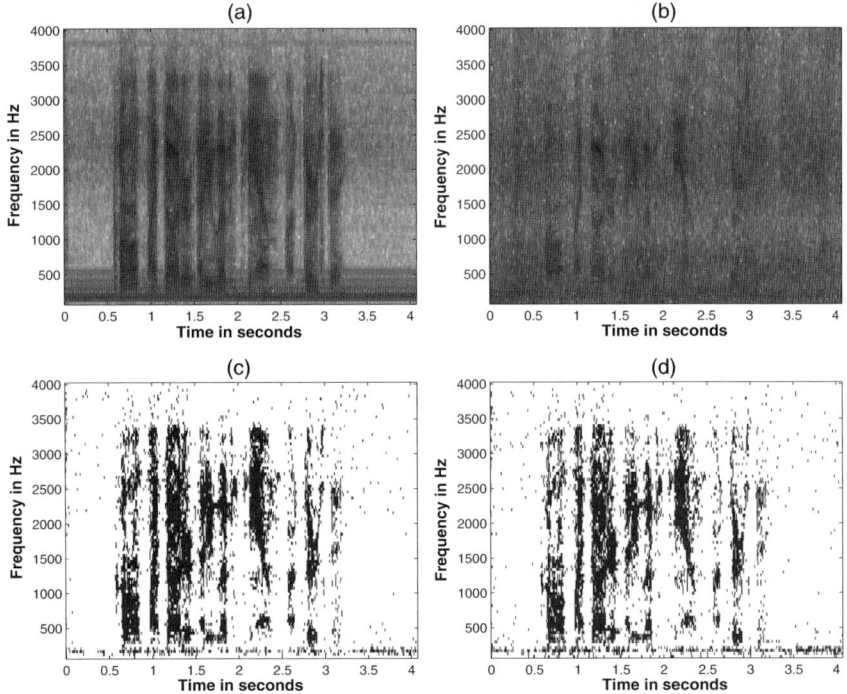

Fig. 14.2. Comparison between estimated and ideal T-F binary masks for a mixture of speech utterance presented in the median plane and an interference signal presented at 30° (redrawn from [53]). (a) Spectrogram of the clean speech utterance. (b) Spectrogram of the mixture. (c) The ideal binary mask. (d) The estimated binary mask.

Fig. 14.2 shows a comparison between an ideal and an estimated T-F binary mask. Figs. 14.2(a) and (b) show the spectrograms of a clean speech utterance and the noisy mixture, respectively. The mixture is obtained using the spatial configuration in Fig. 14.1 and a factory noise as interference. The SNR is 0 dB. The algorithm described above is applied and the T-F binary mask obtained is shown in Fig. 14.2(d). Fig. 14.2(c) shows the corresponding ideal binary mask. As seen in Sec. 14.5, evaluations across a range of SNRs show that the estimated masks approximate very well the ideal binary mask under noisy but anechoic conditions. Similar results are obtained in [49] where the processing of ITD/IID statistics follows the description above but an auditory filterbank is used for frequency decomposition.

In reverberant conditions, the anechoic modeling of time delayed and attenuated mixtures is inadequate. Since the binaural cues of ITD and IID are smeared by reflections, the system performance of ITD/IID based binaural systems degrades considerably. A model of precedence effect is typically employed to improve the robustness against these smearing effects. The system proposed in [43] includes a delayed inhibition circuit which gives more weight to the onset of a sound in order to detect reliable spectral regions that are not contaminated by interfering noise or echoes. Speech recognition is then performed in the log spectral domain by employing missing data ASR. In order to account for the reverberant environment, a spectral energy normalization is employed before recognition. Similarly, the system proposed in [13] uses the interaural coherence to identify the T-F regions that are dominated by the direct sound. Soft masks are derived using probability distributions estimated from histograms of ITD/ILD estimates. The soft masks are then used as front-ends to a modified missing-data ASR. Under mildly reverberant conditions, the authors show that these techniques can improve ASR performance considerably.

14.4 Reverberant Binaural Segregation

We present here an alternative strategy to the binaural processors described previously which can deal more effectively with multiple interfering sources under reverberant conditions. The system proposed is a two-stage model that combines target cancellation with a nonlinear processing stage in order to estimate the ideal binary mask [50]. As seen in Fig. 14.3, an adaptive filter is applied in the first stage to the mixture signal, which contains both target and interference, in order to cancel the target signal. The adaptive filter is trained for simplification in the absence of noise. In the second stage, the system labels 1 only the T-F units that have been largely attenuated in the first stage since those units are likely to have originated from the target source; and labels 0 the other units.

The signal model in Fig. 14.3 assumes that a desired speech source $s(n)$ has been produced in a reverberant enclosure and recorded by two microphones

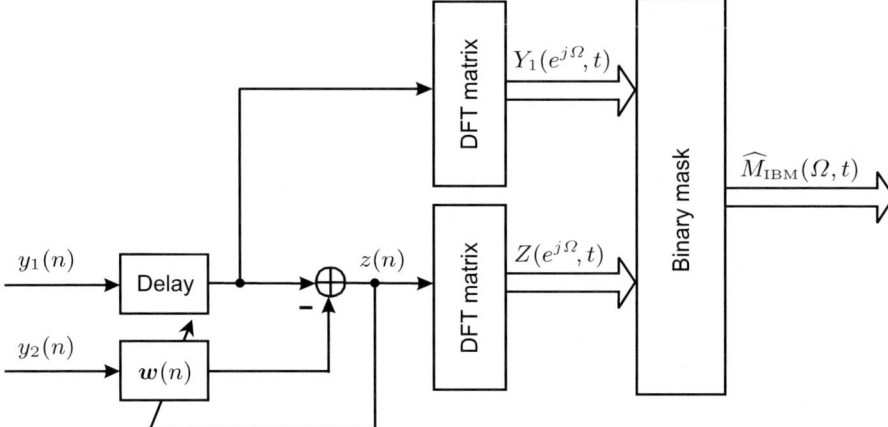

Fig. 14.3. Schematic diagram of the proposed model. The input signal is a mixture of reverberant target sound and acoustic interference. At the core of the system is an adaptive filter for target cancellation. The output of the system is an estimate of the ideal binary mask.

to produce the signal pair $[x_1(n), x_2(n)]$. . The transmission path from target location to microphones is a linear system and is modeled as:

$$x_1(n) = h_1(n) * s(n), \qquad (14.5)$$
$$x_2(n) = h_2(n) * s(n), \qquad (14.6)$$

where $h_i(n)$ corresponds to the room impulse response for the i'th microphone. The challenge of source separation arises when an unwanted interference pair $[n_1(n), n_2(n)]$ is also present at the input of the microphones resulting in a pair of mixtures $[y_1(n), y_2(n)]$:

$$y_1(n) = x_1(n) + n_1(n), \qquad (14.7)$$
$$y_2(n) = x_2(n) + n_2(n). \qquad (14.8)$$

The interference is a combination of multiple reverberant sources and additional background noise. Here, the target is assumed to be fixed but no restrictions are imposed on the number, location, or content of the interfering sources. In realistic conditions, the interference can suddenly change its location and may also contain impulsive sounds. Under these conditions, it is hard to localize each individual source in the scene. The goal is therefore to remove or attenuate the noisy background and recover the reverberant target speech based only on target source location.

In the classical adaptive beamforming approach with two microphones [24], the filter learns to identify the differential acoustic transfer function of a particular noise source and thus perfectly cancels only one directional noise source. Systems of this type, however, are unable to cope well with multiple noise sources or diffuse background noise. As an alternative, the adaptive filter is used here for target cancellation. The noise reference is then used in a nonlinear scheme to estimate the ideal binary mask. This approach offers a potential solution to the multiple interference problem in reverberation.

In the experiments reported here, we assume a fixed target location and the filter $\boldsymbol{w}(n)$ in the target cancellation module (TCM) is trained in the absence of interference. A white noise sequence of 10 s duration is used to calibrate the filter. We implement the adaptation using the Fast-Block Least Mean Square algorithm [25] with an impulse response of 375 ms length (6000 samples at 16 kHz sampling rate). After the training phase, the filters parameters are fixed and the system is allowed to operate in the presence of interference. Both the TCM output $z(n)$ and the noisy mixture at the primary microphone $y_1(n)$ are analyzed using a short time-frequency analysis. The time-frequency resolution is 20-ms time frames with a 10-ms frame shift and 257 DFT coefficients. Frames are extracted by applying a running Hamming window to the signal.

As a measure of signal suppression at the output of the TCM unit, we define the output-to-input energy ratio as follows:

$$OIR(\Omega, t) = \frac{\left|Z\left(e^{j\Omega}, t\right)\right|^2}{\left|Y_1\left(e^{j\Omega}, t\right)\right|^2}, \qquad (14.9)$$

where $Y_1(e^{j\Omega}, t)$ and $Z(e^{j\Omega}, t)$ are the corresponding Fourier transforms of $y_1(n)$ and $z(n)$, respectively.

Consider a T-F unit in which the noise signal is zero. Ideally, the TCM module cancels perfectly the target source resulting in zero output and therefore $OIR(\Omega, t) \to 0$. On the other hand, T-F units dominated by noise are not suppressed by the TCM and thus $OIR(\Omega, t) \gg 0$. Hence, a simple binary decision can be implemented by imposing a decision threshold on the estimated output-to-input energy ratio. The estimated binary mask $\widehat{M}_{\text{IBM}}(\Omega, t)$ is 1 in those T-F units where $OIR(\Omega, t) > \theta(\Omega)$ and 0 in all the other units.

Fig. 14.4 shows a scatter plot of $R(\Omega, t)$ and $OIR(\Omega, t)$ obtained for individual T-F units corresponding to a frequency bin at 1 kHz. Similar results are seen across all frequencies. The results are extracted from 100 mixtures of reverberant target speech fixed at 0° azimuth mixed with four interfering speakers at $-135°$, $-45°$, $45°$ and $135°$ azimuths. The room reverberation time, T_{60}, is 0.3 s (see Sec. 14.5 for simulation details); T_{60} is the time required for the sound level to drop by 60 dB following the sound offset. The input SNR considering reverberant target as signal is 5 dB. Observe that there exists a correlation between the amount of cancellation in the individual T-F units and the relative strength between target and interference. In order to simplify the estimation of the ideal binary mask we have used in our evaluations a frequency-independent threshold of -6 dB on the output-to-input energy ratio. The -6 dB threshold is obtained when the reverberant target signal and the noise have equal energy in Eq. 14.1. Fig. 14.5 demonstrates the performance of the proposed system for the following male target utterance: "Bright sunshine shimmers on the ocean" mixed with four interfering speakers at different locations. Observe that the estimated mask is able to estimate well the ideal binary mask especially in the high target energy T-F regions.

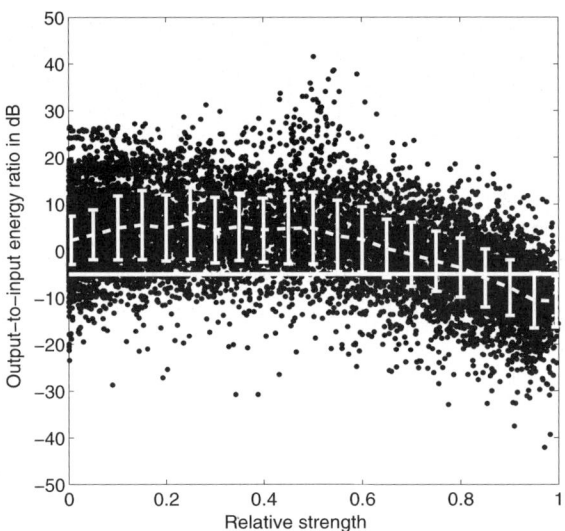

Fig. 14.4. Scatter plot of the output-to-input ratio with respect to the relative strength of the target to the mixture for a frequency bin centered at 1 kHz (from [50]). The mean and the standard deviation are shown as the dashed line and vertical bars, respectively. The horizontal line corresponds to the -6 dB decision threshold used in the binary mask estimation.

14.5 Evaluation

With the emergence of voice-based technologies, current ASR systems are required to deal with adverse conditions including noisy background and reverberation. Conventional ASR systems are constructed as a classification problem which involves the maximization of the posterior probability

$$p\big(W \,\big|\, \boldsymbol{Y}_{\mathrm{sqr}}(t)\big),$$

where $\boldsymbol{Y}_{\mathrm{sqr}}(t)$ is an observed short-term speech spectral power vector

$$\boldsymbol{Y}_{\mathrm{sqr}}(t) = \Big[Y_{\mathrm{sqr}}(\varOmega = 0, t), \ldots, Y_{\mathrm{sqr}}(\varOmega = \pi, t)\Big]^{\mathrm{T}} \qquad (14.10)$$

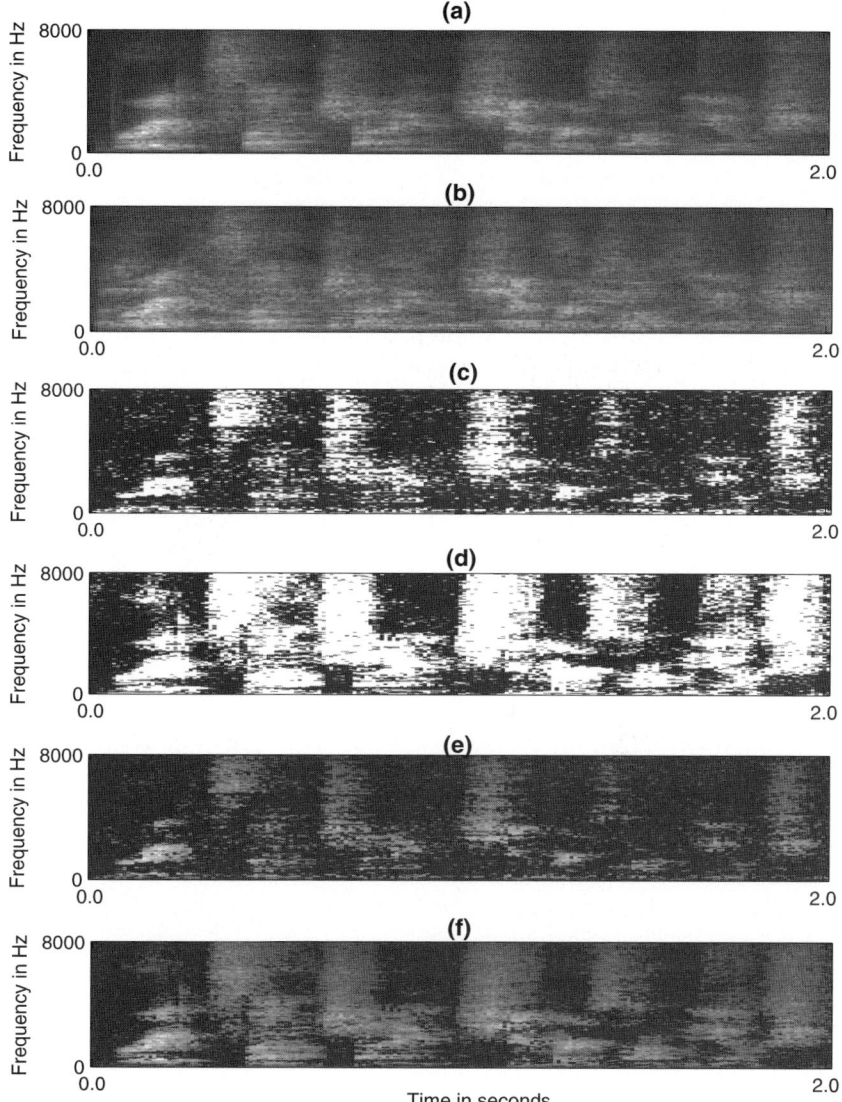

Fig. 14.5. A comparison between the estimated mask and the ideal binary mask for a five-source configuration (from [50]). (a) Spectrogram of the reverberant target speech. (b) Spectrogram of the mixture of target speech presented at 0° and four interfering speakers at locations −135°, −45°, 45° and 135°. The SNR is 5 dB. (c) The estimated T-F binary mask. (d) The ideal binary mask. (e) The mixture spectrogram overlaid by the estimated T-F binary mask. (f) The mixture spectrogram overlaid by the ideal binary mask. The recordings correspond to the left microphone.

with

$$Y_{\text{sqr}}(\Omega, t) = \left|Y\left(e^{j\Omega}, t\right)\right|^2 \tag{14.11}$$

and W is a valid word sequence. The classification is highly sensitive to distortions in the spectral vector $\boldsymbol{Y}_{\text{sqr}}(t)$. The standard approach to improving the ASR robustness is to enhance the target speech in the acoustic input. Given a T-F mask, the signal is resynthesized through the mask to reconstruct the target speech and the output is then fed to a conventional ASR system.

An alternative approach is the missing-data ASR proposed by Cooke et al. [17] which identifies the corrupted T-F spectral regions and treats them as unreliable or missing. In this approach, the spectral vector $\boldsymbol{Y}_{\text{sqr}}(t)$ is partitioned into its reliable and unreliable components as $\boldsymbol{Y}_{\text{sqr,r}}(t)$ and $\boldsymbol{Y}_{\text{sqr,u}}(t)$, where $\boldsymbol{Y}_{\text{sqr}}(t) = \boldsymbol{Y}_{\text{sqr,r}}(t) \cup \boldsymbol{Y}_{\text{sqr,u}}(t)$. The Bayesian decision is then sought using only the reliable components. In the marginalization method, the posterior probability is computed by integrating over the unreliable ones. However, further information about the mixing process can give lower and upper bounds for these unreliable components which can be used in the integral involved in marginalization. Under the assumption of additive and uncorrelated sound sources, the true value of the speech energy in the unreliable parts can be constrained between 0 and the observed spectral energy $\boldsymbol{Y}_{\text{sqr,u}}(t)$. The T-F units indicated as 1 in the binary mask are the reliable units while those indicated as 0 are the unreliable ones. It has been shown that this approach outperforms conventional ASRs with input resynthesized from T-F binary masks. Moreover, ideal binary masks produce impressive recognition scores when applied to the missing-data ASR for a variety of noise intrusions including multiple interfering sources [17, 49].

The binaural system presented in Sec. 14.3 has been evaluated under noisy but anechoic conditions using the missing-data ASR. As in [17], the task domain is speaker independent recognition of connected digits. Thirteen (the number 1-9, a silence, very short pause between words, zero and oh) word-level models are trained using an hidden Markov model (HMM) toolkit, HTK [63]. All except the short pause model have 8 emitting states. The short pause model has a single emitting state, tied to the middle state of the silence model. The output distribution in each state is modeled as a mixture of 10 Gaussians. The grammar for this task allows for one or more repetitions of digits and all digits are equally probable. Both training and testing are performed using the male speaker dataset in the TIDigits database [34]. Specifically, the models are trained using 4235 utterances in the training set of this database. Testing is performed on a subset of the testing set consisting of 461 utterances from 6 speakers, comprising 1498 words. All test speakers are different from the speakers in the training set. The signals are sampled at 20 kHz.

Fig. 14.6 shows recognition results for target source in the median plane and one noise source on the right side at 30° for a range of SNRs from −5 to 10 dB. The noise source is the factory noise from the NOISEX corpus [57]. The factory noise is chosen as it has energy in the formant regions, therefore posing challenging problems for recognition. The binaural mixtures are obtained by convolving the original signals with the HRIRs of the corresponding sound source locations. The binaural processing described in Sec. 14.3 is applied to derive the corresponding T-F binary mask which is then fed to the missing-data ASR. Feature vectors for the missing-data ASR are derived from the 512 DFT coefficients extracted in each time frame. Recognition is performed using log-spectral energy bandlimited to 4 kHz. Hence only 98 spectral coefficients along with delta coefficients in a two-frame delta-window are extracted in each frame. As seen in Fig. 14.6, the estimated masks approximate very well the ideal binary masks resulting in large recognition improvements over the baseline. Similar results have been obtained in multispeaker conditions in [49].

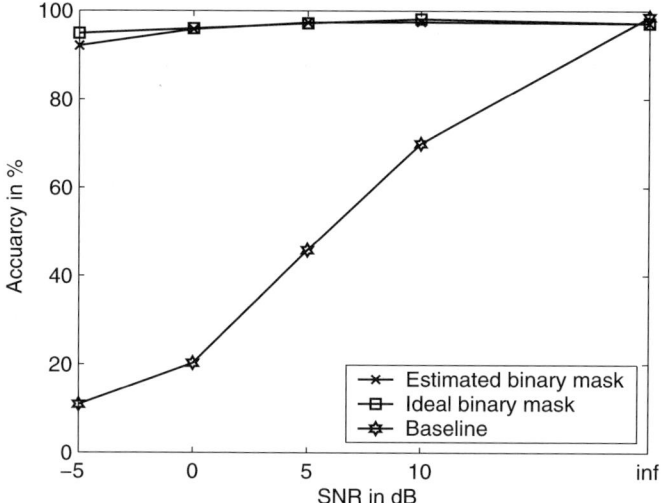

Fig. 14.6. Comparison between estimated and ideal binary masks as front-ends to a missing-data ASR under anechoic conditions (redrawn from [53]). The target source in the median plane is presented with a noise source on the right side at 30° for a range of SNRs from −5 to 10 dB. For comparison, the baseline performance is shown.

To illustrate the binaural system described in Sec. 14.4, we present here systematic recognition results under multi-source reverberant conditions. The reverberation is generated using the room acoustic model described in [43]. The reflection paths of a particular sound source are obtained using the image reverberation model for a small rectangular room (6 m × 4 m × 3 m) [1]. The resulting impulse response is convolved with the same HRIRs as before in order to produce the binaural input to our system. Specific room reverberation times are obtained by varying the absorption characteristics of room boundaries.

The position of the listener was fixed asymmetrically at (2.5 m × 2.5 m × 2 m) to avoid obtaining near identical impulse responses at the two microphones when the source is in the median plane. All sound sources are presented at different angles at a distance of 1.5 m from the listener. For all our tests, target is fixed at 0° azimuth unless otherwise specified. To test the robustness of the system to various noise configurations we have performed the following tests:

- an interference of rock music at 45° (scene 1);
- two concurrent speakers (one female and one male utterance) at azimuth angles of −45° and 45° (scene 2); and
- four concurrent speakers (two female and two male utterances) at azimuth angles of −135°, −45°, 45° and 135° (scene 3).

The initial and the last speech pauses in the interfering utterances have been deleted in conditions scene 2 and scene 3 making them more comparable with condition scene 1. The signals are upsampled to 44.1 kHz and convolved with the corresponding left and right ear HRIRs to simulate the individual sources for the above three testing conditions (scene 1 – scene 3). Finally, the reverberated signals at each ear are summed and then downsampled to 16 kHz. In all our evaluations, the input SNR is calculated at the left ear using reverberant target speech as signal. While in scene 2 and scene 3 the SNR at the two ears is comparable, the left ear is the 'better ear' – the ear with higher SNR – in the scene 1 condition. In the case of multiple interferences, the interfering signals are scaled to have equal energy at the left ear.

While the missing-data approach has shown promising results with additive noise in anechoic conditions, an extension to reverberant conditions has turned out to be problematic (see for example [43]). We therefore adapt here the spectrogram reconstruction method proposed in [46] to reverberant conditions which shows improved performance over the missing-data recognizer. This approach has been suggested in the context of additive noise. In this approach, a noisy spectral vector $\boldsymbol{Y}_{\text{sqr}}(t)$ at a particular frame is partitioned in its reliable $\boldsymbol{Y}_{\text{sqr,r}}(t)$ and its unreliable $\boldsymbol{Y}_{\text{sqr,u}}(t)$ components. The task is to reconstruct the underlying true spectral vector $\boldsymbol{X}_{\text{sqr}}(t)$. Assuming that the reliable features $\boldsymbol{Y}_{\text{sqr,r}}(t)$ are approximating well the true ones $\boldsymbol{X}_{\text{sqr,r}}(t)$, a Bayesian decision is then employed to estimate the remaining $\boldsymbol{X}_{\text{sqr,u}}(t)$ given only the reliable component. Hence, this approach works seamlessly with the T-F binary mask that our speech segregation system produces. Here, the reliable features are the T-F units labeled 1 in the mask while the unreliable features are the ones labeled 0. Although the reliable data in our system contains some reverberation, we train the prior speech model only on clean data. This actually avoids the trouble of obtaining a prior for each deployment condition, and is desirable for robust speech recognition. The signals reconstructed in this way are then used as input to a conventional ASR which employs cepstral features.

The speech prior is modeled empirically as a mixture of Gaussians and trained on the clean database used for training the conventional ASR (see below):

$$p(\boldsymbol{X}_{\text{sqr}}(t)) = \sum_{k=1}^{M} p(k) p(\boldsymbol{X}_{\text{sqr}}(t) \,|\, k), \qquad (14.12)$$

where $M = 1024$ is the number of mixtures, k is the mixture index, $p(k)$ is the mixture weight and $p(X|k) = N(X; \mu_k; \Sigma_k)$.

Previous studies [17, 46] have shown that a good estimate of $\boldsymbol{X}_{\text{sqr,u}}(t)$ is its mean conditioned on $\boldsymbol{X}_{\text{sqr,r}}(t)$:

$$\mathrm{E}\Big\{\boldsymbol{X}_{\text{sqr, u}}(t) \,\Big|\, \boldsymbol{X}_{\text{sqr,r}}(t),\, 0 \le \boldsymbol{X}_{\text{sqr,u}}(t) \le \boldsymbol{Y}_{\text{sqr,u}}(t)\Big\}$$

$$= \sum_{k=1}^{M} p\big(k \,\big|\, \boldsymbol{X}_{\text{sqr,r}}(t),\, 0 \le \boldsymbol{X}_{\text{sqr,u}}(t) \le \boldsymbol{Y}_{\text{sqr,u}}(t)\big)$$

$$\cdot \underbrace{\int_{0}^{\boldsymbol{Y}_{\text{sqr,u}}(t)} X\, p(X \,|\, k, 0 \le X \le \boldsymbol{Y}_{\text{sqr,u}}(t))\, dX}_{\widetilde{\boldsymbol{X}}_{\text{sqr,u}}(t)} \qquad (14.13)$$

where $p(k \,|\, \boldsymbol{X}_{\text{sqr,r}}(t), \ldots)$ is the *a posteriori* probability of the k'th Gaussian given the reliable data and the integral denotes the expectation $\widetilde{\boldsymbol{X}}_{\text{sqr,u}}(t)$ corresponding to the k'th mixture. Note that under the additive noise condition, the unreliable parts may be constrained as $0 \le \boldsymbol{X}_{\text{sqr,u}}(t) \le \boldsymbol{Y}_{\text{sqr,u}}(t)$ [17]. Here it is assumed that the prior can be modeled using a mixture of Gaussians with diagonal covariance. Theoretically, this is a good approximation if an adequate number of mixtures are used. Additionally, empirical evaluations have shown that for the case of $M = 1024$ this approximation results in an insignificant degradation in recognition performance while the computational cost is greatly reduced. Hence, the expected value can now be computed as:

$$\widetilde{\boldsymbol{X}}_{\text{sqr,u}}(t) = \begin{cases} \mu_{\text{u},k}, & 0 \le \mu_{\text{u},k} \le \boldsymbol{Y}_{\text{sqr,u}}(t), \\ \boldsymbol{Y}_{\text{sqr,u}}(t), & \mu_{\text{u},k} > \boldsymbol{Y}_{\text{sqr,u}}(t), \\ 0, & \mu_{\text{u},k} < 0. \end{cases} \qquad (14.14)$$

The a posteriori probability of the k'th mixture given the reliable data is estimated using the Bayesian rule from the simplified marginal distribution $p(\boldsymbol{X}_{\text{sqr,r}}|k) = N(\boldsymbol{X}_{\text{sqr,r}}; \mu_{\text{r},k}, \sigma_{\text{r},k})$ obtained without utilizing any bounds on $\boldsymbol{X}_{\text{sqr,u}}$. While this simplification results in a small decrease in accuracy, it gives substantially faster computation of the marginal.

The same recognition task is used as in the previous evaluation. Training is performed using the 4235 clean signals from the male speaker dataset in the

TIDigits database downsampled to 16 kHz to be consistent with the system described in Sec. 14.4. The HMMs are trained with clean utterances from the training data using feature vectors consisting of the 13 mel-frequency cepstral coefficients (MFCC) including the zeroth order cepstral coefficient, $C_0(n)$, as the energy term together with their first and second order temporal derivatives. MFCCs are used as feature vectors as they are most commonly used in state-of-the-art recognizers [45]. Cepstral mean normalization (CMN) is applied to the cepstral features in order to improve the robustness of the system under reverberant conditions [52]. Frames are extracted using 20 ms windows with 10 ms overlap. A first-order preemphasis coefficient of 0.97 is applied to the signal. The recognition result using clean test utterances is 99 % accuracy. Using the reverberated test utterances, performance degrades to 94 % accuracy.

Testing is performed on a subset of the testing set containing 229 utterances from 3 speakers which is similar to the test used in [43]. The test speakers are different from the speakers in the training set. The test signals are convolved with the corresponding left and right ear target impulse responses and noise is added as described above to simulate the conditions of Scene 1 to Scene 3. Speech recognition results for the three conditions are reported separately in Figs. 14.7, 14.8 and 14.9 at five SNR levels: -5 dB, 0 dB, 5 dB, 10 dB and 20 dB. Results are obtained using the same MFCC-based ASR as the back-end for the following approaches:

- fixed beamforming (delay-and-sum),
- adaptive beamforming,
- target cancellation through adaptive filtering followed by spectral subtraction,
- our proposed front-end ASR using the estimated mask
- and finally our proposed front-end ASR using the ideal binary mask.

Note that the ASR performance depends on the interference type and we obtain the best accuracy score in the two speaker interference. The baseline results correspond to the unprocessed left ear signal. Observe that our system achieves large improvements over the baseline performance across all conditions.

The adaptive beamformer used in evaluations follows the two-stage adaptive filtering strategy described in [56] that improves the classic Griffiths-Jim model [24] under reverberation. The first stage is identical to our target cancellation module and is used to obtain a good noise reference. The second stage uses another adaptive filter to model the difference between the noise reference and the noise portion in the primary microphone in order to extract the target signal. Here, training for the second filter is done independently for each noise condition in the absence of target signal using 10 s white noise sequences presented at each location in the tested configuration. The length of the filter is the same as the one used in the TCM (375 ms). As seen in Fig. 14.7, the adaptive beamformer outperforms all the other algorithms in the case of

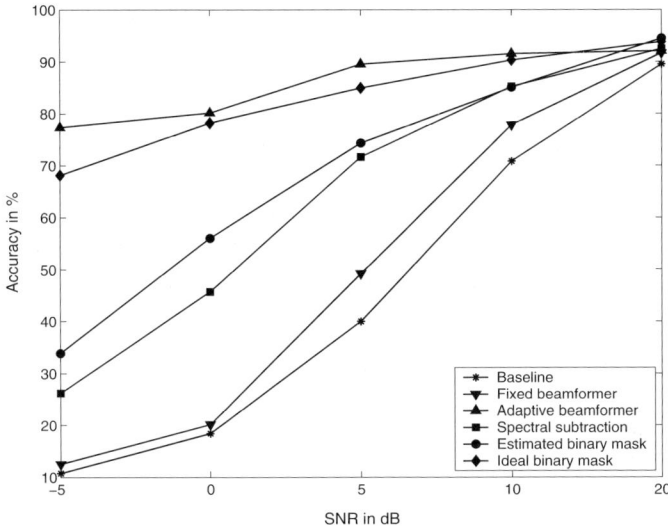

Fig. 14.7. Recognition performance for Scene 1 at different SNR values for the reverberant mixture (∗), a fixed beamformer (▼), an adaptive beamformer (▲), a system that combines target cancellation and spectral subtraction (■), an ASR front-end using the estimated binary mask (●), and an ASR front-end using the ideal binary mask (♦) (from [50]).

a single interference (scene 1). However, as the number of interferences increases, the performance of the adaptive beamformer degrades rapidly and approaches the performance of the fixed beamformer in scene 3. As proposed in [2], we can combine the target cancellation stage with spectral subtraction to attenuate the interference. As illustrated by the recognition results in Figs. 14.8 and 14.9, this approach outperforms the adaptive beamformer in the case of multiple concurrent interferences. While spectral subtraction improves the SNR gain in target-dominant T-F units, it does not produce a good target signal estimate in noise-dominant regions. Note that our ASR front-end employs a better estimation of the spectrum in these unreliable T-F units and therefore results in large improvements over the spectral subtraction method. Although the results using our ASR front-end show substantial performance gains, further improvement can be achieved as can be seen in the results reported with the ideal binary mask.

14.6 Concluding Remarks

In anechoic conditions, there exists a systematic relationship between computed ITD and IID values and local SNR within individual T-F units. This relationship leads to characteristic clustering in the joint ITD-IID feature

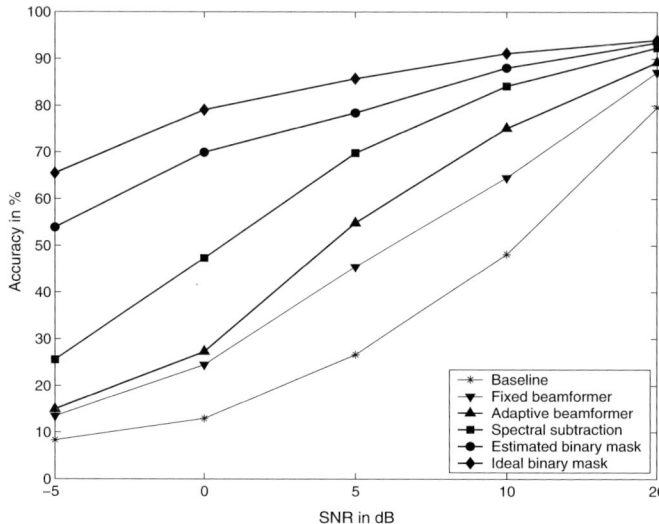

Fig. 14.8. Recognition performance for Scene 2 at different SNR values for the reverberant mixture (∗), a fixed beamformer (▼), an adaptive beamformer (▲), a system that combines target cancellation and spectral subtraction (■), an ASR front-end using the estimated binary mask (●), and an ASR front-end using the ideal binary mask (♦) (from [50]).

space, which enables the effective use of supervised classification to estimate the ideal binary mask. Estimated binary masks thus obtained from mixtures of target speech and acoustic interference have been shown to match the ideal ones very well.

In natural settings, reverberation alters many of the acoustical properties of a sound source reaching our ears, including smearing the binaural cues due to the presence of multiple reflections. This is especially detrimental when multiple sound sources are present in the acoustic scene since the binaural cues are now required to distinguish between the competing sources. Location based algorithms that rely on the anechoic assumption of time delayed and attenuated mixtures are highly affected by these distortions. In this chapter we have described strategies to alleviate this problem as well as a system that integrates target cancellation through adaptive filtering and T-F binary masking which is able to perform well under multi-source reverberant conditions.

Most work in binaural CASA assumes that sound sources remain fixed throughout testing. The system proposed in Sec. 14.4 alleviates somehow the problem; it is insensitive to interference location changes but assumes a fixed target location. None of these are realistic situations since head movement as well as source movement can occur. One way to approach the problem is to add a source tracking component. For example, the system proposed in [48] is able to track the azimuths of multiple acoustic sources using ITD/IID estimates.

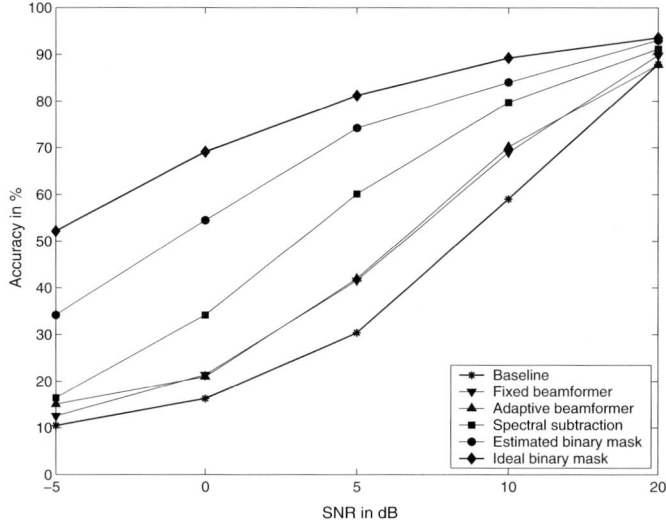

Fig. 14.9. Recognition performance for Scene 3 at different SNR values for the reverberant mixture (∗), a fixed beamformer (▼), an adaptive beamformer (▲), a system that combines target cancellation and spectral subtraction (■), an ASR front-end using the estimated binary mask (●), and an ASR front-end using the ideal binary mask (♦) (from [50]).

Such a system could be coupled with a binaural processor to deal with moving sources. Nix and Hohmann [42] have recently proposed to simultaneously track sound source locations and spectral envelopes using a non-Gaussian multidimensional statistical filtering approach. This strategy could be used to integrate in a robust way different acoustic cues even when they are corrupted by noise, reverberation and motion. Other two-microphone algorithms combine location cues and pitch information or other signal processing techniques to improve system robustness [4, 36, 51].

The computational goal of many CASA algorithms is the ideal binary T-F mask which selects target-dominant spectrotemporal regions. Signals reconstructed from such masks have been shown to be substantially more intelligible for human listeners than the original mixtures [14, 49]. However, conventional ASR systems are extremely sensitive to the distortions produced during resynthesis. Here, we have utilized two strategies that minimize these effects on recognition:

- the missing-data ASR proposed by Cooke et al. [17] that utilizes only the reliable target dominant features in the acoustic mixture
- and a target reconstruction method for the unreliable features proposed by Raj et al. [46].

As seen in our evaluations, the proposed binaural CASA systems coupled with these two strategies can produce substantial ASR improvements over baseline under both anechoic and reverberant multi-source conditions. Recently, a new approach to robust speech recognition has been proposed to additionally take into account the varied accuracy of features derived from front-end preprocessing [18]. Srinivasan and Wang [55] convert binary uncertainty in the T-F mask into real-valued uncertainty associated with cepstral features, which can then be used by an uncertainty decoder during recognition. Such uncertainty-based strategies can be utilized to further improve the performance of current binaural CASA systems when applied to robust speech recognition [54].

14.7 Acknowledgments

This research described in this chapter was supported in part by an AFOSR grant (FA9550-04-01-0117) and an NSF grant (IIS-0534707).

References

1. J. B. Allen, D. A. Berkley: Image method for efficiently simulating small-room acoustics, *JASA*, **65**, 943–950, 1979.
2. A. Álvarez, P. Gómez, V. Nieto, R. Martínez, V. Rodellar: Speech enhancement and source separation supported by negative beamforming filtering, *Proc. ICSP '02*, 342–345, 2002.
3. S. Araki, S. Makino, H. Sawada, R. Mukai: Underdetermined blind separation of convolutive mixtures of speech with directivity pattern based mask and ICA, *Proc. Fifth International Conference on Independent Component Analysis 04*, 898–905, 2004.
4. A. K. Barros, T. Rutkowski, F. Itakura, N. Ohnishi: Estimation of speech embedded in a reverberant and noisy environment by independent component analysis and wavelets, *IEEE Trans. Neural Netw.*, **13**, 888–893, 2002.
5. R. Balan, A. Jourjine, J. Rosca: AR processes and sources can be reconstructed from degenerate mixtures, *Proc. 1st International Workshop on Independent Component Analysis and Signal Separation*, 467–472, 1999.
6. J. Blauert: *Spatial Hearing – The Psychophysics of Human Sound Localization*, Cambridge, MA, USA: MIT press, 1997.
7. M. Bodden: Modeling human sound-source localization and the cocktail-party-effect, *Acta Acoustica*, **1**, 43–55, 1993.
8. M. Brandstein, D. Ward (eds.): *Microphone Arrays: Signal Processing Techniques and Application*, Berlin, Germany: Springer, 2001.
9. A. S. Bregman: *Auditory Scene Analysis*, Cambridge, MA, USA: MIT press, 1990.
10. A. W. Bronkhorst: The cocktail party phenomenon: a review of research on speech intelligibility in multiple-talker conditions, *Acustica*, **86**, 117–128, 2000.
11. G. J. Brown, M. P. Cooke: Computational auditory scene analysis, *Comput. Speech Lang.*, **8**, 297–336, 1994.
12. G. J. Brown, D. L. Wang: Separation of speech by computational auditory scene analysis, in J. Benesty, S. Makino, J. Chen (eds.), *Speech Enhancement*, New York, NY, USA: Springer, 2005, 371–402.

13. G. J. Brown, S. Harding, J. P. Barker: Speech separation based on the statistics of binaural auditory features, *Proc. ICASSP '06*, **5**, 2006.
14. D. S. Brungart, P. S. Chang, B. D. Simpson, D. L. Wang: Isolating the energetic component of speech-on-speech masking with ideal time-frequency segregation, *JASA*, **120**, 4007–4018, 2006.
15. E. C. Cherry: Some experiments on the recognition of speech, with one and with two ears, *JASA*, **25**, 975–979, 1953.
16. M. P. Cooke: *Modeling Auditory Processing and Organization,* Cambridge, U.K.: Cambridge University Press, 1993.
17. M. P. Cooke, P. Green, L. Josifovski,A. Vizinho: Robust automatic speech recognition with missing and unreliable acoustic data, *Speech Commun.*, **34**, 267–285, 2001.
18. L. Deng, J. Droppo, A. Acero: Dynamic compensation of HMM variances using the feature enhancement uncertainty computed from a parametric model of speech distortion, *IEEE Trans. Speech, and Audio Process.*, **13**, 412-421, 2005.
19. P. Divenyi (ed.): *Speech Separation by Humans and Machines,* Norwell, MA, USA: Kluwer Academic, 2005.
20. Y. Ephraim, H. L. Trees: A signal subspace approach for speech enhancement, *IEEE Trans. Speech Audio Process.*, **3**, 251–266, 1995.
21. Y. Ephraim, D. Malah: Speech enhancement using a minimum mean-square error short-time spectral amplitude estimator, *IEEE Trans. Acoust. Speech Signal Process.*, **ASSP-32**(6), 1109-1121, 1984.
22. W. G. Gardner, K. D. Martin: HRTF measurements of a KEMAR dummy-head microphone, *MIT Media Lab Perceptual Computing Technical Report #280*, 1994.
23. J. Garofolo, L. Lamel, W. Fisher, J. Fiscus, D. Pallett, N. Dahlgren: Darpa timit acoustic-phonetic continuous speech corpus, *Technical Report NISTIR 4930*, National Institute of Standards and Technology, Gaithersburg, MD, USA, 1993.
24. L. J. Griffiths, C. W Jim: An alternative approach to linearly constrained adaptive beamforming, *IEEE Trans. Antennas and Propagation*, **30**, 27–34, 1982.
25. S. Haykin: *Adaptive Filter Theory*, 4th ed., Upper Saddle River, NJ, USA: Prentice Hall, 2002.
26. H. Helmholtz: *On the Sensation of Tone*, (A. J. Ellis, Trans.), 2nd English ed., New York, NY, USA: Dover Publishers, 1863.
27. G. Hu, D. L. Wang: Monaural speech segregation based on pitch tracking and amplitude modulation, *IEEE Trans. Neural Netw.*, **15**, 1135–1150, 2004.
28. G. Hu, D. L. Wang: An Auditory Scene Anaylsis Approach to Monaural Speech Segregation, in E. Hänsler, G. Schmidt (eds.), *Topis in Acoustic Echo and Noise Control*, 485–515, Berlin, Germany: Springer, 2006.
29. G. Hu, D. L. Wang: Auditory segmentation based on onset and offset analysis, *IEEE Trans. Audio, Speech and Language Process.*, **15**, 396–405, 2007.
30. X. Huang, A. Acero, H. -W. Hon: *Spoken Language Processing: A guide to theory, algorithms, and system development,* Upper Saddle River, NJ, USA: Prentice Hall PTR, 2001.
31. A. Hyvärinen, J. Karhunen, E. Oja: *Independent component analysis*, New York, NY, USA: Wiley, 2001.
32. L. A. Jeffress: A place theory of sound localization, *Journal of Comparative and Physiological Psychology*, **41**, 35–39, 1948.

33. R. P. Lippman: Speech recognition by machines and humans, *Speech Commun.*, **22**, 1–16, 1997.
34. R. G. Leonard: A database for speaker-independent digit recognition, *Proc. ICASSP '84,*, 111–114, 1984.
35. C. Liu, B. C. Wheeler, W. D. O'Brien, Jr., C. R. Lansing, R. C. Bilger, D. L. Jones, A.S. Feng: A two-microphone dual delay-line approach for extraction of a speech sound in the presence of multiple interferers, *JASA*, **110**, 3218–3230, 2001.
36. H. Y. Luo, P. N. Denbigh: A speech separation system that is robust to reverberation, *Proc. International Symposium on Speech, Image Process. and Neural Netw.*, 339–342, 1994.
37. R. F. Lyon: A computational model of binaural localization and separation, *Proc. ICASSP '83*, 1148–1151, 1983.
38. N. Ma, M. Bouchard, R. Goubran: Perceptual Kalman filtering for speech enhancement in colored noise, *Proc. ICASSP '04*, **1**, 717–720, 2004.
39. E. A. MacPherson: A computer model of binaural localization for stereo imaging measurement, *J. Audio Engineering Soc.*, **39**, 604–622, 1991.
40. R. Martin: Noise power spectral density estimation based on optimal smoothing and minimum statistics, *IEEE Trans. Speech Audio Process.*, **9**, 504–512, 2001.
41. B. C. J. Moore: *An introduction to the Psychology of Hearing*, 5th ed., San Diego, CA, USA: Academic, 2003.
42. J. Nix, V. Hohmann: Combined estimation of spectral envelopes and sound source direction ofconcurrent voices by multidimensional statistical filtering, *IEEE Trans. Audio, Speech and Language Process.*, **15**, 995–1008, 2007.
43. K. J. Palomäki, G. J. Brown, D. L. Wang: A binaural processor for missing data speech recognition in the presence of noise and small-room reverberation, *Speech Commun.*, **43**, 361–378, 2004.
44. M. S. Pedersen, D. L. Wang, J. Larsen, U. Kjems: Two-microphone separation of speech mixtures,*IEEE Trans. Neural Netw.*, in press, 2008.
45. L. R. Rabiner, B. H. Juang: *Fundamentals of Speech Recognition*, 2nd ed., Englewood Cliffs, NJ, USA: Prentice-Hall, 1993.
46. B. Raj, M. L. Seltzer, R. M. Stern: Reconstruction of missing features for robust speech recognition, *Speech Commun.*, **43**, 275–296, 2004.
47. N. Roman, D. L. Wang, G. J. Brown: Speech segregation based on sound localization, *Proc. IJCNN '01*, 2861–2866, 2001.
48. N. Roman, D. L. Wang: Binaural tracking of multiple moving sources, *Proc. ICASSP '03*, **5**, 149–152, 2003.
49. N. Roman, D. L. Wang, G. J. Brown: Speech segregation based on sound localization, *JASA*, **114**, 2236–2252, 2003.
50. N. Roman, S. Srinivasan, D. L. Wang: Binaural segregation in multisource reverberant environments, *JASA*, **120**, 4040–4051, 2006.
51. A. Shamsoddini, P. N. Denbigh: A sound segregation algorithm for reverberant conditions, *Speech Commun.*, **33**, 179–196, 2001.
52. M. L. Shire: Discriminant training of front-end and acoustic modeling stages to heterogeneous acoustic environments for multi-stream automatic speech recognition, Ph. D. dissertation, University of California, Berkeley, 2000.
53. S. Srinivasan, N. Roman, D. L. Wang: Binary and ratio time-frequency masks for robust speech recognition, *Speech Commun.*, **48**, 1486–1501, 2006.
54. S. Srinivasan, N. Roman, D. L. Wang: Exploiting uncertainties for binaural speech recognition, *Proc. ICASSP '07*, **4**, 789–792, 2007.

55. S. Srinivasan, D. L. Wang: Transforming binary uncertainties for robust speech recognition, *IEEE Trans. Audio, Speech, and Language Process.*, **15**, 2130–2140, 2007.
56. D. Van Compernolle: Switching adaptive filters for enhancing noisy and reverberant speech from microphone array recordings, *Proc. ICASSP '90*, 833–836, 1990.
57. A. P. Varga, H. J. M. Steeneken,M. Tomlinson, D. Jones: The NOISEX-92 study on the effect of additive noise on automatic speech recogonition, Technical Report, Speech Research Unit, Defense Research Agency, Malvern, UK, 1992.
58. D. L. Wang: On ideal binary mask as the computational goal of auditory scene analysis, in P. Divenyi (ed.), *Speech Separation by Humans and Machines*, Norwell, MA, USA: Kluwer Academic, 2005, 181–197.
59. D. L. Wang, G. J. Brown: Separation of speech from interfering sounds based on oscillatory correlation, *IEEE Trans. Neural Netw.*, **10**, 684–697, 1999.
60. D. L. Wang, G. J. Brown (eds.): *Computational auditory scene analysis: Principles, algorithms and applications*, IEEE Press/Wiley-Interscience, 2006.
61. T. Whittkop, V. Hohmann: Strategy-selective noise reduction for binaural digital hearing aids,*Speech Commun.*, **39**, 111–138, 2003.
62. O. Yilmaz, S. Rickard: Blind separation of speech mixtures via time-frequency masking, *IEEE Trans. Signal Process.*, **52**(7), 1830–1847, 2004.
63. S. Young, D. Kershaw, J. Odell, V. Valtchev, P. Woodland: The HTK Book (for HTK Version 3.0), Microsoft Corporation, 2000.
64. M. Zibulevsky, B. A. Pearlmutter, P. Bofill, P. Kisilev: Blind source separation by sparse decomposition, in S. J. Roberts, R. M. Everson (eds.), *Independent Component Analysis: Principles and Practice*, Cambridge University Press, 2001.

15

Spatio-Temporal Adaptive Inverse Filtering in the Wave Domain

Sascha Spors[1], Herbert Buchner[1], and Rudolf Rabenstein[2]

[1] Deutsche Telekom Laboratories, Berlin University of Technology, Berlin, Germany
[2] University Erlangen-Nuremberg, Erlangen, Germany

The sound quality for acoustical communication, information, and entertainment is often subject to impairments by room reflections and undesired noise sources. As a remedy, various signal processing techniques have been developed for different applications, like acoustic echo cancelation and active noise control. Starting from the single channel case, these techniques have recently been extended to multiple channels. These extensions increased the intricacy of the original problem by the added complexity of the multichannel case. The effect was that the resulting techniques like multichannel active listening room compensation and multichannel active noise control appeared as unrelated solutions to different kinds of problems. This chapter gives a unifying description of these spatio-temporal adaptive methods on the perspective of sound reproduction by tracing them back to the fundamental problem of inverse filtering. After analyzing the problems of an adaptive solution, eigenspace adaptive filtering is introduced as a concept for decoupling the multichannel problem. Unfortunately, this concept is not straightforward applicable in its pure form, since it requires data-dependent transformations. Therefore, an approximate solution called *wave-domain adaptive filtering* is introduced. It has the advantage of being data independent and still performs a close-to-ideal decoupling. Based on the unifying inverse filtering description, the application of wave-domain adaptive filtering to active listening room compensation and to active noise control is shown.

15.1 Introduction

In many practical situations, signals are impaired by processes which can be regarded as linear systems. Signal processing operations to recover the original signal from such an impaired version are called inverse filtering. In electroacoustics these impairments are typically caused by room reflections and undesired noise sources.

Room reflections may be described as a linear filtering process with the so-called *room impulse response*. Therefore, undoing the effects of room acoustics is equivalent to the inversion of the room impulse response. However, since the propagation of sound waves is a spatially distributed phenomenon, it cannot be represented well by a single-input, single-output system. Therefore the concept of a room impulse response needs some more refinement.

More appropriate is a description of room acoustics as a distributed parameter system in the form of the acoustic wave equation. The corresponding response to a spatio-temporal impulse is then given by the Green's function. However, Green's functions depend on continuous time and continuous space in a fashion which is determined by the boundary conditions of the wave equation, i.e. by the properties of the enclosure. Therefore, Green's functions serve more as a theoretical concept than as a signal processing tool.

A more practical approach is to consider the Green's functions between selected spatial locations like the positions of loudspeakers and the positions of microphones. The time function between a pair of these locations is the corresponding room impulse response. Assembling the room impulse responses between all loudspeaker positions and all microphone positions in matrix form leads to the so-called *room response matrix*. Recovering a signal impaired by room acoustics is therefore equivalent to an inversion of the room response matrix.

Unfortunately, the acoustical properties of an enclosure are not static. Movements of sources, receivers, and other objects in a room as well as variations of the room temperature require to consider the room impulse matrix as time-variant. To cope with such a situation by inverse filtering means to employ adaptive filters. In short, reducing the effects of room acoustics by signal processing requires the inversion of the time-varying room response matrix by adaptive filtering.

Such an endeavor immediately poses a number of problems, like the continuing estimation of the room impulse matrix, the uniqueness of the inversion, and the convergence of the adaptation. So far, these problems have been approached typically in the context of special application fields like multichannel acoustic echo cancelation, room compensation for multichannel reproduction, or active noise control.

This chapter presents a unified representation of spatio-temporal adaptive filtering problems. It encompasses various application areas which are typically seen as independent fields. Two of these are described in the next section: active room compensation (ARC) for multichannel sound reproduction and active noise control (ANC). Room compensation pertains to the inversion of room acoustics as described above, while noise control is related to the reduction of additive noise. Then it is shown that both problems – although quite different in nature – are special cases of the same unified representation.

The remaining sections discuss the solution of the adaptive inverse filtering problem in this unified context. After the formulation of the multichannel adaptation and a discussion of its associated problems, a generic framework for multichannel pre-equalization in the so-called *eigenspace* is presented. Although this framework yields an optimal decoupling of the multichannel problem, it is not yet suitable for implementation. A practical solution is finally obtained by wave-domain adaptive filtering [9, 38]. It is an approximation of the optimal solution based on efficient building blocks.

15.2 Problem Description

In the following, a unified problem description for multichannel inverse adaptive filtering will be developed. For this purpose we review multichannel sound reproduction, active listening room compensation and active noise control. Since we will consider the potential and challenges in conjunction with high channel numbers (>10) we will explicitly refer to such systems as massive multichannel systems in the following.

15.2.1 Nomenclature

The following conventions are used throughout this chapter: Vectors are denoted by lower case boldface, matrices by upper case boldface. The temporal frequency domain is denoted by underlining the quantities, the spatial transform domain (eigenspace, wave domain) by a tilde placed over the respective symbol. The two-dimensional position vector in Cartesian coordinates is given as $\boldsymbol{x} = [x, y]^T$. The Cartesian coordinates are related to polar coordinates by $x = r \cos(\alpha)$ and $y = r \sin(\alpha)$.

15.2.2 Massive Multichannel Sound Reproduction

The solution of the homogeneous wave equation for a bounded region V with respect to inhomogeneous boundary conditions is given by the Kirchhoff-Helmholtz integral [55]

$$\underline{p}(\boldsymbol{x}, \omega) = - \oint_{\partial V} \left\{ \underline{g}(\boldsymbol{x}|\boldsymbol{x}_0, \omega) \frac{\partial}{\partial \boldsymbol{n}} \underline{p}(\boldsymbol{x}_0, \omega) - \underline{p}(\boldsymbol{x}_0, \omega) \frac{\partial}{\partial \boldsymbol{n}} \underline{g}(\boldsymbol{x}|\boldsymbol{x}_0, \omega) \right\} dS_0 , \tag{15.1}$$

where $\underline{p}(\boldsymbol{x}, \omega)$ denotes the pressure field inside a bounded region V surrounded by the border ∂V ($\boldsymbol{x} \in V$), $\underline{g}(\boldsymbol{x}|\boldsymbol{x}_0, \omega)$ a suitable chosen Green's function, $\underline{p}(\boldsymbol{x}_0, \omega)$ the acoustic pressure at the boundary ∂V ($\boldsymbol{x}_0 \in \partial V$) and $\frac{\partial}{\partial \boldsymbol{n}}$ the directional gradient in direction of the inward pointing normal vector $\boldsymbol{n}$ of V. The wave field outside of V is zero and V is assumed to be source-free. Fig. 15.1 illustrates the geometry.

The Green's function $\underline{g}(\boldsymbol{x}|\boldsymbol{x}_0, \omega)$ represents the solution of the inhomogeneous wave equation for an excitation with a spatio-temporal Dirac pulse at

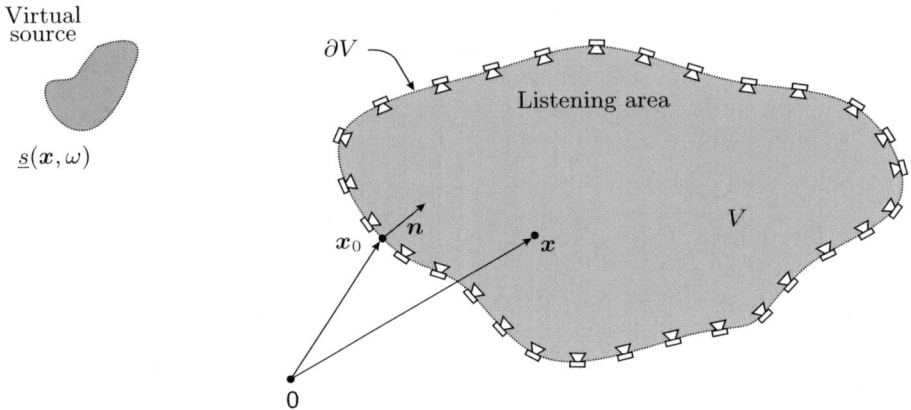

Fig. 15.1. Geometry used for the Kirchhoff-Helmholtz integral.

the position x_0. It has to fulfill the homogeneous boundary conditions imposed on ∂V. Within the context of this chapter we will assume that V is free of any objects and that the border ∂V does not restrict propagation. Hence, we assume free-field propagation within V. The Green's function is then given as the free-field solution of the wave equation and is referred to as free-field Green's function. The free-field Green's function can be interpreted as the field of a monopole source placed at the point x_0 and its directional gradient as the field of a dipole source at the point x_0, whose main axis points towards n.

Eq. 15.1 states that the wave field $\underline{p}(x,\omega)$ inside V is fully determined by the pressure $\underline{p}(x_0,\omega)$ and its directional gradient on the boundary ∂V. Consequently, if we realize the Green's function by a continuous distribution of monopole and dipole sources which are placed on the boundary ∂V, the wave field within V is fully determined by these sources. This principle can be used for sound reproduction as will be illustrated in the following. In this context the monopole and dipole sources on the boundary are referred to as (monopole/dipole) secondary sources.

For authentic sound reproduction it is desired to reproduce the wave field $\underline{s}(x,\omega)$ of a virtual source inside a limited area (listening area) as closely as possible. In the following, the listening area is assumed to be the bounded region V. Concluding the considerations given so far, authentic sound reproduction can be realized if a distribution of secondary monopole and dipole sources on the boundary ∂V of the listening area V are driven by the directional gradient and the pressure of the wave field of the virtual source, respectively. The wave field $\underline{p}(x,\omega)$ inside the listening area V is then equal to the wave field $\underline{s}(x,\omega)$ of the virtual source. Note, that there might be additional near-field contributions present [45].

The Kirchhoff-Helmholtz integral and its interpretation given above lay the theoretical foundation for sound reproduction systems like wave field synthesis

(WFS) and higher-order ambisonics (HOA). However, for a practical realization several simplifications are typically applied. These will be illustrated for the example of WFS. The simplifications that are typically applied for WFS are [20, 33, 43]:

1. elimination of dipole secondary sources,
2. spatial sampling of the continuous secondary source distribution, and
3. usage of secondary point sources for two-dimensional reproduction.

The first simplification is to remove one of the two secondary source types. Typically the dipole sources are removed, since monopole sources can be realized reasonably well by loudspeakers with closed cabinets. The Kirchhoff-Helmholtz integral (Eq. 15.1) takes inherently care that only those secondary sources are driven whose local propagation direction coincides with the wave field to be reproduced. This inherent feature gets lost when using monopoles only. Hence, removing the dipole secondary sources requires to sensibly select the secondary sources used for the reproduction of a particular virtual sound field. Suitable selection criteria have been derived in [44, 46].

The second simplification accounts for the fact that secondary sources (loudspeakers) can only be placed at discrete spatial positions. This implies a spatial sampling of the continuous secondary source distribution in the Kirchhoff-Helmholtz integral. Sampling of the secondary source distribution may lead to spatial aliasing artifacts being present in the reproduced wave field [20, 42]. As a consequence, control over the wave field within the listening area is only possible for a limited bandwidth. However, for a reasonable bandwidth a high number of secondary sources is required. Within the context of this chapter we refer to such systems as massive multichannel sound reproduction systems.

The third simplification is related to the choice of secondary sources. For two-dimensional reproduction line sources would be the appropriate choice as secondary sources. However, two-dimensional WFS utilizes point sources (closed loudspeakers) instead of secondary line sources for reproduction. Please see e. g. [33, 43] for more details.

A number of WFS systems with different shapes, channel numbers and loudspeaker technologies have been successfully realized in the past on the basis of the discussed simplifications. Objective and subjective evaluation has proven the ability of WFS to accurately recreate the impression of a desired virtual source [12, 48, 50, 51]. Other massive multichannel sound reproduction systems, like HOA, are inherently based on the same simplifications.

Common to most approaches for massive multichannel sound reproduction is that they rely on free-field wave propagation and do not consider the influence of reflections within the listening room. Since these reflections may impair the carefully designed spatial sound field, their influence should be minimized by taking appropriate countermeasures. Another kind of impairments are noise sources within the listening room. The next sections will introduce active listening room compensation and active noise control as potential countermeasures for these impairments.

15.2.3 Multichannel Active Listening Room Compensation

The influence of the listening room on the performance of sound reproduction systems is a topic of active research [3–5, 11, 13, 15, 17, 25, 35, 52–54]. Since the acoustic properties of the listening room and the reproduction system used may vary in a wide range, no generic conclusion can be given for the perceptual influence of the listening room. However, it is generally agreed that the reflections imposed by the listening room will have influence on the perceived properties of the reproduced scene. These influences may be, e. g. degradation of directional localization performance or sound coloration. A reverberant listening room will superimpose its characteristics on the desired impression of the recorded room. Listening room compensation aims at eliminating or reducing the effect of the listening room.

Ideally, the desired control over the undesired reflections can be applied within the entire listening area by destructive interference. Advanced reproduction systems like WFS provide control over the reproduced wave field within certain limits provided by the spatial sampling of the secondary source distribution. Hence, in principle these systems allow to compensate for the reflections of the listening room throughout the entire listening area. In order to cope with the limits imposed by spatial sampling of the secondary sources they may be supported easily by passive damping of the listening room for high frequencies.

Listening room compensation requires to analyze the reproduced wave field throughout the entire listening area. Again the Kirchhoff-Helmholtz integral (Eq. 15.1) provides the basis for advanced wave field analysis (WFA) techniques. It states that the wave field within a bounded region is fully determined by the acoustic pressure and its gradient on the boundary of that bounded region. Some simplifications of this fundamental principle have to be applied to arrive at a realizable WFA system [22, 49]. One of these is to spatially sample the analysis positions at the boundary of the region of interest. As for sound reproduction, an adequate analysis of the reproduced wave field within the listening area will require a large number of analysis channels [2, 23, 43].

A suitable combination of both, massive multichannel sound reproduction and WFA, will result in an enlarged compensation zone which is free of the influence of the listening room. Fig. 15.2 illustrates this approach to active listening room compensation.

The wave field reproduced by the (spatially discrete) secondary source distribution on the boundary ∂V_{ls} is analyzed by the (spatially discrete) distribution of microphones on the boundary ∂V_{al}.

Due to the causal nature of listening room reflections, active listening room compensation can be realized by pre-filtering the secondary source

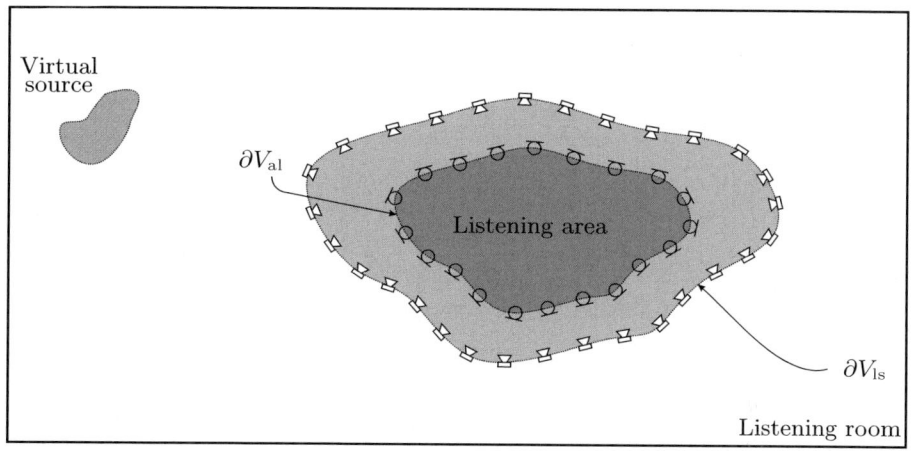

Fig. 15.2. Block diagram illustrating massive multichannel active listening room compensation system based on the Kirchhoff-Helmholtz integral.

(loudspeaker) driving signals with suitable compensation filters. This prefiltering results in cancelation of the room reflections by destructive interference within the listening area.

In general, the listening room characteristics may change over time. For instance as a result of a temperature variation in the listening room the speed of sound and hence the acoustic properties will change [29,32]. This calls for an adaptive computation of the compensation filters on the basis of a continuous analysis of the reproduced wave field.

The block diagram of an adaptive system for multichannel active room compensation (ARC) is shown in Fig. 15.3. The vector of input signals

$$\boldsymbol{d}^{(R)}(n) = \begin{bmatrix} d_1(n), d_2(n), \ldots, d_R(n) \end{bmatrix}^\mathrm{T} \tag{15.2}$$

represents the R secondary source driving signals $d_r(n)$, $r = 1, \ldots, R$ provided by a multichannel spatial rendering system which is based on the assumption of a reflection-free listening room. Throughout this chapter the discrete time index is denoted by n. In Eq. 15.2 r denotes the index of the scalar signals $d_r(n)$ and R is their total number and hence the length of the vector $\boldsymbol{d}^{(R)}(n)$. The same notation scheme is also used for the other signals $\boldsymbol{w}^{(N)}(n)$, $\boldsymbol{a}^{(M)}(n)$, $\boldsymbol{l}^{(M)}(n)$ and $\boldsymbol{e}^{(M)}(n)$ in Fig. 15.3, with with $m = 1, ..., M$ and $p = 1, ..., N$. The matrices of impulse responses $\boldsymbol{C}_{\mathrm{ARC}}(n)$, $\boldsymbol{R}(n)$, and $\boldsymbol{F}(n)$ represent the compensation filters, the listening room impulse responses, and the desired listening room impulses responses, respectively. The elements of the matrices are composed from the n-th element of the impulse response from the respective input to the respective output. For example, the matrix $\boldsymbol{R}(n)$ is defined as

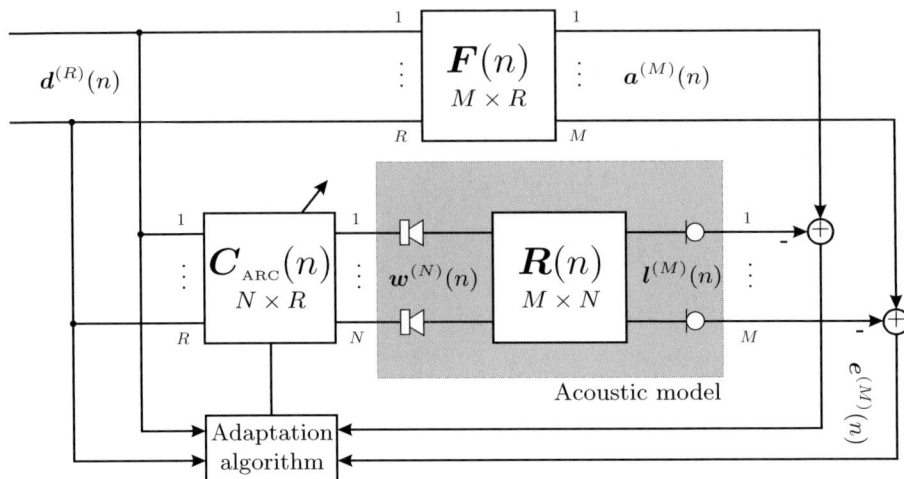

Fig. 15.3. Block diagram illustrating the generic listening room compensation system.

$$\boldsymbol{R}(n) = \begin{bmatrix} r_{1,1}(n) & \cdots & r_{1,N}(n) \\ \vdots & \ddots & \vdots \\ r_{M,1}(n) & \cdots & r_{M,N}(n) \end{bmatrix}, \qquad (15.3)$$

where $r_{m,p}(n)$ denotes the n-th element of the impulse response from the p-th loudspeaker to the m-th microphone. The matrices $\boldsymbol{C}_{\mathrm{ARC}}(n)$ and $\boldsymbol{F}(n)$ are composed from the n-th element of the impulse responses $c_{p,r}(n)$ and $f_{m,r}(n)$.

In order to eliminate the effect of the listening room, free-field conditions are desired for reproduction. Hence, the matrix $\boldsymbol{F}(n)$ is composed from the free-field impulse responses from each loudspeaker to each microphone. The matrix elements $f_{m,r}(n)$ can be derived from the analytical solution of the wave equation for the free-field case. A modeling-delay can be introduced to ensure causal compensation filters. After the compensation filters there are N pre-filtered driving signals for the loudspeakers, that are combined into a column vector $\boldsymbol{w}^{(N)}(n)$. A total of M microphones are used for the analysis of the resulting wave field. Their signals are combined in the vector $\boldsymbol{l}^{(M)}(n)$, representing spatial samples of the desired wave field on which the undesired listening room reflections are superimposed. The $N \times 1$ column vector $\boldsymbol{w}^{(N)}(n)$ and the $M \times 1$ vector $\boldsymbol{l}^{(M)}(n)$ are defined in an analogous way as given by Eq. 15.2. The vector $\boldsymbol{a}^{(M)}(n)$ results from filtering the driving signals $\boldsymbol{d}^{(R)}(n)$ with the idealized $M \times R$ matrix $\boldsymbol{F}(n)$ of free-field impulse responses.

The error $\boldsymbol{e}^{(M)}(n)$ between free-field propagation $\boldsymbol{a}^{(M)}(n)$ and the actual microphone signals $\boldsymbol{l}^{(M)}(n)$ describes the deviation of the rendered wave field from the reflection-free case. It is used to adapt the matrix of compensation filters $\boldsymbol{C}_{\mathrm{ARC}}(n)$. After convergence of $\boldsymbol{C}_{\mathrm{ARC}}(n)$, the response of the pre-filtered reproduction system in the listening room produces the desired free-field sound field.

The shaded region in Fig. 15.3 denotes the acoustic character of the signals and systems between the loudspeakers and the microphones. The other signals/systems are either electrical or digital. The required conversion between analog and digital signals at various locations is not shown explicitly. As mentioned before, a reasonable control and analysis within an enlarged listening area and temporal bandwidth can only be gained with a high number of synthesis and analysis channels. This results in a massive multichannel adaptation problem.

15.2.4 Multichannel Active Noise Control

Acoustic active noise control aims at suppressing an undesired noise source based on the principle of superposition using appropriately driven loudspeakers [26]. The loudspeaker driving signals are typically derived from acoustic measurements and adaptive algorithms in order to cope for arbitrary noise fields and the time varying nature of acoustics. For some applications a large zone where the noise is canceled (quiet zone) is desirable. As for active room compensation, a suitable combination of massive multichannel sound reproduction and WFA will result in an enlarged quiet zone for ANC as will be illustrated.

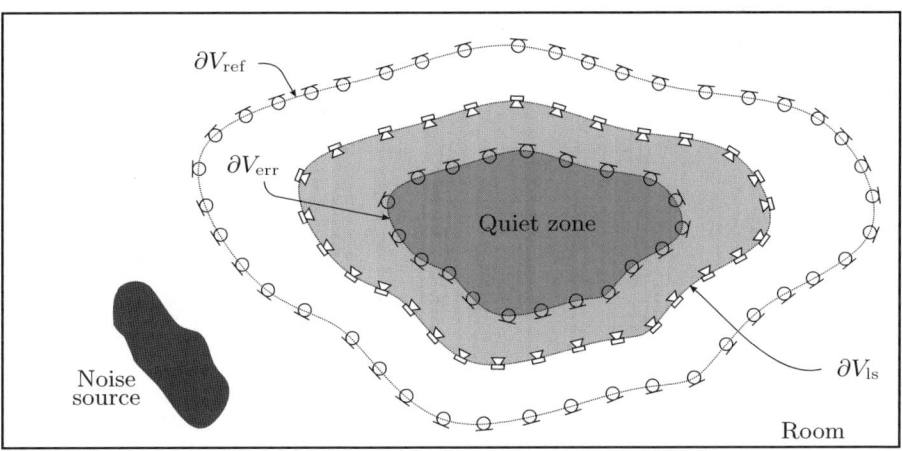

Fig. 15.4. Block diagram illustrating massive multichannel ANC based on the Kirchhoff-Helmholtz integral.

Fig. 15.4 illustrates the setup considered in the following. A distributed noise source located outside the ANC system emits noise that is scattered at the walls of a reverberant room. The resulting complex noise field and its residual (superposition of noise field and the wave field generated by the loudspeakers) is analyzed by reference microphones and error microphones arranged on the closed contours ∂V_{ref} and ∂V_{err}, respectively. The wave field is controlled by the loudspeakers arranged on the contour ∂V_{ls}. For continuous microphone and loudspeaker distributions a perfect analysis and control of the wave field is possible according to the Kirchhoff-Helmholtz integral (Eq. 15.1). The result will be an enlarged quiet zone which fully covers V_{err}. For a practical implementation of the proposed ANC scheme spatial sampling of the continuous microphone and loudspeaker distributions is required. As a consequence, active compensation of the noise field is limited by spatial aliasing artifacts. Therefore, for a reasonable noise suppression and size of the quiet zone a high number of microphones and loudspeakers is required.

As for active listening room compensation, the realization of an ANC system is based on pre-filtering the loudspeaker driving signals with suitable compensation filters. Due to the time-varying nature of the acoustic environment, these filters have to be derived by adaptive algorithms. The discrete time and space block diagram shown in Fig. 15.5 illustrates a generic multichannel ANC system without feedback paths, i.e. we assume that the effects of feedback from the loudspeakers to the reference microphones can be neglected.

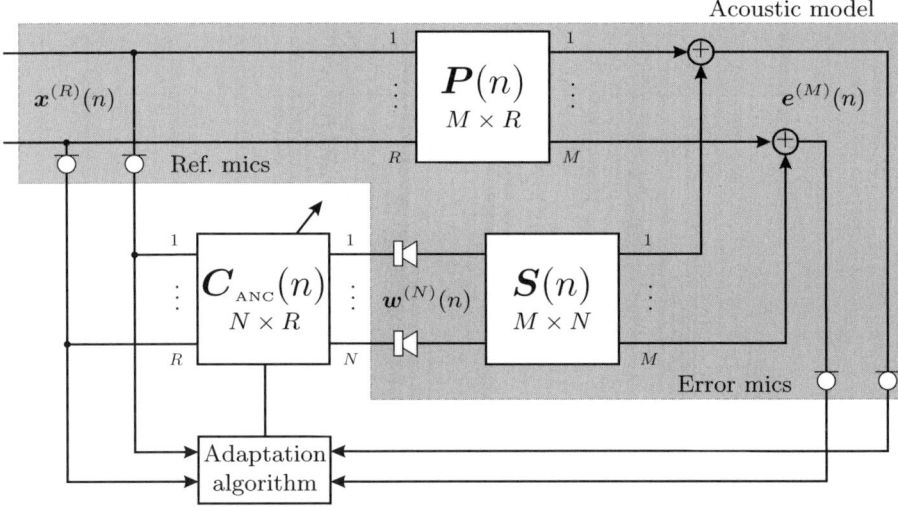

Fig. 15.5. Block diagram illustrating the generic feedforward broadband multichannel ANC system without feedback paths.

The matrices of impulse responses $\boldsymbol{P}(n)$, $\boldsymbol{S}(n)$ and $\boldsymbol{C}_{\text{ANC}}(n)$ characterize the primary paths between the reference microphones and the error microphones, the secondary paths between the secondary sources and the error microphones, and the compensation filters used to generate the secondary source signals from the reference signals. The signals from the reference microphones are combined into the vector $\boldsymbol{x}^{(R)}(n)$. The signals from the error microphones are denoted by the vector $\boldsymbol{e}^{(M)}(n)$ and the secondary source driving signals by $\boldsymbol{w}^{(N)}(n)$. The shaded region denotes again the domain of acoustical signals and systems.

The reference signal is filtered by the compensation filters and reproduced by the secondary sources in order to achieve the desired goal of a quiet zone. The primary path response $\boldsymbol{P}(n)$ and secondary path response $\boldsymbol{S}(n)$ are in general not known a-priori and may change over time due to a varying acoustic environment. This calls for an adaptive computation of the compensation filters. The fundamental problem of adaptive ANC is to compute compensation filters $\boldsymbol{C}_{\text{ANC}}(n)$ on the basis of the reference $\boldsymbol{x}^{(R)}(n)$ and error signals $\boldsymbol{e}^{(M)}(n)$ such that the error signal is minimized with respect to a given cost function. For a high number of analysis and synthesis channels this will be subject to fundamental problems, as outlined in Sec. 15.3.4.

15.2.5 Unified Representation of Spatio-Temporal Adaptive Filtering Problems

The preceding two subsections introduced multichannel adaptive systems for active listening room compensation and active noise control. Inspection of the block diagrams for ARC and ANC shown in Figs. 15.3 and 15.5 reveals a strong similarity between both approaches. The structure of both block diagrams is equal, however the meanings of the signals and systems and the ingredients of the shaded region differ. The following section will derive a unified representation of both multichannel adaptation problems. For this purpose the similarities and differences between both approaches will be discussed first.

Besides the matrix of compensation filters, the block diagrams for ARC and ANC consist of two matrices that model acoustic propagation paths. For active listening room compensation $\boldsymbol{R}(n)$ and $\boldsymbol{F}(n)$ describe the room and desired free-field characteristics, respectively. For ANC, $\boldsymbol{S}(n)$ and $\boldsymbol{P}(n)$ characterize the secondary and primary path characteristics. Both, the room response $\boldsymbol{R}(n)$ and the secondary path $\boldsymbol{S}(n)$ describe the propagation of sound from the secondary sources to the analysis/error microphones. Hence, both matrices represent the same propagation paths and can be treated as equal for a unified description. However, the matrix $\boldsymbol{F}(n)$ for ARC and the matrix $\boldsymbol{P}(n)$ for ANC represent different acoustic paths. The matrix $\boldsymbol{F}(n)$ models the desired propagation characteristics from the loudspeakers to the analysis microphones, while $\boldsymbol{P}(n)$ characterizes the acoustic transmission paths from the reference to the error microphones. Hence, $\boldsymbol{F}(n)$ represents a digital system, while $\boldsymbol{P}(n)$ represents an acoustic system (as depicted by the shaded regions in Figs. 15.3 and 15.5).

Both, ARC and ANC can be classified as pre-equalization problems. In both cases, a multichannel acoustic system is pre-equalized by a multiple-input/multiple-output compensation filter. However, the input signal to the compensation filter is different. For ARC the input signal is the secondary source driving signal $\boldsymbol{d}^{(R)}(n)$, while for ANC the input is the signal of the reference microphones $\boldsymbol{x}^{(R)}(n)$. For active room compensation, $\boldsymbol{a}^{(M)}(n)$ and $\boldsymbol{l}^{(M)}(n)$ represent the desired and the reproduced wave field at the analysis positions/microphones. The error $\boldsymbol{e}^{(M)}(n)$ denotes the difference between both signals. For an ANC system the equivalent signals to $\boldsymbol{a}^{(M)}(n)$ and $\boldsymbol{l}^{(M)}(n)$ are not directly accessible, only the error $\boldsymbol{e}^{(M)}(n)$ can be measured. For active noise control $\boldsymbol{a}^{(M)}(n)$ and $\boldsymbol{l}^{(M)}(n)$ would represent the noise field without active noise control and the anti-noise generated by the loudspeakers at the error microphones. In both approaches, ARC and ANC, the error signals are used to drive the adaptation of the pre-equalization filters $\boldsymbol{C}_{\mathrm{ARC}}(n)$ and $\boldsymbol{C}_{\mathrm{ANC}}(n)$, respectively. However, the construction of the error signal is different in both approaches. For ARC the error signal denotes the difference between $\boldsymbol{a}^{(M)}(n)$ and $\boldsymbol{l}^{(M)}(n)$, while in ANC an acoustic superposition of these signals takes place. In order to stay consistent with the literature on adaptive filtering we will assume that the error $\boldsymbol{e}^{(M)}(n)$ denotes the difference between the signals $\boldsymbol{a}^{(M)}(n)$ and $\boldsymbol{l}^{(M)}(n)$. As a consequence, the pre-filter in the unified representation for ANC will become the ANC pre-filter with negative sign $\boldsymbol{C}(n) = -\boldsymbol{C}_{\mathrm{ANC}}(n)$. This is in conjunction with the literature on ANC [26].

Beside the differences discussed so far, the block diagrams for ARC and ANC are structurally equivalent. In order to derive a unified representation the highlighting of the acoustical systems will be dropped in the remainder of this chapter. Fig. 15.6 illustrates the unified representation of the block diagrams of multichannel active listening room compensation and active noise control.

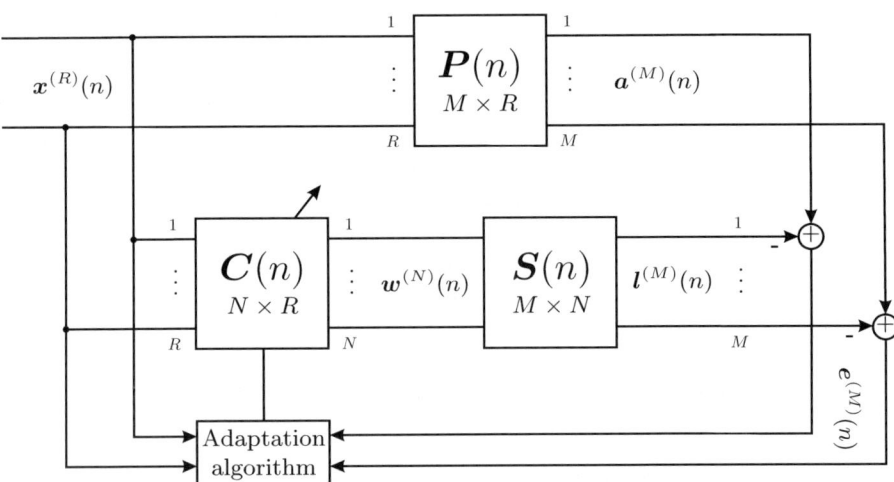

Fig. 15.6. Unified representation of multichannel active listening room compensation and active noise control.

The nomenclature of the ANC block diagram shown in Fig. 15.5 was used for the unified representation. In addition, the signals $\boldsymbol{a}^{(M)}(n)$ and $\boldsymbol{l}^{(M)}(n)$ where introduced since they play a prominent role for active room compensation and the derivation of adaptive algorithms. Note, that the unified scheme does not cover the combination of ARC and ANC, it represents either one of them in a unified manner.

Fig. 15.6 illustrates a generic multichannel pre-equalization or multichannel inverse adaptive filtering problem. The matrices of impulse responses $\boldsymbol{P}(n)$, $\boldsymbol{S}(n)$ and $\boldsymbol{C}(n)$ describe discrete linear multiple-input/multiple-output (MIMO) systems. Hence, the underlying problems of ARC and ANC can be interpreted as MIMO pre-equalization or inverse filtering problems. Due to the time-varying nature of acoustic environments, an adaptive computation of the pre-equalization is favorable as will be discussed in Sec. 15.3.

15.2.6 Frequency-Domain Notation

For the temporal frequency-domain description of the signals and systems used, the discrete time Fourier transform (DTFT) [30] is used. The DTFT and its inverse for the multichannel discrete input signal $x_r(n)$ are defined as

$$\underline{x}_r(\omega) = \sum_{n=-\infty}^{\infty} x_r(n)\, e^{-j\omega n T_s}, \tag{15.4a}$$

$$x_r(n) = \frac{T_s}{2\pi} \int_0^{2\pi/T_s} \underline{x}_r(\omega)\, e^{j\omega n T_s}\, d\omega, \tag{15.4b}$$

where T_s denotes the temporal sampling interval. Frequency-domain quantities are underlined, the temporal frequency is denoted by ω. Analogous definitions apply to the other signals $w_p(n)$, $l_m(n)$, $a_m(n)$ and $e_m(n)$. The vector of input signals $\boldsymbol{x}^{(R)}(n)$ is transformed into the temporal frequency domain by transforming each element $x_r(n)$ separately using the DTFT according to Eq. 15.4a. The resulting vector is denoted by $\underline{\boldsymbol{x}}^{(R)}(\omega)$. Vectors and matrices of frequency-domain signals are underlined in the following. Analogous definitions as for $\boldsymbol{x}^{(R)}(n)$ apply to the other signals used.

The matrix of primary path impulse responses $\boldsymbol{P}(n)$ is transformed into the temporal frequency-domain in the same way as $\boldsymbol{x}^{(R)}(n)$. The resulting matrix in the frequency domain is referred to as *primary path transfer matrix* and denoted by $\underline{\boldsymbol{P}}(\omega)$. The matrices $\boldsymbol{S}(n)$ and $\boldsymbol{C}(n)$ are transformed analogously. The transfer matrix $\underline{\boldsymbol{S}}(\omega)$ is referred to as *secondary path transfer matrix*.

15.3 Computation of Compensation Filters

This section reviews traditional techniques to compute the compensation (pre-equalization) filters for the MIMO inverse filtering problem. First, a rough classification of the known algorithms will be given.

15.3.1 Classification of Algorithms

A wide variety of solutions has been developed in the past for the computation of pre-equalization filters. The solutions can be roughly classified into three classes depending on the algorithms and the underlying preliminaries used to compute the compensation filters:

1. non-iterative algorithms,
2. iterative algorithms, and
3. adaptive algorithms.

The first two classes of algorithms assume that they have direct access to the matrices $\boldsymbol{P}(n)$ and $\boldsymbol{S}(n)$ characterizing the acoustic propagation paths. Typically the acoustic propagation paths have been measured at some fixed time by impulse response measurements. The task of computing the compensation filters is then formulated as a matrix inversion problem which is solved with non-iterative or iterative methods. Hence, the methods inherently assume that the acoustical environment is time-invariant which is not necessarily the case for acoustic systems, as discussed in Sec. 15.2.3. If the compensation filters are not adapted to changes in the acoustic environment, the error introduced might be higher than without using ARC or ANC at all [29,32]. Additionally, explicit measurement of the acoustic transmission paths might not be desirable in all applications e. g. due to the unpleasantness of typical measurement signals.

This calls for an adaptive computation of the compensation filters on basis of an analysis of the wave field. The third class of algorithms, the adaptive ones, needs no direct access to the matrices $\boldsymbol{P}(n)$ and $\boldsymbol{S}(n)$. These algorithms derive suitable compensation filters on basis of the error $\boldsymbol{e}^{(M)}(n)$ and the input signal $\boldsymbol{x}^{(R)}(n)$. They inherently cope with the time-variance of the propagation medium and changes in the acoustic environment. Hence, adaptive algorithms are preferable for ARC and ANC applications.

Please note, that adaptive algorithms can also be used to compute the compensation filters iteratively. However, not all iterative algorithms are suited for adaptive solutions of the pre-equalization problem. In order to gain more insight into the adaptive solution of the pre-equalization problem the idealized case will be regarded first.

15.3.2 Ideal Solution

The non-adaptive solution of the pre-equalization problem depicted in Fig. 15.6 is derived on basis of a frequency-domain description of the respective signals and systems. The error $\boldsymbol{e}^{(M)}(n)$ is defined as the difference between the signals $\boldsymbol{a}^{(M)}(n)$ and $\boldsymbol{l}^{(M)}(n)$. In the frequency domain the error $\underline{\boldsymbol{e}}^{(M)}(\omega)$ can be derived as

$$\begin{aligned}\underline{\boldsymbol{e}}^{(M)}(\omega) &= \underline{\boldsymbol{a}}^{(M)}(\omega) - \underline{\boldsymbol{l}}^{(M)}(\omega) \\ &= \underline{\boldsymbol{P}}(\omega)\,\underline{\boldsymbol{x}}^{(R)}(\omega) - \underline{\boldsymbol{S}}(\omega)\,\underline{\boldsymbol{C}}(\omega)\,\underline{\boldsymbol{x}}^{(R)}(\omega)\,.\end{aligned} \quad (15.5)$$

Optimal pre-equalization is obtained by minimizing the error $\underline{e}^{(M)}(\omega)$ jointly for all M analysis positions $\underline{e}^{(M)}(\omega) \rightarrow \mathbf{0}$. For sufficient excitation $\underline{x}^{(R)}(\omega)$, the least-squares error (LSE) solution to this problem, with respect to the compensation filters, is given as [16, 18]

$$\underline{C}(\omega) = \underline{S}^{+}(\omega) \underline{P}(\omega) , \quad (15.6)$$

where $\underline{S}^{+}(\omega)$ denotes the Moore-Penrose pseudoinverse of $\underline{S}(\omega)$. Successful computation of the pseudoinverse is subject to the conditioning of the matrix $\underline{S}(\omega)$. A regularization might be required. Conditions for the exact solution of the pre-equalization problem in the time-domain are given by the multiple-input/multiple-output inversion theorem (MINT) [27].

15.3.3 Adaptive Solution

The basis of most adaptive algorithms is the so-called *normal equation*. The derivation of the normal equation for the generic pre-equalization problem introduced in Sec. 15.2.5 is briefly reviewed in the following section, as it is somewhat more involved than the post-equalization, i.e., the classical inverse modeling problem [18]. A detailed discussion for acoustic MIMO systems can be found, e.g. in [14, 43]. The solution of the normal equation with the filtered-x recursive least-squares algorithm (X-RLS) is additionally shown in the remaining part of the section.

The adaptation of the compensation filters $\boldsymbol{C}(n)$ is driven by the vector of error signals $\boldsymbol{e}^{(M)}(n)$. Each component $e_m(n)$ is determined by the components of the signal vector $\boldsymbol{x}^{(R)}(n)$ and the respective impulse response matrices as

$$e_m(n) = \sum_{r=1}^{R} \sum_{k} p_{m,r}(k) \, x_r(n-k)$$
$$- \sum_{p=1}^{N} \sum_{r=1}^{R} \sum_{k_1} \sum_{k_2} s_{m,p}(k_1) \, \hat{c}_{p,r}(k_2, n) \, x_r(n - k_1 - k_2) , \quad (15.7)$$

where the error $e_m(n)$ depends, besides the input signal and the impulse responses, on the estimates $\hat{c}_{p,r}(k,n)$ (at time index n) of the ideal compensation filters $c_{p,r}(n)$. With this error signal, the following cost function is defined

$$\xi(\hat{c}, n) = \sum_{\kappa=0}^{n} \lambda^{n-\kappa} \sum_{m=1}^{M} \left| e_m(\kappa) \right|^2 , \quad (15.8)$$

where $0 < \lambda \leq 1$ denotes an exponential weighting factor. The cost function can be interpreted as the time and analysis position averaged energy of the error $e_m(n)$. The optimal filter coefficients in the mean-squared error (MSE) sense are found by setting the gradient of the cost function $\xi(\hat{c}, n)$ with respect to the estimated filter coefficients $\hat{c}(n)$ to zero. The normal equation is derived by expressing the error $e^{(M)}(n)$ in terms of the filter coefficients, introducing the result into the cost function (Eq. 15.8) and calculating its gradient. To express the normal equation as a linear system of equations, the estimated compensation filters are arranged in vector form

$$\hat{c}_{p,r}(n) = \left[\hat{c}_{p,r}(0,n),\, \hat{c}_{p,r}(1,n),\, \cdots,\, \hat{c}_{p,r}(N_c-1,n)\right]^{\mathrm{T}}, \tag{15.9}$$

$$\hat{c}_p^{(R)}(n) = \left[\hat{c}_{p,1}^{\mathrm{T}}(n),\, \hat{c}_{p,2}^{\mathrm{T}}(n),\, \cdots,\, \hat{c}_{p,R}^{\mathrm{T}}(n)\right]^{\mathrm{T}}, \tag{15.10}$$

$$\hat{c}(n) = \left[\left(\hat{c}_1^{(R)}(n)\right)^{\mathrm{T}},\, \left(\hat{c}_2^{(R)}(n)\right)^{\mathrm{T}},\, \cdots,\, \left(\hat{c}_N^{(R)}(n)\right)^{\mathrm{T}}\right]^{\mathrm{T}}, \tag{15.11}$$

where N_c denotes the number of filter coefficients. Typically this number has to be chosen higher than the number of coefficients N_s of the secondary path impulse responses $s_{m,p}(n)$ since we are dealing with an inverse identification problem [27]. The resulting normal equation of the inverse filtering problem is given as

$$\hat{\boldsymbol{\Phi}}_{xx}(n)\, \hat{c}(n) = \hat{\boldsymbol{\Phi}}_{xa}(n)\,. \tag{15.12}$$

The $NRN_c \times N_cRN$ matrix $\hat{\boldsymbol{\Phi}}_{xx}(n)$ is the time and analysis position-averaged auto- and cross-correlation matrix of the filtered input signals

$$\begin{aligned}\hat{\boldsymbol{\Phi}}_{xx}(n) &= \sum_{\kappa=0}^{n} \lambda^{n-\kappa}\, \boldsymbol{X}_S(\kappa)\, \boldsymbol{X}_S^{\mathrm{T}}(\kappa) \\ &= \lambda\, \hat{\boldsymbol{\Phi}}_{xx}(n-1) + \boldsymbol{X}_S(n)\, \boldsymbol{X}_S^{\mathrm{T}}(n)\,,\end{aligned} \tag{15.13}$$

where $\boldsymbol{X}_S(n)$ denotes the matrix of filtered input signals. The matrix of filtered input signals is given by consecutively packing all $N \cdot R$ combinations resulting from filtering the input signals $x_r(n)$ with all inputs of the secondary path responses $s_{m,p}(n)$ for all possible combination of p and r

$$x_{m,p,r}(n) = \sum_k x_r(n-k)\, s_{m,p}(n)\,, \tag{15.14}$$

$$\boldsymbol{x}_{m,p,r}(n) = \left[x_{m,p,r}(n),\, x_{m,p,r}(n-1),\, \cdots,\, x_{m,p,r}(n-N_c+1)\right]^{\mathrm{T}}, \tag{15.15}$$

$$\boldsymbol{x}_{m,p}^{(R)}(n) = \left[\boldsymbol{x}_{m,p,1}^{\mathrm{T}}(n),\, \boldsymbol{x}_{m,p,2}^{\mathrm{T}}(n),\, \cdots,\, \boldsymbol{x}_{m,n,R}^{\mathrm{T}}(n)\right]^{\mathrm{T}}, \tag{15.16}$$

$$\boldsymbol{x}_m^{(N,\,R)}(n) = \left[\left(\boldsymbol{x}_{m,1}^{(R)}(n)\right)^{\mathrm{T}},\, \left(\boldsymbol{x}_{m,2}^{(R)}(n)\right)^{\mathrm{T}},\, \cdots,\, \left(\boldsymbol{x}_{m,N}^{(R)}(n)\right)^{\mathrm{T}}\right]^{\mathrm{T}}, \tag{15.17}$$

$$\boldsymbol{X}_S(n) = \left[\boldsymbol{x}_1^{(N,\,R)}(n),\, \boldsymbol{x}_2^{(N,\,R)}(n),\, \cdots,\, \boldsymbol{x}_M^{(N,\,R)}(n)\right]. \tag{15.18}$$

The complex structure of the matrix $\boldsymbol{X}_S(n)$ and consequently of the correlation matrix $\hat{\boldsymbol{\Phi}}_{xx}(n)$ results from rearranging Eq. 15.7 in order to isolate the filter coefficients $\hat{c}(n)$ in the normal equation 15.12.

The $NRN_c \times 1$ vector $\hat{\boldsymbol{\Phi}}_{xa}(n)$ can be interpreted as the time and analysis position-averaged cross-correlation vector between the filtered input signals and the desired signals which is defined as

$$\hat{\boldsymbol{\Phi}}_{xa}(n) = \sum_{\kappa=0}^{n} \lambda^{n-\kappa} \, \boldsymbol{X}_S(\kappa) \, \boldsymbol{a}^{(M)}(\kappa)$$
$$= \lambda \, \hat{\boldsymbol{\Phi}}_{xa}(n-1) + \boldsymbol{X}_S(n) \, \boldsymbol{a}^{(M)}(n) \, . \tag{15.19}$$

The optimal pre-equalization filter with respect to the cost function (Eq. 15.8) is given by solving the normal equation 15.12 with respect to the filter coefficients $\hat{\boldsymbol{c}}(n)$. The normal equation is typically not solved in a direct manner but by recursive updates of the filter coefficients. A recursive update equation for the filter coefficients can be derived from the normal equation 15.12, the recursive definitions of the correlation matrices given by Eqs. 15.13 and 15.19, and the definition of the error signal $\boldsymbol{e}^{(M)}(n)$ as follows

$$\hat{\boldsymbol{c}}(n) = \hat{\boldsymbol{c}}(n-1) - \hat{\boldsymbol{\Phi}}_{xx}^{-1}(n) \, \boldsymbol{X}_S(n) \, \boldsymbol{e}^{(M)}(n) \, . \tag{15.20}$$

Eq. 15.20 together with the recursive definition of the correlation matrix $\hat{\boldsymbol{\Phi}}_{xx}(n)$ given by Eq. 15.13 constitutes the basis of the X-RLS algorithm. The inverse $\hat{\boldsymbol{\Phi}}_{xx}^{-1}(n)$ of the auto-correlation matrix is typically computed in a recursive fashion by applying the matrix inversion lemma [18].

Inspection of Eq. 15.20 and Eq. 15.13 reveals that the X-RLS algorithm requires access to the input signal $\boldsymbol{x}^{(R)}(n)$, the error signal $\boldsymbol{e}^{(M)}(n)$ and the secondary path responses $s_{m,p}(n)$. The X-RLS algorithm deviates from the standard RLS algorithm by using a filtered version of the input signal for the adaptation. The calculation of the filtered input signals requires knowledge of the secondary path response $\boldsymbol{S}(n)$, which is in general not known a-priori and is potentially time-variant. Hence, the secondary path characteristics have to be identified additionally using a multichannel RLS algorithm. Its normal equation can be derived in the same manner as shown above for the compensation filters. It is given by

$$\hat{\boldsymbol{\Phi}}_{ww}(n) \, \hat{\boldsymbol{s}}(n) = \hat{\boldsymbol{\Phi}}_{wl}(n) \, , \tag{15.21}$$

where $\hat{\boldsymbol{\Phi}}_{ww}(n)$ denotes the auto-correlation matrix of the filtered loudspeaker driving signals $\boldsymbol{w}^{(N)}(n)$, $\hat{\boldsymbol{\Phi}}_{wl}(n)$ the cross-correlation matrix between the filtered loudspeaker driving signals $\boldsymbol{w}^{(N)}(n)$ and the analysis signals $\boldsymbol{l}^{(M)}(n)$, and $\hat{\boldsymbol{s}}(n)$ the estimated coefficients of the secondary path impulse responses.

15.3.4 Problems of the Adaptive Solution

Three fundamental problems of massive multichannel adaptive pre-equalization can be identified from the normal equation 15.12 and the definition 15.13 of the auto-correlation matrix $\hat{\boldsymbol{\Phi}}_{xx}(n)$. These are:

1. non-uniqueness of the solution,
2. ill-conditioning of the auto-correlation matrix $\hat{\boldsymbol{\Phi}}_{xx}(n)$, and
3. computational complexity for massive MIMO systems.

The first problem is related to minimization of the cost function $\xi(\hat{\boldsymbol{c}}, n)$. Minimization of the cost function $\xi(\hat{\boldsymbol{c}}, n)$ may not provide the optimal solution in terms of identifying the inverse system to the secondary path transfer matrix. Depending on the input signals $\boldsymbol{x}^{(R)}(n)$ there may be multiple solutions for $\hat{\boldsymbol{c}}(n)$ that minimize $\xi(\hat{\boldsymbol{c}}, n)$ [6]. This problem is often referred to as *non-uniqueness problem*.

The second and the third fundamental problems are related to the solution of the normal equation 15.12. The normal equation has to be solved with respect to the coefficients $\hat{\boldsymbol{c}}(n)$ of the compensation filter. However, the auto-correlation matrix $\hat{\boldsymbol{\Phi}}_{xx}(n)$ is typically ill-conditioned for the considered multichannel reproduction scenarios [34]. The filtered input signals $\boldsymbol{X}_S(n)$ will contain cross-channel (spatial) correlations due to the deterministic nature of most auralization algorithms. Also temporal correlations may be present for typical virtual source signals. Besides the ill-conditioning also the dimensionality of the auto-correlation matrix $\hat{\boldsymbol{\Phi}}_{xx}(n)$ poses problems. Massive multichannel systems exhibit a high number of reproduction channels and additionally the length N_c of the inverse filter has to be chosen quite long for a suitable suppression of reflections [27]. As a consequence, the derivation of the compensation filters will get computationally very demanding [7].

The same problems as discussed above apply to the identification of the secondary path transfer matrix using a multichannel RLS algorithm [6].

15.4 Eigenspace Adaptive Filtering

Eigenspace adaptive filtering (EAF) provides a generic framework for MIMO pre-equalization which explicitly solves the second problem discussed in Sec. 15.3.4 by utilizing signal and system transformations. The conditioning of $\hat{\boldsymbol{\Phi}}_{xx}(n)$ is then highly alleviated by removing all cross-channel correlations. In the following a brief review of EAF based on [40] will be given. The basic idea is to perform a decoupling of the MIMO systems $\underline{\boldsymbol{S}}(\omega)$ and $\underline{\boldsymbol{P}}(\omega)$ resulting in a decoupling of the MIMO adaptation problem and the correlation matrix $\hat{\boldsymbol{\Phi}}_{xx}(n)$.

15.4.1 Generalized Singular Value Decomposition

The generalized singular value decomposition (GSVD) [16] will be used to derive the desired decoupling. The GSVD for the matrices $\underline{\boldsymbol{S}}(\omega)$ and $\underline{\boldsymbol{P}}(\omega)$ is given as

$$\underline{\boldsymbol{S}}(\omega) = \underline{\boldsymbol{X}}(\omega)\, \underline{\widetilde{\boldsymbol{S}}}(\omega)\, \underline{\boldsymbol{V}}^{\mathrm{H}}(\omega)\,, \tag{15.22a}$$

$$\underline{\boldsymbol{P}}(\omega) = \underline{\boldsymbol{X}}(\omega)\, \underline{\widetilde{\boldsymbol{P}}}(\omega)\, \underline{\boldsymbol{U}}^{\mathrm{H}}(\omega)\,. \tag{15.22b}$$

The matrices $\underline{X}(\omega)$, $\underline{V}(\omega)$ and $\underline{U}(\omega)$ are unitary matrices. The matrix $\underline{X}(\omega)$ is the generalized singular matrix, the matrices $\underline{V}(\omega)$ and $\underline{U}(\omega)$ the respective right singular matrices of $\underline{S}(\omega)$ and $\underline{P}(\omega)$. The matrices $\underline{\widetilde{S}}(\omega)$ and $\underline{\widetilde{P}}(\omega)$ are diagonal matrices constructed from the singular values of $\underline{S}(\omega)$ and $\underline{P}(\omega)$. The diagonal matrix $\underline{\widetilde{S}}(\omega)$ is defined as

$$\underline{\widetilde{S}}(\omega) = \mathrm{diag}\Big\{ \big[\widetilde{S}_1(\omega), \widetilde{S}_2(\omega), \cdots, \widetilde{S}_M(\omega)\big]^\mathrm{T} \Big\}, \qquad (15.23)$$

where $\widetilde{S}_1(\omega) \geq \widetilde{S}_2(\omega) \geq \cdots \geq \widetilde{S}_B(\omega) > 0$ denote the B nonzero singular values $\widetilde{S}_m(\omega)$ of $\underline{S}(\omega)$. Their total number B is given by the rank of the matrix $\underline{S}(\omega)$ with $1 \leq B \leq M$. For $B < M$ the remaining singular values $\widetilde{S}_{B+1}(\omega), \widetilde{S}_{B+2}(\omega), \cdots, \widetilde{S}_M(\omega)$ are zero. Similar definitions as given above for $\underline{\widetilde{S}}(\omega)$ apply to the matrix $\underline{\widetilde{P}}(\omega)$.

The relation given by Eq. 15.22a can be inverted by exploiting the unitary property of the joint and right singular matrices. This results in

$$\underline{\widetilde{S}}(\omega) = \underline{X}^\mathrm{H}(\omega)\,\underline{S}(\omega)\,\underline{V}(\omega) \,. \qquad (15.24)$$

Hence each matrix $\underline{S}(\omega)$ can be transformed into a diagonal matrix $\underline{\widetilde{S}}(\omega)$ using the joint and right singular matrix $\underline{X}(\omega)$ and $\underline{V}(\omega)$. A similar relation as given by Eq. 15.24 can be derived straightforwardly for $\underline{\widetilde{P}}(\omega)$.

The GSVD transforms the matrices $\underline{S}(\omega)$ and $\underline{P}(\omega)$ into their joint eigenspace using the singular matrices $\underline{X}(\omega)$, $\underline{V}(\omega)$ and $\underline{U}(\omega)$. In general, these singular matrices depend on the matrices $\underline{S}(\omega)$ and $\underline{P}(\omega)$. Therefore, the GSVD is a *data-dependent transformation*. The transformation of $\underline{S}(\omega)$ into its diagonal representation $\underline{\widetilde{S}}(\omega)$, as given by Eq. 15.24, can be interpreted as pre- and post-filtering the secondary path transfer matrix $\underline{S}(\omega)$ by the MIMO systems $\underline{V}(\omega)$ and $\underline{X}^\mathrm{H}(\omega)$.

The SVD can be used to define the pseudoinverse $\underline{S}^+(\omega)$ of the matrix $\underline{S}(\omega)$ [18]

$$\underline{S}^+(\omega) = \underline{V}(\omega)\,\underline{\widetilde{S}}^{-1}(\omega)\,\underline{X}^\mathrm{H}(\omega) \,. \qquad (15.25)$$

Eqs. 15.22b and 15.25 can be combined to derive the following result

$$\underline{S}^+(\omega)\,\underline{P}(\omega) = \underline{V}(\omega)\,\underline{\widetilde{S}}^{-1}(\omega)\,\underline{\widetilde{P}}(\omega)\,\underline{U}^\mathrm{H}(\omega) \,, \qquad (15.26)$$

where it is assumed that $\underline{S}(\omega)$ and $\underline{P}(\omega)$ have both full rank. Eq. 15.26 will be utilized in the following to derive a decoupling of the compensation filters.

15.4.2 Eigenspace Adaptive Filtering

The GSVD provides a decomposition of the primary and secondary path, respectively. A decomposition of the compensation filters is derived by expressing the non-adaptive LSE solution $\underline{C}(\omega) = \underline{S}^+(\omega)\underline{P}(\omega)$ using the GSVD [40].

For the ideal solution introduced in Sec. 15.3.2, the decomposition of the compensation filters is given by Eq. 15.26. However, since the primary and secondary path are assumed to be unknown in the adaptive case, we combine the unknown product of the two diagonal matrices $\widetilde{\underline{P}}(\omega)$ and $\widetilde{\underline{S}}(\omega)$ into one unknown diagonal matrix $\widetilde{\underline{C}}(\omega)$ which is then derived by the adaptation algorithm. The decomposition of the compensation filters is given as

$$\underline{C}(\omega) = \underline{V}(\omega)\,\widetilde{\underline{C}}(\omega)\,\underline{U}^{\mathrm{H}}(\omega)\,, \tag{15.27}$$

where $\widetilde{\underline{C}}(\omega)$ denotes the equalization filters in the transformed domain. Introducing Eqs. 15.22 and 15.27 into Eq. 15.5, exploiting the unitary property of the singular matrices and multiplying both sides with $\underline{X}^{\mathrm{H}}(\omega)$ yields

$$\widetilde{\underline{e}}^{(M)}(\omega) = \widetilde{\underline{P}}(\omega)\,\widetilde{\underline{x}}^{(M)}(\omega) - \widetilde{\underline{S}}(\omega)\,\widetilde{\underline{C}}(\omega)\,\widetilde{\underline{x}}^{(M)}(\omega)\,, \tag{15.28}$$

where

$$\widetilde{\underline{e}}^{(M)}(\omega) = \underline{X}^{\mathrm{H}}(\omega)\underline{e}^{(M)}(\omega) \tag{15.29}$$

$$\widetilde{\underline{x}}^{(M)}(\omega) = \underline{U}^{\mathrm{H}}(\omega)\underline{x}^{(R)}(\omega) \tag{15.30}$$

denote the error and reference signal vector in the transformed domain. Eq. 15.28 states that the adaptive inverse MIMO filtering problem is decomposed into M single-channel adaptive inverse filtering problems using the GSVD. The adaptation of the compensation filters is performed independently for each of the transformed components. Introducing the proposed transformations for $\underline{e}^{(M)}(\omega)$, $\underline{x}^{(R)}(\omega)$ and $\underline{C}(\omega)$ into Fig. 15.6 yields Fig. 15.7 which illustrates the application of eigenspace adaptive filtering to the unified massive multichannel inverse filtering problem. Note that $\underline{w}^{(N)}(\omega) = \underline{V}(\omega)\widetilde{\underline{w}}^{(M)}(\omega)$.

15.4.3 Problems

Although EAF provides the optimal solution to massive multichannel ANC with respect to the desired decoupling it has two major drawbacks: (1) the GSVD is computationally quite complex and (2) the optimal transformations depend on the primary and secondary path responses which are potentially time variant. The next section will introduce wave-domain adaptive filtering (WDAF) as a practical solution to these problems.

15.5 Wave-Domain Adaptive Filtering

In the following the concept of WDAF will be introduced followed by an analytic transformation that has proven to be suitable in practical situations.

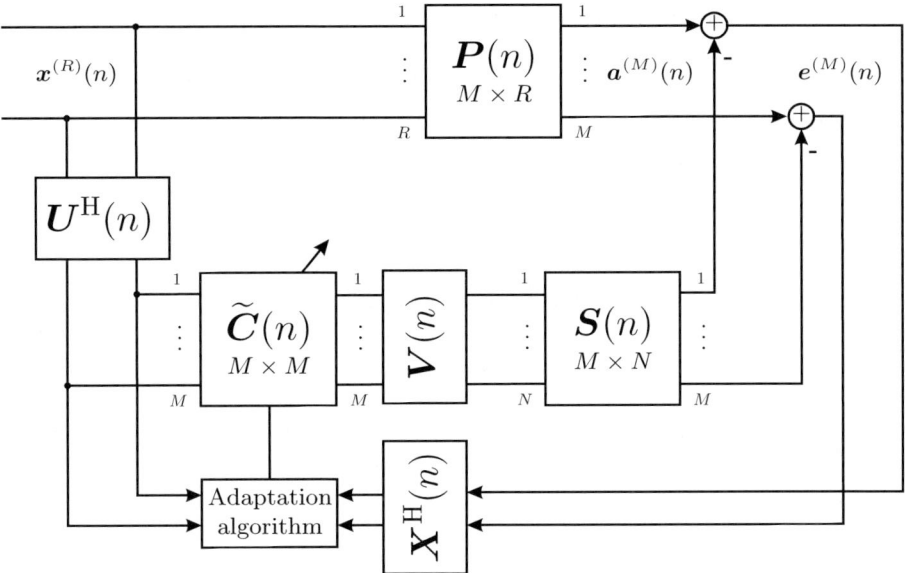

Fig. 15.7. Block diagram illustrating the eigenspace adaptive filtering approach to massive multichannel inverse filtering.

15.5.1 Concept

The concept of wave-domain adaptive filtering is based on two basic ideas:

1. Explicit consideration of the characteristics of the propagation medium for the derivation of suitable transformations, and
2. approximation of the concept of perfect decoupling of the MIMO adaptation problem.

The elements of the primary and secondary path transfer matrices $\boldsymbol{P}(\omega)$ and $\boldsymbol{S}(\omega)$ describe the sound transmission between two points in the listening room with respect to the characteristics of the propagation medium and the boundary conditions imposed by the listening room. Hence these elements have to fulfill the wave equation and the homogeneous boundary conditions imposed by the room. This knowledge can be used to construct efficient transformations for the decoupling of the generic inverse filtering problem. Since these transformations inherently have to account for the wave nature of sound in order to perform well, this approach is referred to as *wave-domain adaptive (inverse) filtering* (WDAF) and the transformed domain as *wave domain*.

The second idea is to approximate the perfect decoupling of the MIMO adaptation problem in favor of generic transformations which are to some degree independent of the listening room characteristics. In order to keep the complexity low, these generic transformations need not to strictly diagonalize the primary and secondary path matrices, but they should represent the MIMO systems with as few paths as possible.

The combination of both ideas allows to derive fixed transformations that provide nearly the same favorable properties as the optimal GSVD-based transformations used for eigenspace adaptive inverse filtering with the benefit of computational efficiency.

Based on the approach of eigenspace adaptive filtering and the above considerations a generic block diagram of the WDAF approach can be developed. Fig. 15.8 displays this generic block diagram. The signal and system transformations are performed by three generic transformations. Their structure is not limited to the MIMO systems derived from the GSVD. Transformation $\mathcal{T}_1$ transforms the driving signals into the wave domain, $\mathcal{T}_2$ inversely transforms the filtered loudspeaker driving signals from the wave domain, and $\mathcal{T}_3$ transforms the signals at the analysis positions into the wave domain. As previously for the eigenspace domain, the signals and transfer functions in the wave domain are denoted by a tilde over the respective variable, since suitable transforms will be based on the idea of a transformation into the eigenspace of the respective systems. The adaptation is performed entirely in the wave domain.

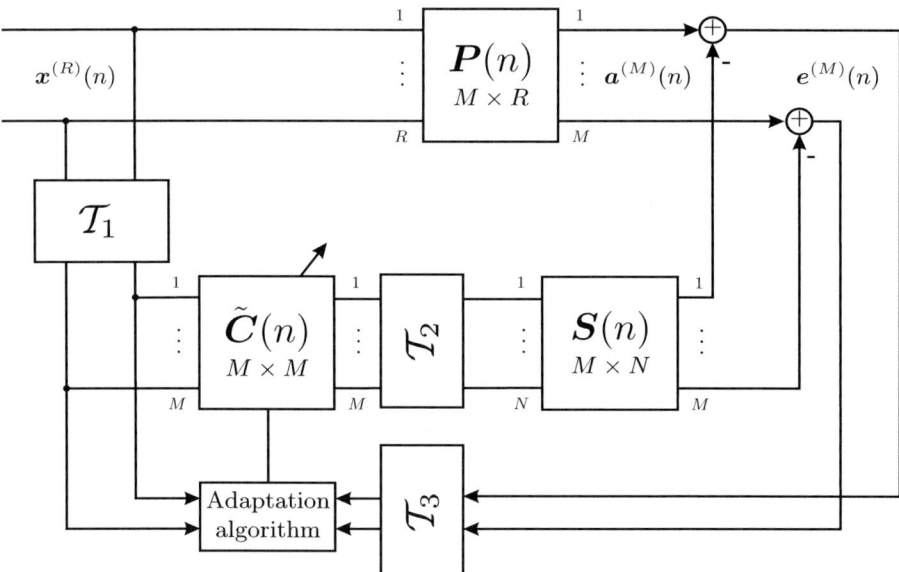

Fig. 15.8. Block diagram illustrating the wave-domain adaptive filtering approach to massive multichannel inverse filtering.

Note, that the generic block diagram depicted by Fig. 15.8 also includes the eigenspace adaptive inverse filtering approach. In this special case the transformations are given as the following MIMO systems (see Sec. 15.4.2): $\mathcal{T}_1 = \underline{\boldsymbol{U}}^\mathrm{H}(\omega)$, $\mathcal{T}_2 = \underline{\boldsymbol{V}}(\omega)$ and $\mathcal{T}_3 = \underline{\boldsymbol{X}}^\mathrm{H}(\omega)$.

The following section introduces a wave field expansion which has proven to be a good candidate as wave-domain transformation. The generic concept of eigenspace adaptive filtering requires no knowledge of the underlying physical system. However, wave field expansions are based on a specific physical scenario. For the ease of illustration we will limit ourselves to two-dimensional wave field representations for the following discussion. Typical sound reproduction systems aim at the reproduction in a plane only. The analysis of the wave fields is then performed in the reproduction plane as well. However, since the reproduction will take place in a three-dimensional environment several artifacts of two-dimensional sound reproduction and analysis have to be considered [39]. A generalization of the proposed decomposition and hence wave-domain adaptive filtering to three-dimensional sound reproduction systems can be derived straightforwardly on the basis of the presented principles.

15.5.2 The Circular Harmonics Decomposition

In order to derive the desired decoupling, the acoustic wave fields should be decomposed with respect to an orthogonal basis. In general, the basis functions required for this basis will depend on the actual characteristics of the propagation medium air with respect to sound transmission and the boundary conditions imposed by the room. However, as concluded before these characteristics are not known a-priori and may change over time.

Of special interest in the following are decompositions which are based on the fundamental free-field solutions of the wave equation. These depend on the underlying coordinate system and its dimensionality. The basis functions connected to spherical, cylindrical and Cartesian coordinates are known as spherical harmonics, cylindrical harmonics and plane waves. An arbitrary wave field can be represented by the expansion coefficients with respect to these basis functions. If two-dimensional wave fields are considered then the polar and Cartesian coordinate systems are common. Here the basis functions are circular harmonics and (two-dimensional) plane waves.

The circular harmonics decomposition of an arbitrary wave field is given by [28, 55]

$$P(\alpha, r, \omega) = \sum_{\nu=-\infty}^{\infty} \left(\check{P}^{(1)}(\nu, \omega) H_\nu^{(1)}\left(\left|\frac{\omega}{c}\right| r\right) e^{j\nu\alpha} + \check{P}^{(2)}(\nu, \omega) H_\nu^{(2)}\left(\left|\frac{\omega}{c}\right| r\right) e^{j\nu\alpha} \right),$$
(15.31)

where $H_\nu^{(1),(2)}(\cdot)$ denotes the ν-th order Hankel function of first/second kind and ν the angular frequency. The infinite sum over ν can be interpreted as Fourier series with respect to the angle α. The coefficients $\check{P}^{(1),(2)}(\nu, \omega)$ are referred to as circular harmonics expansion coefficients in the following and will be denoted by a breve over the respective variable. The Hankel function $H_\nu^{(1)}\left(\left|\frac{\omega}{c}\right|r\right)$ belongs to an incoming (converging) and $H_\nu^{(2)}\left(\left|\frac{\omega}{c}\right|r\right)$ to an outgoing (diverging) cylindrical wave [55]. Thus, the expansion coefficient $\breve{P}^{(1)}(\nu,\omega)$ describes the incoming wave field, whereas $\breve{P}^{(2)}(\nu,\omega)$ describes the outgoing wave field. According to Eq. 15.31 the total wave field is given as a superposition of incoming and outgoing contributions.

In order to get more insight into the circular harmonics expansion a closer look at the basis functions is taken. Each spatial variable has its own basis function. The angular coordinate α has an exponential function as basis. The angular basis functions exhibit a spatial selectivity in the angular coordinate. They can be interpreted as directivity patterns. Hankel functions are the basis functions of the radial coordinate r. See [1] for a detailed discussion of their properties.

Arbitrary solutions of the wave equation can be expressed alternatively as superposition of plane waves traveling into all possible directions. The plane wave expansion coefficients $\bar{P}^{(1)}(\theta,\omega)$ and $\bar{P}^{(2)}(\theta,\omega)$ describe the spectrum of incoming/outgoing plane waves with incidence angle θ. They can be derived by a plane wave decomposition [21, 43]. The plane wave expansion coefficients exhibit a direct link to the expansion coefficients in circular harmonics

$$\bar{P}^{(1)}(\theta,\omega) = \frac{4\pi}{k} \sum_{\nu=-\infty}^{\infty} j^\nu \, \breve{P}^{(1)}(\nu,\omega) \, e^{j\nu\theta} , \qquad (15.32a)$$

$$\bar{P}^{(2)}(\theta,\omega) = \frac{4\pi}{k} \sum_{\nu=-\infty}^{\infty} j^\nu \sum_{\nu=-\infty}^{\infty} j^\nu \, \breve{P}^{(2)}(\nu,\omega) \, e^{j\nu\theta} . \qquad (15.32b)$$

Eq. 15.32 states that the plane wave decomposition of a wave field is, up to the factor j^ν, given by the Fourier series of the expansion coefficients in terms of circular harmonics.

15.5.3 The Circular Harmonics Expansion Using Boundary Measurements

The Kirchhoff-Helmholtz integral implies, as outlined in Sec. 15.2.2, that measurements on the boundary of interest are suitable to characterize the wave field within the entire region. This fundamental principle can also be applied to efficiently calculate the circular harmonics decomposition coefficients from boundary measurements. Due to the underlying geometry, a circular boundary where the measurements are taken provides the natural choice for this task. The circular harmonics decomposition of a wave field could be derived in a straightforward fashion by extrapolation of the boundary measurements using the Kirchhoff-Helmholtz integral and by applying an inverse relation to Eq. 15.31. However, we will present an alternative way of deriving the desired expansion coefficients which is based on a Fourier series expansion of the measured quantities. It is closely related to the work of [21].

An acoustic pressure field can be represented as Fourier series with respect to the angle α

$$P(\alpha, r, \omega) = \sum_{\nu=-\infty}^{\infty} \mathring{P}(\nu, r, \omega) e^{j\nu\alpha} , \qquad (15.33)$$

where the Fourier series expansion coefficients of $P(\alpha, r, \omega)$ are denoted by $\mathring{P}(\nu, r, \omega)$. Comparing this Fourier series representation with the circular harmonics decomposition (Eq. 15.31) at the boundary ($r = R$) provides a relation between the expansion coefficients of these two representations

$$\mathring{P}(\nu, R, \omega) = \check{P}^{(1)}(\nu, \omega) H_\nu^{(1)}(kR) + \check{P}^{(2)}(\nu, \omega) H_\nu^{(2)}(kR) . \qquad (15.34)$$

Unfortunately, Eq. 15.34 does not provide a one-to-one relation between the Fourier series coefficients of the acoustic pressure and the expansion coefficients in terms of circular harmonics. Thus, it cannot be solved to derive the cylindrical harmonics coefficients. This conclusion is not surprising, since the Kirchhoff-Helmholtz integral states that the acoustic pressure and velocity are required on the boundary to describe the wave field within the boundary. Hence, the acoustic velocity $V_r(\alpha, r, \omega)$ in direction of inward pointing radial vector is additionally needed. The radial component of the acoustic velocity for the circular harmonics decomposition can be derived by applying Euler's relation [24] to Eq. 15.31. Expressing the radial component of the acoustic velocity as Fourier series and comparison with the circular harmonics representation provides a similar relationship as Eq. 15.34 for the Fourier series expansion coefficients of the radial particle velocity

$$j\varrho_0 c \mathring{V}_r(\nu, R, \omega) = \check{P}^{(1)}(\nu, \omega) H_\nu^{'(1)}(kR) + \check{P}^{(2)}(\nu, \omega) H_\nu^{'(2)}(kR) , \qquad (15.35)$$

where $\mathring{V}_r(\nu, R, \omega)$ denotes the Fourier series expansion coefficients of the acoustic particle velocity in radial direction and $k = |\omega/c|$ the acoustic wave number. Combination of Eq. 15.34 and Eq. 15.35 together with the definition of the Wronskian for the Hankel functions [1] allows to express the circular harmonics expansion coefficients in terms of the measured quantities as

$$\check{P}^{(1)}(\nu, \omega) = -\frac{\pi k R}{4j} \left(H_\nu^{'(2)}(kR) \mathring{P}(\nu, R, \omega) - H_\nu^{(2)}(kR) j\varrho_0 c \mathring{V}_r(\nu, R, \omega) \right) ,$$
$$(15.36a)$$

$$\check{P}^{(2)}(\nu, \omega) = -\frac{\pi k R}{4j} \left(H_\nu^{'(1)}(kR) \mathring{P}(\nu, R, \omega) - H_\nu^{(1)}(kR) j\varrho_0 c \mathring{V}_r(\nu, R, \omega) \right) .$$
$$(15.36b)$$

Eq. 15.36 provides the basis for the efficient calculation of the circular harmonics expansion coefficients from measurements taken on a circular boundary. Both the acoustic pressure and particle velocity in radial direction have to be measured to allow distinguishing between incoming and outgoing contributions. If it is known that only incoming or outgoing contributions occur in a particular scenario, then measuring only one quantity is sufficient [20, 22].

In general, the measurements will be taken on spatial discrete positions in a practical implementation. This may result in spatial aliasing if anti-aliasing conditions are not met reasonable. For discussion of spatial sampling in this context please refer to [43, 49]. A practical implementation is e.g. presented in [22].

15.6 Application of WDAF to Adaptive Inverse Filtering Problems

Sec. 15.2.5 introduced a unified representation of spatio-temporal adaptive inverse filtering problems. This representation covers amongst other problems both active listening room compensation and active noise control as introduced in Secs. 15.2.3 and 15.2.4. In order to overcome the fundamental problems of filter adaptation in the context of massive multichannel MIMO systems we introduced the versatile concept of WDAF in Sec. 15.5.1 to decouple the adaptation process. This section will specialize the WDAF concept to the problem of active listening room compensation and active noise control.

For this purpose we will first specialize the transformations $\mathcal{T}_1$ through $\mathcal{T}_3$ in Fig. 15.8. The performance of the circular harmonics decomposition to decouple the underlying multichannel adaptation problem for circular microphone arrays and typical acoustic environments has been investigated in [41]. The results presented there show that this decomposition provides a reasonable approximation of EAF. In the following, the transformations $\mathcal{T}_1$, $\mathcal{T}_2$ and $\mathcal{T}_3$ are specialized for a wave-domain representation of the respective signals in circular harmonics decomposition coefficients:

- **Transformation $\mathcal{T}_1$:**
  This transformation transforms the input signal $\boldsymbol{x}^{(R)}(n)$ into its representation in terms of circular harmonics. If the signal $\boldsymbol{x}^{(R)}(n)$ is captured/prescribed on a circular contour a spatially sampled version of Eq. 15.36 can be used for this purpose.
- **Transformation $\mathcal{T}_2$:**
  This transformation generates the loudspeaker driving signals from the filtered driving signals $\breve{\underline{\boldsymbol{w}}}^{(M)}(n)$. Eq. 15.31 together with a suitable loudspeaker selection criterion [44, 46] can be used for this purpose.
- **Transformation $\mathcal{T}_3$:**
  This transformation calculates the circular harmonics decomposition coefficients of the wave field within the listening area/quiet zone from the microphone array measurements. Again a spatially sampled version of Eq. 15.36 can be used for this purpose.

The next two subsections will illustrate the application of WDAF to active listening room compensation and active noise control for massive multichannel systems.

15.6.1 Application of WDAF to Active Listening Room Compensation

Fig. 15.3 shows the block diagram of a generic multichannel listening room compensation system. Comparison with Fig. 15.8 illustrating the WDAF concept for the unified inverse filtering problem reveals that for active listening room compensation the matrix $\boldsymbol{P}(\omega)$ equals the free-field transfer matrix $\boldsymbol{F}(\omega)$, the matrix $\boldsymbol{S}(\omega)$ equals the listening room transfer matrix $\boldsymbol{R}(\omega)$ and the input signals $\boldsymbol{x}^{(R)}(n)$ equal the secondary source driving signals $\boldsymbol{d}^{(R)}(n)$. For active room compensation only the incoming parts (see Sec. 15.5.2) of the respective wave fields are of interest. Reflections inside the listening area (e. g. by listeners) are neglected since they cannot be compensated for actively. The circular harmonics are based upon the fundamental free-field solutions of the wave equation. As a consequence only the free-field transfer matrix $\boldsymbol{F}(\omega)$ will be fully decoupled in this representation. The listening room transfer matrix $\boldsymbol{R}(\omega)$ will not be decoupled in general by this decomposition. However, in practical situations the desired decoupling is approximated reasonably well [41].

For listening room compensation both the free-field transfer matrix $\boldsymbol{F}(\omega)$ and the secondary source driving signals $\boldsymbol{d}^{(R)}(n)$ can be realized directly in the wave domain since only the room transfer matrix $\boldsymbol{R}(\omega)$ represents an actual existing acoustic model. Fig. 15.9 illustrates a simplified structure for the application of WDAF to active listening room compensation.

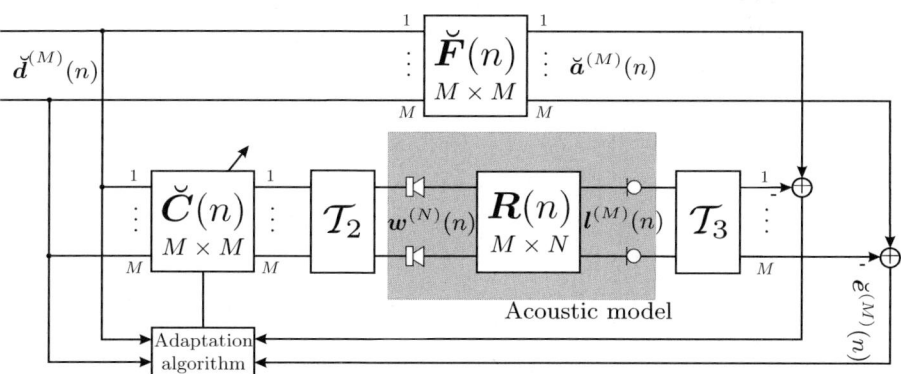

Fig. 15.9. Block diagram illustrating the application of WDAF to active listening room compensation.

The loudspeaker driving signals $\check{\boldsymbol{d}}^{(M)}(n)$ in terms of circular harmonics can be derived in closed form using an appropriate spatial model of the virtual source. Suitable models for the virtual source characteristics are point sources or plane waves. The free-field transfer matrix $\check{\boldsymbol{F}}(n)$ models the free-field propagation in terms of circular harmonics from the secondary sources to the microphone array. In the ideal case this matrix would only model the propagation delay and an additional delay to ensure the computation of causal room compensation filters. However, certain artifacts of WFS systems like e. g. amplitude errors (see [39]) should be considered in the construction of the matrix $\check{\underline{F}}(n)$ to improve the convergence of the compensation filters. A listening room compensation system based on Fig. 15.9 has been simulated for various WFS systems. The derived results prove that the WDAF concept provides fast and stable adaptation for such scenarios. A presentation of detailed results is out of the scope here, but can be found in [36–38].

15.6.2 Application of WDAF to Active Noise Control

Fig. 15.5 shows the block diagram of a generic multichannel active noise control system. Comparison with Fig. 15.8 illustrating the WDAF concept for the unified inverse filtering problem reveals that for ANC the implementation is straightforwardly given by Fig. 15.8. Both the analysis of the reference field and the residual error field can be performed efficiently using circular microphone arrays. The calculation of the circular harmonics decomposition coefficients is the given by Eq. 15.36. If the noise source is located outside of the microphone array analyzing the reference field, only incoming circular harmonics contributions will be present. The circular harmonics are based upon the fundamental free-field solutions of the wave equation. As a consequence both the primary path $\underline{P}(\omega)$ and the secondary path $\underline{R}(\omega)$ matrices will only be decoupled approximately.

Massive multichannel active noise control systems based on Fig. 15.8 have been simulated [31, 47]. The derived results prove that the WDAF concept provides fast and stable adaptation also for massive multichannel ANC. A presentation of detailed results is again out of the scope of this contribution but can be found in the above cited literature.

15.7 Conclusions

This chapter has shown that advanced multichannel adaptive methods like multichannel active listening room compensation and multichannel active noise control can be formulated as special cases of a unifying inverse filtering problem. In this general context, a method for an effective adaptation in the wave domain has been presented. It is close to an optimal solution, which performs an ideal decoupling of the multichannel case by eigenspace adaptive filtering. However, while eigenspace adaptive filtering requires data dependent transformations, wave domain adaptive filtering uses a generic data independent representation. It is obtained from an expansion of the acoustic wave equation into its eigenfunctions for the free field case.

The approach to approximate an ideal but data dependent transformation by a similar transformation with a fixed set of eigenfunction is well known from audio and video coding. There, the ideal transformation which performs an optimal decorrelation is shown to be the Karhunen-Loève transform (under certain assumptions). This optimality is based on a data dependence of the transformation matrix, which renders this approach impractical for most coding and decoding applications. Instead that data independent discrete cosine transformation is often used, which approximates the decoupling properties of the Karhunen-Loève transform sufficiently well.

This general concept presented here does not only allow to describe multichannel listening room compensation and multichannel active noise control in a common framework. It is also suitable to develop applications which suffer from room reflections and undesired noise at the same time. The close relation between multichannel active listening room compensation and multichannel active noise control can be exploited for the design of a system which implements both methods with the same set of hardware. Fig. 15.10 illustrates a combination of massive multichannel listening room compensation and active noise control. In the idealized case, the listening area will be free of contri-

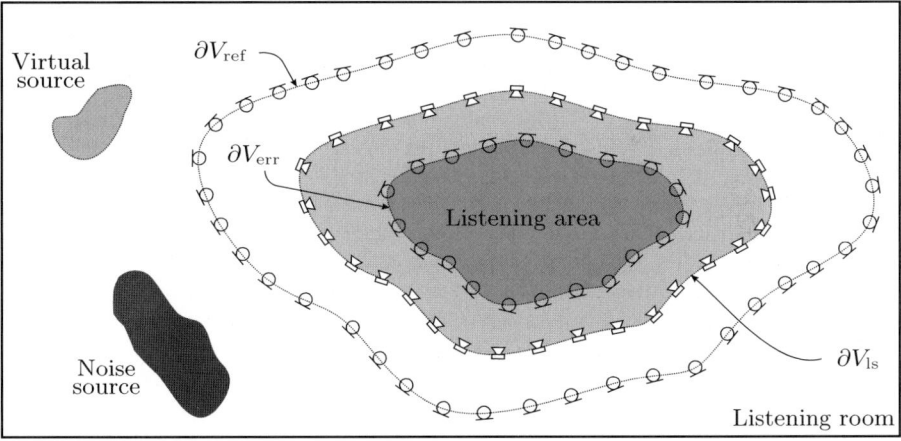

Fig. 15.10. Block diagram illustrating combined massive multichannel active room and noise control based on the Kirchhoff-Helmholtz integral.

butions from the noise source and the reproduction will not suffer from the reflections imposed by the listening room.

The application WDAF to massive multichannel systems is not only limited to inverse filtering problems. Also the adaptation process for identification problems can be highly improved in this context. The application of WDAF to acoustic echo control and interference suppression has been shown in [8–10,19].

References

1. M. Abramowitz, I. A. Stegun: *Handbook of Mathematical Functions,* New York, NY, USA: Dover Publications, 1972.
2. T. Ajdler, M. Vetterli: The plenacoustic function, sampling and reconstruction, *Proc. ICASSP '03,* **5**, 616–619, Hong Kong, 2003.
3. S. Bech: Timbral aspects of reproduced sound in small rooms I, *JASA,* **97**(3), 1717–1726, March 1995.
4. S. Bech: Timbral aspects of reproduced sound in small rooms II, *JASA,* **99**(6), 3539–3549, June 1996.
5. S. Bech: Spatial aspects of reproduced sound in small rooms, *JASA,* **103**(1), 434–445, Jan. 1998.
6. J. Benesty, D. R. Morgan, M. M. Sondhi: A better understanding and an improved solution to the specific problems of stereophonic acoustic echo cancellation, *IEEE Transactions on Speech and Audio Processing,* **6**(2), 156–165, March 1998.
7. M. Bouchard, S. Quednau: Multichannel recursive-least-squares algorithms and fast-transversal-filter algorithms for active noise control and sound reproduction systems, *IEEE Transactions on Speech and Audio Processing,* **8**(5), 606–618, September 2000.
8. H. Buchner, S. Spors, W. Kellermann: Full-duplex systems for sound field recording and auralization based on wave field synthesis, *Proc. 116th AES Convention,* Audio Engineering Society (AES), Berlin, Germany, 2004.
9. H. Buchner, S. Spors, W. Kellermann: Wave-domain adaptive filtering: Acoustic echo cancellation for full-duplex systems based on wave-field synthesis, *Proc. ICASSP '04,* **4**, 117–120, Montreal, Canada, 2004.
10. H. Buchner, S. Spors, W. Kellermann: Wave-domain adaptive filtering for acoustic human-machine interfaces based on wavefield analysis and synthesis, *Proc. EUSIPCO '04,* 1385–1388, Vienna, Austria, 2004.
11. T. Caulkins, O. Warusfel: Characterization of the reverberant sound field emitted by a wave field synthesis driven loudspeaker array, *Proc. 120th AES Convention,* Audio Engineering Society (AES), Paris, France, May 2006.
12. W. de Bruijn: *Application of Wave Field Synthesis in Videoconferencing,* PhD thesis, Delft University of Technology, 2004.
13. M. Dewhirst, S. Zielinski, P. Jackson, F. Rumsey: Objective assessment of spatial localization attributes of surround-sound reproduction systems, *Proc. 118th AES Convention,* Audio Engineering Society (AES), Barcelona, Spain, May 2005.
14. J. Garas: *Adaptive 3D Sound Systems,* Norwell, MA, USA: Kluwer Academic Publishers, 2000.
15. P.-A. Gauthier, A. Berry: Sound-field reproduction in-room using optimal control techniques: Simulations in the frequency domain, *JASA,* **117**(2), 662–678, Feb. 2005.
16. G. H. Golub. C. F. Van Loan: *Matrix Computations,* Baltimore, MD, USA: The Johns Hopkins University Press, 1989.
17. D. Griesinger: Multichannel sound systems and their interaction with the room, *Proc. 16th International Conference on Audio, Acoustics, and Small Places,* Audio Engineering Society (AES), 159–173, Oct./Nov. 1998.
18. S. Haykin: *Adaptive Filter Theory,* Englewood Cliffs, NJ, USA: Prentice-Hall, 1996.

19. W. Herbordt, S. Nakamura, S. Spors, H. Buchner, W. Kellermann: Wave field cancellation using wave-domain adaptive filtering, *Proc. HSCMA '95*, Piscataway, PA, USA, 2005.
20. E. Hulsebos: *Auralization using Wave Field Synthesis,* PhD thesis, Delft University of Technology, 2004.
21. E. Hulsebos, D. de Vries, E. Bourdillat: Improved microphone array configurations for auralization of sound fields by wave field synthesis, *Proc. 110th AES Convention*, Audio Engineering Society (AES), Amsterdam, Netherlands, May 2001.
22. E. Hulsebos, T. Schuurmanns, D. de Vries, R. Boone: Circular microphone array recording for discrete multichannel audio recording, *Proc. 114th AES Convention*, Audio Engineering Society (AES), Amsterdam, Netherlands, March 2003.
23. H. M. Jones, A. Kennedy, T. D. Abhayapala: On dimensionality of multipath fields: Spatial extent and richness, *Proc. ICASSP '02*, **3**, 2837–2840, Orlando, FL, USA, May 2002.
24. L. E. Kinsler, A. R. Frey, A. B. Coppens, J. V. Sanders: *Fundamentals of Acoustics,* 4th ed., New York, NY, USA: Wiley, 2000.
25. B. Klehs, T. Sporer: Wave field synthesis in the real world: Part 1 – In the living room, *114th AES Convention*, Audio Engineering Society (AES), Amsterdam, The Netherlands, March 2003.
26. S. M. Kuo, D. R. Morgan: *Active Noise Control Systems – Algorithms and DSP Implementations,* New York, NY, USA: Wiley, June 1996.
27. M. Miyoshi, Y. Kaneda: Inverse filtering of room acoustics, *IEEE Transactions on Acoustics, Speech, and Signal Processing*, **36**(2), 145–152, February 1988.
28. P. M. Morse, H. Feshbach: *Methods of Theoretical Physics, Part I,* New York, NY, USA: McGraw-Hill, 1953.
29. M. Omura, M. Yada, H. Saruwatari, S. Kajita, K. Takeda, F. Itakura: Compensating of room acoustic transfer functions affected by change of room temperature, *Proc. ICASSP '99*, Phoenix, AZ, USA, March 1999.
30. A. V. Oppenheim, R. W. Schafer: *Discrete-Time Signal Processing,* Englewood Cliffs, NJ, USA: Prentice-Hall, 1999.
31. P. Peretti, S. Cecchi, L. Palestini, F. Piazza: A novel approach to active noise control based on wave domain adaptive filtering, *Proc. WASPAA '07*, New Paltz, USA, Oct. 2007.
32. S. Petrausch, S. Spors, R. Rabenstein: Simulation and visualization of room compensation for wave field synthesis with the functional transformation method, *Proc. 119th AES Convention*, Audio Engineering Society (AES), New York, USA, 2005.
33. R. Rabenstein, S. Spors: Wave field synthesis techniques for spatial sound reproduction, in E. Haensler, G. Schmidt (eds.), *Topics in Acoustic Echo and Noise Control*, 517–545, Berlin, Germany: Springer, 2006.
34. M. M. Sondhi, D. R. Morgan, J. L. Hall: Stereophonic acoustic echo cancellation – an overview of the fundamental problem, *IEEE Signal Processing Letters*, **2**(8), 148–151, August 1995.
35. T. Sporer, B. Klehs: Wave field synthesis in the real world: Part 2 – In the movie theatre, *Proc. 116th AES Convention*, Audio Engineering Society (AES), Berlin, Germany, May 2005.
36. S. Spors, H. Buchner, R. Rabenstein: Adaptive listening room compensation for spatial audio systems, *Proc. EUSIPCO '04*, 1381–1384, Vienna, Austria, 2004.

37. S. Spors, H. Buchner, R. Rabenstein: Efficient active listening room compensation for wave field synthesis, *Proc. 116th AES Convention*, Audio Engineering Society (AES), Berlin, Germany, 2004.
38. S. Spors, H. Buchner, R. Rabenstein: A novel approach to active listening room compensation for wave field synthesis using wave-domain adaptive filtering, *Proc. ICASSP '04*, **4**, 29-32, Montreal, Canada, 2004.
39. S. Spors, M. Renk, R. Rabenstein: Limiting effects of active room compensation using wave field synthesis, *Proc. 118th AES Convention*, Audio Engineering Society (AES), Barcelona, Spain, May 2005.
40. S. Spors, H. Buchner, R. Rabenstein: Eigenspace adaptive filtering for efficient pre-equalization of acoustic MIMO systems, *Proc. EUSIPCO '06*, Florence, Italy, Sept. 2006.
41. S. Spors, R. Rabenstein: Evaluation of the circular harmonics decomposition for WDAF-based active listening room compensation, *Proc. 28th AES Conference: The Future of Audio Technology – Surround and Beyond*, Audio Engineering Society (AES), 134–149, Pitea, Sweden, June/July 2006.
42. S. Spors, R. Rabenstein: Spatial aliasing artifacts produced by linear and circular loudspeaker arrays used for wave field synthesis, *Proc. 120th AES Convention*, Audio Engineering Society (AES), Paris, France, May 2006.
43. S. Spors: *Active Listening Room Compensation for Spatial Sound Reproduction Systems,* PhD thesis, University of Erlangen-Nuremberg, 2006.
44. S. Spors: An analytic secondary source selection criteria for wave field synthesis, *Proc. DAGA '07*, Stuttgart, Germany, March 2007.
45. S. Spors, J. Ahrens: Analysis of near-field effects of wave field synthesis using linear loudspeaker arrays, *Proc. 30th International AES Conference*, Audio Engineering Society (AES), Saariselkä, Finland, March 2007.
46. S. Spors: Extension of an analytic secondary source selection criterion for wave field synthesis, *Proc. 123th AES Convention*, Audio Engineering Society (AES), New York, USA, Oct. 2007.
47. S. Spors, H. Buchner: An approach to massive multichannel broadband feedforward active noise control using wave-domain adaptive filtering, *Proc. WASPAA '07*, New Paltz, USA, Oct. 2007.
48. E. W. Start: *Direct Sound Enhancement by Wave Field Synthesis*, PhD thesis, Delft University of Technology, 1997.
49. H. Teutsch: *Modal Array Signal Processing: Principles and Applications of Acoustic Wavefield Decomposition*, Lecture Notes in Control and Information Sciences 348, Berlin, Germany: Springer, 2007.
50. E. N. G. Verheijen: *Sound Reproduction by Wave Field Synthesis,* PhD thesis, Delft University of Technology, 1997.
51. P. Vogel: *Application of Wave Field Synthesis in Room Acoustics*, PhD thesis, Delft University of Technology, 1993.
52. E. J. Völker: To nearfield monitoring of multichannel reproduction – Is the acoustics of the living room sufficient?, *Proc. Tonmeistertagung*, Hannover, Germany, 1998.
53. E. J. Völker: Home cinema surround sound – Acoustics and neighbourhood, *Proc. 100th AES Convention*, Audio Engineering Society (AES), Copenhagen, Denmark, May 1996.
54. E. J. Völker, W. Teuber, A. Bob: 5.1 in the living room – on acoustics of multichannel reproduction, *Proc. Tonmeistertagung*, Hannover, Germany, 2002.

55. E. G. Williams: *Fourier Acoustics: Sound Radiation and Nearfield Acoustical Holography,* London, GB: Academic Press, 1999.

Part V

Selected Applications

16

Virtual Hearing

Karl Wiklund and Simon Haykin

McMaster University, Canada

With the introduction of digital hearing aids, designers have been afforded a much greater scope in the nature of the algorithms that can be implemented. Such algorithms include designs meant for speech enhancement, interference cancellation, and so on. However, as the range of these algorithms increases, and as greater attention is given to the problems in real acoustic environments, designers are faced with a mounting problem in terms of how to test the algorithms that they have produced. The existing test procedures such as HINT[1] [20] and SPIN[2] [16] do not adequately reflect the problems that many new algorithms were designed to cope with. However, testing under real conditions must involve the problems of reverberation, different signal types (e.g. speech, music, etc.) and multiple spatially distributed interferers. These interferers may also become active and inactive at random intervals.

The acoustic environment however is not the only obstacle that many researchers face. A further variable in their tests exists in the form of the patients themselves. An algorithm that produces a good result according to a common error metric such as the mean-squared error, may not appreciable improve the patients ability to understand speech. Not only must a proposed algorithm be tested against human patients, but it must be tested against a broad range of patients. Hearing impairment after all comes in many different forms and degrees, so it is essential that the performance of an algorithm be determined for different types of patients. Ideally, one ought to be able to tune the algorithm under test so as to offer the best level of performance for a given patient.

Both of the problems outlined above are not insurmountable. It is possible to conduct tests in real acoustic environments, just as it is possible to arrange for human trials. However, such tests can be expensive, time consuming, and may require specialized equipment or even dedicated laboratory space. In

---

[1] *HINT* abbreviates *hearing in noise test.*
[2] *SPIN* stands for *speech perception in noise.*

general, the time involved in such testing also limits the turn-around time between designing, testing, and (if necessary) correcting the design.

16.1 Previous Work

Overcoming these difficulties have motivated us to investigate the development of software environments for testing hearing aids. The initial result of such investigation was the R-HINT-E[3] package, which was described in [30] and [34]. This package was developed after careful consideration both of our needs as researchers and of the methods currently used in both architectural acoustics and in the newer field of virtual acoustics. In our case, we were interested in rendering both the acoustic effects of the room and of the human body. In addition, we wished to present the result to a pair of simulated "ears" which would match the physical configuration of the hearing aid under test. The sound impinging on the device was then to be processed according to some user-defined algorithm.

Our particular needs therefore included both the physical realism of the presentation as well as the flexibility of the software simulation. Different source positions and even cocktail-party situations needed to be incorporated into the range of possible scenarios in order for the software to be an effective testing platform. In addition, the effects of different room acoustics needed to be considered, so the inclusion of multiple acoustic environments became an important requirement of the system's design.

The software model we ultimately followed was chosen on the basis of both our requirements and on the practicality of the system. Unlike multimedia based virtual audio systems [27], there was no need for real-time performance, nor was there a strong concern regarding the storage and retrieval of large amounts of data. This did not mean of course, that there were no time-constraints on the simulation's performance. The simulation methods used in architectural acoustics applications (such as ray-tracing or the image-source method) were detailed, but also very slow. Our need for physical accuracy as well as for reasonable performance times led us to use pre-recorded room impulse response measurements.

The recordings themselves were made using custom-built, flat-response loudspeakers. The source signal was an exponentially swept chirp signal [19], which was swept over a range of 0-22050 Hz and a duration of 1486 ms. By recording the system response at the microphones, it was possible to use the known source signal to obtain an estimate of the acoustic impulse response. This was done by straightforward deconvolution in the manner of

$$H\left(e^{j\Omega}\right) = \frac{Y\left(e^{j\Omega}\right)}{S\left(e^{j\Omega}\right)}. \tag{16.1}$$

---

[3] *R-HINT-E* stands for *realistic hearing in noise test environment*.

In the above equation, $Y(e^{j\Omega})$ is the Fourier transform of the measured response, and $S(e^{j\Omega})$ is likewise the Fourier transform of the input source signal. The acoustic impulse response can then be recovered by taking the inverse Fourier transform of $H(e^{j\Omega})$.

For the sake of convenience, we combined the room impulse responses measurements with the measurements of the head-related transfer function by using a KEMAR[4] dummy as our recording platform (see Fig. 16.1). The dummy was supplied by Gennum Corporation, which also supplied the microphone system used in our experiments. This system used three Knowles FG microphones arranged horizontally in each ear of the mannequin. The combined impulse response was then recovered for each microphone using the method of Eq. 16.1.

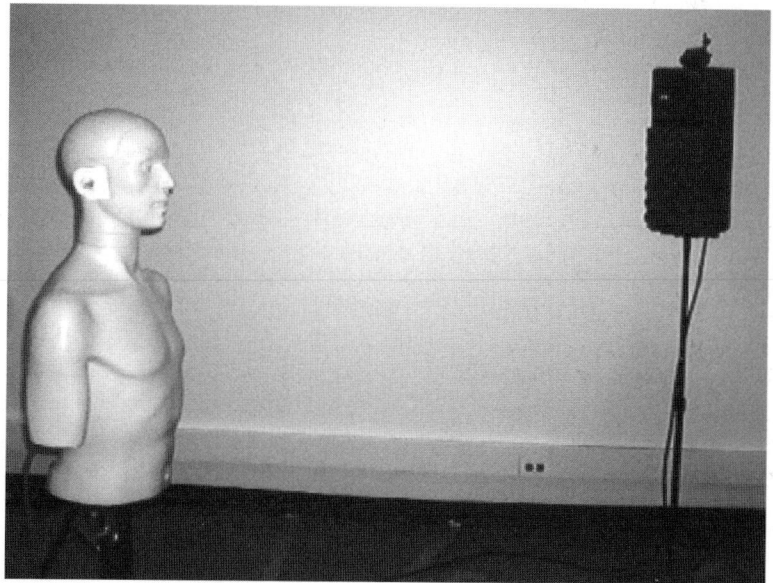

Fig. 16.1. The R-HINT-E recording setup (photo by Ranil Sonnadara).

For the R-HINT-E system, measurements were made from multiple positions in order to approximate realistic acoustic scenarios. In particular, measurements were taken from 12 different angles (0°, 22.5°, 45°, 67.5°, 90°, 135°, 180°, 225°, 270°, 292.5°, 315°, 337.5°), three heights (1'6.5" = 47.0 cm, 4'6" =137.2 cm, 5'5" = 165.1 cm), and from 2 distances (3' = 91.4 cm and 6' = 182.9 cm) for a total of 72 different source-receiver configurations (see Fig. 16.3). These measurements were carried for three different environments representing differing levels of room reverberation. These environments included a small

---

[4] The term *KEMAR* abbreviates *Knowles electronic manikin for acoustic research.*

room with movable velour drapes, as well as a hard-walled, reverberant lecture room. The first two environments were created in the small room by changing the drape positions from open to closed, where the closed drapes were used to reduce the reverberation time of the room.

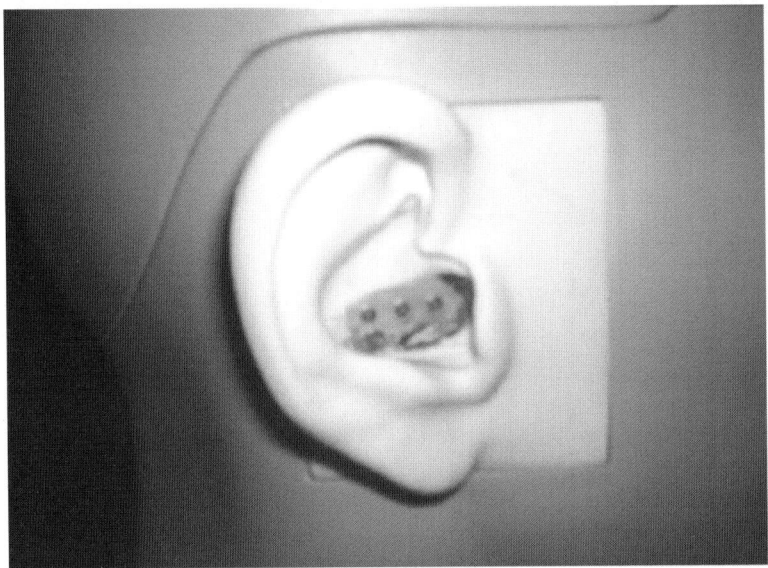

Fig. 16.2. A close up of KEMAR's right ear. The three microphones are placed in the ear and arranged in a horizontal fashion (photo by Ranil Sonnadara).

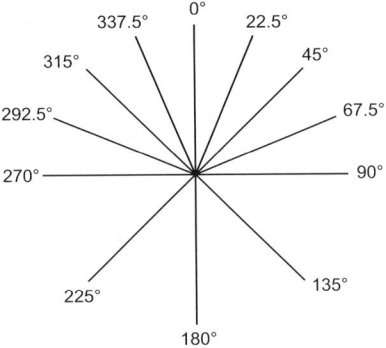

Fig. 16.3. The source positions for the R-HINT-E model.

A GUI[5] front end (see Figs. 16.4 and 16.5) for R-HINT-E was later created using MATLAB, which allowed an experimenter to create custom acoustic scenarios. Using the software we developed, a user could choose the desired room type, source position and source signal in order to create the desired conditions. In addition, multiple signals could be distributed in space to create a 'cocktail party effect' in order to simulate realistic acoustic environments. Additional flexibility was incorporated in the form of allowing the user to test custom hearing aid algorithms within the GUI, as well as allowing the user to provide their own room impulse response (RIR) measurements given a standard format. Using this software also makes it easier to perform large-scale tests involving many different scenarios. This was accomplished through the incorporation of a custom scripting language that could control all of R-HINT-E's major functions.

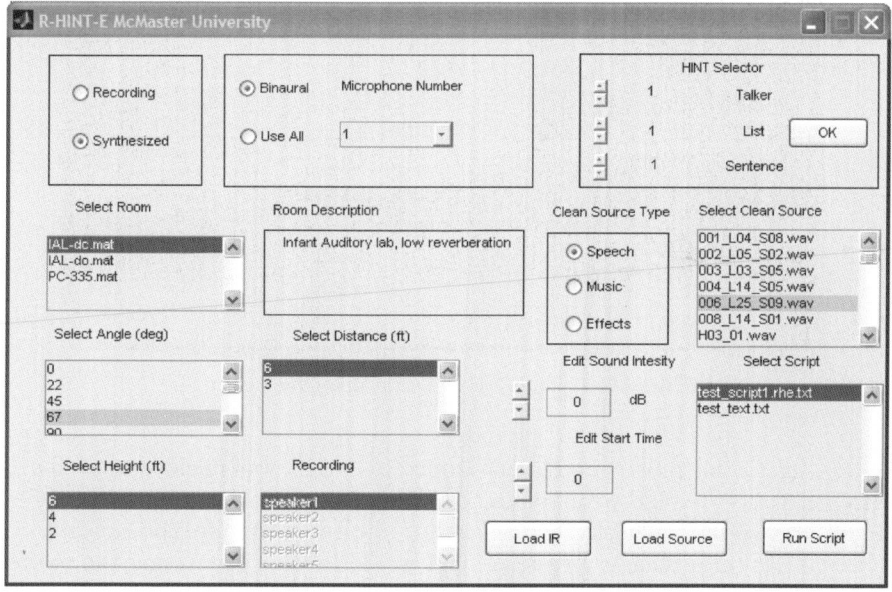

Fig. 16.4. The R-HINT-E main menu.

However, while R-HINT-E is a useful tool with respect to hearing aid research, it is not without its shortcomings. Most important is that by relying on pre-recorded RIRs, we lose a considerable degree of flexibility in the kinds of simulations that can be run. Only a limited number of source-receiver arrangements are possible, and there are even fewer acoustic environments in which one can carry out tests. While it is possible for the user to add new acoustics environments to the R-HINT-E program, this is a time-consuming process,

---

[5] The term *GUI* abbreviates graphical user interface.

and one that requires the right equipment. Moreover, while R-HINT-E does accurately simulate the room acoustics as well as the processing algorithm of interest, it does not simulate the actual patient beyond the level of the head-related transfer functions or HRTFs. In other words, the simulator cannot provide any information on how useful a processing algorithm might be in terms of patient performance given some level of hearing deficiency.

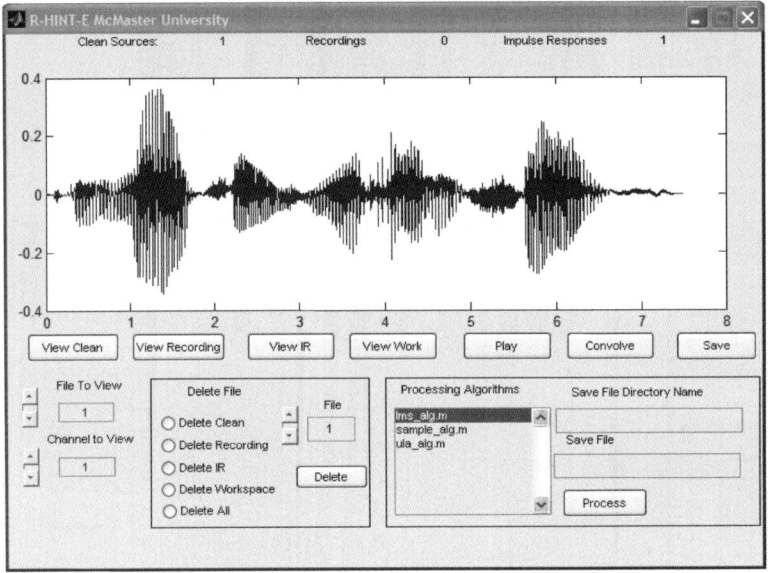

Fig. 16.5. The R-HINT-E processing screen.

Owing to the shortcomings mentioned above, it was decided that the R-HINT-E concept could be extended further to allow not only for greater flexibility in designing acoustic scenarios, but also to encompass the concept of a "virtual patient". Such a virtual patient would involve simulating the pathology of hearing loss in software, which would permit testing and customization of algorithms for specific patterns of hearing deficiency. Ideally, the simulation would also include additional flexibility by incorporating a range of HRTFs, which are known to vary widely from patient to patient.

16.2 VirtualHearing

The VirtualHearing software that we have developed in response to these needs attempts to incorporate all the aforementioned simulation requirements into a single package. That is, it contains modules that allow for the simulation of room acoustics, patient HRTFs, and finally the patient itself. In addition, a further module also makes it possible for the end user to test their own hearing aid algorithms by loading them as pre-compiled binary files. The basic software model thus follows Fig. 16.6.

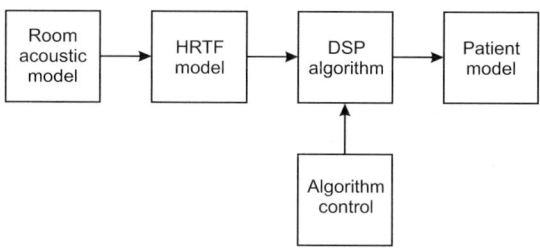

Fig. 16.6. The VirtualHearing software model. If no DSP algorithm is specified, then the output of the HRTF model is sent directly to the patient model.

As Fig. 16.6 implies, this model is a fairly simple one, but allows for individual modules to be specified independently of each other, and to be replaced if necessary in the course of software maintenence. The nature of the individual modules will be discussed in the following sections.

16.3 Room Acoustic Model

The simulation of virtual environments is a fairly well-developed field, and plays an important role in both entertainment and architectural applications. The different requirements of these applications mean that specific implementations will be forced to trade off accuracy vs. speed. Many entertainment applications for example need real-time or close to real-time rendering of acoustic scenes, while on the other hand accuracy is of paramount importance for architects considering the acoustic properties of their designs. The needs of our own software however fall in between these two extremes. In order for the room impulse responses to have a realistic effect on the designers signal processing algorithms, greater accuracy is needed than in the case of multimedia applications where the standard of accuracy is simply perceptual realism. On the other hand, while we do not have the same real-time demands that some multimedia algorithms have, a practical desktop simulator cannot afford to take the time needed to run the most accurate simulations.

In order to meet these constraints, it was decided that a hybrid simulator would be used to implement the room acoustics module. Such an arrangement was chosen because it allowed us to divide the room impulse response into two portions: the early reflections and the late reverberation, which in turn allowed for the use of algorithms best suited for each component. The use of this method was suggested in [27] by Savioja for use in the DIVA[6] multimedia system [15, 26]. Unlike the DIVA system however, which has a greater concern for the architectural acoustics of concert halls, we have chosen to use a much simpler modelling geometry than was implemented in that system. For our purposes, it was deemed sufficient to model rooms as simple rectangular prisms (see Fig. 16.7). The reason for this is that we are solely interested in the

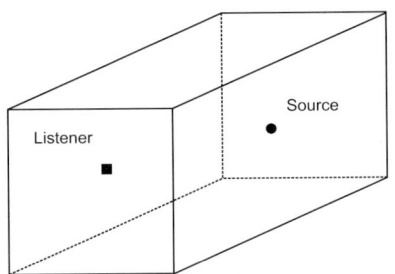

Fig. 16.7. All simulated rooms are modelled as rectangular prisms.

effects of reverberation on hearing, which can be captured sufficiently in simple rooms. Including more complicated geometries would have added nothing to this part of the simulation while greatly increasing the difficulties both in programming and in handling the front-end user interface.

The so-called early reflections are modelled essentially as direct reflections from the source to the receiver. The well known image-source [3] method was used for this effect owing to the simplicity of its implementation and its accuracy. In this method, the reflected sounds are modelled as separate, virtual sources located somewhere outside the bounds of the room (see Fig. 16.8). By adding up the calculated reflection strengths as a series of weighted, time-delayed delta functions, the rooms impulse response can be estimated for any given source-receiver pair.

Working from this model, it is a fairly simple matter to include the attenuation effects resulting from both the distance and the reflection. Additionally, the use of this method also allows us to include the directional effects of the calculated reflections. That is, given the direction from which the virtual source appears to be coming from, it can be convolved with the HRTF for that head direction. Doing this provides a better model for the confusion of binaural cues by room reverberation.

Unfortunately, while the image-source method does provide a good approximation of real room impulse responses, it suffers from a significant problem in that the computational complexity of the algorithm increases exponentially with the order of the reflections to be included in the model. As a result, the use of this algorithm is generally cut off after some more or less arbitrary

---

[6] *DIVA* abbreviates *digital interactive virtual acoustics*.

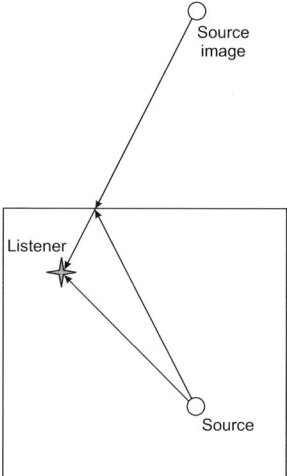

Fig. 16.8. The reflected sound is modelled as a separate source.

reflection order has been passed. Needless to say, such a cut off does not provide a good representation of the real room impulse response, since it ignores the late reverberation.

As a result, a second algorithm was used in conjunction with the image-source method in order to simulate the late reverberation component of the room impulse response. This second algorithm has as a starting point the fact that the late reverberation can essentially be modelled as a diffuse sound field, instead of being made up of a series of discrete reflections. To that end, we investigated the use of various recursive digital filter algorithms [24, 31], and ultimately decided on the one developed by Vaananen et al. [32].

All such algorithms have a roughly similar structure and are required to meet specific criteria relating to the behavior of reverberant sound fields. These criteria are outlined in [27]. In particular, a good digital reverberator algorithm must meet the following criteria:

1. A spectrally dense pattern of reflection with an exponential decay of energy in the time domain.
2. Higher frequencies must have shorter reverberation times than lower frequencies.
3. The late reverberation as perceived by a binaural listener should be partially incoherent.

Of the algorithms we studied, it was found that the Vaananen reverberator not only met these criteria, but it also used less memory than other methods while allowing a much faster growth in reflection density.

The Vaananen algorithm consists of a series of parallel filter blocks and delay lines as shown in Fig. 16.9. The individual filter blocks $H_k(z)$ and $A_k(z)$ are a low-pass filter and an all-pass comb filter respectively:

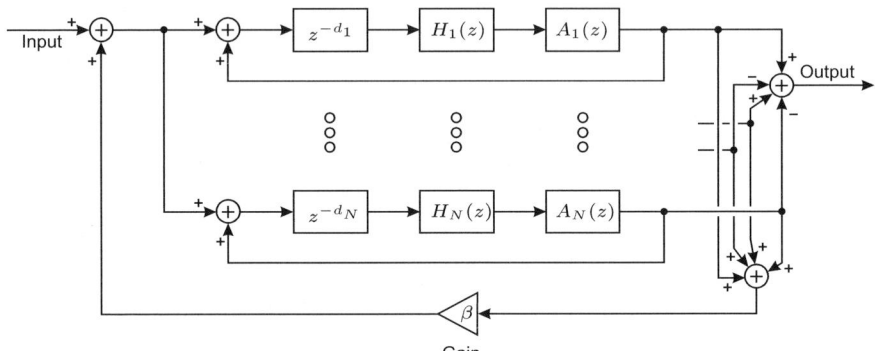

Fig. 16.9. The structure of the Vaananen reverberator consists of a series of parallel delay loops. Each loop contains a lowpass filter and an all-pass comb filter.

$$H_k(z) = \frac{g_k (1 - b_k) z^{-1}}{1 - b_k z^{-1}} \qquad (16.2)$$

$$A_k(z) = \frac{-\frac{1}{2} + z^{-M_k}}{1 - \frac{1}{2} z^{-M_k}} . \qquad (16.3)$$

The purpose of the low-pass filter $H_k(z)$ is to mirror the low-pass characteristics of air, while the all-pass comb filter $A_k(z)$ helps to diffuse the reflections, making them more irregular. These filters are of very simple construction and are shown in Fig. 16.10, for a given kth filter in the loop. These filters are governed by three different parameters, b_k and g_k, and the lengths of the individual delay lines.

These parameters are dependent on the nature of the room; in particular they depend on both the room size and the average reflectivity of the rooms walls. The filter delay lengths for example, are based strictly on the room dimensions, with the smallest value being equal to the largest dimension of the room. The lengths of subsequent delay lines are increased, but in an irregular fashion in order to prevent colouration of the reverberant signal.

The remaining parameters are calculated using the reverberation time T_{60}, which can be estimated for the room using Sabines formula [22]

$$T_{60} = K_{T_{60}} \frac{V}{\sum_i \alpha_i A_i}, \qquad (16.4)$$

where V is the volume of the room, while a_i and A_i are respectively, the reflection coefficient and area of the ith wall. The coefficient $K_{T_{60}}$ is

$$K_{T_{60}} = 0.161 \frac{\text{s}}{\text{m}} . \qquad (16.5)$$

From the reverberation time, the delay line gains g_i can be calculated [31] using Eq. 16.6, while the low-pass coefficients b_i are calculated using Eq. 16.7:

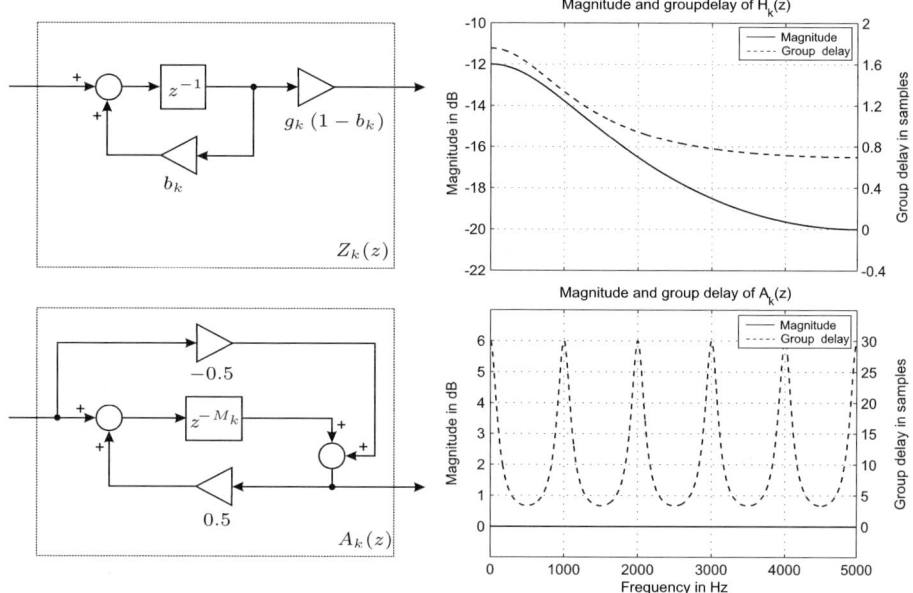

Fig. 16.10. Block diagrams for the low-pass $H_k(z)$ and all-pass diffusive filter $A_k(z)$. $d_k = 1000$, $f_s = 10000$ Hz, $T_{60} = 0.5$ seconds, $\epsilon = 0.6$

$$g_i = 10^{-\frac{3\,d_i}{f_s\,T_{60}}}, \tag{16.6}$$

$$b_i = 1 - \frac{2}{1 + g_i^{\left(1-\frac{1}{\epsilon}\right)}}. \tag{16.7}$$

The quantity d_i in Eq. 16.6 is equal to the delay length of the ith delay line in Fig. 16.9, while f_s is the sampling frequency. In Eq. 16.7, the symbol ϵ represents the ratio of the reverberation time for frequencies $f = f_s/2$ to $f = 0$ Hz. This value cannot be computed directly since it is dependant on the filter coefficients described above; it must instead be specified by the designer of the simulation. For our own purposes, we found that a value of $\epsilon \approx 0.6$ gave results that provided a reasonable approximation the results to be had from several pre-recorded rooms (see Fig. 16.11 and Fig. 16.12).

16.4 HRTF Simulation

The combination of the image-source method with the Vaananen reverberator described above provides an adequate simulation of the acoustics of simple rooms. In addition to this though, the acoustic signal perceived by a human

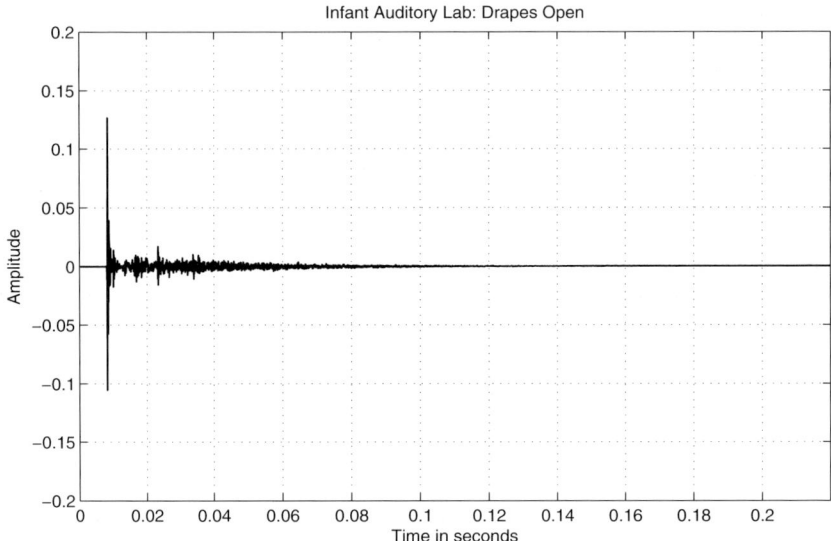

Fig. 16.11. The measured impulse response and HRTF for a real room.

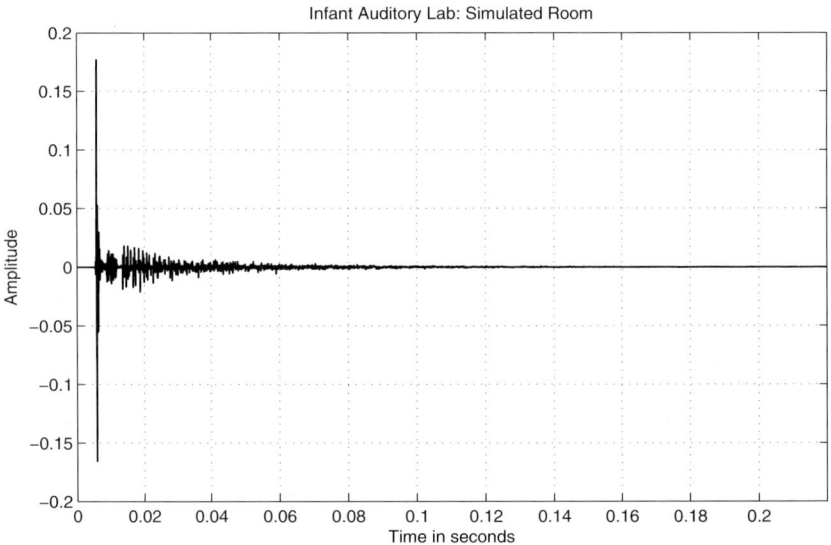

Fig. 16.12. The simulated room impulse response and HRTF using "best guess" parameters to approximate the room used in Fig. 16.11.

listener is also strongly affected by the listeners own HRTFs. These binaural transfer functions arise from the scattering of acoustic energy off of the listeners head, torso and outer ears (pinnae), and are vital in forming the auditory cues needed for the acoustic source localization [1, 28]. The inclusion of HRTF effects is therefore vital for any hearing aid simulation, especially for the development of binaural algorithms.

As it happens though, there is considerable variation among the HRTFs for individual listeners, a fact that arises from the diversity of ear and body shapes [28]. In addition, the HRTF also depends on the direction and range of the sound source impinging on the listener. It was important therefore, during the design of this component of the simulator to ensure that these facts were adequately represented in the software. In practical terms this meant first making sure that it was possible to represent the direction dependent HRTFs in sufficient density to approximate real-life scenarios. In addition, it was desirable to include the possibility of simulating different body and pinna shapes.

In the VirtualHearing software package, this problem was solved by making use of a database of pre-recorded HRTFs, albeit one that was more extensive than that used in our previous R-HINT-E project [30, 34]. This database was compiled by V.R. Algazi et al [2] at the U.C. Davis CIPIC laboratory, and is publicly available online at http://interface.cipic.ucdavis.edu. The database is comprised of HRTF measurements taken for 43 different individuals, plus two KEMAR sets using both large and small pinnae. For each individual, the measurements include 25 different azimuths and 20 different elevations, with each HRTF being 200 samples in length and sampled at rate of $f_s = 44.1$ kHz.

To make use of this database in our software package, a menu can be accessed that allows the user to specify the subject to use as the simulated listener, while at the same time viewing his/her anthropometric data (see Fig. 16.13). Additionally, the KEMAR models may also be selected instead of using the human subjects provided in the database. A separate options screen allows the user to choose whether to use the binaural HRTFs, or the HRTF corresponding to just one ear. This allows for the development of binaural processing algorithms which use the inputs from both ears in order to enhance the input signal, as opposed to most conventional devices today where the hearing aids in either ear operate independently.

Given the wide range of possible positions of course, one cannot expect even a large HRTF database to include all conceivable directions. In order for there to be adequate coverage of the three-dimensional space for the HRTFs, it is necessary to use interpolation in order to approximate the required HRTF from the ones existing in the database. For this application, we have followed the lead of Huopaniemi [14] in implementing Begault's [4] bilinear interpolation scheme. This method uses the four nearest neighbors of the desired point in order to form the approximation, and is accomplished via

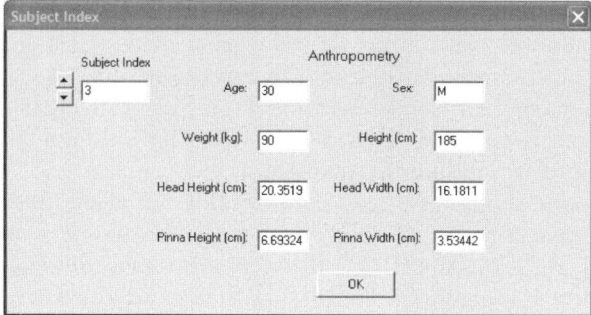

Fig. 16.13. The subject selection window from VirtualHearing allows the user to browse the subjects from whom the HRTFs were selected. Anthropemetric data is also displayed.

$$H_\mathrm{d}\left(e^{j\Omega}\right) = (1-c_\theta)(1-c_\phi)H_1\left(e^{j\Omega}\right) + c_\theta(1-c_\phi)H_2\left(e^{j\Omega}\right)$$
$$+ c_\theta c_\phi H_3\left(e^{j\Omega}\right) + (1-c_\theta)c_\phi H_4\left(e^{j\Omega}\right). \qquad (16.8)$$

In the above formula, the known HRTFs ($H_1(e^{j\Omega})$ through $H_4(e^{j\Omega})$) surround the desired HRTF $H_\mathrm{d}(e^{j\Omega})$, which is situated at the point (θ, ϕ) on the azimuth/elevation grid shown in Fig. 16.14. The interpolation coefficients are computed using

$$c_\theta = \frac{\theta_\mathrm{d} \bmod \theta_\mathrm{grid}}{\theta_\mathrm{grid}}, \qquad (16.9)$$

$$c_\phi = \frac{\phi_\mathrm{d} \bmod \theta_\mathrm{grid}}{\theta_\mathrm{grid}}. \qquad (16.10)$$

16.5 Neural Model

In addition to the acoustic modelling discussed in the previous section, the other major task of the software is to model the expected cochlear response of a listener immersed in the acoustic scenario specified by the user. In order for such a simulation to be useful in the evaluation of hearing aids, it ought to match the responses of a real cochlea subjected to the same stimulus. This is not a trivial task however, given the complexity of the human auditory system, which is both non-linear and time-varying in its responses to acoustic stimuli.

In mammals, the detection of acoustic stimuli and their transduction into neural signals occurs in several stages. Vibrations in the air impinge on the eardrum, inducing movement that is transmitted to the cochlea via the action of several small bones connecting the two structures. Within the cochlea,

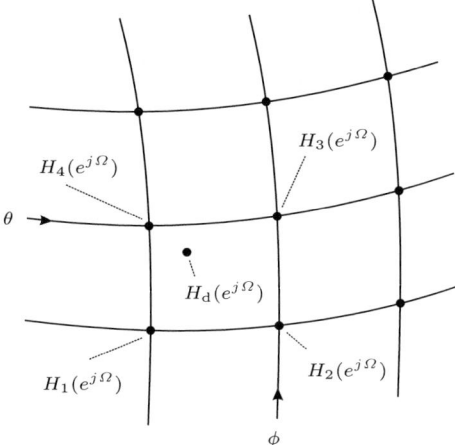

Fig. 16.14. The HRTF for point $H_d(e^{j\Omega})$ must be interpolated from its four nearest neighbours.

the connecting bones induce pressure waves that create small displacements of the basilar membrane [10]. Owing to the mechanical properties of this membrane, local resonances exist that follow a tonotopic pattern [9]. That is, different parts of the basilar membrane respond to different frequencies, with low frequencies causing vibrations at the base of the membrane, and high frequencies at the apex. Detection of these displacements and their conversion into the electrical signals used by the nervous system is carried out by the hair cells that lie along the length of the basilar membrane. As a result of the localized nature of the basilar membrane response, each group of hair cells responds only to a select band of frequencies.

These hair cells themselves may be divided into two types of cell, inner and outer, each of which plays a distinct role in the process of transducing mechanical motions into electrical impulses. The actual transduction in fact, is only carried out by the inner hair cells, while the purpose of the outer hair cells is to act as a feedback mechanism in order to boost the detectability of the incoming acoustic signal, and to sharpen the level of tuning [8]. It is important to note that neither the tonotopic map of the basilar membrane, nor the role of the outer hair cells, are produced by entirely passive mechanisms. The outer hair cells themselves in fact are thought to produce forces that lead to amplification of the stimuli [8, 23]. Also noteworthy is the fact that the ability of the outer hair cells to provide sharp tuning is dependent on the intensity of the stimuli. The tuning is sharpest near the threshold of reception, and broadens as the sound level increases.

The role of the hair cells thus has important implications for modelling hearing loss. Damage to both types of hair cell can occur, but the effects on human auditory perception depend on what amount of damage has been done to each group, and where on the basilar membrane the losses have occurred. Damage to the outer hair cells for example, will result in both broadened tuning curves and elevated reception thresholds. On the other hand, damage to the inner cells will weaken the ability of the cells to transduce the mechanical stimuli, and will therefore raise the reception thresholds.

In order to produce a computational model of the human auditory system and the effects of hair cell loss, it is possible to think in terms of varying levels of simulation detail. At its most basic level, for example the basilar membrane can be thought of as a filterbank, where the individual filters are described by the frequency-domain gamma-tone function

$$G(\omega) \propto \left[\frac{1}{1 + j\tau(\omega - \omega_{\text{CF}})}\right]^\gamma e^{-j\omega\alpha} \qquad (16.11)$$

These auditory filters are model-based approximations to the linear revcor functions which were derived empirically [21] and are parametrized by three main quantities. The first of these is ω_{CF}, which simply represents the centre frequency of the filter. The parameter τ controls the filter's time-decay and bandwidth, while α controls the time delay introduced by the filter, which can also be co-opted to model additional system delays if needed. The remaining parameter, γ is simply an empirically-derived value, and does not directly relate to the more physically meaningful parameters.

A simple model of neural transduction can be also be included by passing the bandpass filtered signals through a half-wave rectifier, which reflects the fact that the hair cells only transduce vibrations in one direction. Such a model is sufficient when only general details about the auditory peripheral system are needed, such as in [7] and [13]. For the sort of detailed modeling of the human auditory system that our "virtual patient" system requires however, these kind of approximations are inadequate. The model described above for example, does not include the non-linear time-varying effects that are known to be present in the human auditory system. These effects include for example, the changes in auditory filter bandwidth with input sound pressure level, or the changes in auditory behavior brought about by hearing impairment. For this reason, we have chosen to employ a more detailed model of the human auditory system, which addresses these needs.

In particular, we have incorporated the Bruce-Zilaney [5, 36] model of the cochlea, which is itself a modification of the model developed first by Carney [6] and later modified by Zhang [35]. This model can be broken down in to five main processing blocks: a middle ear filter, a time-varying gamma-tone filter, an OHC[7] control path, an IHC[8] transduction model, and synaptic model. These blocks are organized in the manner shown in Fig. 16.15, where the input is an acoustic stimulus measured in Pascals. The output of the model

---

[7] *OHC* abbreviates *outer hair cell*.
[8] *IHC* stands for *inner hair cell*.

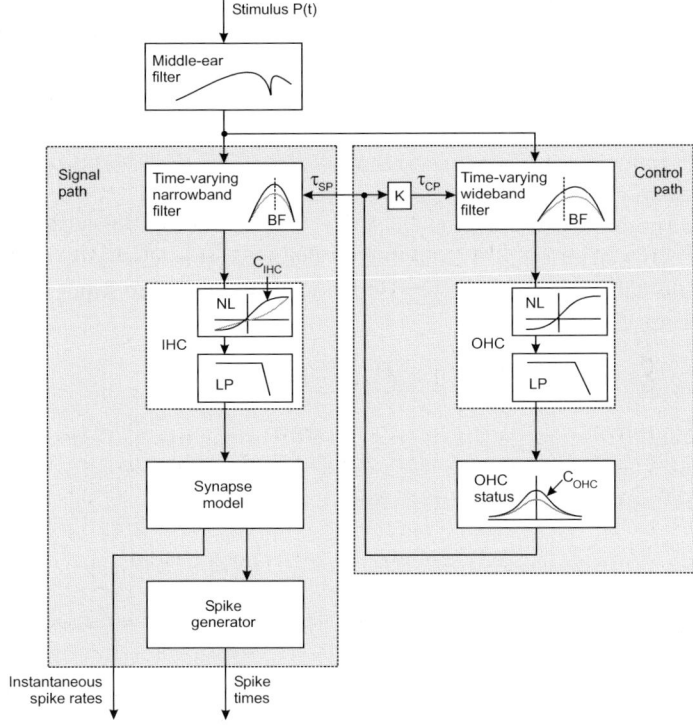

Fig. 16.15. Block diagram of the neural model.

consists of two signals, which include a binary spike train as well as a signal of instantaneous spike rates.

The first actual component of the block, the middle ear filter, is meant to model the filtering aspects of the human ear canal. This is most important for describing wide-band signals, where the relative amplitudes of different frequency components may be changed. The middle ear block may be implemented using a digital filter representation, which is fully described in Appendix A of [5].

Like the simple models discussed earlier in this section, the cochlear model that we make use of also incorporates a bank of gamma-tone filters in order to model the frequency selectivity of the human ear. However, unlike those models, the model of Zilaney and Bruce [36] generates a data-driven control signal that affects the tuning of the gamma-tone filters. This section makes up the control path shown in Fig. 16.15, and models the affects of the OHCs on filter tuning. Specifically, the acoustic input signal is asymmetrically band-pass filtered about the gammatone filter's centre frequency. Afterwards, the resultant signal is rectified and lowpass filtered in a model approximation of the OHC's input/output behavior. An additional nonlinear function maps the

OHC outputs to the gammatone filter's new time constant, which results in a subsequent change in filter bandwidth.

In addition to modeling the time-varying properties of the cochlear filter, this model of the OHC control path is useful in two other significant ways. Firstly, the asymmetric bandpass filtering of the acoustic input allows the incorporation of two-tone suppression effects without needing to directly implement the interactions between separate cochlear filters [35]. In addition, the non-linear function that relates the OHC output to the filter time constants may be easily parametrized in order to describe the effects of OHC impairment. Specifically, all that needs to be done is to scale the function output by some constant C_{OHC} with

$$0 \leq C_{\text{OHC}} \leq 1. \tag{16.12}$$

This parameter represents the degree of OHC impairment. Thus a C_{OHC} value of 1 indicates healthy functionality, while a value of 0 indicates total impairment. This results in the modified time constant output [5]

$$\tau_{\text{sp,impaired}}(n) = C_{\text{OHC}} \left(\tau_{\text{sp}}(n) - \tau_{\text{wide}} \right) + \tau_{\text{wide}}, \tag{16.13}$$

where $\tau_{\text{sp,impaired}}(n)$ is the new time constant, $\tau_{\text{sp}}(n)$ is the non-impaired time constant, and τ_{wide} is a constant value that reflects the time constant of the cochlear filter in the absence of any OHC tuning.

In contrast to the role of the OHCs as a control mechanism, the purpose of the IHCs is to transduce the mechanical stimulus from the basilar membrane into an electrical potential. As a result, the modelling of this portion of the cochlea is rather simpler than the previous section. To simulate the transduction process, a logarithmic function is used as a half-wave rectifier, the output of which is low pass filtered. The rectifier is modelled on a similar rectification process known to exist in the cochlea, while the lowpass filter, which has a cutoff frequency of 3800 Hz, simulates the loss of synchrony capture that occurs in the auditory nerve as the stimulus frequency increases. The modelling of the inner hair cell impairment is also quite simple. The input to the IHC block can simply be scaled by the parameter C_{IHC} [5] with

$$0 \leq C_{\text{IHC}} \leq 1. \tag{16.14}$$

As in the case of OHC impairment, a value of one indicates healthy functioning, while a value of zero indicates a total loss of function.

The remaining portion of the cochlear simulation, the synapse model is based on a discrete-time adaptation [6] of Westerman and Smith's three store diffusion model of the cochlear synapse [33]. This block is followed by a time-varying Poisson discharge generator, which takes as its input the instantaneous rate values provided by the discrete-time synapse model. The spike generator also incorporates the effect of refractoriness by keeping track of the time since the last discharge, and modifying the discharge probabilities accordingly [6].

While this cochlear model is fairly complex, it does capture much of the phenomena related to hearing. In particular, as been shown above, it is capable of modelling the effects of hearing loss beyond simply elevating the reception thresholds. Unfortunately, the parameters associated with the modelling of hearing loss are not directly related to measurable quantities, and new procedures need to be developed before they can be estimated from patient testing. It is likely however, that a combination of auditory threshold measurements combined with loudness growth profiles should be able to characterize the inner and outer hair cell parameters [18]. Further work on this subject needs to be carried out in order for this idea to be realizable.

16.6 The Software and Interface

The software to implement the above modules was written in C++, while a graphical user interface was developed using Borland C++ Builder 5. The interface so created allows the end-user a considerable degree of control in deciding the parameters of the simulation, and the nature of the acoustic scenario to test under. This ensures a high degree of repeatability between experiments, and also allows the user to save and load test scenarios that are commonly used.

The interface screen, which is shown in Fig. 16.16 demonstrates some of this functionality. In particular, this screen deals with the basic elements of the acoustic scenario. From here the user can specify the size of the room, the acoustic reflectivity of the walls, as well as the number and placement of the sources. The placement of the sensor (listener) is also handled here, as well the sensors orientation. For the purposes of orientation, it is always assumed that 0° of azimuth corresponds to the direction of the X-axis, while 0° of elevation corresponds to the subject looking straight ahead in an ordinary upright position.

As Fig. 16.16 shows, the user can load any source file he or she wishes provided it is in the standard PCM[9] .wav format, and sampled at 44.1 kHz. This option allows the user to preview the sound before adding it to the simulation. Additional sources can be added simply by clicking the **Add Source** button after specifying their filenames and positions. The loudness in dB SPL[10] as well as the time-delay of each source can also be chosen, reflecting the fact that sources may start and stop at different times, and be active with different intensities. Sources that have been previously added to the simulation may be viewed and managed by using the **Delete Source** button, which calls up the menu in shown in Fig. 16.17.

The user can also specify whether the source is distributed in space or not (see Fig. 16.18). This may be desirable since most real sound sources are not

---

[9] *PCM* stands for *pulse code modulation*.
[10] *SPL* abbreviates *sound pressure level*.

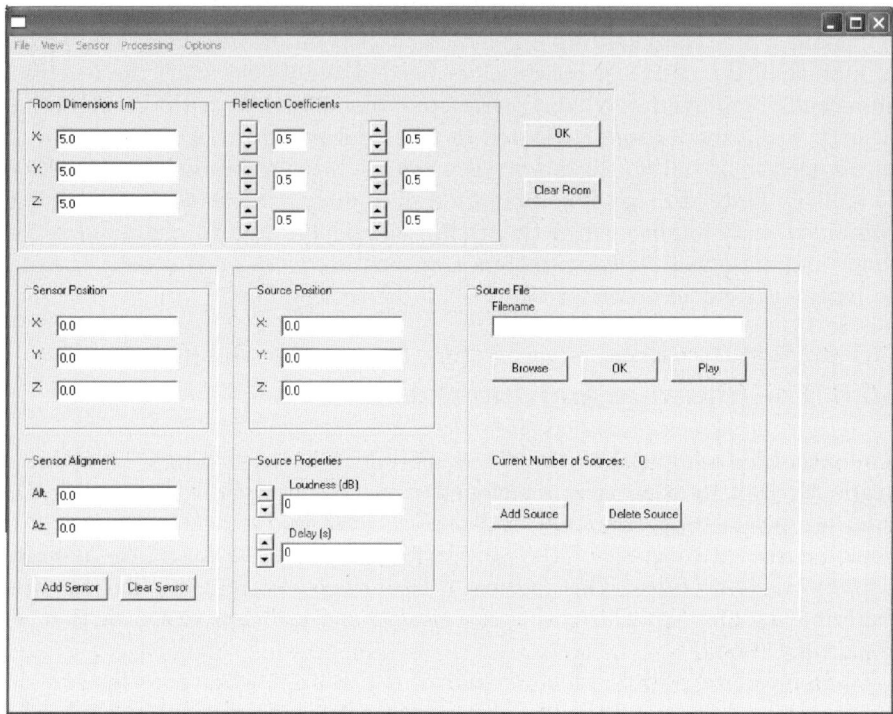

Fig. 16.16. The VirtualHearing main screen. This screen allows the user to specify most of the details regarding the acoustic scenario.

point sources, but are generated over some area or volume. A stereo speaker for example radiates acoustic energy over its whole surface rather than just from a point in its centre. While such a distinction may not matter for many applications, it has been noted that modelling distributed sources as point sources may not allow for proper testing of some spatial processing algorithms. As a result, we have chosen to include the possibility of distributed sources, and to model them as an array of point sources [17]. Such an approach retains both the necessary realism as well as limiting computational complexity.

A source that has been designated as being spatially distributed is centred at the source position previously specified. The user can designate the number of subsources, as well the over all size of the distribution. In addition, the user may also choose from one of several distribution types. At the moment, these include flat surfaces in the XY, XZ, or YZ planes, a cube, or a random cloud, although more types could be added in the future.

With the acoustic environment specified, the user can also choose other simulation parameters to control. In particular, the nature of the hearing loss suffered by the hypothetical patient can be controlled by setting the inner

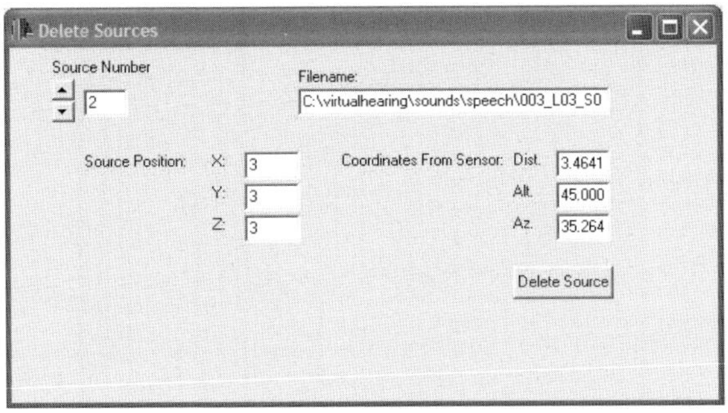

Fig. 16.17. This screen allows the user to browse the selected sources, and to delete some if so desired.

and outer hair cell parameters as shown in Fig. 16.19. For each of the patient centre frequencies, the user can control the hearing impairment levels using the sliding buttons shown above. It is also possible to move through the centre frequencies, and to specify different levels of impairment for each frequency.

The user also has several other simulation options that may be specified (see Fig. 16.20). In particular, it needs to be decided whether or not the simulation should be binaural or monaural. Since most hearing aid algorithms in use today are strictly monaural, this is the default option. In addition, the user is also allowed to specify how many of the early reflections should be considered as being directional. That is, given the placement of the virtual sources, one can decide how many of the impinging virtual sounds need to be convolved with the HRTF for the corresponding angle.

The final simulation option allows the end user to receive the simulation output in terms of a neural spike train in addition to the usual time series of instantaneous spike rates. While this option is only of very limited interest to those developing hearing aids, it is of use to those interested in the

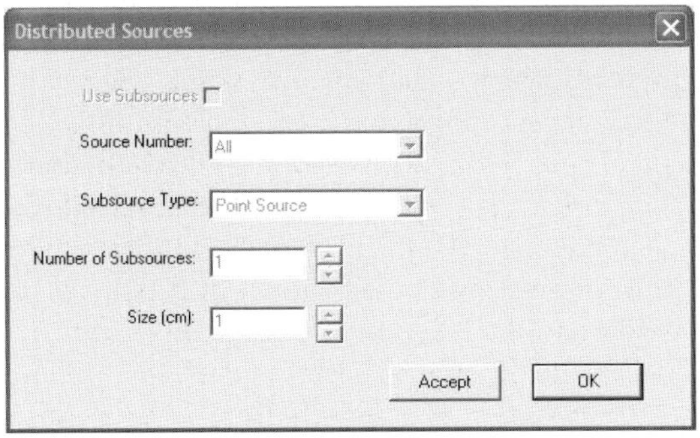

Fig. 16.18. If the user wishes, any individual source may be distributed in space instead of being modelled as a point source.

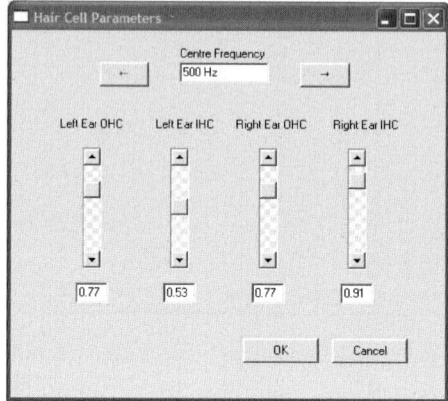

Fig. 16.19. The simulation currently considers seven centre frequencies, which reflect positions on the basilar membrane. The OHC and IHC parameters which specify the level of hearing impairment can be controlled from the menu shown above.

neurobiological processing of sound. Since the real auditory system operates on the basis of spike trains rather than rate signals, study of this system and how it processes sound must be based on realistic inputs. The VirtualHearing simulator is a useful platform in this regard given that it encompasses not only a realistic auditory model, but also offers an easy to use interface for managing signals and environmental effects.

Once the acoustic environment and simulation options are chosen, the simulation itself can be run by clicking on the **Processing** menu and running the appropriate simulation: environmental, DSP, or neural. Naturally, neither of the last two options can be run before the environmental simulation has been

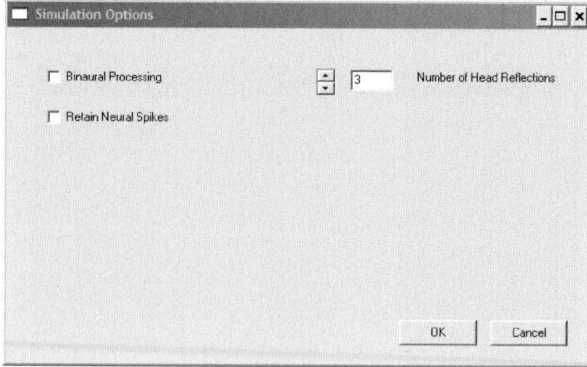

Fig. 16.20. The Simulation menu allows the user to choose several options regarding the simulation output.

completed. In addition, while the neural simulator can be run immediately if the user wishes to do so, the user also has the option of running a custom DSP algorithm that fills the role of the hearing aid processor. User defined algorithms must be in the form of Borland-compatible .dll files. Instructions for creating such files can be found in [12] as well as in the final documentation of our software. A template file will also be provided to assist developers in creating files that will be compatible with the VirtualHearing software. Once created, a custom DSP algorithm can be accessed by clicking File|Load Processor, which calls up the relevant file menu. Similarly, a processor that has already been loaded can be removed by clicking File|Clear Processor, or by clearing the simulation altogether (by clicking File|New).

The results of the simulations can be saved for further analysis depending on the application that the user is interested in. Sound files are saved in PCM .wav format at the standard sampling rate of 44.1 kHz. The other outputs such as the instantaneous spike rates and the spike trains themselves are saved as text files. In this case, there are multiple output files corresponding to the centre frequency as well as the binaural option. For these files, the user is simply asked for a basic file name, to which the particulars of the data are appended. For example, given the base file name out.dat, the program will create several files: out_R0.dat, out_L0.dat, out_R1.dat, and so on. The "R" and "L" suffixes designate the right and left responses, while the numerical entries designate the index (and ultimately the centre frequency) of the particular nerve fibre.

16.7 Software Testing

In order for the software to be useful, it was necessary for us to have confidence that the outputs successfully matched data gathered in real world environments. As a result, it was necessary for us to validate the results. Fortunately, this was unnecessary in the case of the neural model, as extensive comparisons had already been made by the original authors, which indicated its soundness as a model. This meant that only the acoustic modules needed to be examined for accuracy.

The testing procedure we chose to carry out consisted of a listening test given to four volunteers (one male and three females) with unimpaired hearing. For this test, sixty different HINT speech sentences were produced, each filtered by the impulse response of a reverberant room. Thirty of these impulse responses were measured in real rooms, while the remaining thirty were simulated in software using estimates of the room parameters. The sixty speech sentences were presented in random order through a set of headphones to each subject, who asked to rank the "naturalness" of the result on a scale of one to seven. In other words, the subjects were asked to rank their confidence in whether they felt that the reverberation in the signal had been introduced through natural or artificial means. A rank of 1 on this scale meant that the presented sentence sounded wholly artificial, while a rank of 7 indicated that the subjects felt that the reverberation present in the sentence sounded completely natural.

Summed over all participants the results of the listening trial are shown in Tab. 16.1. In this table, we have taken the rankings of all participants and broken them down according to personal preferences with respect to the realism of the presented sound. Thus, the number of times the participants rated either a simulated or a real sound a given "naturalness" score is recorded below.

Table 16.1. These are the rankings summed over all the test subjects.

Type	Rank						
	1	2	3	4	5	6	7
Simulated	1	3	7	20	22	34	33
Real	0	3	33	14	18	28	24

These results indicate a slight preference in belief that the simulated sounds possessed a greater degree of "naturalness" than did the sounds created using the measured impulse responses. This means that there was little perceptual difference between the real and simulated sounds, and thus the simulation is capable of creating perceptually realistic stimuli. It should be noted that the measured room impulse responses used in this trial were also tested against actual room recordings in [30], and no statistically significant preference was found on the part of the subjects. In other words, we could be confident that the measured room impulse responses provided an accurate model of the room acoustics and did not introduce perceptual artifacts.

A further point of comparison between the natural and artificial room responses was the spectrum of the impulse responses. In keeping with the required properties of artificial reverberation discussed in Sec. 16.3, the spectrum should be broad band, although the high frequency parts of the signal should decrease over time. By comparing the windowed FFTs of the room impulse responses, we can see that this property does indeed hold (see Figs. 16.21 -16.24).

16.8 Future Work and Conclusions

While considerable effort has been put into the creation of this simulator as a tool for researchers, future developments could enhance this software's value. In particular, it may be desirable to include objective speech intelligibility

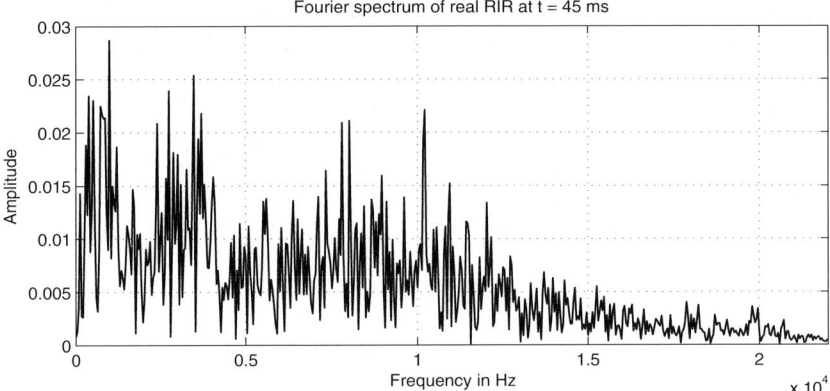

Fig. 16.21. Fourier spectrum of the real room impulse starting at t=45 ms.

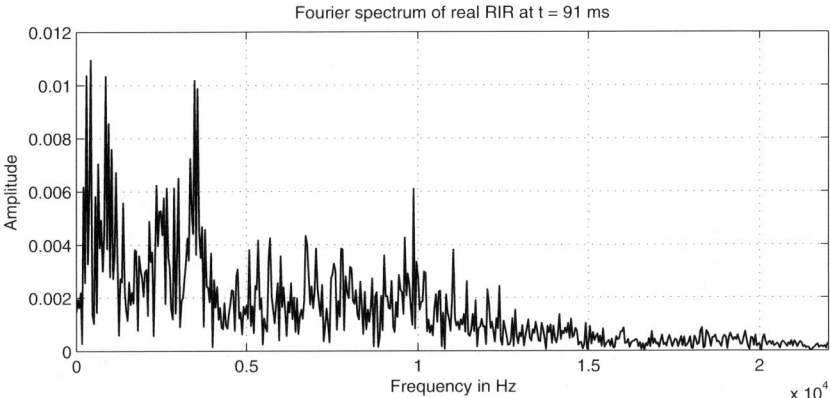

Fig. 16.22. Fourier spectrum of the real room impulse response starting at t=91 ms.

metrics along with automatic comparisons using these metrics. In addition, some thought may also be given to expanding the range of geometries offered in the simulator. Currently, only a binaural simulation is possible. Additional time and effort however, could allow our HRTF models to include multiple microphones, or different receiver geometries. Suggestions from users and potential users may also help shape the development of this software.

Of particular interest to the authors is the development of a hearing test that will allow us to use an audiogram or modified audiogram to estimate the inner and outer hair cell losses. Currently, these parameters are somewhat divorced from the quantities that are measureable on an audiogram. More accurate estimates though could improve algorithm design and fitting by more closely representing the effects of the cochlear damage on the incoming signal.

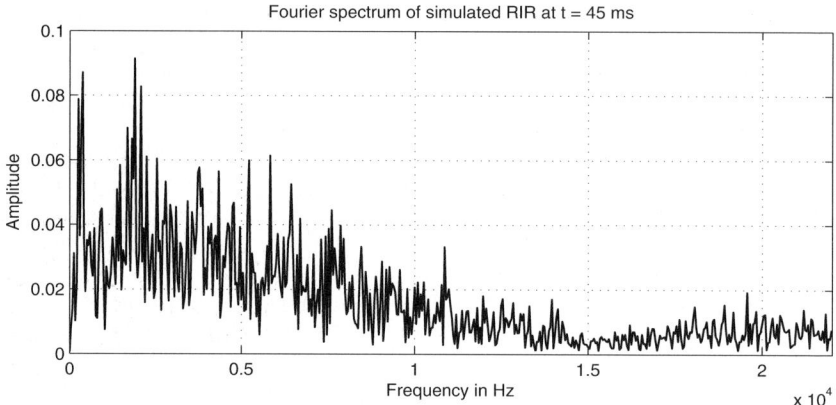

Fig. 16.23. Fourier spectrum of the simulated room impulse response starting at t=45 ms.

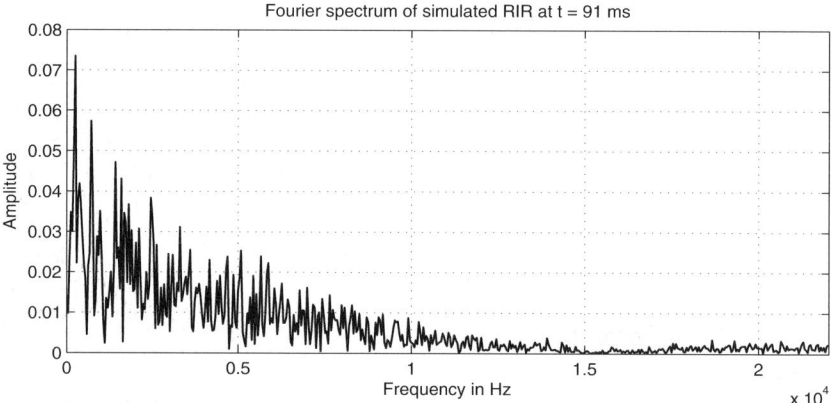

Fig. 16.24. Fourier Spectrum of the simulated room impulse response starting at t=91 ms.

We feel that such an estimation procedure is not out of reach, and that it would be a valuable addition to the tools available to the audiologist.

However, in spite of the possible improvements outline above, we are confident that the VirtualHearing software will prove to be a useful tool for the designers of hearing aid algorithms, as well as those interested in the neurobiological aspects of hearing. Previous experience with the R-HINT-E platform found that it was a useful tool for generating testing data, even if it was somewhat inflexible. The new VirtualHearing software offers a much more flexible simulation environment that encompasses many important problems facing designers, and it does so while ensuring that the simulation is all-inclusive. This encompassing of acoustics, neurobiology and user-defined DSP algorithms will make this software a valuable tool for many researchers.

References

1. V. R. Algazi, R. O. Duda, R. P. Morrison, D. M. Thompson: Structural composition and decomposition of HRTFs, *Proc. WASPAA '01*, 103–106, New Paltz, NY, USA, Oct. 2001.
2. V. R. Algazi, R. O. Duda, D. M. Thompson, C. Avendano: The CIPIC HRTF database, *Proc. WASPAA '01*, 99–102, New Paltz, NY, USA, Oct. 2001.
3. J. B. Allen, D. A. Berkley: Image method for efficiently simulating small-room acoustics, *J. Acoust. Soc. Am.*, **65**(4), 943–950, 1979.
4. D. R. Begault: *3-D Sound for Virtual Reality and Multimedia*, Cambridge, MA, USA: Morgan Kaufmann, 1994.
5. I. C. Bruce, M. B. Sachs, E. D. Young: An auditory periphery model of the effects of acoustic trauma on auditory nerve responses, *J. Acoust. Soc. Am.*, **113**(1), 369–388, 2003.
6. L. H. Carney: A model for the response of low-frequency auditory nerve fibers in cat, *J. Acoust. Soc. Am.*, **93**(1), 401–417, 1993.
7. T. Chi, P. Ru, S. A. Shamma: Multiresolution spectrotemporal analysis of complex sounds, *J. Acoust. Soc. Am.*, **118**(21), 887–906, 2005.
8. P. Dallos: The active cochlea, *Journal of Neuroscience*, **12**, 4575–4585, 1992.
9. R. Fettiplace, C. M. Hackney: The sensory and motor roles of auditory hair cells, *Nature Reviews: Neuroscience*, **7**, 19–29, 2006.
10. S. A. Gelfand: *Hearing: An Introduction to Psychological and Physiological Acoustics*, London, UK: Informa Healthcare, 2004.
11. M. Hauenstein: Application of Meddis inner hair-cell model to the prediction of subjective speech quality, *Proc. ICASSP '98*, **1**, 545–548, Seattle, WA, USA, 1998.
12. J. Hollingworth, et al.: *C++ Builder 5 Developers's Guide,* Indianapolis, IN, USA, Sams Publishing, 2001.
13. G. Hu and D. Wang: Monaural speech segregation based on pitch tracking and amplitude modulation, *IEEE Trans. on Neural Networks*, **15**(5), 1135–1150, September 2004.
14. J. Huopaniemi: *Virtual Acoustics and 3-D Sound in Multimedia Signal Processing*, PhD thesis, Helsinki University of Technology, Helsinky, Finland, 1999.
15. J. Huopaniemi, L. Saioja, T. Tkala: DIVA virtual audio reality system, *Proc. ICAD '96*, 111–116, Palo Alto, CA, USA, 1996.
16. D. N. Kalikow, K. N. Stevens, L. L. Eliot: Development of a test of speech intelligibility in noise using sentence materials with controlled word predictability, *J. Acoust. Soc. Am.*, **61**(5), 1337–1351, 1977.
17. A. L. Lalime, M. E. Johnson: Development of an efficient binaural simulation for the analysis of structural acoustic data, *NASA Technical Report*, CR-2002-211753, 2003.
18. B. C. J. Moore, B. R. Glasberg: A revised model of loudness perception applied to cochlear hearing loss, *Hearing Research*, **188**(1), 70–88, 2004.
19. S. Müller, P. Massarani: Transfer function measurement with sweeps, *Journal of the Audio Engineering Society*, **49**(6), 443–471, June 2001.
20. M. J. Nilsson, S. D. Soli, J. A. Sullivan: Development of a hearing in noise test for the measurement of speech reception thresholds in quiet and in noise, *J. Acoust. Soc. Am.*, **95**(2), 1085–1099, 1994.
21. R. Patterson, et al.: SVOS Final Report: The Auditory Filter Bank, *Technical Report 2341,* Cambridge, UK: MRC Applied Psychology Unit, 1988.

22. A. D. Pierce: Acoustics: *An Introduction to its Physical Principles and Applications*, Woodbury, NY, USA: The Acoustical Society of America, 1991.
23. L. Robles, M. A. Ruggerio: Mechanics of the mammalian cochlea, *Physiological Reviews*, **81**, 1305–1352, 2001.
24. D. Rocchesso, J. O. Smith: Circulant and elliptic feedback delay networks for artificial reverberation, *IEEE Trans. Speech and Audio Process.*, **5**(1), 51–63, 1997.
25. M. B. Sachs, I. C. Bruce, R. L. Miller, E. D. Young: Biological basis of hearing aid design, *Annals of Biomedical Engineering*, **30**(2), 157–168, 2002.
26. L. Savioja, J. Huopaniemi, T. Lokki, R. Vaananen: Virtual environment simulation – Advances in the DIVA project, *Proc. ICAD '97*, 43–46, Palo Alto, CA, USA, 1997.
27. L. Savioja: *Modelling Techniques for Virtual Acoustics*, PhD thesis, Helsinki University of Technology, 1999.
28. R. Shilling, B. Shinn-Cunningham: Virtual auditory displays, in K. M. Stanney (ed.), *Handbook of Virtual Environments: Design, Implementation, and Applications*, 65–92, Mahwah, NJ, USA: Lawrence Erlbaum, 2001.
29. C. Taishih, et al.: Multiresolution spectrotemporal analysis of complex sounds, *J. Acoust. Soc. Am.*, **118**(2), 887–906, 2005.
30. L. J. Trainor, K. Winklung, J. Bondy, S. Gupta, S. Becker, I. C. Bruce, S. Haykin: Development of a flexible, realistic hearing in noise test environment (R-HINT-E), *Signal Processing*, **84**(2), 299–309, 2004.
31. R. Vaananen: *Efficient Modeling and Simulation of Room Reverberation*, Masters Thesis, Helsinki University of Technology, 1997.
32. R. Vaananen, V. Valimaki, J. Huopaniemi, M. Karjalainen: Efficient and parametric reverberator for room acoustics modeling, *Proc. ICMC '97*, 200–203, Thessaloniki, Greece, 1997.
33. L. A. Westerman, R. L. Smith: A diffusion model of the transient response of the cochlear inner hair cell synapse, *J. Acoust. Soc. Am.*, **83**(6), 2266–2276, 1988.
34. K. Wiklund, R. Sonnadara, L. Trainor, S. Haykin: R-HINT-E: a realistic hearing in noise test environment, *Proc. ICASSP '04*, **4**, 5–8, Montreal, Canada, 2004.
35. X. Zhang, M. G. Heinz, I. C. Bruce, L. H. Carney: A phenomenological model for the responses of auditory-nerve fibers: I. Nonlinear tuning with compression and suppression, *J. Acoust. Soc. Am.*, **109**(2), 648–670, 2001.
36. M. S. Zilany, I. C. Bruce: Modeling auditory-nerve responses for high sound pressure levels in the normal and impaired auditory periphery, *J. Acoust. Soc. Am.*, **120**(3), 1446–1466, 2006.

17

Dynamic Sound Control Algorithms in Automobiles

Markus Christoph

Harman/Becker Automotive Systems, Germany

Automobiles comprise acoustical environments far from being considered desirable. Therefore, it is necessary that acousticians conduct a vehicle dependent tuning, in which they try to counteract the deficiencies of the room, the loudspeakers, the way the loudspeakers are embedded in the interior and so on. This is achieved by manipulating the channel dependent phase and magnitude responses. As result they are able to enhance the acoustic quality of the system, providing a more enjoyable sound experience. The whole tuning is done while the car is standing still and the engine is turned off.

Usually passengers listen to any kind of sound source whilst driving. Therefore, they inevitably face background noise of any coloration or intensity. It is the task of a dynamic sound control algorithm to compensate for the dynamic changes of the background noise in order to keep the sound impression like it was after the tuning, which was done without any disturbing background noise.

In this chapter we will give an overview of different dynamic sound control systems. Starting with systems controlling just the volume by utilizing non-acoustical and/or acoustical sensor signals, we will turn to more advanced systems whose task it is to control the whole spectrum. This can be considered as a dynamic equalization control.

Finally, we will describe a spectrum-based dynamic equalization control algorithm, applying a psychoacoustic masking model, in more detail, discussing theoretical as well as practical aspects. The chapter will close with the summary of the discussed approaches and by providing a look in the future of forthcoming systems.

17.1 Introduction

Since a long time the automobile industry asks for acoustical systems, able to counteract against adverse noise conditions. Even before the first digital audio systems emerged in the early nineties, the first attempts had been conducted in the late seventies respectively the early eighties. Systems of that time used a microphone in order to control the volume of analog sound systems [32, 37, 42, 47, 48, 57, 63, 65, 66]. Some of these systems even made their way into commercial products but – due to instability problems, which could not be solved at that time with the available analog technology – they disappeared from the marked rather quickly. Instead of the real noise one started to utilize non-acoustical signals which showed a certain coherence to the background noise measured within the interior of a car, which was most of the time the speed signal [6, 30, 41].

17.1.1 Introduction of Dynamic Volume Control Systems

With the appearance of digital signal processors (DSPs) within acoustic systems the possibilities of such speed dependent sound systems increased dramatically. Now one was able to design any kind of mapping function between the speed respectively the speed difference and the corresponding volume respectively the volume difference. Later research found that the use of a static mapping function did not deliver the desired results, so they included a volume dependency in the systems, leading now from a mapping function to a mapping matrix [67], which increased the intricacy of the tuning effort quite a bit. Further investigation showed that a separate treatment of the low frequencies[1] promised even better results, which of course made it even harder to keep control of the system [7, 16].

It was clear that any kind of dynamic volume control system, based on non-acoustic sensor signals would finally fail to fit the customer's requirements at certain conditions. Thus, some of the original equipment manufacturers (OEMs) turned at that point their research interests again to microphone based volume control systems [2, 8, 43, 44, 50, 62]. Now, after the DSPs made a big leap forward in terms of memory as well as processing power, it was also possible to implement adaptive filters, which one absolutely needs to stabilize the system. By the way, this was the reason why the microphone based analog systems never functioned properly. In the mid nineties one of the first stable and therefore working microphone based volume control systems occurred at the market [2, 50–52]. It worked almost as well as the best speed-dependent volume control systems, but suffered from certain deficiencies as well. One of the major shortcoming was, that the system was unable to react to almost anything else but road-noise. Wind-noise, noise from the fan or the defrost as examples did not trigger the systems at all. The reason behind this drawback was, that this system only used the lower spectra up to $f \approx 160$ Hz. The reason why the inventors limited themselves to this area was on the one hand to get rid of the disastrous effect of voice to the control signal in an easy way [3, 36] and on the other hand to save processing time as well as memory.

---

[1] In the following we will call the low frequency range also *bass range* or simply *bass*.

Other systems tried to counteract towards those deficiencies by combining the microphone signal with other non-acoustical sensor signals like the already known speed-signal, the revolutions per minute (RPM) signal, signals which indicate whether the windows have been opened or not, or for example if the fan is switched on or off and if it is switched on, at which level it is operating among others [13–15, 17, 39]. With the appearance of signal buses like the CAN[2] bus or the MOST[3] bus in modern infotainment[4] systems, it is nowadays possible to get hold on those signals.

Some of the new systems (e.g. [13–15, 17]) already utilized a memory-less smoothing filter known e.g. from [1] in order to get rid of disturbing impulsive noises like voice within the microphone signal.

Besides the fact that the lower frequency range already acquired special treatment in some applications the systems described so far more or less just control the volume and do not utilize the spectral behaviour or the background noise signal. Fig. 17.1 shows a typical behaviour of the background noise power spectral density (PSD) of a car, which has a distinct, lowpass–like spectral shape and varies approximately by ≈ 40 dB over speed.

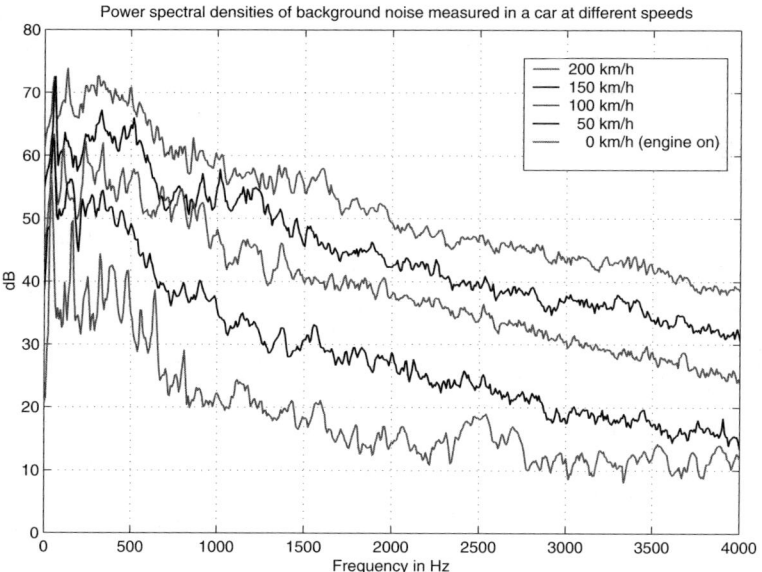

Fig. 17.1. Power spectral densities recorded in a Mercedes S-class at different speed levels.

---

[2] The term *CAN* abbreviates *controller area network*.
[3] The abbreviation *MOST* stands for *media oriented systems transport*.
[4] The term *infotainment* is a newly created word originating from *information* and *entertainment*.

17.1.2 Introduction of Dynamic Equalization Control Systems

After gaining some experience with some dynamic volume control (DVC) algorithms we learned that even with a special treatment of the low frequencies the desired naturalness of the sound coloration could not be attained. Mostly the sound occurred to be too dull, in other words not having enough bass, especially when driving at high speeds. Therefore, some research attempts tried to solve this problem by applying the DVC algorithm in several frequency bands, leading to the first kind of dynamic equalization control (DEC) algorithms [18, 22, 24]. Another idea was to directly use the spectral shape of the power spectral density of the background noise as an equalization function, beside the already applied DVC gain. Therefore one could use the linear predictive coding (LPC) analysis for a direct conversion of the isolated (time domain) background noise signal, respectively its short-term autocorrelation estimate, into an infinite impulse response (IIR) filter coefficient vector (reflection filter coefficients of a prediction error filter), approximating the current shape of the noise PSD [18, 19].

Besides the fact that these time domain based DEC algorithms did show better results than the DVC systems introduced previously, they still suffered from some inadequacies. Impulsive disturbances could not be extracted from the broadband noise signal prior its LPC analysis, which eventually lead to undesired noise PSD estimations, i.e. equalization trajectories. Low update rates of the LPC analysis did soothe this effect but were unable to completely avoid it. Additionally, we noticed that a direct application of the trajectory of the noise PSD as an equalization function did not result in the expected behaviour. Especially tests with artificial narrow band signals, serving as noise, showed that the resulting equalizing did not lead to a natural sound. Hence we were forced to find a different solution.

Turning from the time into the spectral domain opened new possibilities to solve the remaining problems. On the one hand we could now solve the problem accompanied with the impulsive disturbances by applying the memory-less smoothing filter [1] at each frequency bin, without losing the natural shape of the noise PSD. Furthermore, we could now use spectral domain adaptive filter algorithms with its superior properties compared to time domain implementations [11, 35, 58]. In the course of research we finally found that we have to take psychoacoustic properties into account when computing the "acoustically relevant" equalization function. Therefore if we talk about "psychoacoustic" we actually mean masking respectively the masking threshold [69]. There exists several methods of calculating the masking threshold of a signal, e.g. the masking model according to Johnston [40] or one of the standardized ISO/IEC[5] masking models [38]. Due to the fact that both of the previously mentioned masking models almost delivered the same results, we decided to use the masking model according to Johnston because of its

---

[5] The term *ISO* stands for *international standardization organization* and *IEC* stands for *international electrotechnical commission*.

simplicity and its scalability. This model is described in more detail in [56] and [55], whereas in the latter, the ISO/IEC 11172_3 masking model is described as well, but in a far more convenient way as e.g. in [38]. Utilizing the masking model it is thereby not sufficient to apply this model solely to the approximated noise signal in order to get the desired equalizing function. In fact we have to apply it to the source signal as well. For a better explanation: It is clear that we only have to raise those spectral components of the source signal which are masked by the current noise signal. Those spectral parts of it, that already exceed the masking threshold of the background noise need not to be equalized at all. Therefore, we have to use the ratio of the two masking thresholds to get the desired equalization function. In doing so it is logical that we will only apply an equalization which is able to raise the spectral gain – attenuation does not make any sense and thus will be therefore be avoided [20].

The organization of this chapter is as follows: In Sec. 17.2 a review of some dynamic volume control systems will be given which had been or still are employed in current audio systems. Therefore we will introduce the development of non-acoustical sensor based methods, microphone based principles, as well as a mixture of both. In Sec. 17.3 we will describe a possible realization of a spectrum-based dynamic equalization system. This forms the main part of this chapter and will be explained in detail. All systems will be discussed and analyzed more from a practical point of view then from a theoretical one – whereas in some parts the theory will also have its legitimate room.

17.2 Previous Systems – Description and Analysis

17.2.1 Speed Dependant Sound systems

Speed dependent sound systems have been in use for many years in practical signal processing applications to control – in the simplest form solely – the volume $V(n)$, in other designs additionally the bass. Thereby the low frequencies are usually controlled by a second order parametric bandpass filter

$$H_{\text{BP}}(z,n) = \gamma(n) \left[1 - \frac{\alpha(n) + \beta(n)\, z^{-1} + z^{-2}}{1 + \beta(n)\, z^{-1} + \alpha(n)\, z^{-2}} \right], \quad (17.1)$$

with

$$\gamma(n) = 1 + \frac{1}{2}\left(10^{\frac{G(n)}{20}} - 1\right), \quad (17.2)$$

$$\alpha(n) = \frac{1 - \tan\left(\pi f_{\text{cut}}(n)/f_s\right)}{\tan\left(\pi f_{\text{cut}}(n)/f_s\right) + 1}, \quad (17.3)$$

$$\beta(n) = \cos\left(2\pi f_{\text{center}}(n)/f_s\right)\left[\alpha(n) - 1\right]. \quad (17.4)$$

The quantity f_s denotes the sampling frequency. In some cases only the filter gain $G(n)$ and in more advanced cases also the cut-off frequency $f_{\text{cut}}(n)$ respectively the quality factor $Q(n)$, which is defined as

$$Q(n) = \frac{f_{\text{center}}(n)}{2\,f_{\text{cut}}(n)}\,, \tag{17.5}$$

can be controlled. OEM systems will be – in contrast to after market products[6] – adjusted by acousticians in an "online-manner" with the help of a "tuning-tool". This is a program that comes with a certain product and allows the expert to adjust, e.g., the range in which the quality factor $Q(n)$ can be modified and how this value should be changed by the current speed signal. In other words, one is able to set range limits as well as to design mapping functions. Typical range limits are:

$$\begin{aligned}
\text{cut-off frequency} \quad & f_{\text{cut}}(n) \approx 20 \text{ Hz} \ldots 200 \text{ Hz}, \\
\text{quality factor} \quad & Q(n) \approx 0.1 \ldots 3.0, \\
\text{gain} \quad & G(n) \approx -20 \text{ dB} \ldots 10 \text{ dB}, \\
\text{volume} \quad & V(n) \approx -20 \text{ dB} \ldots 10 \text{ dB}.
\end{aligned} \tag{17.6}$$

Usually after market products do not offer such possibilities, but some provide the user at least with the possibility to choose one out of several predefined sensitivity levels. Each level makes the system react more or less aggressively at a certain speed. In some speed dependent volume control systems the acoustician only has the opportunity to adjust a single mapping function. This is far from being ideal, due to the fact that there exists a certain dependency between the mapping function and the volume setting utilized during the tuning session. A better performance can be achieved if one also takes the volume dependency into account during the tuning session. Of course, that this increases the tuning effort but the result is worth the extra work, especially in OEM applications because it only has to be done once for each car-type. Furthermore, such systems are not very demanding, neither in terms of processing power, nor in terms of memory. Therefore, a duplication of the algorithm can, in view of the resources, be achieved easily.

In Fig. 17.2 the signal flow diagram of a speed and (manual) volume depending (automatic) volume control algorithm is shown. At this point it should be clear that one or several adjustments could be optional, such as the cut-off frequency $f_{\text{cut}}(n)$, the filter gain $G(n)$ and the filter quality $Q(n)$ of the parametric bandpass filter $H_{\text{BP}}(z,n)$, as well as the volume dependence of all parameters. In the simplest form only the speed-dependent volume control, which can be adjusted by a simple mapping function between speed and volume gain, will remain.

---

[6] In the automotive business *after market products* mean the addition of non-factory parts, accessories and upgrades to a motor vehicle.

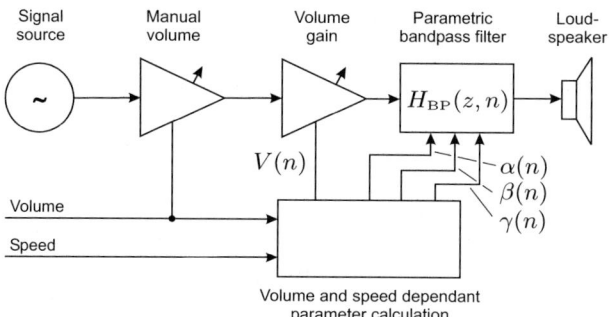

Fig. 17.2. Signal flow diagram of the speed-depending volume control algorithm.

Fig. 17.3 shows screen shots of a typical tuning tool. In the upper plot we see an example how one can adjust the volume gains, depending on the speed as well as on different volume settings. In this example the acoustician has to adjust 10 speed dependent volume gains for each of the 5 separate volume settings, which sums up to an adjustment of 50 individual volume gain values. In the lower part of Fig. 17.3 the possibilities for the adjustment of the parametric bandpass filter $H_{\mathrm{BP}}(z,n)$ can be seen. Therefore we realize that, in this example, the filter-gain $G(n)$ as well as the filter quality $Q(n)$ can be adjusted, in a speed and volume dependant manner – leading to another 100 speed and volume dependant settings. One thing that is missing in the upper example is the speed and volume dependant adjustment of the filter cut-off frequency $f_{\mathrm{cut}}(n)$. In the upper example we face a system that does not support such a speed respectively volume dependant adjustment. Here the value of $f_{\mathrm{cut}}(n)$ could only be adjusted to a fix value, which is settled around $f_{\mathrm{cut}} \approx 60$ Hz.

In Fig. 17.4 one can see a typical, three dimensional representation of a speed and volume dependent volume-gain matrix. The usual dynamic of the volume-gain matrix does not exceed 12 dB, if the system is applied within luxury cars. Otherwise, if applied to economy cars, the dynamic will, of course, increase a bit.

17.2.2 Microphone Based Dynamic Volume Control Sound Systems

As we already mentioned in Sec. 17.1, microphone based control systems constituted one of the first attempts to solve the problem of adjusting sound systems regarding the current background noise. Being restrained – in the early days – to analog solutions it was almost impossible to suppress the influence of the source signal, picked up by the microphone to the resulting control signal(s), e.g. the volume-gain. With the appearance of DSPs in sound systems it was now possible to implement adaptive filters which are necessary

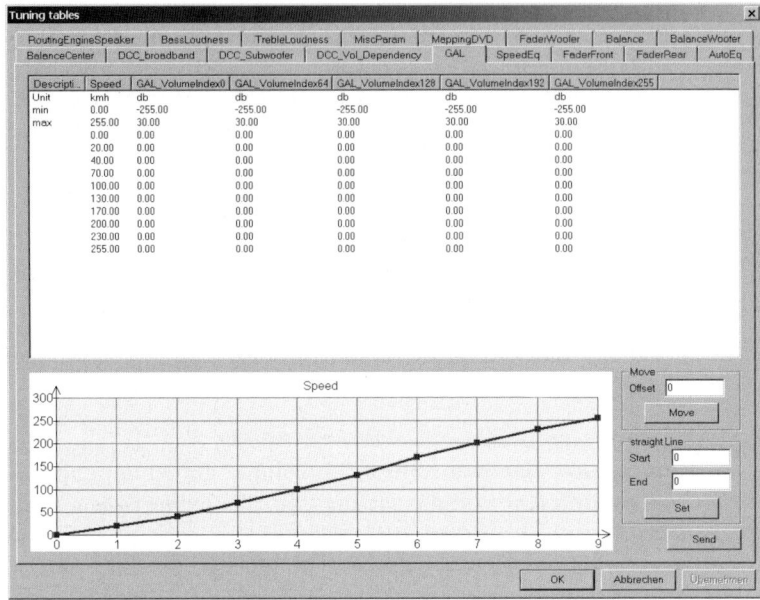

(a) Volume tuning.

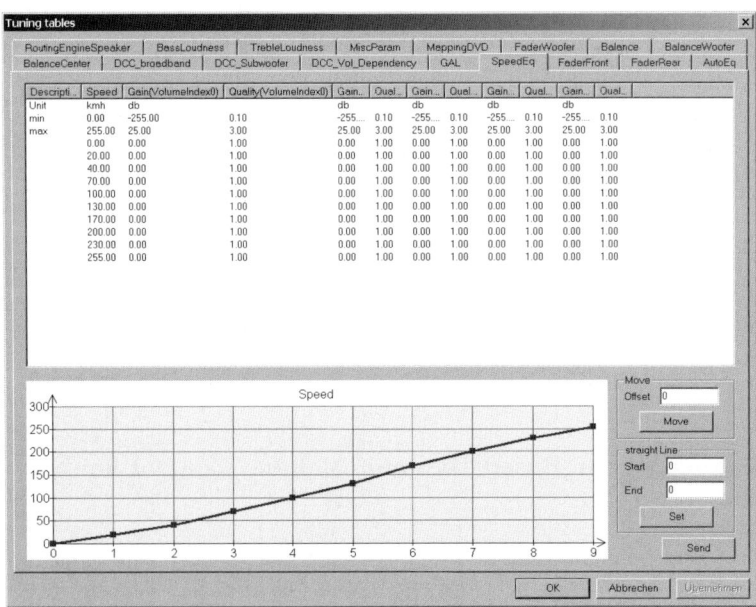

(b) Equalizing tuning.

Fig. 17.3. Tuning example of a multiple volume respectively speed dependent volume and equalization control.

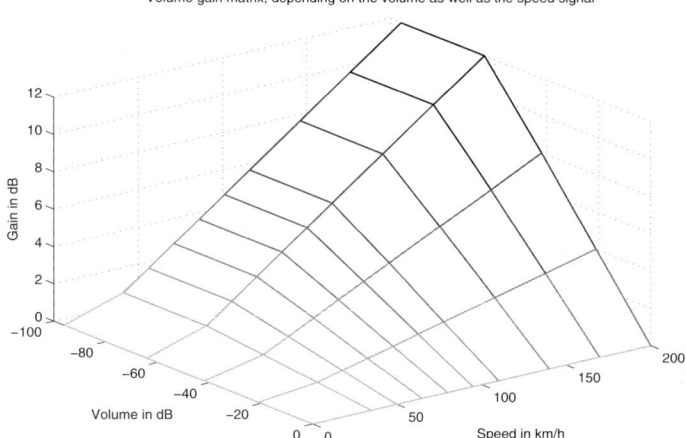

Fig. 17.4. Volume-gain matrix depending on the volume as well as speed signal setting.

to separate the background noise from the source signal, picked up by the microphone.

In this section we will introduce some systems which have been employed in practical sound systems, starting from the simplest form, coming to more advanced solutions.

Fig. 17.5 displays the principle of a microphone based DVC algorithm. Here the volume of the input signal $x(n)$, which represents the source signal, will be controlled statically by the, in Fig. 17.5 not explicitly shown manual volume setting, as well as dynamically regarding the approximated, current estimated background noise signal $\widehat{b}(n)$.

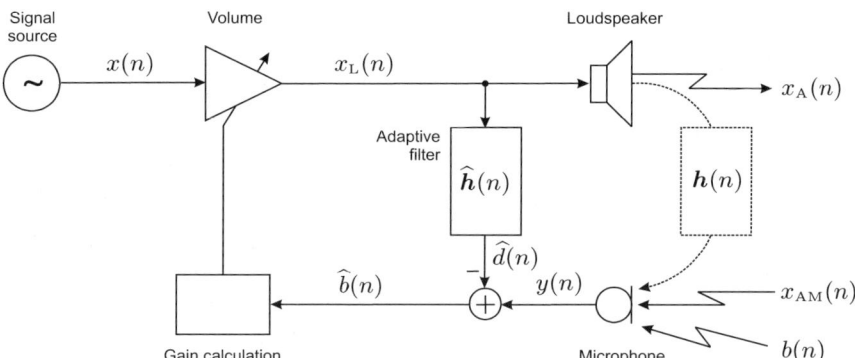

Fig. 17.5. Signal flow diagram of a microphone based dynamic volume control principle.

17.2.2.1 Adaptive Echo Cancellation

In its simplest form $\widehat{b}(n)$ equals the error signal $e(n)$ of the adaptive filter, where the filtered source signal, which drives the loudspeaker $x_\mathrm{L}(n)$ represents the first input signal and the microphone signal $y(n)$ forms the second input of the adaptive filter. Beside the background noise the microphone also receives the by the unknown system $\boldsymbol{h}(n)$ filtered radiated loudspeaker signal $x_\mathrm{AM}(n)$, which corresponds to the desired signal $d(n)$. Usually, the adaptive filter $\widehat{\boldsymbol{h}}(n)$ approximates the unknown system and subtracts the, with $\widehat{\boldsymbol{h}}(n)$ filtered loudspeaker signal, which now represents the estimated desired signal $\widehat{d}(n)$ from the microphone signal to get the error signal $e(n)$ which is the approximation of the background noise signal $\widehat{b}(n)$. If the normalized least mean square (NLMS) algorithm is utilized the adaptive filter can therefore be summarized as follows:

$$\widehat{d}(n) = \widehat{\boldsymbol{h}}^\mathrm{T}(n)\,\boldsymbol{x}_\mathrm{L}(n)\,, \tag{17.7}$$

$$\widehat{b}(n) = e(n) = \widehat{d}(n) - y(n)\,, \tag{17.8}$$

$$\widehat{\boldsymbol{h}}(n+1) = \widehat{\boldsymbol{h}}(n) + \mu \frac{e(n)\,\boldsymbol{x}_\mathrm{L}(n)}{\|\boldsymbol{x}_\mathrm{L}(n)\|^2}\,, \tag{17.9}$$

with

$$\widehat{\boldsymbol{h}}(n) = \left[\widehat{h}_0(n), \widehat{h}_1(n), \ldots, \widehat{h}_{N-1}(N)\right]^\mathrm{T}, \tag{17.10}$$

$$\boldsymbol{x}_\mathrm{L}(n) = \left[x_\mathrm{L}(n), x_\mathrm{L}(n-1), \ldots, x_\mathrm{L}(n-N+1)\right]^\mathrm{T}. \tag{17.11}$$

With the help of the approximated background noise signal $\widehat{b}(n)$ and the loudspeaker signal $x_\mathrm{L}(n)$ we can now calculate a control signal $r_\mathrm{com}(n)$ for a dynamic compressor.

17.2.2.2 Dynamic Compression

Fig. 17.6 demonstrates a simple version of how the ratio $r_\mathrm{com}(n)$ for a dynamic compressor can be calculated. We utilize a dynamic compressor because this simple signal processing tool offers the possibility to, not only adjust the volume gain, respectively the dynamic ratio with respect to the momentarily predominant background noise level, but also in taking the current signal level into account, by forming the final volume gain. Usually the short-term power of $\hat{b}(n)$ varies slowly with time, whereas the dynamic of a typical source-signal $x_\mathrm{L}(n)$, such as music alters much faster. For example: If the music has, at a certain time, a level which exceeds the level of the more or less static background noise, it is clear, that we should not increase the volume at all. But the more the music level drops below the level of $\hat{b}(n)$ the more we have to rise the volume-gain.

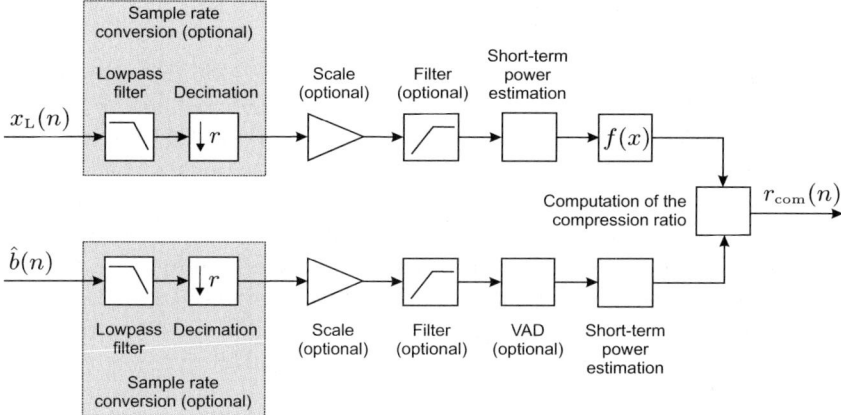

Fig. 17.6. Signal flow diagram of a DVC dynamic compressor.

17.2.2.3 Decimation and Highpass Filtering

Fig. 17.6 also shows many optional signal processing blocks, such as decimating stages with the accompanying downsampling lowpass filter. In our example, the decimation stages only make sense, if the adaptive filter operate at the same sampling rate. Generally, in automobiles by far the strongest part of the background noise lies within the lower frequencies up to approximately 1000 Hz ([33], p. 22ff). Hence, by just using the levels of $x_\mathrm{L}(n)$ and $\widehat{b}(n)$, it is sufficient to calculate the level within a (highly) reduced spectral range, which additionally saves processing time as well. The next optional processing block of Fig. 17.6, worth to be mentioned, is the high pass filter, applied within the $\widehat{b}(n)$ branch, which we call the balancing highpass filter. Its job is to balance the spectrum of the background noise such, that the system will be able to react equally to the very intense road-noise or engine-noise as well as to much less intense disturbances, originating from wind noise coming, e.g. from the fan or an open window. Thereby we have to take the spectral distribution of the different noise sources into account by adjusting the balancing highpass filter adequately.

17.2.2.4 Voice Activity Detection

Last, but not least we also face a block called *VAD*, which stands for *voice activity detection*. The purpose of this block is, as we can already guess by its name, to diminish the influence, especially of voice, but generally of all sorts of impulsive disturbances, to which also the slamming of a door can be counted.

17.2.2.5 The "Gain-Chase" Problem

Despite the application of an adaptive filter it is still possible that the whole system may run into a feedback problem, which we call "gain chase". Therefore the danger of feedback increases as the volume is increased. To avoid such a worse case scenario which, would eventually end in increasing the volume to its maximum, we inserted a function-block denoted with $f(x)$ into the upper branch of Fig. 17.6, which as we see, is not an optional block anymore, due to its importance to the overall stability of the system. The task of the function $f(x)$ is to reduce the influence of the source signal $x_L(n)$, depending on the current volume setting, on the calculation of the final compression-ratio and/or the volume-gain, resulting in a volume depending dynamic range of the DVC algorithm. If we denote the output of the short-term power estimations with $\overline{x_L^2(n)}$ the gain chase function can be realized e.g. as a multiplication with a gain factor that depends on the short-term input power:

$$f\left(\overline{x_L^2(n)}\right) = G_{\mathrm{ldg}}\left(\overline{x_L^2(n)}\right) \overline{x_L^2(n)}. \tag{17.12}$$

An example for a level dependent gain factor is depicted in part (b) of Fig. 17.7. In other words the higher the volume setting of the operator, the lower the remaining dynamic of the DVC system becomes, up to a value where we even deactivate the whole system. At such volume settings the audio signal is considered as loud enough that even the highest background noise levels should not be able to severely mask the audio signal. If we denote the estimated short-term power of the background noise with $\overline{b_{\min}^2(n)}$ we can compute the compression ratio as a function of $\overline{b_{\min}^2(n)}$ and $f(\overline{x_L^2(n)})$

$$r_{\mathrm{com}}(n) = f_r\left(\overline{b_{\min}^2(n)},\, f\left(\overline{x_L^2(n)}\right)\right). \tag{17.13}$$

If we take a look at part (a) of Fig. 17.7 we see what happens in terms of the volume gain at certain volume levels when the level of the background noise $\hat{b}(n)$ varies within a range of 0 dB, ..., 100 dB. The graph also shows that in this example, up to a volume level of approx. -40 dB the full dynamic range of 12 dB will be at the DVC systems command. From there on the maximal dynamic of the DVC system gradually decreases, following therefore the trajectory of the tunable anti gain chase function, shown in part (b) of Fig. 17.7. As soon as a volume level exceeds a level of approx. -17 dB there is no dynamic left any more. This means that at this point the whole DVC system will be deactivated. The anti gain chase function can again be tuned with the help of a table as already shown, e.g., in Fig. 17.3.

17.2.2.6 Estimation of the Background Noise Level

Instead of using a classical VAD algorithm in which the estimation of the background noise will be blocked during speech intervals, we preferred to use

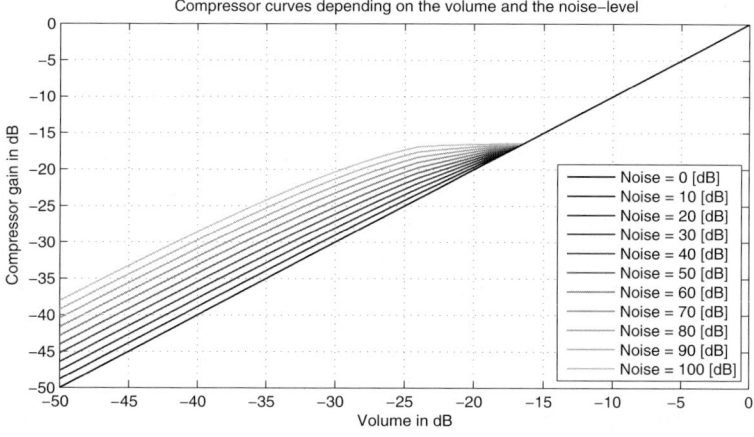

(a) DVC dynamic compressor curves.

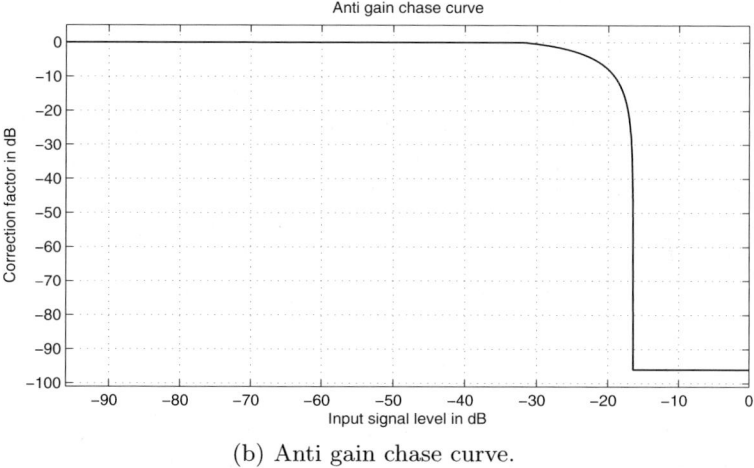

(b) Anti gain chase curve.

Fig. 17.7. DVC compressor curves, corresponding to the volume as well as to the accompanying anti gain chase curve, which is responsible for its actual shape.

a different sort of background noise estimation. This method is able to deliver robust estimation values even during speech periods, to which the *minimum statistic* method, known from [49] can also be counted.

Due to its simplicity and robustness, we, in fact, used a background noise estimation scheme, shown in Fig. 17.8, based on [1], which can be interpreted as a *smoothing filter without memory*. The way in which this scheme estimates the background noise level can be described as follows: As soon as the current, preferably pre-smoothed, estimated short-term power of the background

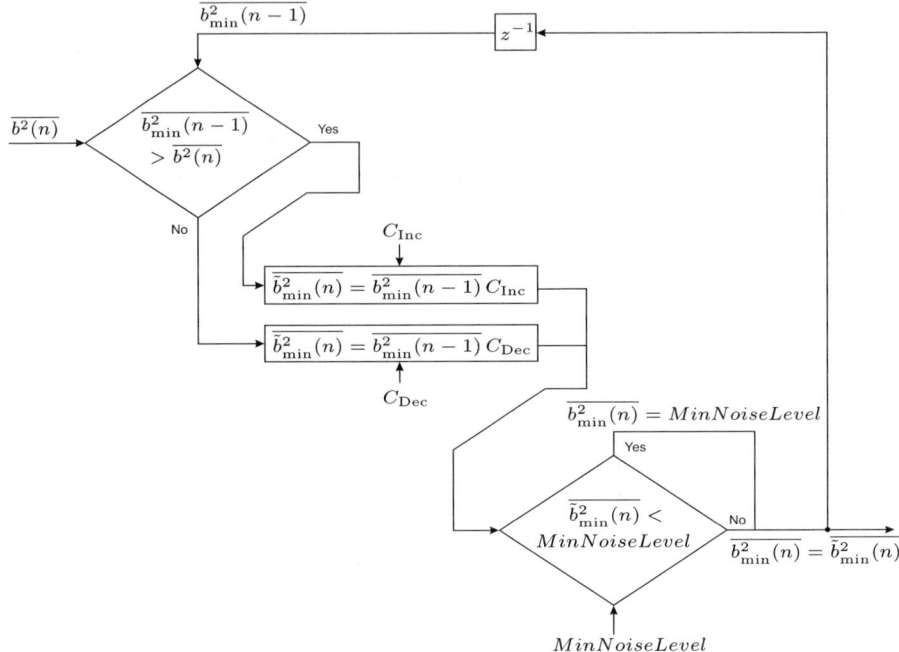

Fig. 17.8. Signal flow diagram of a memoryless smoothing filter.

noise signal $\overline{b^2}(n)$ exceeds the previously estimated background noise level $\overline{b^2_{\min}}(n-1)$, the estimate will be raised by a fix increment C_{Inc}, otherwise it will be decreased by a fix decrement C_{Dec}:

$$\overline{b^2_{\min}}(n) = \begin{cases} \max\left\{MinNoiseLevel,\, C_{\mathrm{Inc}}\,\overline{b^2_{\min}}(n-1)\right\}, & \text{if } \overline{b^2}(n) > \overline{b^2_{\min}}(n-1) \\ \max\left\{MinNoiseLevel,\, C_{\mathrm{Dec}}\,\overline{b^2_{\min}}(n-1)\right\}, & \text{else.} \end{cases}$$

(17.14)

If a car is running e.g. at a speed from 30 to 50 km/h there is no – or just a small – need to do any DVC at all, assuming ordinary operation conditions. Obviously the background noise within the car will not be zero, before the DVC actually starts its work. The threshold of this minimum allowable background noise which does not cause an operation of the DVC system can be adjusted by the parameter $MinNoiseLevel$. But the main reason for a correct adjustment of $MinNoiseLevel$ is to avoid extremely long onset times, which would be encountered if the estimated background noise level $\overline{b^2_{\min}}(n)$ would start at values close to zero, due to the usually applied very small increment values. Fig. 17.8 shows a flow diagram of the background noise estimation scheme.

Typical values for the parameters of the memoryless smoothing filter in DVC applications are:

$$C_{\text{Inc}} \approx 1 \text{ dB/s},$$
$$C_{\text{Dec}} \approx -5 \text{ dB/s}, \quad (17.15)$$
$$MinNoiseLevel \approx -60 \text{ dB}.$$

During several tuning sessions we found out, that the ratio between the linear increment $1 - C_{\text{Inc}}$ and the linear decrement $1 - C_{\text{Dec}}$ remained approximately constant within a range of

$$\frac{1 - C_{\text{Dec}}}{1 - C_{\text{Inc}}} = -4 \ldots -6. \quad (17.16)$$

At the first glance, an increment of about 1 dB/s seems very low, but on the one hand it corresponds quite well with the average noise increment that can be measured in a car when accelerating from 50 km/h, which corresponds to a typical speed when entering an expressway, to 160 km/h, which stands for a considerably high noise disturbance. On the other hand, too high values of C_{Inc} would inevitably lead to a system that reacts on short-term disturbances as they occur, e.g., during pavement changes or shifting of gears.

17.2.2.7 A First Basic System and Some Reflections About It

If we now fit all the components together we would end up in system similar to the one shown in Fig. 17.9. There, the only part that probably still needs some explaining words is the filter entitled with $H_{\text{HP}}(z)$. This filter block can be seen as an option and should be applied if the utilized microphone incorporates a highpass filter, as known as *prewhitening* in acoustic echo cancellation (AEC) applications. Otherwise, if a microphone with a constant transfer function is used, the filter block $H_{\text{HP}}(z)$ need not be inserted.

So far, the DVC system of Fig. 17.9 already solves many practical problems as listed in Tab. 17.1. In the course of the research process we found that there was still enough room for enhancements. For instance, we realized, that the filter coefficient vector of the adaptive filter $\widehat{h}(n)$ diverges whenever an impulsive noise i.e. a short-term noise burst with high energy content, such as voice, appeared. After such a destruction of $\widehat{h}(n)$ it takes a while until the adaptive filter readapts $\widehat{h}(n)$. During this time a lot of energy from the source signal cannot be correctly subtracted from the microphone signal $y(n)$, resulting in an erroneous estimated background noise signal $\widehat{b}(n)$, which can end up in an undesired increase of the volume.

17.2.2.8 Adaptation Control

This instance could be improved by the utilization of an adaptive adaptation step size

$$\mu \longrightarrow \mu(n)$$

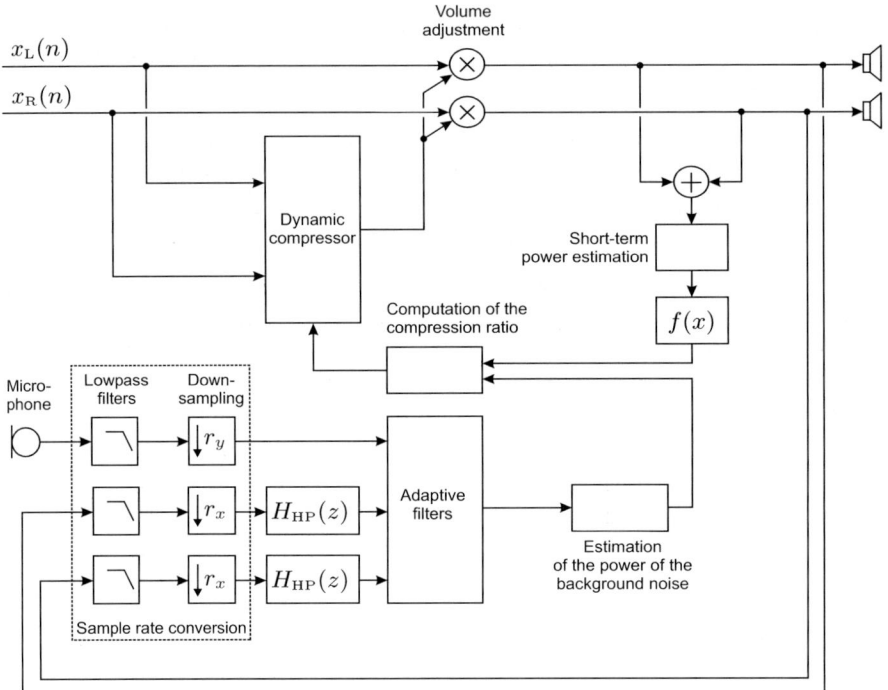

Fig. 17.9. Signal flow diagram of a simple microphone based DVC system.

(see Eqn. 17.9), realized using, e.g., the delay coefficient method [68]. That avoids destructive effects of impulsive interferences like speech signals during the adaptation process. Thus, the approximation of the loudspeaker enclosure microphone (LEM) system remains stable even during such disturbances, presuming the LEM system remains unaltered during this time, which is so in

Table 17.1. Problems solved with the simple DVC system shown in Fig. 17.9.

Problem	Solution
Separation of the source signal and the background noise within the microphone signal	Adaptive filter
Stability of the system, even at high volume settings	Anti gain chase function
Immunity to impulsive disturbances such as voice	Memoryless smoothing filter
Sensitivity to booming disturbances such as road or engine noise	Balancing high pass filter

the vast majority of cases. Step-size control leads to a system reacting much more stable in such situations. Our adaptive adaptation step size $\mu(n)$ takes two measures into account (see Fig. 17.10):

- the current signal-to-noise ratio (SNR), that can be estimated in an easy manner by taking the ratio of the smoothed source signal $\overline{|x(n)|}$ and the smoothed, approximated background noise signal $\overline{|\widehat{b}(n)|}$
- as well as the estimated system distance $Dist(n)$.

The current system distance will be estimated by utilizing the sum of the absolute values of the N_t delayed (filter) coefficients. Therefore the estimation of the system distance can be explained as follows: Due to the fact that we inserted a defined delay element in the microphone branch with a maximum length of N_t, we know how the (at least) N_t leading filter coefficients within $\widehat{\boldsymbol{h}}(n)$ must look like. The sum of the magnitudes of these N_t delayed (filter) coefficients of $\widehat{\boldsymbol{h}}(n)$ can hence be interpreted as a measure for the system distance, telling us how far the adaptive filter has already converged. If we do have to insert a delay line within the microphone branch at all, depends on the distance of the closest loudspeaker to the microphone. If this distance is long enough, such that the wave traveling from the speaker to the microphone exceeds the time represented by the desired N_t delayed coefficients, an insertion of a delay line within the microphone branch can be omitted. Otherwise, at least the difference between the minimum runtime and the N_t delayed filter coefficient have to be considered.

Now after disposing over all necessary signals, we can summarize the generation of our adaptive adaptation step size calculation as shown in Eqs. 17.17 to 17.21:

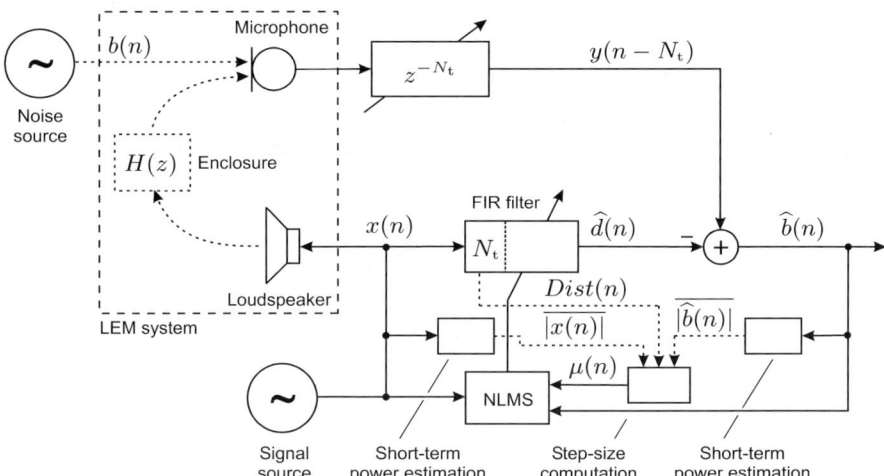

Fig. 17.10. Signal flow diagram of an adaptive filter with an adaptive adaptation step size utilizing the delay coefficient method.

$$\mu(n) = Dist(n)\, SNR(n)\,, \tag{17.17}$$

$$Dist(n) = \frac{1}{N_{\mathrm{t}}}\sum_{i=1}^{N_{\mathrm{t}}} |\widehat{h}_i(n)|\,, \tag{17.18}$$

$$SNR(n) = \frac{\overline{|x(n)|}}{\overline{|\widehat{b}(n)|}}\,, \tag{17.19}$$

$$\overline{|x(n)|} = \alpha_x\,|x(n)| + (1-\alpha_x)\,\overline{|x(n-1)|}\,, \tag{17.20}$$

$$\overline{|\widehat{b}(n)|} = \alpha_{\widehat{b}}\,|\widehat{b}(n)| + (1-\alpha_{\widehat{b}})\,\overline{|\widehat{b}(n-1)|}\,. \tag{17.21}$$

For the constants in the equations above the following settings showed good results:

$$N_{\mathrm{t}} = 5,\ldots,20\,, \tag{17.22}$$
$$\alpha_x = 0.99\,, \tag{17.23}$$
$$\alpha_{\widehat{b}} = 0.999\,. \tag{17.24}$$

17.2.2.9 Gain Mapping Function

Real-time tests showed that the unmodified DVC gain leads to an increasing over-compensation. Hence, we had to introduce a tunable *gain mapping function*, such as the one shown in Fig. 17.11 (a):

$$G_{\mathrm{out}}(n) = f_{\mathrm{GM}}\!\left(G_{\mathrm{in}}(n)\right). \tag{17.25}$$

By applying this gain mapping function we can correct for the undesired behaviour.

An explanation for this phenomenon can be found, e.g., in [69]. There it is described that humans do not mathematically correct sense the loudness of a signal. We rather have a different impression of loudness, depending on the current level respectively on the current volume of the signal, with the effect, that one has to rise its level more if a signal is played back with a small volume, on the one hand and, on the other hand one does not have to increase the level so much if played back with a high volume, to get a psychoacoustically equal increase of loudness.

Additionally we found that we had to tune the *MinNoiseLevel*, already known from Fig. 17.8 anew, every time we changed the volume. Thus, we replaced the single value *MinNoiseLevel* by yet another tunable, volume dependent function like the one shown in Fig. 17.11 (b):

$$MinNoiseLevel \longrightarrow MinNoiseLevel(n) = f_{\mathrm{MNL}}\!\left(V(n)\right). \tag{17.26}$$

The quantity $V(n)$ describes the volume that can be adjusted by the user.

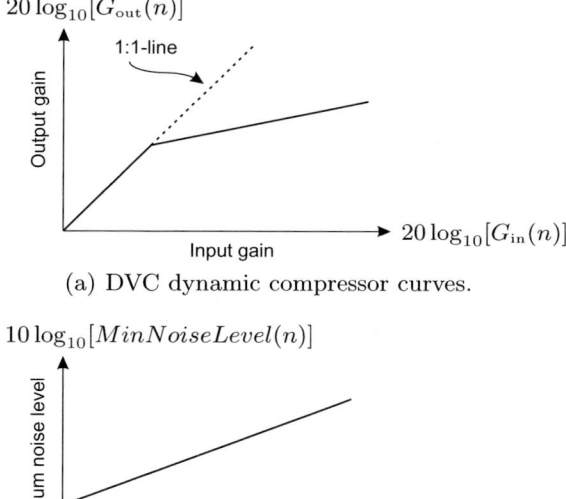

(a) DVC dynamic compressor curves.

(b) Volume dependent $MinNoiseLevel$ function.

Fig. 17.11. Gain mapping and Volume $V(n)$ dependent $MinNoiseLevel$ function.

17.2.2.10 Loudness Filter

Up to now we described DVC systems capable to exclusively modify the volume gain. During several test drives with such DVC systems we recognized that especially at high noise levels the sound became more and more lifeless. This means that the spectral balance of the musical signal had increasingly been deranged, resulting mainly in sounds with partly far too less bass. This problem could not be solved by a pure adjustment of the volume any more. In fact this issue asked for a frequency dependent change of the volume, hence an additional control using an adaptive equalization filter.

As a first and probable simplest solution to counteract this case we thought why not utilize a signal processing tool we have already been using in our audio systems for years, namely the loudness filter $H_{\mathrm{BP}}(z, n)$ as described in Sec. 17.2.1.

The question how we finally succeeded in profitably combining the loudness filter with our DVC system, capable of solely generating a volume gain can easily be answered. As shown in Fig. 17.12, we only had to place the DVC system after the loudness filter, such that volume changes, caused by the DVC system, does not affect the operation of the loudness filter. In other words we separated the volume, set by the operator of the audio system and the volume gain from the DVC system, whereby the loudness filter should only be influenced by the manually and not by the automatically adjusted volume.

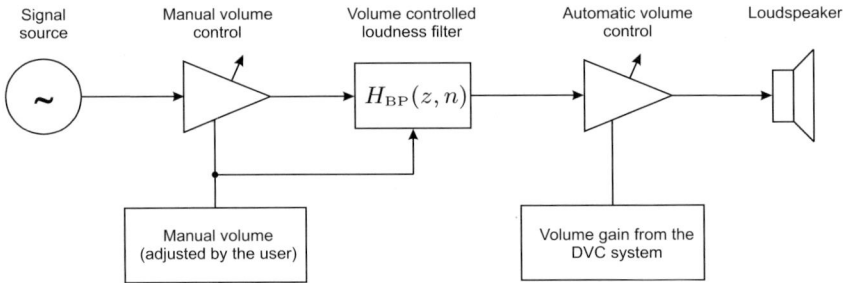

Fig. 17.12. Combination of a DVC system with a conventional loudness filter.

This little trick empowered us to generate something like a first approximation of a desired equalization on top of the DVC system.

17.2.2.11 Multi-Band Dynamic Volume Control Systems

A better performance can be reached if we split the spectrum into several subbands, applying a separate DVC system in every band, leading to a simple form of a *dynamic equalization control* (DEC) algorithm. Therefore, we also have to provide a broadband version of the estimated background noise $\hat{b}(n)$, which cannot be achieved with a single time domain adaptive filter, since the individual reference signals are modified independently. In order to get a broadband estimation of $\hat{b}(n)$ it is sufficient to split the input signal into at least three bands, where the first cut-off frequency may be settled around $f_{\text{Xover,low}} \approx 700$ Hz and the the second one around $f_{\text{Xover,high}} \approx 5000$ Hz. As an alternative one could also use the specified cut-off frequencies of the loudspeakers if two or three way speakers are installed in the car. Fig. 17.13 shows how such a time-domain multi-band DVC system with individual adaptive filters may look like.

Now after introducing several improvements of the previous DVC system, like the one in Fig. 17.9 it is time to collect the – so far achieved – interim results again (see Tab. 17.2).

Despite the fact that we already tried to give the current DVC system something like a "spectral touch" by combining it with the loudness filter, it is still a method which purely generates one or, if applied in several bands, more levels out of the estimated background noise signal $\widehat{b}(n)$. This volume gain need not be applied to all speakers respectively channels in the same manner. We also have the freedom to map the control signal individually for each channel, if desired. This can be beneficial, if we face a situation where we do have different background noise levels at different parts within

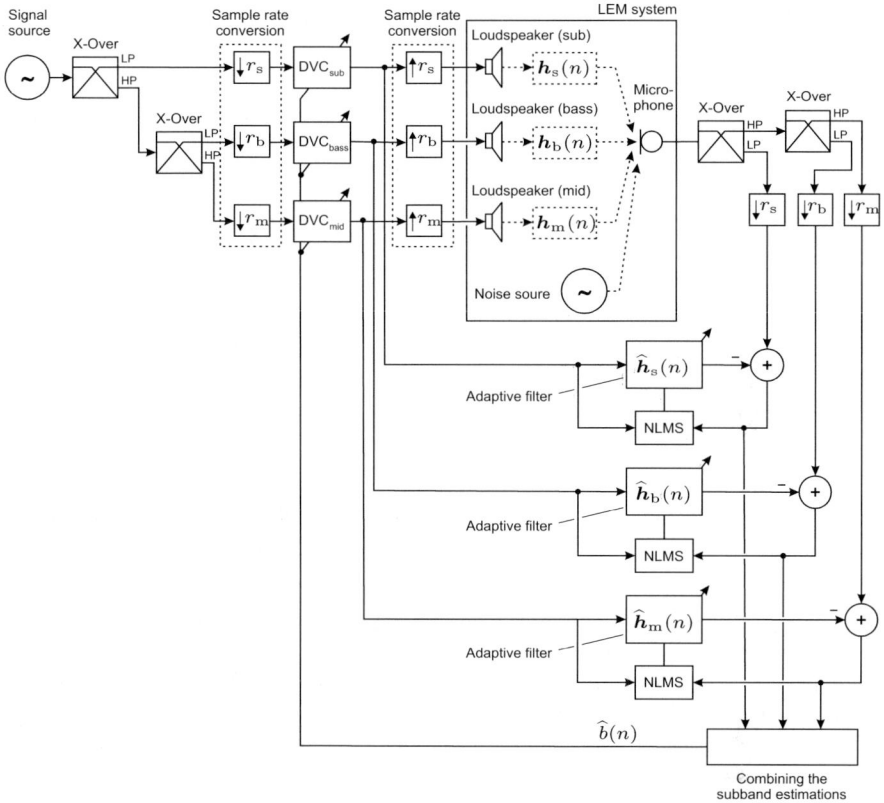

Fig. 17.13. Multi-band adaptive DVC system with adaptive filters.

the vehicular compartment, caused e.g. by a sunroof. In such cases it can be advantageous to provide the front passengers with more control dynamic than the rear passengers.

17.2.2.12 Multi-Channel Systems

Without any doubt, a much better performance could be achieved if every possible seating position within the passengers compartment had its own DVC microphone, resulting in an individualized DVC (=IDVC), respectively an individualized DEC (=IDEC) system, such as displayed in Fig. 17.14.

After elaborating the ability to get a broadband estimation of $\widehat{b}(n)$ we thought about the possibilities to realize a true DEC system, meaning that we want to design a system that operates close to what its name stands for. Such a system should be able to equalize the source signal $x(n)$ depending on the spectral shape of the estimated background noise signal $\widehat{b}(n)$.

Table 17.2. Modification and improvements for DVC systems.

Modification	Improvement
Adaptive adaptation step size	Enhancement of the adaptive filter by making it more robust against burst-like disturbances
Gain mapping function	Considering the psychoacoustically justified, level dependent loudness impression within the volume gain calculation
Volume dependent $MinNoiseLevel$ function	Increasing the initial response time of the DVC system
Combination of the DVC system with a loudness filter	Taking the spiritless low frequency performance of the DVC system into account, especially when high volume gains are required

17.2.2.13 Linear Prediction Based Equalization Systems

An efficient method of calculating such an equalization filter in the time domain, directly out of the estimated background noise signal $\widehat{b}(n)$ uses the *linear predictive coding* (LPC) method. Therefore the LPC analysis, known e.g. from [33], delivers the reflection, so-called PARCOR or LPC coefficients which, inserted into a prediction filter approximates the PSD of the current, estimated background noise signal $\widehat{b}(n)$. After computing the coefficients a filter should be applied to the source signal $x(n)$ as one can see in Fig. 17.15.

Besides a nonlinear frequency resolution, the human ear as well comprises of a level dependent sensitivity which has its maximum at a frequency from around 1 to 2 kHz. This property of the human ear should be accounted for in

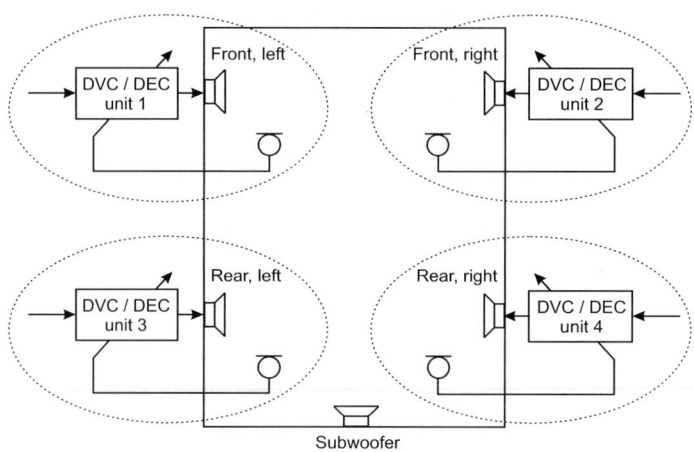

Fig. 17.14. Multi-channel DVC/DEC system.

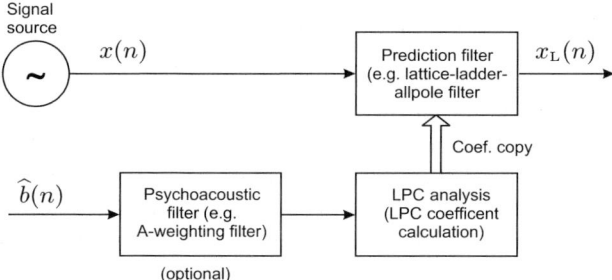

Fig. 17.15. Signal flow diagram how one can utilize an LPC analysis to get a noise dependent equalization function.

Fig. 17.15 by the optional block named *psychoacoustic filter*. As an example this can be an A-weighting filter or an ordinary highpass filter serving as a first approximation of the latter.

A problem, which has not been discussed yet is the influence of potential impulsive disturbances, like voice within $\hat{b}(n)$, on the LPC analysis. For sure, we do not want the LPC analysis being disrupted by such disturbances, hence we have to find a way to make it robust or if possible even immune against them. An easy way making the LPC analysis at least more robust to short-term burst like disturbances within $\hat{b}(n)$ is to enlarge its update time.

An efficient method how the LPC coefficients can, in an iterative fashion, be calculated is given by the so-called *gradient adaptive lattice* (GAL) argorithm as introduced, e.g., in [45]. On the one hand with the GAL we do have the possibility to slow down its update rate simply by enlarging its iteration parameter μ_{step}. On the other hand we do save in comparison to the otherwise often used Levinson-Durbin recursion quite a bit of processing power. A filter structure with which the GAL can be realized in an easy way, is known as the adaptive lattice predictor (ALP).

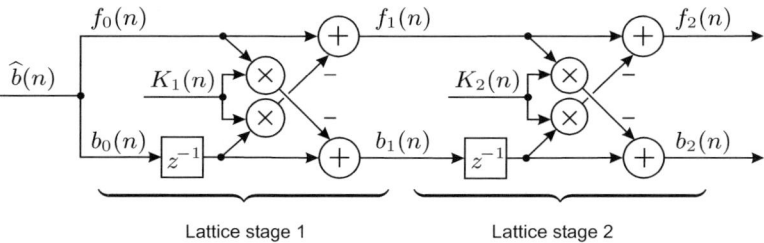

Fig. 17.16. Second order adaptive lattice predictor.

According to the ALP filter structure, where an example of an second order ALP filter is shown in Fig. 17.16, the GAL algorithm can therefore be summarized as follows:

$$K_i(n+1) = K_i(n) + \frac{\mu_{\text{step}}}{P_i(n)} \left(f_{i-1}(n) b_i(n) + b_{i-1}(n-1) f_i(n) \right) \quad (17.27)$$

with

$$P_i(n) = (1-\alpha) P_i(n-1) + \alpha \left(f_{i-1}^2(n) + b_{i-1}^2(n-1) \right). \quad (17.28)$$

Eqs. 17.27 and 17.28 have to be computed for $i = 1, \cdots, N$, whereas N denotes the filter order. It was shown, that this algorithm delivers a good performance for nonstationary signals, when $\mu_{\text{step}} = \alpha$, where α is the smoothing factor used in Eq. 17.28. To disentangle the problem of the naming, Eq. 17.27 is called the GAL algorithm, with which the LPC coefficients can be calculated. The ALP filter, such as shown in Fig. 17.16, forms the foundation on which the GAL algorithm is based on, i.e. where the GAL algorithm obtains its signals from.

As soon as we have derived the LPC coefficients from $\widehat{b}(n)$ we have to insert them into a predictor, respectively, into an allpole filter to get an approximation of the PSD trajectory of $b(n)$ as equalization filter.

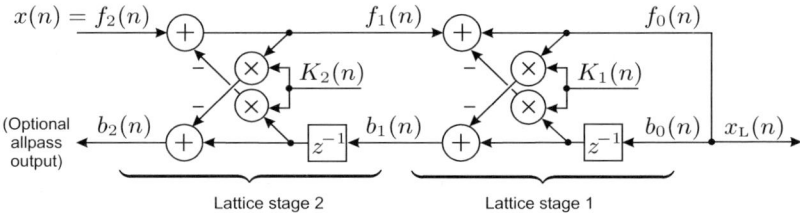

Fig. 17.17. Second order lattice allpole filter.

The lattice allpole filter – as an example, a second order version is displayed in Fig. 17.17 – is an easy way how the LPC coefficients, calculated by the GAL algorithm, can directly be utilized to implement the equalization of the source signal. Depending on the chosen order of the GAL algorithm, the LPC coefficients reflect more or less exact the PSD of the estimated background noise signal $\widehat{b}(n)$ but do not provide any information about its magnitude. Usually the magnitude of the predictor filter, fed with the LPC coefficients, is much too big, hence we have to scale it accordingly. As a simple solution to calculate a scaling factor $\widetilde{G}_{\text{L}_1}(n)$, capable to fit the maximum of the prediction filter to 0 dB, we decided to use the L_1 norm scaling, which can be described as

$$\widetilde{G}_{\text{L}_1}(n) = \frac{1}{\sum_{i=0}^{\infty} \left| h_{\text{LPC},i}(n) \right|}. \quad (17.29)$$

Theoretically we would need infinite samples of the impulse response of the prediction filter to get the exact scaling value. However, after a limited number of samples we can stop the calculation of the scaling factor without losing too much precision. We utilize this value as a new input for a first order IIR lowpass smoothing filter which delivers the scaling factor $G_{L_1}(n)$ that will finally be applied:

$$G_{L_1}(n) = \alpha_G \widetilde{G}_{L_1}(n) + (1 - \alpha_G) G_{L_1}(n-1). \qquad (17.30)$$

The smoothing helps to keep the variance of the scaling low and can be applied due to the fact that the statistics of the background noise usually does not change abruptly. After the normalization of the prediction filter to 0 dB, we have to assess to what extent the equalization – realized by the prediction filter – should be allowed to modify the source signal $x(n)$. This decision mainly depends on the current level of the background noise $b(n)$, from which we already have an estimation, represented by the DVC output signal, i.e. the volume gain $G_{\text{DVC}}(n)$. Hence, it makes sense to couple the functionalities of the predictor filter, representing the DEC system with those of the DVC system. A structure showing how such a combination could be implemented can be found in Fig. 17.18.

If we take a closer look at the DEC kernel, as shown in Fig. 17.19, we discover a similarity to the mode of operation of a parametric filter, known e.g. from [70], which indeed had been adopted in the realization of the DEC part.

The first parameter, denoted as $G_{L_1}(n)$, cares for the normalization of the parametric filter to 0 dB. The task of the second parameter, denoted as $G_{\text{DEC}}(n)$, is to decide how much equalization should be applied. This parameter depends on the estimated current background noise $\widehat{b}(n)$. It is clear that more equalization will be applied, if the background noise level and thus also the parameter $G_{\text{DEC}}(n)$ rises. The parameter $G_{\text{DEC}}(n)$ does not have to be equal to the gain parameter $G_{\text{DVC}}(n)$ that was already calculated by the DVC system. More likely it will be scaled or differently modified to fit the

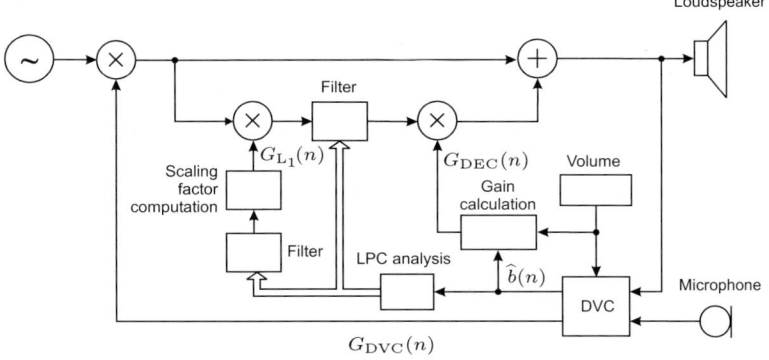

Fig. 17.18. Signal flow diagram showing a possible combination of a DVC and a DEC system.

DEC system such that we will eventually get a comfortable overall control effect.

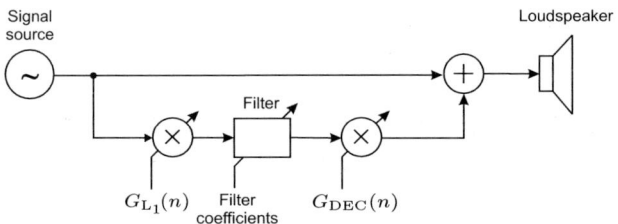

Fig. 17.19. Signal flow diagram of the DEC kernel.

Putting all the things together will lead to a DVC/DEC system as shown e.g. in Fig. 17.20.

Details within Fig. 17.20 that have not been explained yet are, e.g., the second anti gain chase function $f_2(x)$ for the control of the prediction filter, the smoothing of the (manual) volume $V(n)$ as well as the limitation of the maximally allowed system dynamic, adjustable by the parameter $D_{\max}$. Additionally, in contrast to the DVC system of Fig. 17.9 the one in Fig. 17.20 does not use a dynamic compressor for the calculation of the volume gain. It rather uses the smoothed, manual volume setting instead, which makes the whole system more robust against fluctuations of the dynamics within the source signal $x(n)$.

After performing several test rides with an audio system, incorporating such a system combining DVC and DEC functionalities, we actually could denote a certain kind of improvement, compared, for example, to the one known from Fig. 17.9. However, we still found ourselves confronted with some undesired system behaviors. First of all we made the experience that a simple reduction of the update rate of the GAL algorithm made the whole process more robust against impulsive disturbances. On the other hand this improvement was not sufficient for a practical application. Hence, we had to search for different solutions making the GAL algorithm even more robust against voice signals.

17.2.2.14 Systems with Beamforming

First of all, we came up with the idea to use a beamformer either to isolate or deliberately block signals originating from a certain location within the interior. The picture on the upper part of Fig. 17.21 therefore shows an example how a beamformer could isolate a speech signal $s(n)$. The isolated speech signal $s(n)$ can then be combined with the, also undesired source signal $x(n)$, before its mixture could then be blocked by the adaptive filter, leaving ideally only the background noise signal $b(n)$ behind. With the second option, shown

Dynamic Sound Control Algorithms in Automobiles 641

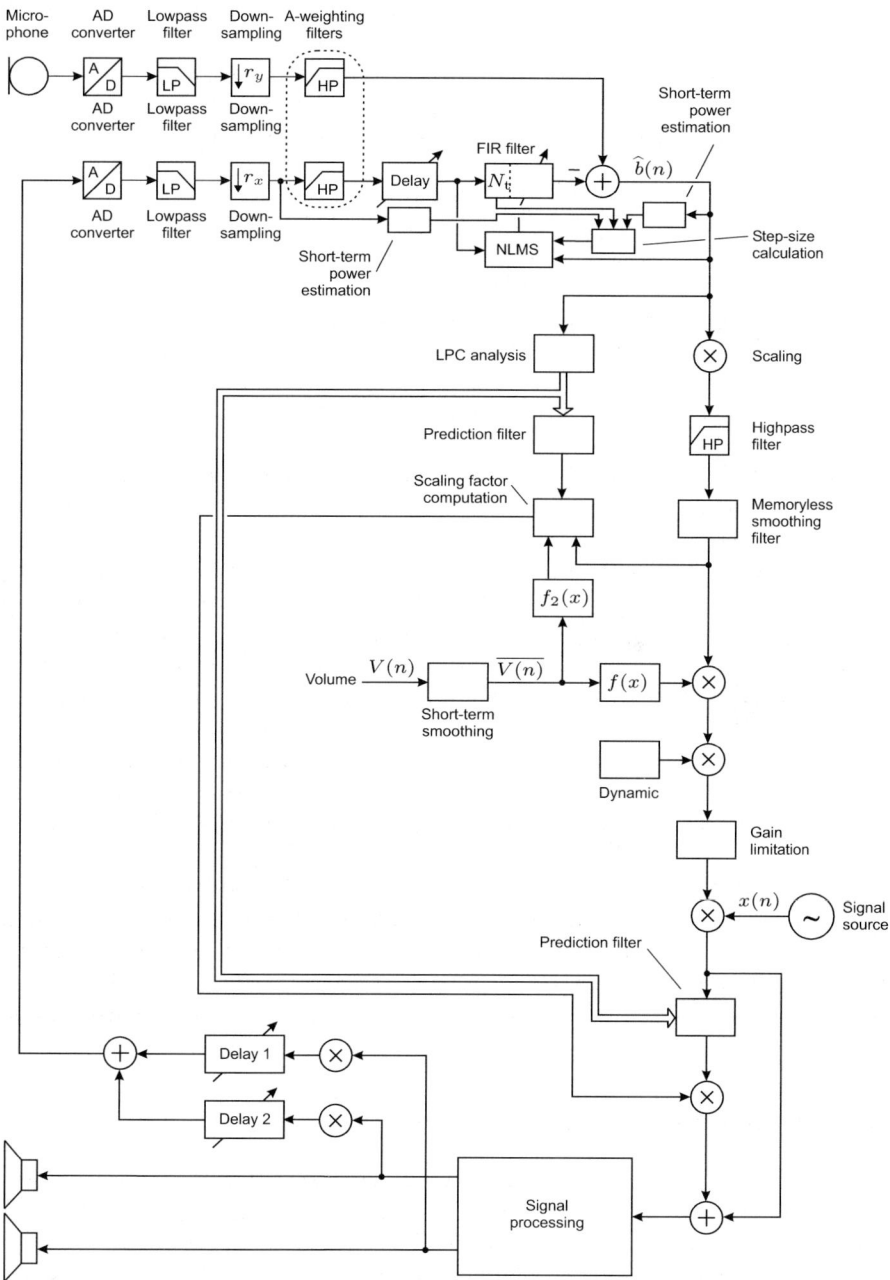

Fig. 17.20. Signal flow diagram of a DVC/DEC system.

in the lower part of Fig. 17.21 we try to get rid of the undesired speech signal $s(n)$ by blocking it with the help of a blocking beamformer, i.e. a beamformer suppressing signals originating from a defined direction. This could be done in automotive applications due to a more or less fix mounting of the seats.

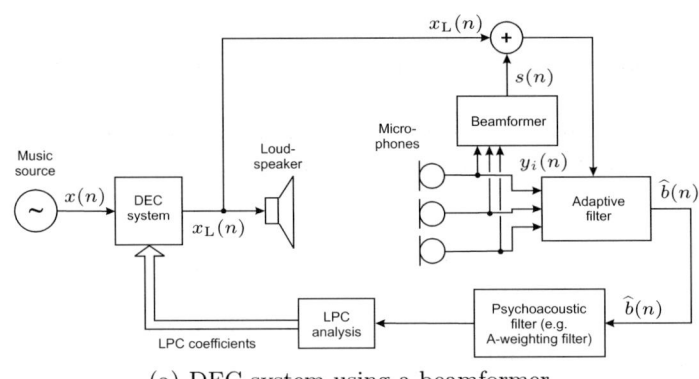

(a) DEC system using a beamformer.

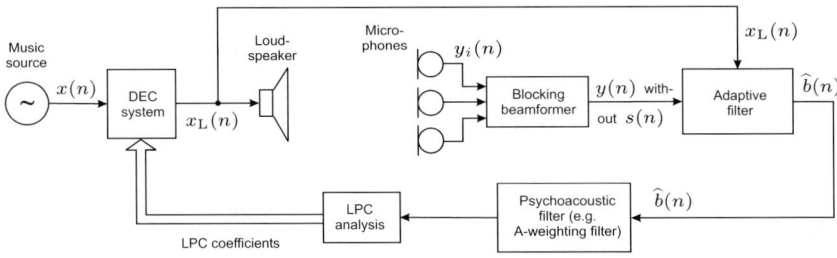

(b) DEC system using a blocking beamformer.

Fig. 17.21. DEC algorithms utilizing different types of beamformers to block undesired speech signals.

17.2.2.15 Psychoacoustic Weighting

However, the more elegant such solutions may seem the less they have to do with a practical realization. Bearing in mind that customers want to spend no or just a minimum amount of money for features such as a DVC or DEC system we know that currently we will not have a chance to install several microphones, necessary to design a beamformer. However, this could certainly be a solution to further enhance the system in the future. Another option to block the speech signal within $\widehat{b}(n)$ would be to convert $\widehat{b}(n)$ into the spectral domain and apply then at every bin the memoryless smoothing filter, known from Fig. 17.15.

If we do so, it is easier if we can use the estimated short-term power spectral density of the background noise directly and apply the psychoacoustic weighting $W_{\mathrm{pa}}(e^{j\Omega_m}, n)$ directly in the DFT-domain. Afterwards, we can get an estimate of the (short-term) autocorrelation function $\hat{s}_{\hat{b}\hat{b}}(l, n)$ by applying an inverse DFT to the weighted noise estimate:

$$\hat{s}_{\hat{b}\hat{b}}(l, n) = \mathrm{IDFT}\left\{\left|W_{\mathrm{pa}}\left(e^{j\Omega_m}, n\right)\right|^2 \overline{B_{\min}^2\left(e^{j\Omega_m}, n\right)}\right\}. \quad (17.31)$$

The estimated autocorrelation $\hat{s}_{\hat{b}\hat{b}}(l, n)$ function can be used within the Levinson-Durbin recursion in order to obtain the coefficients for the DEC shaping filter.

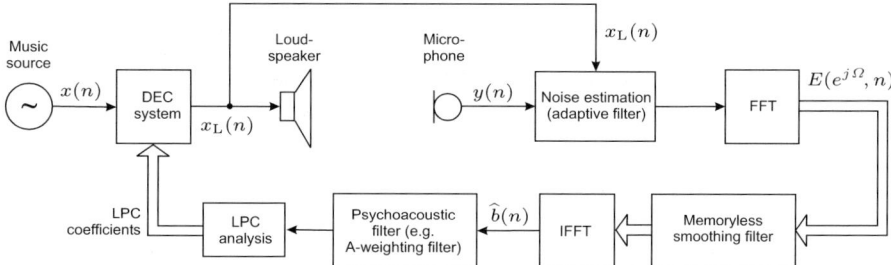

Fig. 17.22. Memoryless smoothing in the spectral domain to suppress impulsive disturbances, without changing the spectral content of the smoothed signal.

With a method as presented in Fig. 17.22 we would be able to reliably block all sorts of burst like disturbances within the signal $\hat{b}(n)$, without destroying its spectral shape, necessary for the following LPC analysis. But it would also imply a transition from the – up to now favored – processing in the time domain to processing in the frequency domain. This offers more flexibility but also means higher costs both in terms of memory as well as processing power.

17.2.3 Non-Acoustic Sensor Based Sound Systems

Before we close the first section of the chapter we will show, by means of an example, how one can, in a beneficial way, combine a microphone signal with several non-acoustic sensor signals.

In the example shown in Fig. 17.23 the microphone based part plays – in contrast to the systems presented before – a minor part. With the reduction of the spectral content of the microphone part up to not more than $f \approx 160$ Hz we do not have, on the one hand, major problems with voice anymore. On the other hand, we give up all the possibilities involved with the measured real background noise. Since the microphone records now only the very low frequency range it could more accurately be described now as a vibration

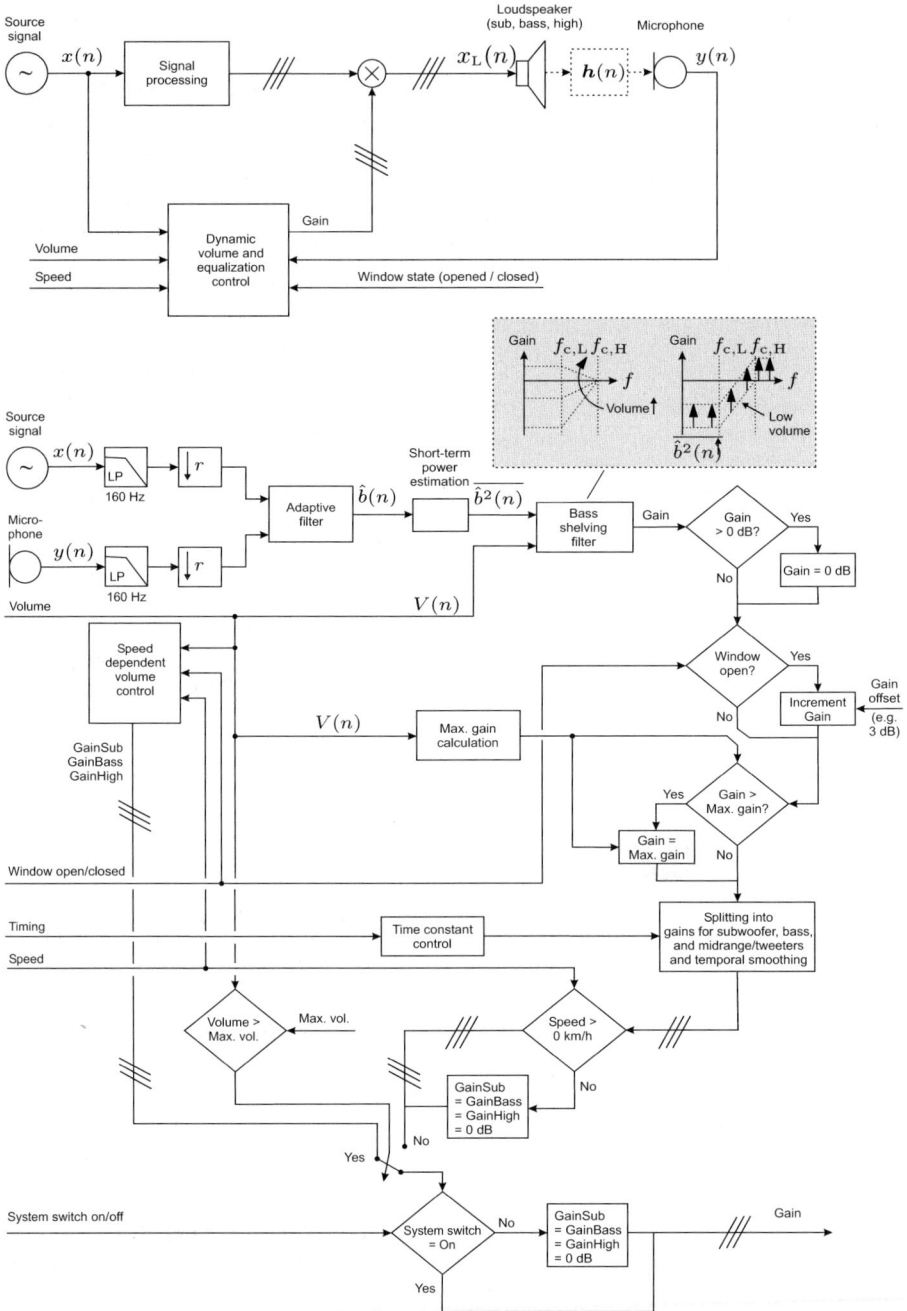

Fig. 17.23. DVC/DEC system based on a microphone as well as several non-acoustic sensor signals.

sensor, whose task it is to react to different environmental conditions like different pavements or tires. All the rest will be controlled by non-acoustic sensor signals such as the speed signal, the window signal, the volume, as well as a signal indicating whether the system should be switched on or off. Additionally, the system operates in three different spectral ranges, this is the sub[7], bass, and mid/high range. For every range a separate control strategy in terms of gain adjustment and timing will be applied, in which we call the time dependent transition from a previous to the current gain value as timing. The gain values will be controlled mainly by the volume setting. Therefore the volume adjusts the gain of a low-cut shelving filter which defines the spectral shape of the gain function. This gain function will then be modified, respectively shifted by the gain value originating from the microphone based part of the system. Also only positive gain values will be allowed. The dynamic will be controlled by the volume, following the volume dependent max-gain function, which also cares about the stability of the system. This is comparable to the previously introduced anti gain chase function. In this example the window sensor signal only controls a simple gain offset, but it would be much better to make this offset speed and/or volume dependent as well. However, one thing is for sure: the faster one drives with an open window the higher the disturbance.

With this short description of such a system we would like to close the introductory part of this chapter and we will turn over to a more natural way of solving the problem how a DEC system should be realized - at least according to our point of view.

17.3 Spectrum-Based Dynamic Equalization Control

A far more important matter – confessed by our test drives – was that an equalization solely driven by the PSD of the background noise $b(n)$, as shown in several systems within Sec. 17.2.2, did not reflect the ideal equalization. This assumption became clear as soon as we confronted our DEC algorithm known e.g. from Fig. 17.20 with artificial noise signals such as sinusoids. The equalization methods described so far as DEC systems failed because they would only be able to rise exactly at the frequency location where the sinusoid had actually been inserted.

The experiences collected so far with the DVC/DEC systems showed that we had to find a different way of considering the psychoacoustic properties of the human auditory system if we want to calculate a better and thus more natural equalizing filter. This leads to the forthcoming DEC system.

We found that such a DEC system can be implemented in a much easier way in the spectral domain. This is mainly due to the availability of psychoacoustic models in the spectral domain but also due to the higher degree of

---

[7] The term *sub* describes the very low frequency range, e.g from 120 Hz down to 20 Hz.

flexibility offered by transferring the signal processing units into the frequency domain. Of course, several parts of the overall system, such as the adaptive filter, could remain in the time domain, but in order to get a closed and compact DEC system we decided to shift the calculation of the adaptive filter into the spectral domain, too. Another advantage of doing the calculations in the frequency domain has already been mentioned at the end of Sec. 17.2.2. There, we mentioned an elegant and easy way of getting rid of impulsive noise disturbances by applying the already known memoryless smoothing filter in the spectral domain, separately for every single FFT bin. In doing so we can, on the one hand, reduce the disturbing influence of impulsive noise, like speech, in a robust and easy way. On the other hand we get an estimation of the spectral shape of the background noise. Fig. 17.24 displays a signal flow diagram, combining all necessary processing blocks within the spectral domain.

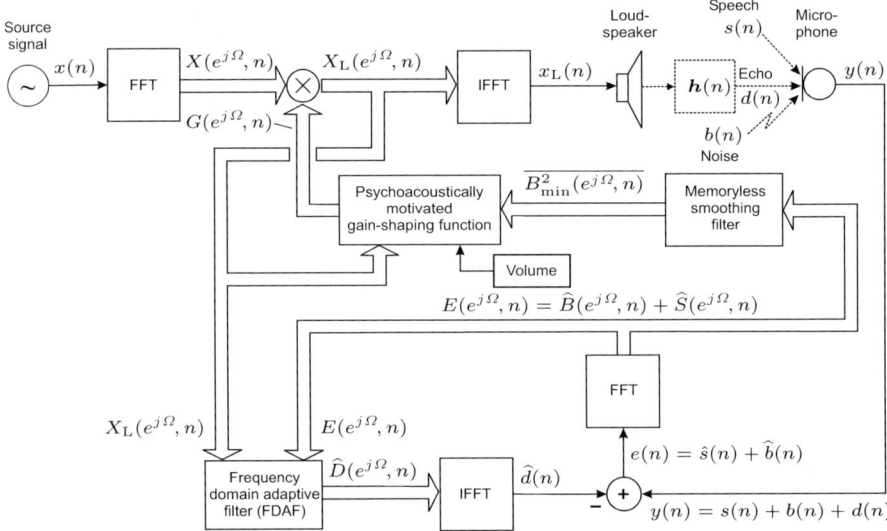

Fig. 17.24. Signal flow diagram, showing the principle of a psychoacoustically motivated, spectral domain DEC system.

17.3.1 Frequency Domain Adaptive Filter

For reasons to decrease the processing load we decided to leave the calculation of the error signal $e(n)$ in the time domain. As an example, we used a *frequency domain adaptive filter* (FDAF) within Fig. 17.24, known e.g. from [21, 29, 58], representative of the obligatory adaptive filter.

Fig. 17.25 shows the signal flow of such a FDAF algorithm in more detail. Thereby the FDAF is nothing but the block LMS algorithm, realized in the frequency domain, with the only difference that the FDAF comprises a time and frequency dependent adaptation step size $\mu(e^{j\Omega_m}, n)$, increasing its rate of convergence by orthogonalizing the input signal. This leads to an adaptive filter, able to converge faster than its time domain counterpart, i.e. the block LMS algorithm. This is true if a colored input signal is used as reference, which is, in practical applications most of the time the case.

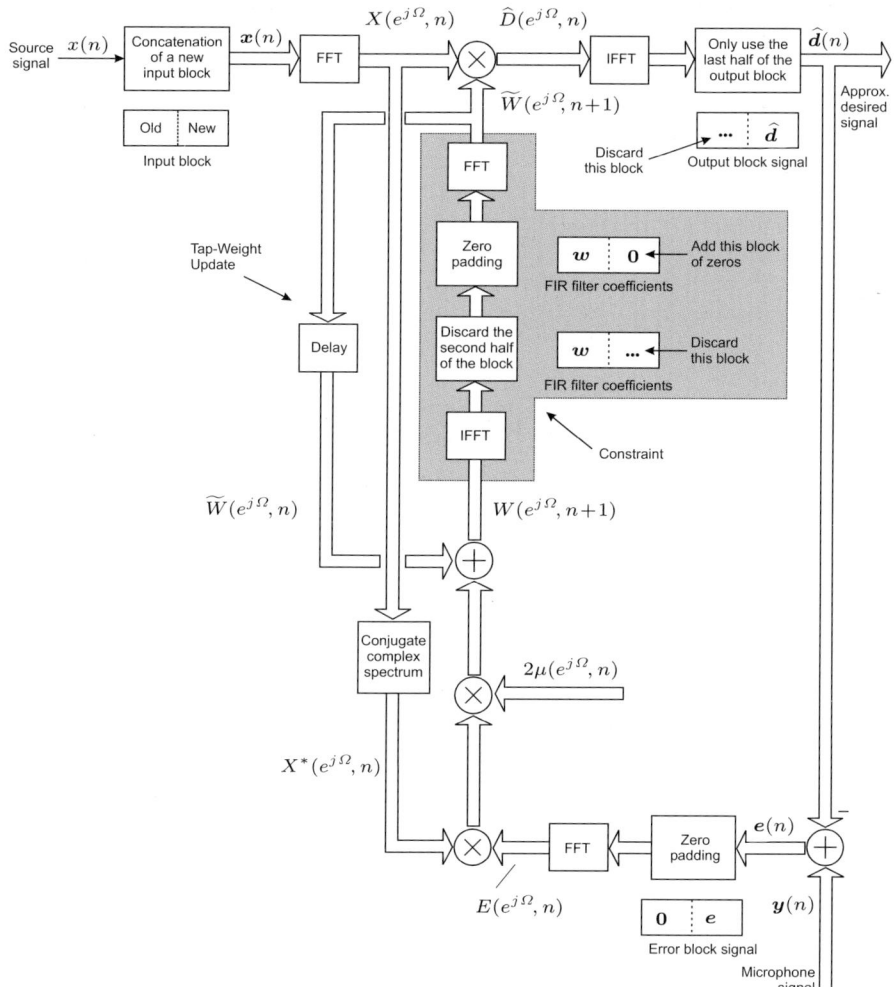

Fig. 17.25. Signal flow diagram of an overlap-save frequency domain adaptive filter.

Tab. 17.3 summarizes the complete FDAF algorithm. One thing worth to be mentioned is that we could also skip the last item within Tab. 17.3 called "Tap weight constraint". This leads to the *unconstraint* FDAF, which is, on

Table 17.3. Summary of the frequency domain adaptive filter (FDAF).

Algorithmic part	Corresponding equations

Filtering:

$$\boldsymbol{X}\left(e^{j\Omega}, n\right) = \text{FFT}\{\boldsymbol{x}(n)\},$$

$$\widehat{\boldsymbol{d}}(n) = \text{IFFT}\{\boldsymbol{X}(e^{j\Omega}, n)\,\widetilde{\boldsymbol{W}}\left(e^{j\Omega}, n\right)\},$$

with

$$\boldsymbol{x}(n) = \left[x(nL - N + 1), \ldots, x(nL + L - 1)\right]^{\text{T}},$$

$$L = \text{Block length},$$

$$N = \text{Filter length}.$$

Error estimation:

$$\boldsymbol{e}(n) = \boldsymbol{y}(n) - \widehat{\boldsymbol{d}}(n),$$

with

$$\boldsymbol{y}(n) = \left[y(nL), \ldots, y(nL + L - 1)\right]^{\text{T}}.$$

Step size:

$$\hat{\sigma}_{X,m}^2(n) = \alpha\,\hat{\sigma}_{X,m}^2(n-1) + (1-\alpha)\left|X\left(e^{j\Omega_m}, n\right)\right|^2,$$

$$\mu_m(n) = \frac{\mu(n)}{\hat{\sigma}_{X,m}^2(n)},$$

for $m = 0, \ldots, M - 1$,

$$\boldsymbol{\mu}(n) = \left[\mu_0(n), \mu_1(n), \ldots, \mu_{M-1}(n)\right]^{\text{T}},$$

with

$$\mu(n) = \text{Not normalized step size},$$

$$M = N + L - 1.$$

Filter adaptation:

$$\boldsymbol{E}\left(e^{j\Omega}, n\right) = \text{FFT}\left\{\begin{bmatrix}\boldsymbol{0}\\ \boldsymbol{e}(n)\end{bmatrix}\right\},$$

$$\boldsymbol{W}\left(e^{j\Omega}, n+1\right) = \widetilde{\boldsymbol{W}}\left(e^{j\Omega}, n\right)$$
$$+ 2\,\text{diag}\{\boldsymbol{\mu}(n)\}\,\text{diag}\{\boldsymbol{X}^*(e^{j\Omega}, n)\}\,\boldsymbol{E}\left(e^{j\Omega}, n\right).$$

Filter constraint:

$$\overline{\boldsymbol{w}}(n+1) = \text{First } N \text{ elements of IFFT}\{\boldsymbol{W}\left(e^{j\Omega}, n+1\right)\},$$

$$\widetilde{\boldsymbol{W}}\left(e^{j\Omega}, n+1\right) = \text{FFT}\left\{\begin{bmatrix}\overline{\boldsymbol{w}}(n+1)\\ \boldsymbol{0}\end{bmatrix}\right\}.$$

the one hand, less demanding but, on the other hand, suffers from a reduced rate of convergence and a limited steady state performance.

In Tab. 17.3 as well as in the following tables all frequency domain vectors are denoted as

$$\boldsymbol{X}\left(e^{j\Omega},n\right) = \left[X\left(e^{j\Omega_0},n\right), X\left(e^{j\Omega_1},n\right), \ldots, X\left(e^{j\Omega_{M-1}},n\right)\right]^{\mathrm{T}} \quad (17.32)$$

with

$$\Omega_m = \frac{2\pi}{M}m. \quad (17.33)$$

Surely, the FDAF algorithm, especially in its unconstraint form, reflects probably the most efficient way how one can implement an adaptive filter in the spectral domain, but it also got its disadvantages. Firstly the delay of the FDAF corresponds to its block length, which, if implemented in its most efficient form, matches half of the FFT length. This is very often considered as too much, especially if used in hands-free telephone systems, in which the ITU, ETSI, and VDA recommendations are quite strict, especially in terms of totally allowed delay times. Secondly, even if realized in its constraint version, the convergence rate of the FDAF sometimes is not considered as fast enough.

17.3.2 Generalized Multidelay Adaptive Filter

An adaptive filter being more flexible then the FDAF algorithm can be found in the *generalized multidelay adaptive filter* (GMAF), also known, e.g., from [4, 5, 9, 10, 12, 27, 31, 46, 53, 54, 59–61] as the *generalized multidelay filter* (GMDF), the *partitioned block frequency domain adaptive filter* (PBFDAF), the *multidelay filter* (MDF) or the *extended multidelay filter* (EMDF), just to name a few, all able to compensate for the deficiencies of the FDAF.

The structure of the GMAF can be seen in Fig. 17.26, whereas Tab. 17.4 shows the corresponding summary of the the algorithm. The GMAF captivates by its ability to freely scale its block length, on the one hand as well as by its flexible adjustment of the delay time by splitting the desired length of the adaptive filter in partitions. Now, the remaining delay solely depends on the effective partition length and not on the overall filter length any more. The block length can again be chosen up to half of the partition length, which corresponds to half of the FFT length, as well. But it can also be much shorter than that, down to an effective block length of only one sample, which marks the lower bound of the limit.

Obviously with the GMAF one is able to freely scale between an effective implementation, by utilizing the maximal block length of half the FFT length and a very fast converging algorithm, by gradually reducing its block length down to just one sample feed. Also the "tap weight constraint" of Tab. 17.4 can be skipped, just as in the FDAF algorithm, leading again to a more efficient implementation, incorporating the same deficiencies, as previously described

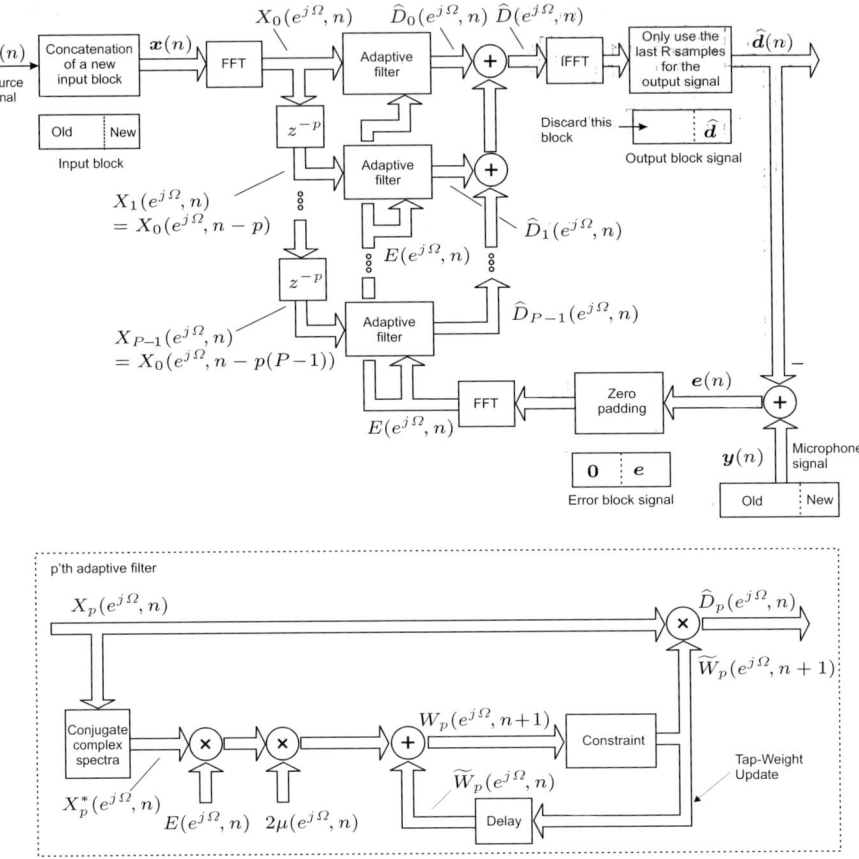

Fig. 17.26. Signal flow diagram of a generalized multidelay adaptive filter.

in the FDAF algorithm. But now we do have to constrain not just one adaptive filter as in the FDAF algorithm, but P partitions, which of course means a higher effort. A strategy, able to tremendously reduce the effort, by worsening the convergence properties of the system, only a bit, at the same time, exist in calculating the constraint of only one partition per block, such as described e.g. in [23, 25, 26, 64].

An additional advantage of the GMAF algorithm is that – due to its scalability – we are now able to design a system that makes use of the available system resources to a much higher extend. For example, usually DSPs do not dispose over enough internal RAM, because much internal RAM means a bigger die, which requires more silicon, which itself corresponds to higher costs. Hence DSPs usually do not possess much internal RAM[8], but often do have

---
[8] The terms *RAM*, *SRAM*, and *DRAM* abbreviate *random access memory*, *static random access memory*, and *dynamic random access memory*, respectively.

Table 17.4. Summary of the generalized multidelay adaptive filter (GMAF).

Algorithmic part	Corresponding equations

Filtering:

$$\boldsymbol{X}_0\left(e^{j\Omega},n\right) = \text{FFT}\left\{\boldsymbol{x}_0(n)\right\},$$

$$\widehat{\boldsymbol{d}}(n) = \text{Last } M \text{ elements of } \widehat{\boldsymbol{d}}_{\text{pre}}(n)$$

with

$$\boldsymbol{x}_0(n) = \left[x(nL-R+1), \ldots, x(nL+L-1)\right]^{\text{T}},$$

$$L = \text{Block length},$$

$$R = \text{Partition length},$$

$$N = PR = \text{Filter length},$$

$$M = R+L-1,$$

$$\widehat{\boldsymbol{d}}_{\text{pre}}(n) = \text{IFFT}\left\{\sum_{p=0}^{P-1}\boldsymbol{X}_p(e^{j\Omega},n)\,\widetilde{\boldsymbol{W}}_p\left(e^{j\Omega},n\right)\right\},$$

$$\boldsymbol{X}_p\left(e^{j\Omega},n\right) = \boldsymbol{X}_{p-1}\left(e^{j\Omega},n-P\right) \text{ for } p=1,\ldots,P-1.$$

Error estimation:

$$\boldsymbol{e}(n) = \boldsymbol{y}(n) - \widehat{\boldsymbol{d}}(n),$$

with

$$\boldsymbol{y}(n) = \left[y(nL-R+1), \ldots, y(nL+L-1)\right]^{\text{T}}.$$

Step size:

$$\hat{\sigma}^2_{X,m}(n) = \alpha\,\hat{\sigma}^2_{X,m}(n-1) + (1-\alpha)\left|X_0\left(e^{j\Omega m},n\right)\right|^2,$$

$$\mu_m(n) = \frac{\mu(n)}{\hat{\sigma}^2_{X,m}(n)},$$

for $m = 0, \ldots, M-1$,

$$\boldsymbol{\mu}(n) = \left[\mu_0(n), \mu_1(n), \ldots, \mu_{M-1}(n)\right]^{\text{T}},$$

with

$$\mu(n) = \text{Not normalized step size}.$$

Filter adaptation:

$$\boldsymbol{E}\left(e^{j\Omega},n\right) = \text{FFT}\left\{\begin{bmatrix}\mathbf{0}\\ \boldsymbol{e}(n)\end{bmatrix}\right\},$$

$$\boldsymbol{W}_p\left(e^{j\Omega},n+1\right) = \widetilde{\boldsymbol{W}}_p\left(e^{j\Omega},n\right)$$
$$+2\,\text{diag}\{\boldsymbol{\mu}(n)\}\,\text{diag}\{\boldsymbol{X}^*_p(e^{j\Omega},n)\}\,\boldsymbol{E}\left(e^{j\Omega},n\right).$$

Filter constraint:

$$\overline{\boldsymbol{w}}_p(n+1) = \text{First } M \text{ elements of IFFT}\left\{\boldsymbol{W}_p\left(e^{j\Omega},n+1\right)\right\},$$

$$\widetilde{\boldsymbol{W}}_p\left(e^{j\Omega},n+1\right) = \text{FFT}\left\{\begin{bmatrix}\overline{\boldsymbol{w}}_p(n+1)\\ \mathbf{0}\end{bmatrix}\right\}.$$

a fast interface to external SRAM or DRAM, which are much cheaper. It is often necessary to exchange data to and from the external storage, especially if we talk about huge amounts of data and are, at the same time keen in keeping the internal data amount to a minimum. If one is confronted with such a practical situation the GMAF can often help a lot in utilizing the given system resources, more efficiently.

17.3.3 Step-Size Control

Furthermore, an adaptive adaptation step size can also be realized in the spectral domain, providing more opportunities as its counterpart in the time domain utilizing, e.g., the delayed coefficients method. Examples how one can effectively realize such an adaptive adaptation step size can e.g. be found in [28, 34].

Tab. 17.5 summarizes the algorithm of [28] which does not solely deliver the desired adaptive adaptation step size $\mu(e^{j\Omega_m}, n)$, but also – in a statistical

Table 17.5. Summary of the calculation of the adaptive adaptation step size and the adaptive post filter.

Algorithmic part	Corresponding equations
Step size:	$\mu\left(e^{j\Omega_m}, n\right) = \dfrac{\left\|G\left(e^{j\Omega_m}, n\right)\right\|^2 \left\|X\left(e^{j\Omega_m}, n\right)\right\|^2}{\left\|E\left(e^{j\Omega_m}, n\right)\right\|^2},$ $\boldsymbol{\mu}\left(e^{j\Omega}, n\right) = \left[\mu\left(e^{j\Omega_0}, n\right), \ldots, \mu\left(e^{j\Omega_{M-1}}, n\right)\right]^{\mathrm{T}}.$
Post filter:	$\boldsymbol{H}\left(e^{j\Omega}, n\right) = \mathbf{1} - \boldsymbol{\mu}\left(e^{j\Omega}, n\right),$ with $\mathbf{1} = \left[1, \ldots, 1\right]^{\mathrm{T}}.$
Convergence state:	$\left\|G\left(e^{j\Omega_m}, n+1\right)\right\|^2 = \left\|G\left(e^{j\Omega_m}, n\right)\right\|^2 \left(1 - \mu\left(e^{j\Omega_m}, n\right)\right)$ $+ \Delta\left(e^{j\Omega_m}, n\right),$ with $\Delta\left(e^{j\Omega_m}, n\right) = C\left\|W\left(e^{j\Omega_m}, n\right)\right\|^2,$ $C =$ Constant which has to be adjusted according the current overlap, with $0 < C \ll 1.$

point of view – an optimal adaptive post filter $H(e^{j\Omega_m}, n)$. This filter is able to suppress residual echoes as well. An advantage of this purely statistical method of calculating the adaptive adaptation step sizes $\mu(e^{j\Omega_m}, n)$ in the frequency domain is that it intrinsically decrees, not only over a method to protect an already converged adaptive filter tap weight vector against impulsive disturbances, but also avoids a blocking of the adaptation process due to abrupt changes of the LEM system, after the adaptive filter has already converged close to its stationary point. Especially the adaptive post filter, which cares for sufficient echo reduction turned out to be of great benefit in our DEC system. This was mainly due to the fact that a typical audio system, in which the DEC system will finally be embedded, usually comprises not only a single loudspeaker but rather a plurality of speakers. Each contributes its own part to the overall microphone signal $y(n)$. Hence we would need as many adaptive filters as we actually have LEM paths in the real system, which would, theoretically be possible, but would also ask for completely uncorrelated reference signals feeding each loudspeaker in the system, to avoid the ambiguity problem, known e.g. from *stereo acoustic echo cancelation* (SAEC).

17.3.4 Multi-Channel Systems

Stereo signals are by far the most applied reference signals in audio systems. Knowing that they are usually sufficiently decorrelated from each other, we decided to use a SAEC instead of a mono AEC algorithm to further increase the echo reduction and so the stability of the DEC system. Of course there are also multi-channel audio formats, such as DTS or AC3, to name just a few, available, but until now, they still play a minor role within automobile entertainment systems. Thus, we generally cannot reckon with more than

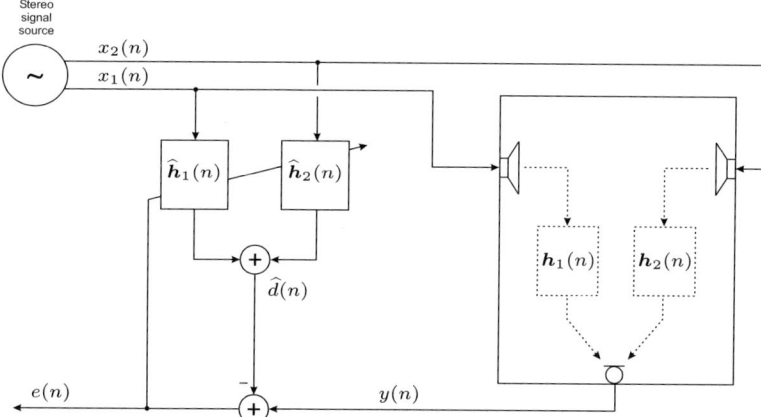

Fig. 17.27. Signal flow diagram showing the principle of a stereo acoustic echo canceller.

two, sufficiently decorrelated channels, even if talking about high end audio systems, limiting us to a SAEC systems, as shown in Fig. 17.27, but without being confronted with the ambiguity problem as some may face in pure SAEC applications. It is possible to artificially decorrelate the rest of the loudspeaker signals such that we could also adapt to the rest of the individual LEM system, but this is strictly forbidden due to the fact that we would then inevitably modify the audio signals in an undesired way, which would then influence some of the audio algorithms such as surround sound algorithms negatively. This, of course, has to be avoided.

Tab. 17.6 shows the collected formulae necessary to implement a SAEC, based on the overlap and save FDAF algorithm. The parameter K has to be set to $K = 2$ in order to correctly reflect the realized stereo version of the SAEC. Again, the constraint can – if desired – be omitted, or as previously described can be done selectively, i.e. one constraint per block, and not as usually desired and described in Tab. 17.6, all constraints per block.

Naturally, the calculation of the adaptive adaptation step size as well as of the adaptive post filter changes with the introduction of the SAEC algorithm. For this reason we again collected the necessary formulae, in Tab. 17.7 to show the interested reader how to do the necessary calculations in this case.

17.3.5 Estimating the Power Spectral Density of the Background Noise

As already mentioned in the beginning of this section, we can use the memoryless smoothing filter, as shown in Fig. 17.15 almost unaltered in the spectral domain as well to exclude undesired impulsive disturbances within the estimated background noise signal $\widehat{b}(n)$, without changing its spectral shape. Better results could be achieved if we pre-process the error signal $E(e^{j\Omega_m}, n)$ originating from the SAEC filter by firstly estimating its PSD, secondly smoothing the same over time with different smoothing parameters for falling and rising signal edges and thirdly smoothing the resulting signal in the frequency domain as well, by applying again different smoothing parameters for the increase and decrease case. Furthermore, doing the same smoothing from the low to the high frequencies and then vice versa, again, with the same parameters, to avoid spectral shifts of the, in this way frequency smoothed signal.

17.3.5.1 Adaptive Increment Parameter $C_{\text{Inc}}(\Omega, n)$

Fig. 17.28 shows the signal flow of the memoryless smoothing filter in the spectral domain, which additionally comprises a frequency dependent, adaptive increment parameter $C_{\text{Inc}}(\Omega, n)$. The reason why we introduced such an adaptive increment parameter is because the algorithm with just a fix increment is not able to react fast enough to abrupt risings of the background noise signal, happening e.g. when one opens a window while driving at a

Table 17.6. Summary of the multichannel acoustic echo canceller, based on the FDAF. For the stereo case we have to set $K = 2$.

Algorithmic part	Corresponding equations

Filtering:
$$\boldsymbol{X}_k\left(e^{j\Omega}, n\right) = \text{FFT}\left\{\boldsymbol{x}_k(n)\right\},$$
$$\widehat{\boldsymbol{d}}(n) = \text{Last } L \text{ elements of } \widehat{\boldsymbol{d}}_{\text{pre}}(n)$$
with
$$\boldsymbol{x}_k(n) = \left[x_k(nL - N + 1), \ldots, x_k(nL + L - 1)\right]^{\text{T}},$$
$$L = \text{Block length},$$
$$N = \text{Filter length},$$
$$K = \text{Number of input channels},$$
$$M = N + L - 1,$$
$$\widehat{\boldsymbol{d}}_{\text{pre}}(n) = \text{IFFT}\left\{\sum_{k=0}^{K-1} \boldsymbol{X}_k(e^{j\Omega}, n)\, \widetilde{\boldsymbol{W}}_k\left(e^{j\Omega}, n\right)\right\}.$$

Error estimation:
$$\boldsymbol{e}(n) = \boldsymbol{y}(n) - \widehat{\boldsymbol{d}}(n),$$
with
$$\boldsymbol{y}(n) = \left[y(nL), \ldots, y(nL + L - 1)\right]^{\text{T}}.$$

Step size:
$$\hat{\sigma}^2_{X,m}(n) = \alpha\, \hat{\sigma}^2_{X,m}(n-1) + (1-\alpha) \frac{1}{K} \sum_{k=0}^{K-1} \left|X_k\left(e^{j\Omega_m}, n\right)\right|^2,$$
$$\mu_m(n) = \frac{\mu(n)}{\hat{\sigma}^2_{X,m}(n)},$$
for $m = 0, \ldots, M - 1$,
$$\boldsymbol{\mu}(n) = \left[\mu_0(n), \mu_1(n), \ldots, \mu_{M-1}(n)\right]^{\text{T}},$$
with
$$\mu(n) = \text{Not normalized step size}.$$

Filter adaptation:
$$\boldsymbol{E}\left(e^{j\Omega}, n\right) = \text{FFT}\left\{\begin{bmatrix} \boldsymbol{0} \\ \boldsymbol{e}(n) \end{bmatrix}\right\},$$
$$\boldsymbol{W}_k\left(e^{j\Omega}, n+1\right) = \widetilde{\boldsymbol{W}}_k\left(e^{j\Omega}, n\right)$$
$$+ 2 \operatorname{diag}\left\{\boldsymbol{\mu}(n)\right\} \operatorname{diag}\left\{\boldsymbol{X}_k^*\left(e^{j\Omega}, n\right)\right\} \boldsymbol{E}\left(e^{j\Omega}, n\right).$$

Filter constraint:
$$\overline{\boldsymbol{w}}_k(n+1) = \text{First } N \text{ elements of IFFT}\left\{\boldsymbol{W}_k\left(e^{j\Omega}, n+1\right)\right\},$$
$$\widetilde{\boldsymbol{W}}_k\left(e^{j\Omega}, n+1\right) = \text{FFT}\left\{\begin{bmatrix} \overline{\boldsymbol{w}}_k(n+1) \\ \boldsymbol{0} \end{bmatrix}\right\}.$$

Table 17.7. Summary of the calculation of the adaptive adaptation step size and the adaptive post filter for the SAEC algorithm.

Algorithmic part	Corresponding equations
Step size:	$$\mu\left(e^{j\Omega_m}, n\right) = \frac{\left|G(e^{j\Omega_m}, n)\right|^2 \frac{1}{K} \sum_{k=0}^{K-1} \left|X_k\left(e^{j\Omega_m}, n\right)\right|^2}{\left|E\left(e^{j\Omega_m}, n\right)\right|^2},$$ $$\boldsymbol{\mu}\left(e^{j\Omega}, n\right) = \left[\mu\left(e^{j\Omega_0}, n\right), \ldots, \mu\left(e^{j\Omega_{M-1}}, n\right)\right]^{\mathrm{T}},$$ with K = Number of input channels.
Post filter:	$$\boldsymbol{H}\left(e^{j\Omega}, n\right) = \mathbf{1} - \boldsymbol{\mu}\left(e^{j\Omega}, n\right),$$ with $$\mathbf{1} = \left[1, \ldots, 1\right]^{\mathrm{T}}.$$
Convergence state:	$$\left|G\left(e^{j\Omega_m}, n+1\right)\right|^2 = \left|G\left(e^{j\Omega_m}, n\right)\right|^2 \left(1 - \mu\left(e^{j\Omega_m}, n\right)\right) + \Delta\left(e^{j\Omega_m}, n\right),$$ with $$\Delta\left(e^{j\Omega_m}, n\right) = \frac{C}{K} \sum_{k=0}^{K-1} \left|W_k\left(e^{j\Omega_m}, n\right)\right|^2,$$ C = Constant which has to be adjusted according the current overlap, with $0 < C \ll 1$.

moderate or high speed. In such situations, the memoryless smoothing filter should act much quicker than the current one. We improved the current memoryless smoothing filter by applying the working principle of the memoryless smoothing filter to the increment parameter as well, leading to the mentioned adaptive increment parameter $C_{\mathrm{Inc}}(\Omega, n)$. Thereby this principle can be described as follows:

- At the beginning the increment parameter $C_{\mathrm{Inc}}(\Omega, n)$ starts at a very low, fix value named $C_{\mathrm{Inc,min}}$, small enough to securely suppress burst like noise like speech, without sustainably deteriorating the long term shape of the background noise PSD $\widehat{S}_{bb}(\Omega, n)$.

- As long as the estimated background noise PSD, which, in fact acts as output signal of the memoryless smoothing filter is lower then the smoothed PSD of the error signal $\overline{S}_{ee}(\Omega, n)$, illustrating the input signal of the algorithm, the current increment parameter $C_{\text{Inc}}(\Omega, n)$ will be increased by a fix increment, denoted as Δ_{Inc}. This continues as long as $\overline{S}_{ee}(\Omega, n) > \widehat{S}_{bb}(\Omega, n-1)$ and $C_{\text{Inc}}(\Omega, n) < C_{\text{Inc,max}}$, where $C_{\text{Inc,max}}$ stands for the upper limit of the increment parameter which need not be exceeded by $C_{\text{Inc}}(\Omega, n)$.
- Otherwise, no matter what the current value of $C_{\text{Inc}}(\Omega, n)$ may be, as soon, as $\overline{S}_{ee}(\Omega, n) \leq \widehat{S}_{bb}(\Omega, n-1)$, $C_{\text{Inc}}(\Omega, n)$ will be, immediately reset to $C_{\text{Inc,min}}$, to ensure a robust behaviour against impulsive disturbances, again.

Of course this principle could also be applied for the decrement parameter C_{Dec}, but tests showed that due to its already high value it has been considered as unnecessary. Thus we left the decrement parameter as fix value, which in contrast to the increment parameter, did not even need to be frequency dependent.

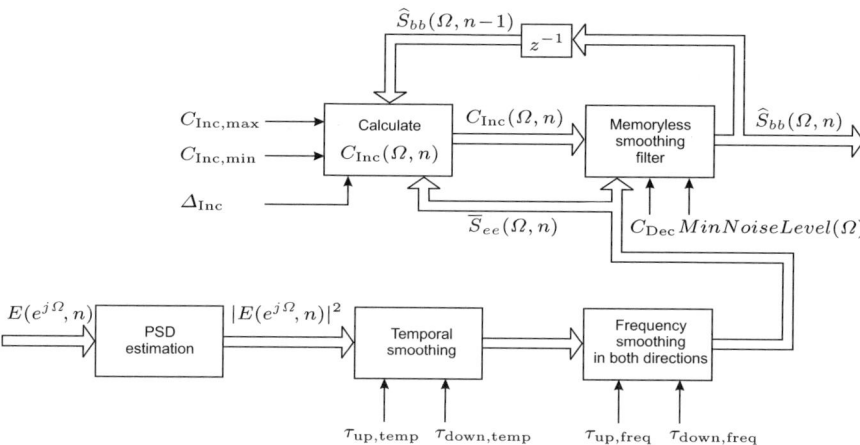

Fig. 17.28. Memoryless smoothing filter with an adaptive increment parameter, realized in the spectral domain.

So far we discussed the adaptive filter block of Fig. 17.24, called *FDAF* and the block responsible for suppressing undesired impulsive disturbances, named *memoryless smoothing filter*. What is still missing is a description of the remaining *psychoacoustically motivated gain-shaping function* block, whose task is to calculate the desired equalizing function $G(e^{j\Omega}, n)$ out of the source signal $X(e^{j\Omega}, n)$ and the estimated background noise signal, depicted as $\overline{B}_{\min}(e^{j\Omega}, n)$. But how do we actually succeed in calculating $G(e^{j\Omega}, n)$ taking psychoacoustic properties of our auditory system into account? First of all, the DEC system calculates with the help of a psychoacoustic masking model the masking threshold of the estimated background noise signal as well as of the input signal respectively the reference signal. If the masking threshold of the reference signal resides completely or partially below the masking threshold of the estimated background noise, the reference signal will be completely or partially masked by the background noise signal. Hence, if we equalize the reference signal by the difference of both masking threshold we can rise the reference signal in such a way, that now all spectral parts of it, masked before would now be located above, or at least at the same level as the masking threshold of the background noise signal. So, it cannot mask the reference signal any more. Frequency parts that has not been masked by the background noise need not be raised, because they are still audible. Thus, it is logical that we again restrict ourself solely to spectral amplifications of the *psychoacoustically motivated gain-shaping function.*

Therefore any masking model able to calculate the masking threshold of a given input signal would do the job but we deliberately decided to use the masking model according to Johnston, which will be described in detail in Sec. 17.3.7, due to its simplicity and scalability. These properties made its integration into our spectral framework possible and easy.

17.3.6 Psychoacoustic Basics

The absolute threshold in quiet of hearing $T_q(f)$ exhibits the least amount of sound pressure needed to make a tone audible in an environment without noise. The sound pressure level L_{SPL} can be converted from the measurable sound pressure p to its level, as shown in Eq. 17.34

$$L_{\text{SPL}} = 20 \log_{10}\left(\frac{p}{p_0}\right), \tag{17.34}$$

with

$$p_0 = 20\,\mu\text{Pa}. \tag{17.35}$$

Analytically the threshold of hearing in quiet $T_q(f)$ can be calculated as depicted in Eq. 17.36:

$$T_q(f) = 3.64 \text{ dB} \left(\frac{f}{1000 \text{ Hz}}\right)^{-0.8} - 6.5 \text{ dB } e^{-0.6\left(\frac{f}{1000 \text{ Hz}} - 3.3\right)^2}$$
$$+ 10^{-3} \text{ dB} \left(\frac{f}{1000 \text{ Hz}}\right)^4. \tag{17.36}$$

The human ear integrates various sound stimuli falling within limited frequency bands, which are called *critical bands* (CB). It combines sounds occurring in certain frequency bands regarding psychoacoustic hearing properties to form a joint hearing sensation. Sound events located in the same critical band influence each other in a different way as sounds residing in different critical bands. For example, two tones having the same level within one critical band are perceived more quietly as when located in different critical bands. Since a test tone within a masker can be heard when the energies are identical and the masker falls into the frequency band which has the frequency of the test tone as center frequency, the required bandwidth of the critical bands can be determined. At low frequencies, the critical bands have a bandwidth of 100 Hz. At frequencies above 500 Hz, the critical bands have a bandwidth of approximately 20 % of the center frequency of the respective critical band, as shown in Tab. 17.8, adopted from [69].

Table 17.8. Bark table, including the upper/lower cut-off frequencies f_{cut}, the center-frequencies of the critical-bands f_c, as well as its bandwidth Δf_b, according to [69].

z	f_{cut}	f_c	z_c	Δf_b	z	f_{cut}	f_c	z_c	Δf_b
Bark	Hz	Hz	Bark	Hz	Bark	Hz	Hz	Bark	Hz
0	0				12	1720			
		50	0.5	100			1850	12.5	280
1	100	150	1.5	100	13	2000	2150	13.5	320
2	200	250	2.5	100	14	2320	2500	14.5	380
3	300	350	3.5	100	15	2700	2900	15.5	450
4	400	450	4.5	100	16	3150	3400	16.5	550
5	510	570	5.5	120	17	3700	4000	17.5	700
6	630	700	6.5	140	18	4400	4800	18.5	900
7	770	840	7.5	150	19	5300	5800	19.5	1100
8	920	1000	8.5	160	20	6400	7000	20.5	1300
9	1080	1170	9.5	190	21	7700	8500	21.5	1800
10	1270	1370	10.5	210	22	9500	10500	22.5	2500
11	1480	1600	11.5	240	23	12000	13500	23.5	3500
12	1720				24	15500			

By lining up all the critical bands over the entire hearing range, a hearing-oriented nonlinear frequency scale is obtained which is called the *equivalent rectangular bandwidth* (ERB). Eq. 17.37 approximates the conversion of the linear to the critical bandwidth resolution, i.e. the ERB scale.

$$BW_c(f) = 25 \text{ Hz} + 75 \text{ Hz} \cdot \left(1 + 1.4 \cdot \left(\frac{f}{1000 \text{ Hz}}\right)^2\right)^{0.69}. \quad (17.37)$$

The distance of one critical bandwidth is commonly referred to as one *Bark*. The function

$$z_{\text{CBR}} = 13\,\text{Bark} \cdot \arctan\left(\frac{0.00076\,f}{\text{Hz}}\right) + 3.5\,\text{Bark} \cdot \arctan\left(\left(\frac{f}{7500\,\text{Hz}}\right)^2\right) \quad (17.38)$$

converts units from the linear frequency scale to the so called *Bark scale*. It represents a nonuniform scaling of the frequency axis such, that critical bands have the same width of exactly 1 Bark at any point.

The nonlinear relationship of frequency and critical-band rate scale has its origin in the frequency-to-location transformation of the basilar membrane. The Bark scale was specified by Zwicker [69] on the basis of masking threshold and loudness investigations. It has been found that just 25 critical bands can be arranged in the auditory frequency band from 0 to approximately 21 kHz so that the associated Bark scale extends from 0 to 24 Bark.

Probably the easiest way to explain the phenomenon of (simultaneously) masking in the spectral domain is by referring to one of the experiments conducted by Zwicker in [69]. It has been found, that a sinusoidal test tone, played at a level approximately 4 dB below the level of a bandpass filtered noise signal, acting as masking signal, will completely be masked by the same. However, this is only true if on the one hand the bandpass filter has a width of 1 Bark and on the other hand the sinusoidal test tone will be adjusted right at the center frequency of the bandpass filter, i.e. of the corresponding Bark. By tuning the sinusoidal test tone off the center frequency we will see that the masking threshold will drop with a slope of about 25 dB/Bark, if $f_{\text{Testtone}} < f_c$ and by approximately -10 dB/Bark, if $f_{\text{Testtone}} > f_c$. This means that the level of the test tone has to be decreased if tuned off the center frequency of the Bark f_c, such that the masker, i.e. the bandpass filtered noise signal, always playing at the same level, is still able to mask the new test tone, now playing at a different frequency. Fig. 17.29 shows the masking thresholds of the sinusoids, gathered by three different maskers, where the center frequency of the masker as well as its bandwidth corresponds to the 2'nd, 8'th and 17'th Bark within the Tab. 17.8. Therefore it is worth mentioning, that the depicted masking thresholds have been found separately, meaning by applying the individual masker one after another and not simultaneously. The typically triangular shape of the masking threshold, depicted in Fig. 17.29 having slopes of approximately 25 dB and -10 dB per Bark, which is also referred to as *spread of masking*, has to be reproduced by the masking model as well, where it is usually called *spreading function*. The formula of Eq. 17.39 analytically represents the spreading function, being able to approximate the triangular shapes of the masking thresholds of Fig. 17.29:

$$SF(i) = 15.81 + 7.5 \cdot (i + 0.474) - 17.5 \cdot \sqrt{1 + (i + 0.474)^2}, \quad (17.39)$$

for

$$i = 1, \ldots, 25.$$

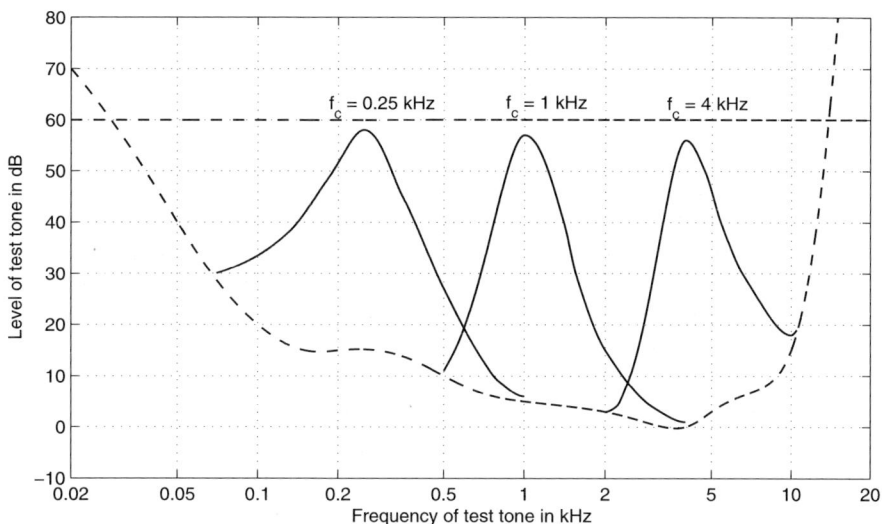

Fig. 17.29. Masking threshold of a sinusoidal test tone, masked by bandpass filtered noise signals, all possessing a width of one Bark, with three different center frequencies of the filtering bandpass, namely $f_c = [0.25, 1.0, 4.0]$ kHz.

17.3.7 The Psychoacoustic Masking Model According to Johnston

Primarily, Johnston created a masking model, incorporating the, up to this point already known and confirmed psychoacoustic phenomenons, especially those related to masking, in order to increase the data compression ratios of coding algorithms, without, decreasing its acoustical performance or quality.

17.3.7.1 Calculation of the Power Spectral Density

Following the masking model after Johnston, we first have to calculate the PSD of the input signal, as displayed in Eqs. 17.40 - 17.42:

$$\widehat{S}_{xx}(\Omega_m, n) = \left| \sum_{k=0}^{N-1} w(k) \cdot x(n-k) \cdot e^{-j\frac{2\pi}{N}km} \right|^2 \tag{17.40}$$

for

$$m = 0, \ldots, \left(\frac{N}{2} + 1\right), \tag{17.41}$$

with

$$w(k) = \frac{1}{2}\left[1 - \cos\left(\frac{2\pi k}{N}\right)\right]. \tag{17.42}$$

The last quantity $w(k)$ represents a windowing function – here a Hamming window – to decrease leakage.

17.3.7.2 Bark Frequency Scale

Bearing the nonlinear frequency resolution of the human auditory system in mind, we will group several bins together in such a way, that we will finally end up in a frequency resolution resembling the Bark frequency scale. Eq. 17.43 depicts how we will have to do the grouping to eventually get the *discrete Bark spectrum* $B(i,n)$.

$$B(i,n) = \sum_{m=b_{\text{low}}(i)}^{b_{\text{high}}(i)} \widehat{S}_{xx}(\Omega_m, n), \qquad (17.43)$$

for

$$i = 1, \ldots, 25.$$

There $b_{\text{low}}(i)$ stands for the bin number, falling into the lower band limit, representing the i'th Bark, whereas $b_{\text{high}}(i)$ marks the upper or high band limit of the i'th Bark. Additionally Eq. 17.43 shows that a total of 25 Barks, representing the whole auditory frequency range of man, will be calculated. The discrete Bark spectrum $B(i,n)$ forms the foundation for the calculation of the Bark related masking effect. Characterizing the input signal in tone-like and noise-like parts, we find ourself confronted not only with the already mentioned *noise-masking-tone* scenario, but also with *tone-masking-noise*, *tone-masking-tone* and *noise-masking-noise* situations, where only the first two are actually of interest. A typical signal normally consists of both signal types, leading to simultaneous masking effects, which of course, have to be considered as well in the masking model.

17.3.7.3 Spreading Matrix

In the Johnston model the concurrent masking effect of neighboring Barks will be accounted for by convolving the discrete Bark spectrum $B(i,n)$ with the spreading function $SF(i)$, summarized in Eqs. 17.44 - 17.48:

$$\boldsymbol{C}(n) = \boldsymbol{S}\,\boldsymbol{B}(n), \qquad (17.44)$$

with

$$\boldsymbol{C}(n) = \begin{bmatrix} C(1,n), C(2,n), \ldots, C(25,n) \end{bmatrix}^{\mathrm{T}}, \qquad (17.45)$$

$$\boldsymbol{B}(n) = \begin{bmatrix} B(1,n), B(2,n), \ldots, B(25,n) \end{bmatrix}^{\mathrm{T}}, \qquad (17.46)$$

$$\boldsymbol{S} = \begin{bmatrix} 10^{\frac{SF(1,1)}{10}} & 10^{\frac{SF(1,2)}{10}} & \cdots & 10^{\frac{SF(1,25)}{10}} \\ 10^{\frac{SF(2,1)}{10}} & 10^{\frac{SF(2,2)}{10}} & \cdots & 10^{\frac{SF(2,25)}{10}} \\ \vdots & \vdots & \ddots & \vdots \\ 10^{\frac{SF(25,1)}{10}} & 10^{\frac{SF(25,2)}{10}} & \cdots & 10^{\frac{SF(25,25)}{10}} \end{bmatrix}, \qquad (17.47)$$

and

$$SF(i,k) = 15.81 + 7.5\left((i-k) + 0.474\right) - 17.5\sqrt{1 + \left((i-k) + 0.474\right)^2}. \tag{17.48}$$

Eq. 17.44 displays the principle in very compact form as vector-matrix product, where $\boldsymbol{S}$ stands for the linear conversion of its logarithmic counterpart $SF(i,k)$, which describes the *spreading matrix*. $SF(i,k)$ of Eq. 17.48 is utilitzed to generate a matrix that has a dimension of 25×25, where i denominates the row and k the column index of the resulting spreading matrix. The output vector $\boldsymbol{C}(n)$ of the spreading process represents, for the first time, something like a global masking.

17.3.7.4 Spectral Flatness Measure

So far, we still have not taken the characteristics of the input signal into account, i.e. whether it can be interpreted more as noise or tone-like signal, by generating the global masking. This has to be done, because in dependance whether the signal is more noise-like or tone-like different masking thresholds have to be generated. In order to include also this dependency into the algorithm we first have to find out whether the input signal can be classified as tone or noise-like. An effective method, delivering a measure allowing such a categorization of the input signals offers the *spectral flatness measure* (SFM):

$$SFM(i,n) = \frac{\sqrt[M(i)]{\prod_{m=b_{\text{low}}(i)}^{b_{\text{high}}(i)} \widehat{S}_{xx}(\Omega_m, n)}}{\frac{1}{M(i)} \sum_{m=b_{\text{low}}(i)}^{b_{\text{high}}(i)} \widehat{S}_{xx}(\Omega_m, n)}, \tag{17.49}$$

with

$$M(i) = b_{\text{high}}(i) - b_{\text{low}}(i) + 1. \tag{17.50}$$

Referring to Eq. 17.49, the spectral flatness measure is calculated by the ratio between the *geometrical mean* and the *arithmetical mean* of the spectral energy per Bark. Thereby a flat spectrum, which can be interpreted as a noise-like spectrum would result in a SFM value close, but less then 1, whereas a peak like spectrum, such as one of a sinusoid, would correspond to a SFM value close, but greater then 0.

Practical implementations of the SFM lead to enormous problems, due to numerical errors. Even PC simulations, disposing over a huge numerical precision suffered under those arithmetically deficiencies, mainly during the calculation of the Bark-wise geometrical mean values, especially if, during the course of its calculation, some of the underlaying PSD values of the input signal $\widehat{S}_{xx}(\Omega_m, n)$ possess low numerical values. Knowing about the eminent effect of a correct spectral flatness measure on the complete masking model, we searched for a way to avoid these numerical problems during the calculation of the SFM, in general and the geometrical mean, in particular. By transferring the SFM calculation from the linear into the logarithmical domain we were able to successfully circumvent these hindering numerical problems:

$$SFM_{\log}(i,n) = 10 \log_{10} \left\{ \frac{\sqrt[M(i)]{\prod_{m=b_{\text{low}}(i)}^{b_{\text{high}}(i)} \widehat{S}_{xx}(\Omega_m, n)}}{\frac{1}{M(i)} \sum_{m=b_{\text{low}}(i)}^{b_{\text{high}}(i)} \widehat{S}_{xx}(\Omega_m, n)} \right\}$$

$$= 10 \log_{10} \left\{ \frac{\sqrt[M(i)]{\frac{1}{N^{2M_i}} \cdot \prod_{m=b_{\text{low}}(i)}^{b_{\text{high}}(i)} \widehat{S}_{xx}(\Omega_m, n)}}{\frac{1}{N^2 M(i)} \sum_{m=b_{\text{low}}(i)}^{b_{\text{high}}(i)} \widehat{S}_{xx}(\Omega_m, n)} \right\}$$

$$= 10 \log_{10} \left\{ \sqrt[M(i)]{\frac{1}{N^{2M(i)}} \prod_{m=b_{\text{low}}(i)}^{b_{\text{high}}(i)} \widehat{S}_{xx}(\Omega_m, n)} \right\}$$

$$- 10 \log_{10} \left\{ \frac{1}{N^2 M(i)} \sum_{m=b_{\text{low}}(i)}^{b_{\text{high}}(i)} \widehat{S}_{xx}(\Omega_m, n) \right\}$$

$$= 10 \log_{10} \left\{ \left(\frac{1}{N^{2M(i)}} \prod_{m=b_{\text{low}}(i)}^{b_{\text{high}}(i)} \widehat{S}_{xx}(\Omega_m, n) \right)^{\frac{1}{M_i}} \right\}$$

$$- 10 \log_{10} \left\{ \left(\frac{1}{M(i)} \sum_{m=b_{\text{low}}(i)}^{b_{\text{high}}(i)} \frac{\widehat{S}_{xx}(\Omega_m, n)}{N^2} \right) \right\}$$

$$= \frac{10}{M(i)} \left[\log_{10} \left\{ \frac{1}{N^{2M(i)}} \right\} + \log_{10} \left\{ \widehat{S}_{xx}\left(\Omega_{b_{\text{low}}(i)}, n\right) \right\} + \ldots \right.$$

$$\left. + \log_{10} \left\{ \widehat{S}_{xx}\left(\Omega_{b_{\text{high}}(i)}, n\right) \right\} \right]$$

$$- 10 \log_{10} \left\{ \frac{1}{M(i)} \sum_{m=b_{\text{low}}(i)}^{b_{\text{high}}(i)} \frac{\widehat{S}_{xx}(\Omega_m, n)}{N^2} \right\}. \tag{17.51}$$

Eq. 17.51 shows the derivation of the logarithmically calculated version of the SFM.

17.3.7.5 Global Masking Threshold

Now, that we have the opportunity to characterize the input signal in noise and tone-like spectral parts, we can use this information and calculate an offset measure $O(i)$, empowering us to consider the different masking properties of the varying spectral types of the input signal by modifying the already available absolute masking vector $\boldsymbol{C}(n)$ to get the *global masking threshold* $T_{\log}(i,n)$:

$$T_{\log}(i,n) = 10 \log_{10}\left\{C(i,n)\right\} - O_{\log}(i,n) \tag{17.52}$$

for

$$i = 1, \ldots, 25,$$

with

$$O_{\log}(i,n) = \left(\alpha(i,n)\,(14.5 + i) + \left(1 - \alpha(i)\right)5.5\right) \text{dB}, \tag{17.53}$$

and

$$\alpha(i,n) = \min\left\{\frac{SFM_{\log}(i,n)}{-30}, 1\right\}. \tag{17.54}$$

With respect to Eq. 17.53 we can perceive how the offset vector $O_{\log}(i,n)$ is calculated in the Johnston model. There $\alpha(i,n)$ denotes the *tonality factor*, of the i'th Bark, which maps the SFM value of the i'th Bark to a range of $0 \leq \alpha(i,n) \leq 1$. In Eq. 17.54, we perceive a normalization to -30 dB of the SFM values during the calculation of $\alpha(i)$, which corresponds to the FFT length N, where the depicted value of -30 dB belongs to an FFT length of $N = 512$.[9] The normalization value is thereby important and has to be adjusted according to the resulting SFM value, occurring after insertion of a tonal input signal into the masking model. Additionally the FFT length N should not be shorter then $N = 512$, to ensure proper SFM values. If N should be chosen any smaller then 512, the SFM values gathered, especially at the low end of the spectrum, compulsorily do not contain enough spectral information any more. In such a setup too less frequency bins would be disposable for a reasonable calculation of the SFM values. If we take a closer look at the calculation of the offset values, as depicted in Eq. 17.53, we recognize that the offset values for a tone-like masker, i.e. tone-like disturbance, where $\alpha(i,n) \approx 1$, rises if occurring in higher Barks, whereas a noise-like masker, with $\alpha(i,n) \approx 0$ results in a frequency independent offset value of $O_{\log}(i,n) = 5.5$ dB.

---

[9] The following considerations are based on a sampling frequency $f_s = 44.1$ kHz.

17.3.7.6 Correction and Normalization

During the application of the spreading matrix to the individual Bark energies, considering the influence of neighboring Barks during the calculation of the global masking threshold, an error has been made. This error appears such that the total energy has been misleadingly increased, whereas the psychophysical process, which we are attempting to model, spreads the energy by dispersing it. For example, examining the behaviour with a hypothetical stimulus with unity energy in each critical band. The actual spreading function of the ear would result in no overall change to the level of energy in any critical band, whereas the here presented spreading function would increase the energy in each band, due to its additive contribution of energy, spread from adjacent critical bands. To simulate this effect, the absolute masking threshold $T_{\log}(i,n)$, found so far, has to be normalized by the vector $\boldsymbol{C}_{\text{err,log}}$, representing the *spread spectrum error*, as shown in Eq. 17.55.

$$\widetilde{\boldsymbol{T}}_{\log}(n) = \boldsymbol{T}_{\log}(n) - \boldsymbol{C}_{\text{err,log}}, \tag{17.55}$$

with

$$\boldsymbol{C}_{\text{err,log}} = \boldsymbol{S}\,\boldsymbol{E}, \tag{17.56}$$

and

$$\boldsymbol{E} = \begin{bmatrix} 1, 1, \ldots, 1 \end{bmatrix}^{\mathrm{T}}. \tag{17.57}$$

Regarding Eq. 17.56 and Eq. 17.57, $\boldsymbol{C}_{\text{err,log}}$ is determined by convolving an artificial stimulus, with unity energy in each Bark $\boldsymbol{E}$ with the spreading matrix $\boldsymbol{S}$. To include the absolute threshold of hearing (see Eq. 17.36) in the estimated, absolute masking threshold $T_{\log}(n)$, it is necessary to relate the digital input signal to its real listening level, i.e. its corresponding sound pressure level. In [56] it is suggested to use a 1 kHz full scale sinusoidal test tone as input signal, whose sound pressure level, measured 1 m away from the speaker, should be 90 dB. This test tone will also be inserted into the masking model, such that we will get its discrete Bark power, after the calculation of Eq. 17.40 and Eq. 17.43. Its resulting discrete Bark power is then used as a reference value $P_{\text{ref}}(i,n)$ in dB which has to be subtracted from the afore adjusted sound pressure level (SPL) of 90 dB to get the value $G_{\text{offset}}(i,n) = (90\,\text{dB} - P_{\text{ref}}(i,n))$, as named in Eq. 17.58.

$$M_{\text{TH,SPL}}(i,n) = \max\left\{ T_{\log}(i,n) - G_{\text{offset}}(i,n),\, T_{\text{q}}(i) \right\}. \tag{17.58}$$

The in this way adjusted absolute masking threshold $T_{\log}(i,n) - G_{\text{offset}}(i,n)$ will then be compared with the absolute threshold in quiet $T_{\text{q}}(i)$, where the maximum of the two finally marks the corresponding sound pressure level of the *global masking threshold* $M_{\text{TH,SPL}}(i,n)$, which can now be utilized for the

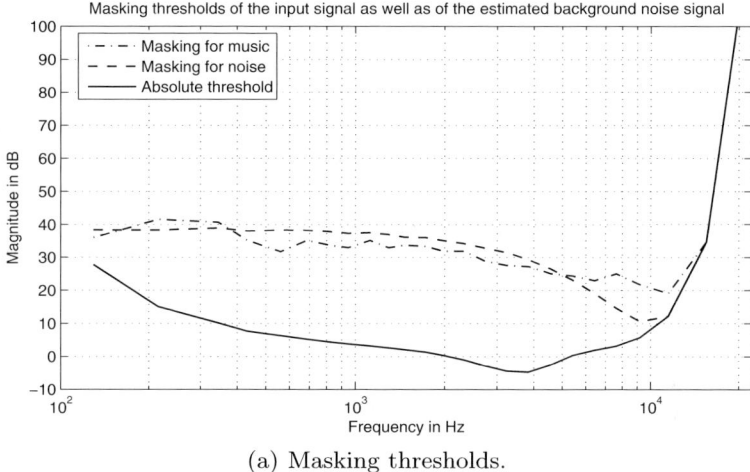

(a) Masking thresholds.

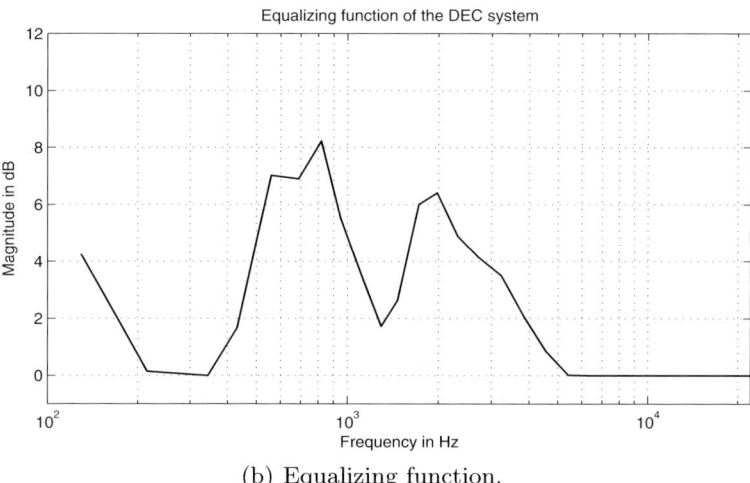

(b) Equalizing function.

Fig. 17.30. Calculation of the DEC gain shaping function from the difference between the masking thresholds of the reference and the estimated background noise signal.

calculation of the desired, psychoacoustically motivated gain-shaping-function or briefly the designated equalization function.

Fig. 17.30 depicts an example which shows how the introduced DEC algorithm calculates the equalization function. The upper part of Fig. 17.30 displays the resulting masking threshold of a typical background noise, recorded in a car, driving at an expressway with a speed of approx. 130 km/h, with deactivated fan and all windows closed, as well as the masking threshold of a typical pop music fraction, played at a moderate volume setting. It is logical, that the resulting masking threshold of the input signal will vary, depending on its current spectral content. In our case the input signal is usually music, which, as we all know can dispose over a high dynamic range, especially if we talk about classical music. In contrast to the input signal, the dynamics of the background noise signal statistically alters much less and thus can almost be considered as quasi-stationary. Hence, the resulting difference between those masking thresholds as, e.g., shown at the picture in the lower part of Fig. 17.30, which realizes the desired equalization function, changes almost in the same rhythm as the masking threshold of the input respectively music signal, compressing therefore its dynamics in dependance of the current level and spectral shape of the background noise signal, or more precisely, the masking threshold of the same. Thus, it is also legal to name our DEC system a *frequency dependant dynamic compressor*.

17.3.7.7 Conversion of the DEC Filter

Now after having the equalization function at our disposal, we faced a different, more practically related problem, namely: "How can we apply the equalizing to all necessary channels?" To answer this question we should first clarify about how many channels we actually talk about. In a high-class car it is not unusual to have 10 channels, which can be e.g. the front left, front right, center, side left, side right, rear left, rear right, woofer left, woofer right, and the subwoofer channel. Furthermore, we also have to deal with several different signal sources like the ordinary stereo input signal, a so-called 5.1 input signal, delivered by, e.g., a DTS[10] or an AC3[11] algorithm, the navigation, telephone, speech dialog, or the chimes signal. In total it is reasonable to say, that we have to filter as much as 15 different channels with the generated noise-shaping-function, which may very well burst the available memory and/or processing load of the DSP. On the one hand we could transform the equalizing function from the frequency into the time domain and may there filter each channels with the corresponding FIR filter, which inevitably would result in an extreme increase of the processing load. On the other hand we could also remain in the spectral domain and filter there by utilizing the very efficient fast convolution, but this would exceed the available memory reservoir. Thus, we were forced to find a different solution for this problem.

A possible remedy with which we came up is displayed in Fig. 17.31. There a solution is disclosed, based on the already mentioned LPC analysis, which

---

[10] The term *DTS* abbreviates *digital theater systems* which is a surround sound format for synchronized film sounds.
[11] The abbreviation *AC3* stands for *adaptive transform coder 3* and describes a multi-channel audio coding scheme for up to six discrete channels of sound, with five channels for normal-range speakers (20 Hz - 20 kHz) (right front, center, left front, right rear and left rear) and one channel (20 Hz - 120 Hz) for the subwoofer.

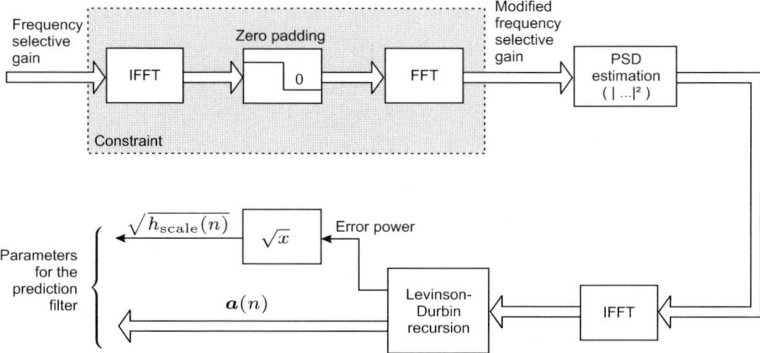

Fig. 17.31. Signal flow diagram, displaying the conversion of the spectral gain vector to LPC coefficients.

can be described as follows: First we have to calculate the constraint to effectively avoid the negative effects accompanied with the circular convolution. Then we calculate the PSD of the gain vector, which after all stands for the equalizing function, before we transform it to the time domain again, which eventually depicts the *autocorrelation* of the equalizing filter, which act as basis for the succeeding *Levinson-Durbin recursion* which at the end delivers the filter coefficients for the *prediction filter*.

In contrast to the GAL algorithm, which deliveres reflection coefficients, the Levinson-Durbin algorithm is used such, that it deliberately generates direct form filter coefficients for the prediction filter. Thus, the prediction filter can be realized as a direct form filter, as presented in Fig. 17.32, which can more effectively be implemented as e.g. the lattice filter structure known form Fig. 17.17. Referring to Fig. 17.32, the implementation cost of the prediction filter in direct form is only marginally higher as if implementing a short FIR filter. Compared to the pure realization in the time domain by utilizing a long FIR filter, as previously described, the processing load can now, by using this method, based on the LPC analysis, be reduced to approximately one tenth of the former version. Assuming we use a FFT with a length of $N = 1024$

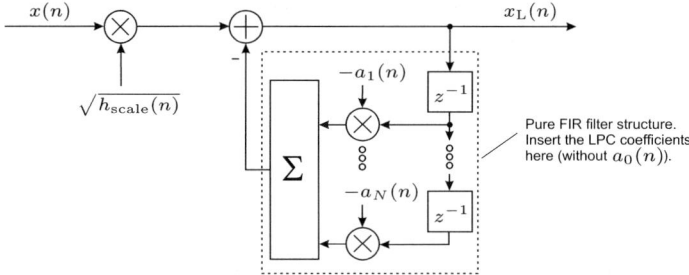

Fig. 17.32. Signal flow diagram of a prediction filter, realized in the direct form.

we would have to realize an FIR filter with a length of 512 coefficients, which can now, with the help of the LPC method, be approximated, with sufficient accuracy by using a prediction filter with a length of only 64 filter coefficients. This enables us to save enough processing power as well as memory, such that we are now able to even filter as much as 15 independent channels with the desired gain respectively equalizing functions.

17.3.7.8 Example of a Psychoacoustically Motivated, Spectral Domain Based DEC-System

Combining all the signal processing blocks, discussed so far, we may end up in a DEC system as depicted in Fig. 17.33. There we see how the stereo echo canceller can be efficiently embedded, where the memoryless smoothing filter should be inserted, which signals should be used as input for the masking models, how the equalizing function should be post-processed to finally feed the DEC filter of the system.

However, at two blocks we should take a closer look, that is on the one hand the volume controlled block named *anti-gain-chase function*, which particularly fulfills the same task as the one shown in the lower picture of Fig. 17.7. On the other hand also the block called *Conversion Bark to linear* should be mentioned, which realizes the conversion of the equalizing function, basically calculated in the Bark domain into the "normal", i.e. linear spectral domain, as usually available after transforming a signal from the time into the spectral domain, e.g. with the help of an FFT.

17.4 Conclusion and Outlook

In this chapter we showed, from a practical point of view, how the problem to realize an automatic volume, or more advanced, equalization control can be realized, so far. We started from established systems that are based on non-acoustical sensor signal(s). The most important one of these sensors signals is the speed signal. It is available nowadays in almost every car on the market. By using it we are able to not only adjust the volume or better the dynamics of the input signal, but also the bass content of the audio signal. Afterwards we turned to microphone based systems, providing a more natural way of tackling the given challenge. Even the first deployed DVC systems utilizing a microphone to actually measure the real background noise suffered from the same problems as known from AEC algorithms. I.e., they had to sufficiently suppress the music signal, picked up by the microphone, acting, from the AEC point of view, as echo signal, to avoid erroneous estimations of the background noise (level), which, in turn would result in feedback problems. To circumvent those stability problems, we had to take counter measures such as the employment of an adaptive filter with all its improvements such as the insertion of an

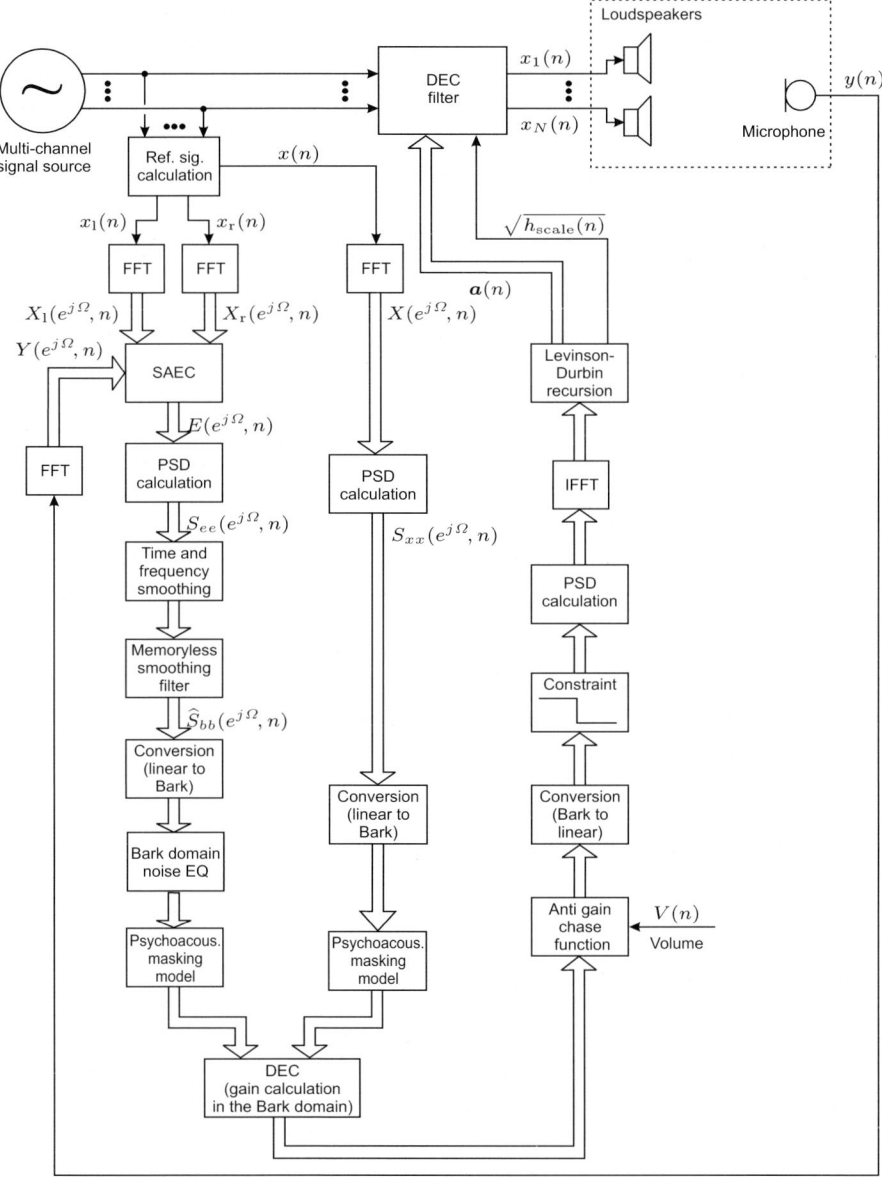

Fig. 17.33. Signal flow diagram of a psychoacoustically motivated, spectral domain DEC system.

adaptive adaptation step size to avoid misleading destructions of the filter coefficient set of the already converged adaptive filter, by impulsive disturbances presented e.g. by voice. Also the possibility that the whole adaptation process could freeze during abrupt changes of the LEM system had to be considered, just like in AEC system. Due to these similarities it is obvious that we could combine, at least parts of an AEC algorithm with a microphone based DVC system to more efficiently use the limited resources of the DSP. Furthermore, we have to estimate the background noise, or at least its level, as well, without being disrupted by residual echoes, i.e. remaining parts of the music signal which could not be reduced by the AEC, or other signal parts which should not particularly be attributed to the background noise, such as speech signals or other burst like disruptions. This problem is also known. Noise reduction algorithms which, most of the time accompanying AEC algorithms, do have exactly the same task to perform, namely to estimate the background noise in a robust way, without considering speech, e.g., as part of the background noise. Therefore *noise reduction algorithms* (NR) usually dispose over VAD methods to block those signal parts during the estimation of the background noise. Here we could as well use synergy effects by sharing algorithmic parts of DVC and NR systems. One open problem consists of the placement of the microphones within the interior of the car, which, mostly lies solely in the responsibility of the customer and not in the hand of the developer. This may lead to the introduction of a so called "noise equalizing function" whose job is to compensate for the deficiencies of the placement of the microphone, by equalizing its signal in such a way that the resulting, estimated background noise may better reflect the real noise situation close to the head of the passengers, preferably the driver. Having several microphones installed in the car, able to be combined as beamformer to isolate or deliberately block signals originating from a certain direction could also be used to solve the placement problem of the microphone. Therefore it is, on the one hand imaginable to combine the output of a classical beamformer with the DVC system or on the other hand use a different set of beamforming filters which, e.g., generate a beam in a direction where the background noise is similar to the desired place, which is, as we already know close to the head of the driver, without receiving undesired speech signals originating from the locations where passengers could possibly sit. Even if we do have more microphones in the car that cannot profitably be combined to a beamformer, such as in *in-car communication systems* (ICC), or *active noise control systems* (ANC) we could, at least use them to create a multi channel, respectively individualized DVC or DEC system, with the option to combine the individual background noise estimations such as to soothe the placement problem as well. Of course it is also supposable to share parts of the signal processing, in creating a more efficient, complete system, with those systems, too. The introduced systems utilizing a microphone and several non-acoustical sensors, as well as the first combined DVC/DEC systems marked only an intermediate step to a psychoacoustically motivated, spectral based DEC system. There the same problems that we were confronted with during the development of the DVC systems, realized in the time domain, had to be solved. However, the spectral domain offered more flexible and powerful solutions. The assignment of the psychoacoustic masking model according to Johnston enabled us to create an equalization function as desired, without extensively increasing neither the processing load nor the memory requirements, due to the fact that we were able to conduct most of the adherent calculations in the Bark domain. This means that as less as 25 different Bark values suffice to implement most of the masking model, which of course does not constitute much effort. The discussion about the SAEC demonstrated how the spectral DEC system could be improved by utilizing a stereo reference signal instead of a mono signal. Following this basic line it is imaginable to further enhance the DEC system by applying the psychoacoustic masking model to both, or even more reference signals to eventually get individual gain-shaping-functions applicable to one or more loudspeakers. As an example, having two separate reference signals, as already described, we would also get two separate equalizing functions were each loudspeaker, usually fed with the left audio signal would be equalized with the corresponding left gain-shaping-function, whereas the loudspeakers at the right side would be weighted with the right equalizing function. Mono loudspeakers such as the center speaker, normally supplied with a mono signal

$$x_{\text{mono}}(n) = \frac{1}{2}\left(x_{\text{left}}(n) + x_{\text{right}}(n)\right) \tag{17.59}$$

could then also be equalized with the mean of the two equalizing functions. Hopefully, in the near future the available and cost efficient DSPs of the market will comprise enough memory or at least a sufficiently fast interface to external memory such as SRAM or DRAM to allow the complete realization of the audio signal processing in the spectral domain. Then the work-around with the LPC analysis to reduce the processing load as well as the memory consumption during the filtering of a multitude of channels will be a relict of the past. This implies, that we could use the more efficient fast convolution, able to do the filtering process with the original filter coefficients, resulting in more accurate results, without rising the number of cycles. An improvement of the whole system can possibly be achieved if the psychoacoustical masking model would be realized, more precisely, which of course would also imply an increase of the FFT length. However, this should only be taken into account, if we have enough system resources to our disposition.

As we see, there is still enough room for advancements, but we are of the opinion that the core problems, involved in generating a system, able to deliver a subjectively constant audio performance, even if disrupted by an arbitrary background noise, have already been solved, especially by considering psychoacoustic properties.

17.5 Acknowledgement

I will herewith thank Gerhard Schmidt firstly for the faith in my abilities and secondly for providing the opportunity to participate in this book. I am also indebted to my company, especially to Gerhard Pfaffinger and other colleagues of our department for their encouragement and understanding in writing this chapter.

References

1. L. Arslan, A. McCree, V. Viswanathan: New methods for adaptive noise suppression, *Proc. ICASSP '95*, **1**, 812–815, Detroit, Mi, USA, 1995.
2. D. Barschdorff, D. Wtzlar, R. Klockner, Blaupunkt Werke GmbH: Process and circuitry to automatically control sound level depending on ambient noise, European Patent Application, *EP0319777*, published June 14'th, 1989.
3. F. J. Op De Beek, J. Kemna, Philips Corp.: Noise-dependent volume control having a reduced sensitivity to speech signals, United States Patent Application, *US4677389*, published June 30'th, 1987.
4. Y. Bendel, D. Burshtein, O. Shalvi, E. Weinstein: Delayless frequency domain acoustic echo cancellation, *IEEE Trans. on Acoustics, Speech and Signal Processing*, **9**(5), July 2001.
5. J. Benesty, Y. Huang: *Adaptive signal processing applications to real-world problems*, Berlin, Germany: Springer, 2003.
6. A. G. Bose, J. L. Veranth, Bose Corp.: Speed controlled amplifying, United States Patent Application, *US4944018*, published July 24'th, 1990.
7. A. G. Bose, Bose Corp.: Speed controlled amplifying, United States Patent Application, *US5034984*, published July 23'th, 1991.
8. J.-P. Boyer, Philips Electronics: Device for automatic control of sounds with fuzzy logic, European Patent Application, *EP0632586*, published January 4'th, 1995.
9. H. Buchner, J. Benesty, W. Kellermann: An extended multidelay filter: Fast low-delay algorithms for very high-order adaptive systems, *Proc. ICASSP '03*, Hong Kong, PRC, April 2003.
10. H. Buchner, J. Benesty, T. Gänsler, Walter Kellermann: An outlier-robust extended multidelay filter with application to acoustic echo cancellation, *Proc. IWAENC '03*, Kyoto, Japan, September 2003.
11. H. Buchner, J. Benesty, W. Kellermann: Generalized multichannel frequency-domain adaptive filtering: efficient realization and application to hands-free speech communication, *Signal Processing*, **85**(3), 549–570, September 2005.
12. K. S. Chan, B. Farhang-Boroujeny: Lattice PFBLMS: Fast converging structure for efficient implementation of frequency-domain adaptive filters, *Signal Processing*, **78**(1), 79–89, October 1999.
13. M. Christoph, Harman/Becker Automotive Systems GmbH: Method and apparatus for dynamic sound optimization, International Patent Application, *WO0180423*, published October 25'th, 2001.
14. M. Christoph, Harman/Becker Automotive Systems GmbH: Dynamic sound optimization in the interior of a motor vehicle or similar noisy environment, a monitoring signal is split into desired-signal and noise-signal components which are used for signal adjustment, German Patent Application, *DE10035673*, published March 7'th, 2002.
15. M. Christoph, Harman/Becker Automotive Systems GmbH: Method and apparatus for dynamic sound optimization, United States Patent Application, *US6529605*, published March 4'th, 2003.

16. M. Christoph, Harman/Becker Automotive Systems GmbH: Device for the noise- dependent adjustment of sound volumes, United States Patent Application, *US2004076302*, published April 22'th, 2004.
17. M. Christoph, Harman/Becker Automotive Systems GmbH: Method and apparatus for dynamic sound optimization, United States Patent Application, *US2004125962*, published July 1'st, 2004.
18. M. Christoph, Harman/Becker Automotive Systems GmbH: Audio enhancement system and method, United States Patent Application, *US2005207583*, published September 22'th, 2005.
19. M. Christoph, Harman/Becker Automotive Systems GmbH: Audio enhancement system and method, European Patent Application, *EP1720249*, published November 8'th, 2006.
20. M. Christoph, Harman/Becker Automotive Systems GmbH: Audio enhancement system and method, European Patent Application, *EP1619793*, published January 25'th, 2006.
21. G. A. Clark, S. R. Parker, S. K. Mitra: A unified approach to time and frequency domain realization of FIR adaptive digital filters, *IEEE Trans. on Acoustics, Speech and Signal Processing*, **ASSP–31**(5), October 1983.
22. D. St. D'Arc De Costemore: Automatic process of adjustment of the volume of sound reproduction, United States Patent Application, *US5530761*, published June 25'th, 1996.
23. R. M. M. Derx, G. P. M. Egelmeers, P. C. W. Sommen: New constraint method for partitioned block frequency-domain adaptive filters, *IEEE Trans. on Signal Processing*, **50**(9), September 2002.
24. M. A. Dougherty: Automatic noise compensation system for audio reproduction equipment, United States Patent Application, *US5907622*, published May 25'th, 1999.
25. G. P. M. Egelmeers, P. C. W. Sommen: A new method for efficient convolution in frequency domain by nonuniform partitioning for adaptive filtering, *IEEE Trans. on Signal Processing*, **44**(12), 3123–3128, December 1996.
26. K. Eneman, M. Moonen: The iterated partitioned block frequency-domain adaptive filter for acoustic echo cancellation, *Proc. IWAENC '01*, Darmstadt, Germany, September 2001.
27. K. Eneman: Subband and frequency-domain adaptive filtering techniques for speech enhancement in hands-free communication, *Katholieke Universiteit Leuven*, PhD thesis, Leuven, Belgium: March 2002.
28. G. Enzner, P. Vary: Robust and elegant, purely statistical adaptation of acoustic echo canceller and postfilter, *Proc. IWAENC '03*, Kyoto, Japan, September 2003.
29. B. Farhang-Boroujeny: *Adaptive filters – Theory and applications*, New York, NY, USA: John Wiley & Sons Inc., October 2000.
30. H. Fischlmayer, Grundig EMV: Circuit arrangement to control the sound volume in a motor car in dependence upon the speed of it, European Patent Application, *EP0212105*, published March 4'th, 1987.
31. L. Freund, M. Israel, F. Rouseau, J.M. Berge, M. Auguin, C. Belleudy, G. Gogniat: A codesign experiment in acoustic echo cancellation: GMDF, *Proc. IEEE international conference on system synthesis*, IEEE computer science press, 1996.

32. H. Germer, J. Käs=er, Blaupunkt Werke GmbH: Method of adapting the sound volume of a loudspeaker in accordance with the ambient noise level, European Patent Application, *EP0141129*, published Mai 15'th, 1985.
33. E. Hänsler, G. Schmidt: *Acoustic echo and noise control – A practical approach*, New York, NY, USA: John Wiley & Sons Inc., 2004.
34. M. Heckman, J. Vogel, K. Kroschel: Frequency selective step-size control for acoustic echo cancellation, *Proc. EUSIPCO '00*, Tampere, Finland, September 2000.
35. W. Herbordt, H. Buchner, W. Kellermann: Computationally efficient frequency-domain combination of acoustic echo cancellation and robust adaptive beamforming, *Proc. of EUROSPEECH '01*, **2**, 1001–1004, Aalborg, Danmark, September 2001.
36. J. Höllermann, Blaupunkt Werke GmbH: Device for the noise dependent volume control of a car radio, European Patent Application, *EP0623995*, published November 9'th, 1994.
37. A. Holly, C. R. Culbertson, D. L. Ham, Walkaway Technologies Inc.: Constant power ratio automatic gain control, United States Patent Application, *US4953221*, published May 28'th, 1990.
38. ISO/IEC CD 11172-3:Coding of moving pictures and associated audio for digital storage media at up to about 1.5 MBit/s: Part 3 audio, *Standard*, March, 28'th 1996.
39. K. Juji, T. Chiba, K. Yamada, Matsushita Electric Ind. Co. Ltd.: Sound reproduction apparatus, United States Patent Application, *US5796847*, published August 18'th, 1998.
40. J. D. Johnston: Estimation of perceptual entropy using noise masking, *Proc. ICASSP '88*, **5**, 2524–2527, New York, NY, ULA, April 1988.
41. B. Jun Jae, Hyundai Mobis Co. Ltd.: Volume control device automatically tuned by speed of vehicle and operating method thereof, Korean Patent Application, *US4944018*, published October 16'th, 2004.
42. M. Kato, S. Kato, F. Tamura, Pioneer Electronic Corp.: Mobile automatic sound volume control apparatus, United States Patent Application, *US4864246*, published September 5'th, 1989.
43. M. Kato, S. Kato, F. Tamura, Pioneer Electronic Corp.: On-vehicle automatic loudness control apparatus, United States Patent Application, *US5081682*, published January 14'th, 1992.
44. S. Kato, H. Kihara, S. Mori, F. Tamura, Pioneer Electronic Corp.: On-board vehicle automatic sound volume adjusting apparatus, United States Patent Application, *US5208866*, published May 4'th, 1993.
45. S. M. Kuo, D. R. Morgan: *Active noise control systems - Algorithms and DSP implementations*, New York, NY, USA: John Wiley & Sons Inc., 1996.
46. J. Lariviere, R. Goubran: GMDF for noise reduction and echo cancellation, *IEEE Signal Processing Letters*, **7**(8), August 2000.
47. K. Leyser, Blaupunkt Werke GmbH: Circuit arrangement for automatic adapting the volume of a loudspeaker to an interfering noise level prevailing at the loudspeaker location, German Patent Application, *DE3320751*, published December 13'th, 1984.
48. K. Leyser, R. Albert, J. Käsßer, Blaupunkt Werke GmbH: Circuit arrangement for automatic adapting the volume of a loudspeaker to a disturbing noise level prevailing at the loudspeaker location, German Patent Application, *DE3338413*, published Mai 2'nd, 1985.

49. R. Martin: An efficient algorithm to estimate the instantaneous SNR of speech signals, *Proc. EUROSPEECH '93*, **3**, 1093–1096, Berlin, Germany, 1993.
50. T. Menzel, H.-U. Aust, Blaupunkt Werke GmbH: Regulation of noise in passenger compartment in motor vehicle - subtracting integrated loudspeaker and microphone signals and detecting lowest value using comparator for regulation of audio end stage amplifier, German Patent Application, *DE4204385*, published October 19'th, 1993.
51. T. E. Miller, K. L. Kantor, J. Barish, D. K. Wise, Jensen Int. Inc.: Method and apparatus for dynamic sound optimization, United States Patent Application, *US5434922*, published July 18'th, 1995.
52. T. E. Miller, K. L. Kantor, J. Barish, D. K. Wise, Jensen Int. Inc.: Method and apparatus for dynamic sound optimization, United States Patent Application, *US5615270*, published March 25'th, 1997.
53. E. Moulines, O. Ait Amrane, Y. Grenier: The generalized multidelay adaptive filter: Structure and convergence analysis, *IEEE Trans. on Signal Processing*, **43**(1), January 1995.
54. S. S. Narayan, A. M. Peterson, M. J. Narasimha: Transform domain LMS algorithm, *IEEE Trans. on Acoustics, Speech and Signal Processing*, **ASSP–31**(3), 609–615, June 1983.
55. P. Pietarila: A frequency-warped front-end for a subband audio codec, *Department of Electrical Engineering, University of Oulu*, Oulu, Finland, Diploma Thesis, May 2001.
56. D. J. M. Robinson: Perceptual model for assessment of coded audio, *Department of Electronic Systems Engineering, University of Essex*, Essex, Great Brittain, PhD Thesis, March 2002.
57. S. Defence: Gain control in telecommunications systems, Great Britain Patent Application, *GB2013051*, published August 1'st, 1979.
58. J. J. Shynk: Frequency-domain and multirate adaptive filtering, *IEEE Signal Processing Magazine*, **9**(1), 14–37, January 1992.
59. J.-S. Soo, K. K. Pang: A multistep size (MSS) frequency domain adaptive filter for stationary and nonstationary signals, *Proc. ICASSP '89*, Glasgow, Great Britain, May 1989.
60. J.-S. Soo, K. K. Pang: Multidelay block frequency domain adaptive filter, *IEEE Trans. on Acoustics, Speech and Signal Processing*, **38**(2), February 1990.
61. J.-S. Soo, K. K. Pang: A multistep size (MSS) frequency domain adaptive filter, *IEEE Trans. on Signal Processing*, **39**(1), January 1991.
62. O. Tadatoshi, T. Ken-Ichi, Mitsubishi Electric Corp.: Automatic volume controlling apparatus, United States Patent Application, *US5450494*, published September 12'th, 1995.
63. A. Tokumo, M. Kato, Pioneer Electronic Corp.: Automatic sound volume control device, United States Patent Application, *US4476571*, published October 9'th, 1984.
64. V. Turbin, A. Gilloire: A frequency domain implementation of the combined acoustic echo cancellation system, *Proc. of the 8th IEEE DSP workshop*, Bryce Canyon, USA: August 1998.
65. W. Vössing, H. Wellhausen, Blaupunkt Werke GmbH: Circuit arrangement for the automatic volume control of a loudspeaker in dependence upon an interference noise level prevailing at the loudspeaker's location, European Patent Application, *EP0027519*, published April 29'th, 1981.

66. K. Wiedemann, Blaupunkt Werke GmbH: Apparatus for matching the sound output of a radio receiver to the ambient noise level, United States Patent Application, *US4247955*, published January 27'th, 1981.
67. F. Wolf: Optimierte Lautstärkeanpassung für Audiosignale im Fahrzeug in Abhängigkeit verschiedener Eingangsgrößen, Internal Document, *Harman/Becker Automotive Systems GmbH*, October 15'th, 2001.
68. S. Yamamoto, S. Kitayama, J. Tamura, H. Ishigami: An adaptive echo canceller with variable step gain method, *Trans. IECE of Japan*, **E 65**(1), 1–8, 1982.
69. E. Zwicker: *Psychoakustik*, Berlin, Germany: Springer, 1982 (in German).
70. U. Zölzer: *Digitale Audiosignalverarbeitung*, Stuttgart, Germany: B.G. Teubner, 1997 (in German).

Towards Robust Distant-Talking Automatic Speech Recognition in Reverberant Environments

Armin Sehr and Walter Kellermann

Multimedia Communications and Signal Processing
University of Erlangen-Nuremberg, Germany

In distant-talking scenarios, automatic speech recognition (ASR) is hampered by background noise, competing speakers and room reverberation. Unlike background noise and competing speakers, reverberation cannot be captured by an additive or multiplicative term in the feature domain because reverberation has a dispersive effect on the speech feature sequences. Therefore, traditional acoustic modeling techniques and conventional methods to increase robustness to additive distortions provide only limited performance in reverberant environments.

Based on a thorough analysis of the effect of room reverberation on speech feature sequences, this contribution gives a concise overview of the state of the art in reverberant speech recognition. The methods for achieving robustness are classified into three groups: Signal dereverberation and beamforming as preprocessing, robust feature extraction, and adjustment of the acoustic models to reverberation. Finally, a novel concept called reverberation modeling for speech recognition, which combines advantages of all three classes, is described.

18.1 Introduction

Even for difficult tasks, current state-of-the-art ASR systems achieve impressive recognition rates if a clean speech signal recorded by a close-talking microphone is used as input [51, 52]. In many applications however, using a close-talking microphone is either impossible or unacceptable for the user.

As an example, for the automatic transcription of meetings or lectures [2, 11], equipping each speaker with a close-talking microphone would be very inconvenient. Instead, distant microphones, e.g., placed at the meeting table, are used. Voice control of medical systems allows a surgeon to work with both hands while controlling diagnostic instruments or assistance devices.

Telephone-based speech dialogue systems for information retrieval or transactions, like telephone-based flight information desks or telephone banking systems, need to cope with users calling from hands-free telephones. Further applications of distant-talking ASR are dictation systems, information terminals, and voice-control of consumer electronics, like television sets or set-top boxes.

In all these scenarios, the distance between speaker and microphone is in the range of one to several meters. Therefore, the microphone does not only pick up the desired signal, but also additive distortions like background noise or competing speakers, and reverberation of the desired signal. While significant progress has been achieved over the last decades in improving the robustness of ASR to additive noise and interferences, the research on reverberation-robust ASR is still in its infancy. This contribution focuses on robust ASR in reverberant environments.

The chapter is structured as follows: The distant-talking ASR scenario is discussed in Sec. 18.2 and the different properties of additive distortions and reverberation are emphasized. Sec. 18.3 outlines how the measures for increasing robustness to reverberation are embedded into ASR systems and explains the basics of ASR which will be needed for describing these measures. The effect of reverberation on speech feature sequences is investigated in Sec. 18.4. The known approaches to achieve robust ASR in reverberant environments are classified into three groups:

- first, signal dereverberation and beamforming as preprocessing (Sec. 18.5),
- second, usage of robust features which are insensitive to reverberation or feature-domain compensation of reverberation (Sec. 18.6),
- third, adjustment of the acoustic models of the recognizer to reverberation by training or adaptation (Sec. 18.7).

A novel approach called reverberation modeling for speech recognition, which combines advantages of all three classes, is discussed in Sec. 18.8. It uses a statistical reverberation model to perform feature-domain dereverberation within the recognizer. Sec. 18.9 summarizes and concludes this contribution.

18.2 The Distant-Talking ASR Scenario

Fig. 18.1 shows a typical *distant-talking* ASR scenario. Compared to the close-talking scenario, the gain of the microphone amplifier has to be increased because of the greater distance between the desired speaker and the microphone. Therefore, the microphone does not only pick up the desired signal but also background noise, interfering speakers and the reverberation of the desired signal. The reverberation results from the fact that the desired signal does not only travel along the direct path from the speaker to the microphone, but is also reflected by walls and other obstacles in the enclosure. Therefore, the microphone picks up many delayed and attenuated copies of the desired signal

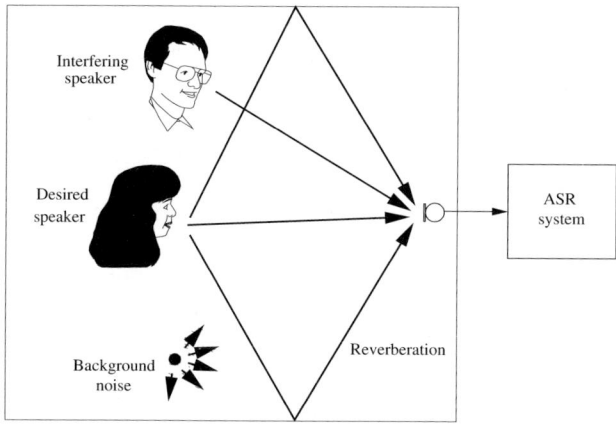

Fig. 18.1. Distant-talking ASR scenario.

which are perceived as reverberation. In the time domain, reverberation can be very well modeled by convolving the signal $s(n)$ of the desired speaker with the impulse response $h(n)$ describing the acoustic path between speaker and microphone [37]. Additive distortions, like background noise and interfering speakers, are modeled by the signal $b(n)$ so that the microphone signal $y(n)$ is given as

$$y(n) = h(n) * s(n) + b(n) . \tag{18.1}$$

The corresponding block diagram for the distant-talking signal capture is depicted in Fig. 18.2.

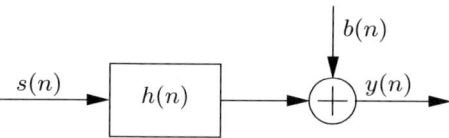

Fig. 18.2. Block diagram of distant-talking signal capture.

As the additive distortions $b(n)$ and the desired speech signal $s(n)$ result from different sources, they can be modeled as statistically independent random processes. Therefore, very effective methods for reducing additive distortions in the microphone signal, and for adjusting speech recognizers to additive distortions have been developed in the last decades. See [7, 8, 24, 26] and [33, 35] for overviews.

In contrast to that, the reverberation is strongly correlated to the desired signal and cannot be described by an additive term. Therefore, the approaches developed for additive distortions are not appropriate to increase robustness against reverberation.

As we will focus on reverberation in this chapter, we neglect the signal $b(n)$ in the following treatment so that the microphone signal is given as

$$y(n) = h(n) * s(n) \,. \tag{18.2}$$

Fig. 18.3 shows a typical *room impulse response* (RIR) measured in a lecture room. After an initial delay of approximately 12.5 ms, which is caused by the time the sound waves need to travel from the speaker to the microphone (here roughly 4 m), the first peak in the RIR is caused by the direct sound. Following the direct sound, several distinct peaks corresponding to prominent reflections can be observed. With increasing delay, more and more reflections overlap so that no distinct peaks but rather an exponentially decaying envelope characterizes the last part of the RIR.

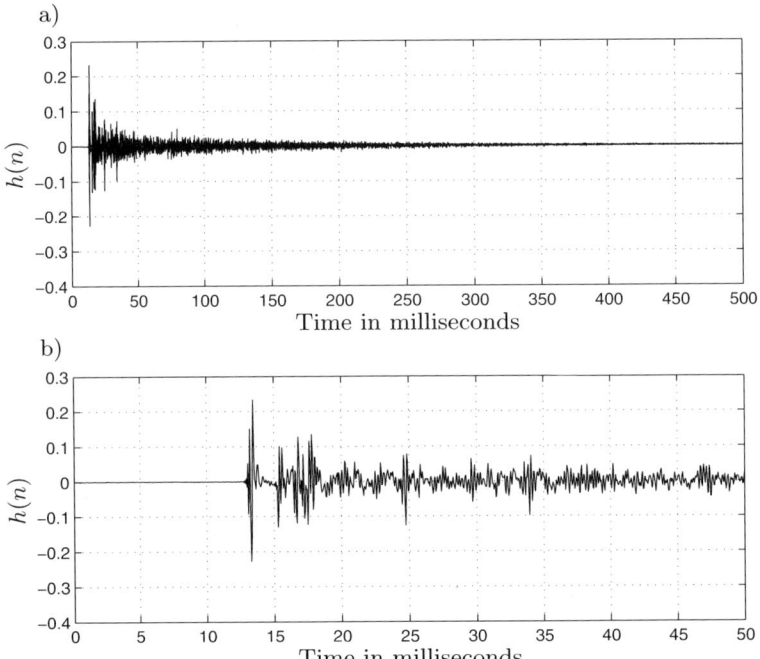

Fig. 18.3. RIR $h(n)$ of a lecture room in the time domain, a) complete RIR, b) first section of RIR.

The time needed for a 60 dB decay in sound energy is called the *reverberation time* T_{60}. Typical reverberation times are in the range of 20-100 ms in cars and 200-800 ms in offices or living rooms. In large lecture rooms, concert halls or churches, the reverberation time is often significantly longer than 1 s.

The *signal-to-reverberation ratio* (*SRR*) compares the energy of the direct sound to the energy of the reverberation and is defined as

$$SRR = 10 \log_{10} \left\{ \frac{\sum_{n=0}^{N_d-1} h^2(n)}{\sum_{n=N_d}^{N_h-1} h^2(n)} \right\},$$

where the first part of the RIR from $0 \ldots N_d - 1$ is considered as direct sound and the second part of the RIR from $N_d \ldots N_h - 1$ is considered as reverberation. The RIR is strongly time-variant. Already small changes in the position of the speaker or the microphone, movements of other objects, like doors, windows or persons, or variations in temperature change the details of the RIR significantly. However, its overall characteristics, like the reverberation time, the SRR and even the envelope of the time-frequency pattern corresponding to the RIR, are hardly affected by such changes.

Please note the distinction between reverberation and acoustic echoes. In this contribution, reverberation is used for multiple delayed copies of the desired signal, while in acoustic echo cancellation [9], the term echo is used to describe multiple delayed copies of interfering signals originating from loudspeakers.

18.3 How to Deal with Reverberation in ASR Systems?

This section discusses how the different approaches to increase robustness against reverberation can be embedded into an ASR system. For this purpose, the general task of ASR is formulated first, and the options for increasing robustness to reverberation are explained using a generic ASR block diagram.

The task of a speech recognizer can be formulated as finding the best estimate $\hat{W}$ of the true word sequence W_t corresponding to a certain utterance, given the respective speech signal $s(n)$. Usually, the recognizer does not use the speech signal itself but rather speech feature vectors $s(k)$ derived from the speech signal. Denoting the sequence of all observed speech feature vectors $s(1) \ldots s(K)$ as S, where K is the length of the sequence, the recognition problem can be expressed as maximizing the posterior probability $P(W|S)$ over all possible word sequences W

$$\hat{W} = \underset{W}{\mathrm{argmax}} \left\{ P(W|S) \right\}. \tag{18.3}$$

Equivalently, the product of the likelihood and the prior probability can be maximized

$$\hat{W} = \underset{W}{\mathrm{argmax}} \left\{ P(S|W) \cdot P(W) \right\}. \tag{18.4}$$

The exact determination of both $P(S|W)$ and $P(W)$ is very difficult in real-world systems. Therefore, the likelihood $P(S|W)$ of observing the feature sequence S given the word sequence W is approximated by some acoustic score $A(S|W)$, which is modeled by the *acoustic model*. The prior probability

$P(W)$ of the word sequence is approximated by some language score $L(W)$ and is modeled by a *language model* so that the recognition problem can be expressed as

$$\hat{W} = \underset{W}{\operatorname{argmax}} \left\{ A(\boldsymbol{S}|W) \cdot L(W) \right\}. \tag{18.5}$$

In a distant-talking scenario, the clean-speech feature sequence $\boldsymbol{S}$ is not available. Instead, the feature sequence $\boldsymbol{Y}$ derived from the reverberant microphone signal $y(n)$ has to be used so that the distant-talking recognition problem is given as

$$\hat{W} = \underset{W}{\operatorname{argmax}} \left\{ A(\boldsymbol{Y}|W) \cdot L(W) \right\}. \tag{18.6}$$

Robustness to reverberation is achieved, if the solution to the problem described in Eq. 18.6 is approaching the solution to the problem of Eq. 18.5. That is, the accuracy of the transcription determined from the reverberant sequence $\boldsymbol{Y}$ approaches the accuracy determined from the clean-speech sequence $\boldsymbol{S}$.

Fig. 18.4 shows a generic block diagram of an ASR system which is used to solve the problem described in Eq. 18.5. The speech signal is preprocessed in order to reduce distortions and then transformed into speech feature vectors. Before the recognizer can be used to determine the word sequence or transcription of unknown utterances, both its acoustic model and its language model have to be trained using training data with known transcriptions.

The function blocks where measures to increase robustness against reverberation can be embedded into the ASR system so that it can effectively solve problem 18.6 are marked by areas shaded in dark gray in Fig. 18.4. The attached labels point out the section where these measures are discussed.

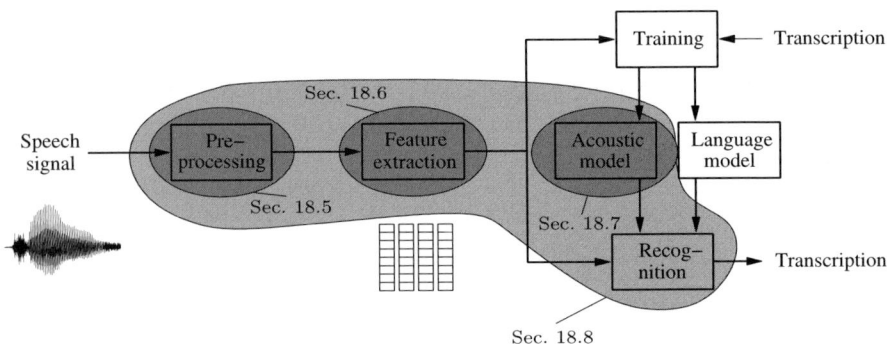

Fig. 18.4. Block diagram of a speech recognition system.

The novel concept of reverberation modeling for speech recognition implements the idea of preprocessing directly in the feature domain using an improved acoustic model to dereverberate the speech features during recognition (Sec. 18.8). In this way, robustness is achieved by utilizing four main function blocks instead of only one as indicated by the area shaded in light gray. Therefore, the advantages of all three classes of approaches are utilized by the novel concept.

In the following, the blocks of the speech recognition system according to Fig. 18.4 are explained in more detail. The goal of the feature extraction is to reduce the dimension of the input data roughly by one order of magnitude (e.g., from 256 samples to 25 features). The features should concentrate all information of the speech signal which is necessary for the classification of different phones and words and all information irrelevant for speech recognition should be removed.

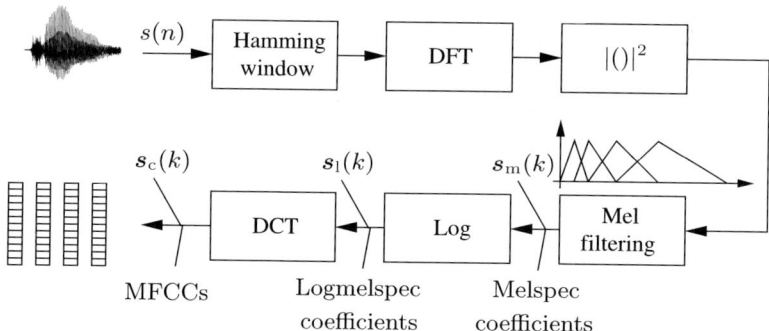

Fig. 18.5. Block diagram of the feature extraction for MFCCs.

Currently, the most popular speech features are the so-called *mel-frequency cepstral coefficients* (MFCCs) [12]. Their calculation is illustrated in Fig. 18.5. In the first step of the feature extraction, a short-time spectrum analysis is performed by windowing overlapping frames of the speech signal with a Hamming window $w(n)$ and applying an F-point discrete Fourier transform (DFT)

$$S(f, k) = \sum_{n=0}^{F-1} w(n)\, s(kN + n)\, e^{-j\frac{2\pi}{F} n f}, \qquad (18.7)$$

where f is the index of the DFT bin, k is the frame index and $N \leq F$ is the frame shift. The magnitude square of the DFT coefficients $S(f, k)$ is filtered by a mel filter bank $C(l, f)$ to obtain the mel-spectral (*melspec*) coefficients

$$s_{\mathrm{m}}(l, k) = \sum_{f=0}^{F/2} C(l, f) |S(f, k)|^2, \qquad (18.8)$$

where the subscript m denotes "melspec domain" and l is the index of the mel channels. Due to the symmetry of the DFT, it is sufficient to calculate the sum over $f = 0 \ldots F/2$. Like the human auditory system, the mel filter bank has a better frequency resolution for low frequencies than for high frequencies.

This is commonly realized by triangular weighting functions $C(l, f)$ for the mel channels as depicted in Fig. 18.6. The widths of these weighting functions increase with the channel number [33] and approximate a logarithmic spectral resolution similar to the human hearing. The feature vector $\boldsymbol{s}_\mathrm{m}(k)$ holds all melspec coefficients of frame k

$$\boldsymbol{s}_\mathrm{m}(k) = \left[s_\mathrm{m}(1,k), \ldots, s_\mathrm{m}(L,k)\right]^\mathrm{T},$$

where T denotes matrix transpose and L is the number of mel channels.

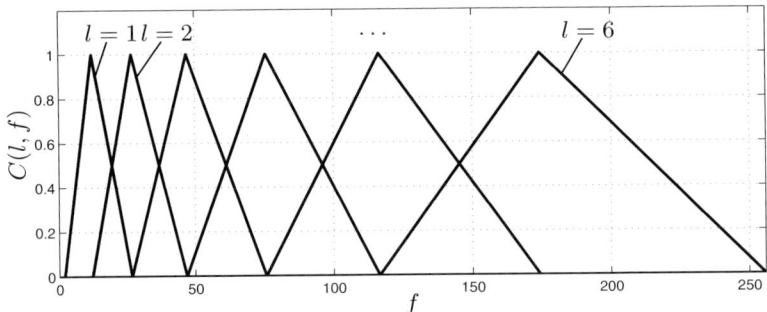

Fig. 18.6. Triangular weighting functions $C(l, f)$ of the mel filter bank for $F = 512$, $L = 6$.

Calculating the logarithm of the melspec coefficients, the logarithmic melspec (*logmelspec*) coefficients are obtained

$$\boldsymbol{s}_\mathrm{l}(k) = \log\left\{\boldsymbol{s}_\mathrm{m}(k)\right\}, \qquad (18.9)$$

where the logarithm is performed element-wise and the subscript l denotes "logmelspec domain".

Due to the spectral overlap of the channels in the mel filter bank, both melspec and logmelspec coefficients are strongly correlated. Performing a discrete cosine transform (DCT) on the logmelspec features, the elements of the feature vectors are largely decorrelated and the MFCCs are obtained. For speech recognition, only the first $I \leq L$ MFCCs are important, and we have

$$\boldsymbol{s}_\mathrm{c}(k) = \boldsymbol{B}\,\boldsymbol{A}\,\boldsymbol{s}_\mathrm{l}(k), \qquad (18.10)$$

where the subscript c denotes "cepstral domain" (MFCC),

$$\boldsymbol{A} = \{a_{il}\} \text{ with } a_{il} = \sqrt{2/L} \cdot \cos\left(\pi/L \cdot i \cdot (l + 0.5)\right)$$

is the $L \times L$ DCT matrix, and the $I \times L$ selection matrix $\boldsymbol{B} = [\mathbf{1}_{I \times I}\ \mathbf{0}_{I \times (L-I)}]$ selects the first I elements of a $L \times 1$ vector by left multiplication. $\mathbf{1}_{I \times I}$ is the $I \times I$ identity matrix, and $\mathbf{0}_{I \times (L-I)}$ is an $I \times (L - I)$ matrix of zeros.

Note that $s(k)$ is used in the following to denote the current clean-speech vector for relationships that hold regardless of the feature kind. Whenever we want to describe relations which are only valid for a certain feature kind, $s(k)$ is replaced by $s_\mathrm{m}(k)$, $s_\mathrm{l}(k)$, or $s_\mathrm{c}(k)$. The corresponding reverberant feature vector $y(k)$ is derived in the same way using the reverberant speech signal $y(n)$.

Most state-of-the-art recognizers use *hidden Markov models* (HMMs) to describe the acoustic score $A(S|W)$. The reasons for the prevalent use of HMMs are the efficient training and recognition algorithms available for HMMs and their ability to model both temporal and spectral variations. See [33, 34, 55] for comprehensive introductions to HMMs.

HMMs can be considered as finite state machines controlled by two random experiments. Fig. 18.7 shows a typical HMM topology used in speech recognition consisting of five states. The first random experiment controls the transition from the previous state $q(k-1)$ to the current state $q(k)$ according to the *state transition probabilities*

$$a_{ij} = P\big(q(k) = j | q(k-1) = i\big) \ .$$

In this way, different phoneme durations can be modeled. Only transitions from left to right are allowed, and we assume that the HMM starts in state 1 at frame 1 and ends in the last state J at the final frame K. The second random experiment determines the output feature vector according to the *output density* $f_\lambda(q(k), s(k))$ of the current HMM state $q(k)$ so that spectral variations in the pronunciation can be captured. In summary, an HMM λ is defined by its transition probabilities a_{ij}, its output densities $f_\lambda(q(k), s(k))$ and the initial state probabilities. Since we always assume that the HMM starts in state 1 at frame 1, the initial state probabilities will be neglected in the following.

The HMM is based on two fundamental assumptions [33]:

- The (first-order) *Markov assumption* implies that the current state $q(k)$ depends only on the previous state $q(k-1)$.
- The *conditional independence assumption* implies that the current output feature vector $s(k)$ depends only on the current state $q(k)$ and not on previous states or previous output feature vectors.

Based on these two assumptions, very effective algorithms for training and recognition could be derived [4, 5].

The clean-speech feature sequence $s(k)$ can be considered as a realization of the vector-valued random process $\mathcal{S}(k)$. The HMM λ models this random process as a non-stationary random process. Because of the conditional independence assumption, two statistically independent random vectors $\mathcal{S}_{k_1}$ and $\mathcal{S}_{k_2}$ are obtained if $\mathcal{S}(k)$ is observed at different frames $k_1 \neq k_2$. Since in the real world, neighboring feature vectors of a clean-speech feature sequence exhibit some statistical dependence, the conditional independence assumption

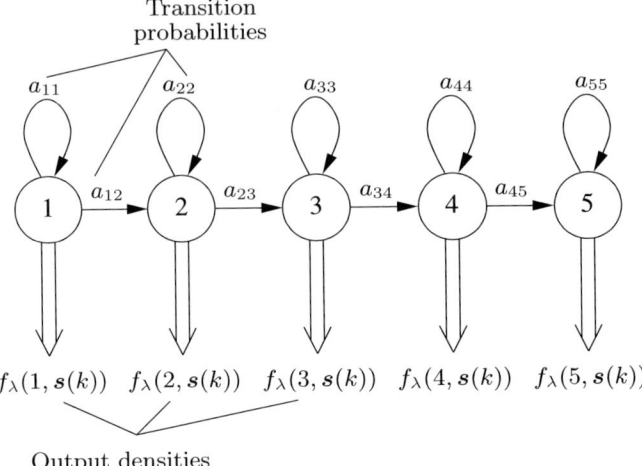

Fig. 18.7. Typical HMM topology used in ASR.

is already a simplification if the HMM is used to model clean-speech feature sequences. In practice however, it has turned out that HMM-based recognizers achieve remarkable recognition rates for clean speech despite this simplification. If HMMs are used to model reverberant feature sequences with much stronger statistical dependencies between frames (see Sec. 18.4), the conditional independence assumption becomes a severe limitation of the model's capability to describe $A(\boldsymbol{Y}|W)$.

The output density $f_\lambda(q(k), \boldsymbol{s}(k))$ of the current HMM state describes the conditional density of the random process $\mathcal{S}(k)$ given the current state $q(k)$

$$f_\lambda\big(q(k), \boldsymbol{s}(k)\big) = f_{\mathcal{S}(k)|q(k)}\big(\boldsymbol{s}(k)\big) \,. \tag{18.11}$$

The task of determining the parameters of an HMM given a set of utterances with known transcription is called training. The parameters of the HMM are chosen so that the probability of observing the feature sequences corresponding to the training utterances is maximized. Usually the Baum-Welch algorithm is used to solve this maximization problem (see e.g. [33, 55]).

To describe the probability $P(\boldsymbol{S}|W)$ or the acoustic score $A(\boldsymbol{S}|W)$ by HMMs, the sequence of words W has to be split into smaller units. For each of these units an HMM is trained. The complexity of this acoustic-phonetic modeling depends on the vocabulary size of the recognition task. For small vocabularies, like e.g., in digit recognition, it is possible to model each word with its own word-level HMM. For large vocabularies, it is more efficient to model subword units like phonemes with HMMs. Due to coarticulation phenomena, the pronunciation of subword units strongly depends on their contexts. Therefore, training different HMMs for the same phoneme with different contexts increases the accuracy of the acoustic-phonetic modeling. Triphones which consider both the previous and the following phoneme are often used in large-vocabulary recognizers (see e. g. [33]).

For continuous speech recognition, HMM networks are constructed incorporating the grammar of the recognition task (see Fig. 18.8) and the pronunciation dictionaries, specifying the subword units which make up a word (see e. g. [78]).

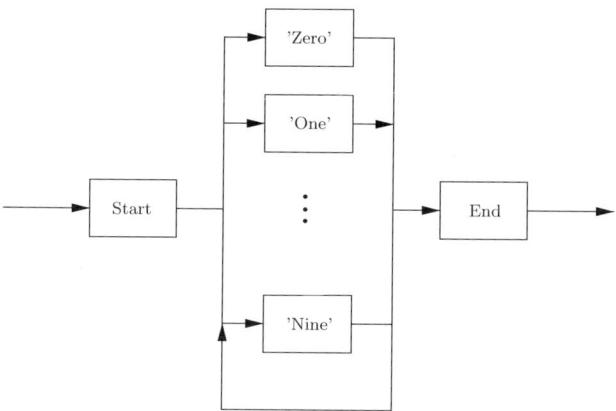

Fig. 18.8. Task grammar for connected digit recognition, 'start' and 'end' is associated with start and end of the utterance.

The information given by a language model can also be included. These recognition networks can be considered as large HMMs. Fig. 18.9 shows a very simple HMM network $\mathcal{N}_\lambda$ for connected digit recognition which can be considered as one of the simplest examples of continuous speech recognition.

Given the feature sequence of the utterance to be recognized, the recognizer searches for the most likely path through the recognition network and records the words along this path so that the most likely transcription can be determined. The *Viterbi algorithm* can be used to find the most likely path through the HMM network.

As we will focus on the determination of the acoustic score, we use a notation similar to [49] to separate the acoustic score and the language score. A large number of search algorithms exists for solving the resulting search problem 18.5, see [33, 49] for overviews. To simplify the search, the acoustic score $A(S|W)$ approximates the probability $P(S|W)$ by only considering the most likely state sequence through the HMM sequence Λ describing W. If, for example, the recognition task is connected digit recognition based on word-level HMMs and the word sequence W is "three, five, nine", the HMM sequence Λ corresponding to W is the concatenation of the word-level HMMs $\lambda_{\text{'three'}}$, $\lambda_{\text{'five'}}$, and $\lambda_{\text{'nine'}}$. Then, the acoustic score $A(S|W)$ can be expressed as

$$A(S|W) = \max_Q \{P(S, Q|\Lambda)\}, \qquad (18.12)$$

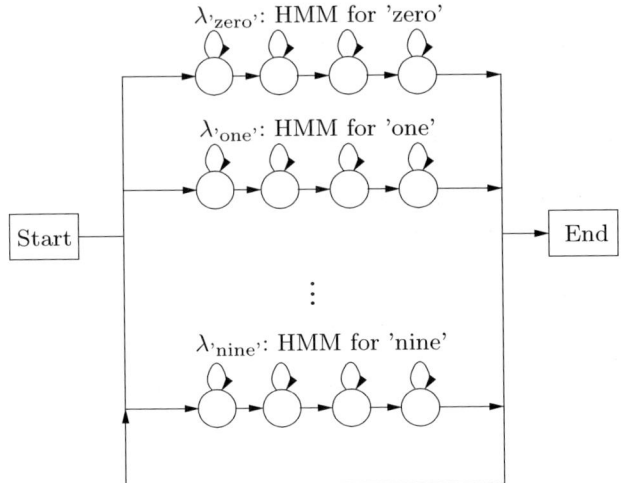

Fig. 18.9. HMM network $\mathcal{N}_\lambda$ for connected digit recognition.

where the maximization is performed over all allowed state sequences Q through Λ.

To calculate the acoustic score $A(\boldsymbol{S}|W) = A(\boldsymbol{S}|\Lambda)$, the Viterbi algorithm, defined by the following equations, is commonly used. Note that it is assumed that the HMM starts in state 1 and ends in the last state J at the final frame K of the sequence $\boldsymbol{S}$.

Initialization:

$$\gamma_1(1) = f_\Lambda\bigl(1, \boldsymbol{s}(1)\bigr) ,$$
$$\gamma_j(1) = 0 \quad \forall j = 2 \ldots J ,$$
$$\psi_j(1) = 0 \quad \forall j = 1 \ldots J .$$

Recursion:

$$\gamma_j(k) = \max_i \{\gamma_i(k-1) \cdot a_{ij}\} \cdot f_\Lambda(j, \boldsymbol{s}(k)) , \tag{18.13}$$
$$\psi_j(k) = \operatorname*{argmax}_i \{\gamma_i(k-1) \cdot a_{ij}\} .$$

Termination:

$$A(\boldsymbol{S}|W) = \gamma_J(K), \quad q(K) = J .$$

Backtracking:

$$q(k) = \psi_{q(k+1)}(k+1) \quad \forall k = K-1, \ldots, 1 .$$

Here, i indexes all considered previous states leading to the current state j, $\gamma_j(k)$ is the Viterbi metric for state j at frame k. The greater the Viterbi metric $\gamma_j(k)$, the more likely is the corresponding partial sequence up to frame k ending in state j. $f_\Lambda(j, \boldsymbol{s}(k))$ is the output density of state j of the HMM sequence Λ describing W evaluated for the clean-speech vector $\boldsymbol{s}(k)$. The backtracking pointer $\psi_j(k)$ refers to the previous state and allows backtracking of the most likely state sequence.

The Viterbi algorithm can be illustrated by a trellis diagram as depicted in Fig. 18.10 for the HMM of Fig. 18.7. The vertical axis represents the states (from 1 to 5 in the given example) and the horizontal axis represents the frames. Each dot in the diagram corresponds to the Viterbi metric $\gamma_j(k)$ of state j for frame k and each arc between the dots illustrates the non-zero transition probability between the respective states. The Viterbi scores are calculated from left to right by multiplying the score of the possible predecessor states with the corresponding transition probability, selecting the maximum over all predecessors, and then multiplying the output density of the current state for the current feature vector as given in Eq. 18.13.

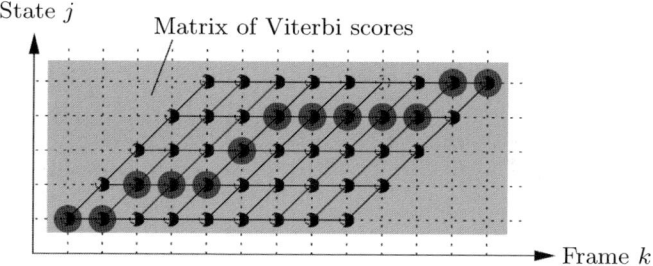

Fig. 18.10. Trellis diagram for the HMM of Fig. 18.7.

In this way, the Viterbi algorithm fills the matrix of Viterbi scores (see Fig. 18.11) with the elements $\gamma_j(k)$. At the same time, the backtracking matrix is filled with elements $\psi_j(k)$. As we assume that the HMM ends in the last state J, the final acoustic score is obtained by reading the Viterbi score $\gamma_J(K)$ of the dot in the upper right corner of the trellis diagram. Using the backtracking matrix, the most likely path through the HMM as indicated by the large dots in Fig. 18.10 for the current utterance is reconstructed.

18.4 Effect of Reverberation in the Feature Domain

This section investigates the effect of reverberation on the speech feature sequences used in ASR. Based on the exact description in the time domain (see Sec. 18.2), an approximative relationship between clean and reverberant feature vectors is derived.

The time-domain convolution of Eq. 18.2 is transformed to a multiplication in the frequency domain if the discrete-time Fourier transform is employed

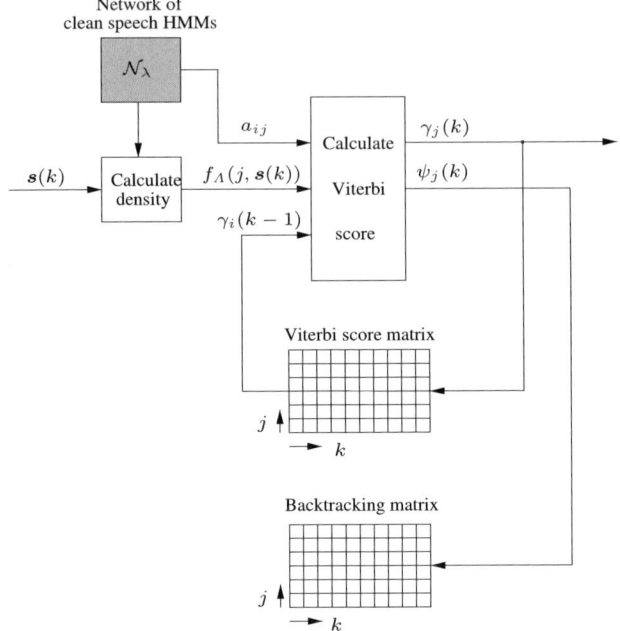

Fig. 18.11. Illustration of the Viterbi algorithm.

$$Y\left(e^{j\Omega}\right) = H\left(e^{j\Omega}\right) \cdot S\left(e^{j\Omega}\right), \qquad (18.14)$$

were $Y\left(e^{j\Omega}\right)$, $H\left(e^{j\Omega}\right)$ and $S\left(e^{j\Omega}\right)$ are the discrete-time Fourier transforms of the complete sequences $y(n)$, $h(n)$, and $s(n)$, respectively. However, common feature extraction schemes as described in Sec. 18.3 use short-time spectral analysis, like the DFT, performing the transform on short windows of the time-domain signal. If these time windows are shorter than the sequences to be convolved, the time-domain linear convolution cannot be expressed as a multiplication in the frequency domain anymore. The overlap-save or overlap-add methods [53] can be used to perform the linear convolution in the short-time spectral domain, if the DFT length is larger than the length of the impulse response.

In most environments, the length of the impulse response (200-800 ms in offices or living-rooms) is significantly longer than the DFT length used for feature extraction (typically 10-40 ms). In this case, partitioned convolution methods can be used. These methods were first introduced in [67] and are successfully used for the implementation of long adaptive filters (e. g. [63, 64]) and the efficient convolution of very long sequences [71]. We use the partitioned overlap-save method to describe the effect of reverberation on speech features.

For feature calculation, the reverberant speech signal $y(n)$ is split into overlapping frames which are weighted with a suitable window function $w(n)$. By calculating an F-point DFT, the short-time frequency-domain representation

$$Y(f,k) = \sum_{n=0}^{F-1} w(n)\, y(kN+n)\, e^{-j\frac{2\pi}{F}nf} \qquad (18.15)$$

of the reverberant speech signal is obtained. Note that for the following analysis, the frame shift N between neighboring frames needs to fulfill $N \leq F/2$. The RIR $h(n)$ is partitioned into M non-overlapping partitions of length N. These partitions are zero-padded to length F so that an F-point DFT yields the short-time frequency-domain representation

$$H(f,k) = \sum_{n=0}^{F-1} w_h(n)\, h(kN+n)\, e^{-j\frac{2\pi}{F}nf} \qquad (18.16)$$

of the RIR, where $w_h(n) = 1 \ \forall\ 0 \leq n < N;\ w_h(n) = 0 \ \forall\ N \leq n < F$ is the window function used for the impulse response. Using the short-time frequency-domain representation $S(f,k)$ of the clean speech signal $s(n)$ (see Eq. 18.7), we obtain

$$Y(f,k) = \sum_{m=0}^{M-1} \text{constraint}\{H(f,m) \cdot S(f,k-m)\}, \qquad (18.17)$$

where the constraint-operation removes the time-aliasing effects due to the circular convolution performed by the multiplication of two DFT sequences (see e.g. [53]). A common constraint operation foresees an inverse DFT, setting the first $F - N$ points in the time domain to zero, and performing a DFT [65].

If the constraint operation is neglected, the relationship between $S(f,k)$ and $Y(f,k)$ is given as

$$Y(f,k) \approx \sum_{m=0}^{M-1} H(f,m) \cdot S(f,k-m). \qquad (18.18)$$

Applying the mel filter bank to the magnitude square of $Y(f,k)$, we obtain the melspec representation

$$y_m(l,k) = \sum_{f=0}^{F/2} C(l,f) |Y(f,k)|^2 \qquad (18.19)$$

$$\approx \sum_{f=0}^{F/2} C(l,f) \left| \sum_{m=0}^{M-1} H(f,m) \cdot S(f,k-m) \right|^2 \qquad (18.20)$$

of the reverberant microphone signal. A simpler approximation is obtained if we exchange the order of the mel-filtering operation and the convolution

$$y_m(l,k) \approx \sum_{m=0}^{M-1} \left(\sum_{f=0}^{F/2} C(l,f)|H(f,m)|^2 \right) \cdot \left(\sum_{f=0}^{F/2} C(l,f)|S(f,k-m)|^2 \right) \quad (18.21)$$

$$= \sum_{m=0}^{M-1} h_m(l,m) \cdot s_m(l,k-m). \quad (18.22)$$

Note that the squared magnitude of the sum in Eq. 18.20 is replaced by the sum of squared magnitudes in Eq. 18.21. In vector notation, Eq. 18.22 reads

$$\boldsymbol{y}_m(k) \approx \sum_{m=0}^{M-1} \boldsymbol{h}_m(m) \odot \boldsymbol{s}_m(k-m), \quad (18.23)$$

where $\odot$ denotes element-wise multiplication. The *melspec convolution* (as described in Eq. 18.23) will be used throughout this contribution to describe the relationship between the clean feature sequence $\boldsymbol{s}(k)$ and the reverberant feature sequence $\boldsymbol{y}(k)$. The approximations included in Eq. 18.23 compared to the exact relationship according to Eq. 18.17 can be summarized as follows:

- The constraint which had to be applied to realize an exact linear convolution by the overlap-save method [53] is neglected.
- Due to the squared magnitude operation in the feature extraction, the phase is ignored.
- Because of the mel-filtering, the frequency resolution is reduced.
- Since the order of convolution and feature extraction is reversed, the squared magnitude of a sum is replaced by a sum of squared magnitudes.

Fig. 18.12 shows that Eq. 18.23 is nevertheless a good approximation of Eq. 18.17. The figure compares three different melspec feature sequences corresponding to the utterance "four, two, seven". The clean sequence (subfigure a)), exhibits a short period of silence before the plosive /t/ in "two" (around frame 52) and a region of low energy for the lower frequencies at the fricative /s/ in "seven" (around frame 78). These are filled with energy from the preceding frames in the reverberant case (subfigure b)). This illustrates that the reverberation has a dispersive effect on the feature sequences: the features are smeared along the time axis so that the current feature vector depends strongly on the previous feature vectors. We believe that this contradiction to the conditional independence assumption of HMMs (compare Sec. 18.3), namely that the current feature vector depends only on the current state, implies a major performance limitation of HMM-based recognizers in reverberant environments.

Comparing the true reverberant feature sequence in subfigure b) and the approximated reverberant feature sequence according to Eq. 18.23 in subfigure c) reveals that the approximation does not capture the exact texture of the time-frequency pattern (time-mel-channel pattern) of the original sequence. However, the envelope of the time-frequency pattern is very well approximated.

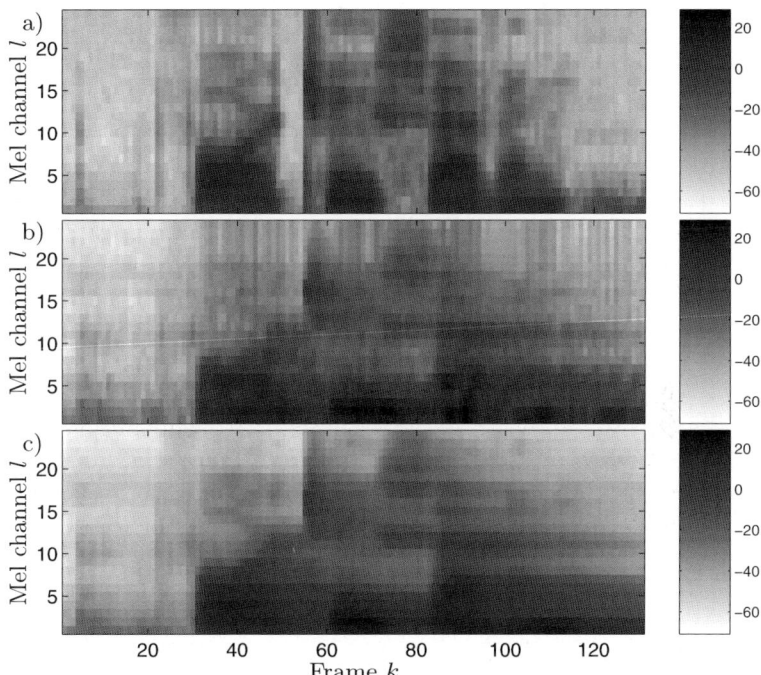

Fig. 18.12. Melspec feature sequences of the utterance "four, two, seven" in dB gray scale a) clean utterance, recorded by a close-talking microphone, b) reverberant utterance, recorded by a microphone four meters away from the speaker, c) approximation of the reverberant utterance by melspec convolution.

Fig. 18.13 b) illustrates the melspec representation of the RIR (frame shift 10 ms) for a very short RIR with a length of only 100 ms and the relationship to its time-domain representation. This picture underlines that even a short RIR extends over several frames in the feature domain. Therefore, the effect of reverberation cannot be modeled by a simple multiplication or addition in the feature domain. A much more accurate approximation is obtained by the melspec convolution of Eq. 18.23.

18.5 Signal Dereverberation and Beamforming

Robust distant-talking ASR can be achieved by dereverberating the speech signal before the feature vectors are calculated. For dereverberation, the convolution of the clean speech signal with the RIR has to be undone by inverse filtering. Since RIRs are in general non minimum-phase, an exact causal inverse filter is not stable [48]. Therefore, only approximations of inverse filters can be determined. As many zeros of the RIR are located close to the unit circle, the inverse of the RIR is usually even longer than the RIR itself so that

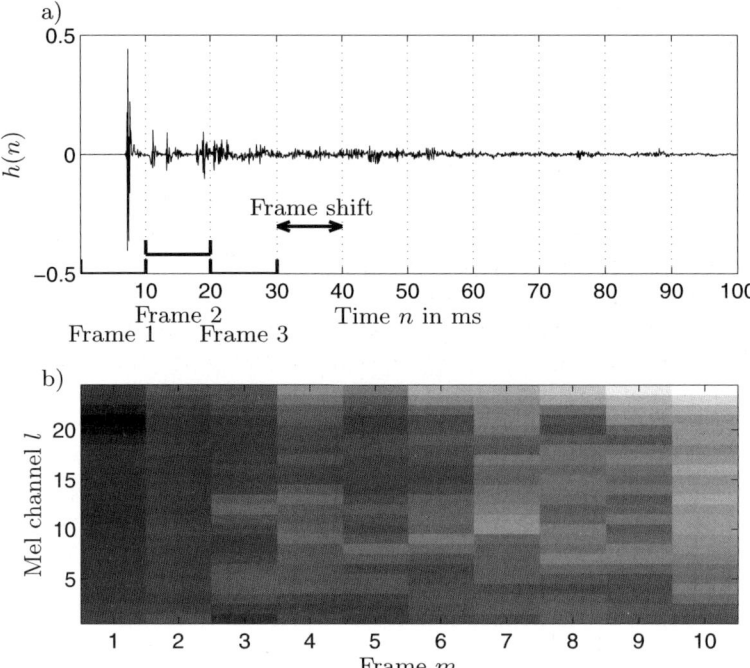

Fig. 18.13. RIR a) $h(n)$ in the time domain b) $\boldsymbol{h}_\mathrm{m}(k)$ in the melspec domain using dB gray scale.

an extremely large number of coefficients is necessary to model the inverse by an FIR filter.

Miyoshi and Kaneda show in [42] that multi-channel recordings allow for an exact realization of the inverse filter if the RIRs of all channels are known and do not exhibit common zeros (multiple input/output inverse theorem, MINT). The inverse filters are obtained by inverting the multi-channel convolution matrix which describes the single-input multiple-output system between speaker and microphones. The lengths of the resulting inverse filters are smaller than those of the RIRs. If the RIRs can only be estimated, small deviations from the true RIRs lead to large deviations from the optimum solution [54] so that, in practice, it is still very difficult to implement robust dereverberation algorithms based on MINT.

In [14], Furuya et al. suggest to use the inverse of an estimated correlation matrix of the reverberant speech signal for the calculation of the inverse filters. This approach is equivalent to MINT if

a) it is known which of the microphones is closest to the speaker,
b) the estimation of the correlation matrix is sufficiently accurate, and
c) the source signal is white.

Therefore, whitening filters are applied to the microphone signals to remove the correlation introduced by the speech production from the correlation matrix. A recursive time-averaging is suggested in [15] for the estimation of the correlation matrix in order to track changes of the RIRs between speaker and microphones.

Eigenvector-based multi-channel blind system identification [6, 22] to estimate the RIRs and subsequent inversion based on the MINT theorem is used in [29]. A major problem of this approach is that the order of the RIRs is usually not known. Therefore, an appropriate size of the correlation matrix can hardly be determined, and the accuracy of the blind system identification is significantly reduced. Hikichi et al. suggest in [29] to overestimate the lengths of the RIRs and to employ a post-processing scheme to compensate for the common part which is introduced into the RIRs because of the overestimation.

An alternative approach to estimating the RIRs by blind system identification and subsequent inversion by MINT is to estimate the inverse filters directly. In [10], a versatile framework for multi-channel blind signal processing is proposed, which can be used for blind dereverberation. In a second-order version of the approach, multi-channel filters are adapted to obtain a desired correlation matrix, where the entries along the main diagonal and close to the main diagonal are unchanged while all other elements are minimized. In this way, the clean speech signal is hardly distorted, since the correlation caused by the vocal tract is concentrated around the main diagonal. In contrast, the correlation due to room reverberation extends across the entire autocorrelation matrix. In this way, partial dereverberation can be achieved.

Single-channel approximate dereverberation of the microphone signal can be accomplished by modifying the linear prediction residual. Yegnanarayana et al. show in [75] that the residual of clean speech exhibits one distinct phonation impulse per pitch period in voiced segments, while the residual of reverberant speech exhibits many impulses. By attenuation of the impulses due to reflections compared to the phonation impulse, a dereverberation effect is achieved. In [75], a weighting factor for the residual based on the entropy and the short-time energy contour is suggested. In [76], the approach is extended to multi-channel recordings by coherent summation over the residuals of the individual microphone channels. Further approaches aiming at speech dereverberation by enhancement of the prediction residual are described in [17, 18, 20].

Nakatani et al. propose to use the short-term harmonicity of voiced speech segments for dereverberation (Harmonicity-based dEReverBeration, HERB) [45]. Based on an estimate of the pitch period, the harmonic part of speech is extracted by adaptive filtering and used as initial estimate of the dereverberated speech signal. Averaging the quotient of the Fourier transforms of the harmonic part and the reverberant part, respectively, over numerous training utterances, a dereverberation filter is determined which reduces reverberation both in voiced and unvoiced speech segments.

An implementation of HERB for the dereverberation of single-word utterances achieves a significant reduction of reverberation [47] as indicated by the resulting reverberation curves. Using HERB as preprocessing for a speaker-dependent isolated word recognition system which employs HMMs trained on clean speech, a decisive increase in word accuracy is achieved. However, the recognition rate is still significantly lower than the clean-speech performance because of changes in the spectral shapes of the dereverberated signals [47]. Using HMMs trained on utterances dereverberated by HERB to recognize dereverberated speech, recognition rates which are very close to the clean-speech performance are achieved even for strongly reverberant speech signals ($T_{60} = 1$ s).

The main problems for the implementation of the approach are the large DFT length required (10.9 seconds in [47]) and the averaging operation necessary for the calculation of the inverse filter. However, the number of utterances needed for the averaging operation has been decreased considerably by several improvements of the approach [36, 46].

Beamforming methods, which use microphone arrays to achieve spatial selectivity, are also potential candidates for signal dereverberation. Steering the main lobe of the beamformer towards the direct sound of the desired source and attenuating reflections arriving from different directions, a dereverberating effect can be achieved. However, several aspects limit the dereverberation capability of beamformers.

By compensating for the delays due to different sound propagation times, the delay-and-sum-beamformer achieves a coherent addition of the signals arriving from the desired direction and an incoherent addition of the signals arriving from other directions. In this way, a relative attenuation of the undesired signals in relation to the desired signals is achieved. The delay-and-sum-beamformer is very robust, but due to the limited spatial selectivity of the apertures of typical microphone arrays, only limited dereverberation can be achieved. Nevertheless, slight improvements of the recognition rates are reported in [50] when delay-and-sum-beamformers are used as preprocessing unit for ASR systems.

Adaptive beamformers [73] are established as powerful approaches for attenuating distortions which are uncorrelated to the desired signal. An adaptive filter is used for each sensor signal and is adapted according to some optimization criterion. For example, the variance of the output signal can be minimized subject to the constraint that the signal arriving from the desired direction passes the filter undistorted (minimum variance distortionless response (MVDR) beamformer). For implementing adaptive beamformers, the structure of the *generalized sidelobe canceler* (GSC) [21] has turned out to be very advantageous. To achieve a robust GSC implementation for broad-band speech signals, restrictions of the filter coefficients have to be enforced [31]. The performance of the GSC for speech signals can be further improved by controlling the adaptation in individual DFT bins instead of using a single broad-band control [26, 27].

A remaining problem for the use of adaptive beamformers in signal dereverberation is the correlation between the desired signal and its reverberation.

Therefore, it is very difficult to completely avoid cancellation of the desired signal and the gain of adaptive beamformers compared to a fixed delay-and-sum-beamformer is reduced.

In [61], Seltzer proposes to integrate beamforming and speech recognition into one unit. The coefficients of a filter-and-sum-beamformer are adapted in order to maximize the probability of the correct transcription, which is estimated by an initial recognition iteration (unsupervised version) or is known for an initial training utterance (supervised version). The probability is determined based on the HMMs of the speech recognizer.

Using the speech features and the acoustic model of the ASR system for the adaptation of the filter coefficients ensures that those speech properties are emphasized which are crucial for recognition. In [61], a noticeable increase of the recognition performance compared to using only a single microphone or using a delay-and-sum-beamformer is reported for both additive noise and moderate reverberation. For strong reverberation, a subband version of the approach [62] is more suitable. The performance of the supervised version is limited if the RIRs change significantly between calibration and test. The performance of the unsupervised version is limited by the accuracy of the initial transcription estimate. In strongly reverberant environments, this initial estimate can be very inaccurate so that, then, hardly any gain can be achieved with the approach.

The adaptation of the filter coefficients is very challenging. Because of the nonlinear relationship between the filter coefficients and the cost function, in general, the error surface exhibits local minima so that the convergence to a satisfying solution is not assured. The reduced data rate of the speech features compared to the speech signal samples (see Sec. 18.3) implies that a large number of filter coefficients has to be adapted with only little training data. Thus, for a given duration of the utterance used for adaptation, the number of adjustable filter coefficients is limited or the optimum coefficients cannot be identified.

18.6 Robust Features

A simple way to alleviate the limitations of the conditional independence assumption (see Sec. 18.3) is to extend the speech feature vector by so-called *dynamic features* like Δ and $\Delta\Delta$ coefficients [13, 25]. These features can be considered as the first (Δ) and second ($\Delta\Delta$) derivative of the static features and are usually approximated by simple differences or by linear regression calculations. In this way, the dynamic features capture the temporal changes in the spectra across several frames (2 to 10 frames) and thus enlarge the temporal coverage of each feature vector. Nevertheless, for strongly reverberant environments, the limited reach of the Δ and $\Delta\Delta$ features is not sufficient to cover the dispersive character of the reverberation so that the recognition performance is still limited.

RASTA (RelAtiveSpecTrA)-based speech features [28] are largely insensitive to a convolution with a short time-invariant impulse response. The key steps in calculating RASTA-based features are the following: The speech signal is divided into sub-bands (e. g., similar to the critical bands) and a nonlinear compressing transform is performed on each sub-band signal. Each compressed sub-band signal is then filtered by a bandpass filter (passband from 0.26 Hz to 12.8 Hz) which removes the very low and high modulation frequencies.

If a logarithmic transform is used, the convolution in the signal domain, which approximately corresponds to a multiplication in the sub-band domain, is transformed into an addition in the compressed sub-band domain so that the impulse response is represented by an additive constant in each sub-band, which is removed by the respective bandpass filters. Therefore, a convolution of the time-domain signal with a short time-invariant impulse response has hardly any influence on RASTA-based speech features.

In virtually all reverberant environments, the RIR is significantly longer than the frames used for feature calculation. Therefore, the time-domain convolution (Eq. 18.2) cannot be represented by a multiplication in the sub-band domain, but rather by a convolution in each sub-band as discussed in Sec. 18.4. Consequently, the time-domain convolution cannot be represented by additive constants in the compressed sub-band domain and will not be removed by the bandpass filters. Therefore, the RASTA-based features are not insensitive to long reverberation.

Cepstral mean subtraction (CMS) (see, e. g., [3] and [33], Sec. 10.6.4) is another way of alleviating convolutional distortions. A convolution with an impulse response in the time domain is transformed into an addition of the cepstral representation of the impulse response in the cepstral domain, if the frame length of the analysis window is long compared to the length of the impulse response. Thus, convolutive effects characterized by a short impulse response result in an addition of the cepstral representation of the impulse response. This representation of the impulse response can be estimated by calculation of the linear mean across the utterance and can be removed by subtraction. If the utterances are long enough so that the cepstral representation of the impulse response can be estimated reliably, the robustness of the recognizer to short convolutional effects can be significantly increased. For long reverberation with a typical duration from 200 to 800 ms in offices and living-rooms compared to the cepstral analysis window length of typically 10 to 40 ms, CMS yields only limited gains.

18.7 Model Training and Adaptation

ASR systems perform best if the acoustical conditions of the environment where the training data have been recorded match the acoustical conditions of the environment where the recognizer is applied. Therefore, using training data recorded in the application environment results in models which are well suited for the application environment. However, recording a complete set of training data for each application environment requires tremendous effort and is therefore unattractive for most real-world applications.

Giuliani et al. [19] generate reverberant training data by convolving clean-speech training utterances with RIRs measured in the application environment. In this way, the data collection effort is considerably reduced. In [66], Stahl et al. show that the performance of HMMs trained on artificially reverberated training data is significantly degraded relative to that of HMMs trained on data recorded in the application environment. On the other hand, the recognition performance based on artificial reverberation is significantly improved compared to models trained on clean speech. However, training the recognizer for each application environment still implies a huge computational load and is quite inflexible.

Therefore, Haderlein et al. [23] use RIRs recorded at different loudspeaker and microphone positions in the application environment to generate artificially reverberated data which allow the training of HMMs suitable for different speaker and microphone positions in the application environment. These HMMs show good performance also in different rooms with similar reverberation characteristics.

To reduce the effort required by a complete training with reverberant data, well-trained clean-speech models can be used as starting point for *model adaptation*. Using only a few utterances recorded in the target environment, the clean-speech models are adapted to the acoustic conditions in the application environment. Numerous adaptation schemes have been proposed for adaptation to additive distortions (e. g. background noise) and channel effects characterized by impulse responses shorter than the frame length of the feature extraction analysis window (e. g. to compensate for different frequency responses of the microphones used for training and test): Maximum a posteriori estimation [38, 39], parallel model combination [16], vector Taylor series (VTS) [43], and HMM composition [44, 68, 69].

These approaches rely on the assumption that the observed features result from the addition of the clean features and a noise term or a channel distortion term, respectively. In the case of room reverberation, the relation between clean-speech features and the feature-domain representation of the RIR is not additive as shown in Sec. 18.4. Therefore, the above-mentioned model adaptation approaches are not appropriate for reverberant environments.

In [56] and [30], two model adaptation techniques tailored to long reverberation are proposed. Based on information of the reverberation characteristics of the application environment, the means of the output density of the current state are adapted taking into account the means of the previous states. This adaptation is performed for all states in the HMM network before recognition. Therefore, the average time of remaining in each state has to be considered. In this way, a performance approaching that of reverberant training is achieved for isolated digit recognition in [56] and for connected digit recognition in [30].

However, both the model adaptation approaches [30, 56] and the training with reverberant data [19, 23, 66] suffer from the conditional independence assumption which limits the capability of the HMMs to accurately model reverberant feature vector sequences.

In [70] a frame-by-frame adaptation method is suggested which overcomes the limitation of the conditional independence assumption. The reverberation of the previous feature vectors is modeled by a first-order linear prediction and is added to the means of the clean-speech HMM at decoding time. This implies an approximation of the reverberation by a strictly exponentially decaying function and achieves slightly lower recognition rates than matched reverberant training [70].

18.8 Reverberation Modeling for Speech Recognition

A novel concept for robust distant-talking ASR in reverberant environments, called **RE**verberation **MO**deling for **S**peech recognition (REMOS), is discussed in this section. The concept was first introduced in [57] and has been extended in [58–60]. The acoustic model of a REMOS-based recognizer is a combination of a clean-speech HMM network $\mathcal{N}_\lambda$ and a statistical *reverberation model* η as depicted in Fig. 18.14. This combined acoustic model allows for very accurate and flexible modeling of reverberant feature sequences without the limitations of the conditional independence assumption.

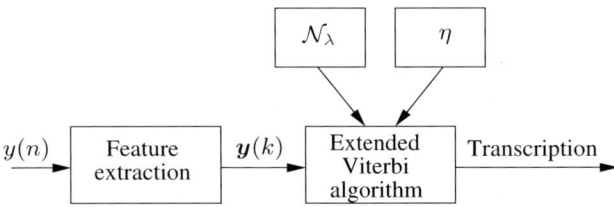

Fig. 18.14. Block diagram of the REMOS concept.

During the recognition process, the improved acoustic model is used to estimate the most likely clean-speech feature sequence directly in the feature domain. This kind of dereverberation follows the idea of preprocessing. Performing the dereverberation directly in the feature domain makes the approach less sensitive to variations of the spectro-temporal details of the acoustic path between speaker and microphone and allows for a more efficient implementation. The calculation of the acoustic score is based on this clean-speech feature estimate. In this way, the REMOS concept combines advantages of all three previously described classes of robust approaches: signal preprocessing, feature compensation and improved acoustical modeling (see Fig. 18.4).

We introduce the REMOS concept from the perspective of feature production. In particular, we show how the combination of the clean-speech HMM network and the statistical reverberation model describes the reverberant feature sequence. For the actual speech recognition, however, the combined model will be employed to find the most likely transcription for a given reverberant input feature sequence. Before deriving a solution for the decoding of the combined model, a detailed description of the reverberation model and its training is given.

18.8.1 Feature Production Model

The idea of modeling reverberation directly in the feature domain is based on the following observation: While the spectro-temporal details of the acoustic path between speaker and microphone are very sensitive to changes like small movements of the speaker, the spectro-temporal envelopes are hardly affected by such changes (see also Sec. 18.4). As the speech features used for ASR only capture the envelopes, a good feature-domain model for describing reverberation in a certain room can be obtained without detailed information on speaker and microphone positions.

We assume that the sequence of reverberant speech feature vectors $\boldsymbol{y}(k)$ is produced by a combination of a network $\mathcal{N}_\lambda$ of word-level HMMs λ describing the clean-speech and a reverberation model η as illustrated in Fig. 18.15. The word-level HMMs λ may be composed of subword HMMs. The task grammar and the language model can be embedded into the network of HMMs to represent the actual recognition task.

The reverberation model is completely independent of the recognition task and describes the reverberation of the room where the recognizer will be used. The strict separation of the task information incorporated into the network of HMMs and the information about the acoustic environment reflected by the reverberation model yields a high degree of flexibility when the recognition system has to be adapted to new tasks or new acoustic environments.

The REMOS concept can be applied to any kind of speech features which allow the formulation of an appropriate relation between the sequence $\boldsymbol{s}(k)$ of output feature vectors of the HMM network, the sequence $\boldsymbol{H}(k)$ of the reverberation model output matrices (see Sec. 18.8.2) and the sequence $\boldsymbol{y}(k)$ of reverberant speech feature vectors.

Based on the melspec convolution described in Eq. 18.23, the feature-dependent *combination operator* in Fig. 18.15 is given in generic form and then for melspec features $\boldsymbol{y}_\mathrm{m}(k)$, logmelspec features $\boldsymbol{y}_\mathrm{l}(k)$, and MFCC features $\boldsymbol{y}_\mathrm{c}(k)$ in the following:

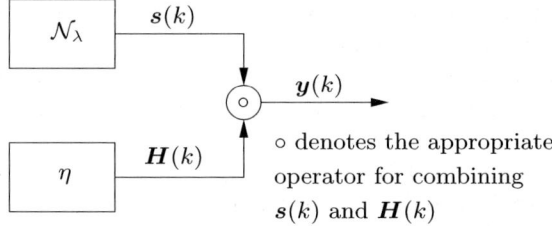

Fig. 18.15. Feature production model of the REMOS concept.

$$y(k) = H(k) \circ s(k) \quad \forall\, k = 1 \ldots K + M - 1, \tag{18.24}$$

$$y_\mathrm{m}(k) = \sum_{m=0}^{M-1} h_\mathrm{m}(m,k) \odot s_\mathrm{m}(k-m), \tag{18.25}$$

$$y_\mathrm{l}(k) = \log\left(\sum_{m=0}^{M-1} \exp\left(h_\mathrm{l}(m,k)\right) \odot \exp\left(s_\mathrm{l}(k-m)\right)\right), \tag{18.26}$$

$$y_\mathrm{c}(k) = B\,A \cdot \log\left(\sum_{m=0}^{M-1} \exp\left(A^{-1}B^\mathrm{T} h_\mathrm{c}(m,k)\right)\right.$$
$$\left. \odot \exp\left(A^{-1}B^\mathrm{T} s_\mathrm{c}(k-m)\right)\right). \tag{18.27}$$

Here, $\odot$ denotes element-wise multiplication, the vector $h(m,k)$ is a realization of the reverberation model for frame delay m and frame k, while M and K are the lengths of the reverberation model and the clean utterance, respectively. The matrices A and B were introduced in Eq. 18.10. The logarithm and the exponential function are applied element-wise. The dependency of $h(m,k)$ on the current frame k results from fact that the reverberation model allows the feature-domain representation of the RIR to change each frame (see Sec. 18.8.2). Note that all combination operators (Eqs. 18.25, 18.26, and 18.27) are equivalent, since they use the same model of reverberation, namely the feature vector convolution in the melspec domain (Eq. 18.23).

The reverberant features sequence $y(k)$ can be considered as a realization of the vector-valued random process $\mathcal{Y}(k)$. The combined acoustic model according to Fig. 18.15 describes $\mathcal{Y}(k)$ as a non-stationary random process with statistical dependencies between neighboring frames characterized by the reverberation model η.

18.8.2 Reverberation Model

The reverberation model represents the acoustic path between speaker and microphone in the feature domain. As the acoustic path can be modeled sufficiently well by an RIR, the reverberation model basically represents the RIR in the feature domain. As shown in Fig. 18.13 b), the feature-domain representation of the RIR can be considered as a matrix, where each column corresponds to a certain frame and each row corresponds to a certain mel channel. Each matrix element in Fig. 18.13 b) has a fixed value as illustrated by the gray level in the image.

Since the exact RIR is usually not known and since the combination operation provides only an approximation of the exact relationship between the clean and the reverberant feature sequences (see Sec. 18.4), we do not use a fixed feature-domain representation of a single RIR as the reverberation model. Instead, we use a statistical model where each matrix element is modeled by an independent identically distributed (IID) random process. For simplification, each element of the matrix is assumed to be statistically independent from all other elements and is modeled by a shift-invariant Gaussian density. Therefore, the reverberation model is completely described by the matrices of the means and variances of the Gaussian distributions. Fig. 18.16 illustrates the reverberation model.

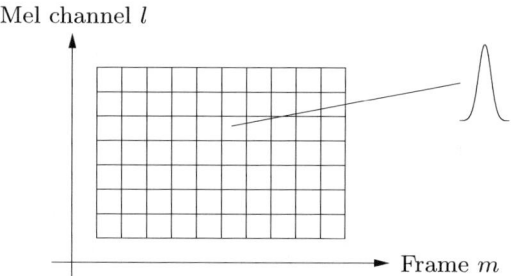

Fig. 18.16. Reverberation model η.

In summary, the reverberation model describes an IID matrix-valued Gaussian random process $\mathcal{H}(k)$. The sequence of the feature-domain RIR representations $\boldsymbol{H}(k)$ is a realization of this random process as illustrated in Fig. 18.17. The IID property of the random process implies that all elements of the random process at frame k_1 are statistically independent from all elements of the random process at frame k_2 as long as $k_1 \neq k_2$. Because the random process $\mathcal{H}(k)$ is strict-sense stationary, its probability density is not time-dependent and is denoted as $f_{\mathcal{H}(k)}(\boldsymbol{H}(k)) = f_\eta(\boldsymbol{H}(k))$.

18.8.3 Training of the Reverberation Model

The training of the reverberation model is based on a number of measured or hypothesized feature-domain RIR representations $\hat{\boldsymbol{H}}(k)$. Using these RIR representations, the mean matrix μ_η and the variance matrix σ_η^2 of the reverberation model η are estimated

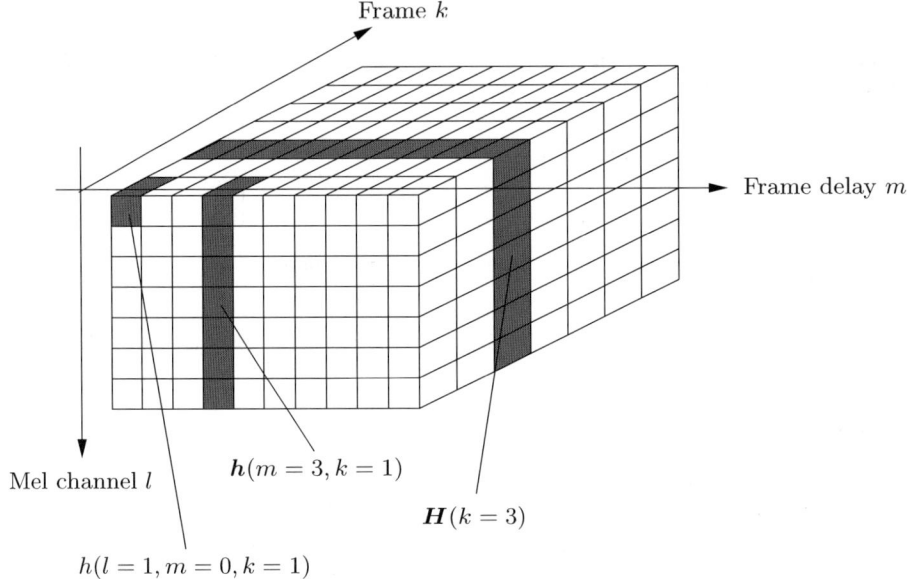

Fig. 18.17. Sequence of feature-domain RIR representations $\boldsymbol{H}(k)$ as a realization of the random process $\mathcal{H}(k)$ described by the reverberation model η.

$$\mu_\eta = \frac{1}{P} \sum_{k=1}^{P} \hat{\boldsymbol{H}}(k) , \qquad (18.28)$$

$$\sigma_\eta^2 = \frac{1}{P-1} \sum_{k=1}^{P} \left(\hat{\boldsymbol{H}}(k) - \mu_\eta \right)^2 , \qquad (18.29)$$

where P is the number of RIR representations $\hat{\boldsymbol{H}}(k)$.

There are two ways of obtaining a set of RIR representations $\hat{\boldsymbol{H}}(k)$: Either time-domain RIRs are transformed to the feature domain, or $\hat{\boldsymbol{H}}(k)$ is estimated directly in the feature domain.

Training of the Reverberation Model using Time-Domain RIRs

A set of time-domain RIRs for different microphone and loudspeaker positions of the room where the ASR system will be applied can be used for calculating a set of realizations $\hat{\boldsymbol{H}}(k)$. These RIRs can either be measured before using the recognizer, estimated by blind system identification approaches or modeled, e.g., using the image method as described in [1]. To train the reverberation model, the RIRs are time-aligned so that the direct path of all RIRs appears at the same delay. Calculation of the features yields a set of realizations $\hat{\boldsymbol{H}}(k)$ which are used to estimate the means and the variances of the reverberation model according to Eq. 18.28 and Eq. 18.29. A block diagram of the training based on time-domain RIRs is given in Fig. 18.18.

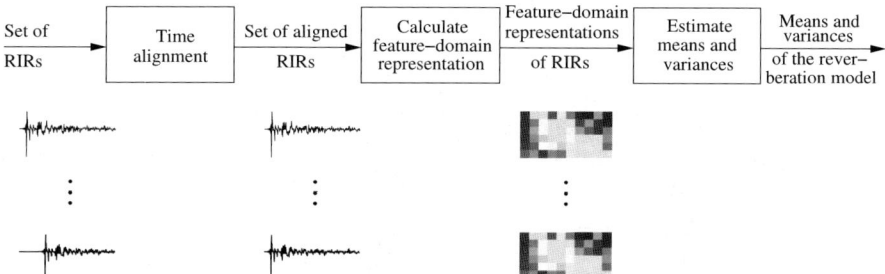

Fig. 18.18. Training of the reverberation model using time-domain RIRs.

Estimation in the Feature Domain

The realizations $\hat{\boldsymbol{H}}(k)$ can also be obtained directly in the feature domain. For example, maximum likelihood (ML) estimation based on a few training utterances with known transcription as depicted in Fig. 18.19 can be employed. Using the reverberant feature sequence $\boldsymbol{y}(k)$, a set of clean-speech HMMs, and the correct transcription of the training utterance corresponding to $\boldsymbol{y}(k)$, the optimum state sequence through the HMM representing the correct transcription is obtained by forced alignment [78]. Using this state sequence and the clean-speech HMMs, a joint density of the clean-speech feature sequence $f_{\mathcal{S}}(\boldsymbol{S})$ is estimated.

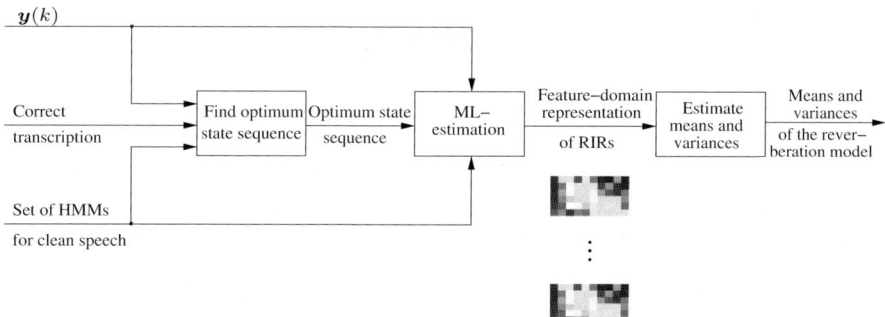

Fig. 18.19. Block diagram for the feature-domain training of the reverberation model based on maximum likelihood estimation.

To obtain the corresponding conditional Gaussian density $f_{\mathcal{Y}|\boldsymbol{H}(k)}(\boldsymbol{Y})$ of the reverberant feature sequence given $\boldsymbol{H}(k)$, the means $\mu_{\mathcal{S}(k)}$ are combined with $\boldsymbol{H}(k)$ to get the means $\mu_{\mathcal{Y}(k)|\boldsymbol{H}(k)}$

$$\mu_{\mathcal{Y}(k)|\boldsymbol{H}(k)} = \boldsymbol{H}(k) \circ \mu_{\mathcal{S}(k)} \ .$$

For simplification, the variances $\sigma^2_{\mathcal{Y}(k)|\boldsymbol{H}(k)}$ are assumed to be equal to the clean-speech variances $\sigma^2_{\mathcal{S}(k)}$ as suggested in [56]. The ML estimate $\hat{\boldsymbol{H}}_{\mathrm{ML}}(k)$ is obtained by maximizing the conditional density of the reverberant feature sequence with respect to $\boldsymbol{H}(k)$

$$\hat{\boldsymbol{H}}_{\mathrm{ML}}(k) = \underset{\boldsymbol{H}(k)}{\operatorname{argmax}} \left\{ f_{\mathcal{Y}|\boldsymbol{H}(k)}(\boldsymbol{Y}) \right\}.$$

A more detailed description of this approach including the derivation of the ML estimate in the melspec domain and corresponding experimental results can be found in [60].

18.8.4 Decoding

So far, we introduced the REMOS concept from the perspective of feature production, describing how reverberant speech features are generated given the model. For speech recognition, however, the opposite task has to be solved. Given a reverberant utterance, a recognition network of clean-speech HMMs and a reverberation model, the task of the recognizer is to find the path through the network yielding the highest probability for the given feature sequence in connection with the reverberation model.

Independently of the acoustic-phonetic modeling, the distant-talking speech recognition search problem has been formulated in Sec. 18.3 as finding the word sequence $\hat{W}$ maximizing the product of the acoustic score $A(\boldsymbol{Y}|W)$ of $\boldsymbol{Y}$ given the word sequence W and the language score $L(W)$

$$\hat{W} = \underset{W}{\operatorname{argmax}} \left\{ A(\boldsymbol{Y}|W) \cdot L(W) \right\}. \tag{18.30}$$

For conventional HMMs, the acoustic score based on the most likely state sequence is given as (see Sec. 18.3)

$$A(\boldsymbol{Y}|W) = \underset{Q}{\max} \left\{ P(\boldsymbol{Y}, Q|\Lambda) \right\}.$$

For the combined acoustic model consisting of a clean-speech HMM network and the reverberation model according to Fig. 18.15, the acoustic score is given as

$$A(\boldsymbol{Y}|W) = \underset{Q,\boldsymbol{S},\boldsymbol{H}}{\max} \left\{ P(Q, \boldsymbol{S}, \boldsymbol{H}|\Lambda, \eta) \right\} \quad \text{subject to Eq. 18.24}$$

$$= \underset{Q}{\max} \left\{ P(Q|\Lambda) \cdot \underset{\boldsymbol{S},\boldsymbol{H}}{\max} \left\{ P(\boldsymbol{S}, \boldsymbol{H}|\Lambda, \eta, Q) \right\} \right\}$$

subject to Eq. 18.24 .

As only the calculation of the acoustic score is different in the REMOS concept compared to conventional HMMs, the same search algorithms as for conventional HMMs can be used to solve the problem described in Eq. 18.30 by the

REMOS concept if a few extensions are added which account for the modified acoustic score calculations. These extensions will be derived in the following.

In the proposed approach, the acoustic score $A(\boldsymbol{Y}|W)$ can be calculated iteratively by an extended version of the Viterbi algorithm, where we assume that the HMM starts in state 1 and ends in state J.

Initialization:

$$\gamma_1(1) = \max_{s(1),h(0,1)} \left\{ f_\Lambda(1, s(1)) \cdot f_\eta(h(0,1)) \right\},$$

subject to $x(1) = s(1) \circ h(0,1)$

$\gamma_j(1) = 0 \quad \forall j = 2 \ldots J$,

$\psi_j(1) = 0 \quad \forall j = 1 \ldots J$.

Recursion:

$$\gamma_j(k) = \max_i \left\{ \gamma_i(k-1) \cdot a_{ij} \cdot O_{ij}(k) \right\}, \tag{18.31}$$

$$\psi_j(k) = \operatorname*{argmax}_i \left\{ \gamma_i(k-1) \cdot a_{ij} \cdot O_{ij}(k) \right\},$$

$$O_{ij}(k) = \max_{s(k),H(k)} \left\{ f_\Lambda(j, s(k)) \cdot f_\eta(H(k)) \right\}, \tag{18.32}$$

subject to $y(k) = H(m,k) \circ s(k)$, \hfill (18.33)

$$\forall j = 1 \ldots J, \quad k = 2 \ldots K + M - 1, \ .$$

Termination:

$$A(Y|W) = \gamma_J(K+M-1), \quad q(K+M-1) = J.$$

Backtracking:

$$q(k) = \psi_{q(k+1)}(k+1) \quad \forall k = K+M-2, \ldots, 1.$$

As in the conventional Viterbi algorithm, i indexes all considered previous states leading to the current state j, $\gamma_j(k)$ is the Viterbi metric for state j at frame k. $f_\Lambda(j, s(k))$ is the output density of state j of the HMM sequence Λ describing W evaluated for the clean-speech vector $s(k)$. $f_\eta(H(k))$ is the probability density of the reverberation model η evaluated for the feature-domain representation $H(k)$ of the RIR (see Sec. 18.8.2). The backtracking pointer $\psi_j(k)$ refers to the previous state and allows backtracking of the most likely state sequence.

The result $O_{ij}(k)$ of the optimization in Eq. 18.32, which is referred to as *inner optimization*, is obtained by varying the vector of the current clean-speech frame $s(k)$ and the matrix of the current feature-domain RIR representation $H(k)$ in order to maximize the product of their probability densities subject to the constraint described in Eq. 18.33. That is, the combination of $H(k)$ and $s(k)$ needs to be equal to the current reverberant feature vector $y(k)$. The subscript ij in $O_{ij}(k)$ indicates that this term is based on the optimum partial state sequence $\hat{Q}_{ij}(k)$ from frame $k-M+1$ to frame k with current state j and previous state i (see Fig. 18.21) given by

$$\hat{Q}_{ij}(k) = \hat{q}_{ij}(k-M+1), \ldots, \hat{q}_{ij}(k-2), \ \hat{q}_{ij}(k-1) = i, \ \hat{q}_{ij}(k) = j \ .$$

Comparing the update equation 18.13 of the conventional Viterbi algorithm to the update equation 18.31 of the extended Viterbi algorithm, we observe two differences. The first difference is that the output density $f_\Lambda(j, s(k))$ of the current HMM state in (18.13) is replaced by the term $O_{ij}(k)$ in Eq. 18.31. This term can be considered as the output density of the combined model according to Fig. 18.15 and is calculated by solving the inner optimization problem (Eq. 18.32) subject to Eq. 18.33. The second difference is that $O_{ij}(k)$ is included in the maximization over all possible state sequences Q in Eq. 18.31 while $f_\Lambda(j, s(k))$ is not included in the corresponding maximization in Eq. 18.13. Therefore, the inner optimization has to be performed for each frame k, each state j and each possible predecessor state i. The inner optimization is the main extension compared to the conventional Viterbi algorithm and will be discussed in more detail in the following section.

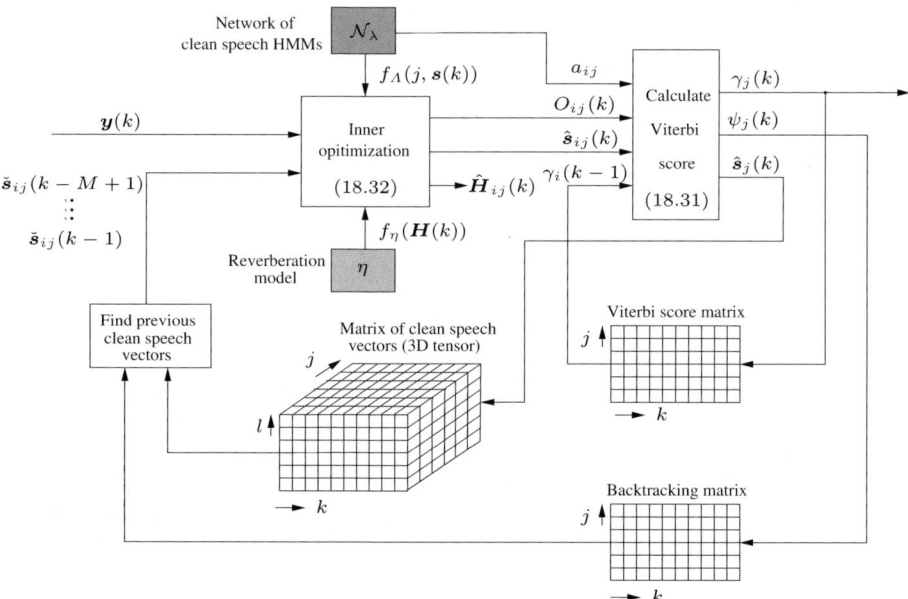

Fig. 18.20. Illustration of the extended Viterbi algorithm.

Fig. 18.20 illustrates the extended Viterbi algorithm. To calculate the current Viterbi score $\gamma_j(k)$, the previous Viterbi score $\gamma_i(k-1)$, the transition probability a_{ij} and the output density $O_{ij}(k)$ of the combined model have to be maximized according to Eq. 18.31. In order to obtain $O_{ij}(k)$, the inner optimization according to Eq. 18.32 has to be solved. Therefore, the optimum contributions $\hat{s}_{ij}(k)$ and $\hat{H}_{ij}(k)$ of the current HMM state and the reverberation model to the current reverberant observation vector $y(k)$ are estimated by maximizing the product of the HMM output density $f_\Lambda(j, s(k))$ and the reverberation model output density $f_\eta(H(k))$ subject to the constraint that the combination of $s(k)$ and $H(k)$ yields $y(k)$. In this way, $O_{ij}(k)$, $\hat{s}_{ij}(k)$, and $\hat{H}_{ij}(k)$ are obtained.

To solve the inner optimization based on one of the combination operators described in Eqs. 18.25, 18.26, or 18.27, all clean-speech feature vectors $s(k - M + 1) \ldots s(k - 1)$ are necessary. These true clean-speech feature vectors are replaced by estimates determined in previous iterations of the extended Viterbi algorithm for the frames $k' < k$ and the states j'. The clean-speech feature vector estimates are calculated as follows.

The inner optimization for frame k', state j', and each possible predecessor state i' yields a clean-speech feature estimate $\hat{s}_{i'j'}(k')$ for each i'. By maximizing over i' in the Viterbi recursion (Eq. 18.31), the most likely predecessor state

$$\hat{i}' = \underset{i'}{\operatorname{argmax}} \left\{ \gamma_{i'}(k' - 1) \cdot a_{i'j'} \cdot O_{i'j'}(k') \right\} \tag{18.34}$$

is determined. Using, $\hat{i}'$, the most likely clean-speech feature estimate among all estimates $\hat{s}_{i'j'}(k')$ is selected according to

$$\hat{s}_{j'}(k') = \hat{s}_{\hat{i}'j'}(k') . \tag{18.35}$$

For each frame k' and each state j', the most likely clean-speech feature estimate $\hat{s}_{j'}(k')$ is stored in a matrix of clean-speech vectors (3D tensor) as depicted in Fig. 18.20.

Since the matrix of clean-speech vectors is filled up to column $k - 1$ by the previous iterations, before the recursions for frame k start, the estimated clean-speech vectors can be obtained from this matrix using the optimum partial path $\hat{Q}_{ij}(k)$. The states corresponding to $\hat{Q}_{ij}(k)$ are determined by tracing back the path from frame $k - 1$ and state i using the backtracking pointers ψ as follows

$$\hat{q}_{ij}(k) = j , \tag{18.36}$$
$$\hat{q}_{ij}(k - 1) = i , \tag{18.37}$$
$$\hat{q}_{ij}(\kappa) = \psi_{\hat{q}_{ij}(\kappa+1)}(\kappa + 1) \quad \forall \kappa = k - 2, \ldots, k - M + 1 . \tag{18.38}$$

Fig. 18.21 illustrates the two optimum partial paths $\hat{Q}_{i_1j}(k)$ and $\hat{Q}_{i_2j}(k)$ for frame k, state j and the two possible predecessor states i_1 and i_2 for the HMM topology according to Fig. 18.7.

Now the clean-speech feature estimates corresponding to $\hat{Q}_{ij}(k)$ are determined by selecting the corresponding vectors from the matrix of clean-speech vectors as follows

$$\check{s}_{ij}(\kappa) = \hat{s}_{\hat{q}_{ij}(\kappa)}(\kappa) \quad \forall \kappa = k - 1, \ldots, k - M + 1 . \tag{18.39}$$

Note that the clean-speech estimate $\check{s}_{ij}(\kappa)$ extracted from the matrix of clean-speech vectors is in general different from the initial clean-speech estimate

$\hat{\mathbf{s}}_{ij}(\kappa)$ obtained from the inner optimization. Now, all input data required for the inner optimization are available, and Eq. 18.32 can be solved.

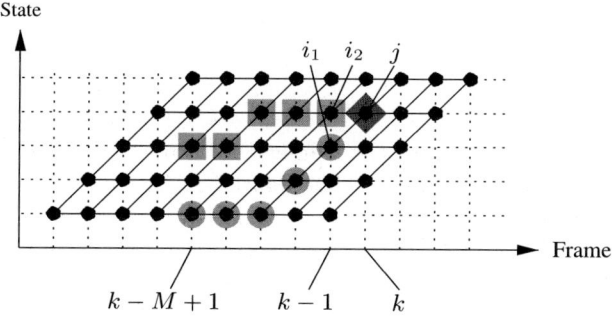

Fig. 18.21. Illustration of two optimum partial paths $\hat{Q}_{i_1j}(k)$ (indicated by the large dots) and $\hat{Q}_{i_2j}(k)$ (indicated by the squares) corresponding to the two possible predecessor states i_1 and i_2 in the trellis diagram of the HMM according to Fig. 18.7 for $k = 10$, $j = 4$, $i_1 = 3$, $i_2 = 4$, $M = 6$.

After each iteration, the Viterbi score $\gamma_j(k)$ and the backtracking pointer $\psi_j(k)$ are stored in the corresponding matrices. After all iterations are finished, these two matrices are used to determine the final acoustic score and to find the optimum path through the HMM network which enables the reconstruction of the most likely word sequence W corresponding to the feature sequence $\mathbf{Y}$.

Note that the decoding of the combined acoustic model described above exhibits some similarities to the HMM decomposition approach proposed in [72] for additive noise. Indeed, REMOS can be considered as a generalization of the HMM decomposition approach to a convolutive combination of the model outputs if the reverberation model is considered as a one-state HMM with matrix-valued output. However, there is a significant difference in the evaluation of the output density of the combined model. The HMM decomposition approach proposes to integrate over all possible combinations of the outputs of the individual models to calculate the output probability of the combined feature vector. We propose to search for the most likely combination to calculate the probability of the reverberant feature vector. While both approaches are feasible for simple combinations like addition, the method proposed here provides significant computational savings for more complex combinations like convolution.

18.8.5 Inner Optimization

To find the best combination of the HMM output and the reverberation model output, the extended Viterbi algorithm performs an inner optimization in each iteration. In this inner optimization, the joint density of the current HMM state and the reverberation model has to be maximized subject to the constraint that the combination of $s(k)$ and $H(k)$ yields the current reverberant feature vector $y(k)$ as described by Eq. 18.32 and Eq. 18.33.

Instead of maximizing the objective function

$$f_{\Lambda\eta} = f_\Lambda(j, s(k)) \cdot f_\eta(H(k))$$

directly, equivalently, the logarithm of the objective function $\log(f_{\Lambda\eta})$ can be maximized, since the logarithm is a monotone function. Therefore, the inner optimization problem can be expressed as

$$\tilde{O}_{ij}(k) = \max_{s(k), H(k)} \{\log\{f_{\Lambda\eta}\}\} \quad \text{subject to Eq. 18.33}. \quad (18.40)$$

The objective function $f_{\Lambda\eta}$ of the inner optimization problem depends on the output density of the current HMM state $f_\Lambda(j, s(k))$ and the output density of the reverberation model $f_\eta(H(k))$. If single Gaussian densities are used both in the HMM and in the reverberation model, $\log\{f_{\Lambda\eta}\}$ is a quadratic function with a single global maximum. If mixtures of Gaussians are used in the HMMs and/or in the reverberation model, $\log\{(f_{\Lambda\eta}\}$ is a sum of weighted quadratic functions and, in general, exhibits several local maxima.

The constraint of the inner optimization problem depends on the kind of features used, since the combination operation of the HMM output features and the reverberation model output features is feature-dependent and is given for melspec features, logmelspec features, and MFCCs in Eqs. 18.25, 18.26, and 18.27. For all three kinds of features, the constraint is a non-linear function. Note that the independent variables to be optimized are $s(k-0), h(0, k), \ldots, h(M-1, k)$. The terms $s(k-M+1), \ldots, s(k-1)$ are known from previous iterations, since they are given by the clean-speech feature estimates $\check{s}_{ij}(k-M+1), \ldots, \check{s}_{ij}(k-1)$.

The discussion above shows that the complexity of the inner optimization problem depends both on the output densities of the HMM and the reverberation model, and the kind of features used in the recognizer. In general, numerical optimization methods have to be employed for the solution of the inner optimization problem.

If single Gaussian densities are used and the constraint is linearized, a closed-form solution of the inner optimization problem can be found in the melspec domain. This solution is derived in the following section as an example of how to solve the inner optimization problem.

18.8.6 Solution of the Inner Optimization Problem in the Melspec Domain for Single Gaussian Densities

In the melspec domain, the inner optimization problem can be expressed as

$$\tilde{O}_{ij}(k) = \max_{\boldsymbol{s}_{\mathrm{m}}(k), \boldsymbol{H}_{\mathrm{m}}(k)} \{\log\{f_{\Lambda\eta}\}\} \tag{18.41}$$

subject to $\quad \boldsymbol{y}_{\mathrm{m}}(k) = \boldsymbol{h}_{\mathrm{m}}(0,k) \odot \boldsymbol{s}_{\mathrm{m}}(k) + \sum_{m=1}^{M-1} \boldsymbol{h}_{\mathrm{m}}(m,k) \odot \check{\boldsymbol{s}}_{\mathrm{m},ij}(k-m) \,.$

$$\tag{18.42}$$

Using single Gaussian densities both in the HMMs and the reverberation model, the objective function $\log\{f_{\Lambda\eta}\}$ becomes a multivariate quadratic function. If the constraint in the melspec domain (Eq. 18.42) is linearized, an optimization problem with a quadratic cost function and a linear constraint is obtained, which exhibits a single unique solution. The determination of this solution using the method of Lagrange multipliers is described in the following.

We introduce a simplified notation which neglects the subscript m indicating "melspec domain", the dependencies on the frame index k and the partial state sequence $Q_{ij}(k)$ as follows: $\boldsymbol{s}_{\mathrm{m}}(k-0) := \boldsymbol{s}_0$, $\check{\boldsymbol{s}}_{\mathrm{m},ij}(k-m) := \check{\boldsymbol{s}}_m$, $\boldsymbol{y}_{\mathrm{m}}(k) := \boldsymbol{y}$, $\boldsymbol{h}_{\mathrm{m}}(m,k) := \boldsymbol{h}_m$. That is, $\boldsymbol{y}$ is the current reverberant feature vector, $\boldsymbol{s}_0$ is the current clean-speech feature vector and $\check{\boldsymbol{s}}_m$ is the estimated clean-speech vector for frame $k-m$. $\boldsymbol{h}_m$ is the m-th column of the current melspec RIR representation (see Fig. 18.13 for illustration).

With this simplified notation, the constraint of Eq. 18.42 can be written as

$$\boldsymbol{y} = \underline{\boldsymbol{h}_0} \odot \underline{\boldsymbol{s}_0} + \sum_{m=1}^{M-1} \underline{\boldsymbol{h}_m} \odot \overline{\check{\boldsymbol{s}}_m} \,, \tag{18.43}$$

where the <u>underlined vectors</u> are unknown realizations of multivariate Gaussian random vectors with diagonal covariance matrix and the $\overline{\text{overlined vectors}}$ are known from previous iterations.

To linearize the constraint, we approximate the generally non-Gaussian random vector $\tilde{\mathcal{Y}}_0 = \mathcal{H}_0 \odot \mathcal{S}_0$ describing the realizations $\tilde{\boldsymbol{y}}_0 = \boldsymbol{h}_0 \odot \boldsymbol{s}_0$ by a Gaussian random vector $\mathcal{Y}_0$ with the same mean and variance as $\tilde{\mathcal{Y}}_0$. The realizations of $\mathcal{Y}_0$ are denoted $\boldsymbol{y}_0$. Thus we obtain the following linear constraint

$$\boldsymbol{y} = \underline{\boldsymbol{y}_0} + \sum_{m=1}^{M-1} \underline{\boldsymbol{h}_m} \odot \overline{\check{\boldsymbol{s}}_m} \,. \tag{18.44}$$

Based on this constraint, a two-step closed-form solution of the inner optimization problem can be derived as follows:

First step: Find $\boldsymbol{y}_0$ and $\boldsymbol{h}_{m'}$.

We apply the method of Lagrange multipliers (see e.g. [41], appendix B.2) to

$$\max_{\boldsymbol{y}_0, \boldsymbol{h}_1, \ldots, \boldsymbol{h}_{M-1}} \{f_{\mathcal{Y}_0}(\boldsymbol{y}_0) \cdot f_\eta(\boldsymbol{h}_1) \cdot \ldots \cdot f_\eta(\boldsymbol{h}_{M-1})\} \quad \text{subject to Eq. 18.44}\,, \tag{18.45}$$

where $f_{\mathcal{Y}_0}(\boldsymbol{y}_0)$ is the probability density of $\mathcal{Y}_0$ evaluated at $\boldsymbol{y}_0$, $f_\eta(\boldsymbol{h}_m)$ is the probability density of the m-th column of the reverberation model evaluated at $\boldsymbol{h}_m$. Since the columns of the reverberation model are assumed to be statistically independent as described in Sec. 18.8.2,

$$f_\eta(\boldsymbol{h}_0) \cdot \ldots \cdot f_\eta(\boldsymbol{h}_{M-1}) = f_\eta(\boldsymbol{H}(k)) \, .$$

Using the negative logarithm of the densities to be maximized and neglecting irrelevant constants, the Lagrangian function $\mathcal{L}_1$ is obtained as

$$\mathcal{L}_1 = \frac{(\boldsymbol{y}_0 - \boldsymbol{\mu}_{\boldsymbol{y}_0})^2}{2\,\sigma^2_{\boldsymbol{y}_0}} + \sum_{m=1}^{M-1} \frac{(\boldsymbol{h}_m - \boldsymbol{\mu}_{\boldsymbol{h}_m})^2}{2\,\sigma^2_{\boldsymbol{h}_m}}$$

$$+ \nu_1 \cdot \left(\boldsymbol{y} - \boldsymbol{y}_0 - \sum_{m=1}^{M-1} \boldsymbol{h}_m \odot \boldsymbol{\check{s}}_m \right), \qquad (18.46)$$

where the squaring and the division operations are performed element-wise (as for the remainder of this section), ν_1 is the Lagrange multiplier, and $\boldsymbol{\mu}_{\boldsymbol{h}_m}$ and $\sigma^2_{\boldsymbol{h}_m}$ denote the mean and the variance vector of $\boldsymbol{h}_m$, respectively, and likewise for the other variables.

Setting the derivatives of the Lagrangian $\mathcal{L}_1$ with respect to $\boldsymbol{y}_0$, $\boldsymbol{h}_1$, $\ldots$, $\boldsymbol{h}_{M-1}$, and ν_1 to zero and solving the resulting system of equations, we obtain $\boldsymbol{\hat{y}}_0$ and $\boldsymbol{\hat{h}}_{m'}$, for $m' = 1, \ldots, M-1$, as solutions

$$\boldsymbol{\hat{y}}_0 = \frac{\sum_{m=1}^{M-1} \boldsymbol{\check{s}}^2_m \odot \sigma^2_{\boldsymbol{h}_m}}{\sigma^2_{\boldsymbol{y}_0} + \sum_{m=1}^{M-1} \boldsymbol{\check{s}}^2_m \odot \sigma^2_{\boldsymbol{h}_m}} \odot \boldsymbol{\mu}_{\boldsymbol{y}_0}$$

$$+ \frac{\sigma^2_{\boldsymbol{y}_0}}{\sigma^2_{\boldsymbol{y}_0} + \sum_{m=1}^{M-1} \boldsymbol{\check{s}}^2_m \odot \sigma^2_{\boldsymbol{h}_m}} \odot \left(\boldsymbol{y} - \sum_{m=1}^{M-1} \boldsymbol{\check{s}}_m \odot \boldsymbol{\mu}_{\boldsymbol{h}_m} \right), \qquad (18.47)$$

$$\boldsymbol{\hat{h}}_{m'} = \frac{\sigma^2_{\boldsymbol{y}_0} + \sum_{\substack{m=1 \\ m \neq m'}}^{M-1} \boldsymbol{\check{s}}^2_m \odot \sigma^2_{\boldsymbol{h}_m}}{\sigma^2_{\boldsymbol{y}_0} + \sum_{m=1}^{M-1} \boldsymbol{\check{s}}^2_m \odot \sigma^2_{\boldsymbol{h}_m}} \odot \boldsymbol{\mu}_{\boldsymbol{h}_{m'}}$$

$$+ \frac{\boldsymbol{\check{s}}^2_{m'} \odot \sigma^2_{\boldsymbol{h}_{m'}}}{\sigma^2_{\boldsymbol{y}_0} + \sum_{m=1}^{M-1} \boldsymbol{\check{s}}^2_m \odot \sigma^2_{\boldsymbol{h}_m}} \odot \frac{1}{\boldsymbol{\check{s}}_{m'}} \odot \left(\boldsymbol{y} - \boldsymbol{\mu}_{\boldsymbol{y}_0} - \sum_{\substack{m=1 \\ m \neq m'}}^{M-1} \boldsymbol{\check{s}}_m \odot \boldsymbol{\mu}_{\boldsymbol{h}_m} \right). \qquad (18.48)$$

Second step: Find $\boldsymbol{h}_0$ and $\boldsymbol{s}_0$ given $\boldsymbol{\hat{y}}_0$.

Applying the method of Lagrange multipliers to

$$\max_{s_0, h_0} \{ f_\Lambda(j, s_0) \cdot f_\eta(h_0) \} \quad \text{subject to} \quad \overline{\hat{y}_0} = \underline{h_0} \odot \underline{s_0} , \quad (18.49)$$

replacing the densities with their negative logarithm, and neglecting irrelevant constants, we obtain the following Lagrangian function

$$\mathcal{L}_2 = \frac{(s_0 - \mu_{s_0})^2}{2\sigma_{s_0}^2} + \frac{(h_0 - \mu_{h_0})^2}{2\sigma_{h_0}^2} + \nu_2 \cdot (\hat{y}_0 - h_0 \odot s_0) . \quad (18.50)$$

Setting the derivatives of the Lagrangian $\mathcal{L}_2$ with respect to h_0, s_0, and ν_2 to zero and solving the resulting system of equations, we obtain the following fourth-order equation to be fulfilled by the desired vector h_0

$$\sigma_{s_0}^2 \odot h_0^4 - \mu_{h_0} \odot \sigma_{s_0}^2 \odot h_0^3 + \mu_{s_0} \odot \sigma_{h_0}^2 \odot \hat{y}_0 \odot h_0 - \hat{y}_0^2 \odot \sigma_{h_0}^2 = 0 , \quad (18.51)$$

where the exponents denote element-wise powers. It can be shown that this equation has a pair of complex conjugate solutions, one real-valued positive and one real-valued negative solution. As only the real-valued positive solution achieves the maximization of the desired probability, we obtain exactly one vector $\hat{h}_0$ and thus exactly one vector $\hat{s}_0$

$$\hat{s}_0 = \frac{\hat{y}_0}{\hat{h}_0} . \quad (18.52)$$

In this way, $\hat{s}_{ij}(k)$ and $\hat{H}_{ij}(k)$ are obtained so that $O_{ij}(k)$ can be calculated as

$$O_{ij}(k) = f_\Lambda(j, \hat{s}_{ij}(k)) \cdot f_\eta(\hat{H}_{ij}(k)) .$$

18.8.7 Simulations

To investigate the effectiveness of the REMOS concept, simulations of a connected digit recognition (CDR) task using melspec features and single Gaussian densities are performed. The performance of the proposed approach is compared to that of conventional HMM-based recognizers trained on clean and reverberant speech, respectively.

The REMOS concept is implemented by extending the functionality of HTK [32] with the inner optimization as described in Sec. 18.8.6. HTK employs Viterbi beam search implemented by the so-called *token passing paradigm* as continuous speech recognition search algorithm [77].

The CDR task is chosen for evaluation, since it can be considered as one of the easiest examples of continuous speech recognition. Furthermore, the probability of the current digit can be assumed to be independent of the preceding digits so that a language model is not required. Therefore, the recognition rate is solely determined by the quality of the acoustic model, making the CDR task well suited for the evaluation of the REMOS concept, which aims at improving the acoustic model.

The simulations are performed using RIRs measured in three different rooms. Room A is a lab environment, room B a studio environment and room C a lecture room. The details of the room characteristics are summarized in Tab. 18.1. Note that room A is a moderately reverberant environment while room B and room C are highly reverberant environments. A set of RIRs is measured for different loudspeaker and microphone positions in each room. In room C, three RIR sets with different loudspeaker/microphone-distances are measured which are denoted C1, C2 and C4, where the number corresponds to the distance in meter. Each set of RIRs is split into two disjoint sets, one used for training and the other used for test (see Tab. 18.1 for detailed numbers). In this way, a strict separation of test and training data is achieved.

Table 18.1. Summary of room characteristics: T_{60} is the reverberation time, d the distance between speaker and microphone and SRR is the signal-to-reverberation-ratio.

	Room A	Room B	Room C1	Room C2	Room C4
Type	Lab	Studio	Lecture rooms		
T_{60}	300 ms	700 ms	900 ms	900 ms	900 ms
d	2.0 m	4.1 m	1.0 m	2.0 m	4.0 m
SRR	4.0 dB	-4.5 dB	7.4 dB	2.9 dB	-1.5 dB
Number of training RIRs	36	18	36	72	44
Number of test RIRs	18	6	18	36	22
Length of rev. model M	20	50	70	70	70

The used feature vectors are calculated in the following way: The speech signal, sampled at 20 kHz, is decomposed into overlapping frames of length 25 ms with a frame shift of 10 ms. After applying a first-order pre-emphasis (coefficient 0.97) and a Hamming window, a 512-point DFT is computed. From the DFT representation, 24 melspec coefficients are calculated. Only static features and no Δ and $\Delta\Delta$ coefficients are used.

A 16-state left-to-right model without skips over states is trained for each of the 11 digits ('0'-'9' and 'oh'). Additionally, a three-state silence model with a backward skip from state 3 to state 1 is trained. The output densities are single Gaussians with diagonal covariance matrices. All HMMs are trained according to the following procedure: First, single Gaussian MFCC-based HMMs are trained by 10 iterations of Baum-Welch re-estimation [33].

Then the melspec HMMs are obtained from the MFCC HMMs by single-pass retraining [74]. In this way, more reliable models are obtained than by training melspec models from scratch.

For the training, 4579 connected digit utterances corresponding to 1.5 hours of speech from the TI digits [40] training data are used. For the training with reverberant speech, the clean training data are convolved with measured RIRs randomly selected from the training set of the corresponding room. A uniform distribution is employed for the random selection so that a balanced use of all RIRs is ensured. The HMMs trained on clean data are denoted λ_{clean}, the HMMs trained on data convolved with RIRs from room A are denoted λ_A and so on. The corresponding HMM networks are denoted $\mathcal{N}_{\lambda_{\mathrm{clean}}}$, $\mathcal{N}_{\lambda_A}$ and so on. For the conventional HMM-based clean recognizer and for the REMOS-based recognizer, identical HMM networks are used. The HMM network of the conventional reverberant recognizers for each room shares the same structural parameters and the same training procedure but differs with respect to the training data.

For the recognition, a silence model is added in the beginning and at the end of the HMM network consisting of the 11 digit-HMMs connected in a loop similar to Fig. 18.9. As test data, 512 test utterances randomly selected from the TI digits test set are used. To obtain the reverberant test data, the clean test data are convolved with RIRs randomly selected from the test set of the corresponding rooms.

To train the reverberation models for each room, the measured RIRs from the corresponding training set are used according to the procedure described in Sec. 18.8.3. The reverberation models are denoted according to the rooms where the RIRs have been measured. E. g., the reverberation model of room A is denoted η_A. In addition to the reverberation models η_{C1}, η_{C2}, η_{C4}, a universal model for room C is trained using all training RIRs measured in room C. This model is denoted η_{C124}.

In a first test series, the performance of REMOS is compared to conventional HMM-based recognizers. Tab. 18.2 shows the word accuracies achieved with conventional HMM-based recognizers and with the REMOS concept for the connected digit recognition task in the rooms described above. The relatively low accuracy of 82 % achieved by applying the conventional HMM-based recognizer using clean HMMs to the clean test data (clean-speech performance) results from the fact that melspec features cannot be modeled very accurately by single Gaussian densities. With increasing reverberation, the accuracy decreases significantly if HMMs trained on clean speech are used in the conventional HMM-based recognizers. The accuracy is improved to some extent if HMMs trained with reverberant data from the corresponding rooms are used.

The lower recognition rate in room B compared to room C4 for the clean HMM-based recognizer can be explained by the strong low-pass characteristic of the transfer functions corresponding to the RIRs measured in room B. Therefore, the mismatch between the clean training data and the reverberant test data is larger in room B than in room C4. As the low-pass characteristic can be modeled very well by the reverberant training, the performance increase between clean and reverberant training is higher in room B than in room C4.

The word accuracy achieved by the REMOS concept is significantly higher than that of the reverberant HMM-based recognizers in all three rooms. In room A, the recognition rate of REMOS even approaches the clean-speech performance. The performance gain compared to the reverberant training increases with growing reverberation from 10.8% absolute in room A to 21.6% absolute in room C4. These results confirm that the REMOS concept is much more robust to reverberation than conventional HMM-based recognizers, even if the latter use HMMs trained on reverberant data.

Table 18.2. Comparison of word accuracies of a conventional HMM-based recognizer and of the proposed REMOS concept in the melspec domain using single Gaussian densities.

Test data	Recognizer						
	Conventional HMM-based				REMOS concept		
	Clean training		Reverberant training				
	HMM	Acc.	HMM	Acc.	HMM	Rev. model	Acc.
Clean	$\mathcal{N}_{\lambda_{clean}}$	82.0%	-	-	-	-	-
Room A	$\mathcal{N}_{\lambda_{clean}}$	51.5%	$\mathcal{N}_{\lambda_A}$	66.8%	$\mathcal{N}_{\lambda_{clean}}$	η_A	77.6%
Room B	$\mathcal{N}_{\lambda_{clean}}$	13.4%	$\mathcal{N}_{\lambda_B}$	54.6%	$\mathcal{N}_{\lambda_{clean}}$	η_B	71.6%
Room C4	$\mathcal{N}_{\lambda_{clean}}$	25.9%	$\mathcal{N}_{\lambda_{C4}}$	46.0%	$\mathcal{N}_{\lambda_{clean}}$	η_{C4}	67.6%

In a second test series, the sensitivity of the REMOS concept to a mismatch between the set-up in the target environment and the reverberation model is investigated. Therefore, the reverberation models η_{C1}, η_{C2}, η_{C4}, η_{C124} are applied to the test data of the scenarios C1, C2 and C4. The word accuracies for all possible combinations are summarized in Tab. 18.3. The results for scenario C1 are similar for all reverberation models, while significant differences between different reverberation models are observed for the set-ups C2 and C4. For all of the tested loudspeaker/microphone-distances, the matched model (e. g., η_{C2} for scenario C2) achieves the best results among all models or is at least close to the best result. Using a reverberation model with higher SRR than the test conditions (e. g., η_{C1} for scenario C2), decreases the recognition rate much more than using a reverberation model with lower SRR (e. g., η_{C4} for scenario C2).

The reverberation model η_{C124} trained on RIRs with different loudspeaker/microphone-distances performs very well for all scenarios C1, C2 and

C4. For the test data with a loudspeaker/microphone-distance of 4 m (scenario C4) it even outperforms the matched model. In summary, we can conclude that using RIRs measured at various loudspeaker and microphone positions with various distances in the target environment enables the training of a reverberation model which achieves a good performance in the target environment regardless of the loudspeaker/microphone-distance.

Table 18.3. Word accuracy of the REMOS concept for test data with different loudspeaker/microphone-distances in room C and different reverberation models.

Test data	Reverberation model			
	η_{C1}	η_{C2}	η_{C4}	η_{C124}
Room C1	73.9 %	74.5 %	73.0 %	73.2 %
Room C2	58.8 %	71.7 %	68.0 %	71.4 %
Room C4	45.6 %	46.9 %	67.6 %	70.2 %

The performance of the REMOS concept as a function of the reverberation model length M is investigated in a third test series in room C4. Therefore, the model η_{C4} with an original length of $M = 70$, covering a reverberation time of 700 ms is truncated to the lengths given in Tab. 18.4. For all tests in this series, the test data of scenario C4 are used.

Table 18.4. Word accuracy of the REMOS concept for room C4 and different lengths of the reverberation model η_{C4}.

Length M of rev. model	1	2	3	4	6	8	10
Accuracy	21.3 %	27.5 %	31.4 %	36.5 %	39.8 %	43.7 %	45.8 %
Length M of rev. model	15	20	30	40	50	60	70
Accuracy	48.3 %	51.1 %	58.7 %	62.5 %	66.4 %	67.5 %	67.5 %

Tab. 18.4 and Fig. 18.22 show that the word accuracy increases monotonically with increasing length M of the reverberation model. At the first glance, it might be surprising that for $M = 1$, the recognition rate of the REMOS concept is slightly lower than that of the clean HMM-based recognizer. Even with a one-frame reverberation model, REMOS can compensate for differences in the transfer function of training and test data. However, the energy of the reverberation model is reduced by the truncation so that the

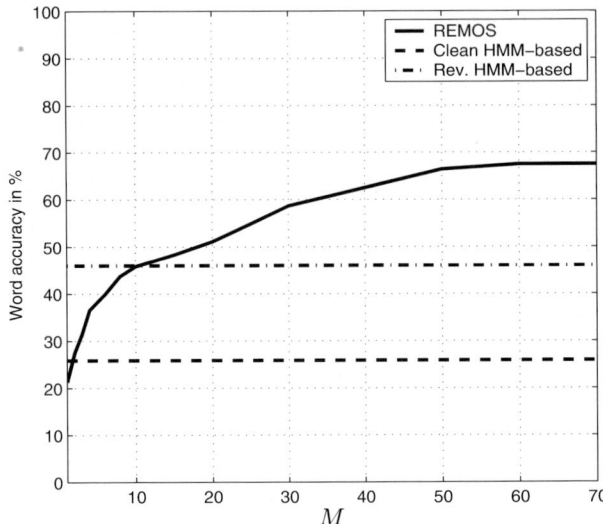

Fig. 18.22. Word accuracy of REMOS in room C4 as a function of the length M of the reverberation model η_{C4}.

resulting mismatch in the signal energy between the test sequence and the model causes the slight decrease in recognition rate.

Already with a length of $M = 10$, the REMOS concept achieves the same recognition rate as the conventional HMM-based recognizer trained on reverberant data. A further increase in the reverberation model length M leads to further significant gains in the recognition rate until a saturation can be observed for lengths larger than $M = 60$. This curve confirms that by modeling the effect of reverberation not simply by a multiplication in the feature domain but rather by a feature-domain convolution, REMOS has the capability to significantly outperform HMM-based recognizers, even if they are trained on reverberant data. If context-dependent sub-word HMMs (e.g. triphones) are used instead of word HMMs, the context of the HMMs is reduced and the gain of REMOS compared to reverberantly trained HMM-based recognizers is expected to increase further.

18.9 Summary and Conclusions

In this contribution, the progress towards robust distant-talking speech recognition in reverberant environments has been reviewed and a novel concept has been described. Since the length of the RIR describing the acoustic path between speaker and microphone is significantly larger than the frame length used for short-time spectrum analysis in the ASR feature extraction, the RIR extends over several frames. Therefore, reverberation has a dispersive effect on the feature vector sequences used for ASR so that the current feature vector strongly depends on the previous feature vectors. This contradiction to the conditional independence assumption of HMMs, which are state-of-the-art in acoustic-phonetic modeling, has been identified as the main performance limitation of HMM-based recognizers in reverberant environments.

The numerous approaches to improve the ASR performance in reverberant environments have been classified into three groups according to the function block of the ASR system they are applied to. Preprocessing algorithms like blind dereverberation and beamforming aim at removing or at least reducing the reverberation of the input signal before the feature vectors are calculated. Robust speech features and feature-domain compensation techniques try to remove the effect of reverberation at the feature level. Alternatively, the acoustic model of the ASR system can be adjusted to reverberation. This can be performed either by training the HMMs with reverberant data or by adapting well-trained clean-speech HMMs using a few calibration utterances recorded in the target environment.

Finally, a novel concept based on reverberation modeling for speech recognition (REMOS) has been discussed. A combination of an HMM network and a feature-domain reverberation model is used to determine the acoustic score. During recognition, an optimization problem is solved in each iteration of the extended Viterbi algorithm to find the most likely contribution of the HMM network and the reverberation model to the current reverberant observation vector. The complexity of this inner optimization depends both on the kind of features and the output densities used in the HMM and the reverberation model. In general, it has to be solved by numerical optimization algorithms.

For melspec features and single Gaussian densities a closed form solution is possible. Based on this solution, simulations of a connected digit recognition task have been performed in three different rooms. These simulations confirm that the REMOS concept, which explicitely models the dispersive character of reverberation, achieves significantly better recognition rates than conventional HMM-based algorithms, even if the latter are trained on reverberant data. Future work on the REMOS concept includes incorporation of more powerful speech features, like MFCCs, and more accurate output densities, like mixtures of Gaussians, as well as the application of REMOS to more complex tasks, such as large-vocabulary continuous speech recognition for more natural human/machine speech interfaces.

References

1. J. B. Allen, D. A. Berkley: Image method for efficiently simulating small-room acoustics, *JASA*, **65**(4), 943–950, April 1979.
2. AMI project: "Webpage of the AMI project," http://corpus.amiproject.org.
3. B. Atal: Effectiveness of linear prediction characteristics of the speech wave for automatic speaker identification and verification, *JASA*, **55**(6), 1304–1312, 1974.

4. L. E. Baum, J. A. Eagon: An inequality with applications to statistical estimation for probabilistic functions of Markov processes and to a model for ecology, *Bulletin of American Mathematical Society*, **73**, 360–363, 1967.
5. L. E. Baum, et al.: A maximization technique occurring in the statistical analysis of probabilistic functions of Markov chains, *Annals of Mathematical Statistics*, **41**, 164–171, 1970.
6. J. Benesty: Adaptive eigenvalue decomposition algorithm for passive acoustic source localization, *Journal of the Acoustical Society of America*, **107**(1), 384–391, Jan. 2000.
7. J. Benesty, S. Makino, J. Chen (eds.): *Speech Enhancement*, Berlin, Germany: Springer, 2005.
8. M. Brandstein, D. Ward (eds.): *Microphone Arrays*, Berlin, Germany: Springer, 2001.
9. C. Breining, P. Dreiseitel, E. Hänsler, A. Mader, B. Nitsch, H. Puder, T. Schertler, G. Schmidt, J. Tilp: Acoustic echo control. An application of very-high-order adaptive filters, *IEEE Signal Process. Mag.*, **16**(4), 42–69, 1999.
10. H. Buchner, R. Aichner, W. Kellermann: TRINICON: A versatile framework for multichannel blind signal processing, *Proc. ICASSP '04*, **3**, 889–892, Montreal, Canada, 2004.
11. CHIL project: "Webpage of the CHIL project," http://chil.server.de.
12. S. Davis, P. Mermelstein: Comparison of parametric representations for monosyllabic word recognition in continuously spoken sentences, *IEEE Trans. Acoust. Speech Signal Process.*, **ASSP-28**(4), 357–366, 1980.
13. S. Furui: On the role of spectral transition for speech perception, *JASA*, **80**(4), 1016–1025, 1986.
14. K. Furuya, S. Sakauchi, A. Kataoka: Speech dereverberation by combining MINT-based blind deconvolution and modified spectral subtraction, *Proc. ICASSP '06*, **1**, 813–816, Toulouse, France, 2006.
15. K. Furuya, A. Kataoka: Robust speech dereverberation using multichannel blind deconvolution with spectral subtraction, *IEEE Trans. Audio Speech Language Process.*, **T-ASLP-15**(5), 1579–1591, 2007.
16. M. J. F. Gales, S. J. Young: Robust continuous speech recognition using parallel model combination, *IEEE Trans. Speech Audio Process.*, **T-SAP-4**(5), 352–359, 1996.
17. N. D. Gaubitch, P. A. Naylor, D. B. Ward: On the use of linear prediction for dereverberation of speech, *Proc. IWAENC '03*, 99–102, Kyoto, Japan, 2003.
18. B. W. Gillespie, L. E. Atlas: Strategies for improving audible quality and speech recognition accuracy of reverberant speech, *Proc. ICASSP '03*, **1**, 676–679, Hong Kong, 2003.
19. D. Giuliani, M. Matassoni, M. Omologo, P. Svaizer: Training of HMM with filtered speech material for hands-free recognition, *Proc. ICASSP '99*, **1**, 449–452, Phoenix, AZ, USA, 1999.
20. S. M. Griebel, M. S. Brandstein: Microphone array speech dereverberation using coarse channel modeling, *Proc. ICASSP '01*, **1**, 201–204, Salt Lake City, UT, USA, 2001.
21. L. Griffiths, C. Jim: An alternative approach to linearly constrained adaptive beamforming, *IEEE Trans. on Antennas and Propagation.*, **30**(1), 27–34, 1982.
22. M. I. Gürelli, C. L. Nikias: EVAM: an eigenvector-based algorithm for multichannel blind deconvolution of input colored signals, *IEEE Trans. on Signal Processing*, **T-SP-43**(1), 134–149, 1995.

23. T. Haderlein, E. Nöth, W. Herbordt, W. Kellermann, H. Niemann: Using Artificially Reverberated Training Data in Distant Talking ASR, in *Proc. TSD '05*, V. Matoušek, P. Mautner, T. Pavelka (eds.), 226–233, Berlin, Germany: Springer, 2005.
24. E. Hänsler, G. Schmidt (eds.): *Topics in Acoustic Echo and Noise Control: Selected Methods for the Cancellation of Acoustical Echoes, the Reduction of Background Noise, and Speech Processing*, Berlin, Germany: Springer, 2006.
25. B. Hanson, T. Applebaum: Robust speaker-independent word recognition using static, dynamic and acceleration features: Experiments with lombard and noisy speech, *Proc. ICASSP '90*, **2**, 857–860, Albuquerque, NM, USA, 1990.
26. W. Herbordt: *Sound Capture for Human/Machine Interfaces – Practical Aspects of Microphone Array Signal Processing*, Heidelberg, Germany: Springer, 2005.
27. W. Herbordt, H. Buchner, S. Nakamura, W. Kellermann: Multichannel bin-wise robust frequency-domain adaptive filtering and its application to adaptive beamforming, *Trans. Audio Speech Language Process.*, **T-ASLP-15**(4), 1340–1351, 2007.
28. H. Hermansky, N. Morgan: RASTA processing of speech, *IEEE Trans. Speech Audio Process.*, **T-SAP-2**(4), 578–589, 1994.
29. T. Hikichi, M. Delcroix, M. Miyoshi: Blind dereverberation based on estimates of signal transmission channels without precise information of channel order, *Proc. ICASSP '05*, **1**, 1069–1072, Philadelphia, PA, USA, 2005.
30. H.-G. Hirsch, H. Finster: A new HMM adaptation approach for the case of a hands-free speech input in reverberant rooms, *Proc. INTERSPEECH '06*, 781–783, Pittsburgh, PA, USA, 2006.
31. O. Hoshuyama, A. Sugiyama, A. Hirano: A robust adaptive beamformer for microphone arrays with a blocking matrix using constrained adaptive filters, *IEEE Trans. Signal Process.*, **T-SP-47**(10), 2677–2684, 1999.
32. HTK: "HTK webpage," http://htk.eng.cam.ac.uk.
33. X. Huang, A. Acero, H.-W. Hon: *Spoken Language Processing: A Guide to Theory, Algorithm, and System Development*, Upper Saddle River, NJ, USA: Prentice Hall, 2001.
34. F. Jelinek: *Statistical Methods for Speech Recognition*, Cambridge, MA, USA: MIT Press, 1998.
35. J.-C. Junqua: *Robustness in Automatic Speech Recognition*, Boston, MA: Kluwer Academic Publishers, 1996.
36. K. Kinoshita, T. Nakatani, M. Miyoshi: Fast estimation of a precise dereverberation filter based on speech harmonicity, *Proc. ICASSP '05*, **1**, 1073–1076, Philadelphia, PA, USA, 2005.
37. H. Kuttruff: *Room Acoustics*, 4th ed., London, UK: Spon Press, 2000.
38. C.-H. Lee, C.-H. Lin, B.-H. Juang: A study of speaker adaptation of continuous density HMM parameters, *Proc. ICASSP '90*, **1**, 145–148, Albuquerque, NM, USA, 1990.
39. C. J. Leggetter, P. C. Woodland: Speaker adaptation of continuous density HMMs using multivariate linear regression, *Proc. ICSLP '94*, **2**, 451–454, Yokohama, Japan, 1994.
40. R. G. Leonard: A database for speaker-independent digit recognition, *Proc. ICASSP '84*, 42.11.1–42.11.4, San Diego, CA, USA, 1984.

41. D. G. Manolakis, V. K. Ingle, S. M. Kogon: *Statistical and Adaptive Signal Processing: Spectral Estimation, Signal Modeling, Adaptive Filtering and Array Processing*, Boston, MA: McGraw-Hill, 2000.
42. M. Miyoshi, Y. Kaneda: Inverse filtering of room acoustics, *IEEE Trans. Acoust. Speech Signal Process.*, **ASSP-36**(2), 145–152, February 1988.
43. P. J. Moreno, B. Raj, R. M. Stern: A vector taylor series approach for environment independent speech recognition, *Proc. ICASSP '96*, **2**, 733–736, Atlanta, GA, USA, 1996.
44. S. Nakamura, T. Takiguchi, K. Shikano: Noise and room acoustics distorted speech reognition by HMM composition, *Proc. ICASSP '96*, **1**, 69–72, Atlanta, GA, USA, 1996.
45. T. Nakatani, M. Miyoshi: Blind dereverberation of single channel speech signal based on harmonic structure, *Proc. ICASSP '03*, **1**, 92–95, Hong Kong, 2003.
46. T. Nakatani B.-H. Juang, K. Kinoshita, M. Miyoshi: Speech dereverberation based on probabilistic models of source and room acoustics, *Proc. ICASSP '06*, **1**, 821–824, Toulouse, France, 2006.
47. T. Nakatani, K. Kinoshita, M. Miyoshi: Harmonicity-based blind dereverberation for single-channel speech signals, *IEEE Trans. Audio Speech Language Process.*, **T-ASLP-15**(1) 80–95, Jan. 2007.
48. S. Neely, J. Allen: Invertibility of a room impulse response, *JASA*, **66**(1), 165–169, July 1979.
49. H. Ney, S. Orthmanns: Dynamic programming search for continuous speech recognition, *IEEE Signal Process. Mag.*, **16**(5), 64–63, 1999.
50. M. Omologo, M. Matassoni, P. Svaizer, D. Giuliani: Microphone array based speech recognition with different talker-array positions, *Proc. ICASSP '97*, **1**, 227–230, Munich, Germany, 1997.
51. D. S. Pallett, J. G. Fiscus, W. M. Fisher, J. S. Garofolo, B. S. Lund, A. Martin, M. A. Przybocki: The 1994 benchmark tests for the ARPA spoken language program, *Proc. Spoken Language Technology Workshop*, 5–38, Austin, TX, USA, 1995.
52. D. S. Pallett: A look at NIST's benchmark ASR tests: past, present, and future, *Proc. ASRU '03*, 483–488, St. Thomas, Virgin Islands, 2003.
53. J. G. Proakis, D. G. Manolakis: *Digital Signal Processing: Principles, Algorithms, and Applications*, Upper Saddle River, NJ, USA: Prentice Hall, 1996.
54. W. Putnam, D. Rocchesso, J. Smith: A numerical investigation of the invertibility of room transfer functions, *Proc. WASPAA '95*, 249–252, Mohonk, NY, USA, 1995.
55. L. R. Rabiner: A tutorial on hidden markov models and selected applications in speech recognition, *Proc. IEEE*, **77**(2), 257–286, 1989.
56. C. K. Raut, T. Nishimoto, S. Sagayama:Model adaptation for long convolutional distortion by maximum likelihood based state filtering approach, *Proc. ICASSP '06*, **1**, 1133–1136, Toulouse, France, 2006.
57. A. Sehr, M. Zeller, W. Kellermann: Hands-free speech recognition using a reverberation model in the feature domain, *Proc. EUSIPCO '06*, Florence, Italy, 2006.
58. A. Sehr, M. Zeller, W. Kellermann: Distant-talking continuous speech recognition based on a novel reverberation model in the feature domain, *Proc. INTERSPEECH '06*, 769 – 772, Pittsburgh, PA, USA, 2006.

59. A. Sehr, W. Kellermann: A new concept for feature-domain dereverberation for robust distant-talking ASR, *Proc. ICASSP '07*, **4**, 369–372, Honolulu, Hawaii, 2007.
60. A. Sehr, Y. Zheng, E. Nöth, W. Kellermann: Maximum likelihood estimation of a reverberation model for robust distant-talking speech recognition, *Proc. EUSIPCO '07*, 1299-1303, Poznan, Poland, 2007.
61. M. L. Seltzer, B. Raj, R. M. Stern: Likelihood-maximizing beamforming for robust hands-free speech recognition, *IEEE Trans. Speech Audio Process.*, **T-SAP-12**(5), 489–498, 2004.
62. M. L. Seltzer, R. M. Stern: Subband likelihood-maximizing beamforming for speech recognition in reverberant environments, *Trans. Audio Speech Language Process.*, **T-ASLP-14**(6), 2109–2121, 2006.
63. P. C. W. Sommen: Partitioned frequency domain adaptive filters, *Proc. 23rd Asilomar Conference on Signals Systems and Computers*, 676–681, Pacific Grove, CA, USA, 1989.
64. J. S. Soo, K. K. Pang: Multidelay block frequency domain adaptive filter, *IEEE Trans. Acoust. Speech Signal Process.*, **ASSP-38**(2), 373–376, 1990.
65. J. S. Soo, K. K. Pang: A multistep size (MSS) frequency domain adaptive filter, *IEEE Trans. Signal Process.*, **T-SP-39**(1), 115–121, 1991.
66. V. Stahl, A. Fischer, R. Bippus: Acoustic synthesis of training data for speech recognition in living-room environments, *Proc. ICASSP '01*, **1**, 285–288, Salt Lake City, UT, USA, 2001.
67. T. G. Stockham: High-speed convolution and correlation, *Proc. AFIPS '66*, **28**, 229–233, 1966.
68. T. Takiguchi, S. Nakamura, Q. Huo, K. Shikano: Model adaption based on HMM decomposition for reverberant speech recognition, *Proc. ICASSP '97*, **2**, 827–830, Munich, Germany, 1997.
69. T. Takiguchi, S. Nakamura, K. Shikano: HMM-separation-based speech reognition for a distant moving speaker, *IEEE Trans. Speech Audio Process.*, **T-SAP-9**(2), 127–140, 2001.
70. T. Takiguchi, M. Nishimura, Y. Ariki: Acoustic model adaptation using first-order linear prediction for reverberant speech, *IEICE Trans. Information and Systems*, **E89-D**(3), 908–914, 2006.
71. A. Torger, A. Farina: Real-time partitioned convolution for ambiophonics surround sound, *Proc. WASPAA '01*, 195–198, Mohonk, NY, 2001.
72. A. P. Varga, R. K. Moore: Hidden Markov model decomposition of speech and noise, *Proc. ICASSP '90*, **2**, 845–848, Albuquerque, NM, USA, 1990.
73. B. van Veen, K. Buckley: Beamforming: A versatile approach to spatial filtering, *IEEE ASSP Magazine*, **5**(2), 4–24, 1988.
74. P. C. Woodland, M. J. F. Gales, D. Pye: Improving environmental robustness in large vocabulary speech recognition, *Proc. ICASSP '96*, **1**, 65–68, Atlanta, GA, USA, 1996.
75. B. Yegnanarayana, P. Satyanarayana Murthy: Enhancement of reverberant speech using LP residual signal, *IEEE Trans. Speech Audio Process.*, **T-SAP-8**(3), 267–281, 2000.
76. B. Yegnanarayana, S. R. Mathadeva Prasanna, K Sreenivasa Rao: Speech enhancement using excitation source information, *Proc. ICASSP '02*, **1**, 541–544, Orlando, FL, USA, 2002.

77. S. J. Young, N. H. Russel, J. H. S. Thornton: Token passing: a simple conceptual model for connected speech recognition systems, CUED technical report, Cambridge University Engineering Department, 1989.
78. S. Young, G. Evermann, D. Kershaw, G. Moore, J. Odell, D. Ollason, D. Povey, V. Valtchev, P. Woodland: *The HTK Book (for HTK Version 3.2)*, Cambridge, UK: Cambridge University Engineering Department, 2002.

Index

Absolute category rating, 344
Absolute threshold in quiet, 666
Absolute threshold of hearing, 666
Absolute-category rating (ACR), 297
Acoustic echo canceller, 188
Acoustic model, 683
Acoustic score, 683
Acoustic systems, 616
Acoustic-phonetic modeling, 688
Active noise control (ANC), 552
Active noise control system, 672
Active room compensation, 552
Adaptation control, 457, 509, 515
Adaptive beamformer, 444
Adaptive lattice predictor, 637
Adaptive mixer, 124
Adaptive noise cancellation (ANC), 229
Adaptive target cancellation, 528, 533
Aliasing distortions, 17, 23, 24
All-pole
  Filter, 142
  model, 160
Allpass transformation, 20, 41
Ambiguity problem, 653
Anechoic binaural segregation, 529
Anti gain chase function, 626
Arithmetical mean, 663
Attributes, 292
Audiogram, 611
Autocorrelation, 669
Automatic speech recognition (ASR), 679

Background noise, 89
  estimation, 100
  maximum attenuation, 93
  overestimation, 93
Background noise simulation, 351
Bandwidth extension, 135
Bark scale, 21, 313, 660
Bark-spectral distance (BSD), 314
Beamformer, 640
Beamforming, 418, 698
Beampattern, 422
Bias removal, 493
  multi-channel, 497
Binaural hearing, 525
Blind source separation
  broadband, 486
  convolutive, 469
  instantaneous, 485, 497
  narrowband, 485
Broadside array, 420

Car noise, 64
Cepstral coefficients, 143, 164
  linear predictive, 144
Cepstral distance, 311
Cepstral distance (CD), 51
Cepstral mean subtraction (CMS), 700
Cocktail-party effect, 525
Codebook, 112
Codebook approach, 160, 169
Combination operator, 703
Comparison-category rating (CCR), 297, 345
Compensation filter, 73

Composite source signal (CSS), 356
Computational auditory scene analysis (CASA), 525
Conditional independence assumption, 687
Connectivity of a TDOA graph, 405
Consistent graph, 398
Continuity, 326
Control unit, 99
Controller area network (CAN), 617
Conversation test, 294, 345
Correlation method, 484
Covariance method, 484
Critical band, 313
Critical bands, 658
Critical distance, 200
Cross-coupled paired filter adaptive noise cancellation structure, 242
Crosstalk, 230, 232
Crosstalk resistant adaptive noise cancellation, 239

Data-dependent transformation, 569
Decision-directed estimator, 194
Decoding, 708
Deconvolution, 472
Degradation category rating (DCR), 345
Degradation mean opinion score (DMOS), 345
Degradation-category rating (DCR), 297
Delay compensation, 235, 244
Delay requirements, 355
Delay-and-sum beamformer, 420
Dereverberation, 472
Diffuse sound field, 491, 505
Direct path TDOA, 384
Directivity, 422
Directness, 324
Discrete Bark spectrum, 662
Discrete Fourier transform (DFT), 19
Discrete wavelet transform (DWT), 16
Dispersive effect, 694
Distance measures, 147
    L_p-norm, 148
    cepstral, 148
    CMOS, 177, 180
    Euclidean, 148
    likelihood ratio, 169
    logarithmic spectral distortion, 177
    Minkowski, 148
    objective, 177
    SDM, 178
    subjective, 180
Distant-talking scenario, 680
Distributed source, 606
Double talk, 277
Double talk test, 346
Double-talk, 190
DSP implementation, 280
Dynamic compressor, 624
Dynamic equalization control (DEC), 618, 634
Dynamic features, 699
Dynamic sound control, 615
Dynamic volume control (DVC), 618
Dynamic volume control system, 616, 619

E-model (ETSI network-planning model), 293
Early reflections, 593
Echo loss requirements, 355
Echo path TDOA, 384
Echo replica, 269
Eigenspace, 553
Eigenspace adaptive filtering, 568
Endfire array, 420
Engine noise, 66
Equivalent rectangular bandwidth, 659
Evaluation of hands-free terminals, 339
Excitation
    extension, 152, 155
    signal, 142, 153, 155
Expert test, 350
Extended Viterbi algorithm, 709

False detection, 392
Fast attack and slow decay, 276
Fast Fourier transform (FFT), 19
Feature classification, 99
Filter
    Mth-band, 31
    auto-regressive (AR), 46
    direct form, 35
    linear-phase, 33
    low delay, 46

minimum-phase, 46
moving-average (MA), 46
prototype, 17, 29, 32
transposed direct form, 35
Filter-and-sum beamformer, 420
Filter-bank
allpass transformed, 20
analysis-synthesis (AS FB), 15
low delay, 26
modulated, 17
oversampled, 16, 18
paraunitary, 18
polyphase network (PPN), 18
summation method (FBSM), 28
tree-structured, 16
Filter-bank equalizer (FBE), 29, 30
allpass transformed, 41
direct form, 38
polyphase network (PPN), 37
transposed direct form, 38
Filtering ambiguity, 472
Finite impulse response (FIR), 18
Fixed beamformer, 440
Formants, 140, 161
Frequency Content, 324
Frequency dependant dynamic compressor, 668
Frequency domain adaptive filter (FDAF), 646
Frequency response characteristics, 355
Frequency warping, 21, 41
Fricatives, 139
Frozen-time transfer function, 38
Function generators, 158
sine generators, 158
Fundamental frequency, 105, see Pitch

Gain chase, 626
Generalized cross-correlation (GCC), 390
Generalized discrete cosine transform (GDCT), 34
Generalized discrete Fourier transform (GDFT), 32
Generalized linear phase response, 42
Generalized multidelay adaptive filter (GMAF), 649
Generalized sidelobe canceler (GSC), 698

Generalized sidelobe canceller, 425
Generalized singular value decomposition, 568
Generic pitch impulse, 119
Geometrical mean, 663
Global masking threshold, 665, 666
Glottis, 139
Gradient adaptive lattice (GAL) algorithm, 637
Green's function, 553

Hair-cell model, 319
Half-way rectification, 95
Hamilton cycle, 397
Hamilton graph, 397
Hands-free telephone, 89
Harmonicity-based dereverberation (HERB), 697
Head and torso simulator (HATS), 352
Head related impulse response (HRIR), 530
Head related transfer function (HRTF), 597
Hearing aid, 587
Hearing impairment, 587
Hearing in noise test (HINT), 587
Hidden Markov model (HMM), 687
Higher-order ambisonics, 555
Higher-order statistics, 471, 479
HMM networks, 689
Homogeneous wave equation, 553
Human-robot communication, 247

Ideal binary mask (IBM), 528
IIR smoothing, 100
Image-source method, 594
Imaging, 149
In-car communication system, 672
In-service non-intrusive measurement devices (INMD), 326
Independent component analysis, 471
Independent subgraph, 398
Independent triple, 399
Individual pitch impulses, 121
Individualized DEC (IDEC), 635
Individualized DVC (IDVC), 635
Infotainment system, 617
Inner optimization, 710, 713
Input-to-noise ratio, 101

732   Index

Instrumental attribute measurement, 320
Integral quality, 292
Intelligibility, 289
Interaural intensity difference (IID), 527
Interaural time difference (ITD), 527
Interference, 230
Intermediate reference system, 288

Kandinski test, 346
Kirchhoff's 2nd law, 398

Language model, 684
Language score, 684
Late reverberation, 593
Lattice allpole filter, 638
LBG algorithm, 171
   centroid, 171
   quantization, 171
   splitting, 171
Levinson-Durbin recursion, 669
Lifting scheme, 26
Line of sight, 383
Line spectral frequencies, 146, 164
Linear mapping, 160, 163
   piecewise, 166, 173
Linear periodically time-variant (LPTV) system, 17
Linear prediction, 310
Linear predictive coding (LPC), 636
Linear time-invariant (LTI) system, 18
Listening test, 239, 296, 347
Listening-only test, 348
Log spectral amplitude estimator, 191
   optimally-modified, 192
Log-area (ratio) distance, 312
Log-likelihood ratio, 311
Logarithmic distance, 116
Logmelspec coefficients, 686
Lombard effect, 368, 458
Lombard speech, 161
Loudness, 298, 322
Loudness rating, 355
Loudness transformation, 313
Loudspeaker distortion, 267
Loudspeaker enclosure microphone system, 188
LPC coefficient, 636

Magnitude-squared coherence, 491
Markov assumption, 687
Masking, 618
   overlap-masking, 195
   self-masking, 195
Masking model, 618, 658
Masking model according to Johnston, 658, 661
Masking threshold, 618, 658
Mean opinion score (MOS), 239, 279, 344
Mean-opinion score (MOS), 295
Media oriented systems transport (MOST), 617
Mel scale, 125
Mel-frequency cepstral coefficient (MFCC), 685
Melspec coefficients, 685
Melspec convolution, 694
Memoryless smoothing filter, 627
Microphone based volume control system, 616, 623
Microphone calibration, 436
Microphone mismatch, 434
Microphone sensitivity, 433
Minimum mean-square error short-time spectral amplitude (MMSE-STSA) estimation, 274
Minimum statistics, 497, 509, 513, 627
Minimum tracker, 101
Minimum variance distortionless response (MVDR) criterion, 424
Mirrored microphone, 401
Misleading consistency, 400
Miss detection, 392
Model adaptation, 701
Model-based speech enhancement, 89
Modulation techniques, 152, 155
Multi-dimensional scaling (MDS), 302
Multi-stage adaptive noise canceller, 232
Multichannel active listening room compensation, 556
Multipath ambiguity, 384
Multipath propagation, 384
Multiple input/output inverse theorem (MINT), 696
Multiple source ambiguity, 384
Multiple-input multiple-output (MIMO), 470

Multivariate Gaussian PDF, 482
Multivariate Gaussian probability density function, 479
Multivariate probability density function, 475, 479, 486
Multivariate score function, 478–480, 482, 486
Musical noise, 33
Mutual information, 474

Natural gradient, 477, 481, 483
Neural model for human auditory perception, 600
Neural network, 160, 166
  activation function, 167
  bias, 167
Noise cancellation, 229
Noise generator, 118
Noise reduction algorithm, 672
Noisiness, 329
Non-linear characteristics, 149, 153, 157
  cubic, 157
  full-way rectification, 158
  half-way rectification, 158
  quadratic, 157
  saturation characteristic, 158
  tanh characteristic, 158
Non-model-based algorithms, 149
Non-uniqueness problem, 568
Nongaussianity, 471, 475
Nonlinear adaptive filter, 267
Nonlinear echo, 267
Nonlinearity, 93
Nonstationarity, 471, 475, 483
Nonwhiteness, 471, 475, 483
Normal equation, 565
Normalized least mean square (NLMS) algorithm, 189
Notch filter, 68

Octave-band analysis, 16
Optimal step-size, 78
Output density, 687
Overlap-add method, 20
Oversampling, 149

Paired filter structure, 233
PARCOR coefficient, 636
Partial spectral reconstruction, 89

Partitioned convolution, 692
Perceptual evaluation of speech quality (PESQ), 315
Periodic signals, 386
Permutation ambiguity, 472
PESQ^{TM}, 356
Phantom source, 385
Phase equalizer, 25, 42
  allpass, 25
  least-squares (LS), 25
Phase transform (PHAT), 390
Pitch
  detection, 157
  frequency, 139, 142
  structure, 152
Pitch frequency
  estimation, 105
  reliability, 106
Pitch impulse prototype, 97
Pitch period, 120
Pitch trajectories, 131
Pitch-specific impulses, 120
Point source, 606
Post-processing, 498
Postfilter, 267
Pre-processing, 493
Prediction filter, 669
Predictor
  coefficients, 143, 164, 169
  error filter, 154, 158
  inverse error filter, 155
Primary path transfer matrix, 563
Psychoacoustic basics, 658
Psychoacoustics, 618

Quality assessment of hands-free terminals, 340
Quality of speech signals, 289
Quality pie, 366
Quasi-stationary, 140

RASTA, 700
Raster condition, 388
Raster matching, 389
Reconstruction
  phase, 121
  unvoiced speech, 118
  voiced speech, 119
Reflection coefficient, 636

Regularization, 485
Relative approach, 361
Residual echo, 186, 269, 270, 273
Reverberant binaural segregation, 533
Reverberation, 195, 680
  early reverberation, 195
  late reverberation, 195
Reverberation cancellation, 196
Reverberation model, 702, 704
Reverberation modeling for speech recognition (REMOS), 702
Reverberation suppression, 196
Reverberation time, 187, 596, 682
Rhyme test, 290
Robust automatic speech recognition, 527
Room acoustic model, 593
Room impulse response, 552
Room impulse response (RIR), 682
Room response matrix, 552
Roughness, 300, 323

Second-order statistics, 482
Secondary path transfer matrix, 563
Segmental noise attenuation, 51
Self-calibration, 449
Semantic differential (SD), 304
Sharpness, 299, 323
Short conversational test, 346
Signal cancellation effect, 426
Signal delay, 51
Signal distortion, 238
Signal reconstruction
  near-perfect, 16
  perfect, 16, 23, 31
Signal-to-noise ratio, 103
Signal-to-noise ratio (SNR), 308
  a priori, 34, 50
  segmental, 50
Signal-to-noise ratio (SNR) estimation, 234
Signal-to-reverberation ratio (SRR), 682
Single talk, 277
Sound quality, 290
Source TDOA vector, 383
Source-filter model, 139, 142, 153
Spectral distance, 309

Spectral envelope, 112, 140, 143, 147, 169
  broadband, 154, 155, 159
  narrowband, 155
Spectral flatness measure, 663
Spectral refinement, 109
Spectral shifting, 151
  adaptive, 152, 157
  fixed, 151
Spectrum-based dynamic equalization control, 619, 645
Speech activity detection, 130
Speech apparatus, 139
Speech enhancement
  based on nonlinear processing, 93
  based on speech reconstruction, 97
  conventional, 91
Speech intelligibility outside vehicles, 372
Speech perception in noise (SPIN) test, 587
Speech quality, 176
Speech recognition, 458
Speech recogniton, 256
Speech reconstruction, 110
Speech segmentation, 258
Speech-dialog system, 89
Speech-quality evaluation tool (SQET), 318
Speed dependent sound system, 616, 619, 620
Spherically invariant random process (SIRP), 479
Spherically symmetric multivariate Laplacian PDF, 481
Spread of masking, 660
Spread spectrum error, 666
Spreading function, 660
Spreading matrix, 663
Star graph, 403
State, 687
Statistical reverberation model, 199
Statistical room acoustics, 199
Steering vector, 419
Step-size control, 75, 234
Subjective evaluation, 239
Subjective performance evaluation, 343
Subsampling
  critical, 16

non-critical, 16
Superdirective beamforming, 420
Susceptibility, 423
Sylvester
  matrix, 476
  structure, 473
Sylvester constraint, 476
  column, 477
  row, 477

Talking test, 347
TDOA ambiguity, 384
TDOA disambiguation, 387
TDOA estimation, 383
TDOA graph, 397
Telecommunication objective speech-quality assessment (TOSQA), 317
Telephone-speech quality, 287
Time alignment, 122
Time difference of arrival (TDOA), 383
Time potential, 398
Tire noise, 65
TMOS, 356
Toeplitz structure, 484
Tolerance function of raster match, 393
Tolerance function of triple match, 402
Tolerance width of raster match, 393
Tolerance width of triple match, 402
Tonality factor, 665
TOSQA 2001 mean opinion score, 356
Training
  data, 160
  database, 160
  parameter, 146
Transition probabilities, 687
Trellis diagram, 691
Triangular inequality, 390

TRINICON, 469, 474, 486

Unconstraint FDAF, 647
Univariate probability density function, 479
Unvoiced excitation, 118
Unvoiced sounds, 139, 142, 149

Vaananen reverberator, 595
Vector quantizer, 171
Virtual patient, 592
VirtualHearing software, 592
Viterbi algorithm, 689
Viterbi metric, 690, 710
Vocal cords, 139
Vocal tract, 140
  transfer function, 169
  transfer function estimation, 159
Voice activity detection, 626
Voice activity detection (VAD), 408
Voiced excitation, 120
Voiced sounds, 139, 142
Voiced/unvoiced classification, 103

Warped discrete Fourier transform (WDFT), 27
Wave field synthesis, 555
Wave-domain adaptive filtering, 551, 570
Wideband, 136
Wiener filter, 501
Wind noise, 65

Yule-Walker equation, 143
Yule-Walker equations, 46

Zero cyclic sum condition, 389, 397
Zero cyclic sum matching, 389